Conic Sections

Circle:

$(x - h)^2 + (y - k)^2 = r^2$

Ellipse:

$\dfrac{(x - h)^2}{a^2} + \dfrac{(y - k)^2}{b^2} = 1$

Hyperbola:

$\dfrac{(x - h)^2}{a^2} - \dfrac{(y - k)^2}{b^2} = 1$

Hyperbola:

$\dfrac{(y - k)^2}{a^2} - \dfrac{(x - h)^2}{b^2} = 1$

Parabola:

$y - k = a(x - h)^2$

Parabola:

$x - h = a(y - k)^2$

Sequences and Series

Arithmetic Sequence: $a_n = a_1 + (n - 1)d$

$$S_n = \frac{n}{2}(a_1 + a_n) = \frac{n}{2}[2a_1 + (n - 1)d]$$

Geometric Sequence: $a_n = a_1 r^{n-1}$

$$S_n = \frac{a_1(1 - r^n)}{1 - r} \quad \text{if } r \neq 1$$

$$= na_1 \qquad \text{if } r = 1$$

Infinite Geometric Series: $S = \dfrac{a_1}{1 - r}, \quad |r| < 1$

A PRIMER FOR CALCULUS

6TH EDITION

Leonard I. Holder
Gettysburg College

PWS PUBLISHING COMPANY
Boston

PWS
Publishing Company

20 Park Plaza
Boston, Massachusetts 02116

I(T)P ™

International Thomson Publishing
The trademark ITP is used under license

Copyright © 1993 by PWS Publishing Company.

PWS Publishing Company is a division of Wadsworth, Inc.

Library of Congress Cataloging-in-Publication Data

Holder, Leonard Irvin, 1923–
    A primer for calculus / Leonard I. Holder. — 6th ed.
      p. cm.
    Includes index.
    ISBN 0-534-17748-4
    1. Mathematics. I. Title.
QA39.2.H63   1993
510—dc20   92-30286
                                         CIP

Printed in the United States of America.
93 94 95 96 97—10 9 8 7 6 5 4 3 2

Mathematics Editor: Anne Scalan-Rohrer
Senior Editorial Assistant: Leslie With
Permissions Editor: Robert Kauser
Production: Greg Hubit Bookworks
Print Buyer: Diana Spence
Text and Cover Designer: Vargas/Williams/Design
Cover Photo: Melvin L. Prueitt, Los Alamos National Laboratory
Copy Editor: Charles D. Cox
Compositor: Polyglot Pte Ltd
Printer: R. R. Donnelley & Sons Company

CONTENTS

BASIC ALGEBRAIC CONCEPTS   1

EQUATIONS AND INEQUALITIES OF THE FIRST AND SECOND DEGREE   63

7 THE TRIGONOMETRIC FUNCTIONS   303

8 TRIGONOMETRIC IDENTITIES AND EQUATIONS   349

9 THE SOLUTION OF TRIANGLES   385

10 FURTHER APPLICATIONS OF TRIGONOMETRIC FUNCTIONS   421

11 SYSTEMS OF EQUATIONS AND INEQUALITIES   451

12 THE CONIC SECTIONS   517

13 SEQUENCES AND SERIES   557

PREFACE

In the first edition of this book, published in 1978, I stated that many calculus students had difficulty not so much because of failure to understand new calculus concepts, but because of inadequate preparation in precalculus mathematics. The only thing different today is that the number of students in this category has increased. Some of these students simply have not been exposed to a good precalculus course. Others have been exposed but for one reason or another have not learned the material adequately. A common complaint among calculus instructors is that many students are weak in basic algebraic skills.

My goal from the beginning has been to write a book that addresses these diverse needs in a "reader friendly," yet mathematically sound way, so that on completion of the course students will have a strong background for studying calculus. With each revision I believe that I have come closer to achieving this goal. For those students with weak algebraic skills, the first two chapters should be studied thoroughly. Some students think of algebra as a "bag of tricks." By showing that all of the so-called rules of algebra follow logically from the basic properties of the real number system, I hope to dispel this misconception. I do not, however, carry this deductive approach to an extreme. In fact, I avoid for the most part a formal "theorem–proof" format, preferring to write in a readable, narrative style.

One of the concepts I have found virtually all calculus students have difficulty understanding is that of a function. For this reason I place particular emphasis on this key concept. Beginning with Chapter 3 and throughout the next six chapters, the notion of function is dominant, first in general terms and then followed by specific functions. This I believe to be the heart of a precalculus course. The remaining chapters, although containing important and interesting material, can be thought of as supplementary. It is doubtful that any one-semester course will include all of these chapters, but they provide for flexibility and enrichment.

Calculators are now so commonplace that I assume students using this book have access at least to one with basic scientific functions. Problems occur in many places where it makes sense to use a calculator. In some cases I call attention to this, but in many cases the need seems so evident that I do not feel any special indication is necessary. The calculator is a tool and should be used when it is helpful, but the basic ideas of the course can—and should be—learned without dependence on it. Graphing calculators (some of which blur the line between calculators and computers) are becoming more affordable, but I believe it is still premature to assume students have access to one of these. Nevertheless, I have included some references to ways graphing calculators can be used, and in a number of exercise sets I have included problems specifically calling for a graphing calculator. These are clearly labeled as "C" exercises and always occur at the end of the exercise sets, so that instructors who wish to omit them may easily do so.

As in earlier editions, I have written with the students in mind, although I do not talk down to them. I believe they *can* read this book with understanding, and I hope they *will* do so. In the note to students that follows this preface I give some suggestions that might help in this regard. The notes of caution that occur from time to time throughout the text have proven to be helpful, and I have added a few new ones in this edition. Also, I have continued the use of chapter openers taken from calculus texts. Each of these illustrates one or more of the concepts or techniques to be studied in the chapter. The idea is to show students that what they are about to study really is useful in calculus. Students should simply observe these, without any attempt to understand all of the details. When they finish the chapter, however, they might find it interesting to look back and see how much they are then able to follow.

Problem sets certainly constitute one of the most important parts of any mathematics textbook. Mine have been cited by many users of previous editions as being especially well chosen with respect to variety and scope. One especially noteworthy feature of the exercise sets is the inclusion of many problems that come directly from calculus. I have retained most of the problems from the fifth edition and have added quite a few more, especially some of intermediate difficulty. As in the past, the problems are divided into A and B categories, with the A problems designed to give sufficient practice to fix ideas in students' minds, and the B problems the more thought-provoking ones. Some of the B problems also extend the theory. I have changed some of the examples and have added some where they seemed to be needed.

Many of the changes I have made from the fifth edition are the result of suggestions by users, and I am most appreciative of these. Among these changes are the following:

- In Chapter 1 the section on exponents and radicals has been separated into two sections, one on integral exponents and one on radicals and fractional exponents.

- The distance and midpoint formulas have been moved to the first section of Chapter 3, and the material on parallel and perpendicular lines has been incorporated into Section 4.2, thereby eliminating the need for the old Section 4.3.

- The introduction to conic sections has been restored as Section 4.5, and this is now reflected in the chapter title. Several users of the fifth edition indicated that time did not permit their getting to Chapter 12, where the conics are taken up in detail, but that they would have liked to have had the brief introduction to them similar to that given in Chapter 4 of earlier editions. At a minimum I feel that students should learn to recognize the different conic sections in standard position from their equations and be able to draw them. (Many calculus students cannot even identify a circle from its equation!)

■ The material in Chapter 7 on trigonometric identities and equations has been separated out into a separate chapter (Chapter 8). I agreed with those who felt that the old Chapter 7 was too long and contained too many concepts.

■ Chapter 11 has been eliminated. Most of the ideas from it have been incorporated into Chapter 5, and some (intercepts and symmetry) into Chapter 3. This relocation places the material where it comes up naturally, and provides a more efficient coverage (e.g., asymptotes are studied in one section, rather than in two widely separated ones).

■ A section on the locus of a point has been added to Chapter 12. This section had been deleted from the fifth edition, and a number of users recommended that it be restored.

The primary focus of this book is clearly on preparing students for calculus. It should be emphasized, however, that the material in it is important in its own right, and students can profit from the course whether or not they follow it with calculus.

I wish to acknowledge the support and help provided by all of those persons at Wadsworth Publishing Company associated with this project, and in particular, the mathematics editor, Anne Scanlan-Rohrer. I also wish to thank my typist, Donna Cullison, who has been with me from the beginning and has always done an outstanding job.

SUPPLEMENTS

- **Student Solutions Manual**   An all-new Student Solutions Manual written by Fred Safier of City College of San Francisco features detailed solutions to one-half of the odd-numbered exercises.

- **Instructor's Manual/Test Bank**   A printed and bound test bank provides sample tests and a battery of test questions for constructing exams. The manual also contains answers to all even-numbered problems.

- **Computerized Testing Programs**   All of the test items are also available on computerized testing programs. EXP Test is available for IBM PCs and compatibles and allows fast and easy creation of tests, with all technical symbols and fonts appearing on the screen as they will appear when printed. Exam Builder is a comparable system for the Macintosh. Available upon adoption.

- **TeMath**   TeMath software for the Macintosh is a set of numeric and graphic tools for exploring mathematics. For further information, contact your local representative.

- *Algebra Facts*   Created by Theodore Szymanski of Tompkins Cortland Community College, this pocket-sized reference book contains essential algebra concepts, formulas, and procedures in a concise and readable format and is available for student purchase.

- *Precalculus in Context: Functioning in the Real World*   By Marsha Davis, Judy Flagg Moran, and Mary Murphy, this laboratory manual for precalculus assumes the use of a graphing calculator or software. A series of twelve labs encourage students to explore concepts, work in groups and provide solutions in writing. A cross-reference chart relating this manual to *Primer for Calculus* follows.

Section in *Precalculus in Context*	Section in *A Primer for Calculus*
1. Linear Functions, Graphs, Constant Rates of Change	3.1, 3.2, 4.3
2. Quadratic Functions, Parabolas, Changing Rates of Change	4.4
3. Transforming a Graph: Shifting, Reflecting, Stretching, Compressing, Symmetry, and Absolute Value	3.3, 3.4, 4.2

ACKNOWLEDGMENTS

Most of the changes I have made in this new edition were suggested by users of the fifth edition, and I am indebted to them. Further useful suggestions were made by reviewers of the manuscript for this edition, and I have incorporated many of their ideas. I would like to thank W. Dwayne Collins, Hendrix College; Gary Grainger, University of Scranton; Yvonne Hunter, Pennsylvania State University; Carl B. Hurd, Pennsylvania State University–Altoona; William Kabeisman, University of South Dakota; Parviz Khalili, Christopher Newport College; Joan Kravchvck, Hillsborough Community College; Eric J. Nelson, Chabot College; Luis Ortiz-Franco, Chapman College; Robert O. Stanton, St. John's University; Janet M. Winter, Pennsylvania State University–Berks; and Ulf Wostner, City College of San Francisco. I would also like to thank the following instructors who checked the accuracy of the examples, exercises, and answers in this edition: Gary Grainger, Diana Gerardi, Terri Bittner, and Groff Bittner.

It would be difficult to recognize by name all persons who have contributed to this book as it has evolved, but I do wish to list the reviewers of the previous edition and thank them all for the important role they have played in making the text a successful teaching tool: James Arnold, University of Wisconsin, Milwaukee; Barbara Bender, Pennsylvania State University–Berks; E. Ray Bobo, Georgetown University; Edward S. Boylan, Rutgers University; Otha Britton, University of Tennessee, Martin; Regina Brunner, Cedar Crest College; Anand M. Chak, University of West Virginia; James F. Chew, North Carolina A&T University; Robert Collings, North Harris County College; Romae Cormier, Northern Illinois University; Edwin Creasy, State University of New York A&T, Canton; Roger L. Creech, East Carolina University; Robert Dahlin, University of Wisconsin, Superior; Robert Dean, Central Washington University; Felix A. Defino, Morris College; Marvin Eargle, Appalachian State University; Wade Ellis, Jr., West Valley College; Richard Fast, Mesa College; Gregory D. Foley, The Ohio State University; Karl Ray Gentry, University of North Carolina, Greensboro; Stuart Goldenberg, California Polytechnic State University; James E. Hall, University of Wisconsin, Madison; Ferdinand Haring, North Dakota State University; Charles Heuer, Concordia College; David Hoak, University of California, Los Angeles; Louis Hoelzle, Bucks County Community College; John Hornsby, University of New Orleans; Paul Hutchens, St. Louis Community College; Glenn E. Johnston, Morehead State University; Gail Jones, Delgado Junior College; William Jones, Xavier University; Charles Kinzek, Los Angeles Valley College; Kurt Lewandowski, Gavilan College; Michael Lewis, Clark College; James McKim, University of Central Arkansas; Jerold C. Mathews, Iowa State University; Deborah Mirdamadi,

Pennsylvania State University, Mont Alto Campus; Jane Morrison, Thornton Community College; Robert Morrow, Central Piedmont Community College; Thomas Mowry, Diablo Valley College; Umesh Nagarkatte, Medgar Evers College, CUNY; K. L. Nielsen, Butler University; Ellard Nunnally, University of Miami; Gary R. Penner, Richland College; Juanita Peterson, College of Alameda; John Petro, Western Michigan University; Edward Pettit, Augusta College; Anne Prial, Hofstra University; Glenn Prigge, University of North Dakota; Elaine Rhoades, Illinois State University, Normal; Fred Safier, City College of San Francisco; Venice Scheurich, Del Mar College; Edwin Schulz, Elgin Community College; Donald Smith, Louisiana State University, Shreveport; Johnny S. Smith, Jacksonville State University; Sybil Smith, Rutgers University; John Spellmann, Southwest Texas State University; Lettie Stallings, Gordon Junior College; E. Genevieve Stanton, Portland State University; Rita Stout, Texas A&M University; Jan Vandever, South Dakota State University; John Vangor, Housatonic Community College; John L. Whitcomb, University of North Dakota; Bruno Wichnoski, University of North Carolina, Charlotte; W. R. Wilson, Central Piedmont Community College; and Charles T. Wolf, Millersville University.

TO THE STUDENT

HOW TO READ THE TEXT

A few words of advice at the outset may help you in your understanding of this text and in mastering its subject matter. First of all, *read* the text *before* attempting to do the exercises. Read it with pencil and paper at hand, writing out the steps of examples and proofs, trying to supply any missing steps. Reading a mathematics text passively, as you would a novel, is likely to be more confusing than enlightening. It is also a good idea to make a list of the definitions and important formulas and other results derived in the section you are studying. *Writing these out helps to fix them in your mind.*

RELEVANCE TO CALCULUS

Most of the topics covered have relevance in calculus. To help emphasize this, each chapter is introduced with an excerpt from some calculus textbook, illustrating a use in calculus of one or more of the concepts to be studied in that chapter. As you begin a chapter, just observe the chapter opener and read the comments about it, without trying to understand all of the mathematics. Most of the excerpts will contain symbols and words that you will learn about only when you take calculus, and you will not yet have studied the precalculus material contained in them. When you finish a chapter, you might want to look at the calculus excerpt again. You should then be able to follow the part dealing with precalculus.

HIGHLIGHTED ITEMS

As you read the text, you will see certain words and phrases in color. This is for emphasis, and you should pay particular attention to these. In some cases these highlighted items constitute a definition, and sometimes they amount to a statement of a mathematical proposition, or *theorem*. Definitions and theorems of special importance are set apart, however, and are identified as such. Also certain special results or formulas are put in boxes, for emphasis and for easy reference. Pay attention also to the words of caution that occur from time to time. They can help you avoid some common pitfalls.

EXERCISE SETS

The exercises are divided into A and B categories. The A exercises are designed to fix ideas in your mind and so contain quite a few "drill type" problems. The B exercises tend to be more challenging and sometimes call for proofs of results stated in the text or otherwise extend the theory. Do not assume, though, that the B exercises are too difficult. You should always attempt any of these that are assigned. Do not give up too quickly. Struggling with a problem and eventually getting it can give you a real sense of accomplishment. If your instructor permits it, I encourage you to team up with one or two other students to work together on homework. If all members of the team contribute and you talk together about the problems and the theory behind them, this can be a good learning experience. Students sometimes seem to shut down their brain function when they see a "word" problem. But these are the most interesting problems, and they more closely resemble "real-world" applications than those that call primarily for manipulative skills. Here again, working with a group to decide on the appropriate mathematical formulation, that is, on the *mathematical model*, can be a big help. Whether you work with a team or alone, it is essential that you do your homework, and do it when it is assigned. Getting behind can be fatal. Mathematics is not a spectator sport. You learn it by doing it.

IMPORTANCE OF ALGEBRA

Algebra is a necessary tool—not only in calculus but in virtually all of mathematics. Contrary to popular opinion, it is not just a collection of arbitrary rules that someone thought up. In fact, all of these so-called rules can be derived from a surprisingly small set of basic assumptions (called *axioms*), in much the same way that plane geometry is developed from its axioms (although the way algebraic results are derived differs in general from the way geometry proofs are constructed). The complete logical development is not carried out in this text, but some of the main results are derived. You should spend some time and effort following these derivations (or *proofs*) and do some of them yourself. Besides understanding where the rules come from, you need to be able to apply them correctly, and you will have plenty of practice doing this.

EMPHASIS ON FUNCTIONS

Probably the most important thing you will study in this course is the concept of a *function*. So you should put special emphasis on Chapter 3, where this concept is first introduced. You may need to refer back to the definitions there many times. In succeeding chapters you will study the most important functions, namely linear and quadratic functions, higher-degree polynomial functions, rational functions, exponential and logarithmic functions, and trigonometric functions, together with their inverses. These functions form the basis for many mathematical models in applications, and they are used repeatedly in calculus, as well as higher-level courses. One of the best ways to understand a function is to see its graph, and you will find much emphasis on graphs in this text. Learning how to obtain the graphs is an important part of the course, and you will need to devote considerable time to this.

In all probability you have seen some of the subject matter of this course previously. I encourage you, even so, to approach it as if you were seeing it all for the first time. You are likely to get new insights and a greater understanding. Devoting the time and effort to learning the material in this course can be rewarding in itself, and it will pay big dividends later.

A NOTE ON CALCULATORS

Scientific calculators have become essential tools for students in mathematics as well as many other fields. In this text calculators are needed in many places but particularly in Chapters 5, 6, and 9. Sometimes their use is explicitly called for, but in most cases the nature of the problem implies their use. Because it is assumed that all students have access to a scientific calculator, tables are not included in the text.

Scientific calculators currently may be divided into at least four levels of sophistication, as follows:

1. Basic scientific calculator, such as the Texas Instruments TI-30 SLR + solar-powered calculator.
2. Elementary programmable calculator, such as the TI-60.
3. Graphics calculator, such as the TI-81, the Casio fx-7000G, and the Sharp EL-5200.
4. Symbol-manipulating calculator, such as the Hewlett-Packard HP-28C. This also has graphics display.

These are listed in increasing order of sophistication, as well as increasing price levels. Calculators of the fourth category, especially, blur the line between computers and calculators. For this text a basic scientific calculator is sufficient, but students with graphing calculators will find many places where the graphing capabilities can be put to good use.

From time to time in the text a keystroke display is given, using keys from a typical basic scientific calculator. Students may need to modify these for their own calculators.

Courtesy Texas Instruments

TI-60 programmable calculator

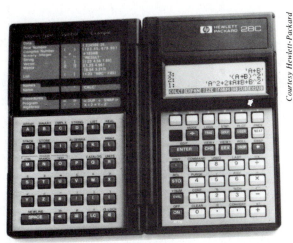

Courtesy Hewlett-Packard

HP-28C symbol-manipulating calculator

1

BASIC ALGEBRAIC CONCEPTS

At the beginning of each chapter we will give an excerpt from a calculus textbook illustrating the use of one or more of the precalculus concepts to be studied in that chapter. *You are not expected to follow all of the details but simply to observe the precalculus concepts that are involved.*

A mastery of the techniques of elementary algebra is essential to success in calculus. Open any calculus book to any page and you are likely to find algebra employed in one form or another. To illustrate one such use, consider the following example taken directly from a calculus textbook.*

Example Simplification by factoring out the least powers

$$\text{Differentiate } f(x) = x^2\sqrt{1 - x^2}$$

Solution

$$f(x) = x^2(1 - x^2)^{1/2} \qquad\qquad\qquad \text{Rewrite}$$

$$f'(x) = x^2\frac{d}{dx}\left[(1 - x^2)^{1/2}\right] + (1 - x^2)^{1/2}\frac{d}{dx}[x^2] \qquad \text{Product Rule}$$

$$= x^2\left[\frac{1}{2}(1 - x^2)^{-1/2}(-2x)\right] + (1 - x^2)^{1/2}(2x) \qquad \text{Power Rule}$$

$$= -x^3(1 - x^2)^{-1/2} + 2x(1 - x^2)^{1/2}.$$

$$= x(1 - x^2)^{-1/2}[-x^2(1) + 2(1 - x^2)] \qquad\qquad \text{Factor}$$

$$= \frac{x(2 - 3x^2)}{\sqrt{1 - x^2}} \qquad\qquad\qquad\qquad \text{Simplify}$$

▲

   * Roland E. Larson, Robert P. Hostetler, and Bruce H. Edwards, *Calculus*, 4th. ed. (Lexington, MA: D. C. Heath and Company, 1990), p. 152. Reprinted by permission.

This is not all there is to the calculus problem, but it is enough to illustrate the point. You should not be concerned with the symbols $f'(x)$ or $\dfrac{d}{dx}$, which are used in calculus for the process called **differentiation,** but you should observe the algebraic operations employed in the last two lines. You will study such operations in this chapter.

THE REAL NUMBER SYSTEM

The **real number system** is fundamental in the study of algebra, as well as in the study of many other branches of mathematics. So it is appropriate that we begin with a review of the basic properties of this most important number system. From this basic list, all other properties of real numbers can be derived. And from these properties, in turn, the so-called rules of algebra can be justified. We will not attempt such a complete deductive approach here, since to do so would constitute a course in itself, but we will indicate how the most important algebraic properties follow from those of the real numbers. Before summarizing their basic properties, we describe real numbers informally.

The simplest real numbers are the **natural numbers,** which are the ordinary counting numbers 1, 2, 3, ... Natural numbers are also called **positive integers,** and they, together with 0 and their negatives (the **negative integers,** -1, -2, -3, . . .), comprise the **integers.** The **rational numbers** are numbers that can be expressed in the form of a ratio, m/n, where m and n are integers, with $n \neq 0$. Rational numbers include ordinary fractions, such as $\frac{3}{4}$, and they also include the integers themselves, since any integer m can be written $m/1$. Real numbers that are not rational are called **irrational.** Some examples are $\sqrt{2}, \pi, \sqrt[3]{7}$.

One convenient way to visualize real numbers is to use a **number line,** as shown in Figure 1. After selecting a point as the **origin** to correspond to 0 and another point to correspond to 1, both a scale and a direction are established, and points corresponding to both positive and negative integers can then be determined. At least theoretically, we can see how to locate a unique point for every rational number. For example, the point corresponding to $\frac{13}{25}$ could be found by dividing the unit interval from 0 to 1 into 25 equal parts and locating the right end point of the thirteenth interval, starting from 0. It is more difficult to locate irrational numbers on the line, but they are there (in abundance!). Irrationals are rather elusive, but they can always be approximated to any desired degree of accuracy by rationals. For example, the well-known number we designate by π is typically approximated by 3.1416 ($\frac{31,416}{10,000}$), and $\sqrt{2}$ is

FIGURE I

approximately 1.414. We should keep in mind that these are approximations only; in fact, we can never write in decimal form the exact value of any irrational number.

When working with a number line, we will often not distinguish between numbers and the corresponding points on the line. For example, we might say "the point 2" rather than the more precise "the point corresponding to 2." A significant feature of the real numbers is that they occupy *every* position on the number line. There are no gaps. This characteristic would not be true if we limited consideration to rational numbers only, for while there are infinitely many rationals and they are dense on the line (that is, between any two points, however close together, there are rationals), they do not occupy all of the line. In fact, in a certain sense the gaps (that is, the irrationals) are more numerous than the rational points.

The rational numbers have two possible forms when expressed as decimal expansions: (1) terminating, such as $\frac{5}{4} = 1.25$, or (2) repeating, such as $\frac{2}{3} = 0.666\ldots$. One way of showing that a repeating decimal represents a rational number is suggested in Problem 6, Exercise Set 1. All decimal expansions that neither terminate nor repeat are irrational numbers. For example, $\sqrt{3} = 1.73205\ldots$.

We will frequently have occasion to refer to a **set** of things, by which we mean a collection. Sets are frequently designated by using braces, { }, and inside the braces either the members of the set are listed, or a rule is given that describes the set. The members of a set are called its **elements,** and we use the symbol $a \in A$ to indicate that a is an element of the set A. If every element of a set B is also an element of the set A, then B is said to be a **subset** of A, designated $B \subseteq A$. It is also convenient to be able to speak of the set that has no elements. We call this set the **empty set** and designate it by $\varnothing$. All other sets are said to be **nonempty.** We designate the set of all real numbers by **R,** the rationals by **Q,** the irrationals by **I,** the integers by **J,** and the natural numbers by **N.**

Since the notion of equality occurs throughout mathematics, it is useful at the outset to say precisely what this means. In mathematics the symbol $A = B$ means A and B are two names for the same thing. Thus, we may replace A by B or B by A in any expression involving either symbol. Following are the fundamental properties of equality as it relates to real numbers.*

---

* The same properties hold for complex numbers, which we will study in Chapter 2, and we will use them there without further discussion.

Basic Properties of Equality

1. $a = a$ Reflexive property
2. If $a = b$, then $b = a$.   Symmetric property
3. If $a = b$ and $b = c$, then $a = c$.   Transitive property
4. If $a = b$, then $a + c = b + c$.   Addition property
5. If $a = b$, then $ac = bc$.   Multiplication property

By the real number *system* we mean the set R of real numbers, together with the two binary operations of addition and multiplication, that satisfy the eight basic properties listed below.

1.   The set R is **closed** with respect to addition and multiplication. That is, **both $a + b$ and $a \cdot b$ are in R.**

Here, the word *closed* means that any number that results from the addition or multiplication of members of R must also be a member of R.

2.   Addition and multiplication in R are **commutative.** That is, $a + b = b + a$ **and** $a \cdot b = b \cdot a$.

Thus, the order of adding or multiplying two real numbers is immaterial.

3.   Addition and multiplication in R are **associative.** That is, $a + (b + c) = (a + b) + c$ **and** $a \cdot (b \cdot c) = (a \cdot b) \cdot c$.

Without this property we would not know, for example, how to interpret $2 + 3 + 4$. The associative property asserts that we may add like this: $2 + (3 + 4) = 2 + 7 = 9$; or like this: $(2 + 3) + 4 = 5 + 4 = 9$. Similar comments apply to multiplication.

4.   Multiplication is **distributive** over addition. That is, $a(b + c) = ab + ac.$

For example, we may evaluate $13(12 + 8)$ either this way: $13(12 + 8) = 13(20) = 260$; or by distributing the multiplication: $13(12 + 8) = 13(12) + 13(8) = 156 + 104 = 260$.

5.   The number **0 is the additive identity** in R, and the number **1 is the multiplicative identity** in R. That is, $a + 0 = a$ **and** $a \cdot 1 = a$.

In simpler terms, when 0 is added to any number, this leaves the number unchanged, and when a number is multiplied by 1, the number is unchanged.

6.   For each element a in R there is a unique **additive inverse** in R, designated by $-a$, and for each $a \neq 0$, there is a unique **multiplicative inverse** in R, designated by a^{-1}. Thus, $a + (-a) = 0$ **and** $a \cdot a^{-1} = 1$.

So the additive inverse of a number is that number which must be added to it to give 0, and the multiplicative inverse of a number is that number by which it must be multiplied to give 1. For example, the additive inverse of 2 is -2, and the additive inverse of $-\frac{2}{3}$ is $\frac{2}{3}$; that is, $2 + (-2) = 0$ and $-\frac{2}{3} + \frac{2}{3} = 0$. The multiplicative inverse of 2 is $\frac{1}{2}$, and the multiplicative inverse of $-\frac{2}{3}$ is $-\frac{3}{2}$; that is, $2 \cdot 2^{-1} = 2 \cdot \frac{1}{2} = 1$ and $\left(-\frac{2}{3}\right) \cdot \left(-\frac{2}{3}\right)^{-1} = \left(-\frac{2}{3}\right)\left(-\frac{3}{2}\right) = 1$.

According to property 6, every real number *except 0* has a multiplicative inverse. That 0 fails to have a multiplicative inverse can be seen by the fact that there is no number whose product with 0 gives 1. (See Problem 22, Exercise Set 1.)

7. The **positive real numbers,** designated by R^+, constitute a subset of R with the following properties:
   a. **If a and b are in R^+, so are $a + b$ and $a \cdot b$.** That is, R^+ is closed with respect to addition and multiplication.
   b. **Every real number falls into exactly one of three distinct categories: It is in R^+, it is zero, or its additive inverse is in R^+.**

Property 7a says that if you add or multiply two positive real numbers, the result is positive. Property 7b divides R into three disjoint classes: (1) the positive real numbers, (2) zero, and (3) the negative real numbers (that is, numbers with additive inverses that are positive).

Before stating the final basic property, it is necessary to give some definitions. You may have wondered why the properties stated so far do not mention subtraction or division. This is because these operations are defined in terms of addition and multiplication.

DEFINITION 1   **The difference between a and b is defined by**

$$a - b = a + (-b)$$

DEFINITION 2   **If $b \neq 0$, the quotient of a by b is defined by**

$$\frac{a}{b} = a \cdot b^{-1}$$

The notation $a \div b$ is also used for the quotient of a by b.

A Word of Caution. Since 0 has no multiplicative inverse, division by 0 is undefined. So **the denominator of a fraction cannot be 0.** Thus, the symbol $\frac{a}{0}$ is not a real number. It is undefined.

DEFINITION 3   **The number a is said to be less than b, written $a < b$, provided $b - a$ is positive. Alternately, we may say b is greater than a, and write $b > a$. If a is either less than or equal to b, we write $a \leq b$. Alternately, we say b is greater than or equal to a, and write $b \geq a$.**

When a number line is directed positively to the right, we can interpret $a < b$ geometrically to mean that a is to the *left* of b. So, for example, $-5 < -2$, $-1 < 0$, and $2 < 4$. In each case you can test to see that Definition 3 is satisfied.

DEFINITION 4   **A subset S of R is said to be bounded above if there is a real number k such that every number in S is less than or equal to k. Such a number k is called an upper bound for S. If no upper bound of S is less than k, then k is said to be the least upper bound of S.**

EXAMPLE I

The set $\{1, 1\frac{1}{2}, 1\frac{3}{4}, 1\frac{7}{8}, 1\frac{15}{16}, \ldots\}$ is bounded above by 4, 5, 100, and many other numbers. But 2 is the least upper bound. ▲

We can now state the final basic property of the real numbers:

8.   Every nonempty subset of R that is bounded above has a least upper bound in R. This is called the **completeness property** of R.

This property is admittedly rather difficult to understand, and in this course it will not be necessary to explore its meaning in depth. But you should be aware that this property guarantees that every point on the number line corresponds to some real number and also that every infinite decimal represents a real number.

The eight basic properties we stated characterize completely the real number system, and for this reason they can be thought of as **axioms** for R. Other number systems possess some of these properties, but the real number system is the only one having all eight.

A host of properties can be derived from the basic ones. We list some of these below for reference.

Further Properties of Real Numbers

1.   The right-hand distributive property: $(a + b)c = ac + bc$.
2.   The extended associative properties: The result is the same regardless of the order of adding or of multiplying any finite collection of real numbers. For example, consider

$$2 + 3 + 4 + 5$$

We may perform this addition in any of the following ways:

$$(2 + 3) + (4 + 5) = 5 + 9 = 14$$
$$[2 + (3 + 4)] + 5 = [2 + 7] + 5 = 9 + 5 = 14$$
$$2 + [3 + (4 + 5)] = 2 + [3 + 9] = 2 + 12 = 14$$

and so on.

3. When 0 is multiplied by any number, the result is 0, that is: $a \cdot 0 = 0$.
4. $-a = (-1)a$
5. $(-a)(b) = -(ab)$
6. $(-a)(-b) = ab$
7. If $ab = 0$, then either $a = 0$ or $b = 0$.

More properties will be given after introducing more definitions.

EXAMPLE 2 |||||||||||||||||||||||||||||||||||||||||||||||||||||||||||||||||||||||||||||||||||||||||||||||||||||||||

Show that $(-a)(b) = -(ab)$.

Note. The instruction "show that" is equivalent to "prove that."

Solution   In the proof we will indicate the justification for each step to the right of the step.

$$a + (-a) = 0 \qquad \text{Definition of additive inverse}$$
$$[a + (-a)] \cdot b = 0 \cdot b \qquad \text{Equality property 5}$$
$$ab + (-a)(b) = 0 \cdot b \qquad \text{Right-hand distributive property}$$
$$ab + (-a)(b) = 0 \qquad \text{Property of 0 (see Problem 22, Exercise Set 1)}$$

The last equation says that $(-a)(b)$ is a number which, when added to ab, gives 0, and since additive inverses are unique (see Problem 23, Exercise Set 1), it follows that $(-a)(b)$ is the additive inverse of ab, that is,

$$(-a)(b) = -(ab) \qquad \blacktriangle$$

EXERCISE SET I

A

In Problems 1 and 2 express the numbers as repeating or terminating decimals.

1. a. $\dfrac{3}{8}$ b. $\dfrac{4}{3}$ c. $-\dfrac{3}{11}$ d. $\dfrac{13}{5}$ e. $\dfrac{16}{27}$

2. a. $\dfrac{13}{16}$ b. $\dfrac{5}{9}$ c. $-\dfrac{45}{8}$ d. $\dfrac{20}{7}$ e. $-\dfrac{12}{37}$

3. Sometimes $\frac{22}{7}$ is used for π. Can this be the exact value? Explain. If not exact, how good is the approximation?

4. Consider the decimal expansion 1.21221222122221.... Does this represent a rational or an irrational number?

5. State which of the following are rational and which are irrational.

   a. -4 b. $\dfrac{2}{\sqrt{25}}$ c. $\sqrt{7}$

   d. $\sqrt[3]{-27}$ e. $\dfrac{4}{5} + \dfrac{2}{3}$

6. Let $x = 2.363636....$ By considering $100x - x$, express x in the form m/n, where m and n are integers.

In Problems 7–9 use a procedure similar to that in Problem 6 to express each number as the ratio m/n of integers.

7. a. $1.666...$ b. $0.111...$

8. a. 0.121212...   **b.** 0.0272727...

9. a. 0.132132132...   **b.** 2.837837837...

10. What can you conclude about $x + y$ and $x \cdot y$ if
- **a.** x and y are rational?
- **b.** x and y are irrational?
- **c.** x is rational and y is irrational?

11. State the basic property of R used in each of the following:
- **a.** $2(3 + 4) = 2 \cdot 3 + 2 \cdot 4$
- **b.** $4 + (-4) = 0$
- **c.** $(3 \cdot 3^{-1}) \cdot 6 = 1 \cdot 6$
- **d.** $1 + 0 = 1$ **e.** $1 \cdot 0 = 0$
- **f.** $(-2 + 3) + 5 = -2 + (3 + 5)$
- **g.** $2 \cdot (3 \cdot 4) = (2 \cdot 3) \cdot 4$
- **h.** $(3 + 2) \cdot 5 = (2 + 3) \cdot 5$
- **i.** $(5 + 7) \cdot 6 = 6 \cdot (5 + 7)$
- **j.** $(2 + 0) \cdot 7 = 2 \cdot 7$

12. a. Give the additive inverse of each of the following:

$$3, -5, 1.2, \frac{4}{3}, -\frac{3}{5}$$

b. Give the multiplicative inverse of each of the following:

$$2, -4, 0.1, \frac{3}{4}, -\frac{5}{3}$$

13. Evaluate each of the following:
- **a.** $3(-4)$ **b.** $(-5)(-2)$ **c.** $-(-3)$
- **d.** 5^{-1} **e.** $2 + [4 + (3 + 5)]$

14. Evaluate each of the following:
- **a.** $1 - 0$ **b.** $0 - 1$ **c.** $\dfrac{0}{1}$
- **d.** $(-1)^{-1}$ **e.** $\left(-\dfrac{2}{3}\right)^{-1}$

15. a. What is the additive inverse of 0?
- **b.** What is the multiplicative inverse of 1?
- **c.** What is the multiplicative inverse of -1?
- **d.** Are there any other "self-inverses"?

16. Does the commutative property hold in R for
- **a.** Subtraction?   **b.** Division?

Justify your answers.

17. Does the associative property hold in R for
- **a.** Subtraction?   **b.** Division?

Justify your answers.

18. Is addition in R distributive over multiplication? Justify your answer.

19. Determine the least upper bound of each of the following sets:
- **a.** $\{3, -1, -10, 5, 4\}$
- **b.** $\left\{-1, -\dfrac{1}{2}, -\dfrac{1}{3}, -\dfrac{1}{4}, \ldots\right\}$
- **c.** $\{0.3, 0.33, 0.333, 0.3333, \ldots\}$
- **d.** $\{x: \ x < 7, x \in R\}$
- **e.** $\left\{\dfrac{n-1}{n}: \ n \in N\right\}$

In Problems 20 and 21 a, b, and c denote real numbers.

20. Prove that if $a = b$, then $a - c = b - c$.

21. Prove that if $ac = bc$ and $c \neq 0$, then $a = b$.

22. State the basic property employed in each step of the following proof that $a \cdot 0 = 0$:

$$a \cdot (0 + 0) = a \cdot 0$$

$$a \cdot 0 + a \cdot 0 = a \cdot 0$$

$$-(a \cdot 0) + (a \cdot 0 + a \cdot 0) = -(a \cdot 0) + (a \cdot 0)$$

$$-(a \cdot 0) + (a \cdot 0 + a \cdot 0) = 0$$

$$[-(a \cdot 0) + a \cdot 0] + a \cdot 0 = 0$$

$$0 + a \cdot 0 = 0$$

$$a \cdot 0 = 0$$

23. The following proof shows that there can be only one additive inverse of a given real number; that is, additive inverses are unique. State the basic property of R used in each step of the proof.*

Let $a \in R$. By basic property 6 we know there is an additive inverse of a, which we designate by $-a$. Suppose b is also an additive inverse of a, so that

$$a + b = 0$$

---

* We included uniqueness in the statement of the basic property of additive inverses. This problem shows that uniqueness need not be assumed but is a consequence of the other properties.

We add $-a$ to both sides and proceed as follows:

$$-a + (a + b) = -a + 0$$
$$-a + (a + b) = -a$$
$$(-a + a) + b = -a$$
$$0 + b = -a$$
$$b = -a$$

So b is the same as $-a$, which shows there is only one additive inverse of a.

24. Use basic properties of R to prove the right-hand distributive law:

$$(a + b)c = ac + bc$$

B

25. a. Prove that 0 is the only additive identity in R.
   Hint. Assume some number z is also an additive identity so that for any $a \in R$, both $a + 0 = a$ and $a + z = a$. Show how it follows that $z = 0$.
   b. Use a procedure similar to that in part **a** to show that 1 is the only multiplicative identity in R.

26. Prove that $1 \neq 0$.
   Hint. Assume $1 = 0$ and for any nonzero number $a \in R$, consider $a \cdot 1$ and $a \cdot 0$ to arrive at a contradiction.

27. Follow a procedure similar to that used in Problem 23 to prove that the multiplicative inverse of a nonzero real number is unique.

28. Show that R is closed with respect to subtraction. Is R closed with respect to division? Show why or why not.

29. Show that $a(b - c) = ab - ac$.
30. Show that $-(a + b) = -a - b$.
31. Show that if $b \neq 0$, then $1/b = b^{-1}$.
32. Show that if $b \neq 0$, then $0/b = 0$.
33. Show that if $a \neq 0$ and $b \neq 0$, then $(ab)^{-1} = a^{-1}b^{-1}$.
34. Use properties of R to prove that for any two real numbers a and b,

$$(a + b)^2 = a^2 + 2ab + b^2$$

Note. a^2 means $a \cdot a$.

35. Use the result of Problem 34 to prove that for any two real numbers a and b,

$$(a - b)^2 = a^2 - 2ab + b^2$$

36. A **lower bound** of a set S of real numbers is a number that is less than or equal to every element of S, and the **greatest lower bound** of S is a lower bound of S that is greater than or equal to all other lower bounds of S. Show that the number m is the greatest lower bound of a set S of real numbers provided the following two conditions are satisfied: (1) $x \geq m$ for all $x \in S$ and (2) if $x \geq k$ for all $x \in S$, then $k \leq m$.

37. Find the greatest lower bound of each of the following sets (see Problem 36):

   a. $\{2, -1, -5, 0, 1\}$ **b.** $\left\{1, \dfrac{1}{2}, \dfrac{1}{3}, \dfrac{1}{4}, \dfrac{1}{5}, \ldots\right\}$

   c. $\{x: \ x > 7, x \in R\}$

   d. $\left\{\dfrac{n + 1}{n}: \ n \in N\right\}$

   e. $\{2 + y: \ y \in R^+\}$

38. The point corresponding to $\sqrt{2}$ on the number line can be located geometrically by constructing a square on the unit interval from 0 to 1 and, with a compass, describing an arc of a circle with center at 0 and with radius equal to the diagonal of the square. The point at which the arc crosses the number line is the desired point.
   a. Explain the basis for this procedure.
   b. How can $\sqrt{5}$ be located in a similar way?
   Hint. For this problem you will need the Pythagorean theorem, which says that in a right triangle with legs a and b and hypotenuse c, $a^2 + b^2 = c^2$.

39. Carry out the details outlined here to show that $\sqrt{2}$ is irrational. Begin by *assuming* $\sqrt{2}$ is rational. Then we can write $\sqrt{2} = m/n$, where m and n have no common factor other than 1. Square both sides of this equation and solve for m^2. Observe that m^2 is an even number, and explain why it follows that m itself must be even. From this conclude that n^2 is even, and also therefore that n is even. Why is this a contradiction? What do you conclude about the original assumption?

ORDER PROPERTIES AND ABSOLUTE VALUE

The following properties of **inequality** can be deduced as consequences of the properties of R given in Section 1.

1. $a > 0$ **if and only if**   **a is positive.**

So, henceforth, the statements "a is positive" and "a is greater than 0" may be used interchangeably.

2.   If $a \neq 0$,   then $a^2 > 0$.

In particular, since $1 \neq 0$, we have $1^2 > 0$. But $1^2 = 1$, so $1 > 0$. Note also that $a^2 > 0$ even if $a < 0$. For example, $(-2)^2 = (-2)(-2) = 4 > 0$.

3.   If $a < b$,   then $a + c < b + c$.

This says that the same number may be added to both sides of an inequality without changing the sense of the inequality (that is, the symbol $<$ is unchanged in direction).

4.   If $a < b$,   then
$$\begin{cases} ac < bc & \text{if} \quad c > 0 \\ ac > bc & \text{if} \quad c < 0 \end{cases}$$

The situation in multiplication is more complicated. Multiplying by a positive number leaves the sense of an inequality unchanged, but multiplying by a negative number reverses the sense.

5.   If a is any real number, then precisely one of the following is true: $a > 0$, $a = 0$, $a < 0$.

This is the property of **trichotomy.**

6.   If $a < b$ and $b < c$,   then $a < c$.

This is the property of **transitivity.**

7.   If $a < b$ and $c < d$,   then $a + c < b + d$.

Note the distinction between this and property 3. It says that as long as both inequalities are of the same type (both "less than" as shown), we may add the respective sides and retain the same sense of inequality.

8.   If $a < b$ and $a \cdot b > 0$,   then $1/a > 1/b$.

Thus, taking reciprocals reverses the sense of the inequality *provided* both numbers are positive or both numbers are negative.

These properties provide the basis for working with more complicated inequalities in algebra, as we will see in Chapter 2. There are times when it is useful to write a combined inequality of the form

$$a < b < c$$

This means that both $a < b$ *and* $b < c$. It is convenient sometimes to read from the middle and say "b is less than c and greater than a."

Sets of the form $\{x: \ a < x < b\}$* and $\{x: \ a \le x \le b\}$ occur frequently, and we give them the following special names and symbols:

Set	Name	Symbol
$\{x: \ a < x < b\}$	**Open interval**	(a, b)
$\{x: \ a \le x \le b\}$	**Closed interval**	$[a, b]$

Figure 2 shows how we illustrate open and closed intervals graphically, using the examples $(0, 2)$ and $[4, 6]$.

Open interval $(0, 2)$ Closed interval $[4, 6]$

FIGURE 2

A set of the form $\{x: \ a < x \le b\}$ is designated $(a, b]$ and is called **half-open** (or **half-closed**). Similarly, the half-open interval $\{x: \ a \le x < b\}$ is designated $[a, b)$. Note that in all cases *a parenthesis indicates the endpoint is not included, whereas a bracket indicates it is.*

We can also use interval notation for unbounded sets by introducing the infinity symbol ∞, read "infinity." It is important to realize that ∞ is not a number. It can be interpreted as meaning "beyond all bound." There are five types of unbounded intervals, as follows:

Set	Symbol	Graph
$\{x: \ x > a\}$	(a, ∞)	
$\{x: \ x \ge a\}$	$[a, \infty)$	
$\{x: \ x < a\}$	$(-\infty, a)$	
$\{x: \ x \le a\}$	$(-\infty, a]$	
All of R	$(-\infty, \infty)$	

* We will understand x to be real in all set designations of the form $\{x: \ _____\}$ unless otherwise specified.

We next introduce the concept of the **absolute value** of a number, which is, roughly speaking, the magnitude of the number without regard to sign. The precise definition follows.

DEFINITION 5   **The absolute value of a real number a is equal to a itself if $a \geq 0$ and is equal to $-a$ if $a < 0$. The absolute value of a is designated by $|a|$.**

From this we see that $|a|$ is always nonnegative. For example, $|2| = 2$ and $|-3| = -(-3) = 3$. Note that when a is negative, $|a| = -a$, and since a itself is negative, $-a$ is positive (despite appearances).

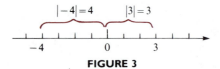

FIGURE 3

Geometrically, the absolute value of a number a can be interpreted as the distance between a and 0 on a number line, as illustrated in Figure 3.

From this geometric interpretation of absolute value as the distance from the origin, we can see that for $a > 0$, the inequality $|x| < a$ will be satisfied provided x is less than a units from the origin, on either side. That is,

$$|x| < a \qquad \text{is equivalent to} \qquad -a < x < a$$

Similarly, $|x| > a$ provided x is more than a units away from the origin, on either side:

$$|x| > a \qquad \text{is equivalent to} \qquad x > a \quad or \quad x < -a$$

These relationships are illustrated in Figure 4. For example, an equivalent way of writing the set $S = \{x: \ |x| < 2\}$ is $S = \{x: \ -2 < x < 2\}$. Similarly, the set $T = \{x: \ |x| > 2\}$ can be written $T = \{x: \ x < -2 \quad or \quad x > 2\}$.

FIGURE 4

Sets that consist of two or more separate parts, called **disjoint** subsets, such as the set T above, can be written more compactly using the notation for the **union** of two sets. The union of two sets A and B is designated $A \cup B$ and

is defined as

$$A \cup B = \{x: \quad x \in A \quad or \quad x \in B\}$$

Consider the set $T = \{x: \quad x < -2 \quad or \quad x > 2\}$ again. Using the idea of the union and interval notation, we can write $T = (-\infty, -2) \cup (2, \infty)$. It should be noted that the definition of the union of sets A and B in general does not require that they be disjoint. A point x is in $A \cup B$ if it is in either A or B, and so if x is in both A and B, it satisfies this requirement. The **intersection** of two sets, written $A \cap B$, is defined by

$$A \cap B = \{x: \quad x \in A \quad and \quad x \in B\}$$

You might wish to verify that the set S above, defined as $\{x: \quad |x| < 2\}$, can be written as

$$S = (-\infty, 2) \cap (-2, \infty)$$

EXERCISE SET 2

A

In Problems 1 and 2 replace each question mark with an appropriate strict inequality sign ($>$ or $<$).

1. a. $0 ? -2$ **b.** $-5 ? -2$ **c.** $2 ? -100$
    d. $-1 ? -10$ **e.** $-20 ? 0.1$

2. a. $\dfrac{3}{4} ? \dfrac{5}{8}$ **b.** $-\dfrac{2}{3} ? -\dfrac{1}{2}$ **c.** $\pi ? \dfrac{22}{7}$

    d. $-\sqrt{3} ? -1.7$ **e.** $2.8 ? \dfrac{17}{6}$

3. Use Definition 3 to show that each of the following is true.
   a. $2 < 5$ **b.** $-3 < -1$ **c.** $-4 < 0$
   d. $7 > 3$ **e.** $-2 > -4$

In Problems 4 and 5 write each statement using inequality symbols.

4. a. x is less than 3.
   b. x is greater than or equal to 4.
   c. x is negative.
   d. x is nonnegative.
   e. x is not positive.

5. a. x is greater than 1 and less than 3.
   b. x is less than or equal to 2 and greater than -1.
   c. x is at least 2 but does not exceed 6.
   d. x is nonnegative and less than 5.
   e. x is negative but not less than -6.

6. Given that $a < b$, replace each question mark with the correct inequality sign.

   a. $-3a ? -3b$ **b.** $\dfrac{a}{4} ? \dfrac{b}{4}$

   c. $a + 2 ? b + 3$ **d.** $a - 3 ? b - 1$

7. In each of the following cases place the correct inequality sign between $1/a$ and $1/b$.
   a. $0 < b < a$ **b.** $a < b < 0$
   c. $b < 0 < a$ **d.** $a < 0 < b$

8. Write each of the following sets in interval notation and show it on a number line.
   a. $\{x: \quad -1 < x < 2\}$ **b.** $\{x: \quad 1 \le x \le 4\}$
   c. $\{x: \quad 0 < x \le 2\}$ **d.** $\{x: \quad x > 2\}$
   e. $\{x: \quad x \le 0\}$

9. Write each of the following intervals in set notation and show it on a number line.
   a. $[2, 5]$ **b.** $(-2, 3]$ **c.** $(3, 5)$
   d. $(-\infty, 2)$ **e.** $[1, \infty)$

10. Write each of the following without absolute value signs.
   a. $|-3|$
   b. $|-x|,\quad x < 0$
   c. $|x - y|,\quad x < y$
   d. $|3.14 - \pi|$
   e. $|x^2|$

11. Write the following inequalities without using absolute value signs.
   a. $|x| < 3$
   b. $|y| > 2$
   c. $|t| \le 1$
   d. $|w| \ge 3$

12. Write each of the following in interval notation and show it on a number line.
   a. $\{x:\ |x| \le 2\}$
   b. $\{x:\ |x| > 2\}$
   c. $\{x:\ |x| \ge 1\}$
   d. $\{x:\ |x| < 3\}$

13. Write each of the following in set notation using absolute values.
   a. $(-3, 3)$
   b. $[-4, 4]$
   c. $(-\infty, -2) \cup (2, \infty)$
   d. $(-\infty, -3] \cup [3, \infty)$

16. Prove property 3.
17. Prove property 4.
18. Prove property 6.
19. Prove property 7.
   Hint. Use properties 3 and 6.
20. Prove that if $a > 0$, then $a^{-1} > 0$.
   Hint. Consider the product of a and a^{-1} and suppose a^{-1} is not positive.
21. Use the result of Problem 20 to give a rule for dividing both sides of an inequality by the same nonzero number, and justify your result.
22. Prove inequality property 8.
   Hint. Multiply both sides of $a < b$ by $a^{-1}b^{-1}$ and explain why this is valid based on Problem 20.
23. Prove or disprove:
   If $a < b$ and $c < d$, then $a - c < b - d$.
24. Prove that for any $a \in R$,
   a. $|-a| = |a|$
   b. $-|a| \le a \le |a|$
25. Prove that for any two real numbers a and b, $|ab| = |a||b|$.
   Hint. Consider various cases.

B

Problems 14–19 refer to the properties of inequalities given at the beginning of this section.

14. Prove property 1.
15. Prove property 2.

3   THE ARITHMETIC OF RATIONAL NUMBERS

All of the familiar rules of integer arithmetic can be derived from the basic properties of real numbers, but we will omit the details and assume that the results are well known. It is instructive, however, to review the arithmetic of rational numbers, since this is the basis for much subsequent work in algebra.

Recall that a/b and c/d are rational numbers provided a, b, c, and d are integers, with $b \ne 0$ and $d \ne 0$. Their product is found as follows.

$$\frac{a}{b} \cdot \frac{c}{d} = (ab^{-1})(cd^{-1}) = (ac)(b^{-1}d^{-1})$$

$$= (ac)(bd)^{-1} = \frac{ac}{bd}$$

You will be asked to supply reasons for these steps in Problem 33, Exercise Set 3. The result,

$$\frac{a}{b} \cdot \frac{c}{d} = \frac{ac}{bd} \tag{1}$$

says that "the product of two fractions is the product of their numerators over the product of their denominators."

An important special case of (1) occurs when $d = c$, where $c \neq 0$. Then, since $cc^{-1} = 1$ and also $cc^{-1} = c/c$, we obtain

$$\frac{a}{b} = \frac{a}{b} \cdot 1 = \frac{a}{b} \cdot \frac{c}{c} = \frac{ac}{bc}$$

Or, stated in reverse,

$$\frac{ac}{bc} = \frac{a}{b} \tag{2}$$

This is sometimes called the **fundamental property of fractions.**

In a product such as $a \cdot b$, a and b are called *factors* of the product. Using this terminology, the fundamental property says that the numerator and denominator can each be multiplied or divided by the same nonzero factor, called a **common factor.** When the only common integer factor is 1 or -1, we say the fraction is in **lowest terms.** *Fractions occurring as final answers should always be reduced to lowest terms* by making use of the fundamental property.

A Word of Caution. The fundamental property (2) is sometimes called the **cancellation law,** since we can cancel like this:

$$\frac{a\cancel{c}}{b\cancel{c}} = \frac{a}{b} \qquad \text{Correct}$$

Observe carefully, however, that this is valid only when the numerator and denominator are **products** containing a common factor. Students are sometimes tempted to "cancel" when **sums or differences** are involved, like this:

$$\frac{a + \cancel{c}}{b + \cancel{c}} = \frac{a}{b} \qquad \text{Incorrect}$$

That this is not true in general can be shown by an example, such as

$$\frac{2 + 3}{5 + 3} = \frac{5}{8} \quad not \quad \frac{2}{5}$$

So you must resist this temptation.

Consider now the sum of two rational numbers. If they have the same denominator, the result is easy: add their numerators and write their sum over the common denominator. In symbols,

$$\frac{a}{c} + \frac{b}{c} = \frac{a + b}{c} \tag{3}$$

This is justified as follows:

$$\frac{a}{c} + \frac{b}{c} = a \cdot c^{-1} + b \cdot c^{-1} \qquad \text{Definition of division}$$

$$= (a + b) \cdot c^{-1} \qquad \text{Right-hand distributive property}$$

$$= \frac{a + b}{c} \qquad \text{Definition of division}$$

Now suppose two fractions have different denominators. The first step is to write them in equivalent forms in which their denominators are the same, using the fundamental property (2). The result can be shown as follows:

$$\frac{a}{b} + \frac{c}{d} = \frac{ad}{bd} + \frac{bc}{bd} \qquad \text{by (2)}$$

$$= \frac{ad + bc}{bd} \qquad \text{by (3)}$$

Thus,

$$\frac{a}{b} + \frac{c}{d} = \frac{ad + bc}{bd} \tag{4}$$

While this always yields the correct result, it is better to learn the procedure than to apply the formula directly. That is, you should first write each fraction with the same denominator and then add as in (3). It is best to write the fractions with the smallest possible common denominator, called the **least common denominator (LCD)**. The next two examples illustrate how to do this.

EXAMPLE 3 ||||||||||||||||||||||||||||||||||||||||||||||||||||||||||||||||||||||||||||||||||||||||||||||||||||||||||||||||||||||

Find the following sums.

a. $\dfrac{3}{4} + \dfrac{-7}{6}$ **b.** $\dfrac{-5}{36} + \dfrac{3}{28}$

Solution   **a.** We mentally determine that 12 is the smallest number having 4 and 6 as factors. So the LCD is 12.

$$\frac{3}{4} + \frac{-7}{6} = \frac{3 \cdot 3}{4 \cdot 3} + \frac{(-7) \cdot 2}{6 \cdot 2} = \frac{9}{12} + \frac{-14}{12} = \frac{9 + (-14)}{12} = \frac{-5}{12}$$

b. Since $36 = 4 \cdot 9$ and $28 = 4 \cdot 7$, the LCD $= 4 \cdot 7 \cdot 9$. Thus,

$$\frac{-5}{36} + \frac{3}{28} = \frac{(-5) \cdot 7}{(4 \cdot 9) \cdot 7} + \frac{3 \cdot 9}{(4 \cdot 7) \cdot 9} = \frac{-35 + 27}{4 \cdot 7 \cdot 9} = \frac{-\overset{2}{8}}{4 \cdot 7 \cdot 9} = \frac{-2}{63}$$

Notice that in the final step we used the fundamental property to divide the numerator and denominator by 4.   ▲

For larger denominators it is best to write each denominator as a product of **primes.** A natural number greater than 1 is a prime if it cannot be written as a product of natural numbers other than itself and 1. Examples are 2, 3, 5, 7, 11, ... Every natural number $n > 1$ that is not a prime is called **composite.** A composite number can be written uniquely (except for the order of the factors) as a product of primes. This result is known as the **Fundamental Theorem of Arithmetic.** The resulting product is called the **prime factorization of the number.** For example, the prime factorization of 72 is $2 \cdot 2 \cdot 2 \cdot 3 \cdot 3 = 2^3 \cdot 3^2$. When two (or more) denominators have been expressed in their prime factorizations, the LCD is found by writing the product of the highest powers of all of the primes occurring in either factorization. The next example illustrates this.

EXAMPLE 4   ꟷꟷꟷꟷꟷꟷꟷꟷꟷꟷꟷꟷꟷꟷꟷꟷꟷꟷꟷꟷꟷꟷꟷꟷꟷꟷꟷꟷꟷꟷꟷꟷꟷꟷꟷꟷ

Find the sum:

$$\tfrac{7}{780} + \tfrac{11}{504}$$

Solution   To obtain the prime factorizations of the denominators, we use repeated divisions by successive primes, starting with 2, as shown below.

$$
\begin{array}{r|r} \quad\quad 2 & 780 \\ 2 & 390 \\ 3 & 195 \\ 5 & 65 \\ & 13 \end{array}
\qquad
\begin{array}{r|r} 2 & 504 \\ 2 & 252 \\ 2 & 126 \\ 3 & 63 \\ 3 & 21 \\ & 7 \end{array}
$$

Thus, $780 = 2^2 \cdot 3 \cdot 5 \cdot 13$ and $504 = 2^3 \cdot 3^2 \cdot 7$. So the LCD $= 2^3 \cdot 3^2 \cdot 5 \cdot 7 \cdot 13$. To bring each denominator up to this LCD, we multiply by the primes in the LCD missing from that denominator.

$$
\begin{aligned}
\frac{7}{780} + \frac{11}{504} &= \frac{7 \cdot (2 \cdot 3 \cdot 7)}{780 \cdot (2 \cdot 3 \cdot 7)} + \frac{11 \cdot (5 \cdot 13)}{504 \cdot (5 \cdot 13)} \\[2mm]
&= \frac{294 + 715}{32{,}760} \\[2mm]
&= \frac{1{,}009}{32{,}760}
\end{aligned}
$$

▲

The technique of Example 4 will always work, but with small denominators the complete prime factorization is usually not necessary, as illustrated in Example 3.

Remark. If m and n are natural numbers, the smallest number having m and n as factors is called their **least common multiple (LCM).** We see then that the LCD of two fractions is the same as the least common multiple of their denominators.

In adding or subtracting fractions, the handling of signs can often be made easier by the following considerations. For $b \neq 0$, we have, by (4),

$$\frac{a}{b} + \frac{-a}{b} = \frac{a + (-a)}{b} = \frac{0}{b} = 0(b^{-1}) = 0$$

It follows by the uniqueness of the additive inverse that $(-a)/b$ is the additive inverse of a/b. That is,

$$-\frac{a}{b} = \frac{-a}{b}$$

Furthermore, by the fundamental property,

$$\frac{a}{-b} = \frac{a(-1)}{-b(-1)} = \frac{-(a \cdot 1)}{b \cdot 1} = \frac{-a}{b}$$

and

$$\frac{-a}{-b} = \frac{a(-1)}{b(-1)} = \frac{a}{b}$$

Thus we have the chain of equalities

$$-\frac{a}{b} = \frac{-a}{b} = \frac{a}{-b} = -\frac{-a}{-b}$$

So we can write the difference of two fractions in either of the forms

$$\frac{a}{c} - \frac{b}{c} = \frac{a}{c} + \frac{-b}{c} \quad \text{or} \quad \frac{a}{c} - \frac{b}{c} = \frac{a}{c} + \frac{b}{-c} \tag{5}$$

We illustrate this in the next example.

EXAMPLE 5 ||||||||||||||||||||||||||||||||||||||||||||||||||||||||||||||||||||||||||||||||||||||||||||||||||||||||||||||||

Perform the indicated operations and reduce answers to lowest terms.

    a. $-\dfrac{7}{15} - \dfrac{5}{-12}$ **b.** $\dfrac{2}{3}\left(\dfrac{4}{5} - \dfrac{-3}{4} + \dfrac{1}{-2}\right)$

Solution   **a.**   Since $15 = 3 \cdot 5$ and $12 = 3 \cdot 4$, the LCD is $3 \cdot 4 \cdot 5 = 60$. So we have, by (5),

$$-\frac{7}{15} - \frac{5}{-12} = \frac{-7}{15} + \frac{5}{12} = \frac{-28 + 25}{60}$$

$$= \frac{-3}{60} = \frac{-1}{20}$$

b. We first combine the fractions inside the parentheses and then perform the multiplication.

$$\frac{2}{3}\left(\frac{4}{5} - \frac{-3}{4} + \frac{1}{-2}\right) = \frac{2}{3}\left(\frac{4}{5} + \frac{3}{4} + \frac{-1}{2}\right)$$

$$= \frac{2}{3}\left(\frac{16 + 15 - 10}{20}\right) = \frac{2}{3}\left(\frac{21}{20}\right) = \frac{7}{10} \qquad \blacktriangle$$

To find a formula for the quotient of two rational numbers, we first observe that if $c \neq 0$ and $d \neq 0$, then

$$\left(\frac{c}{d}\right)\left(\frac{d}{c}\right) = \frac{cd}{cd} = 1$$

so that $(c/d)^{-1} = d/c$. Thus, by Definition 2,

$$\frac{a}{b} \div \frac{c}{d} = \frac{\dfrac{a}{b}}{\dfrac{c}{d}} = \left(\frac{a}{b}\right)\left(\frac{c}{d}\right)^{-1} = \frac{a}{b} \cdot \frac{d}{c} = \frac{ad}{bc}$$

The result,

$$\frac{\dfrac{a}{b}}{\dfrac{c}{d}} = \frac{ad}{bc} \qquad (6)$$

is the familiar rule: *To divide fractions, invert the denominator and multiply.*

A fraction in which the numerator or denominator, or both, contains one or more fractions, is called a **complex fraction.** These fractions can be simplified based on the rules we have developed, as shown in the next three examples.

EXAMPLE 6 |||||||||||||||||||||||||||||||||||||||||||||||||||||||||||||||||||||||||||||||||||||||||||||||||||||||||||||

Simplify: $\dfrac{\frac{2}{3} + \frac{3}{4}}{\frac{5}{6} + \frac{1}{2}}$

Solution In general, the best approach to this sort of problem is first to multiply the numerator and denominator of the main fraction by the LCD of all **minor denominators.** This is justified by rule (2). In this case, the LCD is 12. So we have

$$\frac{(\frac{2}{3} + \frac{3}{4})}{(\frac{5}{6} + \frac{1}{2})} \cdot \frac{12}{12} = \frac{8 + 9}{10 + 6} = \frac{17}{16}$$

The omitted steps can be done mentally. It is important to note that when multiplying the numerator and denominator by 12, *every* term must be multiplied. Notice, too, that in multiplying we have divided out common factors *before* multiplication. For example, the operations done mentally on the numerator are shown in more detail as

$$\left(\frac{2}{3}+\frac{3}{4}\right)12 = \frac{2}{3}\cdot 12 + \frac{3}{4}\cdot 12$$

$$= \frac{2}{3}\cdot\frac{\overset{4}{\cancel{12}}}{1} + \frac{3}{4}\cdot\frac{\overset{3}{\cancel{12}}}{1} = 8 + 9$$

As an alternative way of doing this problem, we could perform the additions on the numerator and denominator separately and then use rule (6). Doing it this way we get

$$\frac{\frac{2}{3}+\frac{3}{4}}{\frac{5}{6}+\frac{1}{2}} = \frac{\frac{8}{12}+\frac{9}{12}}{\frac{5}{6}+\frac{3}{6}} = \frac{\frac{17}{12}}{\frac{8}{6}} = \frac{17}{\cancel{12}_2}\cdot\frac{\overset{1}{\cancel{6}}}{8} = \frac{17}{16}$$

The first method is usually easier, but there are exceptions. ▲

EXAMPLE 7 ‖‖‖

Simplify: $\dfrac{\frac{2}{3}-\frac{1}{4}\left(\frac{5}{6}\right)}{\frac{5}{12}-\frac{-3}{2}}$

Solution The minor denominators are 3, 24, 12, and 2, so that the desired LCD is 24.

$$\frac{\frac{2}{3}-\frac{1}{4}\left(\frac{5}{6}\right)}{\frac{5}{12}-\frac{-3}{2}}\cdot\frac{24}{24} = \frac{16-5}{10+36} = \frac{11}{46}$$ ▲

$16 - 6.20 \quad 79$

EXAMPLE 8 ‖‖‖

Simplify: $\dfrac{2}{1-\dfrac{1}{2-\frac{3}{4}}}$

Solution We do this in stages, beginning with the complex fraction in the denominator.

$$\frac{2}{1-\dfrac{1}{2-\frac{3}{4}}} = \frac{2}{1-\dfrac{1}{2-\frac{3}{4}}\cdot\frac{4}{4}} = \frac{2}{1-\dfrac{4}{8-3}}$$

$$= \frac{2}{1-\frac{4}{5}}\cdot\frac{5}{5} = \frac{10}{5-4} = \frac{10}{1} = 10$$ ▲

EXERCISE SET 3

A

In Problems 1–12 perform the indicated operations and reduce answers to lowest terms.

1. a. $\dfrac{2}{3} + \dfrac{1}{2}$ **b.** $\dfrac{7}{10} - \dfrac{3}{8}$

2. a. $\dfrac{7}{8} + \dfrac{5}{12}$ **b.** $\dfrac{5}{6} - \dfrac{3}{4}$

3. a. $\dfrac{3}{8} - \dfrac{-5}{12}$ **b.** $-\dfrac{7}{12} + \dfrac{5}{-9}$

4. a. $3 - \dfrac{-2}{-5}$ **b.** $\dfrac{4}{-3} - 1$

5. a. $\dfrac{3}{4} - \dfrac{2}{3} + \dfrac{1}{2}$ **b.** $\dfrac{5}{6} - \dfrac{4}{3} + \dfrac{7}{12}$

6. a. $2 - \dfrac{-5}{6} - \dfrac{2}{3}$ **b.** $\dfrac{5}{4} - \dfrac{1}{-2} - 1$

7. a. $\dfrac{3}{4}\left(-\dfrac{8}{9}\right)$ **b.** $\left(-\dfrac{7}{3}\right)\left(-\dfrac{5}{14}\right)$

8. a. $(-4)\left(\dfrac{7}{12}\right)$ **b.** $\left(-\dfrac{7}{16}\right)(-2)$

9. a. $\dfrac{2}{5}\left(\dfrac{5}{8} - \dfrac{3}{4}\right)$ **b.** $\dfrac{3}{11}\left(-\dfrac{5}{6} - \dfrac{-3}{8}\right)$

10. a. $\left(-\dfrac{5}{6}\right)\left(\dfrac{3}{4} - \dfrac{-4}{15}\right)$ **b.** $\left(-\dfrac{3}{4}\right)\left(\dfrac{3}{2} - \dfrac{4}{5}\right)$

11. a. $\dfrac{8}{27} - \dfrac{7}{45} - \dfrac{5}{36}$ **b.** $\dfrac{7}{120} - \dfrac{9}{56} + \dfrac{11}{96}$

12. a. $\dfrac{3}{140} + \dfrac{5}{126}$ **b.** $\dfrac{7}{540} - \dfrac{17}{324}$

In Problems 13–22 express each complex fraction as a simple fraction in lowest terms.

13. $\dfrac{\frac{2}{3} + \frac{5}{6}}{\frac{3}{4} - \frac{7}{12}}$ **14.** $\dfrac{\frac{3}{8} - \frac{2}{3}}{\frac{5}{12} - \frac{1}{6}}$

15. $\dfrac{\frac{5}{8} - 2}{\frac{2}{3} + 1}$ **16.** $\dfrac{\frac{4}{15} - \frac{3}{5}}{1 - \frac{7}{3}}$

17. $\dfrac{\frac{3}{2} + \left(-\frac{4}{3}\right)}{-\frac{5}{6} - 1}$ **18.** $\dfrac{\frac{3}{5} - \frac{-9}{4}}{\frac{-7}{2} + \frac{6}{5}}$

19. $\dfrac{\frac{5}{6} + \frac{1}{2}}{2 - \left(\frac{2}{3}\right)\left(\frac{1}{4}\right)}$

20. $\dfrac{\frac{3}{8} - \frac{3}{2}\left(\frac{5}{6}\right)}{\left(\frac{2}{3}\right)\left(\frac{9}{4}\right) - 1}$

21. $2 - \dfrac{3}{2 - \frac{1}{2}}$

22. $\dfrac{1}{\dfrac{1}{2 - \frac{1}{3}} - \dfrac{2}{5}}$

23. If a, b, and c are natural numbers such that $a = bc$, then both b and c are said to be **divisors** of a. The **greatest common divisor (GCD)** of two natural numbers is the largest natural number that is a divisor of both numbers. Find the GCD of each of the following pairs:

    a. 42, 70   **b.** 52, 78   **c.** 72, 108

    d. 320, 288   **e.** 396, 594

B

24. A definition of subtraction that is valid for the natural numbers N is as follows: "If a and b are natural numbers, then $a - b$ equals that natural number c, if it exists, which when added to b gives a." Show that when a, b, and c are elements of R, this definition is equivalent to Definition 1.

25. Prove each of the following:

    a. The set N is not closed with respect to subtraction, but J is.

    b. The set J is not closed with respect to division by numbers other than 0, but Q is. (See page 3 for the definitions of N, J, and Q.)

26. A definition of division applicable to the integers J is as follows: "If a and b are integers with $b \neq 0$, then $a \div b$ equals that integer c, if it exists, which when multiplied by b gives a." Show that when a, b, and c are elements of R, this definition is equivalent to Definition 2.

27. Suppose the condition $b \neq 0$ is omitted from the definition in Problem 26. Prove the following:

    a. If $a \neq 0$, $\frac{a}{0}$ does not exist because there is *no* integer c satisfying the definition.

b. $\frac{0}{0}$ is not defined because *every* integer c satisfies the definition, so that there is no single correct answer.

In Problems 28–32 perform the indicated operations.

28. a. $3(-\frac{2}{5})(-\frac{1}{4}) + \frac{5}{8}(-\frac{2}{3} + 6)$
b. $-(-\frac{2}{3})(-\frac{7}{8}) - 2(\frac{5}{-6} + \frac{-3}{8}) - 1$

29. a. $\frac{5}{32} - \frac{3}{5}(\frac{7}{-8}) + \frac{2}{3}(3 - \frac{8}{5})$
b. $1 - \frac{9}{14} + \frac{2}{3}(\frac{-5}{7}) - (-\frac{2}{7})(\frac{5}{12})$

30. a. $-\frac{5}{16} - \frac{3}{4}(\frac{2}{3} - \frac{5}{6}) + \frac{5}{2}(-\frac{3}{8})$
b. $(-\frac{3}{5})(\frac{-7}{8} - \frac{5}{6}) - 2 + \frac{7}{12}(-\frac{4}{5} + \frac{3}{2})$

31. $\dfrac{\frac{10}{9}(\frac{1}{3}) - \frac{\frac{25}{81}}{\frac{13}{3}}}{2(\frac{5}{9}) - 1}$

32. $\dfrac{-\dfrac{3}{2} - \left(\dfrac{-27}{8} - 1\right) \cdot \dfrac{1}{\frac{21}{4}}}{\frac{9}{4}}$

33. Justify each step of the derivation of rule (1) for the product of two rational numbers.

4 INTEGRAL EXPONENTS

Positive integral exponents are used as a shorthand for repeated multiplication of a number by itself. For example, $2^3 = 2 \cdot 2 \cdot 2 = 8$ and $(-3)^2 = (-3)(-3) = 9$. In general, if a is any real number and n is a positive integer, then

$$a^n = \underbrace{a \cdot a \cdot a \cdots \cdot a}_{n \text{ factors}}$$

In particular $a^1 = a$. The number a is called the **base** and n the **exponent.**

A Word of Caution. The exponent applies only to the base and not to any coefficient of it. For example,

$$2x^2 = 2(x^2) \quad not \quad (2x)^2, \text{ which means } (2x)(2x) = 4x^2$$

and

$$-3^2 = -(3^2) = -9 \quad not \quad (-3)^2, \text{ which means } (-3)(-3) = 9$$

To multiply two numbers with the same base, we add their exponents, as the following calculation shows.

$$a^m \cdot a^n = \overbrace{(a \cdot a \cdots \cdot a)}^{m \text{ factors}}\overbrace{(a \cdot a \cdots \cdot a)}^{n \text{ factors}}$$

$$= \overbrace{a \cdot a \cdot a \cdot a \cdots \cdot a}^{m + n \text{ factors}} = a^{m+n}$$

For example,

$$2^3 \cdot 2^5 = 2^8$$

For division of numbers with the same nonzero base and different exponents, we subtract the smaller exponent from the larger, but the result depends on whether the numerator or denominator has the larger exponent. For $m > n$,

$$\frac{a^m}{a^n} = \frac{\overbrace{a \cdot a \cdots a \cdot a \cdot a \cdots a}^{m - n \text{ factors}}}{a \cdot a \cdots a} = a^{m-n} \qquad (a \neq 0)$$

whereas, if $m < n$,

$$\frac{a^m}{a^n} = \frac{a \cdot a \cdots a}{\underbrace{a \cdot a \cdots a \cdot a \cdot a \cdots a}_{n - m \text{ factors}}} = \frac{1}{a^{n-m}} \qquad (a \neq 0)$$

For example,

$$\frac{2^5}{2^3} = 2^2 \quad \text{and} \quad \frac{2^3}{2^5} = \frac{1}{2^2}$$

Of course, if the exponents are the same, the quotient is 1. That is,

$$\frac{a^m}{a^m} = 1 \qquad (a \neq 0)$$

These results are summarized as follows.

$$a^m \cdot a^n = a^{m+n} \tag{7}$$

$$\frac{a^m}{a^n} = \begin{cases} a^{m-n} & \text{if} \quad m > n \\ \dfrac{1}{a^{n-m}} & \text{if} \quad m < n \\ 1 & \text{if} \quad m = n \end{cases} \tag{8}$$

Rule (7) can be used to obtain the formula $(a^m)^n = a^{mn}$. To illustrate, consider $(2^3)^4$. We have, by definition of the outer exponent,

$$(2^3)^4 = 2^3 \cdot 2^3 \cdot 2^3 \cdot 2^3$$

and by extending rule (7),

$$2^3 \cdot 2^3 \cdot 2^3 \cdot 2^3 = 2^{12}$$

So, $(2^3)^4 = 2^{12}$. More generally,

$$(a^m)^n = \overbrace{a^m \cdot a^m \cdots a^m}^{n \text{ factors}} = a^{mn} \tag{9}$$

since the sum of n m's is $nm = mn$.

Another property of exponents is that $(a \cdot b)^n = a^n b^n$. This follows from the definition and from commutativity and associativity:

$$\mathbf{(ab)^n} = (ab)(ab)(ab) \cdots (ab)$$
$$= (a \cdot a \cdot a \cdots \cdot a)(b \cdot b \cdot b \cdots \cdot b) = \mathbf{a^n b^n} \tag{10}$$

In a similar way, for $b \neq 0$, we have

$$\left(\frac{a}{b}\right)^n = \frac{a}{b} \cdot \frac{a}{b} \cdots \frac{a}{b} = \frac{a^n}{b^n} \tag{11}$$

This would be about all we could say concerning exponents if consideration were limited to exponents that are positive integers. Their usefulness is greatly extended, however, by giving meaning to negative and zero exponents, and even more by permitting rational and, finally, irrational exponents. We will do this in stages, deferring irrational exponents to Chapter 6.

Recall that we have used the symbol a^{-1} to mean the unique multiplicative inverse of the nonzero number a, and by Definition 2 it follows that

$$\frac{1}{a} = 1 \cdot a^{-1} = a^{-1}$$

We now use this relationship to define a to the power -1. That is, we now can interpret the -1 in the equation $a^{-1} = 1/a$ as an exponent. For consistency with rule (9), we then have for n a positive integer and $a \neq 0$,

$$a^{-n} = a^{(-1)n} = (a^{-1})^n = \left(\frac{1}{a}\right)^n = \underbrace{\frac{1}{a} \cdot \frac{1}{a} \cdots \frac{1}{a}}_{n \text{ factors}} = \frac{1}{a^n}$$

so

$$a^{-n} = \frac{1}{a^n}, \qquad a \neq 0$$

To see how we might reasonably define a^0, consider a^m/a^m with $a \neq 0$. This clearly equals 1. However, if we extend the first part of rule (8) to $m = n$, we obtain $a^m/a^m = a^{m-m} = a^0$. So we are led to the definition

$$a^0 = 1, \qquad a \neq 0$$

It can now be shown that rules (7)–(11) remain true with m and n positive, negative, or zero integral exponents, so long as we avoid division by zero or raising zero to a nonpositive power. With these restrictions understood, we summarize these rules below. They are often referred to as **laws of exponents.**

Laws of Exponents

1. $a^m \cdot a^n = a^{m+n}$

2. $\dfrac{a^m}{a^n} = a^{m-n}$

3. $(a^m)^n = a^{mn}$

4. $(ab)^n = a^n b^n$

5. $\left(\dfrac{a}{b}\right)^n = \dfrac{a^n}{b^n}$

Note that because zero and negative exponents are permitted, the three forms in the quotient rule (8) can be combined into one.

A special case involving a negative power should be mentioned. Consider $(a/b)^{-n}$. Since $(a/b)(b/a) = ab/ab = 1$, it follows that $(a/b)^{-1} = b/a$. Thus,

$$\left(\frac{a}{b}\right)^{-n} = \left[\left(\frac{a}{b}\right)^{-1}\right]^n = \left(\frac{b}{a}\right)^n$$

This result is easy to remember and should be used whenever a fraction is raised to a negative power. So rather than writing, for example,

$$\left(\frac{2}{3}\right)^{-2} = \frac{1}{(\frac{2}{3})^2} = \frac{1}{\frac{4}{9}} = \frac{9}{4}$$

we write immediately

$$\left(\frac{2}{3}\right)^{-2} = \left(\frac{3}{2}\right)^2 = \frac{9}{4}$$

The following examples illustrate how the laws of exponents can be used to simplify certain complex expressions. In all cases answers are written with positive exponents only. We assume that all letters represent numbers different from zero whenever they occur in a denominator or are raised to a nonpositive power.

EXAMPLE 9

Simplify:

a. $(3)^{-2}$ b. $\left(\dfrac{2}{3}\right)^0$ c. $\left(\dfrac{1}{2}\right)^{-3} \cdot 4^{-2}$ d. $\dfrac{2^4 \cdot 3^{-2}}{2^2 \cdot 3^{-5}}$ e. $(4x^3)^2$

f. $\dfrac{-2^2}{(-4)^2}$

Solution   a. $\dfrac{1}{3^2} = \dfrac{1}{9}$ b. 1 c. $2^3 \cdot \dfrac{1}{4^2} = \dfrac{8}{16} = \dfrac{1}{2}$ d. $2^2 \cdot 3^3 = 4 \cdot 27 = 108$

e. $(4x^3)^2 = 4^2 x^6 = 16x^6$ f. $\dfrac{-2^2}{(-4)^2} = \dfrac{-(2^2)}{(-4)(-4)} = \dfrac{-4}{16} = -\dfrac{1}{4}$ ▲

EXAMPLE 10

Simplify: $\dfrac{2^0 \cdot a^2 \cdot b^{-3} \cdot c^7}{3^2 \cdot a^5 \cdot b \cdot c^{-2}}$

Solution

$$\dfrac{2^0 \cdot a^2 \cdot b^{-3} \cdot c^7}{3^2 \cdot a^5 \cdot b \cdot c^{-2}} = \dfrac{c^9}{9a^3 b^4}$$

Note that the form of the quotient rule to be used is determined by which produces a final answer with positive exponents, since this is preferred to an answer with negative exponents. Thus, in solving the problem we wrote

$$\dfrac{a^2}{a^5} = \dfrac{1}{a^3} \qquad \dfrac{b^{-3}}{b} = \dfrac{1}{b^{1-(-3)}} = \dfrac{1}{b^4} \qquad \dfrac{c^7}{c^{-2}} = c^{7-(-2)} = c^9 \qquad ▲$$

EXAMPLE 11

Simplify: $\dfrac{x^{-2} \cdot y^{-3} \cdot z^4 \cdot x^5}{y^{-2} \cdot z^{-3} \cdot x^8}$

Solution

$$\dfrac{x^{-2} \cdot y^{-3} \cdot z^4 \cdot x^5}{y^{-2} \cdot z^{-3} \cdot x^8} = \dfrac{x^3 y^{-3} z^4}{y^{-2} z^{-3} x^8} = \dfrac{z^7}{x^5 y} \qquad ▲$$

In dealing with very large or very small numbers it is often convenient to use what is called **scientific notation.** A number is expressed in scientific notation when it is written as a number between 1 and 10, multiplied by a power of 10. For example,

$$2{,}500{,}000{,}000 = 2.5 \times 10^9$$

(It is customary to use " $\times$ " as the multiplication symbol in scientific notation.) The exponent of 10, when it is positive, indicates how many places to the *right* the decimal should go. A negative exponent indicates how many places to the *left* the decimal should go:

$$0.00000000003 = 3 \times 10^{-11}$$

Scientific notation, combined with the laws of exponents, can often greatly simplify complicated arithmetic problems. The next example illustrates this.

EXAMPLE 12 ||||||||||||||||||||||||||||||||||||||||||||||||||||||||||||||||||||||||||||||||||||||||||||||||||||||||||

Evaluate: $\dfrac{(72,000,000)(0.0000000036)}{(0.12)(27,000)}$

Solution

$$\frac{(72,000,000)(0.0000000036)}{(0.12)(27,000)} = \frac{(7.2 \times 10^7)(3.6 \times 10^{-9})}{(1.2 \times 10^{-1})(2.7 \times 10^4)}$$

$$= \frac{(7.2)(3.6)}{(1.2)(2.7)} \times \frac{10^{-2}}{10^3} = 8.0 \times 10^{-5}$$

$$= 0.000080$$

▲

Most scientific calculators have scientific notation capability, which is automatically employed when the size of the number would otherwise exceed the capacity of the calculator. For example, if you enter the number 23,200 and then multiply by 4,500,000, the answer would be shown on most calculators as

$$\boxed{1.044 \quad 11} \qquad \text{or} \qquad \boxed{1.044 \quad E \quad 11}$$

which is to be interpreted as 1.044×10^{11}. Numbers can also be entered in scientific notation and calculations can be performed on them. The key for multiplying by a power of 10 varies, and you will need to check your own calculator to see which key this is. On some calculators the key is labeled $\boxed{EE}$. Using this, we could enter 7.2×10^7 as

$$7.2 \quad \boxed{EE} \quad \boxed{7}$$

and 3.6×10^{-9} as

$$3.6 \quad \boxed{EE} \quad \boxed{9} \quad \boxed{+/-}$$

The $\boxed{+/-}$ key changes the sign of a number. We could do Example 12 on the calculator as follows,

$$7.2 \quad \boxed{EE} \quad \boxed{7} \quad \boxed{\times} \quad 3.6 \quad \boxed{EE} \quad \boxed{9} \quad \boxed{+/-} \quad \boxed{\div} \quad \boxed{(} \quad 1.2 \quad \boxed{EE} \quad \boxed{1} \quad \boxed{+/-}$$

$$\boxed{\times} \quad 2.7 \quad \boxed{EE} \quad \boxed{4} \quad \boxed{)} \quad \boxed{=}$$

The answer would be displayed as 0.00008.

EXERCISE SET 4

Note. Throughout this exercise set all letters represent *positive* real numbers.

A

In Problems 1–9 carry out the indicated operations and simplify the result, expressing the answer with positive exponents only.

1. a. 2^5 **b.** 3^{-1} **c.** $(-3)^3$

  d. $(-10)^0$ **e.** $\left(\dfrac{2}{3}\right)^{-2}$

2. a. $3^4 \cdot 3^2$ **b.** $\dfrac{2^9}{2^5}$ **c.** $\dfrac{7^3}{7^5}$

  d. $\dfrac{2^3}{2^{-5}}$ **e.** $\dfrac{5^{-3}}{5^{-4}}$

3. a. $x^4 \cdot x^5$ **b.** $\dfrac{x^8}{x^5}$ **c.** $(x^2)^5$

  d. $(x^3 y^2)^3$ **e.** $(3x^5)^0$

4. a. $(4x^3)^2$ **b.** $(3x^2)(4x^5)$

  c. $(3^0 x^{-3})(2x^5)$ **d.** $\left(\dfrac{3}{2x}\right)^{-3}$

  e. $\dfrac{x^{-3}}{x^2}$

5. a. $(-2a^4)^3$ **b.** $(3a^{-2})(4a^{-3})$

  c. $(3^2 a^{-3})(3^{-4} a^5)$ **d.** $\left(\dfrac{2a^{-2}}{3b^3}\right)^{-2}$

  e. $\dfrac{18a^{-2}}{24a^3}$

6. a. $\dfrac{a^3 b^{-2}}{a^{-2} b}$ **b.** $\dfrac{a^{-3} b^5 c^0}{a^{-7} b^{-3} c^4}$

  c. $\dfrac{9s^{-3} t^{-2}}{12s^5 t^{-6}}$ **d.** $\dfrac{x^{3a} y^{4b}}{x^a y^{-2b}}$

7. a. $\dfrac{x^{-4} y^3}{x^3 y^{-5}}$ **b.** $\dfrac{r^{-2} s^{-1} t^4}{r^{-5} s^0 t}$

  c. $\dfrac{9a^{-6} n^3}{15a^{-3} b^{-5}}$

  d. $\dfrac{18u^{3s-2t}}{45u^{s+t}}$, $s > \tfrac{3}{2} t$

8. a. $\dfrac{8a^4 b^{-5}}{12a^{-3} b^{-2}}$ **b.** $\dfrac{(-2)^3 x^{-4} y^{-2}}{8x^3 y^{-1}}$

  c. $\dfrac{3^{-2} a^3 b^{-5}}{3^{-1} a^{-6} b^{-3}}$ **d.** $\dfrac{-2^5 a^{3m} b^{-3n}}{a^{-2m} b^{2n}}$

9. a. $\left(\dfrac{a^2}{b}\right)^{-2}\left(\dfrac{c^{-3}}{d^2}\right)^{-3}$

  b. $\dfrac{\left(\dfrac{x^3}{y^2}\right)^3}{\left(\dfrac{x^2}{y}\right)^{-2}}$

  c. $\left(\dfrac{a^{3n}}{b^{2n}}\right)^{2k}\left(\dfrac{b^{-n}}{a^{2n}}\right)^{-k}$

10. Substitute $x = 2$, $y = 3$ in the following expression and simplify the result. Check your result with a calculator.

$$\dfrac{3x^{-1} - 2y^{-1}}{x^{-2} - y^{-2}}$$

11. Substitute $a = 3$, $b = 2$ in the following expression and simplify the result. Check your result with a calculator.

$$\dfrac{ab^{-2} - a^{-1} b^2}{3a^{-2} - b^{-2}}$$

12. Substitute $x = \tfrac{2}{3}$, $y = -\tfrac{1}{2}$ in the following expression and simplify the result. Check your result with a calculator.

$$\dfrac{x^{-1} y^{-2} - 2x^{-2} y^{-1}}{2x^{-1} y - (3y)^{-1}}$$

13. Express in scientific notation:
  a. 3,000,000,000   **b.** 0.000005
  c. 250,000   **d.** 0.234
  e. 3,568,000

14. Express in decimal notation:
  a. 3.2×10^4 **b.** 2×10^{-5}
  c. 5.27×10^2 **d.** 8.92×10^{-3}
  e. 2.01×10^6

In Problems 15–17 perform the indicated calculations. Check your results with a calculator.

15. a. $(2.3 \times 10^5)(4.0 \times 10^{-3})$

 b. $\dfrac{4.8 \times 10^5}{1.2 \times 10^2}$

16. a. $5.7 \times 10^3 + 2.3 \times 10^4$
 b. $4.3 \times 10^4 - 9.6 \times 10^3$
 Hint. First express each term with the same power of 10.

17. a. $7 \times 10^5 + 3 \times 10^3$
 b. $3.4 \times 10^{-3} - 1.2 \times 10^{-2}$
 (See the Hint for Problem 16.)

18. Use scientific notation to calculate

$$\frac{(0.00009)(12{,}000{,}000)}{(4{,}000{,}000{,}000)(0.0000003)}$$

19. The nearest star is approximately 4 light-years away. Light travels at approximately 186,000 miles per second. Taking 365 as the number of days in a year, find the distance in miles to the nearest star. Use a calculator.

20. a. The mass of a hydrogen atom is approximately 0.00000000000000000000000001673 gram. Express this number in scientific notation.
 b. There are approximately 6.023×10^{23} atoms in a mole (a unit of measure used in chemistry). Express this number as an integer.

In Problems 21–26 m and n are integers.

21. Prove that the formula $a^m \cdot a^n = a^{m+n}$ holds true under the following conditions:
 a. When m is positive and n is negative
 b. When both m and n are negative
 c. When $n = 0$
 Hint. For part **a** let $n = -p$, where p is positive.

22. Prove that $(a^m)^n = a^{mn}$ under the following conditions:
 a. When m is positive and n is negative
 b. When m is negative and n is positive
 c. When m and n are both negative
 d. When $n = 0$

23. Prove that for $a > 0$ and $b > 0$,

$$\left(\frac{a}{b}\right)^{-m/n} = \left(\frac{b}{a}\right)^{m/n}$$

where m and n are positive integers.

24. Prove that in general $(a + b)^n \neq a^n + b^n$. What are the exceptions?

25. Is the following equation true?

$$\frac{1}{a^{-n} + b^{-n}} = a^n + b^n$$

Prove or disprove your answer.

26. Compare 2^{n^2} with 2^{2^n}.

 # RADICALS AND FRACTIONAL EXPONENTS

For $a \geq 0$, the symbol $\sqrt{a}$ means the nonnegative real number whose square is a. This is called the **principal square root** of a.

The symbol $\sqrt{}$ is called a **radical** sign. It is also true that $-\sqrt{a}$ is a square root of a, but it is not the principal square root. It is a common mistake to suppose that $\sqrt{a}$ means both the principal square root and its negative, but such an interpretation would lead to ambiguity and confusion. Thus, $\sqrt{4} = 2$, but $\sqrt{4} \neq -2$. More generally, consider $\sqrt{x^2}$, where x is an unspecified real number. Is the answer x, or is it $-x$? Our definition requires that the result be nonnegative. So $\sqrt{x^2} = x$ if $x \geq 0$, and $\sqrt{x^2} = -x$ if $x < 0$. That is,

$$\sqrt{x^2} = |x|$$

The fact that square roots of all nonnegative numbers exist can be proved using the basic properties of R, but the proof is beyond the level of this course.

If $a < 0$, then $\sqrt{a}$ does not exist in R (however, it does exist in the complex number system, which we will discuss briefly in Chapter 2).

In a similar way, $\sqrt[n]{a}$, where n is any natural number greater than 1 and $a \geq 0$, is defined to mean the (unique) nonnegative real number whose nth power is a. If $a < 0$ and n is an odd natural number greater than 1, then $\sqrt[n]{a}$ is the negative real number whose nth power is a. For example, $\sqrt[3]{-8,} = -2$, since $(-2)^3 = (-2)(-2)(-2) = -8$. If n is even and $a < 0$, then $\sqrt[n]{a}$ does not exist in R. The number designated by $\sqrt[n]{a}$ is called the **principal nth root of a.** In the symbol $\sqrt[n]{a}$, a is called the **radicand** and n is called the **index** of the radical.

From the laws of exponents we see that if $a \geq 0$ and $b \geq 0$, then

$$(\sqrt[n]{a}\,\sqrt[n]{b})^n = (\sqrt[n]{a})^n(\sqrt[n]{b})^n = ab$$

It follows that

$$\sqrt[n]{ab} = \sqrt[n]{a}\,\sqrt[n]{b} \qquad (a \geq 0,\, b \geq 0) \tag{12}$$

This can be extended to arbitrary finite products. When n is odd, the restriction that a and b be nonnegative can be removed.

By using equation (12) we can often simplify expressions involving radicals. The object is to write the radicand as a product of the largest possible perfect nth power, times whatever is left. The nth root of the perfect nth power can be written without the radical sign. This is what we will mean by the instruction "simplify" in connection with problems involving radicals.

EXAMPLE 13

Simplify:

a. $\sqrt{50}$ **b.** $\sqrt{32} - \sqrt{18}$ **c.** $\sqrt[3]{81} + \sqrt[3]{24} + \sqrt[3]{192}$

Solution   **a.** $\sqrt{50} = \sqrt{25 \cdot 2} = \sqrt{25}\sqrt{2} = 5\sqrt{2}$

b. $\sqrt{32} - \sqrt{18} = \sqrt{16 \cdot 2} - \sqrt{9 \cdot 2} = 4\sqrt{2} - 3\sqrt{2} = \sqrt{2}$

c. $\sqrt[3]{81} + \sqrt[3]{24} + \sqrt[3]{192} = \sqrt[3]{27 \cdot 3} + \sqrt[3]{8 \cdot 3} + \sqrt[3]{64 \cdot 3}$
$$= 3\sqrt[3]{3} + 2\sqrt[3]{3} + 4\sqrt[3]{3}$$
$$= 9\sqrt[3]{3}$$

▲

EXAMPLE 14

Simplify the following, where $x \geq 0$ and $y \geq 0$:

a. $\sqrt{112x^3y^6}$ **b.** $\sqrt{6xy^3}\sqrt{2xy}$ **c.** $\sqrt[4]{8x^3y^2}\sqrt[4]{4x^2y^6}$

Solution   **a.** $\sqrt{112x^3y^6} = \sqrt{16 \cdot 7x^2 \cdot x(y^3)^2} = 4xy^3\sqrt{7x}$

b. $\sqrt{6xy^3}\sqrt{2xy} = \sqrt{12x^2y^4} = \sqrt{3 \cdot 4x^2y^4} = 2xy^2\sqrt{3}$

c. $\sqrt[4]{8x^3y^2}\sqrt[4]{4x^2y^6} = \sqrt[4]{32x^5y^8} = \sqrt[4]{16 \cdot 2x^4 \cdot x(y^2)^4} = 2xy^2\sqrt[4]{2x}$

▲

If $b \neq 0$ and $\sqrt[n]{a}$ and $\sqrt[n]{b}$ exist, then the following formula for division, analogous to rule (12), holds true:

$$\sqrt[n]{\frac{a}{b}} = \frac{\sqrt[n]{a}}{\sqrt[n]{b}} \tag{13}$$

You will be asked to prove this in Problem 30, Exercise Set 5. We can use this to remove the radical sign from the denominator, as we show in the next example. This process is called **rationalizing the denominator.**

EXAMPLE 15

Rationalize the denominators in each of the following:

a. $\sqrt{\dfrac{5}{18}}$ **b.** $\dfrac{3}{\sqrt{5}}$ **c.** $\sqrt[3]{\dfrac{2}{9}}$ **d.** $\dfrac{\sqrt[4]{2x^2}}{\sqrt[4]{27xy^5}}$ $(x > 0, y > 0)$

Solution   We can make each denominator a perfect power by multiplying the numerator and denominator by an appropriate number. Then rule (13) can be applied.

a. $\sqrt{\dfrac{5}{18}} = \sqrt{\dfrac{5}{18} \cdot \dfrac{2}{2}} = \sqrt{\dfrac{10}{36}} = \dfrac{\sqrt{10}}{\sqrt{36}} = \dfrac{\sqrt{10}}{6}$

b. $\dfrac{3}{\sqrt{5}} = \dfrac{3}{\sqrt{5}} \cdot \dfrac{\sqrt{5}}{\sqrt{5}} = \dfrac{3\sqrt{5}}{5}$

c. $\sqrt[3]{\dfrac{2}{9}} = \sqrt[3]{\dfrac{2}{9} \cdot \dfrac{3}{3}} = \sqrt[3]{\dfrac{6}{27}} = \dfrac{\sqrt[3]{6}}{\sqrt[3]{27}} = \dfrac{\sqrt[3]{6}}{3}$

d. $\dfrac{\sqrt[4]{2x^2}}{\sqrt[4]{27xy^5}} = \dfrac{\sqrt[4]{2x^2}}{\sqrt[4]{27xy^5}} \cdot \dfrac{\sqrt[4]{3x^3y^3}}{\sqrt[4]{3x^3y^3}} = \dfrac{\sqrt[4]{6x^5y^3}}{\sqrt[4]{81x^4y^8}} = \dfrac{x\sqrt[4]{6xy^3}}{3xy^2} = \dfrac{\sqrt[4]{6xy^3}}{3y^2}$ ▲

Remark. Whether to rationalize the denominator in general depends on the context. As a final answer to a problem it is generally considered good form to avoid fractions as radicands. Parts **a** and **c** of the preceding example illustrate the preferred way of writing the answer in such cases. In part **b** the answer does not appear to be any simpler than the original form. However, if this were part of a larger problem where, for example, it was necessary to find a common denominator, then having the denominator free of radicals might be desirable.

We are now in a position to give a definition of fractional exponents. First consider $a^{1/n}$, where n is a natural number. To be consistent with the laws of exponents, we must have

$$(a^{1/n})^n = a^{(1/n)n} = a^1 = a$$

So we are led to the definition

$$a^{1/n} = \sqrt[n]{a}$$

The restriction that $a \geq 0$ when n is even must be observed. Also, for any integer m,

$$a^{m/n} = a^{(1/n)m} = (a^{1/n})^m = (\sqrt[n]{a})^m \tag{14}$$

Alternatively, we can write

$$a^{m/n} = a^{m(1/n)} = (a^m)^{1/n} = \sqrt[n]{a^m} \tag{15}$$

Again, if n is even, we must have $a \geq 0$. In words, the fractional (rational) exponent m/n means: "Take the nth root and raise to the mth power, in either order." Taking the root first, as in (14), is more desirable than raising to the power first, as in (15), whenever a is a perfect nth power. For example, using (14), we have

$$(27)^{4/3} = (\sqrt[3]{27})^4 = 3^4 = 81$$

whereas using (15),

$$(27)^{4/3} = \sqrt[3]{(27)^4} = \sqrt[3]{531{,}441}$$

This again yields the answer 81, but it is obviously much more difficult to calculate.

When we use a calculator, this distinction between raising to a power first or taking a root first disappears. For example, with a calculator we can evaluate $(27)^{4/3}$ as follows:

27 $\boxed{y^x}$ $\boxed{(}$ $\boxed{4}$ $\boxed{\div}$ $\boxed{3}$ $\boxed{)}$ $\boxed{=}$ Displayed answer 81

In this case it is easier without the calculator, but with $(26)^{4/3}$, for example, the calculator would be almost essential.

All the laws of exponents continue to hold true for rational exponents, but the details of showing this are tedious and will be omitted here.

EXAMPLE 16

Give an equivalent expression that does not involve radicals, and simplify.

a. $\sqrt[4]{4x^6y^3}$ $(x, y \geq 0)$ **b.** $\sqrt[6]{(8a^4b^3)^2}$ $(a, b \geq 0)$

Solution   **a.** $\sqrt[4]{4x^6y^3} = (2^2x^6y^3)^{1/4} = 2^{2/4}x^{6/4}y^{3/4} = 2^{1/2}x^{3/2}y^{3/4}$

b. $\sqrt[6]{(8a^4b^3)^2} = (2^3a^4b^3)^{2/6} = (2^3a^4b^3)^{1/3} = 2a^{4/3}b$ ▲

EXERCISE SET 5

Note. All variables represent positive real numbers unless otherwise specified.

A

Simplify each expression in Problems 1–12.

1. a. $\sqrt{304}$ **b.** $\sqrt[3]{-448}$

2. a. $\sqrt{48} + \sqrt{75}$ **b.** $2\sqrt{162} - 5\sqrt{50}$

3. a. $2\sqrt[3]{-81} + 4\sqrt[3]{24}$

   b. $2\sqrt{45} - 3\sqrt{80} + \sqrt{125}$

4. a. $5\sqrt[3]{108} - 2\sqrt[3]{-32}$

   b. $4\sqrt[4]{162} - 5\sqrt[4]{32}$

5. a. $2\sqrt[3]{375} - 4\sqrt[3]{-192} + 5\sqrt[3]{-81}$

   b. $4\sqrt[5]{-64} + 2\sqrt[5]{486}$

6. a. $\sqrt{75x^6y^9}$ **b.** $\sqrt{72a^8b^{-4}}$

7. $\sqrt{\dfrac{1000a^{-2}b^{-3}c^5}{a^{-4}bc^{-1}}}$ **8.** $\sqrt{\dfrac{256x^{-5}y^7z^{-3}}{8x^{-1}y^{-2}z^{-11}}}$

9. $\sqrt[3]{\dfrac{216a^{-5}b^0c^4}{a^{10}b^{-6}c^{-8}}}$ **10.** $\sqrt[3]{\dfrac{-54x^5y^8z}{x^{-2}y^3z^{-4}}}$

11. $\sqrt{7x^3y}\,\sqrt{14x^5y^3}$ **12.** $\sqrt[3]{9a^4b^2}\,\sqrt[3]{192a^2b}$

In Problems 13–16 rationalize the denominator and simplify.

13. a. $\sqrt{\dfrac{5}{3}}$ **b.** $\sqrt{\dfrac{2}{5}}$ **c.** $\dfrac{3}{\sqrt{7}}$

   d. $\sqrt[3]{\dfrac{3}{4}}$ **e.** $\sqrt{\dfrac{27}{32}}$

14. a. $\sqrt{\dfrac{7}{8}}$ **b.** $\sqrt{\dfrac{3}{7}}$ **c.** $\sqrt{\dfrac{5}{63}}$

   d. $\dfrac{3}{\sqrt{72}}$ **e.** $\sqrt[3]{\dfrac{7}{16}}$

15. a. $\sqrt{\dfrac{2a^5}{3b}}$ **b.** $\sqrt{\dfrac{8a}{27b^3}}$

16. a. $\sqrt{\dfrac{x^3}{2yz^4}}$ **b.** $\sqrt[3]{\dfrac{16a^5}{3b^4c}}$

Simplify the expressions in Problems 17–25, writing answers without zero or negative exponents.

17. a. $(3)^{2/3} \cdot (3)^{1/2}$ **b.** $(2)^{-1/2} \cdot (2)^{4/3}$

   c. $(4)^{-3/2} \cdot \left(\dfrac{8}{27}\right)^{-2/3}$ **d.** $(16)^{-3/4} \cdot (32)^{-3/5}$

   e. $(x^{-3a}y^{6b})^{-2/3}$

18. a. $(9)^{-3/2} \cdot (27)^{2/3}$ **b.** $(-8)^{-2/3} \cdot (16)^{-3/2}$

   c. $\left(\dfrac{9}{4}\right)^{-1/2} \cdot (-2)^{-2}$

   d. $\left[4^2 \cdot 3^0 + \left(\dfrac{1}{27}\right)^{-2/3}\right]^{1/2}$

   e. $(a^{-2n}b^{4m})^{-3/2}$

19. a. $(2x^{1/3})(3x^{1/2})$ **b.** $\dfrac{x^{3/4}y^{1/2}}{x^{2/3}y^{1/3}}$

   c. $(8x)^{1/3}(4x)^{1/2}$

20. a. $(x^{2/3})^{3/4}$ **b.** $(x^{-3/2})^{-4/3}$

   c. $(4x^{4/3}y^{8/9})^{3/2}$

21. a. $(a^{3/2})^{4/3}$ **b.** $(4a^{-2/3})^{3/2}$

   c. $(-8a^{3/2}b^{-6})^{2/3}$

22. a. $(8a^6)^{1/3}$ **b.** $(16a^4b^{-2})^{1/2}$

   c. $(-125x^{-9}y^{12})^{1/3}$

23. a. $(8x^{3/4}y^{-9/2})^{-2/3}$ **b.** $(16a^{4/3}b^{-2/3}c^{4/9})^{-3/2}$

   c. $(32x^{-5/4}y^{10/3})^{-2/5}$

24. a. $\left(\dfrac{8a^{-2}b^{-5}c^0}{27a^{-5}b}\right)^{-2/3}$ **b.** $\left(\dfrac{-32a^3b^{-2}}{243a^{-7}b^3}\right)^{3/5}$

25. a. $\left(\dfrac{9x^{4/3}}{4y^{-2/3}}\right)^{3/2}$ **b.** $\left(\dfrac{-27a^{9/2}b^{-3/4}}{8c^6}\right)^{-4/3}$

In Problems 26 and 27 give an equivalent expression that does not involve radicals.

26. a. $\sqrt[3]{27a^2b^4}$

   b. $(\sqrt{3a^5b^3})^3$

   c. $\sqrt[3]{(2a^2b^4)^2}$

27. a. $\sqrt{9x^3y^5}$

   b. $\sqrt{(3xy^4)^3}$

   c. $(\sqrt[3]{4x^2y^5})^2$

Simplify the expressions in Problems 28 and 29. The variables represent real numbers that are not necessarily positive.

28. a. $\sqrt{(-2xy^2)^2}$ **b.** $\sqrt{25x^4y^6}$
29. a. $(9x^2y^4)^{1/2}$ **b.** $(-8x^6y^9)^{1/3}$

B

30. Prove rule (13).

31. Prove that for any natural number k, $\sqrt[kn]{a^{km}} = \sqrt[n]{a^m}$, where $a \geq 0$, and m and n are natural numbers greater than or equal to 2.

32. Use the result of Problem 31 to simplify each of the following.

  a. $\sqrt[4]{9}$ **b.** $\sqrt[6]{27x^9}$
  c. $\sqrt[8]{4x^{12}y^6}$ **d.** $\sqrt[4]{36a^8b^2c^{10}}$

33. Prove that $\sqrt[m]{\sqrt[n]{a}} = \sqrt[mn]{a}$, where $a \geq 0$ and m and n are natural numbers greater than or equal to 2.
Hint. Use fractional exponents.

34. Use the results of Problems 31 and 33 to write each of the following in simplest form with only one radical sign.

  a. $\sqrt{\sqrt{50}}$ **b.** $\sqrt[3]{\sqrt{64a^9}}$
  c. $\sqrt{8\sqrt[3]{16a^8b^4}}$ **d.** $\sqrt[3]{16\sqrt{8a^{10}y^6}}$

35. Prove that for $a > 0$ and $b > 0$,

$$\left(\frac{a}{b}\right)^{-m/n} = \left(\frac{b}{a}\right)^{m/n}$$

where m and n are positive integers.

36. Use the result of Problem 35 to simplify the following.

  a. $\left(\dfrac{343x^{12}}{27y^9z^6}\right)^{-2/3}$ **b.** $\left(\dfrac{81a^{12}b^4}{256c^8}\right)^{-3/4}$

In Problems 37–42 simplify and perform the indicated operations. Express answers without radicals in the denominator or fractions as radicands.

37. $\dfrac{3}{\sqrt{2}} - \dfrac{2\sqrt{50}}{3} + \dfrac{3\sqrt{72}}{4}$

38. $\dfrac{2}{3}\sqrt{\dfrac{4}{3}} + \dfrac{3}{4}\sqrt{\dfrac{1}{27}} - \dfrac{\sqrt{300}}{2}$

39. $\sqrt[3]{-54} + \sqrt[3]{\dfrac{125}{4}} + \dfrac{5}{\sqrt[3]{32}}$

40. $\dfrac{3}{\sqrt[3]{2}} - \sqrt[3]{\dfrac{27}{16}} + \dfrac{\sqrt[3]{-32}}{6}$

41. $\sqrt{\dfrac{242x^3y^{-5}z^0}{243x^{-7}y^{-3}z^4}}$

42. $\sqrt[3]{\dfrac{729a^6b^{-1}c^{-1}}{500a^0b^2c^{-4}}}$

43. Express in scientific notation and evaluate. Check your result with a calculator.

$$\frac{(0.0004)^2\sqrt{30,000}\sqrt{0.0003}}{\sqrt[3]{0.000064}}$$

THE ARITHMETIC OF POLYNOMIALS

One of the distinctive features of algebra is the use of letters to represent numbers. Among other things this permits us to state general rules about numbers in a clear and concise way, as we have already seen in the preceding sections. Also, letters are used to represent unknown quantities in equations and inequalities, the objective being to find the correct value, or set of values, for which the relation is true. The point to be emphasized is that normally when letters are used in elementary algebra, they are used to represent numbers — usually real numbers. This fact enables us to apply the properties of real numbers to algebraic expressions. Unless otherwise stated, we will assume in what follows that all letters represent real numbers.

If a letter can be assigned a value from a given set of numbers, it is called a **variable.** Letters may also be used to designate fixed, but unspecified, numbers called **constants.** For example, in the expression $ax + b$, the letters a and b designate constants, and the letter x designates a variable. Constants may also be specified, as in $2x + 3$. Usually the context will make clear whether a letter represents a constant or a variable. Frequently, letters from the latter part of the alphabet, such as x, y, or z, are used for variables, and letters such as a, b, or c, from the first part, are used for constants.

We use the term **algebraic expression** to mean any combination of variables and constants that is formed using a finite number of the operations of addition, subtraction, multiplication, division, raising to a power, or taking roots. Thus,

$$ax^3 \qquad \frac{x^{-1} - y^{-1}}{x^{-2} - y^{-2}} \qquad \frac{-10\sqrt{x - y}}{(u + v)^8}$$

are examples of algebraic expressions. Of particular interest in this section are algebraic expressions called **polynomials.** Examples are

$$2x^3 - 3x^2 + 5x - 8 \qquad x^{10} - y^{10} \qquad 4x^2y^3z - 3xyz^3 + 5x^3yz^5$$

A **term** of a polynomial consists of a constant called the **coefficient,** multiplied by one or more variables each raised to a nonnegative integral power. So a term of a polynomial in one variable has the form ax^m, and in two variables ax^my^n, and so on. A polynomial consists of the sum of finitely many such terms. This includes differences since we may write, for example,

$$2x^3 - 3x^2 + 5x - 8 = 2x^3 + (-3)x^2 + 5x + (-8)$$

The **degree of a term** of a polynomial is the sum of the exponents on the variables in that term. For example, the degree of $2x^3$ is 3, and the degree of $4x^2y^3z$ is $2 + 3 + 1 = 6$ (remember that $z = z^1$). The **degree of a polynomial** is the largest of the degrees of its terms. A polynomial of degree n in one variable x can always be written in the form

$$a_nx^n + a_{n-1}x^{n-1} + \cdots + a_1x + a_0 \qquad (a_n \neq 0)$$

The coefficient, a_n, of the highest power of x is called the **leading coefficient.** If $n = 0$ and $a_0 \neq 0$, the polynomial is a **constant polynomial** and has degree 0. The constant 0 has no degree. A polynomial having only one term is called a **monomial;** one with two terms, a **binomial;** and one with three terms, a **trinomial.** We consider now how to add, subtract, multiply, and divide polynomials.

ADDITION AND SUBTRACTION

Polynomials are added term-by-term where **like terms** are added—that is, where terms involving the same variable (or variables) raised to the same power are added. For example,

$$(2x^2 + 3x + 5) + (3x^2 - x + 7) = 5x^2 + 2x + 12$$

This is justified by use of the commutative, associative, and distributive properties of real numbers, as can be seen by the following calculations:

$$(2x^2 + 3x + 5) + (3x^2 - x + 7) = (2x^2 + 3x^2) + (3x - x) + (5 + 7)$$
$$= (2 + 3)x^2 + (3 - 1)x + (5 + 7)$$
$$= 5x^2 + 2x + 12$$

Subtraction is handled similarly, as in the following. (You should supply all steps and justify them based on real number properties.)

$$(2x^2 + 3x + 5) - (3x^2 - x + 7) = -x^2 + 4x - 2$$

MULTIPLICATION

Multiplication of polynomials is carried out by repeated use of the distributive property of real numbers. If one factor is a monomial, only one application of this property is required. For example,

$$3x^2(4x^5 + 7x - 9) = 3x^2(4x^5) + 3x^2(7x) + 3x^2(-9)$$
$$= 12x^7 + 21x^3 - 27x^2$$

Now consider a product of the form

$$(a + b)(c + d)$$

We first treat $(a + b)$ as a single term and distribute the multiplication of it over the sum $(c + d)$:

$$(a + b)(c + d) = (a + b)c + (a + b)d$$

Now we can apply the right-hand distributive law to each of the products on the right to obtain

$$(a + b)(c + d) = ac + bc + ad + bd$$

Observe that **the result is the sum of all possible products involving one term from the first factor and one term from the second.** This result holds true in general. For example,

$$(a + b)(c + d + e) = ac + ad + ae + bc + bd + be$$

and similarly for any number of terms in either of the factors.

If we apply this result to two binomials in x, say, $(2x + 3)(4x + 5)$, we have

$$(2x + 3)(4x + 5) = 8x^2 + 12x + 10x + 15$$
$$= 8x^2 + 22x + 15$$

In this case the combining of like terms is possible, and this situation is typical. This combining of terms can be anticipated when multiplying two *like* binomials, for example, both binomials in x or in x and y. The result is generalized by the following:

$$(ax + b)(cx + d) = acx^2 + (ad + bc)x + bd \qquad (16)$$

This should not be memorized as a formula, but the process of arriving at the answer is important. The diagram below illustrates how each term on the right is obtained.

First term: $(ax + b)(cx + d)$

First times First $= acx^2$

First times Last $= adx$

Middle term: $\{(ax + b)(cx + d)$ Combine: $adx + bcx = (ad + bc)x$

Last times First $= bcx$

Last term: $(ax + b)(cx + d)$

Last times Last $= bd$

All this can be done mentally in some problems. For example,

$$(2x + 3)(4x + 5) = 8x^2 + 22x + 15$$

Here are some other examples:

$$(3x + 2)(5x - 4) = 15x^2 - 2x - 8$$

(Notice that the middle term is the sum of $-12x$ and $10x$.)

$$(2 + 3t)(4 + t) = 8 + 14t + 3t^2$$

$$(x - 2y)(x + 3y) = x^2 + xy - 6y^2$$

Two special cases of products that should be memorized are

$$(a + b)(a - b) = a^2 - b^2 \qquad (17)$$

and

$$(a + b)^2 = a^2 + 2ab + b^2 \qquad (18)$$

The first of these is immediately seen to be true, since the middle term drops out ($-ab + ba = 0$). The second, called a **perfect square,** is verified by writing

$$(a + b)^2 = (a + b)(a + b) = a^2 + (ab + ba) + b^2 = a^2 + 2ab + b^2$$

A Word of Caution. Some students make the mistake of writing $(a + b)^2 = a^2 + b^2$. It would be nice if this were true, but it is not, as can be seen by substituting values for a and b—say, for example, $a = 3$, $b = 2$. The left-hand side then is $5^2 = 25$ and the right-hand side is $3^2 + 2^2 = 13$. The important point is that *you must not forget the middle term $2ab$ in formula (18)*.

Similarly, we have

$$(a - b)^2 = a^2 - 2ab + b^2 \tag{19}$$

It is useful to express these formulas in words. We can express equation (17) by saying: "The sum of two numbers times their difference is the difference of their squares." For the perfect square we can say: "The square of the sum (difference) of two numbers is the square of the first, plus (minus) twice the product of the two, plus the square of the last."

These results are illustrated by the following examples:

Product of sum and difference of two terms:

$$(x + 2)(x - 2) = x^2 - 4$$
$$(2x - 3)(2x + 3) = 4x^2 - 9$$
$$(4s + 5t)(4s - 5t) = 16s^2 - 25t^2$$

Perfect square:

$$(x + 5)^2 = x^2 + 10x + 25$$
$$(2x - 3)^2 = 4x^2 - 12x + 9$$
$$(3x^2 - 8y)^2 = 9x^4 - 48x^2y + 64y^2$$

DIVISION

Division of polynomials is carried out in much the same way as long division in arithmetic. For example, consider the problem $(2x^4 - 3x^3 + 3x^2 - 4) \div (x^2 - 3x + 4)$.

Divisor, Dividend, Quotient, Remainder labels on the long division:

$$
\begin{array}{r}
2x^2 + 3x + 4 \quad\longleftarrow \text{ Quotient}\\
x^2 - 3x + 4 \overline{)2x^4 - 3x^3 + 3x^2 \qquad - 4}\\
\underline{2x^4 - 6x^3 + 8x^2}\\
3x^3 - 5x^2\\
\underline{3x^3 - 9x^2 + 12x}\\
4x^2 - 12x - 4\\
\underline{4x^2 - 12x + 16}\\
- 20 \quad\longleftarrow \text{ Remainder}
\end{array}
$$

Note that a space is left for the missing power of x in the dividend. The successive terms of the quotient are obtained by dividing the leading term of the divisor (x^2) into the terms that are circled. Each time we get a term of the quotient, it is multiplied by the divisor to get the next line of work, which is then subtracted from the previous line. We may write the answer as

$$2x^2 + 3x + 4, \quad \text{Remainder} \ -20$$

Note. For this procedure to work, it is essential that both divisor and dividend be written in descending powers of x and that a place be held for each missing power of x.

We conclude this section with a formula for raising a binomial to an arbitrary power n, where n is a natural number. We know that $(a + b)^1 = a + b$ and $(a + b)^2 = a^2 + 2ab + b^2$. Multiplication of this last result by $a + b$ gives

$$(a + b)^3 = (a + b)(a + b)^2 = (a + b)(a^2 + 2ab + b^2)$$
$$= a^3 + 2a^2b + ab^2 + ba^2 + 2ab^2 + b^3$$
$$= a^3 + 3a^2b + 3ab^2 + b^3$$

To get $(a + b)^4$, we could multiply $(a + b)^3$ by $a + b$. The result (verify this) is

$$(a + b)^4 = a^4 + 4a^3b + 6a^2b^2 + 4ab^3 + b^4$$

This process can be continued, but it becomes tedious very quickly. We might hope instead to discover a pattern that would enable us to write the value of $(a + b)^n$ for any natural number n. The powers of a and b in the successive terms on the right *can* be easily anticipated. For example, we would expect that in the expansion of $(a + b)^5$ the successive powers of a and b on the right would be

$$a^5, \quad a^4b, \quad a^3b^2, \quad a^2b^3, \quad ab^4, \quad b^5$$

But what are the coefficients? If you do not already know the answer, you may wish to carry out a few more multiplications and try to conjecture the general pattern.

The brilliant French mathematician Blaise Pascal (1623–1662) discovered the following scheme for the coefficients in the expansion of $(a + b)^n$. This is known as **Pascal's triangle.**

		Coefficients		

	Coefficients
$n = 0$	1
$n = 1$	1 1
$n = 2$	1 2 1
$n = 3$	1 3 3 1
$n = 4$	1 4 6 4 1
$n = 5$	1 5 10 10 5 1
$n = 6$	1 6 15 20 15 6 1

Do you see how to get the next row? Think about it before you read on. Each time, the leading 1 is written one space farther to the left. Then, each succeeding coefficient through the next-to-last one is obtained by adding the two above. For example, we get the coefficients for $n = 7$ from those for $n = 6$ as shown:

$$n = 6 \qquad 1 \quad 6 \quad 15 \quad 20 \quad 15 \quad 6 \quad 1$$
$$n = 7 \qquad 1 \quad 7 \quad 21 \quad 35 \quad 35 \quad 21 \quad 7 \quad 1$$

Using Pascal's triangle, we get, for example,

$$(a + b)^5 = a^5 + 5a^4b + 10a^3b^2 + 10a^2b^3 + 5ab^4 + b^5$$

Pascal's triangle is interesting and useful, but it does have its limitations. If we wanted a large power, say $(a + b)^{20}$ (which admittedly is not very likely), we would need to construct a rather large triangle. Moreover, there are times when we do need a formula for $(a + b)^n$ for n arbitrary. The formula is stated below. We defer its proof to a later chapter.

THE BINOMIAL THEOREM

For any positive integer n

$$(a + b)^n = a^n + na^{n-1}b + \frac{n(n-1)}{1 \cdot 2} a^{n-2}b^2 + \frac{n(n-1)(n-2)}{1 \cdot 2 \cdot 3} a^{n-3}b^3$$
$$+ \cdots + nab^{n-1} + b^n \tag{20}$$

Equation (20) is called the **binomial formula.** As we have already observed, the key to writing it is to know how to form the coefficients, since the pattern for the powers of a and b is easy to remember. Rather than memorizing how these coefficients are formed from the formula, you might prefer to learn the following procedure (which can be verified from the formula). Observe that the first coefficient is 1 and the second is n. Succeeding coefficients can be determined as follows: **Multiply the coefficient of the preceding term by the exponent of a in that term and then divide by one more than the exponent of b in that term.** For example, having written the first two terms of $(a + b)^{10}$ as $a^{10} + 10a^9b$, we calculate the next coefficient as

$$\frac{10 \cdot 9}{2} = 45$$

So the first three terms are

$$a^{10} + 10a^9b + 45a^8b^2$$

(Now calculate the next coefficient. You should get 120.)

Remark. In general, $(a + b)^n \neq a^n + b^n$. Can you see some exceptions?

EXAMPLE 17

Expand $(x - 2y)^6$.

Solution   We write $(x - 2y)^6 = [x + (-2y)]^6$ and take $a = x$, $b = (-2y)$, $n = 6$. Using either Pascal's triangle, the binomial formula, or the rule stated above, we get

$$[x + (-2y)]^6 = x^6 + 6x^5(-2y) + 15x^4(-2y)^2 + 20x^3(-2y)^3 + 15x^2(-2y)^4$$
$$+ 6x(-2y)^5 + (-2y)^6$$

Now we simplify each term to get the final answer,

$$(x - 2y)^6 = x^6 - 12x^5y + 60x^4y^2 - 160x^3y^3 + 240x^2y^4 - 192xy^5 + 64y^6$$

EXERCISE SET 6

A

In all problems perform the indicated operations and simplify the results.

1. $(3x^2 + 2x + 4) + (2x^2 + 5x + 3)$
2. $(4x^2 + x - 5) + (3x^2 + 7x + 2)$
3. $(2x^2 - 3x + 7) + (3 - 2x + 4x^2)$
4. $(4 - 2y^2 - 3y) + (5y - 3 + 2y^2)$
5. $(2t^2 - 5t + 8) - (t^2 - 3t - 5)$
6. $(3u^4 - 2u^2 - 3) - (2 + u^2 - 4u^4)$
7. $2(3x^2 - 4xy + 5y^2) + 3(2x^2 + 5xy - 3y^2)$
8. $3(7u^2 + 2uv - 5v^2) - 2(8u^2 - 4uv + 3v^2)$
9. $5(x^2 + 2x - 3) - 3(2x^2 - 5x + 8) + 6(3 - 2x - x^2)$
10. $3(1 - 2x^2) - (2x^2 + x - 5) - 4(8x - 3x^2)$
11. $(x + 2)(x + 3)$
12. $(x - 3)(x - 1)$
13. $(x + y)(x + 3y)$
14. $(x + 2y)(x - 3y)$
15. $(2 - 3x)(3 + 4x)$
16. $(4 - 3x)(2 - x)$
17. $(5y - 3)(3y + 4)$
18. $(3t - 5)(7t + 9)$
19. $(2x^2 - 3)(x^2 - 1)$
20. $(3x^2 + 1)(x^2 + 2)$
21. $(x + 2)(x - 2)$
22. $(2x - 1)(2x + 1)$
23. $(2a - 3b)(2a + 3b)$
24. $(3s - 4t)(3s + 4t)$
25. $(\sqrt{x} - 2)(\sqrt{x} + 2)$
26. $(5 - 2\sqrt{x})(5 + 2\sqrt{x})$
27. $(x^n - y^n)(x^n + y^n)$
28. $(2x^k - 1)(2x^k + 1)$
29. $(x^{k/2} - y^{k/2})(x^{k/2} + y^{k/2})$
30. $(a^x - b^y)(a^x + b^y)$
31. $(\sqrt{h + 1} - 1)(\sqrt{h + 1} + 1)$
32. $(\sqrt{3h + 4} - 2)(\sqrt{3h + 4} + 2)$
33. $(\sqrt{x + h} - \sqrt{x})(\sqrt{x + h} + \sqrt{x})$
34. $(\sqrt{2h + 3} - \sqrt{3})(\sqrt{2h + 3} + \sqrt{3})$

35. $(2x + 3y)^2$
36. $(3 - 2t)^2$
37. $(a^2 - 2b^2)^2$
38. $(2u^2 + v^2)^2$
39. $(x^n + y^n)^2$
40. $(\sqrt{x} - \sqrt{y})^2$
41. $1 - x[1 - (1 - x^2)]$
42. $[(x + 1)^2 - x^2]^2$
43. $(x + 2)(3 - 2x) + 2(1 - x)^2$
44. $(3a - 2)^2 - (a + 3)(a - 3)$
45. $(x^3 - 3x^2 + 4) \div (x - 2)$
46. $(2x^4 + 3x^2 - x + 3) \div (x + 4)$
47. $(x^3 - 3x^2 + 4x - 5) \div (x^2 + x - 2)$
48. $(3x^4 - 4x^3 + 8x) \div (x^2 - 3x + 4)$
49. $(x + y)^6$
50. $(x - y)^7$
51. $(2x - 3y)^3$
52. $(3x + 2y)^4$
53. $(x^2 - y)^8$
54. $(2x + 3y^2)^5$
55. $\left(x + \dfrac{1}{x}\right)^5$
56. $\left(\dfrac{3}{2} - x\right)^4$
57. $(x^n + 1)^3$
58. $(2 - x^k)^4$
59. $\sqrt{(x^2 + 1)^2 - (x^2 - 1)^2}$
60. $\sqrt{(x + 1)^2 + (3x + 1)(x - 1)}$

B

61. $(x + 2)(x^2 - x + 3)$
62. $(x - 3)(2x^2 + 3x - 1)$
63. $(x^2 + x - 2)(2x^2 - x + 1)$
64. $(x^2 - 2x - 3)(3x^2 + 4x - 2)$
65. $(x + y + 2)(x + y - 2)$
66. $(2x - 3y - 4)(2x - 3y + 4)$
67. $(x + y - 1)(x - y + 1)$
68. $(3 - m + 2n)(3 + m - 2n)$

69. $(\sqrt{2(x + h) + 3} - \sqrt{2x + 3})$
   $\times (\sqrt{2(x + h) + 3} + \sqrt{2x + 3})$

70. $(\sqrt{3(x + h) - 2} - \sqrt{3x - 2})$
   $\times (\sqrt{3(x + h) - 2} + \sqrt{3x - 2})$

71. $(x + y + z)^2$ **72.** $(3x - 2y + 4z)^2$

73. $(x + 1)(x - 2)^2$ **74.** $[2x - (x + 1)^2]^3$

75. $(a^k + 1)(a^{2k} - a^k + 1)$

76. $(x^m - y^n)(x^{2m} + x^m y^n + y^{2n})$

77. $(4x^4 - 3) \div (3x + 2)$

78. $(3x^5 - 2x^4 + 5x - 4) \div (2x^2 + 1)$

79. $(x^{1/2} + x^{-1/2})^8$ **80.** $\left(2x - \dfrac{1}{2y^2}\right)^7$

81. $x^5 - x[2 - (2 - x)^2]^2$

7 FACTORING

To **factor** an algebraic expression means to write it as a product. So, factoring is just the reverse of multiplication. One might reasonably ask why it is important to factor anything. Some of the reasons are: (1) to simplify a fraction by dividing out common factors, (2) to help recognize the lowest common denominator in adding fractions, (3) to facilitate the solution of equations and inequalities, and (4) to help analyze the sign of an algebraic expression.

The most common factoring problems fall into one of the categories described below.

COMMON FACTORS

The distributive law provides justification for an important factoring technique. When read from right to left, this law says,

$$ab + ac = a(b + c)$$

Since a appears as a factor of each term on the left, it is referred to as a **common factor.** Applying the distributive law to obtain the right-hand side is referred to as "factoring out the common factor." This technique is illustrated in the following examples.

EXAMPLE 18

$$4xy^2 + 6x^3y = 2xy(2y + 3x^2) \qquad \blacktriangle$$

EXAMPLE 19

$$3a^2b^3 - 6a^4b^2c + 9a^3b^4c^2 = 3a^2b^2(b - 2a^2c + 3ab^2c^2) \qquad \blacktriangle$$

EXAMPLE 20

$$2x^{3/2} - 4x^{1/2} = 2x^{1/2}(x - 2) \qquad \blacktriangle$$

Sometimes common factors consist of more than one term. **Common binomial factors** occur in the next three examples.

EXAMPLE 21

$$(a + 2)b + (a + 2)c = (a + 2)(b + c)$$

The common factor is the binomial $(a + 2)$. It is treated as a unit and factored out just as if it were a single term.   ▲

EXAMPLE 22

$$2x^2 + 4xy + xy^2 + 2y^3 = (2x^2 + 4xy) + (xy^2 + 2y^3)$$
$$= 2x\,(x + 2y) + y^2\,(x + 2y)$$
$$= (2x + y^2)\,(x + 2y)$$

The initial step here is called **grouping.** We grouped the first two terms together and the last two together. After factoring out the common factor $2x$ from the first group and y^2 from the second, we recognize the binomial factor $(x + 2y)$ as being common to the two groups.   ▲

EXAMPLE 23

$$ax + 9by - 3ay - 3bx = ax - 3ay - 3bx + 9by$$
$$= a\,(x - 3y) - 3b\,(x - 3y)$$
$$= (a - 3b)\,(x - 3y)$$

Here it was necessary to group the first and third terms together and the second and fourth together. A common error occurs in problems such as this because of the negative signs. The troublesome step is

$$-3bx + 9by = -3b(x - 3y)$$

The point to remember is that when a negative term is factored out, the signs inside the parentheses must be changed.   ▲

THE PERFECT SQUARE AND THE GENERAL TRINOMIAL

Equations (18) and (19) of Section 5 provide useful formulas for factoring. Written in reverse order, these become

$$a^2 + 2ab + b^2 = (a + b)^2$$
$$a^2 - 2ab + b^2 = (a - b)^2$$

Consider, for example, the problem of factoring $4x^2 + 20x + 25$. This can be written as $(2x)^2 + 20x + (5)^2$. So it could be of the form $(a + b)^2$, with $a = 2x$ and $b = 5$. We check the middle term $2ab = 2(2x)5 = 20x$ and find it is correct. So we see that

$$4x^2 + 20x + 25 = (2x + 5)^2$$

EXAMPLE 24

$$4x^2 + 4x + 1 = (2x)^2 + 2(2x) \cdot 1 + 1^2 = (2x + 1)^2$$ ▲

EXAMPLE 25

$$9x^2 - 12xy + 4y^2 = (3x)^2 - 2(3x)(2y) + (2y)^2 = (3x - 2y)^2$$ ▲

EXAMPLE 26

$$a^4 + 8a^2 + 16 = (a^2)^2 + 2(a^2) \cdot 4 + 4^2 = (a^2 + 4)^2$$ ▲

EXAMPLE 27

$$18x^3y - 60x^2y^2 + 50xy^3 = 2xy(9x^2 - 30xy + 25y^2)$$
$$= 2xy[(3x)^2 - 2(3x)(5y) + (5y)^2]$$
$$= 2xy(3x - 5y)^2$$

Here we first factored out the common factor $2xy$ and then recognized the other factor as a perfect square. Any existing common factor should always be factored out first.   ▲

A perfect square of the sum of three or more terms is harder to recognize than that of a binomial, but a clue can be found from squaring a trinomial, as follows.

$$(a + b + c)^2 = [(a + b) + c]^2 = (a + b)^2 + 2(a + b)c + c^2$$
$$= a^2 + 2ab + b^2 + 2ac + 2bc + c^2$$
$$= a^2 + b^2 + c^2 + 2ab + 2ac + 2bc$$

The result can be stated in words as follows:

The square of the sum of three terms equals the sum of the squares of each of the terms, plus twice all products of pairs of different terms.

An analogous result holds true for squares of four or more terms.

A clue to recognizing the square of a trinomial $(a + b + c)^2$, is the presence of the sum of the squares of the individual terms, $a^2 + b^2 + c^2$. Then, you can check to see if you also have twice the product of all **pairs** of different terms. Notice, too, that there must be six terms in all. This is illustrated in the next example.

EXAMPLE 28

$$x^2 + y^2 + 4z^2 + 2xy - 4xz - 4yz = (x + y - 2z)^2 \qquad \blacktriangle$$

A trinomial that is a perfect square of a binomial is, of course, a special situation. Certain trinomials that are not perfect squares can also be factored. We rely here on the multiplication formula (16):

$$(ax + b)(cx + d) = acx^2 + (ad + bc)x + bd$$

To illustrate, consider $6x^2 + 17x + 12$. To put this into the form $(ax + b)(cx + d)$, there are various possibilities to consider. The product of a and c must be 6, and b times d must be 12. The key is the middle term. In this case, we find (after several attempts) that the correct combination is $(2x + 3)(3x + 4)$. This is a trial-and-error procedure, but the number of trials is finite. Often, trials can be made mentally.

EXAMPLE 29

$$x^2 - 2x - 8 = (x - 4)(x + 2) \qquad \blacktriangle$$

EXAMPLE 30

$$4x^2 + 5x - 6 = (4x - 3)(x + 2) \qquad \blacktriangle$$

EXAMPLE 31

$$2x^2 + xy - 15y^2 = (2x - 5y)(x + 3y) \qquad \blacktriangle$$

EXAMPLE 32

$$10x^4 - 11x^2 - 6 = (5x^2 + 2)(2x^2 - 3) \qquad \blacktriangle$$

DIFFERENCE OF SQUARES

Equation (17) in reverse order reads

$$a^2 - b^2 = (a + b)(a - b)$$

and this is an important factoring formula called the **difference of squares.** The examples below illustrate its use.

EXAMPLE 33 ||||||||||||||||||||||||||||||||||||||||||||||||||||||||||||||||||||||||||||||||||||||||||||||||||

Factor each of the following:

a. $x^2 - 4$ **b.** $4x^2 - 9y^2$

c. $a^4 - b^4$ **d.** $3y^2 - 5$ (Factor over the reals.)

Solution   **a.** $x^2 - 4 = (x - 2)(x + 2)$ **b.** $4x^2 - 9y^2 = (2x - 3y)(2x + 3y)$

c. $a^4 - b^4 = (a^2 - b^2)(a^2 + b^2) = (a - b)(a + b)(a^2 + b^2)$

d. $3y^2 - 5 = (\sqrt{3}\,y + \sqrt{5})(\sqrt{3}\,y - \sqrt{5})$ ▲

Note. A polynomial that cannot be written as a product of two other polynomials of positive degree is said to be **irreducible.** Part d of Example 32 is therefore irreducible over the rationals but is reducible over the reals. In the next chapter we will learn how to tell whether a second degree polynomial is irreducible over the reals.

EXAMPLE 34 ||||||||||||||||||||||||||||||||||||||||||||||||||||||||||||||||||||||||||||||||||||||||||||||||||

Factor:

a. $x^2 - 4x + 4 - y^2$ **b.** $a^2 - b^2 + 2b - 1$

Solution   In both of these it is necessary to group the terms first.

a.
$$x^2 - 4x + 4 - y^2 = (x^2 - 4x + 4) - y^2$$
$$= (x - 2)^2 - y^2$$
$$= [(x - 2) - y][(x - 2) + y]$$
$$= (x - y - 2)(x + y - 2)$$

b.
$$a^2 - b^2 + 2b - 1 = a^2 - (b^2 - 2b + 1)$$
$$= a^2 - (b - 1)^2$$
$$= [a - (b - 1)][a + (b - 1)]$$
$$= (a - b + 1)(a + b - 1)$$ ▲

Note that in Example 34 we grouped three terms that formed a perfect square together in order to recognize the difference of two squares. Contrast this with the way we grouped terms in Examples 22 and 23. There we grouped in pairs and recognized a common binomial factor. So in one case the grouping was 3 and 1, whereas in the other it was 2 and 2. When you have a four-term expression to factor, you should be alert to both possibilities and see if one of them works. (It may be that the expression cannot be factored.)

In the next example we illustrate how certain fourth degree polynomials can be factored as the difference of two squares by adding and subtracting an appropriate term.

EXAMPLE 35

Factor:
a. $x^4 + x^2 + 1$ **b.** $a^4 + 4$

Solution **a.** The given expression would be a perfect square if the middle term were $2x^2$ instead of x^2. So we add x^2 and then compensate by subtracting x^2:

$$x^4 + x^2 + 1 = (x^4 + 2x^2 + 1) - x^2$$
$$= (x^2 + 1)^2 - x^2$$
$$= [(x^2 + 1) + x][(x^2 + 1) - x]$$
$$= (x^2 + x + 1)(x^2 - x + 1)$$

b. Following the same procedure as in part **a**, we add and subtract $4a^2$, since this is what is needed to make the given expression a perfect square:

$$a^4 + 4 = (a^4 + 4a^2 + 4) - 4a^2$$
$$= (a^2 + 2)^2 - 4a^2$$
$$= [(a^2 + 2) + 2a][(a^2 + 2) - 2a]$$
$$= (a^2 + 2a + 2)(a^2 - 2a + 2) \qquad \blacktriangle$$

SUM AND DIFFERENCE OF CUBES

Since the difference of two squares factors in such a nice way, we might be led to consider the difference of two cubes, $a^3 - b^3$. A reasonable guess is that one of the factors is $a - b$. This is, in fact, correct, and the complete factorization is given by

$$a^3 - b^3 = (a - b)(a^2 + ab + b^2)$$

You can verify this by multiplying the factors on the right.
There is a similar result for the sum of cubes:

$$a^3 + b^3 = (a + b)(a^2 - ab + b^2)$$

So while the sum of squares, $a^2 + b^2$, is irreducible, the sum of cubes can be factored.

A Word of Caution. Note carefully that $a^3 - b^3 \neq (a - b)^3$, and $a^3 + b^3 \neq (a + b)^3$.

EXAMPLE 36

Factor:

a. $x^3 - 8$ **b.** $a^3 + 27$ **c.** $8x^3 - 125y^3$ **d.** $x^6 - 1$ **e.** $x^6 + 1$

Solution   **a.** $x^3 - 8 = (x - 2)(x^2 + 2x + 4)$
b. $a^3 + 27 = (a + 3)(a^2 - 3a + 9)$
c. $8x^3 - 125y^3 = (2x)^3 - (5y)^3 = (2x - 5y)(4x^2 + 10xy + 25y^2)$
d. This problem is best handled by first factoring $x^6 - 1$ as the difference of squares, resulting in the difference and sum of cubes:

$$x^6 - 1 = (x^3 - 1)(x^3 + 1) = (x - 1)(x^2 + x + 1)(x + 1)(x^2 - x + 1)$$

e. We treat this as $(x^2)^3 + 1$ to get

$$x^6 + 1 = (x^2 + 1)(x^4 - x^2 + 1)$$

▲

USE OF THE BINOMIAL THEOREM

There are occasions when a polynomial can be recognized as being a perfect nth power of a binomial. For example, we saw in Section 6 that

$$a^3 + 3a^2b + 3ab^2 + b^3 = (a + b)^3$$

and that

$$a^4 + 4a^3b + 6a^2b^2 + 4ab^3 + b^4 = (a + b)^4$$

The best approach here is to be alert to the possibility that a polynomial might be a perfect nth power by observation of the exponents; then check by means of the binomial theorem to see if this is indeed the case.

EXERCISE SET 7

Factor all problems completely with rational coefficients unless otherwise specified.

Problems 1–10 involve common factors.

1. a. $3a^2b + 6ab^2$ **b.** $4a^2b^3 - 6a^4b^2$
2. a. $6x^2y^4z^3 + 9x^3y^2z^4$
    b. $35x^6y^4z - 42x^4y^3$
3. a. $24a^2b^3 - 30a^3b^5c + 48a^4b^2c^2$
    b. $126x^3y^2 - 168x^4y$

4. a. $18x^{1/2} + 27x^{3/2}$ **b.** $2x - 4\sqrt{x}$
5. a. $3x^{4/3} + 6x^{1/3}$ **b.** $32\sqrt[3]{x} + 40\sqrt[3]{x^4}$
6. a. $a(b - 3) + 2(b - 3)$
    b. $y(2x - 3) - 4(2x - 3)$
7. a. $(x + 3)y - (x + 3)$
    b. $(a - 2) - b(a - 2)$
8. a. $xy - y + x - 1$ **b.** $ab - ac + bd - cd$
9. a. $x^2 - xy - x + y$ **b.** $xy^2 - x^2y - 2x + 2y$
10. a. $a^2b + 3ab^2 - 2a - 6b$
    b. $2x^3 + 3xy^2 - 2x^2y - 3y^3$

Problems 11–20 involve perfect squares and general trinomials.

11. a. $x^2 - 6x + 9$ **b.** $a^2 + 4a + 4$
12. a. $t^2 + 10t + 25$ **b.** $16m^2 - 8m + 1$
13. a. $36a^2 + 84ab + 49b^2$
  b. $81s^2 - 72st + 16t^2$
14. a. $x^{2k} - 8x^k + 16$ **b.** $9x^{2n} - 12x^n + 4$
15. a. $x^2 - 5x + 4$ **b.** $a^2 + 8a + 12$
16. a. $x^2 + 3x - 4$ **b.** $2x^2 - x - 1$
17. a. $2x^2 + 7x - 4$ **b.** $3x^2 - 7x - 6$
18. a. $6x^2 - 17x + 12$ **b.** $8x^2 - 6xy - 5y^2$
19. a. $20x^2 + 3x - 9$ **b.** $24x^2 - 49x + 15$
20. a. $x^4 + 6x^2 + 9$ **b.** $4x^4 + 20x^2 + 25$
21. a. $2x^4 + 5x^2 + 2$ **b.** $6x^4 + 17x^2 + 12$
22. a. $x^{2k} + 4x^k - 5$ **b.** $3x^{2m} - 5x^m y^n - 2y^{2n}$

Problems 23–34 involve difference of squares and sum and difference of cubes.

23. a. $9x^2 - 4$ **b.** $25a^2 - 16b^2$
24. a. $x^4 - 4x^2$ **b.** $a^3 - ab^2$
25. a. $16a^4 - 81b^4$ **b.** $x^5 - 256x$
26. a. $x^8 - 1$ **b.** $147a^3 - 3a^5$
27. a. $(x + 3)^2 - y^2$ **b.** $x^2 - (y + 1)^2$
28. a. $x^2 - 4x + 4 - y^2$
  b. $4x^2 - y^2 + 6y - 9$
29. a. $x^2 - 4y^2 - 2x + 1$
  b. $9x^2 - 9 - 4y^2 + 12y$
30. a. $4a^2 - 12ab - 9 + 9b^2$
  b. $25 - 20xy - 25x^2 - 4y^2$
31. a. $x^4 - 3x^2 + 1$ **b.** $a^4 - 8a^2 + 4$
32. a. $4x^4 + 1$ **b.** $x^4 + 2x^2 + 9$
33. a. $x^3 - 27$ **b.** $8a^3 + 1$
34. a. $27x^3 + 8y^3$ **b.** $m^6 - 125n^3$

Problems 35–40 involve use of the binomial theorem.

35. $x^3 - 3x^2 y + 3xy^2 - y^3$
36. $8x^3 + 12x^2 + 6x + 1$
37. $a^4 - 8a^3 + 24a^2 - 32a + 16$
38. $x^4 - 4x^3 y + 6x^2 y^2 - 4xy^3 + y^4$
39. $1 + 5x + 10x^2 + 10x^3 + 5x^4 + x^5$
40. $x^6 - 3x^4 y + 3x^2 y^2 - y^3$

Factor Problems 41–90 in whatever way is appropriate.

41. $9x^3 - 49x$ **42.** $12x^2 - 14x - 40$

43. $x^4 - 13x^2 + 36$ **44.** $9x^3 - 30x^2 + 9x$
45. $a^3 + a^2 b - ab^2 - b^3$
46. $x^4 - 18x^2 + 81$
47. $x^2 - y^2 - 4y - 4$ **48.** $x^2 - y^2 - 4y - 4x$
49. $36x^2 - 37xy - 48y^2$ **50.** $8x^3 + 64$
51. $a^6 - 64$ **52.** $10x^2 + x - 2$
53. $2a^3 b - 8a^2 b^2 + 8ab^3$
54. $9x^4 + 3x^2 + 4$
55. $xy + x - y - 1$
56. $2x^4 - 7x^2 - 4$
57. $20a^2 + 37ab - 18b^2$
58. $8t^4 - 64t$
59. $2x^2 + 3y - xy - 6x$
60. $4a^4 + 8a^2 b^2 + 9b^4$
61. $t^3 - 9t^2 + 27t - 27$
62. $a^2 - 4b - 4b^2 - 1$
63. $42x^3 y - 20x^2 y^2 - 32xy^3$
64. $4xy - 15 + 5x - 12y$
65. $8a^3 - 36a^2 b^2 + 54ab^4 - 27b^6$
66. $(2x - 3)^{4/3} - x(2x - 3)^{1/3}$
67. $x^6 + 1$ **68.** $98x^2 - 84xy + 18y^2$
69. $36x^3 - 58x^2 - 80x$
70. $x^2\sqrt{x^2 + 1} - \sqrt{(x^2 + 1)^3}$

B

71. $36x^3 y^2 - 60x^2 y^3 - 75xy^4$
72. $162x^3 y + 180x^2 y^2 + 32xy^3$
73. $30a^2 bc - 48b^3 c - 36ab^2 c$
74. $256 - x^8$
75. $x^2 + 4y^2 + 9z^2 + 4xy + 6xz + 12yz$
76. $x^2 + 4y^2 + 1 + 4xy - 2x - 4y$
77. $36x^4 - 64x^2 + 25$
78. $27a^3 + 54a^2 b + 36ab^2 + 8b^3$
79. $x^3 - y^3 - x + y$
80. $x^2 - 4xy + 4y^2 - 2x + 4y + 1$
81. $x^2 - 2xy + y^2 - a^2 - 4a - 4$
82. $x^6 - 7x^3 y^3 - 8y^6$ **83.** $x^{12} - y^{12}$
84. $x^5 + 10x^4 + 40x^3 + 80x^2 + 80x + 32$
85. $(x^2 + 2)^{5/2} - 6(x^2 + 2)^{3/2}$
86. $9x^4 - 16$ (Factor over the reals.)
87. $9x^4 + 16$ (Factor over the reals.)
88. $x^2 - y^2 - z^2 + 2yz - 4x + 4$
  Hint. Rearrange and use groups of three.
89. $x(3x^2 + 1)^{4/3} - 4x^3(3x^2 + 1)^{1/3}$
90. $9y^2 - 9x^2 + 12x - 4$

RATIONAL FRACTIONS

Fractions that are quotients of polynomials are called **rational.** Examples are

$$\frac{x + 3}{x^2 - 2x + 5} \qquad \frac{1}{x + 2} \qquad \frac{x^6 - y^6}{x^2 + 2xy + y^2}$$

The arithmetic associated with these is governed by rules already established in Section 3. Factoring plays a key role in carrying out the steps involved. When a common factor appears in the numerator and denominator, we make use of the fundamental property of fractions to reduce the fraction to lowest terms. To "simplify" an expression involving rational fractions means to perform the indicated operations and express the result as a simple fraction in which no common factor (other than 1) exists in the numerator and denominator.

The examples below illustrate the most important techniques.

EXAMPLE 37

Perform the indicated operations and simplify.

a. $\dfrac{x^2 - x - 6}{x^2 - 6x + 9}$ **b.** $\dfrac{x^2 + 4x + 4}{x^2 - 3x - 4} \cdot \dfrac{2x^2 + x - 1}{x^2 - 3x - 10}$

c. $\dfrac{2x^2 - 6x}{2x^2 - x - 10} \div \dfrac{x^2 - 5x + 6}{x^2 - 4}$

Solution

a. $\dfrac{x^2 - x - 6}{x^2 - 6x + 9} = \dfrac{\overset{1}{\cancel{(x - 3)}}(x + 2)}{(x - 3)\overset{1}{\cancel{2}}} = \dfrac{x + 2}{x - 3}$

b. $\dfrac{x^2 + 4x + 4}{x^2 - 3x - 4} \cdot \dfrac{2x^2 + x - 1}{x^2 - 3x - 10} = \dfrac{(x + 2)\overset{1}{\cancel{2}}}{(x - 4)\cancel{(x + 1)}} \cdot \dfrac{(2x - 1)\overset{1}{\cancel{(x + 1)}}}{(x - 5)\cancel{(x + 2)}}$

$$= \dfrac{(x + 2)(2x - 1)}{(x - 4)(x - 5)} = \dfrac{2x^2 + 3x - 2}{x^2 - 9x + 20}$$

c. From Section 3 we know that to divide, we invert the denominator and multiply:

$$\dfrac{2x^2 - 6x}{2x^2 - x - 10} \div \dfrac{x^2 - 5x + 6}{x^2 - 4} = \dfrac{2x^2 - 6x}{2x^2 - x - 10} \cdot \dfrac{x^2 - 4}{x^2 - 5x + 6}$$

$$= \dfrac{2x\overset{1}{\cancel{(x - 3)}}}{(2x - 5)\cancel{(x + 2)}} \cdot \dfrac{\overset{1}{\cancel{(x + 2)}}\overset{1}{\cancel{(x - 2)}}}{\cancel{(x - 3)}\cancel{(x - 2)}}$$

$$= \dfrac{2x}{2x - 5}$$

The basic procedure to follow in adding (or subtracting) rational expressions is to write each fraction with the same denominator (the LCD) and then add (or subtract) their numerators, as in equation (3). To do this, you will usually make use of the fundamental property of fractions in the form

$$\frac{a}{b} = \frac{a \cdot c}{b \cdot c} \qquad (c \neq 0)$$

which says that a fraction is unchanged if both numerator and denominator are multiplied by the same nonzero number. Recall that the justification for this is that $c/c = 1$, and a number multiplied by 1 is unchanged. So

$$\frac{a}{b} = \frac{a}{b} \cdot 1 = \frac{a}{b} \cdot \frac{c}{c} = \frac{ac}{bc}$$

We illustrate this procedure in the next three examples.   ▲

EXAMPLE 38

Perform the indicated operations.

a. $\dfrac{1}{x-1} + \dfrac{3}{x+2}$ **b.** $\dfrac{x}{2x-1} - \dfrac{3}{x+2}$

Solution   **a.**   The LCD is the product $(x-1)(x+2)$. We change each fraction to one having this expression as its denominator as follows:

$$\frac{1}{x-1} + \frac{3}{x+2} = \frac{1}{x-1} \cdot \frac{x+2}{x+2} + \frac{3}{x+2} \cdot \frac{x-1}{x-1}$$

$$= \frac{x+2}{(x-1)(x+2)} + \frac{3(x-1)}{(x+2)(x-1)}$$

Now the fractions can be added to give

$$\frac{x+2+3x-3}{(x-1)(x+2)} = \frac{4x-1}{(x-1)(x+2)}$$

With a little practice the work can be shortened as follows:

$$\frac{1}{x-1} + \frac{3}{x+2} = \frac{(x+2) + 3(x-1)}{(x-1)(x+2)}$$

$$= \frac{x+2+3x-3}{(x-1)(x+2)}$$

$$= \frac{4x-1}{(x-1)(x+2)}$$

b.   The basic idea is the same as in part a.

$$\frac{x}{2x - 1} - \frac{3}{x + 2} = \frac{x}{2x - 1} \cdot \frac{x + 2}{x + 2} - \frac{3}{x + 2} \cdot \frac{2x - 1}{2x - 1}$$

— **Be careful here.**

$$= \frac{x^2 + 2x - 6x + 3}{(2x - 1)(x + 2)}$$

$$= \frac{x^2 - 4x + 3}{(2x - 1)(x + 2)}$$

$$= \frac{(x - 3)(x - 1)}{(2x - 1)(x + 2)}$$

Note. We factored the numerator to see if it had a factor in common with the denominator. In this case it did not, but it is good practice to see if the fraction can be reduced by writing both numerator and denominator in factored form.   ▲

EXAMPLE 39

Combine and simplify: $\dfrac{2}{x + y} - \dfrac{x - y}{x^2 + 3xy + 2y^2} + \dfrac{3}{x + 2y}$

Solution   By factoring the denominator of the second fraction we recognize the LCD as $(x + y)(x + 2y)$. To obtain equivalent fractions having this LCD as the denominator, the numerator of the first fraction must be multiplied by $(x + 2y)$ and the numerator of the third fraction must be multiplied by $(x + y)$. This gives

$$\frac{2}{x + y} - \frac{x - y}{(x + 2y)(x + y)} + \frac{3}{x + 2y} = \frac{2(x + 2y) - (x - y) + 3(x + y)}{(x + y)(x + 2y)}$$

$$= \frac{2x + 4y - x + y + 3x + 3y}{(x + y)(x + 2y)}$$

$$= \frac{4x + 8y}{(x + y)(x + 2y)}$$

$$= \frac{4\overset{1}{\cancel{(x + 2y)}}}{(x + y)\underset{1}{\cancel{(x + 2y)}}}$$

$$= \frac{4}{x + y}$$ ▲

EXAMPLE 40

Combine and simplify: $\dfrac{1}{2x + 4} - \dfrac{2}{2 - x} + \dfrac{4}{3x - 6}$

Solution   The first thing to observe is that the denominator of the second fraction would be easier to deal with if it were $x - 2$ instead of $2 - x$. Since these factors are negatives of each other, this change can be accomplished by using the fact that $a/(-b) = -(a/b)$. Thus,

$$\frac{2}{2 - x} = -\frac{2}{x - 2}$$

so that the problem becomes

$$\frac{1}{2(x + 2)} + \frac{2}{x - 2} + \frac{4}{3(x - 2)} = \frac{3(x - 2) + 12(x + 2) + 8(x + 2)}{6(x + 2)(x - 2)}$$

$$= \frac{3x - 6 + 12x + 24 + 8x + 16}{6(x + 2)(x - 2)}$$

$$= \frac{23x + 34}{6(x^2 - 4)} \qquad \blacktriangle$$

The next three examples involve complex fractions. They are handled just as in Section 3.

EXAMPLE 41

Simplify: $\dfrac{\dfrac{1}{a^2} - \dfrac{2}{ab}}{\dfrac{1}{2ab} - \dfrac{1}{b^2}}$

Solution   We multiply numerator and denominator of the main fraction by the LCD of all minor denominators—in this case $2a^2b^2$.

$$\frac{\left(\dfrac{1}{a^2} - \dfrac{2}{ab}\right) \cdot (2a^2b^2)}{\left(\dfrac{1}{2ab} - \dfrac{1}{b^2}\right) \cdot (2a^2b^2)} = \frac{2b^2 - 4ab}{ab - 2a^2}$$

$$= \frac{2b(b - 2a)}{a(b - 2a)}$$

$$= \frac{2b}{a}$$

Note how we mentally divided out each minor denominator as we multiplied by the LCD. $\blacktriangle$

EXAMPLE 42

Simplify:
$$\dfrac{\dfrac{1}{x-1}+\dfrac{1}{x^2-1}}{x-\dfrac{2}{x+1}}$$

Solution   Here the LCD of the minor denominators is $(x+1)(x-1)=x^2-1$. Proceeding as above, we obtain

$$\dfrac{\left(\dfrac{1}{x-1}+\dfrac{1}{x^2-1}\right)\cdot(x+1)(x-1)}{\left(x-\dfrac{2}{x+1}\right)\cdot(x+1)(x-1)}=\dfrac{\dfrac{1}{x-1}(x+1)(x-1)+\dfrac{1}{x^2-1}(x+1)(x-1)}{x(x+1)(x-1)-\dfrac{2}{x+1}(x+1)(x-1)}$$

$$=\dfrac{x+1+1}{x(x+1)(x-1)-2(x-1)}$$

$$=\dfrac{x+2}{(x-1)(x^2+x-2)}$$

$$=\dfrac{x+2}{(x-1)(x+2)(x-1)}$$

$$=\dfrac{1}{(x-1)^2}$$

Note that in the third step the factors appearing in the terms of the denominator were not multiplied out. This enabled us to recognize the common factor $(x-1)$. If the multiplication had been done, the result would have been x^3-3x+2, and factoring this would have been far more difficult. (Try it!) ▲

EXAMPLE 43

Simplify:
$$\dfrac{\dfrac{x}{x-1}-2}{x-2}$$

Solution

$$\dfrac{\left(\dfrac{x}{x-1}-2\right)\cdot(x-1)}{(x-2)\cdot(x-1)}=\dfrac{x-2(x-1)}{(x-2)(x-1)}$$

$$=\dfrac{x-2x+2}{(x-2)(x-1)}=\dfrac{-x+2}{(x-2)(x-1)}$$

$$=-\dfrac{x-2}{(x-2)(x-1)}=-\dfrac{1}{x-1}$$

Note carefully the handling of signs in this problem. ▲

EXERCISE SET 8

A

Perform the indicated operations. Reduce final answers to simplest form.

1. a. $\dfrac{x^2 - 4}{x^2 - x - 2}$ **b.** $\dfrac{2x^2 + 5x - 3}{x^2 + 6x + 9}$

2. a. $\dfrac{1 - x^2}{x^3 - 2x^2 + x}$ **b.** $\dfrac{3x^2 + 4x - 4}{6x^2 + 5x - 6}$

3. $\dfrac{2x^2 - x - 6}{x^2 - 4} \cdot \dfrac{x^2 + 4x + 4}{2x^2 + x - 3}$

4. $\dfrac{2x^2 + 6x}{x^2 - 4x + 4} \cdot \dfrac{3x^2 - 7x + 2}{3x^3 + 8x^2 - 3x}$

5. $\dfrac{6x^2 + xy - 12y^2}{2x^2 + xy - 3y^2} \cdot \dfrac{x^2 - 2xy + y^2}{3x^2 - xy - 4y^2}$

6. $\dfrac{x^2 - 5x - 6}{x^3 - 1} \div \dfrac{x^2 + 3x + 2}{x^2 + x - 2}$

7. $\dfrac{2x^2 + 3x}{x^2 - 4} \div \dfrac{2x^2 - x - 6}{3x^2 - 2x - 8}$

8. $\dfrac{x^2 + 8x + 16}{x^2 + 3x - 4} \div \dfrac{2x^2 + x - 1}{x^2 - 1}$

9. $\dfrac{2}{x + 2} + \dfrac{3}{x - 4}$ **10.** $\dfrac{x}{x - 2} + \dfrac{2}{x + 1}$

11. $\dfrac{3}{x - 5} - \dfrac{4}{x + 5}$ **12.** $\dfrac{3}{2x - 1} - \dfrac{1}{x + 4}$

13. $\dfrac{x + 1}{3x - 4} - \dfrac{x - 2}{x - 1}$ **14.** $\dfrac{2x - 1}{x + 2} - \dfrac{x - 2}{x + 1}$

15. $\dfrac{3}{x + 3} - \dfrac{2}{x - 3} + \dfrac{12}{x^2 - 9}$

16. $\dfrac{5}{2x - 1} - \dfrac{2}{x + 3} - \dfrac{14}{2x^2 + 5x - 3}$

17. $\dfrac{1}{x - 3} + \dfrac{3}{x + 2} - \dfrac{2x - 1}{x^2 - x - 6}$

18. $\dfrac{3}{a + 1} + \dfrac{2}{a - 1} - \dfrac{a - 5}{x^2 - 1}$

19. $\dfrac{1}{x} - \dfrac{2}{x - 2} + 1$ **20.** $\dfrac{x^2 - 5}{x + 3} - x + 5$

21. $\dfrac{x^2 + 2}{x - 3} - x + 4$ **22.** $1 + \dfrac{3}{x - 2} - \dfrac{1}{2 - x}$

23. $\dfrac{2x - 5}{x^2 + x - 2} - \dfrac{x - 3}{x^2 - 1} + \dfrac{x - 3}{x^2 + 3x + 2}$

24. $\dfrac{3}{x + 3} - \dfrac{x}{x - 3} - \dfrac{2x^2}{9 - x^2}$

25. $\dfrac{5}{2x^2 + 3x - 2} + \dfrac{3}{2 - x - x^2}$

26. $\dfrac{\dfrac{1}{x^2} - \dfrac{1}{9}}{x - 3}$ **27.** $\dfrac{\dfrac{x}{y} - \dfrac{y}{x}}{\dfrac{1}{y^2} - \dfrac{1}{x^2}}$

28. $\dfrac{\dfrac{x}{x - 2} - 2}{x - 4}$ **29.** $\dfrac{\dfrac{8}{x^2 - 1} - \dfrac{1}{x - 1}}{\dfrac{1}{x - 1} - \dfrac{2}{x + 1}}$

30. $\dfrac{\dfrac{1}{x - 1} - \dfrac{1}{x + 1}}{1 + \dfrac{1}{x^2 - 1}}$ **31.** $\dfrac{a - 6 + \dfrac{20}{a + 3}}{a + 3 - \dfrac{16}{a + 3}}$

32. $\dfrac{\dfrac{x^2}{y} + \dfrac{y^2}{x}}{\dfrac{x}{y} + \dfrac{y}{x} - 1}$ **33.** $\dfrac{\dfrac{4 + h}{2 + h} - 2}{h}$

34. $\dfrac{\dfrac{2}{1 + h} - 2}{h}$ **35.** $\dfrac{\dfrac{x - 2}{x + 1} - \dfrac{1}{4}}{x - 3}$

36. $\dfrac{\dfrac{x}{x + 1} - \dfrac{a}{a + 1}}{x - a}$ **37.** $\dfrac{\dfrac{1}{(x + 1)^2} - \dfrac{1}{(a + 1)^2}}{x - a}$

38. $\dfrac{\dfrac{2x - 3}{x^2} + 1}{x + 3}$ **39.** $\dfrac{\left(\dfrac{2 + h}{1 + h}\right)^2 - 4}{h}$

B

40. $\dfrac{x^2 + 3}{x^4 + x} - \dfrac{1}{x^2 - 1}$

41. $\dfrac{x + 2}{2x^2 - 7x + 6} + \dfrac{x - 3}{2x^2 - x - 3} - \dfrac{x - 1}{x^2 - x - 2}$

42. $\dfrac{2(x + 3)^3(x - 2) - 3(x + 3)^2(x - 2)^2}{(x + 3)^6}$

43. $\dfrac{2x(2x-1)^2 - 4(x^2-1)(2x-1)}{(2x-1)^4}$

46. $\dfrac{3x^2 + 10x - 8}{2 - \dfrac{3}{1 - \dfrac{x}{x-2}}}$

47. $\dfrac{1 + x^3}{1 - \dfrac{x}{1 + \dfrac{x}{1-x}}}$

44. $\dfrac{\dfrac{x+h+2}{x+h-1} - \dfrac{x+2}{x-1}}{h}$

45. $\dfrac{\dfrac{x^2}{2-3x} - \dfrac{a^2}{2-3a}}{x-a}$

Note. Problems 33–39 and 42–45 were taken directly from calculus.

9 OTHER FRACTIONAL EXPRESSIONS

All the rules for working with rational fractions continue to hold for fractions involving negative and fractional exponents or radicals. To simplify such expressions means to perform indicated operations, to divide out common factors, and to write the answer with positive exponents.

EXAMPLE 44

Simplify: $\dfrac{x^{-1} - y^{-1}}{x^{-2} - y^{-2}}$

Solution   We replace each term by its equivalent expressed with a positive exponent and then proceed as in Section 8.

$$\frac{x^{-1} - y^{-1}}{x^{-2} - y^{-2}} = \frac{\dfrac{1}{x} - \dfrac{1}{y}}{\dfrac{1}{x^2} - \dfrac{1}{y^2}} \cdot \frac{x^2 y^2}{x^2 y^2} = \frac{xy^2 - x^2 y}{y^2 - x^2}$$

$$= \frac{xy(y - x)}{(y+x)(y-x)} = \frac{xy}{y+x} \quad \blacktriangle$$

EXAMPLE 45

Simplify: $x(1-x)^{-3} + (1-x)^{-2}$

Solution   There are two ways to proceed.

a.   Write with positive exponents first.

$$x(1-x)^{-3} + (1-x)^{-2} = \frac{x}{(1-x)^3} + \frac{1}{(1-x)^2}$$

$$= \frac{x + (1-x)}{(1-x)^3} = \frac{1}{(1-x)^3}$$

b.   Factor out the common factor $(1 - x)^{-3}$. It is important to take out the factor with the negative exponent that is largest in absolute value, so that the remaining factor will have a positive exponent.

$$x(1 - x)^{-3} + (1 - x)^{-2} = (1 - x)^{-3}(x + 1 - x) = \frac{1}{(1 - x)^3} \quad \blacktriangle$$

EXAMPLE 46

Simplify: $\dfrac{(x^2 - 1)^{1/2} - x^2(x^2 - 1)^{-1/2}}{x^2 - 1}$

Solution   Just as with a complex fraction, we multiply the numerator and denominator by the same quantity. This time we choose the factor that will eliminate all negative powers, namely, $(x^2 - 1)^{1/2}$:

$$\frac{(x^2 - 1)^{1/2} - x^2(x^2 - 1)^{-1/2}}{x^2 - 1} \cdot \frac{(x^2 - 1)^{1/2}}{(x^2 - 1)^{1/2}} = \frac{x^2 - 1 - x^2}{(x^2 - 1)^{3/2}} = \frac{-1}{(x^2 - 1)^{3/2}} \quad \blacktriangle$$

EXAMPLE 47

Simplify: $\dfrac{\dfrac{x^2}{\sqrt{x^2 - 1}} - \sqrt{x^2 - 1}}{x^2}$

Solution   We multiply the numerator and denominator by $\sqrt{x^2 - 1}$ to obtain

$$\frac{\dfrac{x^2}{\sqrt{x^2 - 1}} - \sqrt{x^2 - 1}}{x^2} \cdot \frac{\sqrt{x^2 - 1}}{\sqrt{x^2 - 1}} = \frac{x^2 - (x^2 - 1)}{x^2\sqrt{x^2 - 1}} = \frac{1}{x^2\sqrt{x^2 - 1}} \quad \blacktriangle$$

EXAMPLE 48

Rationalize the denominator.

$$\frac{1}{\sqrt{a} - \sqrt{b}}$$

Solution   To *rationalize* means to make the expression rational—that is, to get rid of radicals. We can accomplish this by multiplying numerator and denominator by $\sqrt{a} + \sqrt{b}$.

$$\frac{1}{\sqrt{a} - \sqrt{b}} = \frac{1}{\sqrt{a} - \sqrt{b}} \cdot \frac{\sqrt{a} + \sqrt{b}}{\sqrt{a} + \sqrt{b}} = \frac{\sqrt{a} + \sqrt{b}}{a - b}$$

Note that we used the fact that *the difference of two numbers times their sum equals the difference of their squares.* $\blacktriangle$

EXAMPLE 49

Rationalize the numerator: $\dfrac{\sqrt{x+2}-2}{x-2}$

Solution This type of problem frequently occurs early in the study of calculus. We make the numerator rational by multiplying the numerator and denominator by $\sqrt{x+2}+2$:

$$\frac{\sqrt{x+2}-2}{x-2}\cdot\frac{\sqrt{x+2}+2}{\sqrt{x+2}+2}=\frac{(x+2)-4}{(x-2)(\sqrt{x+2}+2)}$$

$$=\frac{x-2}{(x-2)(\sqrt{x+2}+2)}$$

$$=\frac{1}{\sqrt{x+2}+2}\quad\blacktriangle$$

Remark. The expressions in Examples 45, 46, 47, and 49, as well as those in many of the exercises that follow, are taken directly from calculus, where they arise in the process of differentiation.

EXERCISE SET 9

A

Simplify the expressions in Problems 1–20. Express answers with positive exponents only.

1. $\dfrac{x^{-1}+y^{-1}}{x^{-1}-y^{-1}}$

2. $\dfrac{x^{-2}-y^{-2}}{x^{-1}+y^{-1}}$

3. $\dfrac{b^{-1}-a^{-1}}{a^{-2}-b^{-2}}$

4. $\dfrac{x^{-1}+2y^{-1}}{x^{-2}+4x^{-1}y^{-1}+4y^{-2}}$

5. $\dfrac{x^{-1}-2}{x^{-2}+2x^{-1}-8}$

6. $\dfrac{2x^{-2}-x^{-1}-3}{1+x^{-1}}$

7. $x^{-1}-x^{-2}(x+4)$

8. $2x^{-1}-2x^{-3}(x^2-1)$

9. $2(x-2)(x+2)^{-1}-(x-2)^2(x+2)^{-2}$

10. $x^2(2x+1)^{-1/2}+2x(2x+1)^{1/2}$

11. $\dfrac{x+2}{2\sqrt{x-1}}+\sqrt{x-1}$

12. $\dfrac{1-x^2}{2\sqrt{x}}-2x^{3/2}$

13. $\dfrac{\sqrt{1-2x}+\dfrac{x}{\sqrt{1-2x}}}{1-2x}$

14. $\dfrac{\dfrac{x}{2\sqrt{x+2}}-\sqrt{x+2}}{x+2}$

15. $\dfrac{(x^2+1)^{1/2}-x^2(x^2+1)^{-1/2}}{x^2+1}$

16. $\dfrac{x^2(4-x^2)^{-1/2}+\sqrt{4-x^2}}{x^2}$

17. $(2x-1)^{-1/2}(4-x)^{1/2}-\dfrac{1}{2}(2x-1)^{1/2}(4-x)^{-1/2}$

18. $2(x^2-1)^{-1/2}-2x^2(x^2-1)^{-3/2}$

19. $\dfrac{2x\sqrt{2x+3}-\dfrac{x^2}{\sqrt{2x+3}}}{2x+3}$

20. $\dfrac{\dfrac{x^2 - 1}{2\sqrt{x}} - 2x\sqrt{x}}{(x^2 - 1)^2}$

In Problems 21–26 rationalize the denominator.

21. $\dfrac{4}{\sqrt{5} - \sqrt{3}}$

22. $\dfrac{\sqrt{3} + 2}{\sqrt{3} - 2}$

23. $\dfrac{\sqrt{2} - 2}{7 - 5\sqrt{2}}$

24. $\dfrac{x - a}{\sqrt{x} + \sqrt{a}}$

25. $\dfrac{3}{\sqrt{x + 3} - \sqrt{x}}$

26. $\dfrac{4}{\sqrt{x^2 + 2} + x}$

In Problems 27–32 rationalize the numerator.

27. $\dfrac{\sqrt{x + h} - \sqrt{x}}{h}$

28. $\dfrac{\sqrt{x} - 3}{x - 9}$

29. $\dfrac{\sqrt{3x - 5} - 1}{x - 2}$

30. $\dfrac{\sqrt{7 - 3x} - 4}{x + 3}$

31. $\dfrac{\sqrt{16 + 3h} - 4}{h}$

32. $\dfrac{\sqrt{3(x + h) - 2} - \sqrt{3x - 2}}{h}$

B

In Problems 33–41 simplify the expressions completely. Express answers with positive exponents only.

33. $\dfrac{x^{-3} - y^{-3}}{xy^{-1} + x^{-1}y + 1}$

34. $\dfrac{\sqrt[3]{x} - \sqrt[3]{a}}{x - a}$

(Rationalize the numerator.)

35. $\dfrac{\dfrac{1}{\sqrt{x + h}} - \dfrac{1}{\sqrt{x}}}{h}$

36. $-\dfrac{8x^2}{9}(x^2 - 1)^{-4/3} + \tfrac{4}{3}(x^2 - 1)^{-1/3}$

37. $\dfrac{(3 - x)^{1/2}(2 + x)^{-1/2} + (3 - x)^{-1/2}(2 + x)^{1/2}}{2(3 - x)}$

38. $\dfrac{(1 - 2x)^{3/2} \cdot 2x - x^2 \cdot \tfrac{3}{2}(-2)(1 - 2x)^{1/2}}{(1 - 2x)^3}$

39. $\dfrac{\sqrt[3]{x^2 - 1} - x^2 \cdot \tfrac{2}{3}(x^2 - 1)^{-2/3}}{(x^2 - 1)^{2/3}}$

40. $\dfrac{\tfrac{1}{3}(2x + 3)^{1/4}(x + 1)^{-2/3} - (x + 1)^{1/3} \cdot \tfrac{1}{2}(2x + 3)^{-3/4}}{\sqrt{2x + 3}}$

41. $\dfrac{\tfrac{2}{3}x(x^2 - 4)^{1/2}(x^2 - 1)^{-2/3} - x(x^2 - 1)^{1/3}(x^2 - 4)^{-1/2}}{x^2 - 4}$

REVIEW EXERCISE SET

A

1. Express as a repeating or terminating decimal.
   a. $\tfrac{11}{16}$ **b.** $\tfrac{25}{8}$ **c.** $\tfrac{3}{7}$ **d.** $\tfrac{5}{27}$ **e.** $-\tfrac{13}{6}$

2. Express in the form m/n, where m and n are integers.
   a. $0.111\ldots$ **b.** $0.242424\ldots$
   c. $1.0363636\ldots$ **d.** $0.216216216\ldots$

3. Identify which numbers are rational and which are irrational.

   a. $-\sqrt[3]{27}$ **b.** $\dfrac{5}{\sqrt{121}}$ **c.** $\dfrac{1}{\sqrt{2}}$

   d. $\sqrt[3]{-125}$ **e.** $(\sqrt{3})^6$ **f.** $\sqrt{3} + 2$

   g. $\dfrac{0}{\sqrt{7}}$ **h.** $\left(\dfrac{2}{3}\right)^{-1}$ **i.** π^2

   j. $\sqrt{32}$

4. Write an equivalent expression without using absolute value signs.
   a. $|x| < 7$ **b.** $|x| \le 4$
   c. $|x| > 8$ **d.** $|x| \le 3$

5. Write each of the following sets in interval notation and show it on a number line:
   a. $\{x:\ x > -1\}$ **b.** $\{x:\ 0 < x \le 2\}$
   c. $\{x:\ |x| < 3\}$ **d.** $\{x:\ |x| \ge 2\}$
   e. $\{x:\ x < -2\} \cup \{x:\ x \ge 1\}$

In Problems 6–9 perform the indicated operations. Reduce final answers to lowest terms.

6. a. $\frac{3}{7} + \frac{5}{9}$ **b.** $\frac{2}{3} - \frac{3}{4} + \frac{5}{6}$

7. a. $\frac{5}{96} - \frac{7}{160} + \frac{3}{112}$ **b.** $(-\frac{15}{16})(-\frac{7}{10} + \frac{3}{2})$

8. a. $\dfrac{\frac{2}{3} - \frac{1}{2}}{\frac{5}{6} + \frac{3}{2}}$ **b.** $\dfrac{1 - \frac{3}{4}}{\frac{1}{3} - \frac{1}{2}}$

9. a. $\dfrac{-2 + \frac{3}{8}}{\frac{3}{4} - \frac{5}{6}}$ **b.** $\dfrac{\frac{2}{3} - \frac{1}{4}}{1 + (\frac{2}{3})(\frac{1}{4})}$

10. Write in scientific notation:
  a. 132,000,000   **b.** 0.00023
  c. 207,000   **d.** 0.00000035

11. Evaluate using scientific notation. Check your result with a calculator.
  a. $3.7 \times 10^5 + 2.4 \times 10^4$
  b. $1.5 \times 10^{-7} - 3.6 \times 10^{-5}$
  c. $\dfrac{(260{,}000)(0.0003)}{(0.013)(12{,}000{,}000)}$

In Problems 12–14 simplify the expressions, writing answers without negative exponents.

12. a. $(3x^4y^2)(4x^3y^5)$ **b.** $(5x^3y^{-2}z^0)(2x^{-1}y^5)$

13. a. $\dfrac{3^2x^4y^{-8}z^0}{3^{-4}x^6y^{-5}z^{-2}}$ **b.** $\dfrac{2^{-3}a^{-2}b^5}{2^{-5}a^{-4}b^3}$

14. a. $(81x^{-12}y^4z^{-8})^{3/4}$ **b.** $\left(\dfrac{-8a^2b^{-5}}{27a^{-1}b}\right)^{-2/3}$

Simplify the expressions in Problems 15–17. All variables represent positive numbers.

15. a. $\sqrt{50x^3y^4z^0}$ **b.** $\sqrt[3]{-32a^6b^{-3}}$

16. a. $\sqrt{6xy^3}\sqrt{12x^3y^5}$ **b.** $\sqrt{\dfrac{98x^{-3}y^5}{125x^{-1}y^{-3}}}$

17. a. $\sqrt{\dfrac{2}{3}}$ **b.** $\dfrac{2}{\sqrt{5}}$ **c.** $\sqrt[3]{\dfrac{16}{9}}$ **d.** $\sqrt{\dfrac{3x}{8y}}$

In Problems 18–22 perform the indicated operations.

18. a. $4(2x^3 - 3x^2 + 4x - 8) + 3(x^3 - 5x + 7)$
  b. $2(3t^2 - 4t + 6) - 5(t^2 - 3t + 2)$
19. a. $(2x - 3y)(5x + 4y)$
  b. $(x + 2y)(x - 2y)$
20. a. $(3x - 5)^2$
  b. $(x - 1)(x^3 + 2x^2 - 3x + 4)$
21. a. $(x^3 - 3x^2 + 4x - 5) \div (x - 3)$
  b. $(2x^4 - 5x^2 - 2x + 7) \div (x^2 + 3x - 2)$
22. a. $(x + 2y)^5$ **b.** $(3a - 2b)^4$

In Problems 23–35 factor completely with rational coefficients.

23. a. $3x^2 - 5x - 2$ **b.** $x^3y - 12x^2y^2 + 36xy^3$
24. a. $18x^2 + 33x - 40$ **b.** $a^3 - a^2b - ab^2 + b^3$
25. a. $2x^4 - 5x^2 - 12$ **b.** $x^2 - y^2 + 4x + 4$
26. a. $x^2 - 4y^2 - 2x + 4y$ **b.** $a^3 + 27$
27. a. $24x^3 - 23x^2y - 12xy^2$
  b. $a^3 + 3a^2 - 9a - 27$
28. a. $2x^5y^3 - 16x^2$ **b.** $a^4 - 256$
29. a. $4 - x^2 - y^2 + 2xy$
  b. $6x^4 - 5x^2y - 6y^2$
30. a. $a^2b - 4b - a^2 + 4$
  b. $x^2 - x - y^2 + y$
31. a. $x^6 - \dfrac{y^3}{8}$ **b.** $x^8 - 1$
32. a. $15x^2 - xy - 28y^2$
  b. $16a^4 + 32a^3b^2 + 24a^2b^4 + 8ab^6 + b^8$
33. a. $12a^2 - 36ab + 27b^2$ **b.** $s^3 + 64t^6$
34. a. $x^4 + 4y^4$ **b.** $4a^4 + 81$
35. a. $a^4 + a^2b^2 + b^4$ **b.** $9x^4 - 16x^2 + 4$

In Problems 36–43 perform the indicated operations and simplify the result. Express all answers with positive exponents only.

36. a. $\dfrac{2x^2 + 5x - 12}{x^2 - 25} \cdot \dfrac{3x^2 + 13x - 10}{6x^2 - 13x + 6}$

  b. $\dfrac{3x^2 - 7x - 20}{x^2 - 8x + 16} \div \dfrac{3x^2 + 14x + 15}{2x^2 - 9x + 4}$

37. $\dfrac{1}{x + 2} + \dfrac{3}{2 - x} + \dfrac{4}{x^2 - 4}$

38. $\dfrac{4x + 5}{2x - 3} - \dfrac{x + 7}{x + 4} - \dfrac{25x + 23}{2x^2 + 5x - 12}$

39. $\dfrac{\dfrac{2x}{x - 3} + 4}{x - 2}$

40. $\dfrac{\dfrac{1}{x - 2} - \dfrac{3}{x + 2}}{\dfrac{3x}{x^2 - 4} - \dfrac{2}{x - 2}}$

41. $\dfrac{2\sqrt{x^2 - 4} - \dfrac{x(2x + 3)}{\sqrt{x^2 - 4}}}{x^2 - 4}$

42. $3x^2(1 - 2x)^{-2} + 4x^3(1 - 2x)^{-3}$

43. $\dfrac{\sqrt{1 - x^2} + \dfrac{x^2}{\sqrt{1 - x^2}}}{1 - x^2}$

44. Rationalize the numerator:

a. $\dfrac{\sqrt{2x-3}-1}{x-2}$

b. $\dfrac{\sqrt{x+h-2}-\sqrt{x-2}}{h}$

45. Rationalize the denominator:

a. $\dfrac{1}{\sqrt{x+3}-2}$

b. $\dfrac{\sqrt{a+4}-\sqrt{a-4}}{\sqrt{a+4}+\sqrt{a-4}}$

 B

46. By using the meaning of the additive inverse of a number, together with its uniqueness, show that $-(-a)=a$.

47. Prove that if a, b, and c are real numbers, with $a>b$, then $a-c>b-c$.

48. Sometimes the definition for $a<b$ is given as follows: "$a<b$ means there is a positive real number c such that $b=a+c$." Show that this is equivalent to Definition 3.

49. Show that if $a\cdot b=0$, then either $a=0$ or $b=0$. **Hint.** Suppose $a\ne 0$, so that a^{-1} exists. Multiply both sides by a^{-1}.

50. Simplify:

a. $\sqrt{18}-\dfrac{2\sqrt{200}}{3}+\sqrt{\dfrac{81}{2}}$

b. $\sqrt{\dfrac{2}{3}}-\dfrac{\sqrt{24}}{5}+\sqrt{\dfrac{27}{2}}$

In Problems 51 and 52 evaluate using scientific notation. Check your results with a calculator.

51. $\dfrac{(320{,}000)(0.000025)^{1/2}}{(0.00008)(5{,}000{,}000)}$

52. $\dfrac{\sqrt[5]{243{,}000{,}000{,}000{,}000{,}000}}{\sqrt[3]{0.0002}\ \sqrt[3]{40{,}000}}$

In Problems 53–57 factor completely with rational coefficients.

53. a. x^2-4+4y^2-4xy
b. x^4-2x^3-x+2

54. a. $4x^2-3xy-10y^2-3x+6y$
b. $8x^3-60x^2y+150xy^2-125y^3$

55. $4x^2+9y^2-z^2-12xy-6z-9$

56. $(x-2)^3(2x+1)^{-1/2}+3(x-2)^2\sqrt{2x+1}$

57. $(2x-3)^{-1/3}(x+1)^{-2/3}-2(2x-3)^{-4/3}(x+1)^{1/3}$

In Problems 58–60 perform the indicated operations and simplify the result. Express answers with positive exponents only.

58. $\dfrac{2x-1}{x^2+x-2}-\dfrac{x+5}{x^2-1}-\dfrac{x+3}{x^2+3x+2}$

59. $\dfrac{9(2x-5)^2(3x+2)^2-4(3x+2)^3(2x-5)}{(2x-5)^3}$

60. $\dfrac{2x\sqrt[3]{1-x^2}+\frac{2}{3}x^3(1-x^2)^{-2/3}}{(1-x^2)^{2/3}}$

2

EQUATIONS AND INEQUALITIES OF THE FIRST AND SECOND DEGREE

Inequalities occur almost as frequently in calculus as equations, so you must gain proficiency with both. The excerpt below illustrates a use of inequalities that also involve absolute values.*

Prove that $\lim\limits_{x \to 2} x^2 = 4$.

Solution   To begin, we assume that a number $\varepsilon > 0$ is given. We must show that there exists a number $\delta > 0$ such that

$$|x^2 - 4| < \varepsilon \qquad \text{whenever} \qquad 0 < |x - 2| < \delta$$

We need to establish a connection between $|x^2 - 4|$ and $|x - 2|$. We note that

$$|x^2 - 4| = |(x + 2)(x - 2)| = |x + 2||x - 2|$$

If we can find a number K such that $|x + 2| < K$, the choice of δ is clear, namely, $\delta < \varepsilon/K$. If x is confined to some interval centered about 2, then K can be found. For example, suppose $|x - 2| < 1$—that is, $1 < x < 3$. Then we add 2 to each part to get $1 + 2 < x + 2 < 3 + 2$; and, in particular, $|x + 2| = x + 2 < 5$.

It follows that whenever $|x - 2| < 1$,

$$|x^2 - 4| = |(x + 2)(x - 2)| < 5|x - 2|$$

If $|x - 2| < \varepsilon/5$ also, then

$$|x^2 - 4| < 5|x - 2| < 5\left(\frac{\varepsilon}{5}\right) = \varepsilon$$

---

* Abe Mizrahi and Michael Sullivan, *Calculus and Analytic Geometry*, 3rd ed. (Belmont, Calif.: Wadsworth Publishing Company, © 1990, 1986, 1982, Wadsworth, Inc.), p. 118. Reprinted by permission.

The two Greek letters ε (epsilon) and δ (delta) are often used in calculus in problems of this nature. The main thing for you to concentrate on at this point is the equivalence of the inequalities $|x - 2| < 1$ and $1 < x < 3$. We will soon see why this is true.

INTRODUCTION

The solution of equations lies at the very heart of algebra and its applications. Much of what we have done up to this point has provided tools for solving equations. The importance of equations stems from the fact that they provide a means by which many complicated relationships in real-world problems can be written in a concise and precise form. In the terminology used by mathematicians (and others), equations often provide **mathematical models** of real-world problems.

Equations that arise as mathematical models often involve more than algebraic concepts. Some involve trigonometry, and many involve calculus. Even in these more complicated equations, however, a knowledge of the techniques of algebra is essential. In this chapter we consider algebraic equations involving only one unknown. Equations with more than one unknown will be discussed later, when we consider systems of several equations.

Equations of special importance are **polynomial equations in one unknown.** Some examples are

$$2x - 3 = 7$$
$$x^2 - 3x + 2 = 0$$
$$x^3 - 2x^2 - 3 = 0$$

In general, if n is a positive integer, a polynomial equation of the nth degree in the variable x is an equation that can be written in the form

$$a_n x^n + a_{n-1} x^{n-1} + \cdots + a_1 x + a_0 = 0$$

where $a_0, a_1, \ldots, a_n$ are constants, with $a_n \neq 0$. If $n = 1$, the equation is called **linear or first degree,** and for $n = 2$, it is called **quadratic or second degree.** Higher-degree equations are sometimes given names (for example cubic, quartic, and quintic for $n = 3, 4,$ and 5, respectively), but we usually treat these as a group. In this chapter we consider the two simplest cases—linear and quadratic equations. A distinction should be made between a polynomial in x and a polynomial equation in x. For example, consider the polynomial

$$2x^2 - 3x - 5$$

and the polynomial equation

$$2x^2 - 3x - 5 = 0$$

In the first case, no restriction is placed on the variable x, and in fact, x is simply left unspecified unless further instructions are given. In the polynomial equation, however, we are seeking the specific values of x that make the equation true. For this reason, we often refer to the variable in such an equation as an **unknown**; it is something with a value that is yet to be determined.

By a **solution** to an equation we mean a number that, when substituted for the unknown, makes the equation a true statement. A solution is also referred to as a **root** of the equation. For example, the number 2 is a solution, or root, of $x^2 - 4 = 0$, since, when 2 is substituted for x, we get $4 - 4 = 0$, which is true. The set of all solutions to an equation is called its **solution set**. Since $(-2)^2 = 4$, we see that -2 is also a solution to the equation $x^2 - 4 = 0$, and we will soon see that 2 and -2 are the only solutions. So we can say that the solution set of this equation is $\{2, -2\}$. Two equations are said to be **equivalent** if they have exactly the same solution set. To **solve** an equation means to find its solution set.

We can change a given equation to an equivalent one by either of the following operations:

1. Add the same number to both sides.
2. Multiply both sides by the same nonzero number.

These are justified by the basic properties of equality (Chapter 1, Section 1). Since $a - b = a + (-b)$ and $a/b = a \cdot b^{-1}$, it also follows that subtracting the same number from both sides or dividing both sides by the same nonzero number results in an equivalent equation.

Do you see why, in multiplying both sides, we can use only a nonzero number? If 0 were allowed, we could get, for example, that

$$5 \cdot 0 = 6 \cdot 0$$

but $5 \neq 6$.

LINEAR EQUATIONS

By definition, a **linear equation** is an equation that can be written in the form

$$ax + b = 0 \qquad (a \neq 0)$$

and by adding $-b$ to both sides and dividing by a we obtain

$$x = -\frac{b}{a}$$

Thus, the only possible solution is $x = -b/a$. We can verify that this is the solution by substitution:

$$a\left(-\frac{b}{a}\right) + b = -b + b = 0$$

We could say that the solution set is $\{-b/a\}$.

There is no point, of course, in memorizing the result just obtained. It is simple enough in each situation to carry out the steps leading to the answer.

EXAMPLE 1 ||||||||||||||||||||||||||||||||||||||||||||||||||||||||||||||||||||||||||||||||||||||||||||||||||||||||||||||||||||

Solve the equation: $3x - 4 = x + 7$

Solution   We want to obtain an equivalent equation in which all terms involving x are on one side and all other terms are on the other. This can be accomplished by adding $-x$ to both sides (to eliminate x from the right) and by adding 4 to both sides (to eliminate -4 from the left). So we add $-x + 4$ to both sides and proceed as follows.

$$3x - 4 = x + 7$$
$$2x = 11 \qquad \text{Add } -x + 4 \text{ to both sides}$$
$$x = \tfrac{11}{2} \qquad \text{Divide both sides by 2} \qquad \blacktriangle$$

EXAMPLE 2 ||||||||||||||||||||||||||||||||||||||||||||||||||||||||||||||||||||||||||||||||||||||||||||||||||||||||||||||||||

Solve the equation: $2(x - 3) = 5x - 8$

Solution
$$2(x - 3) = 5x - 8$$
$$2x - 6 = 5x - 8 \qquad \text{Perform multiplication on left}$$
$$2 = 3x \qquad \text{Add } -2x + 8 \text{ to both sides}$$
$$x = \tfrac{2}{3} \qquad \text{Divide by 3 and interchange right and left sides} \quad \blacktriangle$$

In solving equations we make repeated use of the properties of equality given in Section 1 of Chapter 1. In particular, we may interchange the two sides of an equation as in Example 2, using the symmetric property. We may add the same quantity to both sides (this includes subtraction, since this is the same as adding a negative quantity), and we may multiply by the same *nonzero* quantity.

EXAMPLE 3 ||||||||||||||||||||||||||||||||||||||||||||||||||||||||||||||||||||||||||||||||||||||||||||||||||||||||||||||||||

Solve the equation: $\dfrac{2x}{3} - \dfrac{5}{4} = \dfrac{1}{2}\left(\dfrac{x}{4} + 1\right)$

Solution   When an equation involves fractions, it is best to clear it of fractions by multiplying both sides by the LCD. In this problem we need first to multiply out the right-hand side in order to recognize the LCD. This gives

$$\frac{2x}{3} - \frac{5}{4} = \frac{x}{8} + \frac{1}{2}$$

Now we see that the LCD is 24. It is essential that we multiply *every* term on both sides by this LCD. We show the details as follows.

$$24\left(\frac{2x}{3} - \frac{5}{4}\right) = 24\left(\frac{x}{8} + \frac{1}{2}\right)$$

$$\frac{\overset{8}{\cancel{24}}}{1} \cdot \frac{2x}{\cancel{3}} - \frac{\overset{6}{\cancel{24}}}{1} \cdot \frac{5}{\cancel{4}} = \frac{\overset{3}{\cancel{24}}}{1} \cdot \frac{x}{\cancel{8}} + \frac{\overset{12}{\cancel{24}}}{1} \cdot \frac{1}{\cancel{2}}$$

$$16x - 30 = 3x + 12$$

$$13x = 42$$

$$x = \tfrac{42}{13}$$

With care, the multiplications can be done mentally.   ▲

EXAMPLE 4

Solve for *t*: $at + d = b - ct$

Solution   Equations such as this are called **literal** equations since letters are used as coefficients rather than specific numbers. The procedure is essentially the same as in the other examples.

$$at + d = b - ct$$

$$at + ct = b - d \qquad\qquad \text{Add } ct - d \text{ to both sides}$$

$$(a + c)t = b - d \qquad\qquad \text{Factor out } t \text{ on the left}$$

$$t = \frac{b - d}{a + c} \qquad (c \neq -a) \qquad \text{Divide by } a + c$$ ▲

EXAMPLE 5

Solve the equation: $(x - 1)^2 - 3(x + 2) = x^2 - 4$

Solution   This does not appear to be a linear equation, but when it is simplified, the second degree terms drop out.

$$(x - 1)^2 - 3(x + 2) = x^2 - 4$$

$$x^2 - 2x + 1 - 3x - 6 - x^2 + 4 = 0$$

$$-5x - 1 = 0$$

$$-5x = 1$$

$$x = -\frac{1}{5}$$ ▲

EXERCISE SET 2

A

Find the complete solution set for each of the following equations:

1. $3x - 2 = 5$

3. $3x - 4 = 7 + 2x$

5. $3 - x = x + 5$

7. $3(x - 2) = 4x + 7$

9. $2(3x - 4) = 5$

11. $\dfrac{x}{3} + \dfrac{1}{2} = 1$

13. $\dfrac{2x}{3} - \dfrac{3}{4} = \dfrac{5}{6}$

15. $\dfrac{3}{4} - \dfrac{5x}{8} = \dfrac{1}{2}$

17. $\dfrac{3x + 2}{4} = \dfrac{2x - 5}{7}$

19. $\dfrac{3s + 2}{8} - \dfrac{2 - 5s}{6} = 1$

21. $\dfrac{1}{4}\left(\dfrac{x}{6} - 1\right) = \dfrac{2x}{3} - \dfrac{3}{4}$

22. $\dfrac{2}{3}\left(\dfrac{1}{2} - \dfrac{s}{5}\right) = \dfrac{5s}{6} - \dfrac{11}{15}$

23. $\dfrac{3}{4}\left(2t - \dfrac{1}{3}\right) = \dfrac{5}{4} + \dfrac{t}{6}$

25. $\dfrac{8}{3}\left(\dfrac{5 - 2z}{4}\right) - \dfrac{7}{12} = \dfrac{3z}{2} + \dfrac{7}{4}$

26. $\dfrac{3}{2}\left(\dfrac{1}{6} - \dfrac{5t}{8}\right) = \dfrac{t - 4}{12} + \dfrac{7}{16}$

27. $(x - 2)^2 + 3x = x^2 + x - 4$

28. $2x^2 - 5x - (x - 1)^2 = x(x + 3) + 8$

In Problems 29–32 use a calculator to solve the equation.

29. $0.62(200 - w) + 0.775w = 150$

30. $2.43(y - 1.67) + 5.86 = 1.88 - 3.75y$

2. $4x + 3 = 10$

4. $5x + 3 = 11 - 2x$

6. $5 - x = 2x + 3$

8. $4 - 3x = 6(2x + 3)$

10. $2(5 - x) = 3(2x - 1)$

12. $\dfrac{3x}{4} - \dfrac{1}{3} = \dfrac{x}{2}$

14. $\dfrac{2x + 7}{3} = 1 - \dfrac{x}{2}$

16. $3 - \dfrac{7x}{10} = \dfrac{2x}{5} - \dfrac{1}{2}$

18. $\dfrac{1 - 2t}{3} - \dfrac{1}{2} = \dfrac{2t + 1}{4}$

20. $\dfrac{3}{2}\left(\dfrac{m}{4} - \dfrac{1}{3}\right) = \dfrac{3 - m}{6}$

24. $\dfrac{3y - 2}{15} + 1 = \dfrac{2}{5} - y$

31. $0.372t - (1.267 - 0.598t) = 4.833$

32. $3.42[5.38 - (2.73 - 4.87x)] = 9.06x$

B

In Problems 33 and 34 express the answer in exact form with a rational denominator.

33. $4(1 - t) = 3\sqrt{2}t - \sqrt{8}$

34. $\sqrt{5}(m - \sqrt{5}) = 2(m + \sqrt{5}) - 8$

In Problems 35–42 solve for the indicated unknown.

35. x: $ax + b = cx - d$

36. t: $rs - at = bt - s$

37. d: $S = \dfrac{n}{2}[2a_1 + (n - 1)d]$

38. h: $S = \pi r^2 + 2\pi rh$

39. $\bar{Y}$: $\bar{Y}r = \rho\left(\bar{Y} - \dfrac{A}{s}\right)$

40. t: $\rho V = \rho_0 V_0(\beta t + 1)$

41. I: $RI + \dfrac{q}{C} = E$

42. v^2: $mgy = \dfrac{1}{2}mv^2 + \dfrac{Iv^2}{2R}$

43. Find k so that $x = 3$ is a solution of

$$\dfrac{3x}{4} + \dfrac{2k}{3} = \dfrac{3}{2}\left(\dfrac{5k}{6} - \dfrac{x}{2}\right)$$

44. Find k so that $x = -2$ is a solution of

$$\dfrac{3x - 2k}{5} = \dfrac{2}{3}\left(k - \dfrac{x}{3}\right)$$

In Problems 45 and 46 show that the equation reduces to a linear equation, and find its solution.

45. $(2 - x)^3 + x(x^2 - 1) = 6x^2 + 1$

46. $(x - 1)(2x^2 - x + 3) = 2(x + 1)^3 - (3x - 2)^2$

QUADRATIC EQUATIONS

3

A **quadratic equation** is an equation that can be expressed in the form

$$ax^2 + bx + c = 0 \qquad (a \neq 0)$$

There are three principal techniques for solving them: (1) by factoring, (2) by completing the square, and (3) by the quadratic formula. In this section we consider the solution by factoring, and in Section 4 we will study the other two methods.

The following property of real numbers is of fundamental importance in the solution by factoring. We stated this in Chapter 1, Section 1 but did not prove it.

If a and b are real numbers, then

$$a \cdot b = 0 \quad \text{if and only if } a = 0 \text{ or } b = 0$$

To prove this, suppose first that $a \cdot b = 0$. If $a \neq 0$, we may multiply both sides by a^{-1} to obtain

$$a^{-1}(ab) = a^{-1} \cdot 0$$
$$(a^{-1}a)b = 0$$
$$1 \cdot b = 0$$
$$b = 0$$

So if $a \neq 0$, then b must be 0. On the other hand, if either $a = 0$ or $b = 0$, then the equation $ab = 0$ is satisfied. So the proof is complete.

A Word of Caution. The reasoning used in the above argument is not valid if the right-hand side is any number other than 0. For example, consider the following *fallacious argument*:

$$x^2 - 2x = 4$$
$$x(x - 2) = 4$$
$$x = 4 \quad \text{or} \quad x - 2 = 4, \quad \text{so that} \quad x = 6 \qquad \textbf{\textcolor{red}{Incorrect}}$$

A check of these results reveals that they are wrong. The trouble lies in the fact that if $ab = 4$, we cannot infer anything about the values of a and b.

The next three examples illustrate how to solve a quadratic equation by factoring.

EXAMPLE 6

Solve the equation: $x^2 - x - 2 = 0$

Solution
$$x^2 - x - 2 = 0$$
$$(x - 2)(x + 1) = 0$$

Now, from what we have just shown we know that the product $(x - 2)(x + 1)$ will be 0 if and only if either $x - 2 = 0$ or $x + 1 = 0$, that is, if $x = 2$ or $x = -1$. So these two numbers are solutions, and they are the only solutions. Therefore, the solution set is $\{2, -1\}$. ▲

EXAMPLE 7

Solve: $\dfrac{2x^2}{3} - x - \dfrac{5}{3} = 0$

Solution   We first clear of fractions by multiplying both sides by 3.

$$2x^2 - 3x - 5 = 0$$
$$(2x - 5)(x + 1) = 0$$
$$2x - 5 = 0 \quad \text{or} \quad x + 1 = 0$$
$$x = \tfrac{5}{2} \quad \text{or} \quad x = -1$$

So the solution set is $\{\tfrac{5}{2}, -1\}$. Again, either of these numbers satisfies the original equation, but no other numbers do so. ▲

EXAMPLE 8

Solve: $x(x - 3) = 4$

Solution   As we have seen, it is essential for one side of the equation to be 0. So we add -4 to both sides. This gives

$$x(x - 3) - 4 = 0$$
$$x^2 - 3x - 4 = 0$$
$$(x - 4)(x + 1) = 0$$
$$x - 4 = 0 \quad \text{or} \quad x + 1 = 0$$
$$x = 4 \quad \text{or} \quad x = -1$$

So the solution set is $\{4, -1\}$. ▲

EXERCISE SET 3   2-20 odd

Solve Problems 1–30 by factoring.

1. $x^2 - 4x + 3 = 0$ **2.** $x^2 - 2x - 8 = 0$

3. $t^2 + 3t - 10 = 0$ **4.** $s^2 + 7s + 12 = 0$

5. $2y^2 + 5y - 3 = 0$ **6.** $3r^2 - 10r + 8 = 0$

7. $\dfrac{2}{3}x^2 - \dfrac{x}{3} = 1$ **8.** $3x^2 - \frac{4}{3}x = 0$

9. $\dfrac{x}{3}(x - 5) = 2$ **10.** $t^2 - \frac{3}{2}t = \frac{5}{2}$

11. $9r^2 - 25 = 0$ **12.** $9r^2 - 25r = 0$

13. $6x^2 + x - 12 = 0$ **14.** $20x^2 - 7x - 6 = 0$

15. $t(4 - t) = 4$ **16.** $m^2 - 6m + 9 = 16$

17. $18x^2 + 15x = 18$ **18.** $4 - 9t^2 = 0$

19. $t\left(t + \dfrac{5}{4}\right) = \dfrac{3}{2}$ **20.** $2x^2 - \dfrac{x}{4} = \dfrac{5}{8}$

21. $x(x - 5) = 14$ **22.** $s(s + 1) = 20$

23. $\dfrac{x^2 - 1}{3} = \dfrac{x^2 + x}{4}$ **24.** $\dfrac{2x}{3}\left(\dfrac{x - 1}{4}\right) = 1$

25. $6(2 + 3x^2) = 35x$ **26.** $x(x + 1) = 2(x + 3)$

27. $x(x + 1) = 3(2 - x) + 6$

28. $(x - 1)(2x - 1) = 1$

29. $(2s + 1)(s - 1) + s - 1 = 0$

30. $3t(t - 2) = 2t(t + 1) + 20$

31. $\sqrt{3}r^2 + \sqrt{12} = 5r$ **32.** $x(\sqrt{2}x - 3) = \sqrt{8}$

In Problems 33 and 34 show that the equation reduces to a quadratic, and solve it.

33. $x^3 + (2 - x)^3 = 26$

34. $(x + 1)^4 - 4(x + 1)^3 = x^4 - 11$

In Problems 35–40 solve for the indicated unknown.

35. x: $8a^2x^2 - 6abx - 27b^2 = 0$

36. s: $12s^2t^2 + 8rst - 15r^2 = 0$

37. n: $\dfrac{n(3m - n)}{m} = \dfrac{2m + n}{2}$

38. t: $4(k^2t^2 - 3s^2) = 13kst$

39. Find k such that $x = 2$ is a solution of $3(k - 2) + 4kx + 10k^2 = 0$.

40. Find k such that $x = -\frac{2}{3}$ is a solution of $3k^2x^2 - (2k - 3)x - 3k = 0$.

COMPLETING THE SQUARE AND THE QUADRATIC FORMULA

The technique of factoring a quadratic polynomial to solve an equation is useful only when the factors can be readily determined. There is a more general technique by which *all* quadratic equations can be solved, which leads to the **quadratic formula.** This is the technique known as **completing the square.** To illustrate it, we need to recall the form of the square of a binomial $(x + a)$. This is

$$(x + a)^2 = x^2 + 2ax + a^2$$

Now, suppose we are confronted with an expression such as

$$x^2 + 6x$$

and we wish to determine what should be added to this to make it a perfect square. We see that $2a = 6$, so that $a = 3$. Therefore, $a^2 = 9$, which is the correct term to add, yielding

$$x^2 + 6x + 9 = (x + 3)^2$$

More generally, since the coefficient of x in the expansion of $(x + a)^2$ is $2a$, it follows that **the correct constant term to add is the square of one-half of the coefficient of x.** Here are some examples.

EXAMPLE 9

Determine what number should be added to each of the following to make it a perfect square:

a. $x^2 + 4x$ **b.** $x^2 + 3x$ **c.** $x^2 - 5x$ **d.** $x^2 - \frac{5}{3}x$

Solution   **a.** $\left(\frac{1}{2} \cdot 4\right)^2 = 2^2 = 4$ **b.** $\left(\frac{1}{2} \cdot 3\right)^2 = \frac{9}{4}$

c. $\left[\frac{1}{2} \cdot (-5)\right]^2 = \frac{25}{4}$ **d.** $\left[\frac{1}{2} \cdot \left(-\frac{5}{3}\right)\right]^2 = \left(\frac{-5}{6}\right)^2 = \frac{25}{36}$ ▲

One additional fact is needed before proceeding to the solution of equations by this method. This is that $a^2 = b^2$ if and only if $a = b$ or $a = -b$ (or, more briefly, $a = \pm b$). If either $a = b$ or $a = -b$, then squaring both sides shows that $a^2 = b^2$. On the other hand, if $a^2 = b^2$, we have

$$a^2 - b^2 = 0$$

$$(a - b)(a + b) = 0$$

Now we use the fact that the product is zero if and only if one of the factors is zero.

$$a - b = 0 \quad \text{or} \quad a + b = 0$$

$$a = b \quad \text{or} \quad a = -b$$

The next two examples illustrate the technique of solution by completing the square.

EXAMPLE 10

Solve: $x^2 + 3x - 2 = 0$

Solution   We determine that there are no rational factors and proceed as follows: Add 2 to both sides:

$$x^2 + 3x = 2$$

Complete the square on the left and add the same number on the right:

$$x^2 + 3x + \tfrac{9}{4} = 2 + \tfrac{9}{4}$$

Factor the left side and simplify the right:

$$\left(x + \frac{3}{2}\right)^2 = \frac{17}{4}$$

Use the fact that $a^2 = b^2$ if and only if $a = \pm b$:

$$x + \frac{3}{2} = \pm \frac{\sqrt{17}}{2}$$

Solve for x:

$$x = -\frac{3}{2} \pm \frac{\sqrt{17}}{2}$$

we can combine the fractions to get

$$x = \frac{-3 \pm \sqrt{17}}{2} \qquad \blacktriangle$$

EXAMPLE 11

Solve: $2x^2 - 8x + 3 = 0$

Solution We will carry out steps similar to those in Example 10 but will omit the explanation. The fact that the coefficient of x^2 is not 1 is a slight complication, but this is handled by dividing both sides of the equation by 2. In solving a quadratic equation by completing the square, this step of dividing both sides of the equation by the coefficient of x^2, if it is not 1 to begin with, should always be done first.

$$2x^2 - 8x + 3 = 0$$
$$x^2 - 4x + \tfrac{3}{2} = 0$$
$$x^2 - 4x = -\tfrac{3}{2}$$
$$x^2 - 4x + 4 = -\tfrac{3}{2} + 4$$
$$(x - 2)^2 = \tfrac{5}{2}$$
$$x - 2 = \pm \sqrt{\tfrac{5}{2}}$$
$$x = 2 \pm \sqrt{\frac{5}{2} \cdot \frac{2}{2}}$$
$$x = 2 \pm \frac{\sqrt{10}}{2} \qquad \blacktriangle$$

We now apply this technique to a general quadratic equation, that is, one in which we do not specify the coefficients. Such an equation may be written in the form

$$ax^2 + bx + c = 0 \qquad (a \neq 0) \tag{1}$$

We will refer to equation (1) as the **standard form** of a quadratic equation.

Remark. In defining a polynomial equation of degree n we used $a_0, a_1, a_2, \ldots, a_n$ for the coefficients because to use different letters of the alphabet would be complicated by not knowing how many letters are needed. But in the present case we know we need three coefficients, so we choose to use a, b, and c rather than a_0, a_1, and a_2.

We proceed as in Example 10:

$$ax^2 + bx + c = 0$$

Divide both sides by a:

$$x^2 + \frac{b}{a}x + \frac{c}{a} = 0$$

Add $(-c)/a$ to both sides:

$$x^2 + \frac{b}{a}x = -\frac{c}{a}$$

Complete the square by adding $[\frac{1}{2} \cdot (b/a)]^2 = b^2/(4a^2)$ to both sides:

$$x^2 + \frac{b}{a}x + \frac{b^2}{4a^2} = -\frac{c}{a} + \frac{b^2}{4a^2}$$

Factor the left side and collect terms on the right:

$$\left(x + \frac{b}{2a}\right)^2 = \frac{b^2 - 4ac}{4a^2}$$

Use the fact that $X^2 = N$ if and only if $X = \sqrt{N}$ or $X = -\sqrt{N}$:

$$x + \frac{b}{2a} = \frac{\pm\sqrt{b^2 - 4ac}}{2a}$$

Finally, solve for x, thus obtaining the **quadratic formula:**

The Quadratic Formula

$$x = \frac{-b \pm \sqrt{b^2 - 4ac}}{2a}$$

By committing this formula to memory, you can at once write the solutions to any quadratic equation. It should be emphasized, however, that factoring is usually easier when the factors can be readily found.

It is essential before attempting to apply the quadratic formula that the equation be written precisely in the standard form given in equation (1) so that the correct values of a, b, and c can be determined.

EXAMPLE 12 ıı

Solve the equation $2x^2 - 3x - 7 = 0$ by the quadratic formula.

Solution   We think of the equation in the equivalent form $2x^2 + (-3)x + (-7)$, so that $a = 2$, $b = -3$, and $c = -7$. So

$$x = \frac{-(-3) \pm \sqrt{(-3)^2 - 4(2)(-7)}}{2 \cdot 2} = \frac{3 \pm \sqrt{9 + 56}}{4}$$

$$= \frac{3 \pm \sqrt{65}}{4} \qquad \blacktriangle$$

An alert reader might have noticed a difficulty. The quantity $(b^2 - 4ac)$ appearing under the radical in the quadratic formula could be negative, in which case there is no solution in the set of real numbers. As we have seen, if k is any nonzero real number, then $k^2 > 0$, so that there is no square root of a negative number in R. To overcome this difficulty, it is necessary to consider a wider class of numbers called the **complex numbers.** These will be discussed in the next section.

  EXERCISE SET 4   18-34 odd

A

Solve Problems 1–16 by completing the square.

1. $x^2 + 2x - 4 = 0$ **2.** $x^2 - 4x + 2 = 0$
3. $5 - 4x - x^2 = 0$ **4.** $9 + 6x - x^2 = 0$
5. $y^2 - 3y - 5 = 0$ **6.** $t^2 + t - 3 = 0$
7. $x^2 + 5x = 15$ **8.** $2x^2 - 6x + 3 = 0$
9. $2t^2 - 3t - 2 = 0$ **10.** $12 - 2s - 3s^2 = 0$
11. $3r^2 - 8r - 6 = 0$ **12.** $4t^2 - 2t - 5 = 0$
13. $2x(x - 2) = 3$ **14.** $3x^2 = 2(x + 4)$
15. $4m^2 - 6 = 3m$ **16.** $2y(3 - y) = 3$

Solve Problems 17–34 by the quadratic formula.

17. $x^2 + 2x - 4 = 0$ **18.** $x^2 - 3x - 3 = 0$
19. $t^2 - 9t + 5 = 0$ **20.** $s^2 + 4s - 6 = 0$

21. $2x^2 + 3x - 4 = 0$ **22.** $2x^2 - 5x = 2$
23. $3x^2 = x + 1$ **24.** $3x(2 - x) = 2$
25. $5y(y + 2) + 2 = 0$ **26.** $4t^2 = 3(t + 1)$
27. $3x + 2 = 4x^2$ **28.** $8x = 4x^2 - 3$
29. $x(5 - 2x) = 1$
30. $5x(x + 1) = 3(x + 2)$
31. $(t + 1)(2t - 1) = 1$ **32.** $(2t - 3)(2t + 1) = 2$
33. $r^2 = 2(\sqrt{2}r - 1)$ **34.** $t(2t + 3\sqrt{5}) = 10$

Solve Problems 35–42 for the indicated unknown. Assume all letters represent positive real numbers.

35. r: $S = \pi r(r + 2l)$ **36.** R: $v^2 = \dfrac{R^2 g}{R + h}$

37. x: $(a - x)(b - x) = k$

38. t: $-\frac{1}{2}gt^2 + v_0t + s_0 = 0$

39. r: $Lr^2 + Rr + \dfrac{1}{C} = 0$

40. Q: $(Q - m)(\alpha - \beta Q) = r$

41. s: $2s^2 + 3\sqrt{3}\,ks - 15k^2 = 0$

42. t: $2t^2 + 3kt - 3k = 2$

In Problems 43 and 44 solve for x with the aid of a calculator.

43. $3.24x^2 - 6.83x + 3.05 = 0$

44. $4.238 - 0.3621x - 0.02473x^2 = 0$

5 COMPLEX NUMBERS

We have already noted that certain quadratic equations cannot be solved in the domain of real numbers. Even the simple equation $x^2 + 1 = 0$ falls into this category. If x is any real number, its square is nonnegative, so adding it to the positive integer 1 results in a positive number, which is therefore greater than zero. This difficulty persists in higher-degree equations. The resolution of the problem lies in expanding our number system.

Number systems have evolved over thousands of years, and each step has met with resistance from segments of the intellectual community. The introduction of negative and irrational numbers were certainly two of the most controversial stages. As a matter of fact, all numbers are creations of the human mind (even the integers, Kronecker notwithstanding*). Numbers do not exist in nature waiting to be discovered. They exist only in the realm of ideas. We are free, then, to invent new numbers if a need exists. To make them useful, additions should be consistent with our known number systems, and should, in some sense, include these as subsystems. The real numbers, for example, include the rational numbers as a subsystem, which include the integers, which in turn include the natural numbers.

The extension of the real number system that will provide for solutions of all quadratic equations, as well as many other types of equations that cannot be solved in R, is contained in the following definition.

DEFINITION I   **The set C of complex numbers is defined by**

$$C = \{(a + bi):\quad a \in R,\ b \in R,\quad \text{and}\quad i^2 = -1\}$$

The complex number $i = 0 + 1i$ is called the imaginary unit.

These complex numbers exist just as surely as any other numbers do; there is nothing imaginary about them. It is an unfortunate fact of history that the terms *real* and *imaginary* are used, because of their connotations. One might

---

* Leopold Kronecker (1823–1891) said that "God made the integers, and all the rest is the work of man."

better refer to real numbers as one-dimensional numbers and complex numbers as two-dimensional numbers. This terminology reflects the fact that real numbers can be associated with points on a line, whereas complex numbers can be made to correspond to points in a plane (as we will see in Chapter 10).

The real numbers are included among the complex numbers, since a real number a may be thought of as being of the form $a + 0i$. If $b \neq 0$, the complex number $a + bi = bi$ with $b \neq 0$ is called an **imaginary number.** Complex numbers of the form $0 + bi = bi$ with $b \neq 0$ are called **pure imaginary numbers.**

Two complex numbers $a + bi$ and $c + di$ **are equal if and only if** $a = c$ **and** $b = d$. For example, if $x + yi = 3 + 4i$, then we must have $x = 3$ and $y = 4$.
Addition is defined by

$$(a + bi) + (c + di) = (a + c) + (b + d)i$$

and multiplication by

$$(a + bi)(c + di) = (ac - bd) + (ad + bc)i$$

The latter formula is the result of multiplying as with two real binomials and then substituting $i^2 = -1$:

$$(a + bi)(c + di) = ac + bic + adi + bdi^2 = ac + i(bc + ad) + bd(-1)$$
$$= (ac - bd) + (ad + bc)i$$

Multiplication is usually carried out in this way rather than by simply memorizing the definition. For example,

$$(2 + 3i)(4 - 7i) = 8 - 2i - 21i^2 = 8 - 2i + 21 = 29 - 2i$$

The identity elements are $0 = 0 + 0i$ and $1 = 1 + 0i$. The additive inverse of $a + bi$ is seen to be $(-a) + (-b)i$. When a and b are not both equal to 0 (that is, when $a + bi \neq 0$), we can show that the number $a + bi$ has a multiplicative inverse within the complex number system. To do this, we must show that $(a + bi)^{-1}$ is a number of the form $A + Bi$, where A and B are real. The following calculations show that this is the case, with $A = a/(a^2 + b^2)$ and $B = -b/(a^2 + b^2)$:

$$(a + bi)^{-1} = \frac{1}{a + bi} = \frac{1}{a + bi} \cdot \frac{a - bi}{a - bi} = \frac{a - bi}{a^2 - b^2 i^2} = \frac{a - bi}{a^2 + b^2}$$

$$= \frac{a}{a^2 + b^2} + \frac{-b}{a^2 + b^2} i$$

Rather than memorizing this result, it is better to learn the procedure. For example,

$$\frac{1}{2 + 3i} = \frac{1}{2 + 3i} \cdot \frac{2 - 3i}{2 - 3i} = \frac{2 - 3i}{4 - 9i^2}$$

$$= \frac{2 - 3i}{4 + 9} = \frac{2}{13} + \frac{-3}{13} i$$

We define the conjugate of a complex number $a + bi$ to be the number $a - bi$.
So to find the inverse of a complex number, we write its reciprocal and then multiply the numerator and denominator by its conjugate.

Division is accomplished in a similar way. For example:

$$\frac{3 + 2i}{5 - 4i} = \frac{3 + 2i}{5 - 4i} \cdot \frac{5 + 4i}{5 + 4i} = \frac{15 + 22i + 8i^2}{25 - 16i^2}$$

$$= \frac{15 + 22i - 8}{25 + 16} = \frac{7}{41} + \frac{22}{41} i$$

More generally, to divide one complex number by another, multiply numerator and denominator by the conjugate of the denominator, and simplify the result:

$$\frac{a + bi}{c + di} = \frac{a + bi}{c + di} \cdot \frac{c - di}{c - di} = \frac{(ac + bd) + (bc - ad)i}{c^2 + d^2}$$

$$= \frac{ac + bd}{c^2 + d^2} + \frac{bc - ad}{c^2 + d^2} i$$

To indicate the operation of taking the conjugate, a bar is used above the number:

$$\overline{a + bi} = a - bi$$

For example:

$$\overline{2 + 3i} = 2 - 3i$$
$$\overline{5 - 4i} = 5 + 4i$$
$$\overline{-7i} = \overline{0 - 7i} = 7i$$
$$\overline{-2} = \overline{-2 + 0i} = -2 - 0i = -2$$

Often the letter z is used to designate a complex number. Then $\bar{z}$ indicates its conjugate. The following facts about conjugates can be shown:

1. $\overline{z_1 + z_2} = \bar{z}_1 + \bar{z}_2$ Conjugate of a sum = Sum of conjugates
2. $\overline{z_1 \cdot z_2} = \bar{z}_1 \cdot \bar{z}_2$ Conjugate of a product = Product of conjugates
3. $\overline{\left(\dfrac{z_1}{z_2}\right)} = \dfrac{\bar{z}_1}{\bar{z}_2}$ Conjugate of a quotient = Quotient of conjugates
4. $z\bar{z}$ is real
5. If z is real, then $\bar{z} = z$.

Facts 4 and 5 result from the following reasoning: Suppose $z = a + bi$. Then $z\bar{z} = (a + bi)(a - bi) = a^2 - b^2i^2 = a^2 + b^2$, which is real. If $z = a$, where a is real, we have $\bar{z} = \bar{a} = \overline{a + 0i} = a - 0i = a = z$.

It is useful to know the various powers of i:

$$i^1 = i$$

$$i^2 = -1$$

$$i^3 = i^2 \cdot i = (-1)i = -i$$

$$i^4 = i^2 \cdot i^2 = (-1)(-1) = 1$$

Now, higher powers can be readily calculated. For example, $i^{22} = i^{20} \cdot i^2 = (i^4)^5 \cdot i^2 = (1)^5 \cdot i^2 = i^2 = -1$. Also, $i^{35} = i^{32} \cdot i^3 = (i^4)^8 \cdot i^3 = (1)^8 \cdot i^3 = i^3 = -i$. The object is to factor out the highest power of i^4, which is 1, and the remaining factor can be evaluated by using the results just obtained, that is, $i^1 = i$, $i^2 = -1$, $i^3 = -i$. More generally, if $n > 4$, and we write $n = 4k + r$, where $0 \le r \le 3$, then $i^n = i^{4k+r} = (i^{4k}) \cdot i^r = (i^4)^k \cdot i^r = 1^k \cdot i^r = i^r$. Negative powers of i can also be simplified, using the results just obtained and the fact that

$$i^{-1} = \frac{1}{i} = \frac{1}{i} \cdot \frac{i}{i} = \frac{i}{i^2} = \frac{i}{-1} = -i$$

For example,

$$i^{-3} = \frac{1}{i^3} = \frac{1}{-i} = -\frac{1}{i} = -(-i) = i$$

In summary, any integral power of i can be reduced to one of the numbers 1, -1, i, or $-i$.

When a is a nonnegative real number, the symbol $\sqrt{a}$ means the nonnegative real number whose square is a, but when $a < 0$, the symbol $\sqrt{a}$ as yet has no meaning. Within the complex number system we can now assign a meaning in this case, consistent with the case $a \ge 0$. Suppose $a = -p$, where p is a positive real number. Then we define

$$\sqrt{-p} = i\sqrt{p} \tag{2}$$

To see why this is reasonable, we can square the right-hand side to get

$$(i\sqrt{p})^2 = i^2(\sqrt{p})^2 = (-1)p = -p$$

So $i\sqrt{p}$ is a (complex) number whose square is $-p$. As a special case of definition (2), we have for $p = 1$,

$$\sqrt{-1} = i$$

In calculations involving square roots of negative numbers it is important to make use of definition (2) at the outset. To see why, consider $\sqrt{-2}\sqrt{-5}$. If we carelessly applied formula (12) from Chapter 1, without remembering the requirement that a and b both be positive, we would incorrectly conclude that

$$\sqrt{-2} \cdot \sqrt{-5} = \sqrt{(-2)(-5)} = \sqrt{10}$$

This is wrong because the property

$$\sqrt[n]{a}\,\sqrt[n]{b} = \sqrt[n]{ab}$$

is correct only if each of the individual roots exists in the set of real numbers. Instead, we should do as follows in the set of complex numbers.

$$\sqrt{-2} \cdot \sqrt{-5} = (i\sqrt{2})(i\sqrt{5}) = i^2(\sqrt{2} \cdot \sqrt{5}) = -\sqrt{10}$$

Let us return now to the solution of quadratic equations. Recall that by the quadratic formula, the solutions of

$$ax^2 + bx + c = 0 \qquad (a \neq 0)$$

are given by

$$\frac{-b + \sqrt{b^2 - 4ac}}{2a} \qquad \text{and} \qquad \frac{-b - \sqrt{b^2 - 4ac}}{2a}$$

Although the quadratic formula holds true even when a, b, and c are complex, for the present we assume them to be real. The nature of these two solutions depends on the number under the radical. This number, $b^2 - 4ac$, **is called the discriminant of the quadratic equation.** When the discriminant is positive, the square root is a positive real number, so the solutions are real and distinct. When the discriminant is 0, the square root is 0, so the solutions are real and equal. (There is only one solution in this case.) Finally, when the discriminant is negative, the square root is imaginary, so the solutions are imaginary; they are, in fact, conjugates of each other. We summarize these results in the box.

$$\text{If} \quad b^2 - 4ac \begin{cases} > 0 \\ = 0 \\ < 0 \end{cases} \text{the solutions are} \begin{cases} \text{real and unequal} \\ \text{real and equal} \\ \text{imaginary} \end{cases}$$

EXAMPLE 13

Without solving, determine the nature of the solutions of each of the following.

a. $2x^2 - 3x - 4 = 0$ **b.** $9x^2 - 24x + 16 = 0$ **c.** $3x^2 - 2x + 4 = 0$

Solution **a.** $b^2 - 4ac = (-3)^2 - 4(2)(-4) = 9 + 32 = 41 > 0$. So the solutions are real and unequal.

b. $b^2 - 4ac = (-24)^2 - 4(9)(16) = 576 - 576 = 0$. So the solutions are real and equal.

c. $b^2 - 4ac = (-2)^2 - 4(3)(4) = 4 - 48 = -44 < 0$. So the solutions are imaginary. ▲

EXAMPLE 14 ||||||||||||||||||||||||||||||||||||||||||||||||||||||||||||||||||||||||||||||||||||||||||||||||||||

Solve: $x^2 - 6x + 10 = 0$

Solution   By the quadratic formula we obtain

$$x = \frac{6 \pm \sqrt{36 - 40}}{2} = \frac{6 \pm \sqrt{-4}}{2} = \frac{6 \pm i\sqrt{4}}{2}$$

$$= \frac{6 \pm 2i}{2} = 3 \pm i$$ ▲

EXERCISE SET 5

13 – 19 odd

A

1 – 9 odd

In Problems 1–8 perform the indicated operations. Express answers in the form $a + bi$.

1. **a.** $(3 - 2i) + (5 + 4i)$ **b.** $(2 - 6i) + (-3 + 9i)$
   c. $(4 + 7i) - (5 - i)$ **d.** $(5 - 3i) - (2 - i)$
2. **a.** $(9 + 3i) + (7 - 8i)$ **b.** $(6 - 15i) + (4 + 8i)$
   c. $(3 - 5i) - (4 + 7i)$ **d.** $(3i - 4) - (5 - 2i)$
3. **a.** $(2 + 3i)(4 - 2i)$ **b.** $(3 - i)(3 + i)$
   c. $(4i - 7)(5i + 8)$ **d.** $(i + 2)(3 - 4i)$
4. **a.** $(6 + 5i)(5 + 3i)$ **b.** $(7 - 6i)(3 + 2i)$
   c. $(8 - 3i)(3 + 4i)$ **d.** $(3i - 2)(5 + 4i)$

5. **a.** $(2 - 3i)^{-1}$ **b.** $\dfrac{1}{2 + i}$

   c. $(2i - 3)^{-1}$ **d.** $\dfrac{1}{7 - 6i}$

6. **a.** $(3 + 4i)^{-1}$ **b.** $\dfrac{1}{1 - i}$

   c. i^{-1} **d.** $\dfrac{1}{5 + 4i}$

7. **a.** $\dfrac{3 - 2i}{3 + 2i}$ **b.** $\dfrac{4 + 5i}{5 - 6i}$

   c. $\dfrac{1 + i}{1 - i}$ **d.** $\dfrac{2i - 1}{3i + 2}$

8. **a.** $\dfrac{7 - 5i}{5 - 6i}$ **b.** $\dfrac{4 - 5i}{4 + 5i}$

   c. $\dfrac{i}{i - 1}$ **d.** $\dfrac{3i - 2}{4 - 5i}$

In Problems 9–12 simplify the expressions by using the properties of i.

9. **a.** $\sqrt{-9}$ **b.** $\sqrt{-108}$

   c. $\sqrt{-9}\sqrt{-25}$ **d.** $\dfrac{1}{\sqrt{-8}}$?

10. **a.** $\sqrt{-27}$ **b.** $\sqrt{-6}\sqrt{24}$

    c. $\dfrac{\sqrt{-8}}{\sqrt{-32}}$ **d.** $\sqrt{-10}\sqrt{-2}$

    e. $\dfrac{1}{\sqrt{-18}}$

11. **a.** i^6 **b.** i^{11} **c.** i^{-3} **d.** i^{33} **e.** i^{22}

12. **a.** i^{-5} **b.** $\dfrac{1}{i^7}$ **c.** i^{13} **d.** i^{-23} **e.** $\dfrac{1}{i^{15}}$

Solve the quadratic equations in Problems 13–20.

13. **a.** $x^2 + 9 = 0$ **b.** $2x^2 + 3 = 0$
14. **a.** $x^2 = -4$ **b.** $4x^2 + 1 = 0$
15. **a.** $x^2 - 2x + 2 = 0$ **b.** $x^2 + 2x + 4 = 0$
16. **a.** $t^2 + t + 1 = 0$ **b.** $2s^2 - s + 1 = 0$
17. **a.** $m(4 - m) = 5$ **b.** $2k^2 + 3 = 4k$
18. **a.** $3y^2 + 8y + 6 = 0$ **b.** $5r(2 - r) = 8$
19. **a.** $3x^2 = 2(x - 1)$ **b.** $4x(x - 2) = -7$
20. **a.** $9t^2 + 4 = 8t$ **b.** $3s^2 = 12s - 13$

In Problems 21–24 determine whether the solutions are real and unequal, real and equal, or imaginary by using the discriminant. Do not solve.

21. **a.** $5x^2 - 8x + 4 = 0$ **b.** $8x^2 + 9x + 2 = 0$

22. a. $16t^2 = 3(8t - 3)$ **b.** $5s^2 = 14(s - 1)$

23. a. $3x^2 - \sqrt{13}\,x + 1 = 0$

    b. $\sqrt{2}\,y^2 + 4y + \sqrt{8} = 0$

24. a. $8t^2 - 2\sqrt{21}\,t + 3 = 0$

    b. $(t + 3)(2t + 5) = 3$

 B

In Problems 25 and 26 expand and simplify.

25. a. $(1 - i)^6$ **b.** $(2 + i)^5$

26. a. $(3 - 2i)^3$ **b.** $(2 + 3i)^4$

27. Evaluate the polynomial $x^3 - 2x^2 + 3x - 7$ when $x = 1 + i$.

28. Evaluate the polynomial $x^4 + 3x^2 - 2$ when $x = 3 - 2i$.

29. Solve:

    a. $4x^2 + 8ix - 3 = 0$

    b. $ix^2 - 2x + 3i = 0$

30. By writing $z_1 = a + bi$ and $z_2 = c + di$, prove each of the following:

    a. $\overline{z_1 + z_2} = \overline{z_1} + \overline{z_2}$ **b.** $\overline{z_1 z_2} = \overline{z_1}\,\overline{z_2}$

    c. $\overline{\left(\dfrac{z_1}{z_2}\right)} = \dfrac{\overline{z_1}}{\overline{z_2}}$

31. For what values of k will the equation $k^2 x^2 + (k - 1)x + 4 = 0$ have equal solutions?

EQUATIONS THAT ARE CONVERTIBLE TO LINEAR OR QUADRATIC FORMS

Some equations, although they are not linear or quadratic, can nevertheless be solved by using techniques developed in this chapter. We consider first equations that are **quadratic in form.** These are equations that can be written in the form

$$a(\quad)^2 + b(\quad) + c = 0$$

where the same expression appears in each set of parentheses. For example,

$$x^4 - 5x^2 + 4 = 0$$

is quadratic in form, since it can be written

$$(x^2)^2 - 5(x^2) + 4 = 0$$

We could say it is quadratic in the unknown x^2. Another example is

$$3x^{-2} - 4x^{-1} + 1 = 0$$

which can be written

$$3(x^{-1})^2 - 4(x^{-1}) + 1 = 0$$

In the next example we solve these two equations.

EXAMPLE 15

Solve the equations:

    a. $x^4 - 5x^2 + 4 = 0$ **b.** $3x^{-2} - 4x^{-1} + 1 = 0$

Solution   **a.** We solve by factoring, first treating x^2 as the unknown.

$$x^4 - 5x^2 + 4 = 0$$

$$(x^2 - 1)(x^2 - 4) = 0$$

$$x^2 - 1 = 0 \quad \text{or} \quad x^2 - 4 = 0$$
$$x^2 = 1 \quad \text{or} \quad x^2 = 4$$
$$x = \pm 1 \quad \text{or} \quad x = \pm 2$$

So the solution set is $\{\pm 1, \pm 2\}$.

b.
$$3x^{-2} - 4x^{-1} + 1 = 0$$
$$(3x^{-1} - 1)(x^{-1} - 1) = 0$$
$$3x^{-1} - 1 = 0 \quad \text{or} \quad x^{-1} - 1 = 0$$
$$3x^{-1} = 1 \quad \text{or} \quad x^{-1} = 1$$
$$\frac{3}{x} = 1 \quad \text{or} \quad \frac{1}{x} = 1$$
$$x = 3 \quad \text{or} \quad x = 1$$

So the solution set is $\{3, 1\}$.

As an alternative procedure we may make an appropriate substitution so that the equation becomes quadratic. In part **a**, for example, we could let $t = x^2$. Then the equation becomes

$$t^2 - 5t + 4 = 0$$

Solving for t yields $t = 1$ or $t = 4$. Finally, we replace t by x^2 in each case and solve for x:

$$t = x^2 = 1 \quad \text{or} \quad t = x^2 = 4$$
$$x = \pm 1 \quad \text{or} \quad x = \pm 2$$

Similarly in part **b**, we could let $t = x^{-1}$. The equation would become

$$3t^2 - 4t + 1 = 0$$

yielding $t = \frac{1}{3}$ or $t = 1$, from which we would get

$$x^{-1} = \tfrac{1}{3} \quad \text{or} \quad x^{-1} = 1$$
$$\frac{1}{x} = \frac{1}{3} \quad \text{or} \quad \frac{1}{x} = 1$$
$$x = 3 \quad \text{or} \quad x = 1$$

The next type of equation we consider is one involving fractions, where the unknown appears in a denominator. Our procedure will be to clear of fractions by multiplying both sides by the LCD. Now we know that multiplying an equation by a nonzero number results in an equivalent equation. But if the LCD involves the unknown, then we do not know in advance if the number we are multiplying by is nonzero. The way out of this dilemma is to proceed with the clearing of fractions and solve the resulting equation but then to check each answer obtained to see whether it causes the LCD to be 0. If the LCD $\neq 0$, this ensures that multiplying by the LCD resulted in an equivalent equation. If the LCD $= 0$ for a value of x, this value must be discarded. It should be noted that simply showing that the LCD $\neq 0$ for a value of x that has been found as a possible solution shows that our procedure was justified for that value of x, but it does not guarantee that the solution is correct (an arithmetic error may have been made).

EXAMPLE 16

Solve for x: $\quad \dfrac{1}{x-2} - \dfrac{2}{x+1} = \dfrac{7}{x^2 - x - 2}$

Solution   The procedure is to multiply both sides of the equation by the LCD, which in this case is $(x-2)(x+1)$. It is essential to multiply every term on both sides of the equation. In so doing, all denominators cancel.

$$\left(\frac{1}{x-2} - \frac{2}{x+1}\right)(x-2)(x+1) = \left(\frac{7}{x^2 - x - 2}\right)(x-2)(x+1)$$

$$x + 1 - 2(x - 2) = 7$$

$$x + 1 - 2x + 4 = 7$$

$$-x = 2$$

$$x = -2$$

We check $x = -2$ in the LCD and see that it is not 0. So (assuming no arithmetic error) the solution is $x = -2$. ▲

EXAMPLE 17

Solve for x: $\quad \dfrac{3}{x-3} - \dfrac{2}{x-2} = \dfrac{3}{x^2 - 5x + 6}$

Solution   The LCD is $(x-3)(x-2)$. Multiplying both sides by this yields

$$3(x - 2) - 2(x - 3) = 3$$

$$3x - 6 - 2x + 6 = 3$$

$$x = 3$$

Checking the LCD, we find that it is 0 for this value of x. Since this is the only possible solution and it is not admissible, we conclude that the equation has **no solution.** The solution set is therefore the empty set. ▲

EXAMPLE 18 ‖‖

Solve for x: $\dfrac{1}{x-1} + \dfrac{3}{x+3} = 1$

Solution   We multiply both sides by the LCD, $(x-1)(x+3)$, to obtain

$$x + 3 + 3(x-1) = (x-1)(x+3)$$
$$x + 3 + 3x - 3 = x^2 + 2x - 3$$
$$0 = x^2 - 2x - 3$$
$$(x-3)(x+1) = 0$$
$$x = 3 \quad \text{or} \quad x = -1$$

Since the LCD is nonzero for each of these, the solution set is $\{-1, 3\}$.   ▲

EXAMPLE 19 ‖‖

Solve for y and check the results: $\dfrac{2}{3y} + \dfrac{1}{y-1} = \dfrac{1}{2}$

Solution   The LCD is $6y(y-1)$. Multiplying by this yields

$$4(y-1) + 6y = 3y(y-1)$$
$$4y - 4 + 6y = 3y^2 - 3y$$
$$3y^2 - 13y + 4 = 0$$
$$(3y-1)(y-4) = 0$$
$$3y - 1 = 0 \quad \text{or} \quad y - 4 = 0$$
$$3y = 1 \quad \text{or} \quad y = 4$$
$$y = \tfrac{1}{3}$$

The LCD is nonzero for each answer, so we know the equation obtained after multiplication is equivalent to the original. To verify that these actually are solutions, we substitute in the original equation.

Check. $y = \tfrac{1}{3}$

$$\frac{2}{3 \cdot \tfrac{1}{3}} + \frac{1}{\tfrac{1}{3} - 1} = \frac{2}{1} + \frac{3}{1-3} = 2 + \frac{3}{-2}$$

$$= 2 - \frac{3}{2} = \frac{4}{2} - \frac{3}{2} = \frac{1}{2}$$

So $y = \tfrac{1}{3}$ is a solution.

Check. $y = 4$

$$\frac{2}{3\cdot 4} + \frac{1}{4-1} = \frac{2}{12} + \frac{1}{3} = \frac{1}{6} + \frac{1}{3} = \frac{1}{6} + \frac{2}{6} = \frac{3}{6} = \frac{1}{2}$$

So $y = 4$ is also a solution. The solution set is therefore $\{\frac{1}{3}, 4\}$.   ▲

Remark. Multiplying both sides of an equation by an expression involving the unknown may introduce extraneous roots (as in Example 17). On the other hand, dividing both sides by such an expression may result in the loss of solutions. To illustrate the latter, if in the equation $x^2 = 4x$, we divide both sides by x, getting $x = 4$, we lose the solution $x = 0$.

The next three examples involve squaring both sides of an equation. This may or may not lead to an equivalent equation, as the following considerations show: If $a = b$, then clearly $a^2 = b^2$. But conversely, if $a^2 = b^2$, then we have either $a = b$ or $a = -b$. So squaring both sides of an equation results in the possibility of additional solutions not contained in the original (namely, those contained in $a = -b$). So, when squaring is desirable, the appropriate way to proceed is to square both sides, solve the resulting problem, and check all answers in the original equation. The squared equation contains all solutions of the original, but it may contain additional solutions that are not valid.

EXAMPLE 20

Solve for x: $\sqrt{x-1} = 2$

Solution   On squaring, we obtain

$$x - 1 = 4$$
$$x = 5$$

Check. $\sqrt{5-1} = \sqrt{4} = 2$
So the answer is 5.   ▲

EXAMPLE 21

Solve for x: $\sqrt{2x-3} + x = 3$

Solution   Before squaring we wish to isolate the radical in order to eliminate it after squaring. So, we first add $-x$ to both sides:

$$\sqrt{2x-3} = 3 - x$$
$$2x - 3 = 9 - 6x + x^2$$
$$x^2 - 8x + 12 = 0$$
$$(x-2)(x-6) = 0$$
$$x = 2 \quad \text{or} \quad x = 6$$

Check. $x = 2$

$$\sqrt{2 \cdot 2 - 3} + 2 = \sqrt{1} + 2 = 1 + 2 = 3$$

So $x = 2$ is a solution.

Check. $x = 6$

$$\sqrt{2 \cdot 6 - 3} + 6 = \sqrt{9} + 6 = 3 + 6 = 9 \ne 3$$

So $x = 6$ is not a solution. Thus, $x = 2$ is the only solution.   ▲

EXAMPLE 22

Solve for x: $\sqrt{2x + 3} - \sqrt{x + 1} = 1$

Solution   In a situation such as this it is usually best to isolate the more complicated radical, square once, isolate the remaining radical, and square again.

$$\sqrt{2x + 3} = 1 + \sqrt{x + 1}$$
$$(\sqrt{2x + 3})^2 = (1 + \sqrt{x + 1})^2$$
$$2x + 3 = 1 + 2\sqrt{x + 1} + x + 1$$
$$x + 1 = 2\sqrt{x + 1}$$
$$x^2 + 2x + 1 = 4(x + 1)$$
$$x^2 - 2x - 3 = 0$$
$$(x - 3)(x + 1) = 0$$
$$x = 3 \quad \text{or} \quad x = -1$$

Check. $x = 3$

$$\sqrt{2 \cdot 3 + 3} - \sqrt{3 + 1} = \sqrt{9} - \sqrt{4} = 3 - 2 = 1$$

So $x = 3$ checks.

Check. $x = -1$

$$\sqrt{2(-1) + 3} - \sqrt{-1 + 1} = \sqrt{-2 + 3} - 0 = \sqrt{1} = 1$$

So $x = -1$ also checks. The complete solution set is therefore $\{-1, 3\}$.   ▲

A Word of Caution. In the solution of problems such as in Example 22 students often make the mistake of removing radicals by squaring each individual term; after writing the equation $\sqrt{2x + 3} = 1 + \sqrt{x + 1}$, they "square" and get $2x + 3 = 1 + (x + 1)$. This is *not correct*. The right-hand side must be treated as a binomial $a + b$, so that when it is squared, the result is $a^2 + 2ab + b^2$. *Do not forget the middle term.*

EXERCISE SET 6

In Problems 1–60 find the complete solution set.

A   1-14 odd   21-47 odd

1. $x^4 - 5x^2 + 4 = 0$ **2.** $4x^4 - 17x^2 + 4 = 0$
3. $x^4 - 5x^2 = 36$ **4.** $x^4 + 26x^2 + 25 = 0$
5. $x^{-2} - x^{-1} - 2 = 0$
6. $2x^{-2} - 7x^{-1} + 3 = 0$
7. $6t^{-2} - t^{-1} = 15$ **8.** $2v^{-2} - 3v^{-1} = 20$
9. $x^{2/3} - 3x^{1/3} + 2 = 0$
10. $6x^{2/3} - 5x^{1/3} - 6 = 0$
11. $x^{2/3} - 4x^{1/3} + 4 = 0$
12. $2x^{-2/3} + 3x^{-1/3} = 2$
13. $x^3 - 9x^{3/2} + 8 = 0$
14. $8x^3 - 35x^{3/2} + 27 = 0$

15. $\dfrac{2}{x+3} = \dfrac{1}{x-2}$ **16.** $\dfrac{2}{x} - \dfrac{3}{x+1} = 0$

17. $\dfrac{1}{x+2} - \dfrac{3}{x-2} = \dfrac{4}{x^2-4}$

18. $\dfrac{5}{2x-3} - \dfrac{2}{x+1} = \dfrac{3}{2x^2-x-3}$

19. $\dfrac{3}{x} - \dfrac{4}{3-x} + \dfrac{5}{x^2-3x} = 0$

20. $\dfrac{x}{x-2} - \dfrac{3}{x+6} = \dfrac{x^2}{x^2+4x-12}$

21. $\dfrac{2}{x+1} - \dfrac{8}{x-1} = 1$

22. $\dfrac{1}{2x-3} + \dfrac{3}{x+1} = 2$

23. $\dfrac{2}{x} = \dfrac{x}{x+4}$

24. $\dfrac{x}{x+3} = \dfrac{2}{x-3}$

25. $\dfrac{2}{x} - \dfrac{x}{x-1} = 5$

26. $\dfrac{x}{x-3} - \dfrac{4}{x} = \dfrac{9}{x^2-3x}$

27. $\dfrac{2}{x} - \dfrac{1}{x+1} = \dfrac{1}{x^2+x}$

28. $\dfrac{x}{x-2} + \dfrac{1}{x+1} = \dfrac{3x}{x^2-x-2}$

29. $\dfrac{2x}{x-1} - \dfrac{x+1}{x+3} = \dfrac{8}{x^2+2x-3}$

30. $\dfrac{4}{2x-1} + \dfrac{x}{x+2} = \dfrac{10}{2x^2+3x-2}$

31. $\sqrt{3x+2} = 1$ **32.** $\sqrt{x+2} = -1$
33. $\sqrt{2x-1} + 1 = 0$ **34.** $\sqrt[3]{2x-1} + 1 = 0$
35. $\sqrt{x+5} = 2\sqrt{x+2}$
36. $2\sqrt{x} - \sqrt{3(x+3)} = 0$
37. $\sqrt{4x+3} = 2x$
38. $\sqrt{2-x} = x$
39. $\sqrt{5-4x} = 2 - x$
40. $\sqrt{3x-2} = x$
41. $\sqrt{1-3x} = x - 1$
42. $\sqrt{3x-5} = 1 - x$
43. $\sqrt{1-3x} = 1 - x$
44. $\sqrt{5x+4} - 2 = 3x$
45. $8(x^6 - 1) = 63x^3$
46. $x^8 + 16 = 17x^4$
47. $2x^{-4} - 4x^{-2} + 1 = 0$
48. $24x + 11\sqrt{x} - 18 = 0$ **Hint.** Let $t = \sqrt{x}$.

49. $\dfrac{3x-1}{x+3} - \dfrac{x+4}{2x-3} = 1$

50. $\dfrac{1-2x}{x+1} - \dfrac{3x+2}{2x-1} = \dfrac{9x}{2x^2+x-1}$

51. $\dfrac{2x-3}{3x-2} - \dfrac{x+3}{x+2} = x$

52. $\dfrac{2x^2+1}{x^2+1} - \dfrac{3x^2-4}{x^2-1} = 1$

53. $\dfrac{1}{x} - \dfrac{2x}{x+3} + \dfrac{2x-3}{x-4} = 0$

54. $\dfrac{x}{x+2} + \dfrac{5x+4}{x^2+x-2} = \dfrac{3}{x-1}$

55. $2\sqrt{x+3} = \sqrt{2x+5} + 1$
56. $\sqrt{3x+4} - \sqrt{2x+1} = 1$

57. $\sqrt{x + 4} + 2 = \sqrt{3 - 2x}$

58. $\dfrac{1}{\sqrt{x + 2}} = \sqrt{x + 2} - \dfrac{3}{2}$

59. $\sqrt{x + 1} + \sqrt{2x + 3} = \sqrt{6x + 7}$

60. $\dfrac{1}{\sqrt{x + 2}} + \dfrac{2}{\sqrt{2x - 5}} = \dfrac{3}{\sqrt{2x^2 - x - 10}}$

In Problems 61–64 solve for the indicated quantity.

61. e^t: $e^t + e^{-t} = 2$

62. e^t: $2e^t + e^{-t} = 3$

63. $\ln t$: $3(\ln t)^2 - 2(\ln t) = 1$

64. $\ln t$: $\dfrac{1}{\ln t} + \ln t = 2$

7 APPLICATIONS

In the application of mathematics to real-life situations problems are typically stated in words, not mathematical symbols. Learning to translate such verbal statements into mathematical terms is as important an aspect of the study of mathematics as is learning how to solve problems after they are formulated. The examples and exercises in this section provide practice in setting up these so-called word problems that lead to linear or quadratic equations.

Unfortunately, there is no magical road to success in setting up word problems. Guidelines can be given, but each problem has its own unique features. There are, however, general classes into which many problems can be categorized, and learning how to do one problem of a given class will usually help in solving other problems of that class. For example, in physical applications many problems deal with motion in a straight line in which the relation "distance equals rate times time" is applicable (where the rate, or speed, is constant). Thus, the formula Distance = (Rate)(Time), or $d = rt$, is used in all such situations. Another class of problems has to do with the mixing of substances, as in chemistry experiments. These and other classes will be illustrated in the examples.

The following guidelines are suggested:

1. Read the problem through, more than once if necessary, and identify what unknown quantity is to be determined.

2. Introduce some letter to designate the unknown quantity. The letter x is often used for historical reasons, but sometimes it is helpful to use a suggestive letter, such as d for distance or v for velocity. Be very specific about this; for example, a statement like "Let t = the time in hours required to complete the job" is typical.

3. Begin a careful rereading of the problem, phrase by phrase, and write the relevant information that is given, expressing relationships with the unknown when appropriate. Often, a sketch can be useful in looking for relationships.

4. Try to identify in the problem some equation that relates the unknown and the known information and write this in terms of the symbols you have introduced. Here it is useful to observe that the words *is* and *are* often translate as *equals* in an equation.

5. Solve the equation obtained in Step 4, and check your answers to see if they are reasonable in the context of the original problem. Sometimes, one or more solutions must be eliminated because of physical limitations imposed by the problem.

When more than one unknown is involved, obvious modifications in these steps should be made. Also, there are occasions when it is best to introduce as the unknown a quantity related to, but not the same as, the quantity to be determined.

The importance of Steps 2 and 3 should be emphasized. A thorough and careful listing of the relevant given information and a precise identification of the unknown are often the keys to success in solving problems.

EXAMPLE 23

MOTION PROBLEM

A man leaves town A by car and travels at a constant speed of 40 miles per hour toward town B, which is 100 miles away. One hour later a woman leaves town B and travels on the same highway toward town A at a constant speed of 50 miles per hour. Find how much time elapses from the time the man leaves town A until they meet.

Solution   Let

$$t = \text{Number of hours elapsed from time man leaves until they meet}$$

Then

$$t - 1 = \text{Number of hours woman travels}$$

We use the fundamental relationship

$$\text{Distance} = (\text{Average rate})(\text{Time}) \qquad \text{or} \qquad d = rt$$

to determine the distances for the man and woman.

$$40t = \text{Distance man travels before they meet}$$
$$50(t - 1) = \text{Distance woman travels before they meet}$$

Since the total distance is 100, we see from Figure 1 that

$$40t + 50(t - 1) = 100$$
$$40t + 50t - 50 = 100$$
$$90t = 150$$
$$t = \tfrac{150}{90} = \tfrac{5}{3}$$

So the man travels $1\tfrac{2}{3}$ hours, or 1 hour and 40 minutes.

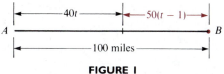

FIGURE I

▲

EXAMPLE 24 ııı

GEOMETRY PROBLEM

A flower bed is in the shape of a rectangle. Its length is twice its width. The bed is surrounded by a walkway 4 feet wide. If the area of the walk is exactly twice the area of the bed, find the dimensions of the bed.

Solution   In problems such as this a sketch such as Figure 2 is definitely helpful.

Let

$$x = \text{Width of bed in feet}$$

Then

$$2x = \text{Length of bed in feet}$$
$$\text{Area of bed} = (\text{Length})(\text{Width}) = (2x)(x) = 2x^2$$
$$\text{Area of walk} = 2 \cdot 4(x + 8) + 2 \cdot 4(2x)$$
$$= 8x + 64 + 16x$$
$$= 24x + 64$$

(This is obtained by dividing the walk into rectangles as shown in Figure 2.) Therefore,

$$24x + 64 = 2(2x^2)$$

or, on dividing both sides by 4 and rearranging,

$$x^2 - 6x - 16 = 0$$
$$(x - 8)(x + 2) = 0$$
$$x - 8 = 0 \quad \text{or} \quad x + 2 = 0$$
$$x = 8 \quad \text{or} \quad x = -2$$

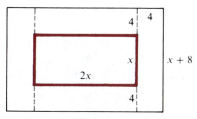

FIGURE 2

Clearly, x cannot be negative, so the width of the bed is 8 feet and the length is $2x = 16$ feet. ▲

EXAMPLE 25 ||||||||||||||||||||||||||||||||||||||||||||||||||||||||||||||||||||||||||||||||||||||||||||||||||||||||||

MIXING PROBLEM

A large tank contains 100 gallons of pure water. How many gallons of a saline solution containing 25% salt must be added to obtain a solution that is 10% salt?

Solution Let

$$x = \text{Number of gallons of 25\% solution that must be added}$$

Then

$$0.25x = \text{Number of gallons of salt added}$$

$$100 + x = \text{Total number of gallons of final solution}$$

$$0.10(100 + x) = \text{Number of gallons of salt in final solution}$$

Since the salt added is the only source of salt, it follows that the amount of salt added equals the amount of salt in the final solution. So we have the equation

$$0.25x = 0.10(100 + x)$$

Decimals can be eliminated by multiplying both sides by 100.

$$25x = 10(100 + x)$$
$$25x = 1,000 + 10x$$
$$15x = 1,000$$
$$x = \frac{1,000}{15} = \frac{200}{3}$$

So $66\frac{2}{3}$ gallons of the solution must be added. ▲

EXAMPLE 26 ||||||||||||||||||||||||||||||||||||||||||||||||||||||||||||||||||||||||||||||||||||||||||||||||||||||||||

MIXING PROBLEM

A solution containing 80% sulfuric acid is to be mixed with a 65% solution to obtain a 75% solution. If 10 gallons of the final solution are desired, how many gallons of each of the original solutions should be used?

Solution Let

$$x = \text{Number of gallons of 80\% solution}$$

Then

$$10 - x = \text{Number of gallons of } 65\% \text{ solution}$$

$$0.80x = \text{Number of gallons of pure sulfuric acid in first solution}$$

$$0.65(10 - x) = \text{Number of gallons of pure sulfuric acid in second solution}$$

$$0.75(10) = \text{Number of gallons of pure sulfuric acid in final solution}$$

So, since the amount of pure sulfuric acid in the first solution plus the amount in the second solution must equal the amount in the final solution, we have

$$0.80x + 0.65(10 - x) = 0.75(10)$$

$$80x + 65(10 - x) = 750$$

$$80x + 650 - 65x = 750$$

$$15x = 100$$

$$x = \tfrac{100}{15} = \tfrac{20}{3}$$

and

$$10 - x = 10 - \tfrac{20}{3} = \tfrac{10}{3}$$

Thus, $6\tfrac{2}{3}$ gallons of the 80% solution and $3\tfrac{1}{3}$ gallons of the 65% solution should be used. ▲

The basic idea in mixture problems is:

$$\begin{pmatrix} \textbf{Amount of pure} \\ \textbf{substance in the} \\ \textbf{first solution} \end{pmatrix} + \begin{pmatrix} \textbf{Amount of pure} \\ \textbf{substance in the} \\ \textbf{second solution} \end{pmatrix} = \begin{pmatrix} \textbf{Amount of pure} \\ \textbf{substance in the} \\ \textbf{final mixture} \end{pmatrix}$$

EXAMPLE 27

FALLING BODY PROBLEM

In calculus it is proved that near the surface of the earth, if air resistance is not considered, the distance s above the earth of a falling body after t seconds is given by the law

$$s = -\tfrac{1}{2}gt^2 + v_0t + s_0$$

where g is the acceleration due to gravity (approximately 32 feet per second per second), v_0 is the initial velocity of the object, and s_0 is the initial distance above the earth. Here, s is measured positively upward from ground level. (This accounts for the negative sign on the first term, since acceleration due to gravity is directed downward.) If a ball is thrown upward from the top edge of a 128 foot building with a velocity of 32 feet per second, find how long it will be before the ball strikes the ground.

Solution   In this problem the unknown t has already been introduced, and in fact the equation relating t to known information is given. All that remains is to interpret the data given in the problem in terms of the constants in the equation of motion. It is evident that $v_0 = 32$ and $s_0 = 128$. Furthermore, at the time the ball strikes the ground $s = 0$; so we have

$$0 = -\tfrac{1}{2}(32)t^2 + 32t + 128$$
$$16t^2 - 32t - 128 = 0$$
$$t^2 - 2t - 8 = 0$$
$$(t - 4)(t + 2) = 0$$
$$t = 4 \quad \text{or} \quad t = -2$$

Clearly, $t = 4$ is the only admissible solution. So the ball will strike the ground after 4 seconds. ▲

EXAMPLE 28

ECONOMICS PROBLEM

A theater has an average daily attendance of 400 with the current ticket price of $2.00. It is estimated that for each 10¢ decrease in the ticket price the average attendance will increase by 40 persons. What price should be charged to increase revenue by $100? What will be the new average attendance?

Solution   This is a situation in which it is best to use as the unknown something related to the ticket price rather than the ticket price itself.

Let

$$x = \text{Number of 10¢ reductions in price needed to achieve desired increase in revenue}$$

Then

$$2.00 - 0.10x = \text{Price of each ticket in dollars}$$
$$40x = \text{Increase in average attendance}$$
$$400 + 40x = \text{New average attendance}$$
$$\text{Revenue in dollars} = \text{Price of each ticket in dollars times number attending}$$
$$\text{Present revenue in dollars} = (2.00)(400) = 800$$
$$\text{Revenue desired in dollars} = 800 + 100 = 900$$

So,

$$900 = (2.00 - 0.10x)(400 + 40x)$$
$$900 = 800 + 40x - 4x^2$$

$$4x^2 - 40x + 100 = 0$$
$$x^2 - 10x + 25 = 0$$
$$(x - 5)^2 = 0$$
$$x = 5$$

So the ticket price should be reduced by 5 increments of 10¢ for a new price of $1.50 per ticket. The new average attendance will be $400 + (40)(5) = 600$. (It can be shown that this is the optimum price; that is, any other price, either higher or lower, would produce less revenue under the stated assumptions. We will see one way of showing this in Chapter 4.)   ▲

EXAMPLE 29   ‖‖

MOTION PROBLEM

An airplane flies from city A to city B in 1 hour and 40 minutes against a headwind of 80 miles per hour. The return trip, with no shift in the wind, requires 1 hour. The pilot flew at the same indicated airspeed going and coming. Find his indicated airspeed.

Note. The indicated airspeed is the speed the plane would move relative to the ground if there were no wind.

Solution   Let

$$v = \text{Indicated airspeed in miles per hour}$$

Then

$$v - 80 = \text{Actual speed going}$$
$$v + 80 = \text{Actual speed returning}$$
$$\text{Time going} = 1 \text{ hour } 40 \text{ minutes} = \tfrac{5}{3} \text{ hours}$$
$$\text{Time returning} = 1 \text{ hour}$$
$$\text{Distance} = (\text{Rate})(\text{Time})$$
$$\text{Distance from } A \text{ to } B = (v - 80)(\tfrac{5}{3})$$
$$\text{Distance from } B \text{ to } A = (v + 80)(1)$$

These distances are equal, so

$$(v - 80)\tfrac{5}{3} = (v + 80)(1)$$
$$5v - 400 = 3v + 240$$
$$2v = 640$$
$$v = 320$$

The indicated airspeed is 320 miles per hour.   ▲

EXAMPLE 30 ||||||||||||||||||||||||||||||||||||||||||||||||||||||||||||||||||||||||||||||||||||||||||||||||||||||||||||||||||

WORK PROBLEM

A pump call fill a reservoir in 12 days. A second pump, operating independently, can fill the same reservoir in 8 days. How long will it take to fill the reservoir if both pumps operate simultaneously?

Solution   Let

x = Number of days required to fill the reservoir with both pumps in operation

$\frac{1}{12}$ = Fractional part of the job first pump does per day

$\frac{1}{8}$ = Fractional part of the job second pump does per day

So the fractional part of the job done by first pump in x days = $x/12$, and the fractional part of the job done by second pump in x days = $x/8$. Working together they do the entire job.

$$\frac{x}{12} + \frac{x}{8} = 1$$
$$8x + 12x = 96$$
$$20x = 96$$
$$x = 4.8$$

Thus, 4.8 days would be required to fill the reservoir with both pumps in operation.   ▲

EXAMPLE 31 ||||||||||||||||||||||||||||||||||||||||||||||||||||||||||||||||||||||||||||||||||||||||||||||||||||||||||||||||||

ECONOMICS PROBLEM

A vendor purchased a shipment of lamps for a total cost of $1,000. Company clerks damaged 4 of the lamps so that they could not be sold. The remaining lamps were sold at a profit of $25.00 each, and a total profit of $200 was realized when all the lamps were sold. How many lamps were purchased?

Solution   Let

x = Total number of lamps purchased

Then

$x - 4$ = Number of lamps sold

$\dfrac{1,000}{x}$ = Cost of each lamp, in dollars

$$\frac{1,200}{x-4} = \text{Revenue from selling each lamp, in dollars}$$

(Since total profit was $200, total revenue was $1,200.)

Profit per lamp = (Revenue per lamp) − (Cost per lamp)

$$25 = \frac{1,200}{x-4} - \frac{1,000}{x}$$

$$25(x-4)x = 1,200x - 1,000(x-4)$$

$$x^2 - 4x = 48x - 40x + 160 \qquad \text{Dividing by 25}$$

$$x^2 - 12x - 160 = 0$$

$$(x-20)(x+8) = 0$$

$$x = 20 \quad \text{or} \quad x = -8$$

So 20 lamps were purchased.

In Example 31 we used the following fundamental relationship:

Profit = (Revenue) − (Cost)

This applies to all problems involving the sale of manufactured or purchased items.

EXERCISE SET 7

In Problems 1–50 carry out these steps:

a. Introduce a variable name for an appropriate unknown quantity, and express all other relevant unknown quantities in terms of this variable.

b. Obtain an equation relating the known and unknown quantities, involving only the variable introduced in part **a**.

c. Solve the equation in part **b**, and answer any other questions posed in the problem.

A

1. A test driver for a new car drove around a 12.2-kilometer course 50 times in 4 hours and 20 minutes. What was his average speed?

2. A driver averaged 55 miles per hour on an interstate highway for the first 70 miles of a 100-mile trip, but averaged only 33 miles per hour for the remainder through a congested area. How long did the trip take?

3. A freight train leaves Chicago and travels at an average speed of 50 miles per hour. Thirty minutes later a passenger train leaves from the same station and travels at an average speed of 80 miles per hour in the same direction on a parallel track. How long will it take for the passenger train to overtake the freight train?

4. Two cars going in opposite directions on a straight highway pass each other at 1:00 P.M. At 1:15 P.M. they are 30 miles apart. If one car is averaging 12 miles per hour more than the other, find the average speed of each car.

5. A cyclist and a jogger leave at the same time and follow the same route for a distance of 4 kilometers. The cyclist arrives 15 minutes before the jogger. If the cyclist goes twice as fast as the jogger, find the speed of the jogger.

6. A girl walked from her home into town at the rate of 4 miles per hour. She decided to return by bus, over the same route. The bus averaged 20 miles per hour, and the entire round trip took 1 hour and 40 minutes, including a 10 minute wait for the bus. How far did the girl walk?

7. A field is in the shape of a rectangle 3 times as long as it is wide. A fence costing $3.00 per foot is to be placed around the field, and two fences costing $1.50 per foot are to be placed across the field so as to divide it into three squares. If the total cost of the fencing is $1,080, find the dimensions of the field.

8. The length of a room is 3 feet more than twice the width, and the perimeter is 60 feet. Find the dimensions of the room.

9. A patio is 6 feet longer than it is wide. One of the long sides is adjacent to the house, and the other three sides are bordered by a flower bed 3 feet wide. If the combined area of the patio and flower bed is 360 square feet, find the dimensions of the patio.

10. A pasture is bounded by a river on one side. The farmer has 600 yards of fencing for enclosing a rectangular area of 40,000 square yards. Only three sides will be fenced, using the river as the boundary for the fourth side. There are two ways to do this. Find them.

11. The sum of three consecutive even numbers is 444. Find the numbers.

12. A student has grades of 78, 66, and 87 on the first three exams in a math course. What grade does the student need on the fourth exam to obtain an average of 80?

13. A student's average grade on the three one-hour exams in a course is 76. If the final exam is weighted as two one-hour exams, what grade would the student have to make on the final to bring the average up to 80?

14. The prices of admission at an amusement park are $2.50 for children, $6.00 for adults, and $4.00 for senior citizens. On a given day total receipts were $80,920 from 21,773 paid admissions. If there were twice as many senior citizen tickets sold as adult tickets, find how many of each type of ticket was sold.

15. At a movie theater the admission for adults is $4.00 and for children $1.50. Popcorn costs $1.00 a bag. On a certain day gross receipts for admission and popcorn combined were $966. If three times as many children as adults attended and half of the persons attending bought popcorn, find how many children attended and how many bags of popcorn were sold.

16. By radar a highway patrolman observes a speeder going 80 miles an hour. The patrolman has difficulty starting his car so that 5 minutes elapse before he is able to begin pursuit. It is 20 miles to the state line. What average speed would the patrolman have to go to overtake the speeder before he crosses the state line?

17. A boy in a town 32 miles from his home got a ride for all but the last 2 miles, which he had to walk. The average speed of the car was 40 miles per hour, and the whole trip took 1 hour and 15 minutes. Find his rate of walking.

18. How many cubic centimeters of a 40% sulfuric acid solution must be added to 400 cubic centimeters of a 15% sulfuric acid solution to obtain a 25% concentration?

19. Coffee costing $6.25 per pound is to be mixed with coffee costing $8.50 per pound to obtain 45 pounds of a blend costing $7.50 per pound. How much of each type should be used?

20. A dairy wishes to obtain 200 gallons of milk that is 3.8% butterfat by mixing lowfat milk, which is 2% butterfat, with half-and-half, which is 11% butterfat. How much of each should be used?

21. A 10-gallon container is filled with a salt solution containing 20% salt. How many gallons should be drained and replaced by pure water to reduce the salinity to 15%?

22. From a helicopter hovering 6,400 feet above the ground a projectile is fired straight down with an initial velocity of 1,200 feet per second. When will the projectile strike the ground? (See Example 27.)

23. After a 25% reduction, a television set was sold for $487.50. What was the original price?

24. The length of a room is 5 feet less than twice its width, and its area is 187 square feet. Find its dimensions.

25. An open-top box is to be made by cutting squares from the corners of a flat 10 × 12-inch piece of cardboard and folding up the sides (see sketch). What size square should be removed to produce a box having a total surface area of 95 square inches?

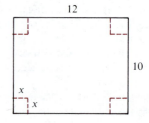

26. A woman invests $10,000 in two accounts, one an ordinary savings account yielding $5\frac{1}{2}\%$ annual interest and the other a certificate of deposit yielding 7% annual interest. At the end of one year the combined interest is $652. How much did she invest in each account?

27. A farmer can buy feeder calves weighing 530 pounds on the average for $280 each. If it costs an average of 42¢ per pound added to fatten a calf, and they can be sold at an average of 62¢ per pound, how much weight would have to be added to each calf on the average to realize a profit of $225 per animal?

28. Find two consecutive positive odd integers whose product is 255.

29. The sum of the reciprocals of two consecutive positive even integers is $\frac{9}{40}$. Find the integers.

30. An electrician charges $15.00 per hour, and his assistant, who is an apprentice, receives $9.00 per hour. On a certain job the assistant worked 30 minutes less than the electrician, and the total bill was $110, including $18.50 for making a house call. How long did the electrician work?

31. In a manufacturing process two machines produce parts of the same type. The newer machine produces 3,000 parts per hour, and the older machine produces 2,000 parts per hour. A total of 30,000 of these parts must be produced each day. On a certain day the older machine broke down before the job was completed, and it took the newer machine two hours more to complete the job. How long did each machine operate?

32. Twice a certain natural number exceeds 8 times its reciprocal by 15. What is the number?

33. The owner of a dress shop paid a total of $1,950 for some dresses of a certain style. She set the selling price of each dress at $25 more than the cost. At the end of one week all but 5 of the dresses had been sold, and the profit was $300. How many dresses were purchased?

34. A boy can paddle his canoe at an average rate of 7 miles per hour in still water. He paddles 5 miles down a river and back in 1 hour and 45 minutes. Find the rate of the current.

35. A video store owner averages 200 rentals of videotapes per day. As an experiment he decides to reduce the rental price by 50¢ per tape on a certain day. On that day 300 tapes are rented, resulting in an increase of $48 in income over

that for an average day. What is the normal rental rate?

36. A pilot flies a distance of 360 miles with a tailwind. The return trip, against the wind, takes 48 minutes longer than the trip going. Her average indicated airspeed (the speed if there were no wind) both going and coming was 240 miles per hour. Find the speed of the wind.

37. A picture is 6 inches taller than it is wide. It is framed with a matting 2 inches wide on the top and sides and 3 inches wide at the bottom. The area of the picture plus the matting is 294 square inches. Find the dimensions of the picture.

38. A group of 40 people charter a bus for an excursion. They are told that if they can get 15 additional people to go, the price of each ticket will be reduced by $7.50 (while the total cost for chartering the bus remains the same). Find what the cost of each ticket will be if they get the additional 15 passengers. What will it be if only the original 40 people go?

39. Bill can mow a large lawn with his rider mower in $3\frac{1}{2}$ hours, and Joanne can do the same job with her hand-operated power mower in 6 hours. How long will it take them working together?

40. A pump can drain all of the water from a tank in 8 hours. If a second, larger pump is used along with the first one, the tank can be emptied in 3 hours. How long would it take the second pump alone to empty the tank?

41. Forty people charter a boat at a cost of $15 each. They are told that for every additional person they can get to go, the price of each ticket will be reduced by 25¢, provided the total does not exceed 60, which is the capacity of the boat. The net income to the boat operator was $625. How many people took the boat ride?

42. A railroad company agrees to run an excursion train for a group under the following conditions. If 200 or fewer people go, the rate is $10 per ticket. The rate for all tickets will be reduced by 2¢ per ticket for each person in excess of 200. If the total intake by the railroad was $2,450, find how many people took the trip.

B

43. It takes a small pump 4 hours longer to fill a tank than a large pump. If both pumps work together,

the tank can be filled in 2 hours and 40 minutes. How long will it take each pump working alone to fill the tank?

44. A certain city bus line charges 50¢ per ticket and has an average daily ridership of 200,000. The capacity is double this amount, and to provide an incentive for people to use public transportation, the company experiments with reducing the fare by 1¢ increments. They find that for each 1¢ fare reduction the number of riders each day increases by an average of 7,500. They also find that up to a certain point their income rises, but further reductions cause the income to decline. Their maximum income exceeds what it was before any fare reduction by $10,200. What fare did they charge to produce this income, and what was the average daily ridership with this fare?

45. The octane rating of gasoline is a number that measures its antiknock value. The octane rating of a fuel consisting of a mixture of normal heptane (which has a decided knocking tendency) and isooctane (which has a decided antiknocking tendency) is the percentage of isooctane in the mixture. Other fuels are rated in comparison with this mixture. For example, gasoline rated at 90 octane has the same antiknock qualities as a mixture of heptane and isooctane that is 90% isooctane. In working with octane ratings, then, we may treat the fuel as if it were a mixture of isooctane and heptane. If two batches of gasoline, one rated at 96 octane and the other at 87 octane, are to be mixed to obtain 200 gallons of 93 octane gasoline, find how many gallons of each should be used.

46. Fifty gallons of 88 octane gasoline are mixed with 75 gallons of 94 octane gasoline. What is the octane rating of the mixture? (See Problem 45.)

47. One pump can fill a reservoir in 8 days and another can do it in 6 days. If both pumps begin working together but the faster pump breaks down at the end of the third day, how long will it take to fill the reservoir?

48. A girl is in a boat on a lake 4 miles from the point nearest to her on the shore (point A). She wants to get to point B, 9 miles along the straight shore from A. She can row at an average of 6 miles per hour and walk at 4 miles per hour. She decides to row directly to a point C between A and B and walk from there. The total trip takes 2 hours and 20 minutes. How far is it from A to C?

49. Point A is on one bank of a river and B is on the other, directly opposite A. The river is 120 feet wide and is essentially straight at the area in question. An underground telephone cable is to go from A to a point C, which is at the river's edge on the same side as B and 150 feet from B. It costs $4 per foot to run the cable under the river and $2 per foot to place it underground. The cable is run under the river from A to a point D between B and C, and from there to C underground. If the total cost is $720, how far is it from B to D?

50. An oil-drilling platform is located 4.8 miles from the nearest point P on the shore. Towns A and B on the shore are on opposite sides of P and are 5 miles apart. A motorboat averaging 11 miles per hour can go from town A to the platform and from there to town B in 1 hour. If town A is closer to P than town B, find the distance from town A to P.

51. A room containing 1,990 cubic feet is originally free of carbon monoxide. Cigarette smoke, containing 2.4% carbon monoxide, is then introduced into the room at the rate of 0.1 cubic feet per minute.

a. Find a general expression for the concentration of carbon monoxide in the room in terms of the elapsed time t in minutes that the smoke has been entering the room.

b. Extended exposure to concentrations of carbon monoxide as low as 0.00012 is harmful to the human body. Find the critical value of t at which this concentration is reached.

 # LINEAR INEQUALITIES

It may come as a surprise to you that inequalities are used in mathematics almost as much as equations. The only difference in appearance of an inequality and an equation is that the equals sign is replaced by an inequality sign ($<$, $>$, $\leq$, or $\geq$). For example, $2x - 3 > 4$ is a linear inequality in one unknown, and $x^2 - 2x - 3 \leq 0$ is a quadratic inequality in one unknown. By the solution set of an inequality we mean the set of all values of the unknown (or unknowns) for which the inequality holds true. In this section and the next we consider the solution of linear inequalities. Quadratic inequalities will be discussed in Section 10, and inequalities involving more than one unknown will be taken up in Chapter 11, when we discuss systems of inequalities.

The order properties given in Section 2 of Chapter 1 provide the basis for solving inequalities of the first degree in one unknown. Briefly, these state that we can add the same number to, or subtract the same number from, both sides of an inequality. Also, we can multiply or divide by any *positive* number and retain the sense (that is, the direction of the inequality symbol) of the inequality. But multiplying or dividing by a *negative* number reverses the sense of the inequality. This last point requires special care, and this feature distinguishes the solution of linear inequalities from that of linear equations. We illustrate the technique with several examples.

EXAMPLE 32

Solve for x: $2x - 3 < 5x + 6$

Solution   We solve just as if it were an equation, collecting the x's together on one side of the inequality sign and the constants on the other. One way to do this is to add 3 to both sides and also to subtract $5x$, obtaining

$$-3x < 9$$

Now to get x we want to multiply by $-\frac{1}{3}$ (or divide by -3), but remember that multiplying by a negative number reverses the sense of the inequality. So we get

$$x > -3$$

We write the solution set as

$$\{x:\quad x > -3\}$$

or, equivalently, as the interval $(-3, \infty)$. The solution consists of not just one value of x but *all* numbers greater than -3. This can be depicted on a number line, as shown in Figure 3.

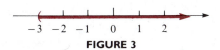

FIGURE 3

Some people prefer to avoid dividing by negative numbers and so collect terms in such a way as to result in a positive coefficient for x. Thus, we could have added $-2x$ and -6 to each member in this example to get:

$$-9 < 3x$$

$$-3 < x$$

which can be turned around to read $x > -3$.   ▲

EXAMPLE 33 ||||||||||||||||||||||||||||||||||||||||||||||||||||||||||||||||||||||||||||||||||||||||||||||||||||||||||||||||||||||||||

Solve the inequality: $\dfrac{2x - 3}{5} \geq \dfrac{x}{2} - 1$

Solution   First, we clear fractions by multiplying by the LCD, which is 10.

$$4x - 6 \geq 5x - 10$$

$$4 \geq x$$

$$x \leq 4$$

So the solution set is $\{x: \;\; x \leq 4\}$ or, in interval notation, $(-\infty, 4]$. This can be depicted graphically as shown in Figure 4.

FIGURE 4   ▲

EXAMPLE 34 ||||||||||||||||||||||||||||||||||||||||||||||||||||||||||||||||||||||||||||||||||||||||||||||||||||||||||||||||||||||||||

Find all x such that $-4 < 2x - 3 \leq 5$.

Solution   This is really the combination of two inequalities, $-4 < 2x - 3$ and $2x - 3 \leq 5$, and we are seeking the values of x that satisfy both inequalities at once. We could solve each inequality individually and find those values of x common to the two solution sets (that is, the intersection of the two solution sets). An easier way, however, is to work with the combined inequality as follows. Add 3 to every member and then divide by 2:

$$-1 < 2x \leq 8$$

$$-\tfrac{1}{2} < x \leq 4$$

The solution set is $\{x: \;\; -\tfrac{1}{2} < x \leq 4\}$. We can read this from the middle: x is greater than $-\tfrac{1}{2}$ and less than or equal to 4. In interval notation the solution set is $(-\tfrac{1}{2}, 4]$. Graphically, we show this in Figure 5.

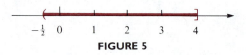

FIGURE 5   ▲

EXERCISE SET 8

A

1–23 odd

In Problems 1–30 find the solution sets and depict these on a number line.

1. $3x + 2 < 7$

2. $4x - 3 < 5$

3. $2x + 5 < x + 9$

4. $3x - 4 < x + 3$

5. $3x - 5 > 5x - 3$

6. $4x + 1 > 6x - 3$

7. $2 - 3x \le 1 - 4x$

8. $3(x - 2) \ge -4$

9. $2(3 - x) \ge 11$

10. $\dfrac{x + 3}{4} \le \dfrac{2}{3}$

11. $\dfrac{x - 3}{2} < \dfrac{3x}{5}$

12. $\dfrac{x}{2} - \dfrac{1}{3} < \dfrac{2x}{3} - \dfrac{5}{6}$

13. $\dfrac{3x - 4}{2} > \dfrac{2 - 3x}{6}$

14. $\dfrac{3}{4} - \dfrac{x}{3} \ge \dfrac{5x}{12} - 1$

15. $\dfrac{x}{2} - \dfrac{3 + x}{6} > \dfrac{2}{3}$

16. $\dfrac{3 - x}{2} \le \dfrac{2x + 5}{3} + \dfrac{5}{6}$

17. $(x - 2)(x + 3) \ge x(x - 4) + 2$

18. $(2x - 1)(x - 2) - (x + 1)(x - 2) < x^2$

19. $\dfrac{4x}{3} - 2 \le \dfrac{5x}{6} + \dfrac{1}{3}$

20. $-1 < 2x - 3 < 4$

21. $3 \le 2x + 1 \le 5$

22. $0 \le 3x - 2 < 7$

23. $-2 < \dfrac{2x + 3}{4} \le 2$

24. $-3 < \dfrac{4x - 5}{2} < 4$

25. $1 \le \dfrac{5 - 2x}{4} \le 3$

26. $-\dfrac{2}{3} \le 3 - 2x \le \dfrac{1}{3}$

27. $0 < \dfrac{4 - 3x}{2} \le \dfrac{3}{4}$

28. $-\dfrac{3}{4} \le \dfrac{7 - 2x}{3} \le \dfrac{5}{6}$

29. $-0.02 \le 2x - 1 \le 0.02$

30. $-0.1 < 1 - 2x < 0.1$

31. A student has grades of 73, 82, and 79 on three tests. How high must she score on the fourth test to have an average of 80 or greater?

32. The temperature in degrees Fahrenheit (F) on a certain South Sea island ranges from 76 to 87. Celsius temperature (C) is related to Fahrenheit temperature by the formula $C = \frac{5}{9}(F - 32)$. What is the Celsius temperature range on this island?

33. Apples keep best in cold storage if they are held above freezing but at no greater than 5°C, that is, $0 < °C \le 5$. Find the corresponding range for the Fahrenheit temperature (°F). (See Problem 32.)

34. A rectangular plate is to be cut from a 10-inch-wide steel strip. If the perimeter of the plate must be at least 57 inches but no more than 60 inches, find the allowable range for the length.

35. A woman drives 18 miles to work each morning. Depending on the traffic, her average speed varies between 36 and 45 miles per hour. Find the range for the time of the trip.

36. A swimming pool is 18 feet wide and 30 feet long. It is 6 feet deeper at one end than at the other, and the depth varies linearly. The water level is never allowed to drop below $2\frac{1}{2}$ feet at the shallow end nor to rise above $3\frac{1}{2}$ feet there. Find the range of values for the volume of water in the pool.

37. Explain why each of the following has no solution.

  a. $-2 < x < -4$

  b. $3 \le 5 - 2x < 0$

  c. $-2 \ge \dfrac{3(2x - 7)}{9} > -1$

B

Solve the inequalities in Problems 38 and 39.

38. $\dfrac{3x - 4}{7} - \dfrac{3 - 2x}{5} \ge \dfrac{2x}{3}$

39. $\dfrac{3}{5} \ge \dfrac{2(4 - 5x)}{9} > -2$

40. Find all values of x that satisfy either $3 - 4x > 9 - 2x$ or $4 + 5x \ge 2(x - 1)$, and depict the solution graphically. Write the solution as the union of two sets.

41. Follow the instructions for Problem 40 with the values of x that satisfy either

$$\dfrac{4 - 2x}{3} \ge \dfrac{2 - x}{4} \qquad \text{or} \qquad \dfrac{6 - x}{4} < \dfrac{2x - 7}{2}$$

In Problems 42 and 43 determine the set of all x values for which the given expression will be real.

42. $\sqrt{4 - 3x} + \sqrt{2x + 1}$

43. $\dfrac{x}{\sqrt{7 - 3x}} - \dfrac{1}{\sqrt{5 + 2x}}$

44. The cost of manufacturing x microcomputers of a certain kind each week is given by $C = 2{,}500 + 35x$, and the revenue from selling these is given by $R = 80x$. How many microcomputers must be produced and sold each week to realize a profit?

9 INEQUALITIES AND ABSOLUTE VALUE

Inequalities often occur in combination with absolute value, such as $|x - 3| < 1$ or $|x - 2| > 3$. To see how to solve these, we first consider some further properties of absolute value.

Recall that when viewed geometrically $|a|$ can be interpreted as the distance between a and 0 on the number line. For example, $|-2| = 2$, and the point corresponding to -2 on the number line is 2 units from the origin. Similarly, $|a - b|$ can be interpreted as the distance between a and b on the number line. You can convince yourself of this by considering several possibilities of a and b on a number line.

The following further properties of absolute value are often useful:

Properties of Absolute Values	
1. $\|-x\| = \|x\|$	2. $\|xy\| = \|x\|\|y\|$
3. $\|x\| = \sqrt{x^2}$	4. $\|x + y\| \le \|x\| + \|y\|$

You can convince yourself of the truth of properties 1 and 2 by considering all of the different cases in which the variables are positive, negative, or zero. We showed property 3 in Section 5 of Chapter 1. To review this, recall that $\sqrt{x^2}$ must be the *nonnegative* number whose square is x^2, namely x if $x \ge 0$ and $-x$ if $x < 0$. But this is precisely the meaning of $|x|$. So $\sqrt{x^2}$ and $|x|$ are equal for all real numbers x. Property 4 is known as the **triangle inequality** and is probably the most used inequality in all of mathematics. We give a proof below and outline an alternative proof in the exercises (Problem 32). First, though, you should note that it is intuitively evident, because when x and y are both positive or both negative, the two sides are the same. For example, $|5 + 2| = |5| + |2|$ and $|(-5) + (-2)| = |-5| + |-2|$, since in each case we get 7 on both sides of the equation. But when x and y are opposite in sign, the left-hand side really involves a subtraction, whereas the right-hand side always involves addition and hence is greater than the left. Thus $|5 + (-2)| = |5 - 2| = 3$, but $|5| + |-2| = 5 + 2 = 7$.

To prove property 4, observe that $-|x| \le x \le |x|$. In fact, this is an extreme inequality in the sense that the quantity in the middle always equals one or the other of the end points. Now we write a similar inequality for y and add:

$$\begin{array}{rcl} -|x| \le & x & \le |x| \\ -|y| \le & y & \le |y| \\ \hline -|x| - |y| \le & x + y & \le |x| + |y| \end{array}$$

or

$$-(|x| + |y|) \le x + y \le |x| + |y| \tag{3}$$

The left-hand inequality can be rewritten after multiplying by -1 as

$$-(x + y) \le |x| + |y|$$

Since, by the right-hand inequality in (3), $(x + y) \le |x| + |y|$, it follows that both $x + y$ and its negative are always less than or equal to $|x| + |y|$. But $|x + y|$ *is either $x + y$ or its negative.* So

$$|x + y| \le |x| + |y|$$

In Chapter 1 we saw equivalent ways of writing inequalities of the form $|x| < a$ and $|x| > a$, where $a > 0$. The results are summarized on the next page.

For $a > 0$,

$$|x| < a \quad \text{is equivalent to} \quad -a < x < a \tag{4}$$

$$|x| > a \quad \text{is equivalent to} \quad x > a \text{ or } x < -a \tag{5}$$

Similar results hold for $|x| \le a$ and $|x| \ge a$.

We now use these results to solve the problems posed at the beginning of this section.

EXAMPLE 35 ||||||||||||||||||||||||||||||||||||||||||||||||||||||||||||||||||||||||||||||||||||||||||||||||||||||||

Find the solution set for the inequality $|x - 3| < 1$.

Solution   In (4) replace x by $x - 3$ and a by 1. This gives that $|x - 3| < 1$ is equivalent to

$$-1 < x - 3 < 1$$

Now add 3 to each member of the inequality, getting

$$2 < x < 4$$

The solution set is therefore the open interval $(2, 4)$. It is shown graphically in Figure 6.

FIGURE 6

▲

EXAMPLE 36 ‖‖

Find the solution set for the inequality $|x - 2| > 3$.

Solution   From (5) we see that $|x - 2| > 3$ is equivalent to $x - 2 > 3$ or $x - 2 < -3$. If we add 2 to each side of these inequalities, we get $x > 5$ or $x < -1$. The solution set can be written as $(5, \infty) \cup (-\infty, -1)$. This is shown graphically in Figure 7.

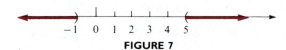

FIGURE 7

▲

EXAMPLE 37 ‖‖

Solve the inequality $9 - 2|5 - 3x| \geq 1$.

Solution   First we isolate the term involving the absolute value by subtracting 1 from each side and adding $2|5 - 3x|$ to each side. This gives

$$8 \geq 2|5 - 3x|, \quad \text{or} \quad \text{equivalently,} \quad 2|5 - 3x| \leq 8$$

Dividing by 2, we get

$$|5 - 3x| \leq 4$$

so that

$$-4 \leq 5 - 3x \leq 4$$
$$-9 \leq -3x \leq -1$$
$$3 \geq x \geq \frac{1}{3}$$

or, equivalently,

$$\frac{1}{3} \leq x \leq 3$$

So the solution set is the closed interval $[\frac{1}{3}, 3]$.

▲

EXERCISE SET 9

A

5-27 odd

1. Verify the four absolute value properties for each of the following:
   a. $x = 5, y = 2$
   b. $x = -8, y = 5$
   c. $x = 11, y = -6$
   d. $x = -7, y = -9$
2. Verify that the distance between a and b on a number line is $|a - b|$ for each of the following:
   a. $a = 2, b = 6$
   b. $a = 3, b = -2$
   c. $a = -4, b = 2$
   d. $a = -1, b = -4$
3. Show that $|a|^2 = a^2$. Is it also true that $|a|^3 = a^3$? Explain why or why not. What can you say in general about $|a|^n$ and a^n?

Solve the inequalities in Problems 4–27.

4. $|x - 3| < 2$
5. $|x - 5| < 3$
6. $|3x + 2| \leq 4$
7. $|2x + 3| \leq 1$
8. $|2x - 1| > 5$
9. $|4x + 5| > 3$
10. $|3 - 2x| \geq 1$
11. $|8 - 5x| \leq 2$
12. $|x - a| < 3$
(a is a constant.)
13. $|x - b| \geq 2$
(b is a constant.)
14. $\left|\dfrac{x + 1}{2}\right| < 3$
15. $\left|\dfrac{2x - 1}{3}\right| \leq 1$
16. $\left|\dfrac{2 - x}{3}\right| \geq 4$
17. $\left|\dfrac{5 - x}{2}\right| < 4$
18. $|x + 2| \geq 0.01$
19. $|3 - x| < 0.05$
20. $|-2(3x - 5)| \leq 4$
21. $|-3(4 - x)| > 6$
22. $2|x - 1| + 3 > 6$
23. $4 - 2|x - 2| \leq 3$
24. $3 - 2|4 - 3x| \leq 0$
25. $5 - 3|2 - x| > 2$
26. $\left|\dfrac{3 - 5x}{2}\right| \leq \dfrac{9}{4}$
27. $\dfrac{7}{3} - \left|\dfrac{1 - x}{4}\right| > \dfrac{5}{6}$

B

28. Show that $|a - b| \leq |a| + |b|$.
   Hint. Write $a - b = a + (-b)$.

29. Show that $|a^{-1}| = |a|^{-1}$.
   Hint. Show that $|a^{-1}|$ fulfills the requirement of being the multiplicative inverse of $|a|$.
30. Use the result of Problem 29 to show that $|a/b| = |a|/|b|$.
31. Show that $|a + b + c| \leq |a| + |b| + |c|$.
   Hint. Write $a + b + c = (a + b) + c$.
32. Justify all steps in the following alternative proof of the triangle inequality.

$$|x + y|^2 = (x + y)^2$$
$$= x^2 + 2xy + y^2$$
$$= |x|^2 + 2xy + |y|^2$$
$$\leq |x|^2 + 2|xy| + |y|^2$$
$$= |x|^2 + 2|x||y| + |y|^2$$
$$= (|x| + |y|)^2$$

Thus, since $|x + y|^2 \leq (|x| + |y|)^2$, it follows that

$$|x + y| \leq |x| + |y|$$

33. a. In the triangle inequality, let $x = a$ and $y = b - a$ to obtain

$$|b - a| \geq |b| - |a|$$

   b. Similarly, let $x = b$ and $y = a - b$ to obtain

$$|a - b| \geq |a| - |b|$$

   c. Use the results of parts **a** and **b** to obtain the inequality

$$|a - b| \geq ||a| - |b||$$

Solve the inequalities in Problems 34–37.

34. $0 < |3x - 2| < 4$
35. $0 < |4 - 3x| \leq 5$
36. $0 < |x - a| < \delta, \quad (\delta > 0)$
37. $k|a - bx| < \varepsilon, \quad (k > 0, b > 0, \varepsilon > 0)$

QUADRATIC AND OTHER NONLINEAR INEQUALITIES

Using the example $x^2 - 3x - 4 < 0$, we will illustrate one technique for solving a quadratic inequality. A second method, involving graphing, will be discussed in Chapter 4.

First we factor, just as if we were solving a quadratic equation, getting

$$(x - 4)(x + 1) < 0 \tag{6}$$

The product on the left will be less than zero, that is, it will be negative, when the factors are opposite in sign. Our procedure provides a systematic way of seeing where this occurs. First we mark on a number line the points $x = 4$ and $x = -1$, where the product equals zero. We refer to these points as **critical numbers.** As Figure 8 shows, these divide the line into the three intervals, $(-\infty, -1), (-1, 4)$, and $(4, \infty)$. The product $(x - 4)(x + 1)$ is of constant sign in each interval, since it can change sign only at one of the critical numbers. All we need to do now is determine the sign of the product in each interval. Since the factor $x - 4$ is positive to the right of 4 and negative to the left, and the factor $x + 1$ is positive to the right of -1 and negative to the left, we have the following:

Interval $(-\infty, -1)$: $(x - 4)(x + 1) = (-)(-) = +$

Interval $(-1, 4)$: $(x - 4)(x + 1) = (-)(+) = -$

Interval $(4, \infty)$: $(x - 4)(x + 1) = (+)(+) = +$

FIGURE 8

We see then that the solution we are seeking is the open interval $(-1, 4)$.

Remark. An alternative way of determining the sign of the product $(x - 4)(x + 1)$ in each interval is to substitute some test value for x in each interval. For example, we could use $x = -2$ in $(-\infty, -1)$, $x = 0$ in $(-1, 4)$, and $x = 5$ in $(4, \infty)$. In each case the objective is to determine the sign of the product.

In our example we were working with a strict inequality $(<)$, so the critical numbers themselves are not included in the solution set. If either $\leq$ or $\geq$ had been involved, we would also have had to check to see if the critical numbers are to be included in the solution set.

In practice much of what we have discussed above can be done mentally, and we can indicate the results on the number line more briefly, as in Figure 9. We refer to this figure as a **sign graph** for the product $(x - 4)(x + 1)$.

FIGURE 9

A Word of Caution. It is tempting to try to solve inequality (6) just as if it were an equation and write

$$(x + 1)(x - 4) < 0$$
$$x + 1 < 0 \quad \text{or} \quad x - 4 < 0$$
$$x < -1 \quad \text{or} \quad x < 4$$

You must resist this temptation, because this procedure *is not correct*. (Do you see why?)

The sign graph technique can be used for any quadratic inequality and also any other inequality in which one side is zero and the other can be written as a product and/or quotient of linear factors. The critical numbers are the numbers that make each of these factors zero. They divide the number line into intervals where we determine the sign of each factor and hence their product (and/or quotient). Finally, we read the answer from the sign graph, depending on the type of inequality involved. (+ corresponds to > 0, − corresponds to < 0). Remember also to check the critical numbers themselves if ≤ 0 or ≥ 0 is involved.

Note. For this procedure to work, it is essential that one side of the inequality be zero.

EXAMPLE 38

Solve the inequality $x^2 - 2x \geq 15$.

Solution   First add -15 to both sides to get 0 on the right, and then factor the left-hand side.

$$x^2 - 2x - 15 \geq 0$$
$$(x - 5)(x + 3) \geq 0$$

The critical numbers are $x = 5$ and $x = -3$, and by examining the sign of each of the factors in the intervals determined by these critical numbers, we get the sign graph in Figure 10. (You should verify this.) Since the inequality sign is $\geq$, we want the intervals where the product is positive or zero. So the solution set is

$$(-\infty, -3] \cup [5, \infty)$$

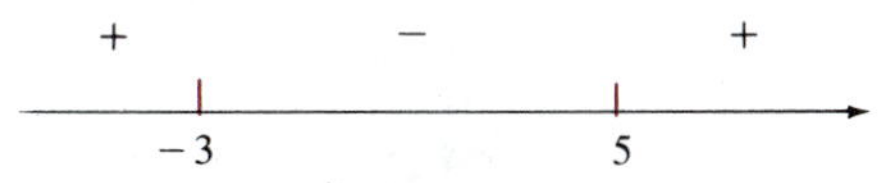

FIGURE 10

EXAMPLE 39

Solve the inequality: $\dfrac{(x-1)^2}{x+2} < 0$

Solution   The critical numbers are $x = 1$ and $x = -2$, and we obtain the sign graph in Figure 11. There is no sign change associated with the critical number $x = 1$ because of the squared factor in the numerator. The quotient is negative (< 0) only in the region $x < -2$. We write the solution set as $(-\infty, -2)$.

FIGURE 11

Remark. As Example 39 illustrates, when a linear factor occurs to an *even* power, the fraction does not change sign as we move through the corresponding critical point. On the other hand, it will change sign if the power is odd. Using this, you can determine the sign graph rapidly. All you really need is the sign in one interval (the rightmost one is usually the easiest to work with). Then signs alternate or stay the same from one interval to the next, depending on whether the factor giving rise to the critical point separating the intervals is raised to an odd or even power.

EXAMPLE 40

Solve the inequality: $\dfrac{3x-4}{x+2} \le 1$

Solution   We cannot simply clear of fractions by multiplying by the LCD as in an equation, because since the LCD involves the unknown x, we do not know if this LCD is positive or negative—and in multiplying both sides of an inequality, it is essential to know the sign of the multiplier, since multiplying by a negative number changes the sense of the inequality. We can, however, combine the terms into one fraction. The first step is to add -1 to both sides, since for our procedure to work, we must have 0 on one side of the inequality:

$$\frac{3x-4}{x+2} - 1 \le 0$$

Then we combine the left-hand side into one fraction:

$$\frac{3x-4-(x+2)}{x+2} \le 0$$

$$\frac{2x - 6}{x + 2} \leq 0$$

$$\frac{2(x - 3)}{x + 2} \leq 0$$

The sign graph is shown in Figure 12. The solution set is the half-open interval $(-2, 3]$. Notice that 3 is included, since the quotient is 0 there, but -2 is not, because the quotient is undefined for this value.

FIGURE 12

EXAMPLE 41

Find the set of values of x for which the following expression will be real:

$$\sqrt{\frac{x^2 - 2x}{(x + 3)^2}}$$

Solution   The expression will be real provided

$$\frac{x^2 - 2x}{(x + 3)^2} \geq 0$$

or, equivalently,

$$\frac{x(x - 2)}{(x + 3)^2} \geq 0$$

The critical numbers are $x = 0$, $x = 2$, and $x = -3$. We obtain the sign graph in Figure 13. The even power in the denominator accounts for the fact that

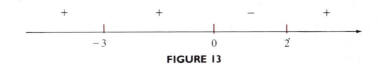

FIGURE 13

there is no sign change at -3. The quotient is either positive or zero if x is in $(-\infty, -3) \cup (-3, 0] \cup [2, \infty)$. Note that the critical numbers 0 and 2 are included, but -3 is not.

EXERCISE SET 10   (1-39 odd)

A

Solve the inequalities in Problems 1–36 by the sign graph technique.

1. $x^2 - 2x - 8 \leq 0$

2. $x^2 - x - 12 > 0$

3. $x^2 - 5x \geq 0$

4. $x^2 - 9 \leq 0$

5. $2x^2 + 3x \leq 2$

6. $6x^2 - x < 2$

7. $x(2x - 1) > 3$

8. $13x > 5x^2 + 6$

9. $3x(8 - 3x) < 16$

10. $x + 6 \leq 12x^2$

11. $x(6x - 7) \leq 20$

12. $4(5x^2 + 3) < 31x$

13. $6x^2 + 17x + 12 \leq 0$

14. $5x(14 - 5x) \geq 49$

15. $\dfrac{x - 4}{x + 3} < 0$

16. $\dfrac{x}{x - 1} > 0$

17. $\dfrac{2 - x}{x + 2} > 0$

18. $\dfrac{2x + 1}{x + 3} \leq 0$

19. $(x - 2)(x + 1)^2(x + 4) \leq 0$

20. $(2x + 3)(x - 3)^3(x - 5)^2 \geq 0$

21. $\dfrac{x^2 - 4}{x^2 - 9} \geq 0$

22. $\dfrac{(x - 4)^2}{x^2 - 1} < 0$

23. $\dfrac{x^2 - 2x + 1}{x^2 - 4} < 0$

24. $\dfrac{x + 2}{x^2 - 2x - 15} \geq 0$

25. $\dfrac{x^2 - x - 2}{x^2 - 4x + 3} \leq 0$

26. $\dfrac{2x^2 - 5x + 2}{x^2 + 5x + 6} < 0$

27. $\dfrac{2x + 1}{x - 1} > 1$

28. $\dfrac{x - 3}{2x + 3} \leq 2$

29. $\dfrac{x}{x + 1} < \dfrac{x + 2}{x - 3}$

30. $\dfrac{x - 1}{x + 2} > \dfrac{x - 2}{x + 1}$

31. $\dfrac{1}{x} - \dfrac{1}{x + 1} \geq \dfrac{1}{2}$

32. $\dfrac{1}{1 - x} + \dfrac{2}{x + 2} \leq 1$

33. $1 + \dfrac{x}{x - 2} < \dfrac{x - 1}{x^2 - 4}$

34. $\dfrac{1}{x} + \dfrac{10}{x^2 + 2x} > \dfrac{x}{x + 2}$

35. $\dfrac{x^2(x - 2)}{x^2 - 5x - 6} \geq 0$

36. $\dfrac{(x + 2)(x - 1)^3}{4x^2 - 4x + 1} < 0$

In Problems 37–40 determine the set of values of x for which the given expression will be real.

37. $\sqrt{x^2 - 4x - 12}$

38. $\sqrt{\dfrac{x + 5}{3 - x}}$

39. $\dfrac{1}{\sqrt{3 - x - 2x^2}}$

40. $\dfrac{x + 1}{\sqrt{2x - x^2}}$

B

41. Find all values of k for which the solutions of the given equation will be real.

$$3kx^2 - 2kx + k = 3$$

42. Find all values of c for which the solutions of the given equation will be imaginary.

$$x^2 - (c - 1)x + c^2 = 0$$

43. Prove that the inequality $(k^2 + 1)x^2 > kx - 1$ is true for all real values of x and k.

44. Prove that if $k > 1$ the inequality

$$x(2 - x) < k$$

is true for all real x.

Solve the inequalities in Problems 45–48.

45. $\dfrac{x - 3}{x^2 + 3x + 2} - \dfrac{2x - 5}{x^2 + x - 2} < \dfrac{x - 3}{1 - x^2}$

46. $\dfrac{x + 2}{x^2 + 2x - 3} - \dfrac{x}{2x^2 + 5x - 3} \geq \dfrac{x - 4}{2x^2 - 3x + 1}$

47. $\frac{2}{3}x^{-1/3}(4 - x)^2 - 2x^{2/3}(4 - x) > 0$

48. $x^{-1/3}(2x + 3)^{-1/3} - x^{2/3}(2x + 3)^{-4/3} \leq 0$

49. For what values of x will the expression below be real?

$$\sqrt{\dfrac{x^2 - 2x - 3}{x^3 - 2x^2}}$$

50. An object is thrown upward from the top edge of a building 96 feet high with an initial velocity of 16 feet per second. The formula for the distance s above the ground of the object at time t is given by $s = -16t^2 + 16t + 96$. (See Example 27.) For what nonnegative values of t will the object be at least 64 feet above the ground?

51. The intensity I, in candlepower, of a certain light is given by $I = 6{,}400/d^2$, where d is the distance in feet from the source. For what range of distances will I satisfy the inequality $900 \le I \le 3{,}600$?

52. The load L that can be safely supported by a wooden beam of length l with rectangular cross-section of width w and depth d is given by the formula

$$L = \frac{kwd^2}{l}$$

where k is a constant depending on the type of wood. A beam is to be 15 feet long and 4 inches

wide, and it is known that for the material being used, $k = 120$. If the maximum load the beam will have to support is 3,200 pounds, find the minimum depth d that should be used.

Note. The given value of k is for both w and d in inches and for l in feet.

In Problems 53 and 54 use a calculator to find the solution set with two decimal places of accuracy.

53. $\dfrac{x^2 - 3x - 2}{x^2 + 2x - 4} < 2$

54. $\dfrac{2}{3x^2 - 2x - 4} - \dfrac{1}{x^2 + 5x + 3} < 0$

REVIEW EXERCISE SET

A

In Problems 1–18 find the solution set.

1. a. $2(3x - 4) = 6 - 7x$

   b. $\dfrac{x}{2} - \dfrac{3}{4} = \dfrac{5(2 - 3x)}{6}$

2. a. $4(x - 3) = 2(3 - 4x)$

   b. $\dfrac{2x}{5} - \dfrac{1}{3} = \dfrac{2(4 - x)}{15}$

3. a. $6x^2 + x - 15 = 0$

   b. $12x^2 - 25x + 12 = 0$

4. a. $x(23 - 6x) = 20$ **b.** $4x^2 - 25 = 0$

5. a. $x^2 - 4x + 3 = 0$ **b.** $3y^2 - 2y - 4 = 0$

6. a. $9x^2 + 16 = 0$ **b.** $10x^2 - 3x = 18$

7. a. $2r^2 - 5r + 4 = 0$ **b.** $3y^2 + 2 = 4y$

8. a. $x^2 = 2(x - 2)$ **b.** $t^2 = \dfrac{4(t - 1)}{5}$

9. a. $2(4 + m) = 15m^2$ **b.** $x^2 = 1 + \dfrac{4x}{3}$

10. a. $\dfrac{2}{x - 2} - \dfrac{3}{x + 1} = 0$

    b. $\dfrac{2}{x} - \dfrac{5}{x + 1} = \dfrac{4}{x^2 + x}$

11. a. $\dfrac{4}{2x - 3} - \dfrac{5}{4 - x} = 0$

    b. $\dfrac{x}{x + 1} - \dfrac{2}{x - 1} = \dfrac{x^2 + 3}{x^2 - 1}$

12. a. $\dfrac{x - 1}{x + 2} = \dfrac{x - 3}{x + 1} + \dfrac{2x}{x^2 + 3x + 2}$

    b. $\dfrac{x}{3x + 4} - \dfrac{5}{2x - 3} = \dfrac{7 - 3x}{6x^2 - x - 12}$

13. a. $\dfrac{x + 1}{x - 2} - \dfrac{x}{2x - 3} = \dfrac{3x + 5}{2x^2 - 7x + 6}$

    b. $\dfrac{x - 1}{x - 3} - \dfrac{2x}{x + 2} = \dfrac{4x}{x^2 - x - 6}$

14. a. $\sqrt{5x - 2} = 3$ **b.** $1 - \sqrt{x - 1} = x$

15. a. $\sqrt{x - 2} = \sqrt{x + 3} - 1$

    b. $\dfrac{1}{\sqrt{2x - 1}} + \sqrt{2x - 1} = 2$

16. a. $\sqrt{2 + x} - \sqrt{5 - 2x} + 3 = 0$

    b. $\sqrt{x - 1} = \sqrt{2x - 1} - 1$

17. a. $9x^4 + 32x^2 - 16 = 0$

    b. $12x^{-2} + x^{-1} - 6 = 0$

18. a. $2x^{2/3} + x^{1/3} - 15 = 0$

    b. $4x^{-4} - 37x^{-2} + 9 = 0$

In Problems 19 and 20 perform the indicated operations and write the answer in the form $a + bi$.

19. a. $(2 + 3i)(4 - 5i)$ **b.** $\dfrac{3 + 2i}{4 - 3i}$

c. $(4 - 5i)^3$

20. a. $(3 + 4i)^{-1}$ **b.** $(3 - \sqrt{-4})^4$ **c.** $\dfrac{2i - 4}{3 + 2i}$

Solve the equations in Problems 21 and 22 by completing the square.

21. a. $2x^2 - 5x + 4 = 0$ **b.** $3t^2 + 2t - 6 = 0$
22. a. $2x^2 + 4x + 3 = 0$ **b.** $4x^2 - 3 = 6x$
23. Without solving the equations below, determine whether the solutions are real and unequal, real and equal, or imaginary.
    a. $3x^2 - 7x + 12 = 0$
    b. $9x^2 - 15x - 13 = 0$
    c. $12x(5 - 3x) = 25$
    d. $8t^2 + 9t + 3 = 0$
24. Evaluate each of the following:

    a. i^{43} **b.** i^{-10} **c.** $\sqrt{-5}\sqrt{-10}$ **d.** $\dfrac{1}{i^{17}}$

    e. $\dfrac{\sqrt{24}}{\sqrt{-6}}$

25. The length of a room is 6 feet more than its width, and its perimeter is 64 feet. Find the length and width.
26. After selling four-fifths of her stock of a certain type of winter jacket at $55 each, a store owner placed the others on sale at $40 each. If all the jackets were sold and the gross amount realized from them was $3,900, how many jackets did she sell at the regular price and at the sale price?
27. Two cars leave from the same point and travel the same route. The first car travels at an average speed of 65 kilometers per hour, and the second car, which leaves 30 minutes after the first, travels at an average speed of 90 kilometers per hour. How many kilometers will each car have gone when the second car overtakes the first?
28. By combining two types of candy, one costing $3.00 per pound and the other $4.00 per pound, a 2-pound box is prepared that costs $6.50. How much of each kind of candy is used?

29. A quantity of brine containing 25% salt is on hand, and it is desired to reduce the salinity to 10% by adding pure water. If 7 gallons of pure water are required, how many gallons of brine were there originally?
30. Two pumps working simultaneously can fill a water tank in 20 hours. The first one alone can fill it in 50 hours. How long would it take the second pump alone?
31. Driver M makes a trip of 220 miles. Driver N makes the same trip in 20 minutes less time, averaging 5 miles per hour faster than driver M. What is the average speed of each driver?
32. Pumps A and B working together can pump all the water out of a reservoir in 20 hours. Pump A could do the job alone 9 hours faster than pump B alone. How long would it take each pump alone?
33. A rain gutter is to be formed from a long, flat piece of tin, 10 inches wide, by bending up at a right angle a certain amount on each side (see the sketch). By calculus it is determined that the greatest amount of water can be accommodated when the cross-sectional area is 12.5 square inches. How many inches on each side should be turned up to produce the gutter?

34. The length of a rectangle is 14 inches greater than the width, and the diagonal is 26 inches. Find the dimensions of the rectangle.
35. How many kilograms of silver alloy containing 35% silver should be mixed with 12 kilograms of an alloy having 43% silver to obtain an alloy with 38% silver?

Find the solution sets for the inequalities in Problems 36–46.

36. a. $\dfrac{2x - 5}{3} \geq \dfrac{x}{4} - 1$ **b.** $\dfrac{4 - 3x}{2} < \dfrac{2x + 5}{3}$

37. a. $-1 \leq \dfrac{x - 2}{3} < 1$ **b.** $0 < \dfrac{2x}{2} - \dfrac{1}{4} \leq 2$

38. a. $9x^2 - 6x - 8 < 0$ **b.** $x(2x + 5) \geq 3$

39. a. $\dfrac{x + 2}{x^2 - 9} \geq 0$ **b.** $\dfrac{x - 4}{3} < \dfrac{x - 6}{x + 4}$

40. a. $\dfrac{3x^2 + 2x - 8}{x^2 - 3x} \geq 0$ **b.** $\dfrac{2}{x + 3} - \dfrac{1}{x - 2} < 1$

41. a. $(x - 2)(x + 4)(3x - 10) \leq 0$

   b. $\dfrac{(x - 1)(x - 2)}{5 - x} < 1$

42. a. $|x - 2| < \frac{1}{2}$ **b.** $|2x + 1| > 3$
43. a. $|3x - 4| \leq 1$ **b.** $|2 - x| > 3$

44. a. $\left|\dfrac{x + 1}{3}\right| \geq 1$ **b.** $\left|\dfrac{1}{2} - \dfrac{x}{3}\right| \leq \dfrac{2}{3}$

45. a. $3|1 - 2x| + 4 > 6$ **b.** $|3(2 - x)| - 4 \leq 1$

46. a. $\left|\dfrac{2x}{5} - \dfrac{3}{4}\right| - \dfrac{1}{2} > \dfrac{2}{3}$ **b.** $1 - \dfrac{|x - 3|}{2} \geq 3$

47. Write in terms of an inequality involving absolute values:
   a. The distance between x and 5 is less than 2.
   b. x is closer to -3 than 1 unit but is not equal to -3.

48. A student has grades of 72, 63, 68, and 75 on the four hour exams in a course. The final exam is to count the equivalent of two hour exams. In what range should the final exam grade fall so that the student's average grade for the course will be greater than or equal to 70? It is possible for the average to be 80 or better?

49. The temperature on a certain day in Los Angeles ranged between 59 and 86°F. What was the range in degrees Celsius?
   Note. $°C = (\frac{5}{9})(°F - 32)$

B

50. Solve for x:
   a. $2a^2x - 3b = b^2x + 4a$
   b. $10a^2x^2 + 9abxy - 36b^2y^2 = 0$
51. a. Solve for y: $y^2 - 2ky + 2k^2 = 1$, $k > 1$
   b. Solve for x: $\dfrac{2}{x} - \dfrac{1}{x - a} = \dfrac{1}{x - b}$
52. Find all values of k for which the equation below will have equal solutions.
$$kx^2 - 2(k - 3)x + 3k + 1 = 0$$

For each value of k so determined, find the solutions of the equation.
53. For what values of k will the solutions of the equation below be real?
$$kx^2 + (k - 2)x + 2 = k$$

54. Use a hand calculator to solve for x:
   a. $34.27x^2 + 28.39x - 52.76 = 0$
   b. $3.258x^2 - 1.237x + 0.965 = 0$

In Problems 55–57 find the solution set.
55. $\sqrt{x + 3} + \sqrt{2x + 5} = 2$
56. $\sqrt{x - 1} + \sqrt{x - 4} = \sqrt{2x - 1}$
57. $\dfrac{x + 3}{2x^2 + x - 3} - \dfrac{3x}{2x^2 + 7x + 6} = \dfrac{5x + 1}{x^2 + x - 2}$
   (Check all answers.)
58. Find the solution set and depict it graphically.
   a. $-\dfrac{3}{2} < \dfrac{3(2x - 4)}{5} \leq 2$
   b. $\dfrac{x - 1}{x + 2} - \dfrac{x - 3}{x + 1} < 1$

59. Solve the inequality.
   a. $\dfrac{2x}{x^2 - 2x - 3} + \dfrac{x + 2}{1 - x^2} \geq \dfrac{6}{x^2 - 4x + 3}$
   b. $\frac{2}{3}x^2(x - 2)^{-1/3} + 2x(x - 2)^{2/3} < 0$
60. Find the values of x for which the expression below will be real.
$$\sqrt{\dfrac{(x - 1)^3(x + 2)}{x^3 + x^2}}$$

61. A group of 25 students rents a bus for a field trip at a cost of $8.00 per person. They learn that the price of each ticket will be reduced by 10¢ for each additional student they can recruit to go on the trip, provided that the total number does not exceed 45 (the capacity of the bus). If the final total cost of the bus was $251.60, find out how many students actually went and the price of each ticket.
62. A container to hold a piece of machinery is to be constructed with a square base and with height 2 feet more than the side of the base. The material for the top and bottom costs 50¢ per square foot, and the material for the sides costs 25¢ per square foot.

If the total cost is $24, find the dimensions of the container.

63. A motorist drives 300 kilometers, and after a 2-hour visit returns over the same route. His average speed on the return trip is 20 kilometers per hour greater than his average speed going. If the total elapsed time from the time he left until he returned was 10 hours and 45 minutes, find his average speed going and coming.

64. An 8-inch by 10-inch picture is to be surrounded by a matting of the same width on all sides. The total area of the picture plus matting is not to exceed 195 square inches. What is the maximum width of matting that may be used?

3

FUNCTIONS
AND GRAPHS

The notion of function plays a central role in calculus. Understanding this important concept and the associated notation are essential prerequisites for studying calculus. The following example from a calculus textbook illustrates a particular way of combining functions, called composition.*

Find $(f \circ g)(x)$ if $f(x) = x^2 + 3$ and $g(x) = \sqrt{x}$.
Solution   The formula for $f(g(x))$ is

$$f(g(x)) = [g(x)]^2 + 3 = (\sqrt{x})^2 + 3 = x + 3$$

Since the domain of g is $[0, +\infty)$ and the domain of f is $(-\infty, +\infty)$, the domain of $f \circ g$ consists of all x in $[0, +\infty)$ such that $g(x) = \sqrt{x}$ lies in $(-\infty, +\infty)$; thus, the domain of $f \circ g$ is $[0, +\infty)$. Therefore,

$$(f \circ g)(x) = x + 3, \quad x \geq 0$$

You will learn how to do problems such as this in this chapter.

_____

* Howard Anton, *Calculus with Analytic Geometry*, 4th ed. (New York: John Wiley & Sons, 1992), p. 86. © 1992 Anton Textbooks, Inc. Reprinted by permission.

THE CARTESIAN COORDINATE SYSTEM AND GRAPHS OF EQUATIONS

The French mathematician and philosopher René Descartes (1596–1650) is usually given credit for the idea of representing points in a plane by ordered pairs of numbers. By an **ordered pair** is meant a pair of numbers (a, b) in which the position of each number is important.* Thus, (2, 3) and (3, 2) are two different ordered pairs. As is so often the case in mathematical discoveries, the same concept was obtained independently at about the same time by Pierre de Fermat (1601–1665), also a Frenchman, and a jurist as well as a mathematician. Descartes' book, *Geometry*, appeared in 1635, and was his only book on mathematics, but it had a profound influence on the subsequent development of mathematics because it provided an essential prerequisite for the invention of calculus. The approach to geometry introduced by Descartes forms the basis of what is now referred to as **analytic geometry.**

The method employed by Descartes makes use of the so-called **rectangular** or **cartesian** (after Descartes) **coordinate system.** Although this system is familiar to many students from high school mathematics, we will review its essential elements. We consider two number lines perpendicular to each other, one vertical and one horizontal, so that the point of intersection corresponds to 0 on each line; this point is called the **origin.** It is customary to have positive numbers to the right on the horizontal line and along the top part of the vertical line. The two lines are called the **horizontal and vertical axes.** It is also customary (but not essential) to name the horizontal axis the x **axis** and the vertical axis the y **axis.** These axes divide the plane into four parts, called **quadrants,** numbered I, II, III, and IV counterclockwise, as shown in Figure 1. Corresponding to each ordered

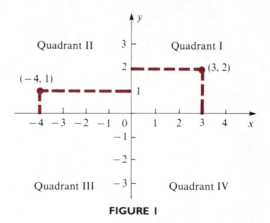

FIGURE I

---

* The context will usually make clear whether (a, b) designates an open interval or a point in a plane.

pair (x, y) of real numbers is a point in the plane determined as follows: Starting from the origin, we move x units along the x axis (right or left, depending on whether x is positive or negative) and from there move y units parallel to the y axis (up or down, depending on whether y is positive or negative). The resulting point is said to have **coordinates** (x, y); for brevity we often say "the point (x, y)." The points having coordinates $(3, 2)$ and $(-4, 1)$ are pictured in Figure 1. The first number of the ordered pair is called the x **coordinate,** or **abscissa,** of the point, and the second number is called the y **coordinate,** or **ordinate.**

If a point in the plane is given, then that point has unique coordinates (x, y) determined by erecting lines through the point and perpendicular to the x and y axes. So the coordinate system provides a one-to-one correspondence between all points in the plane (geometric objects) and all ordered pairs of real numbers (algebraic objects).

A formula for the distance between any two points in the plane can be found by using the Pythagorean theorem, which states that the square of the length of the hypotenuse of any right triangle equals the sum of the squares of the lengths of its legs. Let P_1 and P_2 be any two points in the plane, and designate their coordinates by (x_1, y_1) and (x_2, y_2), respectively. As shown in Figure 2, we introduce the point Q with coordinates (x_2, y_1). Then triangle P_1QP_2 is a right triangle. Its legs are P_1Q, with length $|x_2 - x_1|$, and P_2Q, with length $|y_2 - y_1|$. So if d is the length of the hypotenuse, P_1P_2, we have by the Pythagorean theorem,

$$d^2 = |x_2 - x_1|^2 + |y_2 - y_1|^2 = (x_2 - x_1)^2 + (y_2 - y_1)^2$$

Since distance is a nonnegative quantity, we obtain the following very important formula for the distance d between P_1 and P_2.

The Distance Formula

$$d = \sqrt{(x_2 - x_1)^2 + (y_2 - y_1)^2}$$

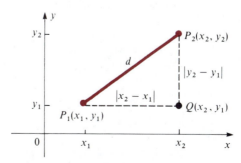

FIGURE 2

For example, the distance between the points $(-1, 3)$ and $(7, 5)$ is

$$d = \sqrt{[7-(-1)]^2 + (5-3)^2} = \sqrt{8^2 + 2^2} = \sqrt{68} = 2\sqrt{17}$$

Note that in applying the formula it makes no difference which point is labeled (x_1, y_1) and which is (x_2, y_2). It should also be noted that although in our derivation we tacitly assumed that P_1 and P_2 were not on the same horizontal or vertical line, the final result holds for these two cases also. For example, if they are on the same horizontal line, then $y_1 = y_2$, and so by the distance formula we derived,

$$d = \sqrt{(x_2 - x_1)^2 + (y_1 - y_2)^2} = \sqrt{(x_2 - x_1)^2} = |x_2 - x_1|$$

which is correct for the distance between P_1 and P_2 in this special case.

Another useful result is the formula for the coordinates of the midpoint of a line segment joining two points. As above, let P_1 and P_2 designate the points, with coordinates (x_1, y_1) and (x_2, y_2), respectively. Let (x, y) denote the unknown coordinates of the midpoint M. As shown in Figure 3, we project each of these three points to the x axis. From plane geometry we know that the projected segments on the x axis are in the same ratio as the segments on the line. Thus, since M divides P_1P_2 into equal parts, the distance from x_1 to x must also be equal to the distance from x to x_2. That is,

$$x - x_1 = x_2 - x$$
$$2x = x_1 + x_2$$
$$x = \frac{x_1 + x_2}{2}$$

By projecting to the y axis, we can show in a similar way that $y = (y_1 + y_2)/2$. So we have the following result.

The Midpoint Formula

$$\text{Midpoint of } P_1P_2 = \left(\frac{x_1 + x_2}{2}, \frac{y_1 + y_2}{2} \right)$$

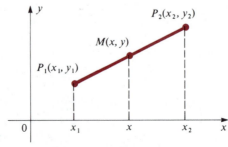

FIGURE 3

For example, the midpoint of the line segment joining $(-1, 3)$ and $(7, 5)$ is

$$\left(\frac{-1 + 7}{2}, \frac{3 + 5}{2}\right) = (3, 4)$$

Observe that each coordinate of the midpoint is the *average* of the corresponding coordinates of the endpoints.

The correspondence between ordered pairs of numbers and points in the plane provides a means of representing certain equations graphically, which we now describe. Equations such as

$$y = 2x + 3 \qquad x^2 + y^2 = 4 \qquad y^2 = \frac{x - 1}{x + 3}$$

are said to be **equations in two variables.** An ordered pair of numbers (a, b) is said **to satisfy such an equation** if, when a is substituted for x and b is substituted for y, **the equation becomes an identity;** that is, the left-hand side and right-hand side are identical. For example, the equation $y = 2x + 3$ is satisfied by the pair $(1, 5)$, since when $x = 1$, we have $y = 2(1) + 3 = 5$. Similarly, this equation is satisfied by $(-2, -1)$, $(-1, 1)$, $(0, 3)$, $(\frac{1}{2}, 4)$, $(2, 7)$, and infinitely many other ordered pairs. Figure 4 shows these points, and they appear to lie on a straight line.

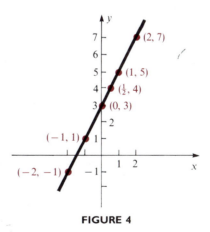

FIGURE 4

The **graph of an equation** in two variables is the collection of all points whose coordinates satisfy the equation. Usually it is impossible to locate (that is, to **plot** every point on the graph, since normally there are infinitely many of them. For equations we consider in this section a reasonably accurate sketch of the graph can be obtained by plotting several points and connecting them with a smooth curve. There are uncertainties associated with this point-plotting procedure, however. How can we be sure, for example, that we have enough points to exhibit all of the main features of the graph? How can we be sure the graph consists of just one connected piece? (Sometimes, in fact, it will consist of two or more such pieces.) How can we be sure the graph is smooth and not jagged? We will answer some of these questions in subsequent chapters for certain types of equations, but calculus is needed to give fully satisfactory answers. If you have a graphing calculator, you are encouraged to use it to check your results.

EXAMPLE 1

Draw the graph of $y = x^2 - 4$.

Solution   We make a table of values by choosing selected values of x and determining corresponding values of y. After plotting the points, we connect them with a smooth curve, as shown in Figure 5. This is an example of a **parabola,** which we will study in Chapter 4.

x	0	1	2	3	−1	−2	−3
y	−4	−3	0	5	−3	0	5

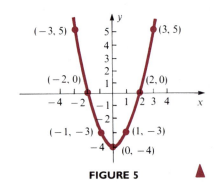

FIGURE 5

Note that in this example a given value of x and its negative produce the same value of y, so we could have shortened the table by using the $\pm$ symbol. For example, for $x = \pm 1$, we would get $y = -3$. We use this shorthand in the next example.

EXAMPLE 2

Draw the graph of $x = \sqrt{1 + y^2}$.

Solution   This time we assign selected real values to y and solve for x, which is always positive. We will use a calculator to approximate the square roots when necessary. The graph, shown in Figure 6, is the right half of a **hyperbola** (as we will see in Chapter 4).

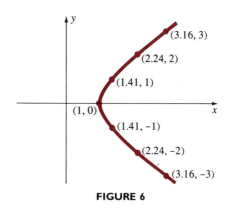

x	1	1.41	2.24	3.16
y	0	± 1	± 2	± 3

FIGURE 6

EXAMPLE 3

Draw the graph of $x + y^3 = 0$.

Solution   Solving for x gives $x = -y^3$, and again we substitute for y to get x. The graph is shown in Figure 7.

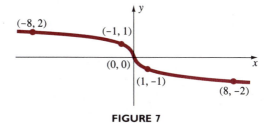

x	0	-1	-8	1	8
y	0	1	2	-1	-2

FIGURE 7

Notice that in this case changing the sign of y also changes the sign of x.

EXAMPLE 4

Draw the graph of $y = |x - 2|$.

Solution   In making a table of values it is important to take some x values for which $x - 2 \geq 0$ and some for which $x - 2 < 0$. The graph is shown in Figure 8.

x	-2	-1	0	1	2	3	4	5	6
y	4	3	2	1	0	1	2	3	4

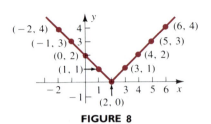

FIGURE 8

The graph appears to consist of parts of two straight lines. In fact, if $x \geq 2$, then $|x - 2| = x - 2$, and the equation is $y = x - 2$. If $x < 2$, then $|x - 2| = 2 - x$, and the equation is $y = 2 - x$. In Chapter 4 we will see that both $y = x - 2$ and $y = 2 - x$ do have straight lines as graphs. ▲

EXAMPLE 5 ||||||||||||||||||||||||||||||||||||||||||||||||||||||||||||||||||||||||||||||||||||||||||||||

Draw the graph of $y = x^3 + x^2 - 2x$.

Solution   Here it is helpful to factor the right-hand side.

$$y = x(x - 1)(x + 2)$$

In this form we see that each of the x values—$x = 0$, $x = 1$, and $x = -2$—causes y to be 0. These x values are called **x intercepts,** since they are the abscissas of points where the graph crosses, or at least touches, the x axis. We now find a few additional points and draw the graph (Figure 9).

x	0	1	2	-1	-2	-3
y	0	0	8	2	0	-12

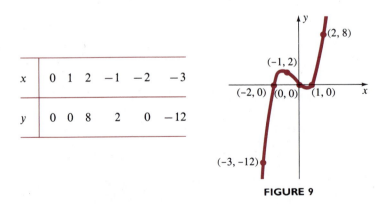

FIGURE 9   ▲

Both **x and y intercepts** can be quite helpful in drawing a graph, and they are often easy to find, as in the preceding example. The x intercepts are found by setting $y = 0$ and solving for x, and the y intercepts are found by setting $x = 0$ and solving for y. This is illustrated in the next example.

EXAMPLE 6 ||||||||||||||||||||||||||||||||||||||||||||||||||||||||||||||||||||||||||||||||||||||||||||||||

Find the x and y intercepts of the graph of

$$y^2 = \frac{2x^2 - 3x - 9}{(x - 1)^3}$$

Solution   A fraction is not defined as a real number if its denominator is zero. Thus, we restrict x so that $x \neq 1$. Setting $y = 0$ gives

$$\frac{2x^2 - 3x - 9}{(x - 1)^3} = 0$$

On multiplying both sides by the nonzero number $(x - 1)^3$, we obtain

$$2x^2 - 3x - 9 = 0$$

$$(2x + 3)(x - 3) = 0$$

$$2x + 3 = 0 \quad \text{or} \quad x - 3 = 0$$

$$x = -\frac{3}{2} \quad \text{or} \quad x = 3$$

Thus, the x intercepts are $-\frac{3}{2}$ and 3.

On setting $x = 0$ we get

$$y^2 = \frac{-9}{(-1)^3}$$

So $y^2 = 9$ and thus $y = 3$ or -3. The y intercepts are therefore 3 and -3.   ▲

The graphs in Examples 1, 2, and 3 exhibit certain *symmetries*, which we now discuss.

DEFINITION I   **The graph of an equation in x and y is said to be symmetric with respect to the x axis if for each point (x, y) on the graph, the point $(x, -y)$ is also on the graph. It is symmetric with respect to the y axis if for each point (x, y) on the graph, the point $(-x, y)$ is also on the graph. It is symmetric with respect to the origin if for each point (x, y) on the graph, the point $(-x, -y)$ is also on the graph.**

Figure 10 illustrates this.

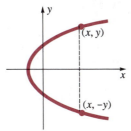
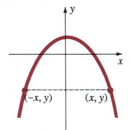
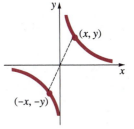

a. Symmetry with respect to the x axis

b. Symmetry with respect to the y axis

c. Symmetry with respect to the origin

FIGURE 10

The graphs in Examples 1, 2, and 3 are symmetric to the y axis, x axis, and the origin, respectively. (The graph in Example 4 has a certain type of symmetry also. How would you describe it?) If a curve is symmetric to the y axis, the part to the left of the y axis is the mirror image of the part to the right. To get the part on the left of the axis, we could *reflect* the part on the right. Similarly, for a curve symmetric to the x axis, the part below the x axis is the reflection through the x axis of the part above it. A curve that is symmetric to the origin can be thought of as the part in any one quadrant together with its combined reflection through both the x and y axes, in succession (in either order).

The following test for symmetry to the two axes and the origin is a direct consequence of Definition 1.

Tests for Symmetry

1. A curve is symmetric with respect to the x axis if replacing y by $-y$ in its equation results in an equivalent equation.
2. A curve is symmetric with respect to the y axis if replacing x by $-x$ in its equation results in an equivalent equation.
3. A curve is symmetric with respect to the origin if replacing both x by $-x$ and y by $-y$ results in an equivalent equation.

For example, the graph of $y^2 = 2x$ is symmetric to the x axis, since replacing y by $-y$ produces no change. Thus, whenever a point (x, y) satisfies the equation, so does the point $(x, -y)$

Some examples of curves symmetric with respect to the origin are those whose equations are

$$y = x^3 - 2x$$

$$xy = 4$$

$$2x^2y - 3y^3 - x^5 = 0$$

$$y = \frac{x^2 - 1}{2x}$$

since in each case replacement of both x and y by their negatives results in an equivalent equation, so that whenever (x, y) is on the graph, so is $(-x, -y)$.

EXAMPLE 7

Test for symmetry with respect to the x axis, y axis, and origin:

a. $y^2 = \dfrac{x - 1}{x + 2}$ b. $xy^2 - 4x^3 + 3x^2y^5 = 0$ c. $y = \dfrac{\sqrt{4 - x^2}}{x^4}$

d. $9x^2 - 4y^2 = 36$ e. $y = \dfrac{2x^2 + 1}{x - 2}$

Solution **a.** If y is replaced by $-y$, the equation is unchanged. So the curve is symmetric to the x axis. If x is replaced by $-x$, the resulting equation is not equivalent to the original. It follows that the curve is not symmetric to the y axis or the origin.

b. If *either x or y* is replaced by its negative, the equation is altered, but if *both x and y* are replaced by their negatives, the result is

$$(-x)(-y)^2 - 4(-x)^3 + 3(-x)^2(-y)^5 = 0$$
$$-xy^2 + 4x^3 - 3x^2y^5 = 0$$

which, on multiplying both sides by -1, becomes

$$xy^2 - 4x^3 + 3x^2y^5 = 0$$

This is exactly the same as the original. Therefore, the curve is symmetric only with respect to the origin. A quicker way to see this is to observe that each term of the polynomial has *odd* degree (remember the degree of a term is the sum of the exponents of the variables in that term), so if the signs of x and y are both changed, *every* term will have its sign changed, and since the right-hand side is 0, this yields an equivalent equation.

c. Since x appears to even powers only, replacing x by $-x$ leaves the equation unchanged. So the curve is symmetric to the y axis. The equation is altered if y is replaced by $-y$, so there is no other symmetry.

d. There is no change if either x or y is replaced by its negative, so the curve is symmetric to the x axis, y axis, and origin.

e. If either x or y is replaced by its negative, the equation is changed. If both x and y are replaced by their negatives, the equation again is changed. So there is no symmetry to either axis or to the origin. ▲

If a curve has any two of the three types of symmetry (x axis, y axis, origin), it will also have the third. (Why?) So if you find that a curve has one type of symmetry, but not a second, then it will not have the third type either.

When attempting to find the graph of an equation, it is helpful to discover in advance if it possesses some sort of symmetry. If it is known that a curve is symmetric, say, with respect to the x axis, it is necessary only to determine its shape above the x axis; then below that axis the curve is the mirror image of the part above.

REMARKS ON GRAPHING CALCULATORS

If you have one of the calculators with graphing capabilities, such as were mentioned in the front pages, then obtaining graphs of equations of the types we have discussed is usually relatively simple, particularly if you can solve for y in terms of x. In fact, the calculator can "draw" with ease graphs that would be very difficult to draw by hand. As you progress through the remainder of this course, graphs will play a major role. *The analysis we present is important, whether or not you have a graphing calculator.* If you do have such a calculator, you are encouraged to use it as a check on your work. Also, in certain exercise sets, some problems occur at the end (labeled C exercises), that are to be done using a graphing calculator. It should be emphasized, however, that your progress in the course will not be adversely affected if you do not have such a calculator and therefore are unable to do these exercises.

When using a graphing calculator, for each graph it is necessary to select an interval for x and an interval for y. These intervals are usually referred to as the **range for x** and the **range for y.** You must select the left and right endpoints for the x range and the lower and upper endpoints for the y range. If you select ranges that are too small, you may cut off from view important features of the graph, such as intercepts, or high points and low points (called **maximum** points and **minimum** points). On the other hand, if you choose ranges that are too large, some important features may not show up clearly. In subsequent chapters you will learn how to analyze certain types of equations, providing a basis for determining appropriate ranges for their graphs. Also, in calculus additional techniques that are helpful in this regard will be studied. For now you will have to try different ranges to see which ones seem to give a satisfactory graph.

EXERCISE SET I

A

For the line segment joining the two points in Problems 1–4 find **(a)** its length and **(b)** its midpoint.

1. $(2, -1), (4, 5)$ **2.** $(3, 2), (-5, -6)$
3. $(-8, 4), (5, 2)$ **4.** $(2, -6), (-3, 1)$
5. The distance between $(5, -2)$ and $(x, -6)$ is 5. Find x. (There are two solutions.)
6. The distance between $(4, y)$ and $(-2, -3)$ is $3\sqrt{5}$. Find y. (There are two solutions.)
7. One end point of a line segment has coordinates $(2, -1)$, and the midpoint has coordinates $(3, 2)$. Find the coordinates of the other end point.
8. The point $(4, 1)$ is the midpoint of the line segment joining $(x, 5)$ and $(-1, y)$. Find x and y.

In Problems 9–16 find the x and y intercepts.

9. $y = \dfrac{x-2}{x+1}$ **10.** $y = \dfrac{4-x^2}{x+3}$

11. $y = 2x^2 - 3x - 5$ **12.** $y = 15 - x - 6x^2$

13. $y = \dfrac{x^2 - 2x + 1}{x^2 - 4}$ **14.** $y = \dfrac{3x^2 - 4x + 1}{2x^2 + 5x - 3}$

15. $x^2 - 2xy + 3y^2 - x = 12$
16. $x^2 y - 2y^2 + 3x - 4y + 6 = 0$

In Problems 17–24 test for symmetry with respect to the x axis, the y axis, and the origin.

17. a. $y = x^2 - 1$ **b.** $y = x^3 - 2x$
18. a. $y^2 = x + 2$ **b.** $2x^2 + 3y^2 = 4$
19. a. $y = \sqrt{x^2 + 4}$ **b.** $x^2 = y^3$

20. a. $xy = 3$ **b.** $y = x + \dfrac{1}{x}$

21. a. $3x^2 - 4y^2 = 7$ **b.** $y = x^2 + x^3$
22. a. $y = x(x^2 - 1)$ **b.** $y = |x|$
23. a. $x^3 + y^3 = 8$ **b.** $x^2 + 2xy + 4y^2 = 4$
24. a. $|x| + |y| = 1$ **b.** $y = x^3(x^2 - 2)$

In Problems 25–49 make a table of values and draw the graph, making use of intercepts and symmetry as appropriate.

25. a. $y = x + 2$ **b.** $y = x - 1$
26. a. $y = 2x$ **b.** $y = 2 - x$
27. a. $y = 2x + 3$ **b.** $y = 4 - 3x$
28. a. $y = 3x - 5$ **b.** $x = 2y + 1$

29. **a.** $x + y = 2$ **b.** $x - y = 1$
30. **a.** $3x + 2y = 4$ **b.** $x - 3y = 2$
31. **a.** $4x - 3y = 8$ **b.** $3x + 5y + 10 = 0$
32. **a.** $x = -3y$ **b.** $x = 3 - 2y$
33. **a.** $x - y = 0$ **b.** $x + y = 0$
34. **a.** $x = 1$ **b.** $y = 2$
35. **a.** $x + 3 = 0$ **b.** $y + 1 = 0$
36. **a.** $y = x^2$ **b.** $y = 1 - x^2$
37. **a.** $y = \sqrt{x}$ **b.** $y = \sqrt{x - 1}$
38. **a.** $y = x^2 + 1$ **b.** $y = \dfrac{1}{x}$
39. **a.** $x = y^2$ **b.** $x = y^2 + 1$
40. **a.** $y = |x|$ **b.** $y = |x - 2|$
41. **a.** $y = |1 - x|$ **b.** $y = 2|x| + 1$

B

42. **a.** $y = \sqrt{x^2}$ **b.** $y = \dfrac{|x|}{x}$
43. **a.** $y^2 = x^2$ **b.** $y = \sqrt{25 - x^2}$
44. **a.** $y = x^3$ **b.** $y = \sqrt[3]{x}$
45. **a.** $y = \dfrac{1}{1 + x^2}$ **b.** $x^2 + y^2 = 4$
46. **a.** $y = \dfrac{x + 1}{x}$ **b.** $y = \dfrac{x}{x - 1}$

47. **a.** $y = x - x^2$ **b.** $y = (x - 1)(x + 2)$
48. **a.** $y = |3x - 2|$ **b.** $y = 2 - |x|$
49. **a.** $|x + y| = 2$ **b.** $|x| + |y| = 2$
50. A farmer estimates that if he plants x pounds of seed in a plot of land, the yield per pound of seed will be $50 - 2x$ pounds of produce. The total yield y from the plot is the number of pounds of seed planted multiplied by the yield per pound. Find the equation for y and graph that portion of the equation for which $x \geq 0$ and $y \geq 0$.

C (Graphing calculator)

Obtain the graphs in Problems 51–58. Then by repeated use of the zooming feature, estimate the x intercepts to two decimal places of accuracy.

51. $y = x^3 - 2x^2 - 3$
52. $y = 3 - 2x + x^2 - x^4$
53. $y = \dfrac{x^3 - 2}{x^3 + 1}$ 54. $y = \dfrac{2x^2 - 3x - 4}{x^3 + x - 1}$
55. $y = \sqrt[3]{4 - 2x - x^3}$
56. $y = \sqrt{|x^2 - 5x + 7|}$
57. $y^2 = \dfrac{x^2 - 3}{x^3 - 1}$ 58. $2x^2 + 3y^2 = 4$

THE FUNCTION CONCEPT

2

The concept of a **function** occupies a central place in most branches of mathematics. A formal definition will be given below, but we will precede it with some examples.

To say that a certain thing is a function of another can be interpreted as meaning the first thing depends on the second. This terminology is often used in nonmathematical situations. For example, one might say, "The size of the wheat crop is a function of the weather." Of course, it is a function of many other things as well. The mathematical use of the phrase "is a function of" is consistent with its everyday use, but the meaning is more precisely defined. For example, the circumference C of a circle is a function of its radius r, the exact nature of the dependence being given by the equation $C = 2\pi r$.

Or, consider a car traveling at a constant speed of 45 miles per hour. The distance d the car travels is a function of the time t elapsed. In this case, $d = 45t$.

As another example, if money is invested at compound interest, the accumulated amount of money A after t years is given by

$$A = P(1 + r)^t$$

where P is the original amount invested (the principal) and r is the annual interest rate (expressed as a decimal). Suppose \$1,000 is invested at compound interest for 5 years. Then we can express A as a function of r by

$$A = 1,000(1 + r)^5$$

It is not always easy, and sometimes it is not possible, to express the dependence given by a functional relationship by means of an equation. Try, for example, to write an equation expressing precisely how the cost of mailing a first class letter is a function of its weight. It is easier in this case to state the relationship in words. Sometimes a table can be used to describe a function. For example, it is known that the number of deaths annually in the United States resulting from highway accidents is a function of the maximum speed limit. A table showing this relationship might look like the following:

Speed limit in miles per hour	50	55	60	65	70	Unlimited
Number of deaths annually	42,000	45,000	52,000	66,000	75,000	85,000

Whether given by an equation, stated in words, or shown by a table, there is always implicit in a functional relationship some sort of rule whereby the value of one quantity is determined by the value of another (or others). Moreover, each of these quantities assumes values ranging over some prescribed set. For example, in the equation relating the circumference of a circle to its radius, $C = 2\pi r$, the value of r may be chosen arbitrarily, but only from the set of positive real numbers. For each choice of r the value of C is uniquely determined, and the totality of values of C as r ranges over all positive real numbers also constitutes a set of positive numbers. In fact, it can be shown that C also assumes every positive real number value.

So the basic ingredients of a function are these: two sets and a rule relating the elements of one of the sets to elements in the other.

DEFINITION 2   **Let A and B designate two nonempty sets. A function from A to B is a rule whereby each element in A is made to correspond to exactly one element in B.**

The set A in this definition is called the **domain** of the function, and the set of elements of B that correspond, under the rule, to elements in A is called the **range** of the function. Note that the range may or may not be all of B. If the range is all of B, we say the function is from A **onto B**. The set B is sometimes called the **codomain** of the function. It is customary to designate functions by letters such as f, g, F, or ϕ. If f is a function from A to B, and x is an element of A, then the element of B that corresponds to x is called the **image of x under f,** and is designated $f(x)$, read "f of x." The set of all such images of elements of A is the range of f. If y designates an element in the range of f, and x is an element in A for which $f(x) = y$, then we say that x is a **preimage** of y. Depending on the nature of the function, an element of the range may have more than one preimage. Figure 11 illustrates these ideas schematically.

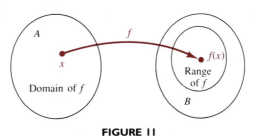

FIGURE 11

It is sometimes helpful to think of a function as a machine. This is suggested in Figure 12. A value of x is taken from the A tray (the domain) and is fed into the machine, which then operates on x according to the defining rule of f, and out comes $f(x)$ into the B tray. This machine analogy suggests the use of the term **input** for x and **output** for $f(x)$, and we will sometimes use these terms independently of the machine analogy. Also, we will think of f as *operating* on x to produce $f(x)$. One of the most common function machines is the calculator. The calculator, in fact, combines many function machines into one unit. For example, you can input a number, say 3, punch the $\boxed{x^2}$ key, so that the squaring function is activated, and the output 9 appears on the display screen.

The term *variable* is appropriate in speaking of functions. If x stands for an arbitrary value in the domain of a function f, and y designates the image of x under f—that is, $y = f(x)$—then x is the **independent variable** and y is the corresponding **dependent variable.** Any number in the domain may be substituted for the independent variable; the corresponding value for the dependent variable is then uniquely determined by means of the function.

To make these ideas more concrete, let us consider the function f with domain being the set of all real numbers, and value at x being $2x^2 + 3$. That is,

$$f(x) = 2x^2 + 3$$

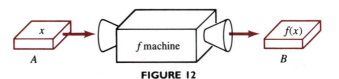

FIGURE 12

This function is now completely defined. The set of images lies in R, and specifically the range of f is the subset of R consisting of those positive real numbers greater than or equal to 3:

$$\text{Range of } f = \{y: \quad y \geq 3\} \qquad \textbf{Why?}$$

We have, for example, $f(0) = 3$, since $f(0) = 2 \cdot (0)^2 + 3 = 2(0) + 3 = 3$. Also, $f(1) = 5$, $f(4) = 35$, and $f(-2) = 11$.

An equation such as $y = 2x^2 + 3$ implicitly defines a function, even though no letter is specified for the function. The equation is the rule, and the domain is assumed to be the largest subset of R such that for x in this subset, the corresponding value of y will be real. In this example, the domain, then, is all of R. Such an assumption about the domain of functions we use will always be made unless the domain is explicitly stated. In the function implied by the equation $y = \sqrt{x}$, then, we would assume the domain to be the set of all nonnegative real numbers.

To reiterate, we may write

$$f(x) = 2x^2 + 3$$

or

$$y = 2x^2 + 3$$

to define the same function. The first way has the advantage that it assigns a letter f to name the function, and images of various values of x are easily indicated, for example, $f(0) = 3$ and $f(1) = 5$. The second way has the advantage that a name, y, is assigned to the dependent variable. Both methods are useful and will be employed throughout the remainder of this text. It should be noted, too, that the particular letters used to designate the function or the variables are not important in themselves; the form is the important part. For example, the function g defined by

$$g(t) = 2t^2 + 3$$

is the very same function as f described previously, because both functions give the same result for each value of the independent variable. For this reason, we can write $f = g$. When the variables represent certain physical quantities, suggestive letters are often used, such as $v = f(t)$ for the velocity v in terms of time t.

EXAMPLE 8

Let $f(x) = \dfrac{x}{x + 1}$. Find the following:

a. $f(2)$ **b.** $f(0)$ **c.** $f(\tfrac{2}{3})$ **d.** $f(x - 1)$ **e.** $f(x + h)$

Solution **a.** $f(2) = \dfrac{2}{2 + 1} = \dfrac{2}{3}$ **b.** $f(0) = \dfrac{0}{0 + 1} = \dfrac{0}{1} = 0$

c. $f(\frac{2}{3}) = \dfrac{\frac{2}{3}}{\frac{2}{3}+1} = \dfrac{2}{2+3} = \dfrac{2}{5}$ **d.** $f(x-1) = \dfrac{(x-1)}{(x-1)+1} = \dfrac{x-1}{x}$

e. $f(x+h) = \dfrac{x+h}{x+h+1}$

Note. Whatever occurs in the parentheses following f is substituted for x wherever x appears. ▲

EXAMPLE 9

Let $g(t) = |1 - t|$. Find the following:

a. $g(0)$ **b.** $g(2)$ **c.** $g(1+h)$ **d.** $g(t^2+1)$ **e.** $g(x + \Delta x)$

Solution **a.** $g(0) = |1 - 0| = |1| = 1$
b. $g(2) = |1 - 2| = |-1| = 1$
c. $g(1 + h) = |1 - (1 + h)| = |-h| = |h|$
d. $g(t^2 + 1) = |1 - (t^2 + 1)| = |-t^2| = t^2$
e. $g(x + \Delta x) = |1 - (x + \Delta x)| = |1 - x - \Delta x|$

Note. The symbol Δx is read "delta x" and has historically been used in calculus. It is treated as a unit and does not mean the product of Δ and x. ▲

EXAMPLE 10

Find the domain and range of the function f.

a. $f(x) = x^2 + 4$ **b.** $f(x) = \dfrac{1}{\sqrt{x-1}}$

Solution **a.** The domain of f is all of R, since $x^2 + 4$ is finite for all real x. Since $x^2 \geq 0$, we see that $x^2 + 4 \geq 4$, and thus $f(x) \geq 4$. Furthermore, any number ≥ 4 is the image of at least one x in the domain. For example, to find an x such that $f(x) = 7$, we solve the equation $7 = x^2 + 4$ for x and get $x = \pm\sqrt{3}$. More generally, for any number $k \geq 4$, putting $f(x) = k$ gives $k = x^2 + 4$, and this has the solutions $x = \pm\sqrt{k-4}$. This proves that the range of f is the set of all numbers ≥ 4; that is,

$$\text{Range of } f = \{y: \quad y \geq 4\}$$

b. For $f(x)$ to be real, we must have $x - 1 > 0$. So the domain of $f = \{x: \quad x > 1\}$. It is clear that $f(x) > 0$ for all x in the domain. To prove that the range of f is the set of all positive real numbers, let k be any positive number and see if the equation $f(x) = k$ can be solved for x:

$$\frac{1}{\sqrt{x-1}} = k$$

$$\sqrt{x-1} = \frac{1}{k}$$

$$x - 1 = \frac{1}{k^2}$$

$$x = 1 + \frac{1}{k^2}$$

Since this is a number in the domain, we have shown that

$$\text{Range of } f = \{y: \quad y > 0\}$$ ▲

EXAMPLE 11

Find the domain of each of the following functions.

a. $f(x) = \sqrt{3 - x - 2x^2}$ **b.** $g(x) = \sqrt{\dfrac{x+3}{x^3 - 4x^2}}$

Solution   **a.**   The domain is the set of all x values for which

$$3 - x - 2x^2 \geq 0$$

We use the procedure of Section 10 of Chapter 2 to solve this inequality:

$$(3 + x)(1 - 2x) \geq 0 \quad [(2+3)(1-x)]$$

The critical numbers are $x = -3$ and $x = \frac{1}{2}$. The sign graph is shown in Figure 13. From it we determine that the domain of f is $[-3, \frac{1}{2}]$.

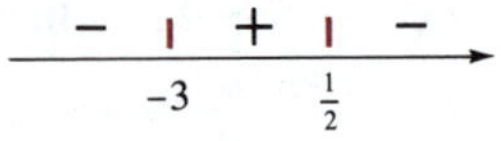

$$\underset{-3 \qquad\quad \frac{1}{2}}{\overline{\quad - \quad | \quad + \quad | \quad - \quad}} \longrightarrow$$

FIGURE 13

b.   Proceeding as in part **a**, we want

$$\frac{x+3}{x^2(x-4)} \geq 0$$

The critical numbers are $x = -3$, $x = 0$, and $x = 4$, giving the sign graph in Figure 14. For $x = -3$ the inequality is satisfied, since the fraction is 0 there; but for $x = 4$ it is not, since the fraction is not defined. Thus, the domain of g is $(-\infty, -3] \cup (4, \infty)$.

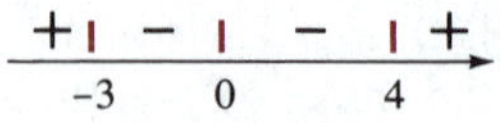

$$\underset{-3 \qquad 0 \qquad\quad 4}{\overline{\quad + | \quad - \quad | \quad - \quad | + \quad}} \longrightarrow$$

FIGURE 14 ▲

EXAMPLE 12 ┃┃┃

Let $f(x) = \sqrt{x}$.   Find

$$\frac{f(x) - f(4)}{x - 4} \qquad (x \neq 4)$$

and write the answer with a rational numerator.

Solution

$$\frac{f(x) - f(4)}{x - 4} = \frac{\sqrt{x} - 2}{x - 4}$$

In calculus, where this type of problem frequently occurs, it is essential that the numerator be rationalized, as follows.

$$\frac{\sqrt{x} - 2}{x - 4} \cdot \frac{\sqrt{x} + 2}{\sqrt{x} + 2} = \frac{\cancel{x - 4}}{\cancel{(x - 4)}(\sqrt{x} + 2)} = \frac{1}{\sqrt{x} + 2}$$

Dividing out the factor $x - 4$ is justified since x is restricted to be different from 4, so that $x - 4 \neq 0$.   ▲

EXAMPLE 13 ┃┃┃

Let $f(x) = \dfrac{1}{2x - 1}$.   Find

$$\frac{f(x + h) - f(x)}{h} \qquad (h \neq 0)$$

and simplify the result.

Solution   This is similar to Example 12 and also is an important type of calculation in calculus. Direct substitution gives

$$\frac{f(x + h) - f(x)}{h} = \frac{\dfrac{1}{2(x + h) - 1} - \dfrac{1}{2x - 1}}{h}$$

Now we simplify by multiplying numerator and denominator by the LCD of the minor denominators, $(2x - 1)[2(x + h) - 1]$. This gives (verify)

$$\frac{2x - 1 - [2(x + h) - 1]}{h(2x - 1)[2(x + h) - 1]} = \frac{2x - 1 - 2x - 2h + 1}{h(2x - 1)(2x + 2h - 1)}$$

$$= \frac{-2\cancel{h}}{\cancel{h}(2x - 1)(2x + 2h - 1)}$$

$$= \frac{-2}{(2x - 1)(2x + 2h - 1)}$$

Dividing out the factor h is justified since $h \neq 0$.   ▲

Sometimes the rule for a function is given by more than one expression, as the next example shows. Such a function is called a **piecewise-defined function.**

EXAMPLE 14

Let

$$f(x) = \begin{cases} -1 & \text{if} & x < 0 \\ x & \text{if} & 0 \le x < 2 \\ 1 - x^2 & \text{if} & x \ge 2 \end{cases}$$

Find:

a. $f(1)$ **b.** $f(2)$ **c.** $f(-2)$ **d.** $f(0)$ **e.** $f(3)$

Solution   **a.**   When $0 \le x < 2$, $f(x) = x$. So $f(1) = 1$.
b.   When $x \ge 2$, $f(x) = 1 - x^2$. So $f(2) = 1 - 4 = -3$.
c.   When $x < 0$, $f(x) = -1$. So $f(-2) = -1$.
d.   Since 0 is in the interval $0 \le x < 2$, for which $f(x) = x$, it follows that $f(0) = 0$.
e.   When $x \ge 2$, $f(x) = 1 - x^2$. So $f(3) = 1 - 9 = -8$.   ▲

EXAMPLE 15

Express the area A of a circle as a function of its radius r.

Solution $$A = \pi r^2$$

460

EXAMPLE 16

A woman invests $1,000 at 5% simple interest. Express the interest earned I as a function of time t in years.

Solution   Using the formula Interest = Principal × Rate × Time for simple interest, we get

$$I = (1,000)(0.05)t = 50t$$

EXAMPLE 17

A rectangle is inscribed in a circle of radius 2. Express the area of the rectangle as a function of its base.

Solution   Here it is helpful to draw a sketch, as shown in Figure 15. Since the radius of the circle is 2, its diameter is 4. Divide the rectangle into two triangles as shown,

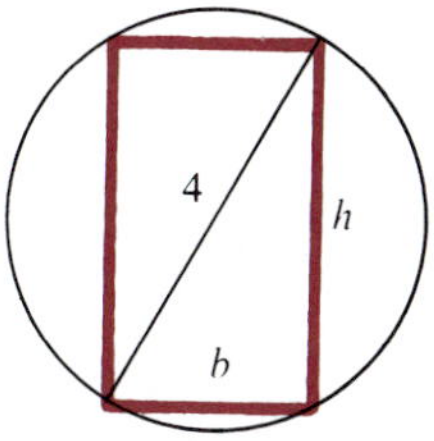

FIGURE 15

by means of a diagonal. This diagonal is a diameter of the circle. Designate the base of the rectangle by b and its height by h. The area is $b \cdot h$, but to obtain this as a function of b only, we need to express h in terms of b. This can be done by means of the Pythagorean theorem. For our triangle, this says that

$$b^2 + h^2 = 4^2$$

So $h^2 = 16 - b^2$, and $h = \sqrt{16 - b^2}$. Finally, then, the area A is given by

$$A = b\sqrt{16 - b^2}$$

▲

Certain functions have the property that for each x in the domain, its negative, $-x$, is also in the domain, and the function values are the same at x and $-x$. That is, $f(-x) = f(x)$. For example, $f(x) = x^2$ has this property. On the other hand, some functions have values that are always exactly opposite in sign at x and $-x$ for each x in the domain, that is $f(-x) = -f(x)$. The function $f(x) = x^3$ is of this type. Functions of these two types are given special names in the following definition.

DEFINITION 3   **Let f be a function whose domain includes $-x$ whenever it includes x. Then f is said to be even if $f(-x) = f(x)$ for every x in its domain, and f is said to be odd if $f(-x) = -f(x)$ for every x in its domain.**

Remark. Even functions and odd functions are special types. Many functions are neither even nor odd. So if a function fails to be even, you cannot conclude that it is necessarily odd, or conversely.

EXAMPLE 18 ||||||||||||||||||||||||||||||||||||||||||||||||||||||||||||||||||||||||||||||||||||||||||||||||||||||||||||||||||||

Classify each of the following functions as even, odd, or neither.

a. $f(x) = \dfrac{x}{x^2 + 1}$ **b.** $g(x) = 2x^4 - 3x^2 + 5$ **c.** $h(x) = x^2 - 3x + 1$

Solution   **a.** $f(-x) = \dfrac{-x}{(-x)^2 + 1} = -\dfrac{x}{x^2 + 1} = -f(x)$

So f is odd.

b. $g(-x) = 2(-x)^4 - 3(-x)^2 + 5$
$\qquad\quad = 2x^4 - 3x^2 + 5 = g(x)$

So g is even.

c. $h(-x) = (-x)^2 - 3(-x) + 1 = x^2 + 3x + 1$

The result is neither $h(x)$ nor $-h(x)$ for all x in the domain. So h is neither even nor odd.   ▲

EXERCISE SET 2   1-13 odd   21-25 $\overset{odd}{}$ 29-33 $\overset{odd}{}$

A

1. If $f(x) = 2x - 3$, find
   a. $f(0)$ b. $f(1)$ c. $f(-2)$ d. $f(4)$
2. If $f(x) = 2$, find
   a. $f(5)$ b. $f(-3)$ c. $f(0)$ d. $f(100)$
3. If $g(x) = \dfrac{x^2 - 1}{x + 2}$, find
   a. $g(1)$ b. $g(-1)$ c. $g(0)$ d. $g(-3)$
4. If $h(x) = \sqrt{3x^2 - 4x + 5}$, find
   a. $h(1)$ b. $h(2)$ c. $h(-2)$ d. $h(-5)$
5. If $F(t) = \dfrac{2t - 1}{t + 3}$, find
   a. $F(\tfrac{1}{2})$ b. $F(\tfrac{2}{3})$ c. $F(-2)$ d. $F(a)$
6. If $f(x) = \sqrt{1 - x^3}$, find
   a. $f(0)$ b. $f(1)$ c. $f(-2)$ d. $f(h)$
7. If $\phi(x) = |x - 3|$, find
   a. $\phi(4)$ b. $\phi(1)$
   c. $\phi(t + 3)$ d. $\phi(3 - x^2)$
8. If $h(u) = \dfrac{u}{|u|}$, find
   a. $h(1)$ b. $h(-1)$ c. $h(|x|)$ d. $h(x^2)$
9. If $f(t) = t^2$, find
   a. $f(2x)$ b. $f(t + 1)$
   c. $f(2 + h)$ d. $f(t + h)$
10. If $g(x) = \dfrac{x}{x + 1}$, find
    a. $g(x + 1)$ b. $g\left(\dfrac{1}{x}\right)$
    c. $g(h - 1)$ d. $g(x + \Delta x)$
11. If $g(x) = \dfrac{1}{x}$, find
    $$\dfrac{g(x) - g(2)}{x - 2} \qquad (x \neq 0, 2)$$
    and simplify the result.
12. If $f(x) = 2x + 3$, find
    $$\dfrac{f(1 + h) - f(1)}{h} \qquad (h \neq 0)$$
    and simplify the result.

13. If $f(t) = \dfrac{t}{t - 3}$, find
    $$\dfrac{f(4 + h) - f(4)}{h} \qquad (h \neq 0; t \neq 3)$$
    and simplify the result.
14. If $\phi(x) = \sqrt{2x - 1}$, find
    $$\dfrac{\phi(x) - \phi(1)}{x - 1} \qquad (x \neq 1)$$
    and simplify by rationalizing the numerator.

In Problems 15–20 find the domain and range of f.

15. a. $f(x) = 3x$ b. $f(x) = 2x^2$
16. a. $f(x) = x^3 + 1$ b. $f(x) = \dfrac{2}{x}$
17. a. $f(x) = |x|$ b. $f(x) = \sqrt{x - 1}$
18. a. $f(x) = 1$
    b. $f(x) = \begin{cases} 1 & \text{if } x \geq 0 \\ -1 & \text{if } x < 0 \end{cases}$
19. a. $f(x) = \sqrt{4 - x^2}$
    b. $f(x) = \begin{cases} x^2 & \text{if } x \geq 0 \\ -x & \text{if } x < 0 \end{cases}$
20. a. $f(x) = \dfrac{1}{|x - 1|}$ b. $f(x) = \dfrac{x - 1}{x}$

In Problems 21–26 find $[f(x + h) - f(x)]/h$ for $h \neq 0$, and simplify the result.

21. $f(x) = 3x + 2$ 22. $f(x) = 1 - x^2$
23. $f(x) = \dfrac{2}{x}$ 24. $f(x) = x^2 - 2x + 5$
25. $f(x) = x^3$
26. $f(x) = \sqrt{x + 2}$ (Rationalize the numerator.)
27. For $f(x) = \begin{cases} 2x - 3 & \text{if } x \geq 0 \\ 1 - x & \text{if } x < 0 \end{cases}$ find
    a. $f(0)$ b. $f(-1)$ c. $f(1)$
    d. $f(4)$ e. $f(|x|)$

28. For $g(x) = \begin{cases} x^2 - 1 & \text{if } x > 1 \\ 2 & \text{if } x = 1 \\ -3 & \text{if } x < 1 \end{cases}$ find

    a. $g(2)$ **b.** $g(0)$ **c.** $g(1)$

    d. $g(-1)$ **e.** $g\left(\dfrac{x^2}{x^2 + 1}\right)$

In Problems 29–34 find the domain of the given function.

29. $f(x) = \dfrac{3x - 2}{2x^2 - 5x - 3}$

30. $f(x) = \dfrac{x^3 - 1}{2x^2 - 7x + 8}$

31. $g(x) = \sqrt{3x - x^2}$

32. $g(x) = \sqrt{x^2 - 3x - 10}$

33. $h(t) = \sqrt{\dfrac{t - 3}{t + 1}}$

34. $h(t) = \sqrt{\dfrac{t}{4 - t^2}}$

35. Express the area of a triangle with altitude 8 as a function of its base b.

36. The length l of a certain rectangle is one less than twice the width w. Express the perimeter as a function of w.

37. Express **(a)** the circumference and **(b)** the area of a circle as a function of its diameter d.

38. The sum of $800 is invested at $6\frac{1}{2}\%$ simple interest for t years. Express the total interest earned as a function of t.

39. Express the amount of money accumulated after t years from investing $12,000 at 8% compounded annually, as a function of t.

40. A can is in the form of a right circular cylinder with height three times the radius of the base. Express each of the following as a function of the base radius r:

    a. Volume

    b. Lateral exterior surface area

    c. Total exterior surface area

41. Express the Fahrenheit temperature F as a function of the Celsius temperature C.

    Hint. Write $F = aC + b$ and find a and b by using the relationships $F = 32$ when $C = 0$ and $F = 212$ when $C = 100$.

42. Let $f(x) = (x - 2)/(1 - 2x)$. Show that if $x \neq 0, \frac{1}{2}, 2$, then

$$f\left(\frac{1}{x}\right) = \frac{1}{f(x)}$$

43. Let $f(x) = ax$. Show that

    a. $f(kx) = kf(x)$ for all real numbers k

    b. $f(x + y) = f(x) + f(y)$

Show by means of an example that neither of these formulas is true for all functions f.

44. For each of the following determine if the function is even, odd, or neither.

    a. $f(x) = 1 - 2x^2$ **b.** $g(x) = x^3 - 5x$

    c. $h(x) = 3x^5 - 2x^3 + 1$

    d. $F(x) = x|x|$ **e.** $f(x) = \dfrac{x^2}{x^2 - 4}$

45. Follow the instructions for Problem 44 for the following functions.

    a. $f(x) = \dfrac{x^3}{x^4 + 1}$ **b.** $g(x) = 2|x| - 3$

    c. $F(x) = (x^2 - 1)^3$ **d.** $h(t) = \dfrac{t - 1}{t + 1}$

    e. $f(x) = (x + 1)^3 + (x - 1)^3$

46. A piece of manufacturing equipment is purchased for $250,000. For tax purposes the value is decreased by $20,000 for each year that elapses after its purchase.

    a. Write the rule for the function f.

    b. Find the domain of f if the value cannot become negative.

47. If the selling price of a certain product is set at x dollars, it is estimated that $1500 - 25x$ units will be sold. Let $R(x)$ be the revenue when the price is x, where revenue is the selling price per unit times the number of units sold.

    a. Write the rule for R.

    b. Find the domain of R if the revenue cannot be negative.

B

In Problems 48 and 49 find the domain of the given function.

48. $f(x) = \sqrt{\dfrac{x^3 - 8}{x^2 + 2x - 3}}$

49. $g(x) = \sqrt{\dfrac{x^3 + x^2 - x - 1}{x^2 - 9}}$

50. Express the area of an equilateral triangle as a function of one of its sides s.

Hint. Use the Pythagorean theorem.

51. An isosceles triangle of base 2 is inscribed in a circle of radius $r > 1$. Express the area of the triangle as a function of r. (There are two solutions.)

Hint. Use the Pythagorean theorem.

52. A right circular cylinder is inscribed in a sphere of radius a. Express the volume of the cylinder as a function of its base radius r.

53. A rectangle is inscribed in a triangle of height 6 and base 4, with one side of the rectangle lying on the base of the triangle. Express the area of the rectangle as a function of its base.

Hint. Use similar triangles.

54. A pyramid has a square base 4 feet on a side, and its height is 10 feet. Express the area of a cross-section, parallel to the base, that is h feet above the base, as a function of h.

55. A storage tank is in the form of a cylinder x feet long and y feet in diameter, with hemispheres at both ends (see sketch).

  a. Express the volume of the tank as a function of x and y.

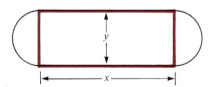

  b. Express the exterior surface area of the tank as a function of x and y.

  c. If $x = 4y$, express the volume and the exterior surface area as functions of x only.

56. An organization issuing credit cards charges the card holder interest of $1\frac{1}{2}\%$ per month for the first $\$1,000$ unpaid balance after 30 days and 1% a month on the unpaid balance above $\$1,000$. Let $f(x)$ be the interest charged per month on an unpaid balance of $\$x$, and find the rule for f.

57. A company deducts 2% of gross salary from each employee's pay as the employee's contribution to the pension fund. However, no employee pays more than $\$400$ into the fund in a single year. Let $f(x)$ be the contribution of an employee with gross salary of $\$x$ a year. Write the rule for f.

58. A right circular cone is inscribed in a sphere of radius a. Express the volume of the cone as a function of its base radius r.

Hint. The volume of a cone is one-third the area of its base times its altitude.

59. Let f be any function whose domain contains the negative of each of its elements. Define g and h as follows:

$$g(x) = \frac{f(x) + f(-x)}{2} \qquad h(x) = \frac{f(x) - f(-x)}{2}$$

  a. Show that g is even and h is odd.

  b. Show that $f(x) = g(x) + h(x)$. (This proves that every such function f can be expressed as the sum of an even function and an odd function.)

60. Prove that the range of the function $f(x) = 1 + \sqrt{x}$ is the set $\{y:\ y \geq 1\}$.

61. Prove that the range of the function $f(x) = 2x/(x^2 + 1)$ is the set $\{y:\ |y| \leq 1\}$.

GRAPHS OF FUNCTIONS

Let f be a function from A to B, where A and B are subsets of R. If x is an element in A, let y denote its image under f, that is, $y = f(x)$. Thus, f establishes a pairing of values (x, y) with x in A and y in B. The totality of such pairs as x varies over all of A can be plotted in a cartesian coordinate system, and the result is called the **graph of f.** If the rule for f is given in the form of an equation, the graph of f is identical to the graph of this equation as defined in Section 1. For example, the graph of the function f for which $f(x) = 2x + 3$ is the same as the graph of the equation $y = 2x + 3$. However, an important distinction must be made. Not all graphs of equations in two variables are graphs of functions. This is because of

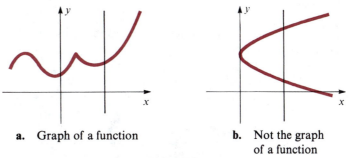

a.   Graph of a function

b.   Not the graph
           of a function

FIGURE 16

the requirement that in the case of a function, each x in the domain determines exactly one value y in the range. **Geometrically, this amounts to the requirement that a vertical line through any point on the x axis strikes the graph of f at no more than one point** (Figure 16).

As a specific example of an equation that does not define y as a function of x, consider $y^2 = x$. If we solve for y, we get $y = \pm\sqrt{x}$, so that each $x > 0$ determines two values of y, which cannot happen in the case of a function. The graph of

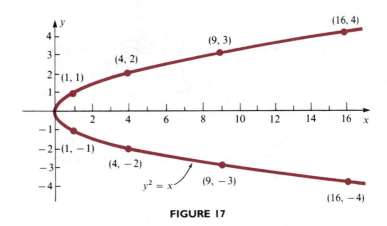

FIGURE 17

$y^2 = x$ (Figure 17) can be obtained by making a table of values, as follows:

x	1	4	9	16
y	± 1	± 2	± 3	± 4

The graph clearly fails the vertical line test. So $y^2 = x$ does not define y as a function of x.

EXAMPLE 19 ||||||||||||||||||||||||||||||||||||||||||||||||||||||||||||||||||||||||||||||||||||||||||||||

Determine which of the graphs in Figure 18 are graphs of functions and which are not.

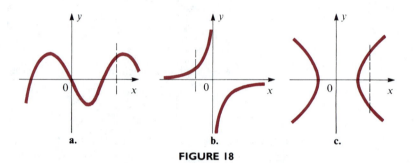

a.   b.   c.

FIGURE 18

Solution   **a.**   This is a function since each vertical line cuts the curve in only one point.
b.   Each vertical line cuts the curve at most in one point, so this is a function.
c.   Some vertical lines cut the curve in two points, so this is not a function.   ▲

EXAMPLE 20 ||||||||||||||||||||||||||||||||||||||||||||||||||||||||||||||||||||||||||||||||||||||||||||||||||

Draw the graph of the function f defined by

$$f(x) = \begin{cases} x & \text{if} \quad x \le 0 \\ 0 & \text{if} \quad 0 < x < 2 \\ 1 & \text{if} \quad x \ge 2 \end{cases}$$

Solution   We set $y = f(x)$ and make a table of values, including points in each of the intervals on which various parts of the function are defined. The graph is given in Figure 19.

x	-3	-2	-1	0	1	2	3
y	-3	-2	-1	0	0	1	1

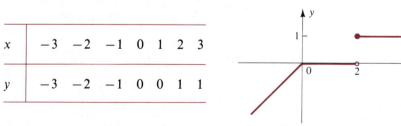

FIGURE 19

Notice the break in the graph at $x = 2$. The function is said to be **discontinuous** there. Even though the graph is made up of three distinct parts, it is still the graph of the single function f.

Note. An open circle at a point on a graph means the point is not included, and a closed circle means it is.   ▲

If f is an even function, we know that $f(-x) = f(x)$. So replacing x by $-x$ in the equation $y = f(x)$ leaves the equation unchanged. Thus, the graph of f is symmetric with respect to the y axis. If f is odd, then $f(-x) = -f(x)$, and so replacing both x and y by their negatives changes $y = f(x)$ to $-y = -f(x)$, which is equivalent to the original equation. This implies that the graph of f is symmetric with respect to the origin.

A function is said to be *increasing* on an interval if for any x_1 and x_2 on the interval, with $x_1 < x_2$, it is true that $f(x_1) < f(x_2)$. If when $x_1 < x_2$, it is true that $f(x_1) > f(x_2)$, then f is *decreasing* on the interval. Figure 20 illustrates increasing and decreasing functions.

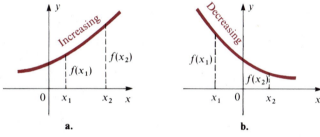

a.   **b.**

FIGURE 20

It is useful to be able to identify from the graph of a function the intervals on which it is increasing and those on which it is decreasing. The next example illustrates this.

EXAMPLE 21 |||||||||||||||||||||||||||||||||||||||||||||||||||||||||||||||||||||||||||||||||||||||||||||||||||||||||||||||||||||||||

Determine the largest intervals on which the function whose graph is shown in Figure 21 is increasing and the largest intervals on which it is decreasing.

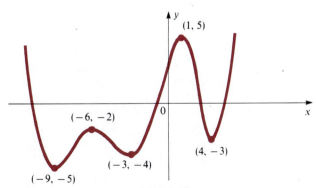

FIGURE 21

Solution   Increasing on $(-9, -6) \cup (-3, 1) \cup (4, \infty)$.
Decreasing on $(-\infty, -9) \cup (-6, -3) \cup (1, 4)$.

▲

We conclude this section by showing how we can infer information about graphs of certain functions from their relationship to some basic function whose graph we know. In particular, suppose we know the graph of some function $y = f(x)$. We will refer to this as the **basic function** and to its graph as the **basic graph.** From this basic graph we can obtain graphs of each of the following:

1. $y = f(x) + k$ 2. $y = f(x - h)$
3. $y = af(x)$ 4. $y = f(bx)$

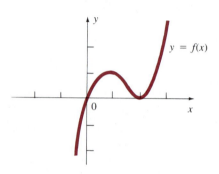

FIGURE 22

Also, we can graph any combination of these four types, such as $y = af(x - h) + k$. We consider these one at a time. To make matters more concrete, we will use the graph in Figure 22 as our basic graph.

1. **$y = f(x) + k$.** Here each y coordinate on the basic graph is shifted vertically by k units, upward if $k > 0$ and downward if $k < 0$. Figure 23(a) illustrates this for $k > 0$. The basic graph is also shown for comparison. This is referred to as a **vertical translation.**

2. **$y = f(x - h)$.** If a point (x_1, y_1) lies on the basic graph, so that $y_1 = f(x_1)$, then the point $(x_1 + h, y_1)$ lies on $y = f(x - h)$, as can be seen by substituting $x_1 + h$ for x. Thus, the effect is to shift the entire graph h units horizontally, to the right if $h > 0$ and to the left if $h < 0$. Figure 23(b) illustrates this for $h > 0$. This shift is referred to as a **horizontal translation.**

3. **$y = af(x)$ $(a \neq 0)$.** First consider $a > 0$. Each ordinate (y value) of the basic graph is multiplied by a. So if $a > 1$, the graph is **stretched,** or **elongated, vertically.** If $0 < a < 1$, the graph is **compressed,** or **shrunk, vertically.** Now suppose $a < 0$. Then each ordinate, in addition to being stretched (if $|a| > 1$)

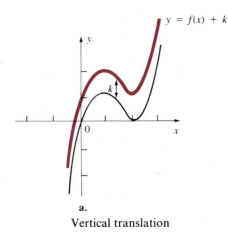

y = f(x) + k

a.

Vertical translation

y = f(x − h)

b.

Horizontal translation

FIGURE 23

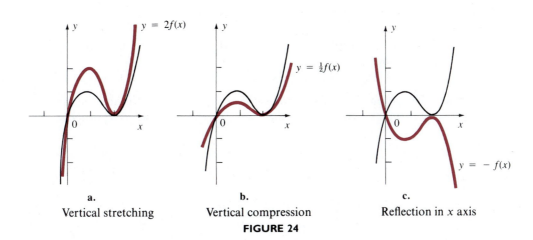

y = 2f(x)

a.

Vertical stretching

y = ½f(x)

b.

Vertical compression

y = − f(x)

c.

Reflection in x axis

FIGURE 24

or compressed (if $0 < |a| < 1$), is changed in sign. So we obtain a **reflection in the x axis.** In particular, if $a = -1$, the basic graph is **reflected.** Figure 24 illustrates vertical stretching, shrinking, and reflection corresponding to particular values of a.

4. $y = f(bx),$ $(b \neq 0)$.   If (x_1, y_1) is on the basic graph, then direct substitution shows that $\left(\dfrac{x_1}{b}, y_1\right)$ is on the new graph. Thus, the basic graph is **compressed horizontally** if $b > 1$ and **stretched horizontally** if $0 < b < 1$. If $b < 0$, in addition to compression (if $|b| > 1$) or stretching (if $0 < |b| < 1$), there is a **reflection in the y axis.** Figure 25 illustrates three particular cases.

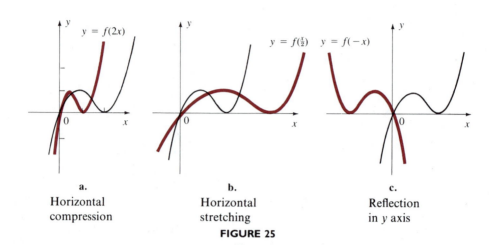

a.
Horizontal
compression

b.
Horizontal
stretching

c.
Reflection
in y axis

FIGURE 25

EXAMPLE 22

Sketch the graph of $f(x) = x^2$ and from it obtain the graph of each of the following:

a. $y = 2x^2$ **b.** $y = 2x^2 - 3$ **c.** $y = 2(x - 1)^2 - 3$

Solution The graph of f is sketched in Figure 26. Note that f is an even function, so that its graph is symmetric to the y axis. It is an example of a *parabola*.

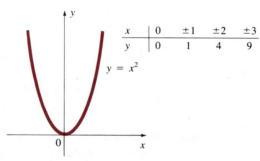

x	0	±1	±2	±3
y	0	1	4	9

$y = x^2$

FIGURE 26

a. The graph of $y = 2x^2$ is obtained from the graph of f by multiplying each y value by 2. So it represents a vertical stretching. Its graph is shown in Figure 27(a).

b. The graph of $y = 2x^2 - 3$ is obtained from the graph of part **a** by a downward translation of 3 units. Its graph is shown in Figure 27(b).

c. The graph of $y = 2(x - 1)^2 - 3$ is obtained from the graph in part **b** by a horizontal translation 1 unit to the right. Note that the equation is of the form $y = af(x - h) + k$, where $a = 2$, $h = 1$, and $k = -3$. So compared to the graph of f, it represents the vertical stretching of part **a,** followed by the downward translation of part **b,** followed by the horizontal translation of part **c.** Its graph is shown in Figure 27(c).

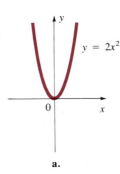

$y = 2x^2$

a.

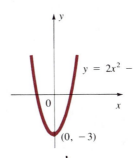

$y = 2x^2 - 3$

$(0, -3)$

b.

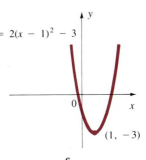

$y = 2(x - 1)^2 - 3$

$(1, -3)$

c.

FIGURE 27

 EXERCISE SET 3 $1-17, 19-23, 27-31$ $abde$

A

In Problems 1–18 make a table of values and draw the graph.

1. $f(x) = x + 1$
2. $f(x) = 2x - 3$
3. $f(x) = 1 - 2x$
4. $g(x) = 3 - 4x$
5. $h(x) = \dfrac{x + 2}{3}$
6. $f(x) = \dfrac{2 - 3x}{4}$
7. $f(x) = 2x - x^2$
8. $g(x) = x^3 + 1$
9. $h(x) = \sqrt{x + 1}$
10. $F(x) = \sqrt{2 - x}$
11. $f(x) = \dfrac{2}{x}$
12. $g(x) = \dfrac{1}{x - 1}$
13. $f(x) = \sqrt{16 - x^2}$
14. $F(x) = \dfrac{2}{x^2}$

15. $g(x) = \begin{cases} x + 1 & \text{if } x \geq 0 \\ 1 & \text{if } x < 0 \end{cases}$

16. $f(x) = \begin{cases} x^2 - 4 & \text{if } x > 2 \\ 1 & \text{if } x = 2 \\ 2 - x & \text{if } x < 2 \end{cases}$

17. $F(x) = \begin{cases} 3 & \text{if } x \geq 1 \\ 1 - x & \text{if } -1 \leq x < 1 \\ x + 1 & \text{if } x < -1 \end{cases}$

18. $h(x) = \begin{cases} \sqrt{4 - x^2} & \text{if } |x| \leq 2 \\ |x| - 2 & \text{if } |x| > 2 \end{cases}$

In Problems 19 and 20 determine which of the graphs are graphs of functions and which are not.

19. a.

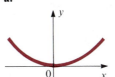

b.

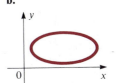

c.

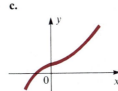

d.

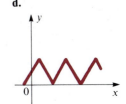

e.

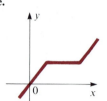

f.

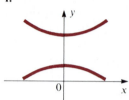

b.

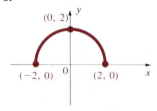

20. a.

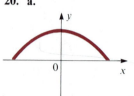

b.

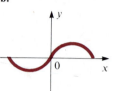

c.

c.

d.

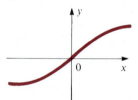

d.

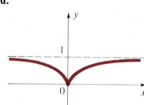

e.

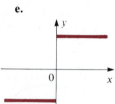

f.

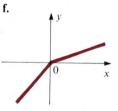

22. a.

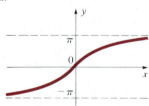

In Problems 21 and 22 determine the domain and range of each function whose graph is given.

21. a.

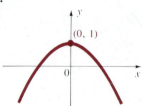

b.

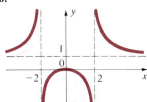

c.

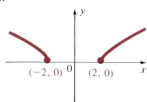

d.

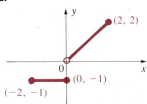

In Problems 23 and 24 determine the intervals on which the function whose graph is shown is increasing and those on which it is decreasing.

23.

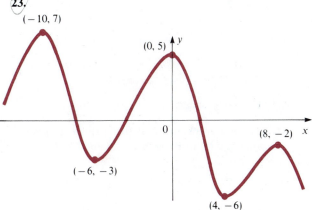

24.

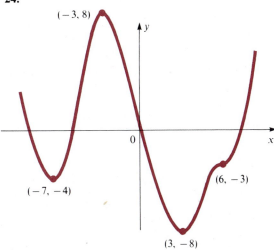

25. The graph for $x \geq 0$ of a function f is given. Complete the graph for $x < 0$ if
   a. f is even   **b.** f is odd

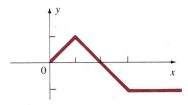

26. Repeat Problem 25 for the graph shown here.

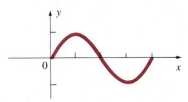

In Problems 27–32 the graph of a function $y = f(x)$ is given. From it obtain the graphs of each of the following:

    a. $y = f(x) + 2$ **b.** $y = f(x - 3)$
    c. $y = 2f(x)$ **d.** $y = -f(x)$
    e. $y = f(-x)$

27.

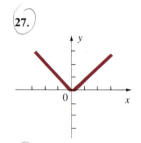

28.

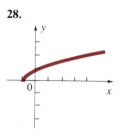

29.

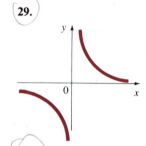

30.

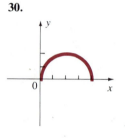

31.

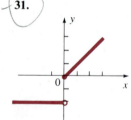

32.

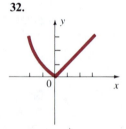

33. Draw the graph of $y = x^3$ and from it obtain the graph of $y = \frac{1}{2}(x - 1)^3 + 1$.

34. Draw the graph of $y = \sqrt{x}$ and from it obtain the graph of $y = \sqrt{1 - x}$.

B

In Problems 35–38 exhibit two functions defined by the given equation, and graph each function.

35. $x^2 + y^2 = 4$

36. $y^2 = 1 - x$

37. $x = \sqrt{1 + y^2}$

38. $|x| + |y| = 1$

  Hint. Consider $y \geq 0$ and $y < 0$ separately.

39. The graph of a function f is given. From it obtain the graph of the function g, defined by $g(x) = 3 - 2f(\frac{1}{2}x - 1)$.

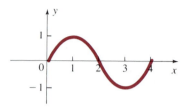

40. Graph the function f defined by

$$f(x) = \begin{cases} x^3 - 1 & \text{if} & x > 1 \\ 1 - x & \text{if} & 0 < x \leq 1 \\ 2 & \text{if} & x = 0 \\ 1 - x^2 & \text{if} & x < 0 \end{cases}$$

41. A model for the operation of a hydroelectric plant determines the function f below, which gives the amount of water that should be allowed to pass through the turbines if the amount of water in the reservoir is x. It is assumed that a steam-generating plant supplying k units of energy is available to supplement the hydroelectric plant. If the total power required in a given period of time is A, then there are water levels a and b so that

$$f(x) = \begin{cases} x & \text{if} & 0 \leq x < A - k \\ A - k & \text{if} & A - k \leq x < a \\ \frac{k}{b - a}(x - b) + A & \text{if} & a \leq x < b \\ A & \text{if} & x \geq b \end{cases}$$

Given that $A = 100$, $k = 70$, $a = 42$, and $b = 74$, graph the function f.

C (Graphing calculator)

42. Use your graphing calculator to check the graphs you obtained in Problems 1–18 and 33–38.

COMBINATIONS OF FUNCTIONS

Functions are often combined in various ways to form new functions. The following definition gives four ways of doing this.

DEFINITION 4   **Let f and g be any two functions, and let D be the intersection of their domains. Assume D is nonempty. Then we define the functions $f + g$, $f - g$, $f \cdot g$, and f/g as follows:**

$$(f + g)(x) = f(x) + g(x) \qquad (f \cdot g)(x) = f(x)g(x)$$

$$(f - g)(x) = f(x) - g(x) \qquad \left(\frac{f}{g}\right)(x) = \frac{f(x)}{g(x)} \quad \text{if} \quad g(x) \neq 0$$

The domain of $f + g$, $f - g$, and $f \cdot g$ is all of D. The domain of f/g is the set of points in D for which $g(x) \neq 0$.

EXAMPLE 23 ||||||||||||||||||||||||||||||||||||||||||||||||||||||||||||||||||||||||||||||||||||||||||||||||||||||||||||||||

Let $f(x) = \sqrt{1 - x}$ and $g(x) = \sqrt{1 + x}$. Find $f + g$, $f - g$, $f \cdot g$, and f/g, and give the domain of each.

Solution   The domain of f is the interval $(-\infty, 1]$, and the domain of g is the interval $[-1, \infty)$. Their intersection is the interval $[-1, 1]$. This is the set D in Definition 4. Using that definition, we have

$$(f + g)(x) = \sqrt{1 - x} + \sqrt{1 + x}, \qquad x \in [-1, 1]$$
$$(f - g)(x) = \sqrt{1 - x} - \sqrt{1 + x}, \qquad x \in [-1, 1]$$
$$(f \cdot g)(x) = \sqrt{1 - x} \cdot \sqrt{1 + x} = \sqrt{1 - x^2}, \qquad x \in [-1, 1]$$
$$\left(\frac{f}{g}\right)(x) = \frac{\sqrt{1 - x}}{\sqrt{1 + x}} = \sqrt{\frac{1 - x}{1 + x}}, \qquad x \in (-1, 1]$$

Note that $x = -1$ is excluded from the domain of f/g. ▲

There is another very important way of combining functions in which the result of applying one function is input into another function. For example, suppose $f(x) = x + 1$ and $g(x) = 3x$. If we first apply f to x, getting $x + 1$, and then apply g to this result, we get $g(f(x)) = 3(x + 1) = 3x + 3$. This two-stage process can be illustrated as follows:

$$x \overset{f}{\to} x + 1 \overset{g}{\to} 3(x + 1) = 3x + 3$$

A function whose value is obtained by combining two other functions in this way is called a **composite** function. This is made precise in the definition below.

DEFINITION 5   **Let f and g be functions. If $f(x)$ is in the domain of g, the function $g \circ f$, called the composition of g with f, is defined by**

$$(g \circ f)(x) = g(f(x))$$

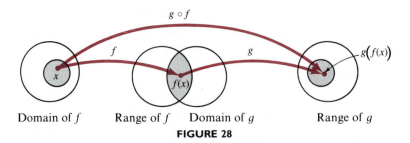

Domain of f Range of f Domain of g Range of g

FIGURE 28

Note that $g \circ f$ is defined when, and only when, $f(x)$ is in the domain of g. So **the domain of $g \circ f$ is the set of all x in the domain of f for which $f(x)$ is in the domain of g.**

In Figure 28 we show a diagram of the actions of two functions f and g, together with the composition $g \circ f$.

If we reverse the roles of f and g in the definition, we get the composite function $f \circ g$. With our same example, where $f(x) = x + 1$ and $g(x) = 3x$, we have

$$(f \circ g)(x) = f(g(x)) = f(3x) = 3x + 1$$

Since we found that for these functions, $(g \circ f)(x) = 3x + 3$, this shows that in general

$$f \circ g \neq g \circ f$$

In this example both f and g have all of R as their domain, and so both $f \circ g$ and $g \circ f$ also are defined for all of R. In the next example this is not the case.

Going back to the machine analogy for functions, the composition function $g \circ f$ can be described as indicated in Figure 29. An input value x is fed into the f machine, which operates on it, producing $f(x)$. Then $f(x)$ is fed into the g machine, producing the final output, $g(f(x))$. If these two machines are enclosed in a large container, we can think of the unit as one big new machine, namely the $g \circ f$ machine. All we see is the input value x and the output value $g(f(x))$, which is $(g \circ f)(x)$.

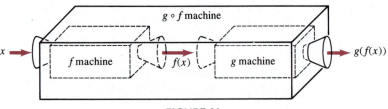

FIGURE 29

EXAMPLE 24

Let $f(x) = \sqrt{x - 4}$ and $g(x) = x^2$. Find $g \circ f$ and $f \circ g$, and give the domain of each.

Solution   For $g \circ f$, we have

$$(g \circ f)(x) = g(f(x)) \qquad \text{by Definition 5}$$
$$= g(\sqrt{x - 4}) \qquad \text{since } f(x) = \sqrt{x - 4}$$
$$= (\sqrt{x - 4})^2 \qquad \text{since } g(x) = x^2$$
$$= x - 4$$

and for $f \circ g$,

$$(f \circ g)(x) = f(g(x)) \qquad \text{by Definition 5}$$
$$= f(x^2) \qquad \text{since } g(x) = x^2$$
$$= \sqrt{x^2 - 4} \qquad \text{since } f(x) = \sqrt{x - 4}$$

The domain of f is the set $\{x: \ x \geq 4\}$, and for all such x, $f(x)$ is in the domain of g, so this is also the domain of $g \circ f$. The domain of g is all of R, but only those values x for which $x^2 \geq 4$ are in the domain of f. Thus, the domain of $f \circ g$ is $\{x: \ |x| \geq 2\} = (-\infty, -2] \cup [2, \infty)$. ▲

Sometimes it is important to write a given function as the composition of two simpler functions. The next example illustrates this.

EXAMPLE 25

Let $F(x) = (2x + 3)^5$. Find two functions f and g such that $F = g \circ f$.

Solution   This can be done in more than one way, but a natural way is to let

$$f(x) = 2x + 3 \qquad \text{and} \qquad g(x) = x^5$$

Then

$$(g \circ f)(x) = g(f(x)) = (2x + 3)^5 = F(x) \qquad ▲$$

 EXERCISE SET 4 $1-9 \qquad 11-20$

A

In Problems 1–10 find $f + g$, $f - g$, $f \cdot g$, and f/g.

1. $f(x) = 3x - 5; \quad g(x) = 2x + 3$

2. $f(x) = 1 - x; \quad g(x) = 1 + x$

3. $f(x) = \dfrac{x - 1}{2}; \quad g(x) = \dfrac{x + 1}{4}$

4. $f(x) = \dfrac{1}{x - 1}; \quad g(x) = \dfrac{1}{x + 1}$

5. $f(x) = \sqrt{x + 4};\quad g(x) = \sqrt{4 - x}$

6. $f(x) = x;\quad g(x) = \sqrt{x - 1}$

7. $f(x) = \dfrac{1}{x - 1};\quad g(x) = \dfrac{x}{x + 2}$

8. $f(x) = \dfrac{1}{x};\quad g(x) = \dfrac{x}{x - 3}$

9. $f(x) = \begin{cases} -1 & \text{if } x < 0 \\ 1 & \text{if } x \geq 0 \end{cases}$;

$g(x) = \begin{cases} 1 & \text{if } x < 0 \\ -1 & \text{if } x \geq 0 \end{cases}$

10. $f(x) = \begin{cases} x & \text{if } x \geq 0 \\ 0 & \text{if } x < 0 \end{cases}$;

$g(x) = \begin{cases} 0 & \text{if } x \geq 0 \\ -x & \text{if } x < 0 \end{cases}$

In Problems 11–20 find $f \circ g$ and $g \circ f$, and give the domain of each.

11. $f(x) = 2x - 1,\quad g(x) = x + 2$

12. $f(x) = 3x,\quad g(x) = x - 1$

13. $f(x) = x^2 + 1,\quad g(x) = 2x - 3$

14. $f(x) = x^2,\quad g(x) = x^3$

15. $f(x) = \dfrac{1}{x - 2},\quad g(x) = \dfrac{1}{x + 3}$

16. $f(x) = \dfrac{x}{x - 1},\quad g(x) = \dfrac{3}{x}$

17. $f(x) = \sqrt{x - 1},\quad g(x) = x^2 - 3$

18. $f(x) = x^3 - 8,\quad g(x) = 2$

19. $f(x) = \sqrt{4 - x},\quad g(x) = x^2$

20. $f(x) = x^2 - 9,\quad g(x) = \dfrac{1}{\sqrt{x}}$

In Problems 21–26 find functions f and g such that $F = g \circ f$.

21. $F(x) = (4 - 3x^2)^6$

22. $F(x) = \dfrac{2}{(x - 3)^5}$

23. $F(x) = \sqrt{1 - x^3}$

24. $F(x) = 3(2x^2 - 1)^{-3}$

25. $F(x) = \left(\dfrac{x}{x + 2}\right)^{2/3}$

26. $F(x) = \sqrt{(x^2 - 1)^3}$

27. What conclusion can you draw with respect to being even or odd for $f + g$, $f - g$, $f \cdot g$, and f/g if
   a. f and g are both even?

b. f and g are both odd?
c. One is even and the other is odd?

B

28. Let $f(x) = |x|$ and $g(x) = |x - 1|$. Find $f + g$ and $f \cdot g$, and draw the graph of each.

29. Let $f(x) = x^2 - 1$, $g(x) = 2x + 3$, and $h(x) = 1/(x + 1)$. Find

$$[f \circ (g \circ h)](x) \quad \text{and} \quad [(f \circ g) \circ h](x).$$

In general, if f, g, and h are arbitrary functions (with suitably restricted domains), what do you conjecture about $f \circ (g \circ h)$ and $(f \circ g) \circ h$?

30. Let $F(x) = \sqrt{1 - \sqrt{1 - x}}$. Find functions f, g, and h such that $F = f \circ (g \circ h)$. Show that for these same functions $F = (f \circ g) \circ h$. What is the domain of F?

31. If a satellite of mass m is at a distance $s(t)$ from the center of the earth at time t, the force of gravity exerted by the earth on the satellite is $F(t) = (h \circ s)(t)$, where $h(x) = gR^2m/x^2$, g is the acceleration due to gravity on the earth's surface, and R is the radius of the earth.
   a. Write an explicit expression for $F(t)$ if $s(t) = -\frac{1}{2}gt^2 + v_0t + s_0$, where v_0 is the initial vertical velocity and s_0 the initial distance from the earth's center.
   b. Suppose the satellite has a mass of 2000 kilograms and is lifted from the earth's surface at an initial velocity of 4000 meters per second. Find the force of the earth's gravity on it after 10 seconds. Take $R = 6.37 \times 10^6$ meters and $g = 9.80$ meters/sec^2. The answer will be in **newtons,** the unit of force in the **mks (meter-kilogram-second) systems.** (Use a calculator.)

32. Prove or disprove:
   a. $f \circ (g + h) = f \circ g + f \circ h$
   b. $(g + h) \circ f = g \circ f + h \circ f$

33. Let $f(x) = \dfrac{1}{x}$. Prove that for functions g and h,

$$f \circ \left(\frac{g}{h}\right) = \frac{f \circ g}{f \circ h}$$

What restrictions must be placed on the domains of g and h?

34. Let $f(x) = x^{2/3}$ and $g(x) = 4x^2 - 9$. Use a calculator to find each of the following correct to five significant figures.

a. $(f + g)(x)$ for $x = 3.0257$
b. $(f \cdot g)(x)$ for $x = 0.023768$

c. $\left(\dfrac{f}{g}\right)(x)$ for $x = 213.82$

d. $(f \circ g)(x)$ for $x = 1.3956$
e. $(g \circ f)(x)$ for $x = -17.248$

35. Let

$$f(x) = \begin{cases} 2x & \text{if } \ 0 \leq x < 3 \\ x^2 + 1 & \text{if } \qquad x \geq 3 \end{cases}$$

Find $(f \circ f)(x)$.

5 ONE-TO-ONE FUNCTIONS AND INVERSES

As we saw in Section 3, the graph of a function has the property that a vertical line intersects the curve in at most one point. It is entirely possible, however, that a horizontal line might intersect the graph in two or more points. Consider, for example, the graph of the function f defined by $f(x) = x^2$, shown in Figure 30. In this case any horizontal line above the x axis intersects the curve in two points. So there are distinct points in the domain having the same function value. For example, $f(2) = f(-2)$. More generally, when a horizontal line intersects the graph of a function in two or more points, we can always find two numbers x_1 and x_2 in the domain for which $f(x_1) = f(x_2)$ but $x_1 \neq x_2$. In this section we study an important class of functions for which this situation cannot occur. These are called *one-to-one* functions (written 1–1).

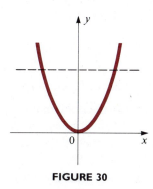

FIGURE 30

DEFINITION 6   **A function f is said to be one-to-one provided that whenever x_1 and x_2 are in the domain of f and $f(x_1) = f(x_2)$, then $x_1 = x_2$.**

Note. An equivalent formulation is to say that f is one-to-one if two distinct points x_1 and x_2 in the domain of f have distinct images, $f(x_1) \neq f(x_2)$. Geometrically, this definition requires that each horizontal line intersect the graph of f in at most one point.

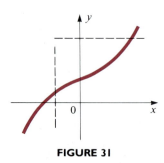

FIGURE 31

Figure 31 illustrates a 1–1 function. We conclude that for a graph to be that of a 1–1 function, it must pass both the vertical and horizontal line tests: *Every vertical line and every horizontal line must intersect the graph in at most one point.*

We cannot rely solely on the graphical method for determining whether a function is 1–1, since the graph may be difficult to draw with accuracy. The next three examples show how to use the definition in making this determination.

EXAMPLE 26

Let $f(x) = 3x + 4$. Show that f is 1–1.

Solution   Suppose $f(x_1) = f(x_2)$, that is, $3x_1 + 4 = 3x_2 + 4$. Then $3x_1 = 3x_2$, and hence $x_1 = x_2$. So f is 1–1.   ▲

EXAMPLE 27

Show that the function g defined by $g(x) = \sqrt{x}$ is 1–1 on its domain.

Solution   Note that the domain consists of all nonnegative real numbers, that is, the set $\{x: \;\; x \geq 0\}$. Suppose $g(x_1) = g(x_2)$, that is, $\sqrt{x_1} = \sqrt{x_2}$. Squaring both sides yields immediately that $x_1 = x_2$. Thus, g is 1–1.   ▲

The next example illustrates how we can sometimes restrict the domain of a function to make it 1–1.

EXAMPLE 28

Let $h(x) = 4 - x^2$. Show that h is not 1–1, but find a suitable restriction on the domain so that it will be 1–1.

Solution   To show that h is not 1–1 we can take any value of x (other than zero) and its negative—for example, $h(1) = 3$ and $h(-1) = 3$. Since $1 \neq -1$, the definition is not satisfied. From Figure 32 we can see that the graph fails the horizontal line test.

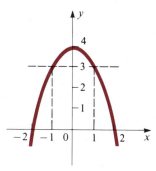

FIGURE 32

If we restrict the domain to nonnegative values of x only, that is, to the set $\{x: \quad x \geq 0\}$, it appears from the graph that the function is then 1–1. In fact if x_1 and x_2 are nonnegative and $h(x_1) = h(x_2)$, we have

$$4 - x_1^2 = 4 - x_2^2$$
$$x_1^2 = x_2^2$$
$$x_1 = x_2 \quad \text{or} \quad x_1 = -x_2$$

but $x_1 \neq -x_2$ since neither x_1 nor x_2 is negative. So by Definition 6, h is 1–1 on this restricted domain. ▲

We now wish to introduce the concept of the **inverse** of a function. We will see that this has a direct relationship with both 1–1 functions and composite functions. Roughly speaking, two functions are inverses of one another if each undoes what the other does. Thus, if $f(x) = 2x$ and $g(x) = x/2$, then f and g are inverses; f doubles any given number in its domain, whereas g halves numbers in its domain. So the effect of first applying f and then g, or vice versa, to a given number x is that there is no net change. This is typical of functions and their inverses.

DEFINITION 7   **Let f be a function with domain A and range B. If there exists a function g with domain B and range A such that**

$$g(f(x)) = x \quad \text{for all} \quad x \in A$$

and

$$f(g(x)) = x \quad \text{for all} \quad x \in B$$

then g is said to be the inverse of f.

Remark. If g is the inverse of f, we can reverse the roles of g and f in the definition to see that f is also the inverse of g. So we can say that f and g are inverses of one another.

Instead of using some other letter, such as g, to designate the inverse of a function f, it is customary to use the symbol f^{-1} for this purpose. This is read "f inverse." By the remark above, we have $(f^{-1})^{-1} = f$ (that is, the inverse of f inverse is f). Note carefully that the -1 appearing as a superscript in the notation f^{-1} is *not* an exponent. So f^{-1} *does not mean* $1/f$.

If f has an inverse, then we have from Definition 7, writing f^{-1} in place of g,

$$f^{-1}(f(x)) = x \quad \text{for all} \quad x \in A$$
$$f(f^{-1}(x)) = x \quad \text{for all} \quad x \in B$$

where, as before, A denotes the domain of f and B denotes its range. Thus, each of the composite functions $f^{-1} \circ f$ and $f \circ f^{-1}$ maps an element of its domain onto itself, and for this reason each of these functions is called an **identity function.**

If we let $y = f(x)$, then we have $f^{-1}(y) = f^{-1}(f(x)) = x$. These relationships are illustrated in Figure 33.

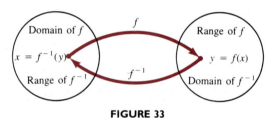

FIGURE 33

For the inverse of a function f to exist, it must be true that each element of the range of f is the image of precisely one element in the domain. If we begin with an x in the domain of f, its image is $f(x)$. Now if f^{-1} exists, it must make $f(x)$ correspond to x and to nothing else:

$$f^{-1}(f(x)) = x$$

But the requirement that each element in the range of f be the image of exactly one element in the domain is just the requirement that f be one-to-one.

Suppose now that f is 1–1. We will show that it has an inverse. All we need to do is to define f^{-1} as that function having domain equal to the range of f and for which

$$f^{-1}(y) = x$$

where $y = f(x)$. This is a valid definition, since each such y uniquely determines an x.

So we have the following result:

A function has an inverse if and only if it is 1–1.

Suppose we determine that a given function does have an inverse. How do we go about actually finding the inverse? Let

$$y = f(x)$$

Then, as we have seen,

$$x = f^{-1}(y)$$

It appears, then, that to find an equation defining f^{-1}, we solve the equation $y = f(x)$ for x in terms of y. That is, we reverse the roles of dependent and independent variables. Consider, for example, the problem of finding the inverse of the function f defined by $f(x) = 2x + 3$. We can show that f is 1–1, so it has an inverse. Now set $y = 2x + 3$ and solve for x:

$$2x = y - 3$$

$$x = \frac{y - 3}{2}$$

So $f^{-1}(y) = (y - 3)/2$. This adequately defines the function f^{-1}; it is the function that subtracts 3 from the input value and then divides the result by 2. The particular letter used in the defining equation for a function is not important. We could just as well write

$$f^{-1}(t) = \frac{t - 3}{2} \quad \text{or} \quad f^{-1}(u) = \frac{u - 3}{2} \quad \text{or} \quad f^{-1}(x) = \frac{x - 3}{2}$$

Each of these in effect tells us what the function f^{-1} does. Since it is customary to use the letter x as the independent and y as the dependent variable in a functional relationship, this suggests the following procedure for finding the inverse of a 1–1 function f:

1. Set $y = f(x)$.
2. Solve the equation for x in terms of y (if possible).
3. Interchange x and y.

The resulting equation gives $y = f^{-1}(x)$.

Again, consider the example $f(x) = 2x + 3$. We write

$$y = 2x + 3$$

$$x = \frac{y - 3}{2}$$

So, on interchanging x and y, we get

$$y = \frac{x - 3}{2} = f^{-1}(x)$$

Two points should be emphasized regarding the procedure given above for finding the inverse. First, before attempting to apply the procedure it should be determined that the inverse actually exists. However, if this is not done in advance, the procedure itself will generally show this. For example, consider $f(x) = x^2 + 1$. We set $y = x^2 + 1$, solve for x:

$$x = \pm\sqrt{y - 1}$$

and interchange x and y:

$$y = \pm\sqrt{x - 1}$$

But the ambiguity of sign shows that no unique inverse exists.

Second, a major problem with the procedure is that it is not always possible to solve for x. For example, try to solve $y = x^5 + x - 1$ for x!

EXAMPLE 29 ||||||||||||||||||||||||||||||||||||||||||||||||||||||||||||||||||||||||||||||||||||||||||||||||||||||

Show that if $f(x) = \sqrt{x - 1}$, then f has an inverse on the domain $\{x: \;\; x \geq 1\}$, and find that inverse.

Solution   We show first that f is 1–1. Suppose $f(x_1) = f(x_2)$, so that $\sqrt{x_1 - 1} = \sqrt{x_2 - 1}$. It follows by squaring both sides that $x_1 - 1 = x_2 - 1$, or $x_1 = x_2$. Thus, f is 1–1 and hence has an inverse. Now set

$$y = \sqrt{x - 1}$$

Squaring and solving for x yields

$$y^2 = x - 1$$
$$x = y^2 + 1$$

Finally, the rule for f^{-1} is

$$y = x^2 + 1$$

that is,

$$f^{-1}(x) = x^2 + 1$$

But we must carefully state the domain of f^{-1}. It is precisely the range of f. Since $f(x) = \sqrt{x - 1}$, the range is contained in the set of nonnegative real numbers; in fact, this is the range of f. If we take any nonnegative real number k, we can find an x for which $f(x) = k$ as follows:

$$\sqrt{x - 1} = k$$
$$x - 1 = k^2$$
$$x = 1 + k^2$$

This value of x does work, since

$$f(x) = f(1 + k^2) = \sqrt{1 + k^2 - 1} = \sqrt{k^2} = |k| = k$$

since $k \geq 0$. Therefore, the domain of f^{-1} is the set of all nonnegative real numbers.   ▲

We can now see the geometrical relationship between a function and its inverse. Let f be a 1–1 function from A onto B, where A and B are subsets of R. In order to obtain the equation $y = f^{-1}(x)$, we solve the equation $y = f(x)$ for x and then interchange the roles of x and y. It follows that if (a, b) is a point on the graph of $y = f(x)$, then (b, a) is on the graph of $y = f^{-1}(x)$. The situation is shown in Figure 34. The graph of $y = f^{-1}(x)$ is the reflection of the graph of $y = f(x)$ in the line through the origin making a 45° angle with the x axis. On this line the x coordinate and the y coordinate of any point are equal, and this holds true only for points on this line. Thus, the equation of the line is $y = x$. We say that the graphs of a 1–1 function and its inverse are *symmetric with respect to the line* $y = x$.

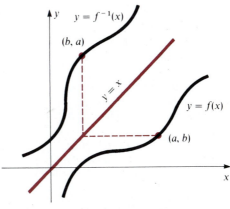

FIGURE 34

EXERCISE SET 5

A

1. Which of the graphs in Problem 19 of Exercise Set 3 are graphs of 1–1 functions?
2. Which of the graphs in Problem 20 of Exercise Set 3 are graphs of 1–1 functions?

In Problems 3–14 use Definition 6 to show that the functions are 1–1 on R.

3. $f(x) = 3x$
4. $f(x) = x - 3$
5. $g(x) = 2x - 3$
6. $h(x) = 4 - 2x$
7. $f(t) = 7t - 9$
8. $F(t) = 5t + 3$
9. $\phi(x) = \dfrac{2x}{3} - \dfrac{3}{4}$
10. $g(x) = x^3 - 1$
11. $f(x) = \frac{3}{5}(2 - 3x)$
12. $h(x) = \dfrac{3 - 5x}{4}$

13. $g(t) = \begin{cases} \dfrac{1}{t} & \text{if } t \neq 0 \\ 0 & \text{if } t = 0 \end{cases}$

14. $f(x) = \begin{cases} 1 + \sqrt{2x} & \text{if } x \geq 0 \\ x & \text{if } x < 0 \end{cases}$

In Problems 15–20 give the domain and range of f, and draw its graph. Determine from the graph whether f is 1–1 on its domain.

15. $f(x) = \dfrac{2}{x - 1}$
16. $f(x) = \sqrt{x - 2}$
17. $f(x) = 2$
18. $f(x) = |x - 3|$
19. $f(x) = \dfrac{x}{|x|}$

20. $f(x) = \begin{cases} x^2 & \text{if } x \geq 0 \\ x^3 & \text{if } x < 0 \end{cases}$

In Problems 21–30 show that the given function has an inverse, and find it.

21. $f(x) = x + 2$

22. $f(x) = 3x - 5$

23. $g(x) = \dfrac{5x - 2}{3}$

24. $F(x) = \dfrac{4 - 7x}{2}$

25. $h(x) = \dfrac{1}{x - 2}$

26. $f(t) = \dfrac{t}{t + 1}$

27. $G(x) = \sqrt{x - 4}$

28. $h(x) = \dfrac{1}{\sqrt{x}}$

29. $F(t) = t^2 - 1, \quad t \geq 0$

30. $G(z) = \sqrt{z^2 - 1}, \quad z \geq 1$

31. Let $f(t) = (3t - 2)/(2t + 3)$. Give the domain of f, and show that f is 1–1. Find f^{-1} and give its domain.

32. Determine if ϕ^{-1} exists for the function ϕ defined by $\phi(x) = x^2 + 2x - 4, \, x \in R$.

33. Let $g(x) = x/(2x - 1)$. Find g^{-1}, or show that it does not exist.

34. Let $h(t) = 3 + \sqrt{1 - t}$. Give the domain and range of h, and show that h is 1–1. Find h^{-1}. What is its domain?

In Problems 35–40 show that each function has an inverse. Without finding the inverse, graph each function and its inverse on the same set of axes.

35. $f(x) = 3x - 2$

36. $f(x) = \dfrac{2x - 3}{5}$

37. $g(x) = x^2 + 1, \quad x \geq 0$

38. $h(x) = \sqrt{x - 1}$

39. $F(x) = x^3$

40. $G(x) = 1 - |x|, \quad x \leq 0$

In each part of Problems 41 and 42 the graph of a 1–1 function is given. On the same set of axes sketch the graph of its inverse.

41. a.

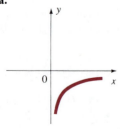

b.

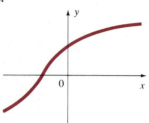

42. a.

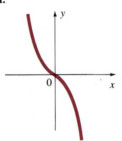

b.

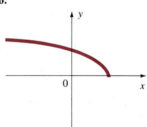

43. Give three distinct values of x that have the same image for the function f defined by $f(x) = 4x^3 - 8x^2 - 9x + 18$. What conclusions can you draw?

44. Let $f(x) = 2x - x^2$. Show that f is not 1–1 on R but that it is on the domain $[1, \infty)$. Find f^{-1} for f on this restricted domain.

45. Show that no even function is 1–1. Show by means of examples that odd functions may or may not be 1–1.

46. Show that if f is either increasing on its domain or decreasing on its domain, then f is 1–1.

47. A function f is said to be **periodic** if there exists a positive real number k such that $f(x + k) = f(x)$

for all x in the domain of f. Show that a periodic function cannot be 1–1.

48. To show that f is a function from A onto B, it is sufficient to show that whatever element y in B is chosen, there is some element x in A for which $f(x) = y$. Show that the function f for which $f(x) = \sqrt[3]{x}$ is a 1–1 function from R onto R.

49. Show that if $g(x) = (3 - 2x)/4$, then g is a 1–1 function from R onto R.

50. Let $f(x) = \sqrt{(x + 1)/(x - 1)}$.
   a. Show that f is 1–1 on its domain.
   b. The set of images of f is contained in the set $\{y:\ y \geq 0\}$. Is f a 1–1 function from its domain onto this set?

51. The notation $[x]$ is defined to mean the greatest integer less than or equal to x. For example, $[2.3] = 2$, $[5\frac{3}{4}] = 5$, $[6] = 6$, and $[-1.5] = -2$. Show that the function f defined by $f(x) = [x]$ from R to R is neither 1–1 nor onto.

In Problems 52–55 show that the function is not 1–1 on its full domain. Then find a suitable restriction on its domain so that it is 1–1, and find the inverse of the restricted function.

52. $f(x) = \dfrac{1}{x^2 + 2}$ **53.** $g(x) = \dfrac{1}{\sqrt{4 - x^2}}$

54. $h(x) = x^2 - 4x + 4$

55. $F(x) = |3 - 2x|$

56. Let $f(x) = (x + a)/(x + b)$, where $a \neq b$. Show that f^{-1} exists, and find it. Give the domain and range of both f and f^{-1}. Verify that $(f^{-1} \circ f)(x) = x$ for all x in the domain of f and that $(f \circ f^{-1})(x) = x$ for all x in the domain of f^{-1}.

57. Prove that if a function has an inverse, the inverse is unique.

Hint. Suppose g and h are both inverses of f. Use the properties of inverses to prove that $g = h$.

VARIATION

The following experiment demonstrates a recurrent type of functional relationship in physical science, as well as in the social and life sciences: An experiment in physics consists of attaching various weights to a spring and measuring the amount by which the spring is stretched; this is called the **elongation.** The object of the experiment is to determine a functional relationship between elongation and weight. Suppose that for a certain spring the measurements in Table 1 were found. From these data we would expect the pattern of each additional pound stretching the spring another 0.5 inch to continue, so long as the elastic limit of

Weight (pounds)	Elongation (inches)
1	0.5
2	1.0
3	1.5
4	2.0

TABLE I

the spring is not exceeded. (At this limit the spring is stretched so much that it will not return to its original shape.)

Let us designate the weight by w and the elongation by x. The functional relationship between w and x can be determined if we extend our table to include

w	x	w/x
1	0.5	2
2	1.0	2
3	1.5	2
4	2.0	2

TABLE 2

the ratio of w to x, as shown in Table 2. Since the ratio is constant for these four values, we assume it remains constant, so that for all x less than the elastic limit of the spring, $w/x = 2$, or $w = 2x$. This is an example of what is called **direct variation,** or **direct proportion.** The fact that w/x is constant is an expression of what is known as **Hooke's law.** The particular constant depends on the spring and is called the **spring constant.**

In other types of situations we may find that instead of the ratio of two quantities being constant, it is their product that is constant, and this leads to what is known as **inverse variation,** or **inverse proportion.** The following experiment in chemistry illustrates this: The experiment consists of measuring the pressure (p) of a gas occupying varying volumes (v) while held at constant temperature. The data are given in Table 3. Note that as the volume decreases, the pressure increases. It appears that for this gas at the given constant temperature, $pv = 6$, or $p = 6/v$.

Pressure, p (millimeters of mercury)	Volume, v (cubic centimeters)	$p \cdot v$
2	3.00	6
3	2.00	6
5	1.20	6
8	0.75	6

TABLE 3

The expressions *varies directly as* and *varies inversely as* are used to describe such situations as these two experiments illustrate. For example, Hooke's law says that the weight necessary to stretch a given spring (within its elastic limit) *varies directly as* the amount it is stretched beyond its natural length. As we saw, this means that $w = kx$. Also, we can say that for a gas at constant temperature the pressure *varies inversely as* the volume. In biology we know from experiment that within certain limits the rate of growth of a bacteria culture *varies directly as* the number of bacteria present.

These concepts are made more precise by the following definition:

DEFINITION 8   **The statement y varies directly as x means there is a real number $k \neq 0$ such that**

$$y = kx$$

The statement y varies inversely as x means there is a real number $k \neq 0$ such that

$$y = \frac{k}{x}$$

The statement *w* **varies jointly as** *x* **and** *y* **means there is a real number** $k \neq 0$ **such that**

$$w = kxy$$

Instead of saying *y* varies directly as *x* we sometimes say *y* is **directly proportional** to *x*, and this is understood to mean the same thing. Similarly, inverse variation and joint variation can be stated using the language of proportionality. If the word *directly* is omitted, it is understood, but *inversely* and *jointly* must always be explicitly stated. There can be combined statements, such as "*y* varies jointly as *x* and the square of *t* and inversely as v^3," which means

$$y = \frac{kxt^2}{v^3}$$

An example of such combined variation is Newton's law of gravitation, which states that the force of attraction between two bodies *varies jointly* as their masses and *inversely* as the square of the distance between them. Here we would write

$$F = \frac{km_1 m_2}{d^2}$$

The real number *k* in each case is called the **constant of variation,** or the **constant of proportionality.** Often, sufficient information is available to find *k*, in which case a completely defined functional relationship is determined and various values of the dependent variable can be found. The examples below should help to make these ideas clearer.

EXAMPLE 30

Translate each of the following into an appropriate mathematical statement:
a.   The temperature *T* of a gas in a container of fixed volume varies directly as the pressure *p*.
b.   The force *F* necessary to stop a body that is moving in a straight line varies directly as the square of the velocity *v*.
c.   The time *t* required for a satellite traveling in a circular orbit around the earth to complete its orbit is inversely proportional to its velocity *v*.
d.   The drag force *F* on an airplane varies jointly as the total cross-sectional area *A* and the square of the velocity *v*.

Solution   a. $T = kp$ b. $F = kv^2$ c. $t = \dfrac{k}{v}$ d. $F = kAv^2$

EXAMPLE 31

It is known that *y* varies directly as *x* and that when $x = 2$, then $y = 10$. Find *y* when $x = 7$.

Solution   The variational statement translates to $y = kx$. Now we know one pair of values satisfying this equation, namely $x = 2$, $y = 10$. So we substitute these to find k:

$$10 = k \cdot 2$$
$$k = 5$$

Therefore, the specific functional relationship is

$$y = 5x$$

From this we can find y for any other value of x. The problem asks for y when $x = 7$:

$$y = 5(7) = 35 \qquad\blacktriangle$$

Note that one set of values of the variables is sufficient to determine k, because k is constant. Having found it for one set of the variables, it is determined once and for all.

The preceding example is typical of most variation problems. There are four steps to their solution:

1. Translate the variational statement into a mathematical equation by using Definition 7.
2. Substitute the given set of values for all the variables involved, and solve for k.
3. Substitute the value of k into the original equation. This gives an explicit functional relationship.
4. Solve for the unknown requested by substituting the corresponding known value (or values).

EXAMPLE 32

If w varies directly as t^2 and inversely as v, and $w = 15$ when $t = 2$ and $v = 3$, find w when $t = 6$ and $v = 8$.

Solution   We have

$$w = k \cdot \frac{t^2}{v}$$

Now we substitute the given values for w, t, and v, and solve for k.

$$15 = k \frac{(2)^2}{3}$$

$$15 = \frac{4k}{3}$$

$$4k = 45$$
$$k = \tfrac{45}{4}$$

So the original equation becomes

$$w = \frac{45}{4} \cdot \frac{t^2}{v}$$

Substituting $t = 6$ and $v = 8$, we get

$$w = \frac{45}{\underset{1}{\cancel{4}}} \cdot \frac{\overset{9}{\cancel{36}}}{8} = \frac{405}{8}$$

▲

EXAMPLE 33 ||||||||||||||||||||||||||||||||||||||||||||||||||||||||||||||||||||||||||||||||||||||||||||||||

The kinetic energy E of a moving object is jointly proportional to its weight w and the square of its velocity v. If an object weighing 32 pounds and moving with a velocity of 60 feet per second has a kinetic energy of 1,800 foot-pounds, find the kinetic energy of an object weighing 96 pounds and traveling at 20 feet per second.

Solution   The given statement translates to

$$E = kwv^2$$

We find k by substituting the known set of values:

$$1,800 = k \cdot (32)(60)^2$$
$$1,800 = k(32)(3,600)$$
$$k = \tfrac{1}{64}$$

So the equation becomes

$$E = \tfrac{1}{64}wv^2$$

Finally, we substitute $w = 96$ and $v = 20$:

$$E = \tfrac{1}{64} \cdot (96)(20)^2 = \tfrac{3}{2} \cdot 400 = 600 \text{ foot-pounds}$$

▲

EXERCISE SET 6

A

In Problems 1–10 translate the given statement into a mathematical equation.

1. u varies directly as v.
2. w varies inversely as t.

3. z varies jointly as x and y.
4. y is directly proportional to x^2.
5. s is inversely proportional to $\sqrt{t}$.
6. w varies directly as x and inversely as y.
7. F is jointly proportional to m_1 and m_2 and inversely proportional to r^2.

8. The force required to stretch a spring x units beyond its natural length is proportional to x.

9. The distance s traveled by a freely falling object varies directly as the square of the time t it has fallen.

10. The volume V occupied by a gas varies directly as the temperature T and inversely as the pressure P.

11. If y varies directly as x, and $y = 4$ when $x = 12$, then find y when $x = 20$.

12. If s varies inversely as r, and $s = 5$ when $r = 8$, then find s when $r = 3$.

13. If u varies jointly as x and y, and $u = 12$ when $x = 6$ and $y = 8$, then find u when $x = 4$ and $y = 12$.

14. If w varies directly as u and inversely as v, and $w = 6$ when $u = 4$ and $v = 2$, then find w when $u = 10$ and $v = 6$.

15. If x is jointly proportional to y and t^2, and $x = 108$ when $y = 4$ and $t = 3$, then find x when $y = 7$ and $t = 4$.

16. The volume of a sphere varies as the cube of its radius. A sphere of radius 3 has a volume of 36π. Find the constant of variation. Write the general formula for the volume of a sphere. What is the volume of a sphere of radius 5?

17. The weight of a body on or above the surface of the earth varies inversely as the square of the distance from the center of the earth. If a body weighs 50 pounds on the earth's surface, how much would it weigh 1,000 miles above the surface of the earth? (Assume the radius of the earth is 4,000 miles.)

18. A printing company has found that for orders between 5,000 and 50,000 the unit cost of printing a college catalogue is inversely proportional to the number printed. If the cost of printing 10,000 of these catalogues is 80¢ per copy, what is the unit cost of printing 25,600 copies?

19. According to Ohm's law the current I in a wire varies directly as the potential E and inversely as the resistance R of the wire. When $R = 20$ ohms and $E = 220$ volts, it is found that $I = 11$ amperes. Find I when $E = 110$ volts and $R = 5$ ohms.

20. The resistance R of a wire varies directly as its length l and inversely as the square of its diameter d. A certain type of wire 60 feet long and with diameter 0.02 inch has a resistance of 15 ohms. What would be the resistance of 90 feet of the same type of wire with diameter 0.01 inch?

21. The volume of a right circular cone varies jointly as the height h and the square of the base radius r.

A cone of base radius 2 and height 6 has volume 8π. Find the constant of variation, and write the general formula for the volume of any right circular cone. What is the volume of a cone of height 10 and base radius 6?

22. The force of the wind on a wall is jointly proportional to the area of the wall and the square of the wind velocity. When the wind is blowing at 20 miles per hour, the force on a wall having area 100 square feet is 180 pounds. Find the force on a wall of area 400 square feet caused by a wind blowing at 50 miles per hour.

23. Newton's law of cooling states that the rate r at which a body cools is proportional to the difference between the temperature T of the body and the temperature T_0 of the surrounding medium. A thermometer registering 70°F is taken outside where the temperature is 40°F, and at that instant the rate of cooling, r, is 18°F per minute. Find the rate of cooling when the thermometer reads 50°F.

24. The vibrating frequency (pitch) of a string varies directly as the square root of the tension in the string. If when the tension is 4 pounds, the frequency of vibration is 250 times per second, find the frequency of vibration when the tension is 16 pounds.

25. One mathematical model of the spread of an infectious disease in a community says that if x is the number of persons already infected and y is the number not infected, then the rate r at which additional persons become infected is jointly proportional to x and y. In a community of 500 persons, when 20 already have a certain infectious disease, the rate of spread is 2 additional cases per day. What will be the rate of spread when 80 persons in the community have the disease? (This model fails to take into consideration such things as quarantine and the duration of the disease.)

B

26. The intensity of light varies inversely as the square of the distance from the source. The intensity of a certain light is 200 candlepower at a point 8 feet from the source. At what distance from the source will the intensity be 100 candlepower?

27. Kepler's third law states that the square of the time it takes a planet to complete its orbit around the sun varies directly as the cube of its mean distance from the sun. The mean distance of Mars from the sun is approximately $1\frac{1}{2}$ times that of the earth.

Find the approximate time (in "earth days") it takes Mars to complete its orbit.

28. The force of attraction between two bodies of masses m_1 and m_2, respectively, is jointly proportional to m_1 and m_2 and inversely proportional to the square of the distance d between them. What will be the effect on the force if m_1 is doubled, m_2 is tripled, and the distance d is cut in half?

29. The weight that can be safely supported by a wooden beam of rectangular cross-section varies jointly as the width and square of the depth of the cross-section and inversely as the length of the beam. A beam of length 12 feet with cross-section 4 inches wide by 8 inches deep will safely support 2,000 pounds. How much weight can be safely supported by a beam of the same material that is 8 feet long and has cross-section 2 inches wide by 4 inches deep?

30. The pressure P of a gas varies directly as its Kelvin temperature T and inversely as its volume V. Gas at pressure 200 pounds per square inch is at a temperature of 47°C and occupies a volume of 500 cubic feet. If the gas is allowed to expand so that its

pressure is reduced to 20 pounds per square inch and its temperature is 17°C, find the volume it occupies.

Note. **Kelvin temperature** is measured through positive values from absolute 0, which is equivalent to −273°C, and one degree on the Kelvin scale equals one degree on the Celsius scale.

31. The mass of a spherical body varies jointly as its density and the cube of its radius. Find the ratio of the mass of Jupiter to that of the earth if the density of Jupiter is $\frac{5}{22}$ that of the earth and its radius is 11 times that of the earth.

32. The time of exposure necessary to photograph an object varies directly as the square of the distance d of the object from the light source and inversely as the intensity of illumination I. When the light source is 10 feet from the object, the correct exposure time for a certain type of film is $\frac{1}{30}$ second. For the same film, if the intensity of the light is doubled and the distance from the light to the object is cut in half (to 5 feet), what will be the correct exposure time?

REVIEW EXERCISE SET

A

In Problems 1–5 make a table of values and draw the graph.

1. **a.** $y = 2x - 1$ **b.** $2x + 3y = 4$
2. **a.** $x - 2y + 3 = 0$ **b.** $y - 3 = 0$
3. **a.** $5x - 2y = 4$ **b.** $x + 2 = 0$
4. **a.** $\dfrac{x + 2y}{4} = 3$ **b.** $\dfrac{y - 3}{2} = x$
5. **a.** $f(x) = 1 - x^2$ **b.** $g(x) = 1 - \sqrt{1 - x}$
6. Let $f(x) = x/(x - 2)$. Find:
    a. $f(1)$ **b.** $f(0)$ **c.** $f(3)$
    d. $f(-5)$ **e.** $f(2.1)$
7. Let $g(t) = 2t/(t - 3)$. What is the domain of g? Find:
    a. $g(1)$ **b.** $g(-2)$ **c.** $g(1/t)$ **d.** $g(t + \Delta t)$
8. Let $g(x) = \sqrt{9 - x^2}$. What is the domain of g? Find:
    a. $g(0)$ **b.** $g(-3)$ **c.** $g(2)$
    d. $g(\frac{12}{5})$ **e.** $g(-\frac{9}{5})$

9. Let
$$h(t) = \begin{cases} 2t^2 - 1 & \text{if} & t \geq 2 \\ 4 - t & \text{if} & 0 \leq t < 2 \\ 0 & \text{if} & t < 0 \end{cases}$$
Find:
    a. $h(5)$ **b.** $h(-1)$ **c.** $h(0)$
    d. $h(2)$ **e.** $h(1)$
10. Let $f(t) = t - (1/t)$. Find:
    a. $f(1)$ **b.** $f\left(-\dfrac{1}{2}\right)$ **c.** $f\left(\dfrac{1}{t}\right)$ **d.** $\dfrac{1}{f(t)}$
11. Let $f(x) = 1/x^2$. Find $[f(x) - f(2)]/(x - 2)$ and simplify your result.
12. Let $f(x) = x/(x + 1)$. Find $[f(x) - f(-2)]/(x + 2)$ and simplify your result.
13. Let $F(x) = \sqrt{x - 1}$. Show that
$$\frac{F(x + h) - F(x)}{h} = \frac{1}{\sqrt{x + h - 1} + \sqrt{x - 1}}$$

14. Find the domain of each of the following:

a. $f(x) = \sqrt{2x^2 - 3x - 5}$

b. $g(x) = \sqrt{\dfrac{x + 1}{x - 3}}$

15. Express the length of the diagonal of a square as a function of one of its sides s.

16. Express the diameter of a circle as a function of its area.

17. The sum of \$5,000 is invested in an account yielding 6% interest compounded annually. Express the total amount accumulated at the end of t years as a function of t.

18. Express the altitude h of an equilateral triangle as a function of one of its sides s.

19. The costs of renting a truck for 1 day are: \$25 rental fee, 20¢ for each mile driven, \$5.50 for insurance, 10¢ for each mile driven for gasoline. Let $C(x)$ be the total cost for driving x miles in a day. Write the rule for $C(x)$.

In Problems 20 and 21 show that each function is 1–1.

20. a. $f(x) = 3x - 7$ **b.** $f(x) = \dfrac{2x + 7}{3}$

21. a. $g(x) = 8 - 5x$ **b.** $h(x) = \dfrac{x + 1}{x}$

22. Show that the function f defined by $f(x) = 4 - x^2$ is not 1–1 on R. Find a suitable restriction on the domain of f so that it will be 1–1.

23. Determine which of the following functions are even, which are odd, and which are neither even nor odd.

a. $f(x) = x^4 - 2x^2 + 1$

b. $g(x) = x(x^2 - 1)$ **c.** $h(t) = t^3 + 2t - 4$

d. $F(x) = x - \dfrac{1}{x}$ **e.** $G(x) = \sqrt{\dfrac{4}{1 - x^2}}$

24. Show that the function $f(x) = |x - 1|$ is not 1–1 on R. Find a suitable restriction on the domain of f so that it will be 1–1.

25. Let $\phi(x) = (2x - 3)/5$ and $\psi(x) = (4 - 2x)/7$. Find $\phi \circ \psi$ and $\psi \circ \phi$.

26. Let $f(x) = x^2$ and $g(x) = \sqrt{x - 4}$. Find $f \circ g$ and $g \circ f$, and determine the domain of each.

27. Let $F(x) = (3 - 4x)^{-3/2}$. Find functions f and g such that $F = f \circ g$.

28. Let $G(t) = \sqrt[3]{(1 - t^2)^2}$. Find functions g and h such that $G = g \circ h$.

29. At temperature x the thermal conductivity of a wall is given by $f(x) = k(1 + ax)$, where k and a are constants (k is the thermal conductivity at temperature $0°$C). Write the rule for the thermal resistance $g \circ f$, given that $g(x) = T/(Ax)$, where T is the thickness of the wall and A is its cross-sectional area.

In Problems 30–37 show that an inverse exists, and find it.

30. $f(x) = 4x - 5$

31. $f(x) = \dfrac{5x + 7}{3}$

32. $g(x) = 1 - \dfrac{2x}{3}$

33. $h(t) = \dfrac{2 - 3t}{5}$

34. $F(z) = 1 - \dfrac{2}{z}$

35. $g(x) = \dfrac{1}{\sqrt{x - 2}}$

36. $f(t) = \dfrac{1}{\sqrt{t}}$

37. $h(x) = \dfrac{x + 2}{x - 3}$

In Problems 38 and 39 determine which graphs are graphs of functions and which are not. For those that are graphs of functions, state whether an inverse exists.

38. a.

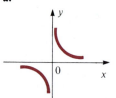

b.

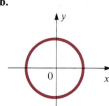

c.

d.

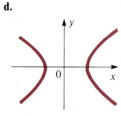

e.

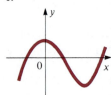

f.

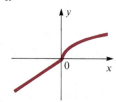

39. a.

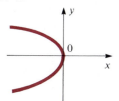

b.

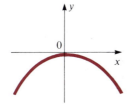

c.

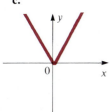

d.

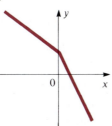

e.

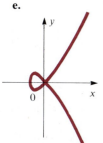

f.

In Problems 40 and 41 draw f and f^{-1} on the same set of axes.

40. $f(x) = \dfrac{3x - 5}{2}$ **41.** $f(x) = \sqrt{x + 1}$

42. Determine the largest intervals on which the function whose graph is shown is increasing and those on which it is decreasing.

a.

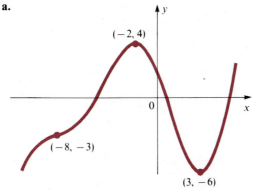

$(-2, 4)$

$(-8, -3)$

$(3, -6)$

b.

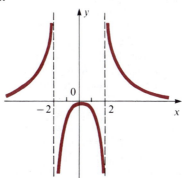

In Problems 43 and 44 the graph of a function f is given. From it obtain the graphs of the specified functions.

43. a. $y = f(x) + 1$ **b.** $y = 2f(x)$
    c. $y = f(x + 1)$ **d.** $y = f(-x)$

    e. $y = -f\left(\dfrac{x}{2} - 1\right)$

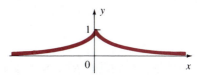

44. a. $y = f(x) - 2$ **b.** $y = f(x - 2)$

    c. $y = f\left(\dfrac{x}{2}\right)$ **d.** $y = -2f(x)$

    e. $y = f(2x + 3)$

45. Let $f(x) = \dfrac{1}{x^2 + x}$ and $g(x) = \dfrac{1}{x^2 - 1}$. Find $(f + g)(x)$, $(f - g)(x)$, $(f \cdot g)(x)$ and $(f/g)(x)$, and give the domain of each.

46. If y varies directly as x and inversely as the square root of t, and $y = 5$ when $x = 3$ and $t = 4$, find y when $x = 8$ and $t = 9$.

47. If a body is dropped and falls freely for t seconds, the distance through which it falls varies directly as the square of t (neglecting air resistance). If after 2 seconds a body has fallen 64 feet, how far will it fall in 10 seconds?

48. The surface area S of a sphere is proportional to the square of its radius. If a sphere of radius 3 has a surface area of 36π, find a formula for the surface area of any sphere. What is the surface area of a sphere of diameter 10?

49. If the temperature is constant, the volume of a gas varies inversely as its pressure. A gas with a pressure of 150 pounds per square inch occupies a volume of 25 cubic feet. What volume will it occupy when the pressure is 45 pounds per square inch, assuming the temperature is held constant?

 B

50. Make a table of values and draw the graph of each of the following:

a. $y^2 = 18 - 2x^2$ **b.** $f(x) = 2 - x - x^2$

51. Find the domain of each of the following functions:

a. $f(x) = \dfrac{2x + 3}{x^3 - 3x^2 - 4x + 12}$

b. $g(x) = \sqrt{\dfrac{4 - x^2}{x^3 + 4x^2}}$

52. Let

$$f(x) = \begin{cases} \sqrt{x^2 - 9} & \text{if} & x \geq 3 \\ 2 - x & \text{if} & 0 \leq x < 3 \\ -1 & \text{if} & x < 0 \end{cases}$$

Draw the graph of f and find:

a. $f(5)$ **b.** $f(-1)$ **c.** $f(0)$
d. $f(3)$ **e.** $f(1)$

53. A rectangle of base b and altitude h is inscribed in an isosceles triangle of altitude 10 and base 6, as shown in the sketch. Express h as a function of b.

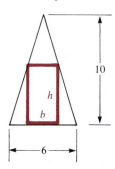

54. A circle is inscribed in an equilateral triangle. Express the area of the circle as a function of the length of a side of the triangle.

55. Let $f(x) = 4x - x^2$. Show that f is not 1–1 on R but that it is 1–1 on the domain $\{x: x \geq 2\}$. Find f^{-1} for this restricted domain.

56. Show that no inverse exists for the function defined by

$$f(x) = \dfrac{4}{(x - 2)^2}$$

but that by a suitable restriction on the domain an inverse does exist. Find this inverse.

57. Let $g(x) = x^2 - 4x - 5$. Show that g is not 1–1 on R but that it is 1–1 on the domain $\{x: \ x \leq 2\}$.

58. The strength of a beam of rectangular cross-section is jointly proportional to the width w and the square of the depth d of the cross-section and inversely proportional to the length l of the beam. What will be the effect on the strength if w and d are each doubled and l is cut in half?

 CUMULATIVE REVIEW EXERCISE SET I (CHAPTERS 1–3)

1. Prove that if $a < b$ and $c < d$, then $ac < bd$ provided b and c are positive. Show by examples that the conclusion is not necessarily true if b and c are not both positive.

2. Simplify the expressions. Answers should not involve negative exponents, and there should be no fractions under the radical. All letters represent positive numbers.

a. $\sqrt{\dfrac{2x^{-3}y^5z^0}{3x^{-2}y^3}}$ **b.** $\sqrt{a^{-1}b - 2 + b^{-1}a}$

3. Let $f(x) = 1/\sqrt{x}$. Find:

$$\dfrac{f(x + h) - f(x)}{h}$$

and express the answer with a rational numerator.

4. A 50% alcohol solution is to be combined with a 25% alcohol solution to obtain 50 cubic centimeters of a solution having 35% alcohol. How much of each solution should be used?

5. a. Express 0.2135135135 ... as the ratio of two integers.

  b. If a, b, and c are integers, with $a \neq 0$, show that regardless of the nature of the roots of the equation $ax^2 + bx + c = 0$, their sum and their product are always rational numbers.

6. Solve for x:

  a. $3x^2 - 8x + 6 = 0$

  b. $(3 + x)(14 - 3x) = 30$

7. Perform the indicated operations and simplify:

$$\frac{x^2 - 2x + 1}{2x^2 + x - 3} \cdot \frac{8x^2 + 6x - 9}{3x^2 + x - 4}$$

$$-\frac{6x^2 + 5x - 4}{4x^2 + 5x - 6} \div \frac{1 - 4x^2}{2x^2 + 5x + 2}$$

8. The volume of oil that flows through a pipe line per day is jointly proportional to the velocity of flow and the square of the diameter of the pipe. If approximately 3 million barrels of oil per day flow through a pipe that is 3 feet in diameter when the velocity is 10 feet per second, how many barrels would flow per day through a pipe 2 feet in diameter if the velocity is 15 feet per second?

9. a. Rationalize the numerator:

$$\frac{\sqrt{2x + 2h - 3} - \sqrt{2x - 3}}{h}$$

  b. Simplify and combine terms. Leave no fraction under a radical and no radical in a denominator.

$$\sqrt{\frac{3}{2} - \frac{1}{3}\sqrt{24}} + \frac{3}{2}\sqrt{\frac{32}{3}} - \frac{5}{2\sqrt{6}}$$

10. Solve for t: $\sqrt{2t + 5} - \sqrt{2 - t} = 1$

11. A flower garden in the shape of a rectangle 8 feet wide by 12 feet long is surrounded by a walkway of uniform width. If the combined area of the bed and walk is 221 square feet, find the width of the walk.

12. a. Expand and simplify: $\left(x^2 - \frac{2}{x} \right)^6$

  b. Find the first four terms of the expansion of $(2x + 3y)^{10}$.

13. Let

$$f(x) = \begin{cases} x^2 - 16 & \text{if} \quad x > 4 \\ \sqrt{x} & \text{if} \quad 0 < x \leq 4 \\ -1 & \text{if} \quad x \leq 0 \end{cases}$$

Draw the graph of f. Find:

  a. $f(4)$ **b.** $f(0)$ **c.** $f(1)$

  d. $f(5)$ **e.** $f(-1)$

14. Solve the inequalities and show the solution sets on a number line.

  a. $x(3x + 14) \geq 24$ **b.** $\left| \frac{3x - 4}{5} - \frac{3x}{4} \right| < \frac{1}{2}$

15. Factor completely with integer coefficients:

  a. $4xy + 3x^2 - x^3y - 12$

  b. $72x^3 + 10x^2y - 48xy^2$

16. Solve for x: $\dfrac{2x}{x - 2} - \dfrac{3x - 1}{x + 1} = 4$

17. a. Solve for s: $F = \dfrac{rs - 1}{r + 2s}$

  b. Solve for t: $s = \frac{1}{2}gt^2 + v_0 t (s \geq 0, t \geq 0)$

18. a. A right circular cone has base radius r and altitude h. Express the area of a cross-section taken x units above the base as a function of x.

  b. Let $f(x) = 1 + |x - 1|$. Give the domain and range of f, and draw its graph. Is f 1–1? Explain why or why not.

19. a. Simplify, and express the answer with positive exponents only:

$$\left(\frac{25a^{-2}b^{4/3}}{\sqrt[3]{16a^{8/3}b^2c^0}} \right)^{-3/2}$$

  b. Evaluate, using scientific notation, and check your answer with a calculator.

$$\frac{(0.000003)\sqrt{250,000,000,000}}{2 \times 10^{-3} + 5 \times 10^{-4}}$$

20. A woman drove a distance of 120 kilometers to do some research in the Library of Congress. She stayed 4 hours and 24 minutes before returning home. On the return trip she encountered rush hour traffic, and as a result her average speed on the return trip was 15 kilometers per hour less than on the trip going. If the total elapsed time from leaving home until returning was 8 hours, find her average speed going and returning.

21. Simplify, and write the answer with positive exponents only:

$$\frac{\sqrt[3]{(x^2-1)^2} - \dfrac{4x^2}{3}(x^2-1)^{-1/3}}{(x^2-1)^{4/3}}$$

22. a. Find the domain of f.

$$f(x) = \sqrt{\frac{2x-1}{x^2-4}}$$

b. Show that the function

$$f(x) = \frac{1-2x}{3}$$

is 1–1 on R, and find f^{-1}.

23. Solve the inequalities.

a. $\dfrac{x}{x-1} - \dfrac{1}{x+3} \leq 1$ **b.** $\dfrac{2}{x-1} \geq \dfrac{x+3}{x+11}$

24. For $a = \pm 1$ and $b = \pm 2$ show that $a + bi$ satisfies the equation $x^4 + 6x^2 + 25 = 0$. Also, solve the equation by the quadratic formula. Verify that the solutions are the same.

25. Let $f(t) = \sqrt{t-1}$ and $g(t) = t^2 - 3$. Find $f \circ g$ and $g \circ f$, give the domain and range of each, and draw their graphs.

26. Perform the indicated operations and simplify:

a. $\dfrac{\dfrac{1}{x-1} - \dfrac{2}{x+2}}{\dfrac{18}{x^2+x-2} - 1}$

b. $\dfrac{x+2}{x} - \dfrac{8}{4x-x^3} - \dfrac{x}{x+2}$

27. A men's clothing store owner had a month-long 20% off sale on suits of a certain type. His gross income from the sale of these suits during the month of the sale was $15,120. During the preceding month, when the suits were full price, his gross income from selling 30 fewer of these suits was $12,600. Find how many of this type of suit he sold during the month of the sale. What was the original price of the suit?

28. Find f and g so that $F = f \circ g$.

a. $F(x) = \sqrt[3]{(1-x)^2}$

b. $F(x) = |3x-4|$

29. Let f and g be defined on R as follows:

$$f \text{ is even and } f(x) = \begin{cases} x & \text{if } 0 \leq x \leq 1 \\ 1 & \text{if } \quad x > 1 \end{cases}$$

g is odd and $g(x) = 2f(x)$ if $x \geq 0$. Find $(f + g)(x)$, $(f - g)(x)$, $(f \cdot g)(x)$, and $(f/g)(x)$, and give the domain of each.

30. From the graph of $f(x) = \dfrac{1}{x^2}$ obtain the graph of

$$g(x) = 3 - \frac{8}{(2x-3)^2}$$

4

LINEAR AND QUADRATIC FUNCTIONS; INTRODUCTION TO CONIC SECTIONS

Linear and quadratic functions are among the most widely used in mathematical models. A recurrent type of problem in calculus is that of finding equations of the tangent line to a curve at a point and the line perpendicular to it, called the **normal line.** The following example illustrates this.*

Write equations for both the tangent line and normal line to the parabola $y = 2x^2 - 3x + 5$ at the point $P(-1, 10)$.

Solution. We use the derivative that we just computed. The slope of the tangent line at $(-1, 10)$ is

$$f'(-1) = 4(-1) - 3 = -7$$

Hence the point-slope equation of the desired tangent line is

$$y - 10 = -7(x + 1)$$

The normal line has slope $m' = -1/(-7) = \frac{1}{7}$, so its point-slope equation is

$$y - 10 = \tfrac{1}{7}(x + 1)$$

We will learn about the slope of a line, the point-slope form of the equation of a line, and the relationship between slopes of perpendicular lines in this chapter. We will also see that quadratic functions have parabolas as graphs.

* C. H. Edwards, Jr. and David E. Penney, *Calculus and Analytic Geometry*, 2/E, © 1986, p. 35. Reprinted by permission of Prentice-Hall, Inc., Englewood Cliffs, NJ.

INTRODUCTION

A function f defined by an equation of the form

$$f(x) = a_n x^n + a_{n-1} x^{n-1} + \cdots + a_1 x + a_0 \qquad (a_n \neq 0)$$

where n is a nonnegative integer, is called a **polynomial function of degree n.**
If $n = 1$, the function is said to be **linear,** and if $n = 2$, it is **quadratic.** We will
study these two cases in some detail in this chapter. Higher-degree polynomial
functions will be taken up in Chapter 5.

As you might expect, a linear function has some relationship to a line. To see
what this relationship is, we will begin by studying straight lines and their
equations. We know from plane geometry (and intuition) that a straight line is
uniquely determined by two points. A line is also determined by one point and the
"inclination," or **slope,** of the line. In the next section we explain the precise
meaning of this concept and show how it is used to obtain an equation of the line.

EQUATIONS OF LINES

Suppose we are given a nonvertical straight line l and know any two points on
the line, say, $P_1(x_1, y_1)$ and $P_2(x_2, y_2)$, as shown in Figure 1. Then we have the
definition below.

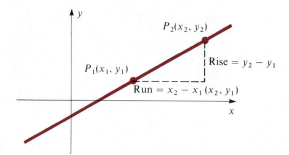

FIGURE I

DEFINITION I   **The slope m of line l is defined as**

$$m = \frac{y_2 - y_1}{x_2 - x_1}$$

(Can you see why we restricted l to be nonvertical?)

The slope is the vertical displacement from P_1 to P_2 divided by the horizontal
displacement (sometimes described as the **rise** divided by the **run**). It does not

matter which point is labeled P_1 and which is labeled P_2, since

$$\frac{y_2 - y_1}{x_2 - x_1} = \frac{y_1 - y_2}{x_1 - x_2} \qquad \textbf{Why?}$$

It can also be shown, using similar triangles, that it does not matter which particular two points on the line are chosen. The value found for the slope will always be the same.

EXAMPLE 1

A line passes through the points (1, 2) and (5, 4). Find the slope of the line.

Solution We will let P_1 be the point (1, 2) and P_2 be (5, 4), but we could equally well reverse these. By Definition 1, we have

$$m = \frac{y_2 - y_1}{x_2 - x_1} = \frac{4 - 2}{5 - 1} = \frac{2}{4} = \frac{1}{2}$$

We can think of this as saying that for every 2 units we go horizontally to the right, the line rises 1 unit vertically, or for every 1 unit we go horizontally to the right, the line rises $\frac{1}{2}$ unit vertically. The graph is shown in Figure 2.

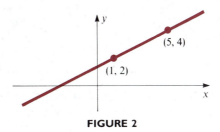

FIGURE 2

An examination of the definition of the slope of a line will show that lines that go upward to the right have positive slopes and those that go upward to the left have negative slopes (Figure 3). Also, a small numerical value for the slope indicates a line that is close to horizontal, whereas a large numerical value indicates a line that is nearly vertical. (What would a slope of 1 indicate? What about -1?) Horizontal lines have a slope of 0 (why?), and slope is not defined for vertical lines.

Let l be a nonvertical line passing through a given point $P_1(x_1, y_1)$ and having slope m. Any other point $P(x, y)$ will lie on l if and only if

$$\frac{y - y_1}{x - x_1} = m$$

If we clear this equation of fractions, we obtain the following equation, known as the **point–slope form** of the equation of a line.

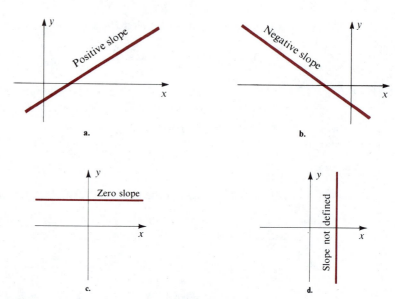

FIGURE 3

Point–Slope Form

$$y - y_1 = m(x - x_1) \tag{1}$$

Note that all points that lie on l, including P_1, have coordinates that satisfy equation (1), and this is true for those points only.

EXAMPLE 2

Find the equation of the line having slope $\frac{3}{4}$ and passing through the point $(-2, 5)$.

Solution   We simply substitute in equation (1), using $x_1 = -2$ and $y_1 = 5$:

$$y - 5 = \tfrac{3}{4}(x + 2)$$
$$4y - 20 = 3x + 6$$
$$3x - 4y + 26 = 0$$

▲

Note. In applying equation (1), x_1 and y_1 represent the coordinates of the given point and so will be specific numbers. On the other hand, x and y will remain in the equation as variables.

EXAMPLE 3 ıı

Find the equation of the line joining $(-6, -2)$ and $(2, 5)$.

Solution   First we must calculate the slope:

$$m = \frac{5 - (-2)}{2 - (-6)} = \frac{7}{8}$$

We may choose either point as (x_1, y_1) for the purpose of substituting in equation (1). Using $(2, 5)$, we get

$$y - 5 = \tfrac{7}{8}(x - 2)$$
$$8y - 40 = 7x - 14$$
$$7x - 8y + 26 = 0$$

You should verify that using $(-6, -2)$ as (x_1, y_1) gives the same result.   ▲

A special case of equation (1) results when the point (x_1, y_1) is taken as the point where the line crosses the y axis. The y coordinate of this point is called the **y intercept** and is usually denoted by b. Substituting $(0, b)$ for (x_1, y_1) in equation (1), we obtain

$$y - b = m(x - 0)$$

When we solve this for y, we obtain another important form of the equation of a nonvertical line, called the **slope–intercept form:**

Slope–Intercept Form

$$y = mx + b \qquad\qquad (2)$$

Note. The slope–intercept form refers to the y intercept b. If the line is not horizontal, it also has an x intercept, usually denoted by a. In Problem 28, Exercise Set 2, you will be asked to obtain a form of the equation of a line that uses both the x intercept and the y intercept.

EXAMPLE 4 ıı

Show that the equation $2x - 3y + 6 = 0$ is the equation of a line, and find the slope and y intercept of the line. Graph the line.

Solution   If we can put the equation in the slope–intercept form, equation (2), we will know it is a line. This is done by solving for y:

$$3y = 2x + 6$$
$$y = \tfrac{2}{3}x + 2$$

Therefore, this is the equation of a line with slope $\frac{2}{3}$ and y intercept 2. We can use the slope to find another point on the line as follows: From (0, 2) if we move 3 units in the x direction and 2 units in the y direction, we must come to another point on the line. This results in the point (3, 4). The line can now be drawn as shown in Figure 4. An easier way to get a second point on the line is to find the x intercept by letting $y = 0$. This gives

$$0 = \tfrac{2}{3}x + 2$$
$$2x = -6$$
$$x = -3$$

So $(-3, 0)$ is another point on the line.

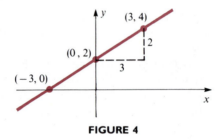

FIGURE 4

In either the point–slope form, $y - y_1 = m(x - x_1)$, or the slope–intercept form, $y = mx + b$, of the equation of a line, if the terms are rearranged, the result can be written in the form

$$Ax + By + C = 0 \qquad (3)$$

But do all equations of this form have straight lines as graphs? The answer is yes, as we show below. Because equation (3) always represents a line, it is called the **general equation** of a line.

We suppose that A and B are not both 0, because if they were, then C would also have to be 0, and the equation would read $0 = 0$, which is true but not of much interest. Now let us consider the case in which $B \neq 0$ (A may or may not be 0). We can solve for y to get

$$y = -\frac{A}{B}x + \left(-\frac{C}{B}\right)$$

This is in the form $y = mx + b$, with

$$m = -\frac{A}{B} \qquad \text{and} \qquad b = -\frac{C}{B} \qquad (4)$$

So this represents a line with slope $-A/B$ and y intercept $-C/B$.

In the special case where $A = 0$ (but $B \neq 0$), so that $m = 0$, we see from the above that

$$y = -\frac{C}{B}$$

or

$$y = b$$

Since this equation places no restriction on x, it follows that all points with coordinates of the form (x, b) satisfy the equation, and therefore the graph is a horizontal line. The slope is 0.

One final case needs to be considered, and that is the case in which $B = 0$ but $A \neq 0$. Equation (3) then has the form

$$Ax + C = 0$$

$$x = -\frac{C}{A}$$

If we let $a = -C/A$, this can be written

$$x = a$$

and this is the equation of a vertical line, since all points with coordinates of the form (a, y), where y can be anything whatsoever, satisfy the equation. Slope is not defined for a vertical line.

We summarize these two special cases of equation (3) below.

Vertical and Horizontal Lines

$x = a$ Vertical line

$y = b$ Horizontal line

Remark. For a horizontal line the slope is 0, so the equation $y = b$ can be considered as a special case of the slope–intercept form (2). But slope is not defined for vertical lines, so the equation $x = a$ cannot be obtained as a special case of either equation (1) or (2). Note also that vertical lines have no y intercept (except for the y axis itself, and in this case there is no unique y intercept).

EXAMPLE 5 ||||||||||||||||||||||||||||||||||||||||||||||||||||||||||||||||||||||||||||||||||||||||||||||||||||||||||

Draw the graphs of the lines:
a. $y = 3$ **b.** $x = -2$

Solution

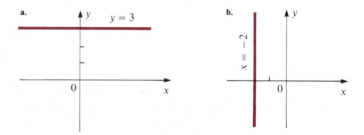

Note. A common mistake is to suppose that $y = 3$ means only the single point $(0, 3)$ or that $x = -2$ means just the point $(-2, 0)$. In the context of two dimensions both of these represent lines, not points.   ▲

The following relationship between slopes of parallel lines is intuitively obvious. (A proof is called for in Problem 64, Exercise Set 2.)

Parallel Lines

Two nonvertical lines are parallel if and only if their slopes are equal.

The relationship between slopes of perpendicular lines is not at all obvious. Of course, if either line is horizontal, the other must be vertical, and conversely. So the only case we need consider is that in which neither line is horizontal (and hence neither is vertical). There is no loss of generality in assuming both lines pass through the origin, for if this were not the case, we could consider lines parallel respectively to the given lines that do pass through the origin. Then, since parallel lines have the same slope, our conclusions about slopes would extend to the original lines.

So let l_1 and l_2, with slopes m_1 and m_2, respectively, be perpendicular to one another and intersect at the origin. By definition of slope, the point $P_1(1, m_1)$ lies on l_1, and $P_2(1, m_2)$ lies on l_2. You can verify this by using the origin together with P_1 in the one case and P_2 in the other, to obtain the slopes m_1 and m_2, respectively. A typical situation is shown in Figure 5. The triangle $P_1 O P_2$ is a right triangle, and so by the Pythagorean theorem $(\overline{P_1 P_2})^2 = (\overline{OP_1})^2 + (\overline{OP_2})^2$. Thus, by the distance formula

$$(m_1 - m_2)^2 = (1^2 + m_1^2) + (1^2 + m_2^2)$$
$$m_1^2 - 2m_1 m_2 + m_2^2 = 2 + m_1^2 + m_2^2$$
$$m_1 m_2 = -1$$

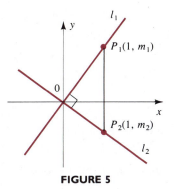

FIGURE 5

That is,

$$m_2 = -\frac{1}{m_1} \tag{5}$$

We say in words that m_2 is the **negative reciprocal** of m_1. Similarly, m_1 is the negative reciprocal of m_2.

In our analysis we assumed that l_2 was perpendicular to l_1 and found equation (5) to be true. But we could begin with equation (5) and reverse all steps, using the converse of the Pythagorean theorem, to conclude that equation (5) implies perpendicularity. Thus, we have the following.

Perpendicular Lines

Two lines, neither of which is horizontal, are perpendicular if and only if their slopes are negative reciprocals of each other.

EXAMPLE 6 |||||||||||||||||||||||||||||||||||||||||||||||||||||||||||||||||||||||||||||||||||||||||||||||||||||||||||||||||

Find the equations of the two lines through the point $(-1, 3)$ parallel and perpendicular, respectively, to the line $2x - 3y + 4 = 0$.

Solution Writing the given line in slope-intercept form

$$y = \tfrac{2}{3}x + \tfrac{4}{3}$$

we see that it has slope $\tfrac{2}{3}$. Thus, the line through $(-1, 3)$ and parallel to this has the equation

$$y - 3 = \tfrac{2}{3}(x + 1)$$
$$2x - 3y + 11 = 0$$

The slope of the perpendicular line is $-\frac{3}{2}$. So its equation is

$$y - 3 = -\tfrac{3}{2}(x + 1)$$
$$3x + 2y - 3 = 0 \qquad \blacktriangle$$

EXAMPLE 7 ‖‖

Find the equation of the perpendicular bisector of the line segment joining $(-2, 3)$ and $(4, 5)$.

Solution   The slope of the line joining the two points is

$$m = \frac{5 - 3}{4 - (-2)} = \frac{1}{3}$$

and the midpoint of the segment is

$$\left(\frac{4 + (-2)}{2}, \frac{3 + 5}{2}\right) = (1, 4)$$

The slope of the perpendicular bisector is therefore -3, and its equation is

$$y - 4 = -3(x - 1)$$
$$3x + y - 7 = 0 \qquad \blacktriangle$$

EXAMPLE 8 ‖‖

Find the equation of the family of all lines that are (a) parallel to and (b) perpendicular to the line $3x - 4y + 3 = 0$.

Note. A **family** of lines is a set of lines satisfying a given condition.

Solution   **a.**   By equation (4) a line with equation $Ax + By + C = 0$ has slope $-A/B$. So the given line has slope $\frac{3}{4}$. We can write the equation of any line parallel to it in the form

$$3x - 4y + C = 0 \qquad (C \text{ arbitrary})$$

since the slope for any real number C will again be $\frac{3}{4}$.

b.   The equation of the family of lines perpendicular to the given line can be written in the form

$$4x + 3y + C = 0 \qquad (C \text{ arbitrary})$$

since all such lines have slope $-\frac{4}{3}$. $\qquad \blacktriangle$

If P and Q are two points on a curve, the line joining them is called a **secant** line for the curve. If P is held fixed and Q comes closer and closer to P, the secant

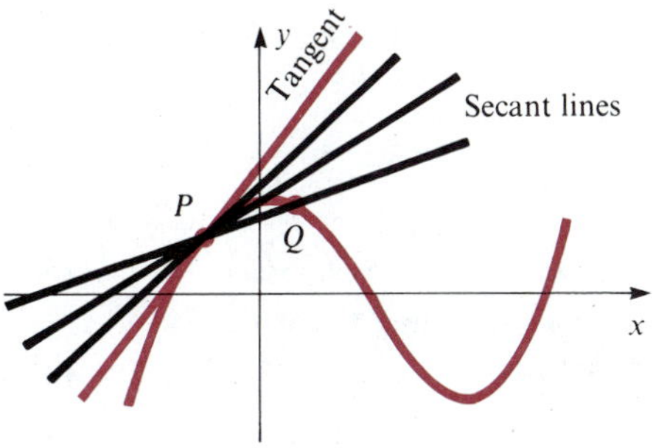

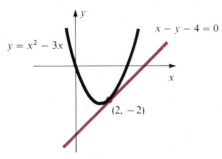

FIGURE 6

lines through P and Q may approach some line as a limit; if so, this line is called the **tangent** to the curve at P. Figure 6 illustrates this.

For any position of Q distinct from P the slope of the secant line can be determined. Then, by the use of a limiting process studied in calculus, the slope of the tangent line at P can be found, if it exists, as the limiting value of the slopes of the secant lines as Q approaches P. The equation of the tangent line can then be obtained by using this slope and the coordinates of P in the point–slope form. We will not pursue these ideas further, but the next example shows that if we are given the results from calculus for the slope of the tangent line, we can find its equation. (See also Problems 71–74.)

EXAMPLE 9 ||||||||||||||||||||||||||||||||||||||||||||||||||||||||||||||||||||||||||||||||||||||||||||||||||||||||

In calculus it is shown that the curve defined by $y = x^2 - 3x$ has a tangent at each point (x, y) whose slope is given by $m = 2x - 3$. Find the equation of the tangent line to this curve at the point for which $x = 2$.

Solution   When $x = 2$, $y = 2^2 - 3(2) = 4 - 6 = -2$. Also, the slope $m = 2(2) - 3 = 1$. So the equation is

$$y + 2 = 1(x - 2)$$
$$x - y - 4 = 0$$

The graph is shown in Figure 7.

FIGURE 7

EXERCISE SET 2

A   10-26 31-40

1. Find the slopes of the lines joining the pairs of points given. Sketch the lines.
   a. $(-1, 2)$ and $(3, 4)$
   b. $(-2, -3)$ and $(7, 3)$
   c. $(5, -3)$ and $(-1, 2)$
   d. $(-4, 3)$ and $(0, -1)$
2. The points $A(3, 1)$, $B(-2, -2)$, and $C(-7, -5)$ all lie on a line. Calculate the slope of the line, using three different pairs of points, and verify that the results are the same.
3. Find the slopes of the sides of the triangle with vertices $A(2, 3)$, $B(-4, -2)$, and $C(-3, 5)$. Draw the triangle.
4. Find the slopes of the two diagonals of the quadrilateral with vertices at the points $(4, 3)$, $(-1, 2)$, $(0, -3)$, and $(7, -1)$. Draw the figure.
5. What is the horizontal displacement from $(-5, -3)$ to $(3, -2)$? What is the vertical displacement?
6. In each of the following a line passes through the given point with the given slope. Find a second point on the line, and sketch the graph.
   a. $(-4, 2)$, slope $\frac{2}{3}$ b. $(-1, -5)$, slope $\frac{-5}{2}$
7. The slope of the line passing through $(4, -3)$ and $(-3, y)$ is $-\frac{3}{2}$. Find y.
8. The slope of a line passing through $(x, 3)$ and $(-1, -5)$ is $\frac{3}{4}$. Find x.
9. A quadrilateral has vertices $A(3, 2)$, $B(7, 5)$, $C(-1, 3)$, and $D(-5, 0)$. Find the slope of each of the sides. Does your result suggest anything special about this quadrilateral? Explain. Sketch the figure.

In Problems 10–19 find equations of the lines satisfying the given conditions.

10. Slope $\frac{2}{3}$, passing through $(-1, 4)$
11. Slope -3, passing through $(-3, 5)$
12. Passing through $(3, -1)$ and $(1, 2)$
13. Passing through $(-2, -3)$ and $(1, 0)$
14. Passing through $(5, -2)$ and $(-1, 2)$
15. Passing through $(0, -4)$ and $(3, 1)$
16. Slope $-\frac{2}{3}$, y intercept 2

17. Slope 3, y intercept -1
18. Slope $\frac{1}{2}$, y intercept $\frac{3}{4}$
19. Slope 0, y intercept -2

In Problems 20–25 find the slope m and the y intercept b. Draw the graph.

20. $y = 2x - 3$
21. $2x - 3y = 4$
22. $3x + 4y + 8 = 0$
23. $3y - 2x + 5 = 0$
24. $y = \dfrac{4 - 2x}{3}$
25. $x = \dfrac{2y - 3}{4}$

In Problems 26 and 27 write the equation of each horizontal or vertical line, as indicated.

26. a. x intercept -3, vertical
    b. y intercept 4, horizontal
    c. Passing through $(5, -9)$, vertical
    d. Passing through $(27, -4)$, horizontal
27. a. The x axis
    b. The y axis
    c. Passing through $(1, 4)$, horizontal
    d. Passing through $(-3, -100)$, vertical
28. Derive the **two-intercept form** of the equation of a line:
$$\frac{x}{a} + \frac{y}{b} = 1$$

   Hint. You know two points on the line.
29. Use the result of Problem 28 to find the equations of the lines having the given intercepts.
    a. x intercept 5, y intercept -2
    b. x intercept $-\frac{2}{3}$, y intercept $\frac{4}{5}$
30. The x intercept of a certain line is twice the y intercept, and the sum of the intercepts is 7. Find the equation of the line.

In Problems 31–40 find the equation of the line satisfying the given conditions.

31. Passing through $(2, -3)$, parallel to $4x - 3y + 17 = 0$
32. Passing through $(-1, 4)$, perpendicular to $x + 2y = 3$
33. Passing through $(2, 0)$, perpendicular to $8x - 3y + 7 = 0$
34. Passing through $(7, -3)$, parallel to $2x - 3y = 8$

35. Parallel to the line joining $(4, -1)$ and $(2, 3)$ and passing through the origin

36. Perpendicular to the line joining $(-2, -3)$ and $(4, 1)$ and having y intercept 3

37. Perpendicular to $x + 3 = 0$ and passing through $(5, -1)$

38. Parallel to $3y - 7 = 0$ and passing through $(13, 2)$

39. Parallel to $4x - 5y = 8$ and whose x intercept is the negative of the x intercept of this line

40. Perpendicular to $5x + 3y = 4$ at the point where this line crosses the y axis

In Problems 41 and 42 use the converse of the Pythagorean theorem, which states that if $c^2 = a^2 + b^2$, where a, b, and c are lengths of the sides of a triangle, then the triangle is a right triangle.

41. Prove that the triangle with vertices $(2, 1)$, $(6, 9)$, and $(-2, 3)$ is a right triangle.

42. An **isosceles triangle** is one in which two sides are equal. Show that the triangle with vertices $(-7, 5)$, $(-3, -2)$, and $(4, 2)$ is an isosceles right triangle.

43. Show that the triangle with vertices $(-4, -3)$, $(-3\sqrt{3}, 4\sqrt{3})$, and $(4, 3)$ is **equilateral** (all sides equal).

44. A right triangle is formed by the x axis, the y axis, and the line $3x - 4y = 24$. Find (a) the length of the hypotenuse, and (b) the midpoint of the hypotenuse.

45. Find the equation of the line joining the point $(3, -1)$ and the midpoint of the segment from $(7, 3)$ to $(1, -1)$.

46. The vertices of a quadrilateral are $(-1, 2)$, $(4, 5)$, $(-3, -4)$, and $(0, -6)$. Prove that the line segments joining successive midpoints of the sides of the quadrilateral form a parallelogram.
Note. It is an interesting fact that this result holds for *any* quadrilateral.

47. By using slopes, show that the points $(-2, 4)$, $(3, -6)$, and $(6, -2)$ are vertices of a right triangle.

48. Show that the points $(3, -1)$, $(5, 4)$, $(-5, 8)$, and $(-7, 3)$ are vertices of a rectangle.

49. Show that the points $(2, 2)$, $(0, -1)$, $(-4, 1)$, and $(-2, 4)$ are the vertices of a parallelogram.

50. Find the equation of the family of lines parallel to the line $7x - 2y + 3 = 0$. Find the equation of the member of this family that goes through the point $(3, -1)$.

51. Find the equation of the family of lines perpendicular to the line $2x + 3y - 4 = 0$. Find the

equation of the member of this family that has the same y intercept as that of the given line.

In Problems 52–55 find the equation of the perpendicular bisector of the line segment joining the two points.

52. $(2, -3)$ and $(5, 1)$ **53.** $(-4, -2)$ and $(0, 6)$

54. $(4, 3)$ and $(1, -2)$ **55.** $(-5, 2)$ and $(11, 3)$

56. An **altitude** of a triangle is a line segment from a vertex perpendicular to the opposite side. Find the equations of the altitudes of the triangle with vertices $(3, -2)$, $(-4, 1)$, and $(2, -5)$.

57. A **median** of a triangle is a line segment from a vertex to the midpoint of the opposite side. Find the equations of the medians of the triangle with vertices $(4, 5)$, $(2, -1)$, and $(0, 7)$.

B

In Problems 58–60 show in two ways that the given points are **collinear** (that is, they lie on the same line).

58. $(1, -2)$, $(2, 3)$, $(-1, -12)$

59. $(0, 2)$, $(2, -1)$, $(-4, 8)$

60. $(1, 2)$, $(-1, \frac{1}{2})$, $(-7, -4)$

61. The midpoints of the sides of a triangle are $(-2, -1)$, $(2, 4)$, and $(1, -3)$. Find the coordinates of the vertices.

62. A line parallel to $3x + 4y = 7$ intersects the positive x axis and positive y axis so that the triangle formed by the line and these axes has area 6. Find the equation of the line.

63. A point $P(x, y)$ moves in the plane so that its distance from $(5, -2)$ is always 4. Find the equation of the path traced out by P. How would you describe this path?

64. Let $y_1 = m_1 x + b_1$ and $y_2 = m_2 x + b_2$. Carry out the details of the argument outlined here to prove that the corresponding lines are parallel if and only if $m_1 = m_2$. We know that two lines are parallel if and only if they do not intersect. If $m_1 = m_2$, show that the lines do not intersect (unless they coincide). If $m_1 \neq m_2$, show that the lines do intersect.

65. A physicist describes the results of an experiment on a plane coordinate system. Two magnets have centers at the points with coordinates $(0, -\frac{1}{2})$ and $(2, \frac{3}{2})$. The path of a particle in the experiment is the perpendicular bisector of the line segment joining the centers of the magnets. Find the equation of the path of the particle.

66. Prove that the points $(-2, 9), (-4, -2), (1, -12)$, and $(3, -1)$ are vertices of a **rhombus** (a parallelogram with all sides equal in length). Show that the diagonals are perpendicular.

67. Coordinates are introduced into a town planner's map. One of the town's boundaries corresponds to the graph of the equation $y = 2x + 1$, while the entrance to the fire station is at the point with coordinates $(3, 4)$. Find the formula, in terms of x, for the distance from the fire station entrance to each point on that boundary of the town.

68. A salesperson works for a fixed base weekly salary of \$$b$, plus a commission of \$$c$ for each unit sold. Write the weekly income y as a function of the number x of units sold. Suppose the salesperson receives \$300 in a week in which 10 units were sold and \$375 in a week in which 15 units were sold. Find the person's base salary b and commission c. What will the person's income be in a week in which 18 units are sold?

69. a. It can be shown by calculus that the slope of the tangent line to the graph of $f(x) = 4 - x^2$ at any point $(x, f(x))$ is given by $m = -2x$. Find the equation of the tangent line at $(1, f(1))$.

b. The line perpendicular to the tangent line of a curve is called the **normal** line. Find the equation of the normal to the curve of part **a** at the given point. Draw the graph of this function f, showing the tangent and normal lines at the given point.

70. The slope of the tangent line to the curve $y = x^3 - 9x + 5$ at any point (x, y) on the curve can be shown by calculus to be given by the formula $m = 3x^2 - 9$. Find the equations of the

tangent and normal lines at the point for which $x = -1$. (See Problem 69.)

71. Let the graph in Figure 6 be the graph of a function $y = f(x)$. Let the fixed point P have coordinates $(a, f(a))$ and the variable point Q have coordinates $(x, f(x))$, where $x \neq a$. Show that the slope of the secant line joining P and Q is given by

$$m_{\text{sec}} = \frac{f(x) - f(a)}{x - a}$$

Now let $f(x) = x^2$ and $a = 2$. Calculate m_{sec} and simplify the result. By letting x come closer and closer to 2 (so that Q approaches P), find the slope of the tangent line at P.

72. Refer to Problem 71. This time let the coordinates of P be $(x, f(x))$ and those of Q be $(x + h, f(x + h))$, where $h \neq 0$. Show that

$$m_{\text{sec}} = \frac{f(x + h) - f(x)}{h}$$

Again let $f(x) = x^2$. Calculate m_{sec} and simplify. By letting h come closer and closer to 0 (so that Q approaches P), find the slope of the tangent line at P. (The answer will be in terms of x).

73. Repeat the process outlined in Problem 71 with $f(x) = 1/x$ at the point $P = (2, f(2))$. Also find the equation of the tangent line at P.

74. Repeat the process outlined in Problem 72 with $f(x) = 1/x$ at the point $P = (x, f(x))$, to find a formula for the slope of the tangent line at any point on the curve.

LINEAR FUNCTIONS

In Section 1 we defined a linear function as a polynomial function of degree 1, that is, a function of the form $f(x) = a_1 x + a_0$, with $a_1 \neq 0$. The reason for calling this *linear* should now be clear. If we put $y = f(x)$, we get

$$y = a_1 x + a_0$$

which is of the form $y = mx + b$, where $m = a_1$ and $b = a_0$. It follows that the graph of every linear function is a straight line. The requirement that $a_1 \neq 0$

ensures that the line is not horizontal. If $a_1 = 0$, the function will be of the form $f(x) = a_0$, which is called a **constant function.** Its graph is the horizontal line having the equation $y = a_0$. So even though the graph is a line, the function is not called linear in this case.

We can now conclude that every linear equation in two variables,

$$Ax + By + C = 0$$

in which neither A nor B is 0, defines a linear function. The rule for the function is obtained by solving for y:

$$y = -\frac{A}{B}x + \left(-\frac{C}{B}\right)$$

or, in functional notation,

$$f(x) = -\frac{A}{B}x + \left(-\frac{C}{B}\right)$$

Polynomial equations of the first degree in more than two variables are also called *linear*, although their graphs are no longer straight lines. For example, the equation $2x - 3y + 4z - 7 = 0$ is a linear equation in three variables. Graphing this equation requires three dimensions. Can you guess what sort of graph this equation would have? Equations such as this will be considered later, when we discuss systems of equations.

EXAMPLE 10

For a certain linear function f, it is known that $f(3) = -1$ and $f(-1) = 1$. Find $f(x)$.

Solution Let $y = f(x)$. Then $y = -1$ when $x = 3$ and $y = 1$ when $x = -1$. That is, the points $(3, -1)$ and $(-1, 1)$ are on the graph of f, and since f is linear, its graph is a straight line. Its slope is

$$m = \frac{1 - (-1)}{-1 - 3} = \frac{2}{-4} = -\frac{1}{2}$$

So, by the point–slope form, using the point $(3, -1)$, we obtain the equation

$$y + 1 = -\frac{1}{2}(x - 3)$$

Now, since we want $f(x)$, and $y = f(x)$, we need to solve for y:

$$y = -\frac{1}{2}x + \frac{1}{2}$$

Finally, we can write $f(x)$ in the form

$$f(x) = -\frac{1}{2}(x - 1)$$

An alternative method to find $f(x)$ is as follows. We know that since f is linear, it has the form

$$f(x) = a_1 x + a_0$$

The problem is to find a_1 and a_0. Using $f(3) = -1$ and $f(-1) = 1$, we obtain the two equations

$$\begin{cases} 3a_1 + a_0 = -1 \\ -a_1 + a_0 = 1 \end{cases}$$

If we now subtract the terms of the second equation from those of the first, we get

$$4a_1 = -2$$

$$a_1 = -\frac{1}{2}$$

We can substitute this value of a_1 into either equation to find a_0. Choosing $-a_1 + a_0 = 1$, we find that

$$a_0 = 1 + \left(-\frac{1}{2}\right) = \frac{1}{2}$$

So

$$f(x) = -\frac{1}{2}x + \frac{1}{2} = -\frac{1}{2}(x - 1)$$ ▲

EXAMPLE 11

A car rental agency charges \$18 per day plus 20¢ per mile for renting a subcompact car. Express the daily cost of renting a car of this type as a function of the number of miles driven. What is the cost of driving 375 miles in one day?

Solution   Let x equal the number of miles driven and let $C(x)$ be the total cost (in dollars) per day. Then

$$C(x) = 0.20x + 18$$

Note that the cost is a linear function of x. The cost of driving 375 miles in one day is

$$C(375) = (0.20)(375) + 18 = \$93$$ ▲

EXAMPLE 12 ||||||||||||||||||||||||||||||||||||||||||||||||||||||||||||||||||||||||||||||||||||||||||||||||||||||||

In economics the **cost function** C, whose value $C(x)$ is the cost of producing x units of some commodity, is often assumed to be linear, say $C(x) = mx + b$. Its slope m is called **marginal cost** and represents the cost of producing one more unit regardless of how many units have been produced. The number b is a **fixed cost,** also called **start-up cost,** in contrast to mx, which is the **variable cost** and depends on the number of units produced.

Suppose that for some commodity the marginal cost is $2.50 and that it costs $400 to produce 10 units. Find $C(x)$. What is the fixed cost?

Solution   Let $C(x) = mx + b$. We are given that $m = 2.50$. So $C(x) = 2.50x + b$. Also from the given information $C(10) = 400$, so

$$(2.50)(10) + b = 400$$
$$b = 375$$

Thus, $C(x) = 2.50x + 375$. The fixed cost is $375.   ▲

Remark. The notion of marginal cost also is applicable for nonlinear cost functions but requires calculus for its computation.

 EXERCISE SET 3

A

1. Let $f(x) = 5x - 3$. Identify and draw the graph of f.
2. Let g be the linear function defined by $g(x) = (3 - 2x)/4$. Draw the graph of g and give its slope.
3. Let $f(x) = (3x - 7)/2$. What are the slope and y intercept of the graph of f? Draw the graph.
4. Graph the constant function defined by $f(x) = 3$.
5. For a certain linear function f we are given that $f(-1) = 3$ and $f(2) = 5$. Find $f(x)$.
6. For a certain linear function f the slope of the graph is -2, and $f(0) = 3$. Find $f(x)$.
7. If g is a linear function with $g(0) = 2$ and $g(-2) = 3$, find $g(x)$.
8. If $g(0) = -\frac{1}{3}$, g is linear, and the slope of the graph of g is 2, find $g(x)$.
9. If h is linear with $h(-2) = 3$ and $h(1) = -1$, find $h(7)$.   ~ 4
10. If f is a linear function with $f(0) = 3$ and $f(2) = 0$, find $f(-5)$.

11. If f is a function such that for all real numbers a and b, $f(b) - f(a) = 2(b - a)$ and $f(1) = -2$, find $f(x)$.
12. For a certain polynomial function f it is known that $f(3) = -4$ and $f(0) = -4$. Can f be linear? Explain. If the degree of f does not exceed 1, find $f(x)$.
13. Suppose the cost of producing 30 units of a certain commodity is $150 and the cost of producing 45 units is $185. Find the cost function $C(x)$ under the assumption that it is linear. What is the marginal cost? What is the start-up cost?
14. A car rental agency charges $32 per day plus a mileage charge of 25¢ per mile for each mile in excess of 200. Express the daily cost of renting this type of car as a function of the number of miles driven. Is this a linear function? Explain. Find the cost of driving 460 miles in one day.
15. Express the amount $A(t)$ accumulated in a savings account after t years from a principal of P dollars

earning simple interest at rate r (expressed as a decimal). Show that this is a linear function of t. What is the slope of its graph? If $P = \$3,000$ and $r = 0.06$, find $A(8)$.

16. According to Hooke's law, the force $F(x)$ required to stretch a spring x inches beyond its natural length is a linear function of x. If, for a certain spring, it takes 30 pounds to stretch it 8 inches, find $F(x)$, and determine the force necessary to stretch the spring 12 inches.

    Note. When $x = 0$, $F(x) = 0$.

17. Just as with cost functions, the **revenue function** $R(x)$, representing gross receipts from the sale of x units of some commodity, is often assumed to be linear, and its slope is called **marginal revenue.** Suppose $R(x)$ is linear, and the marginal revenue is \$65. If revenue of \$3,450 is realized from selling 30 units, find $R(x)$. How much revenue would be expected from selling 50 units?

18. Show that the composition of two linear functions is also a linear function.

    Hint. Let $f(x) = a_1x + a_0$ and $g(x) = b_1x + b_0$, and consider $f \circ g$.

19. Let $f(x) = (3x - 1)/4$ and $g(x) = (2 - x)/3$. Find $f \circ g$ and $g \circ f$. Give the slopes of the graphs of both $f \circ g$ and $g \circ f$, and draw the graphs.

B

20. Show that every linear function is 1–1 from R onto R.

21. What is the relationship between the slope of the graph of a linear function f and that of its inverse f^{-1}?

22. If f and g are linear functions with graphs that are perpendicular, show that the slopes of the graphs of $f \circ g$ and of $g \circ f$ are both -1.

23. A common practice in business and industry is for capital equipment to be **depreciated linearly** over a specified period of years. If we let $V(t)$ represent the value of the equipment t years after purchase, this means that $V(t) = at + b$, where a and b are constants determined by the following conditions: When $t = 0$, $V(t) = C$, the original cost. When $t = N$, $V(t) = 0$, where N is the number of years over which the equipment is being depreciated. Find the constants a and b in terms of C and N. What is the value after 8 years of an item originally costing \$10,000, which is to be depreciated linearly to 0 in 20 years?

24. In Problem 23 it is reasonable to assume that certain items will not depreciate to 0 but will have some residual value (at least as junk) after N years. If the residual value is R, find $V(t)$ in terms of C, N, and R. If the \$10,000 item in Problem 23 has a residual value of \$400 after 25 years, what is its value after 10 years?

25. A certain nonlinear function $L(x)$ is tabulated for x values given to two decimal places between $x = 1.00$ and $x = 9.99$. It is common practice to use **linear interpolation** to approximate $L(x)$ for x expressed to the nearest thousandth, that is, to assume $L(x)$ is linear between consecutive entries in the table. If $L(3.54) = 0.5490$ and $L(3.55) = 0.5502$, use linear interpolation to approximate $L(3.546)$.

26. A nonlinear function $C(x)$ is tabulated so that two consecutive entries in the table are $C(0.4363) = 0.9063$ and $C(0.4392) = 0.9051$. It is desired to find $C(0.4375)$. Use linear interpolation to approximate this. (See Problem 25.)

QUADRATIC FUNCTIONS

As we indicated in Section 1, a polynomial function of degree 2 is called a *quadratic* function. So the general form is $f(x) = a_2x^2 + a_1x + a_0$, or as it is more customarily written,

$$f(x) = ax^2 + bx + c \qquad (a \neq 0) \qquad (6)$$

The simplest case is the one for which $a = 1$ and b and c are both zero:

$$f(x) = x^2$$

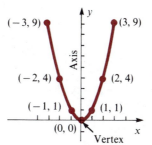

x	0	± 1	± 2	± 3
y	0	1	4	9

FIGURE 8

This is an even function, so its graph is symmetric to the y axis. Its domain is all of R, and its range is the set of all nonnegative real numbers. By plotting selected points and joining them with a smooth curve, we obtain the graph in Figure 8. The curve is called a **parabola.** The low point on the parabola is its **vertex,** and the axis of symmetry, in this case the y axis, is called the **axis** of the parabola.

We will refer to the function $f(x) = x^2$ as the **basic quadratic function** and to its graph in Figure 8 as the **basic parabola.** Using the ideas developed in Section 3 of Chapter 3, we can obtain from this basic curve the graphs of all other quadratic functions. First consider $y = ax^2$ for $a \neq 1$ or 0. Then $y = af(x)$, where $f(x)$ is our basic function. For $a > 0$, we know that the graph is obtained from the basic parabola by a vertical stretching if $a > 1$, or a compression if $0 < a < 1$. If $a < 0$, there is also a reflection in the x axis. So in all cases, the graph is still a parabola, with vertex at the origin and with axis along the y axis. Figure 9 shows the graphs for the cases $a = 2$, $a = \frac{1}{2}$, and $a = -\frac{3}{2}$.

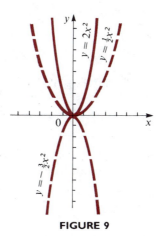

FIGURE 9

Now consider $y = ax^2 + k$. Recall that adding k produces a vertical translation of $|k|$ units, upward if $k > 0$ and downward if $k < 0$. Figure 10 illustrates this with $y = 2x^2 + 3$.

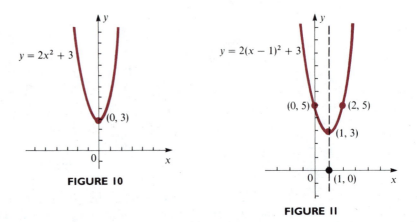

FIGURE 10

FIGURE 11

Finally, if we replace x by $x - h$, getting $y = a(x - h)^2 + k$, we know that the effect is to translate the graph of $y = ax^2 + k$ by $|h|$ units in the horizontal direction, to the right if $h > 0$ and to the left if $h < 0$. Figure 11 shows the graph of $y = 2(x - 1)^2 + 3$. It is the same as the parabola in Figure 10, shifted 1 unit to the right.

Summarizing, we know that the graph of $y = ax^2$ is obtained from the basic parabola by a vertical stretching or shrinking, together with a reflection if $a < 0$. Also, the graph of $y = a(x - h)^2 + k$ is the same as that of $y = ax^2$, translated h units horizontally and k units vertically. Thus, we can conclude the following:

The graph of

$$y = a(x - h)^2 + k \qquad (7)$$

is a parabola with the following properties:

1. Its vertex is at (h, k).
2. Its axis is the line $x = h$.
3. It opens upward if $a > 0$ and downward if $a < 0$.

Now let us return to the general quadratic function (6). Let $y = f(x)$. We factor out the coefficient a from the first two terms and complete the square on x:

$$y = a\left(x^2 + \frac{b}{a}x\right) + c$$

$$= a\left(x^2 + \frac{b}{a}x + \frac{b^2}{4a^2}\right) + c - \frac{b^2}{4a}$$

$$= a\left(x + \frac{b}{2a}\right)^2 + \frac{4ac - b^2}{4a}$$

Note that we added $b^2/4a^2$ inside the parentheses (the square of one-half of the coefficient of x), which had the effect of adding $b^2/4a$ after multiplying by a, so we subtracted this same amount to balance what we added.

If we let

$$h = -\frac{b}{2a} \qquad \text{and} \qquad k = \frac{4ac - b^2}{4a} \tag{8}$$

we can finally write

$$y = a(x - h)^2 + k$$

Therefore, the graph of

$$y = ax^2 + bx + c$$

is a parabola with vertex at (h, k), where h and k are given by equations (8). Its axis is the line $x = h$, and the parabola opens upward if $a > 0$ and downward if $a < 0$.

EXAMPLE 13 ⅠⅠⅠ

Find the vertex and the axis of the parabola with equation $y = x^2 - 2x + 3$ and draw its graph.

Solution   We write

$$y = x^2 - 2x + 3$$
$$= (x^2 - 2x + 1) + 3 - 1$$
$$= (x - 1)^2 + 2$$

We added 1 inside the parentheses to complete the square and subtracted 1 outside the parentheses to balance this off. Comparing this with equation (7), we conclude that the vertex is $(1, 2)$ and the axis is the line $x = 1$. We make a table of values and show the graph (Figure 12).

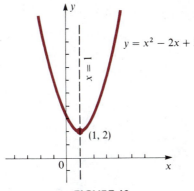

x	0	1	2	3	-1
y	3	2	3	6	6

FIGURE 12

Let us look again at equation (7), this time writing it in the form

$$y - k = a(x - h)^2 \qquad (9)$$

If we define new variables x' and y' by

$$\begin{cases} x' = x - h \\ y' = y - k \end{cases}$$

equation (9) becomes

$$y' = ax'^2$$

which we know is a parabola with vertex at the origin of the $x'y'$ axes, namely, the point (h, k) of the original coordinate system. This illustrates a general result which we now state as a theorem. It generalizes the translations we studied in Section 3 of Chapter 3, since there we were concerned only with graphs of functions, whereas this theorem deals with graphs of equations that may or may not represent functions.

Translation Theorem

If the equation of a curve in an xy coordinate system is known, and the curve is then translated h units horizontally and k units vertically, its new equation is obtained from the old by replacing x by $x - h$ and y by $y - k$.

Figure 13 illustrates this situation. To prove the theorem, we assume that the equation of the curve C is known and that C' is the same curve after translation. If we introduce x' and y' axes as shown with origin at (h, k), then the equation of C' with respect to the x', y' coordinates has exactly the same form as the equation of C with respect to the x, y coordinates. From the figure we see that the coordinates (x', y') of a point P on C' are related to its (x, y) coordinates by

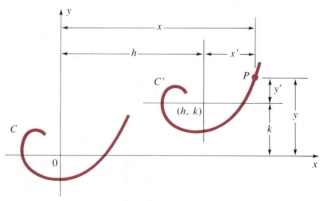

FIGURE 13

the equations $x = x' + h$, $y = y' + k$, or, equivalently,

$$\begin{cases} x' = x - h \\ y' = y - k \end{cases}$$

Then, after writing the equation of C' in terms of x' and y', we replace x' by $x - h$ and y' by $y - k$. Finally, the conclusion of the theorem follows if we observe that we can eliminate the intermediate step of using x', y' coordinates (they were only a means to an end).

Again, we can illustrate this with the parabola $y = ax^2$, whose equation after translation of the vertex to (h, k) becomes, by the theorem, $y - k = a(x - h)^2$, in agreement with equation (9). We will have occasion to use the theorem on translations in the next section.

It is often important to be able to determine the largest or smallest value of a function. In general, calculus is needed to do this, but in the case of a quadratic function, whose graph, as we know, is a parabola, all we need to do is to locate the vertex. The next three examples illustrate this.

EXAMPLE 14

Find the largest value of the function $f(x) = 1 - 8x - 2x^2$. Draw the graph.

Solution   Let $y = f(x)$ and complete the square on x in order to put the equation in the form of equation (7):

$$\begin{aligned} y &= -2x^2 - 8x + 1 \\ &= -2(x^2 + 4x) + 1 \\ &= -2(x^2 + 4x + 4) + 1 + 8 \\ &= -2(x + 2)^2 + 9 \end{aligned}$$

Notice that when we added 4 inside the parentheses to complete the square, we were actually adding -8 to the expression because of the factor -2, and so we had to balance this by adding $+8$.

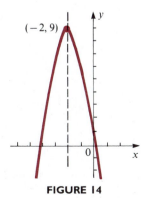

$(-2, 9)$

x	0	-1	-3	-4
y	1	7	7	1

FIGURE 14

The graph is a parabola, with vertex $(-2, 9)$, opening downward since the coefficient of the squared factor is negative. The graph is shown in Figure 14. The highest point on the graph is the vertex $(-2, 9)$, so the largest value of y, that is, $f(x)$, is 9.

▲

Example 14 illustrates a technique for finding **maximum and minimum values** of quadratic functions. Since the graph is a parabola, its vertex is the lowest point on the curve if it opens upward and the highest point if it opens downward. When a curve attains such a highest or lowest point, this is called a **maximum point or minimum point,** respectively, of the curve, and the corresponding y coordinate is the **maximum or minimum value of the function.** More general techniques for finding maximum and minimum values of a much broader class of functions are studied in calculus. The next example illustrates an application of finding the maximum value of a function.

EXAMPLE 15 |||||||||||||||||||||||||||||||||||||||||||||||||||||||||||||||||||||||||||||||||||||||||||||||||||||||||||||||||||||||||||||||

A company finds through experience that the cost function for a certain commodity is approximated by

$$C(x) = 3x^2 - 40x + 2{,}500$$

and that the revenue function is $R(x) = 200x$. Find the number of units that should be produced and sold to maximize profit.

Solution   Let $P(x)$ denote the profit. Then $P(x) = R(x) - C(x)$, since profit is revenue minus cost. So we have

$$P(x) = 200x - (3x^2 - 40x + 2{,}500) = -3x^2 + 240x - 2{,}500$$

The graph of $y = P(x)$ is a parabola opening downward, so $P(x)$ is greatest at the vertex. To find this, we complete the square, after factoring out the coefficient of

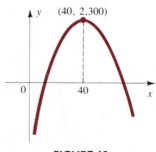

FIGURE 15
(not to scale)

x^2 from the first two terms:

$$y = -3(x^2 - 80x + 1,600) - 2,500 + 4,800$$
$$= -3(x - 40)^2 + 2,300$$

The graph is shown in Figure 15. Since the vertex $(40, 2,300)$ is the high point, we conclude that 40 units produced and sold yield the maximum profit of $2,300.

▲

EXAMPLE 16   ⅠⅠⅠ

The average nightly attendance at a certain movie theater is 100 when the ticket price is $4.00. The manager estimates that for each 10¢ reduction in price, the average attendance will increase by 10. Find the revenue function $R(x)$, where x is the number of 10¢ reductions in price. What ticket price yields the maximum revenue, and what is the maximum revenue?

Solution   The new price after x reductions of 10¢ is $4.00 - 0.10x$, and the new average attendance is $100 + 10x$. The revenue is the ticket price times the number in attendance. So

$$R(x) = (4.00 - 0.10x)(100 + 10x) = -x^2 + 30x + 400$$

Completing the square gives

$$R(x) = -(x^2 - 30x + 225) + 400 + 225 = -(x - 15)^2 + 625$$

The graph of R is a parabola, opening downward, with vertex at $(15, 625)$, as in Figure 16. So the revenue is maximum when $x = 15$. Thus, there should be 15 reductions in price of 10¢ each, or a total reduction of $1.50, making the new price $2.50 per ticket. The average attendance would then be estimated at $100 + 10(15) = 250$, and the revenue would be $625.

▲

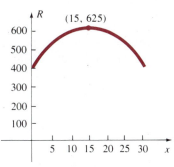

FIGURE 16

For certain purposes it is not necessary to know the exact location of the vertex of a parabola, and all that is needed is a rapid sketch indicating its general appearance. In this case the *intercepts*, especially the x intercepts, are particularly helpful. The x intercepts, if they exist, are found by setting $y = 0$, and the y intercept (there is always just one) is found by setting $x = 0$. The next example illustrates this.

EXAMPLE 17 ||||||||||||||||||||||||||||||||||||||||||||||||||||||||||||||||||||||||||||||||||||||||||||||||||||||||||||||||||||

Make a rapid sketch of the parabola $y = 2x^2 - 5x - 3$, showing where it crosses the x axis.

Solution   The parabola opens upward, since the coefficient of x^2 is positive. To find the x intercepts, we set $y = 0$ and solve for x:

$$2x^2 - 5x - 3 = 0$$
$$(2x + 1)(x - 3) = 0$$
$$x = -\tfrac{1}{2} \quad \text{or} \quad x = 3$$

The y intercept is -3, obtained by setting $x = 0$ in the original equation.

We can now make a sketch as shown in Figure 17. You might note that the axis of the parabola lies midway between the x intercepts.

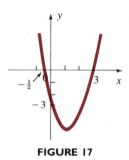

FIGURE 17

A rapid sketch of a parabola such as we have just described provides a useful alternative means of solving quadratic inequalities. The next three examples illustrate this.   ▲

EXAMPLE 18 ||||||||||||||||||||||||||||||||||||||||||||||||||||||||||||||||||||||||||||||||||||||||||||||||||||||||||||||||||||

Solve the inequality $x^2 - 12x < 0$ by graphical means.

Solution   We consider the function

$$y = x^2 - 12x = x(x - 12)$$

We know the graph is a parabola that opens upward, since the coefficient of x^2 is positive. From the factored form of the right-hand side, the x intercepts are seen

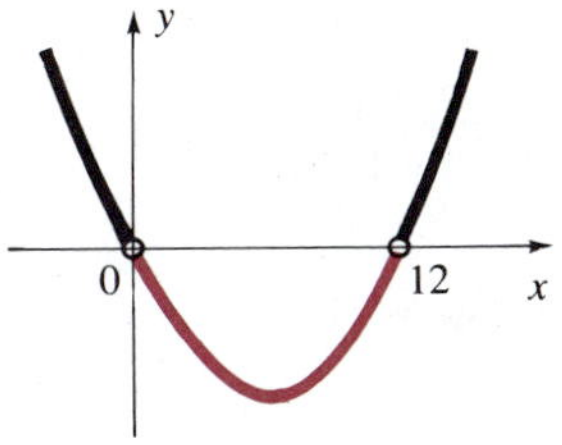

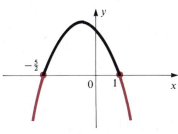

FIGURE 18

to be $x = 0$ and $x = 12$. From this we sketch the parabola, as shown in Figure 18, without attempting to locate the vertex exactly. All we want to know are the x values for which y is negative, that is, for which the curve lies *below* the x axis. We can see from the figure that the answer is those values of x for which $0 < x < 12$. Or we could say the solution set is the open interval $(0, 12)$.   ▲

Note. The graph in this method is a means to an end only. The answer must be expressed either in set notation or in interval notation.

EXAMPLE 19 ıı

Solve by graphical means: $5 - 3x - 2x^2 \le 0$

Solution   Let $y = 5 - 3x - 2x^2$. We will make a sketch of the graph and determine where $y \le 0$. To find the x-intercepts, we set $y = 0$ and factor:

$$5 - 3x - 2x^2 = 0$$
$$(5 + 2x)(1 - x) = 0$$
$$5 + 2x = 0 \quad \text{or} \quad 1 - x = 0$$
$$x = -\tfrac{5}{2} \quad \text{or} \qquad x = 1$$

FIGURE 19

We know the graph is a parabola that opens downward, since the coefficient of x^2 is negative, and we draw it, as shown in Figure 19. So $y \le 0$ when $x \ge 1$ or $x \le -\frac{5}{2}$, that is, the solution set is $(-\infty, -\frac{5}{2}] \cup [1, \infty)$.   ▲

EXAMPLE 20 ||||||||||||||||||||||||||||||||||||||||||||||||||||||||||||||||||||||||||||||||||||||||||||||||||||

Show that the inequality

$$3x^2 - 8x + 7 > 0$$

is satisfied by all real x.

Solution   Let $y = 3x^2 - 8x + 7$. Its graph is a parabola opening upward. To find the x intercepts, we set $y = 0$ and solve, making use of the quadratic formula.

$$3x^2 - 8x + 7 = 0$$

$$x = \frac{8 \pm \sqrt{64 - 84}}{6} = \frac{8 \pm \sqrt{-20}}{6}$$

$$= \frac{8 \pm 2i\sqrt{5}}{6} = \frac{4 \pm i\sqrt{5}}{3}$$

Since the solutions are imaginary, the graph does not cross the x axis. The only way an upward-opening parabola can fail to cross the x axis is for the vertex, and hence the entire parabola, to lie above the x axis. Thus, $y > 0$ for all x. No sketch is needed in this case. A similar analysis applies for a downward-opening parabola. If there are no real zeros, the vertex, and hence the entire graph, lie below the x axis. ▲

Note. The graphical method for solving inequalities is not limited to quadratic functions. If the general shape of the graph of a function f is known, and its x intercepts can be found, then solutions of the inequalities $f(x) < 0$, $f(x) \leq 0$, $f(x) > 0$, and $f(x) \geq 0$ can be read from the graph. For this technique to be practical, however, you should be able to make a rapid sketch of the function. A graphing calculator is very useful for this method.

EXAMPLE 21 ||||||||||||||||||||||||||||||||||||||||||||||||||||||||||||||||||||||||||||||||||||||||||||||||||||

Draw the graph of the function

$$f(x) = |x^2 - 4x + 3|$$

Solution   Since $x^2 - 4x + 3 = (x - 1)(x - 3)$, we see from Figure 20 that $x^2 - 4x + 3 < 0$ on the interval $(1, 3)$, and $x^2 - 4x + 3 \geq 0$ on $(-\infty, 1] \cup [3, \infty)$. Thus, by definition of absolute value,

$$f(x) = \begin{cases} -(x^2 - 4x + 3) & \text{if} \quad 1 < x < 3 \\ x^2 - 4x + 3 & \text{if} \quad x \leq 1 \quad \text{or} \quad x \geq 3 \end{cases}$$

Thus, to get the graph of f, we reflect the portion of the parabola in Figure 20 that is below the x axis in the x axis to obtain Figure 21.

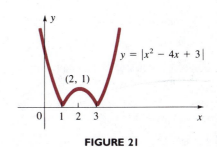

$y = x^2 - 4x + 3$

$(2, -1)$

FIGURE 20

$y = |x^2 - 4x + 3|$

$(2, 1)$

FIGURE 21

Note. The procedure of Example 21 can be used in general to obtain the graph of $f(x) = |g(x)|$, whenever the graph of g is known. The graph of f is the same as that of g wherever $g(x) \geq 0$, and where $g(x) < 0$ its graph is reflected in the x axis to get the graph of f.

EXERCISE SET 4

A

In Problems 1–20 find the vertex and axis of the parabola, and draw the graph.

1. $y = \dfrac{x^2}{4}$

2. $y = -x^2$

3. $y = x^2 + 3$

4. $y = 2 - x^2$

5. $y = (x + 3)^2 + 2$

6. $y = -(x - 1)^2 + 3$

7. $y = \frac{3}{2}(x - 2)^2 - 1$

8. $y = -2(x + 1)^2 - 2$

9. $y = x^2 + 2x + 5$

10. $y = 3 - 4x - x^2$

11. $y = 2x^2 - 6x + 7$

12. $y = 2x^2 + 5x - 4$

13. $y = 2x - x^2$

14. $y = 3x - 2x^2$

15. $y = 4 - 3x - 2x^2$

16. $y = 3x^2 - 4x + 5$

17. $y = \dfrac{x^2}{2} - x - 3$

18. $y = \dfrac{x(3 - x)}{2} + 1$

19. $y = x(3 - 2x) - 4$

20. $y = x(2x + 1) - 7$

In Problems 21–28 find the maximum or minimum value of each function, and state the value of x that gives this maximum or minimum value.

21. $f(x) = 3x - x^2$

22. $f(x) = x^2 - 2x + 5$

23. $g(x) = 2x^2 - 8x + 7$

24. $h(x) = 9 - 5x - 2x^2$

25. $F(x) = 3 - 4x - 5x^2$

26. $G(x) = 4x^2 + 7x - 12$

27. $\phi(x) = 3x^2 - 14x$

28. $\psi(x) = 3x(2 - x) + 11$

29. For a small manufacturing firm the cost $C(x)$ in dollars of producing x items per day is given by

$\sqrt{ertex}$ $C(x) = x^2 - 140x + 5500$

How many items should be produced each day to minimize cost? What is the minimum cost? y

30. A company that produces cellular phones analyzes its production and finds that the cost function C and the revenue function R (both in dollars) are approximated by

$$C(x) = 0.2x^2 + 320x + 25{,}000 \quad \text{and} \quad R(x) = 780x$$

where x is the number of cellular phones produced and sold each month. How many phones should be produced and sold each month to maximize profit? What is the maximum profit?

31. An object is projected vertically upward from the ground at an initial velocity of 192 feet per second. If air resistance is neglected, its distance $s(t)$ in feet above the ground after t seconds is given by $s(t) = 192t - 16t^2$. Find the maximum height the object will rise.

32. An object is propelled vertically upward at time $t = 0$ in such a way that its height y above the

ground at time t is given by

$$y = -16t^2 + \tfrac{247}{6}t + 4$$

Draw the graph of this equation. How long will the object remain in the air?

In Problems 33–46 solve the inequality by using the graphical method.

33. $x^2 - 4x > 0$
34. $x^2 - 2x - 3 \geq 0$
35. $x^2 + 3x - 4 \leq 0$
36. $x^2 - x - 20 < 0$
37. $2 - 5x - 3x^2 \geq 0$
38. $8 - 2x - 3x^2 > 0$
39. $2x^2 - 5x < 3$
40. $12x^2 - 5x \geq 3$
41. $2(2 - x^2) > 7x$
42. $3(4 - x^2) < 7x$
43. $10x(x - 2) \geq 3(x - 4)$
44. $4x(3x - 2) \geq 15$
45. $x^2 - 2x \leq 4$
46. $x(x + 2) \leq 1$

B

47. A rectangular plot of ground is to be fenced on three sides to form a garden, with the fourth side bounded by a brick wall. If there are 50 feet of fencing available, what should be the dimensions of the garden to produce the maximum area?
48. A yacht is rented for an excursion for 60 passengers at $20 per passenger. For each person in excess of 60 that goes, up to a maximum of 40 additional passengers, the fare of each passenger will be reduced by 25¢. Find the total number of passengers that will produce the maximum revenue.
49. A restaurant has a fixed price of $12 for an early evening special complete dinner. The average number of customers for this special is 160. The owner estimates on the basis of prior experience that for each 50¢ increase in the cost of the dinner, there will be 5 fewer customers on the average. What price should be charged to produce the maximum revenue, and what is this maximum revenue?

In Problems 50 and 51 use the graphical method of solving inequalities to find the domain of each function.

50. $f(x) = 3x - \sqrt{18 - 3x - 15x^2}$

51. $g(x) = \dfrac{2x - 3}{\sqrt{12x^2 + x - 20}}$

52. Solve by graphical means:

$$x^3 - x^2 - 4x + 4 \leq 0$$

Hint. Denote the left-hand side by $f(x)$. Factor to find the x intercepts, and sketch the graph with the aid of a few additional points.

In Problems 53 and 54 show that the inequality is satisfied for all real values of x.

53. $2x^2 - 8x + 9 > 0$ 54. $x(1 - x) \leq \tfrac{1}{4}$
55. The revenue from the sale of x units of a certain commodity is 38 times the number of units sold. The cost of producing x units is $105 + 12x + x^2$. Write the profit function and find the range of values of x for which the profit is positive. What value of x produces the maximum profit, and what is that profit?
56. A company that produces cells for solar collectors finds through experience that on the average it can sell x cells per day at a selling price (in dollars) of $p(x) = 100 - 0.05x$ (p is called the **demand function**). The cost function $C(x)$ for producing x cells per day is $C(x) = 4000 + 60x - 0.01x^2$.
   a. Write the revenue function $R(x)$.
   b. Write the profit function $P(x)$.
   c. Find the value of x that maximizes the profit, and find the maximum profit.

In Problems 57–60 draw the graph of f.

57. $f(x) = |x^2 - 4|$ 58. $f(x) = 2 - |3 - x^2|$
59. $f(x) = |3 - 2x - x^2|$
60. $f(x) = |2x - x^2| + 1$

INTRODUCTION TO CONIC SECTIONS

The parabola is one of four curves known as **conic sections.** The other three are the circle, the ellipse, and the hyperbola. The reason they are called conic sections is that each of them is formed from the intersection of a plane and a cone, as shown in Figure 22. In this section we give equations of these curves and their graphs, when they have particular orientations with respect to the axes. For a

Plane perpendicular
to axis of cone;
intersection is a
circle.

Plane parallel
to element of cone;
intersection is a
parabola.

Plane intersecting
both nappes of cone;
intersection is a
hyperbola.

Any other plane
intersecting one nappe;
intersection is an
ellipse.

(a)

(b)

(c)

(d)

FIGURE 22

more detailed treatment, including definitions of each of the conic sections and derivation of their equations, see Chapter 12.

PARABOLAS WITH HORIZONTAL AXES

We saw in Section 4 that the graph of an equation of the form

$$y = ax^2 + bx + c \qquad (a \neq 0) \tag{10}$$

is a parabola with a vertical axis, opening upward if $a > 0$ and downward if $a < 0$. By completing the square we found that we could write the equation in the form $y = a(x - h)^2 + k$, or equivalently,

$$y - k = a(x - h)^2 \tag{11}$$

The vertex is (h, k), and the axis is the line $x = h$.

If in equation (10) we interchange the roles of x and y, we obtain

$$x = ay^2 + by + c \qquad (a \neq 0) \tag{12}$$

which is the equation of a parabola with horizontal axis, opening to the right if $a > 0$ and to the left if $a < 0$. Completing the square yields the following equation, which is analogous to equation (11):

$$x - h = a(y - k)^2 \tag{13}$$

Again the vertex is (h, k). The axis is the horizontal line $y = k$. Equations (11) and (13) are called **standard forms** of equations of parabolas with vertical and horizontal axes, respectively.

EXAMPLE 22

Sketch the graph of the equation $4y^2 - x - 8y + 2 = 0$.

First we solve for x, obtaining an equation in the form of equation (12):

$$x = 4y^2 - 8y + 2$$

We complete the square in y and write the equation in the form of equation (13):

$$x = 4(y^2 - 2y + 1) + 2 - 4$$
$$= 4(y - 1)^2 - 2$$
$$x + 2 = 4(y - 1)^2$$

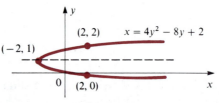

FIGURE 23

So $h = -2$ and $k = 1$. These are the coordinates of the vertex. The axis is the line $y = 1$, and the parabola opens to the right, as shown in Figure 23. ▲

It should be emphasized that an equation of the form (12) or (13) does not define y as a function of x, since there are vertical lines that intersect the graph in more than one point. However, either the upper or lower half of such a parabola is the graph of a function. The next example illustrates this.

EXAMPLE 23

Identify and draw the graph of the function defined by

$$f(x) = \sqrt{x - 2} + 1$$

Solution   The domain of f is the set $\{x: \ x \geq 2\}$, and for x in this domain, $f(x) \geq 1$. Set $y = f(x)$, and write the equation in the form

$$y - 1 = \sqrt{x - 2}$$

If we square both sides, we get

$$x - 2 = (y - 1)^2$$

This is in the form of equation (13), and so its graph is a parabola with vertex $(2, 1)$, axis $y = 1$, and opening to the right. For our original function, however, we had the restriction that $y \geq 1$. So instead of the entire parabola, we have only the upper half, shown in Figure 24.

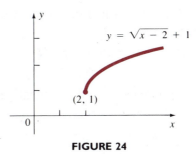

FIGURE 24

CIRCLES

A circle is defined as the set of points in a plane at a fixed distance (the **radius**) from a fixed point (the **center**). Let the center be at (h, k) and the radius be r. Then a point (x, y) lies on the circle if and only if its distance from (h, k) is r (Figure 25). By the distance formula, this condition can be written

$$\sqrt{(x - h)^2 + (y - k)^2} = r$$

or equivalently,

$$(x - h)^2 + (y - k)^2 = r^2 \tag{14}$$

Equation (14) is called the **standard form of the equation of a circle.** If the center is at the origin, it assumes the particularly simple form

$$x^2 + y^2 = r^2 \tag{15}$$

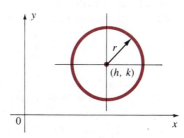

FIGURE 25

EXAMPLE 24

The endpoints of a diameter of a certain circle are $(2, -1)$ and $(4, 3)$. Find the center, radius, and equation of the circle, and draw its graph.

Solution   The center is at the midpoint of the diameter:

$$(h, k) = \left(\frac{2 + 4}{2}, \frac{-1 + 3}{2}\right) = (3, 1)$$

The radius is half the length of the diameter:

$$r = \tfrac{1}{2}\sqrt{(4 - 2)^2 + (3 + 1)^2}$$
$$= \tfrac{1}{2}\sqrt{4 + 16} = \sqrt{5}$$

So by equation (14), the equation of the circle in standard form is

$$(x - 3)^2 + (y - 1)^2 = 5$$

Its graph is shown in Figure 26.

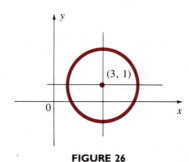

FIGURE 26

If equation (14) is expanded and the terms are rearranged, we obtain an equation of the form

$$x^2 + y^2 + ax + by + c = 0 \tag{16}$$

A natural question is whether every equation of this form has a circle as a graph. As the next example shows, this is not necessarily the case.

EXAMPLE 25

Discuss the graphs of each of the following equations:
a. $x^2 + y^2 - 2x + 4y + 1 = 0$ b. $x^2 + y^2 - 2x + 4y + 5 = 0$
c. $x^2 + y^2 - 2x + 4y + 9 = 0$

Solution   **a.**   We complete the squares in x and y:

$$(x^2 - 2x + ①) + (y^2 + 4y + ④) = -1 + ① + ④$$

The encircled terms were added on the left to complete the squares and on the right to balance what was added on the left. The equation can now be written in the form of equation (14).

$$(x - 1)^2 + (y + 2)^2 = 4$$

Its graph is a circle of radius 2 with center at $(1, -2)$.

b.   The completion of the squares is the same as in part **a.** The only difference is that after collecting terms on the right, we get 0. So the equation becomes

$$(x - 1)^2 + (y + 2)^2 = 0$$

This equation is satisfied by the point $(1, -2)$ and by no other point. The equation is said to represent a **degenerate circle**, since the radius has shrunk to 0.

c.   Again, completing the squares is the same, but the right-hand side now becomes negative:

$$(x - 1)^2 + (y + 2)^2 = -4$$

Since the left-hand side, being the sum of squares, is nonnegative, this equation is not satisfied by any pair (x, y). So there is no graph.   ▲

This example is typical of the way to handle equations in the form (16). By completing the squares, we obtain an equation that is in the form (14), provided the right-hand side is positive, and so its graph is a circle. If the right-hand side is 0, it is a degenerate circle (a point), and if the right-hand side is negative, there is no graph. So we can say that an equation in the form (16) always represents a circle or a degenerate circle, if it has a graph at all. The key to recognizing the circle is the combination $x^2 + y^2$ in equation (16). Contrast this with equations (10) and (12), representing parabolas with vertical and horizontal axes, respectively. In each of these x^2 *or* y^2 appears, but not both. This is the key to recognizing the equation of a parabola of one of these types. When x^2 is the only second degree term, the graph is a parabola with vertical axis, and when y^2 is the only second degree term, the parabola has a horizontal axis.

As with parabolas having horizontal axes, circles cannot be graphs of functions. However, semicircles can be, as the next example illustrates.

EXAMPLE 26   ｜｜｜

Identify and draw the graph of the function f, defined by

$$f(x) = \sqrt{9 - x^2}$$

Solution   The domain of f is the interval $[-3, 3]$, and for x in this domain $f(x) \geq 0$. Set $y = f(x)$ and square both sides:

$$y = \sqrt{9 - x^2}$$
$$y^2 = 9 - x^2$$
$$x^2 + y^2 = 9$$

This is in the form (15) and so represents a circle centered at the origin, with radius 3. However, for our original function y must be nonnegative. So the graph is just the upper half of this circle, as shown in Figure 27.

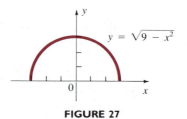

$$y = \sqrt{9 - x^2}$$

FIGURE 27

ELLIPSES

The standard form of the equation of an ellipse centered at the origin is

$$\frac{x^2}{a^2} + \frac{y^2}{b^2} = 1 \tag{17}$$

Note that if $a = b$, this assumes the form of equation (15) of a circle of radius a, after clearing of fractions. The graph of equation (17) is symmetric to both axes and the origin. Its x intercepts are $\pm a$, and the y intercepts are $\pm b$. If $a > b$, the ellipse is elongated on the x axis, as in Figure 28a; and if $b > a$, it is elongated on the y axis, as in Figure 28b. The origin is called the **center** of the ellipse, and the line segments joining the intercepts on each of the axes are called the **axes** of the ellipse. The longer axis is the **major axis** and the shorter one is the **minor axis.** The endpoints of the major axis are called the **vertices.**

If the center of the ellipse is shifted to the point (h, k), the equation becomes, by the translation theorem,

$$\frac{(x - h)^2}{a^2} + \frac{(y - k)^2}{b^2} = 1 \tag{18}$$

The next example shows how to put the expanded form of the equation of an ellipse into the standard form (18).

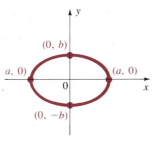

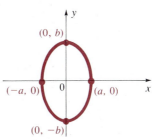

a. $\dfrac{x^2}{a^2} + \dfrac{y^2}{b^2} = 1$ $(a > b)$ **b.** $\dfrac{x^2}{a^2} + \dfrac{y^2}{b^2} = 1$ $(b > a)$

FIGURE 28

EXAMPLE 27 ||||||||||||||||||||||||||||||||||||||||||||||||||||||||||||||||||||||||||||||||||||||||||||||||||||||||||

Identify and draw the graph of the equation

$$4x^2 + 9y^2 + 16x - 18y - 11 = 0$$

Solution   To complete the squares, we first factor out the coefficient of x^2 from terms involving x, and the coefficient of y^2 from those involving y:

$$4(x^2 + 4x + ④) + 9(y^2 - 2y + ①) = 11 + ⑯ + ⑨$$

Notice that we added 4 and 1 inside the parentheses to complete the squares, but this had the effect of adding 16 and 9 on the left-hand side and so had to be balanced off accordingly on the right. Simplifying, we get

$$4(x + 2)^2 + 9(y - 1)^2 = 36$$

which, on dividing by 36, becomes

$$\frac{(x + 2)^2}{9} + \frac{(y - 1)^2}{4} = 1$$

By comparison with the standard form (18) we recognize this as the equation of an ellipse with center $(-2, 1)$ and major axis horizontal. Since $a^2 = 9$ and $b^2 = 4$, we have $a = 3$ and $b = 2$. So the length of the major axis is $2a = 6$ and that of the minor axis is $2b = 4$. The graph is shown in Figure 29.

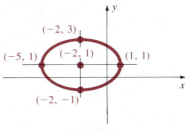

FIGURE 29

HYPERBOLAS

There are two standard forms for equations of hyperbolas centered at the origin. These are

$$\frac{x^2}{a^2} - \frac{y^2}{b^2} = 1 \tag{19}$$

and

$$\frac{y^2}{b^2} - \frac{x^2}{a^2} = 1 \tag{20}$$

In each case the graph is symmetric to the x axis, the y axis, and the origin. The graphs are shown in Figure 30. For the graph of equation (19) in Figure 30(a),

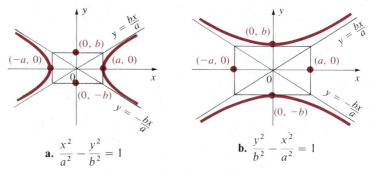

FIGURE 30

the **vertices** are the points $(a, 0)$ and $(-a, 0)$, and the line segment joining these is called the **transverse axis.** The line segment joining $(0, b)$ and $(0, -b)$ is called the **conjugate axis.** For the hyperbola (20) shown in Figure 30(b), the vertices are $(0, b)$ and $(0, -b)$, and the transverse axis is the line segment joining them. The conjugate axis in this case is the segment joining $(a, 0)$ and $(-a, 0)$. In each case the origin is the **center** of the hyperbola. The lines $y = bx/a$ and $y = -bx/a$ are called **asymptotes** of the hyperbola. The hyperbola approaches these lines more and more closely as the curve recedes indefinitely. An easy way to draw the asymptotes is to mark the points $(\pm a, 0)$ and $(0, \pm b)$ and draw a rectangle with vertical and horizontal lines, respectively, through them, as shown in Figure 30. This is called the **fundamental rectangle.** The asymptotes are the diagonals of the rectangle.

If the center is translated to (h, k), then by the translation theorem, the equations become

$$\frac{(x-h)^2}{a^2} - \frac{(y-k)^2}{b^2} = 1 \tag{21}$$

and

$$\frac{(y-k)^2}{b^2} - \frac{(x-h)^2}{a^2} = 1 \tag{22}$$

The asymptotes in this case are the lines

$$y - k = \pm\frac{b}{a}(x - h)$$

and can again be obtained from the fundamental rectangle centered at (h, k).

EXAMPLE 28

Identify and draw the graph of the equation

$$25x^2 - 16y^2 + 400 = 0$$

Solution We proceed as follows to put the equation in one of the standard forms:

$$25x^2 - 16y^2 = -400$$

Now divide by -400 to get a 1 on the right:

$$\frac{y^2}{25} - \frac{x^2}{16} = 1$$

This is of the form (20) and so represents a hyperbola centered at the origin, opening vertically. We draw the fundamental rectangle, using the points $(\pm4, 0)$ and $(0, \pm5)$, and then use the asymptotes to help sketch the graph, as in Figure 31.

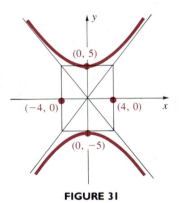

FIGURE 31

EXAMPLE 29

Discuss and sketch the graph of the equation

$$4x^2 - 9y^2 - 16x - 18y - 29 = 0$$

Solution   We complete the squares, as with the ellipse:

$$4(x^2 - 4x + 4) - 9(y^2 + 2y + 1) = 29 + 16 - 9$$

Be careful here

$$4(x - 2)^2 - 9(y + 1)^2 = 36$$

$$\frac{(x - 2)^2}{9} - \frac{(y + 1)^2}{4} = 1$$

We recognize this as a hyperbola centered at $(2, -1)$, with horizontal transverse axis. Since $a^2 = 9$ and $b^2 = 4$, so that $a = 3$ and $b = 2$, we draw the fundamental rectangle by going ± 3 units horizontally and ± 2 units vertically from the center. The result is shown in Figure 32.

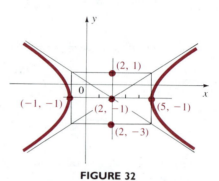

FIGURE 32

EXERCISE SET 5

A

In Problems 1–40 identify the curve and sketch its graph.

1. $x = y^2$
2. $y^2 = x + 2$
3. $x^2 + y^2 = 16$
4. $x^2 + y^2 = 5$
5. $\dfrac{x^2}{9} + \dfrac{y^2}{4} = 1$
6. $\dfrac{x^2}{1} + \dfrac{y^2}{4} = 1$
7. $\dfrac{x^2}{9} - \dfrac{y^2}{4} = 1$
8. $\dfrac{y^2}{4} - \dfrac{x^2}{1} = 1$
9. $x^2 + 4y^2 = 4$
10. $4x^2 + 4y^2 = 25$
11. $9x - 4y^2 = 0$
12. $9x^2 + 4y^2 = 36$
13. $x^2 = 4(y^2 - 1)$
14. $2x = y^2 - 4$
15. $3(x^2 + y^2) = 16$
16. $y^2 = \dfrac{16}{9}(9 - x^2)$

17. $y^2 = \dfrac{16}{25}(x^2 - 25)$
18. $2x = (y + 2)^2$
19. $(x - 3) = -2(y + 2)^2$
20. $(x - 1)^2 + (y + 2)^2 = 4$
21. $\dfrac{(x + 3)^2}{16} + \dfrac{(y - 2)^2}{4} = 1$
22. $(x + 2)^2 = 4(y - 3)$
23. $\dfrac{x^2}{4} - \dfrac{(y - 1)^2}{9} = 1$
24. $\dfrac{y^2}{4} - \dfrac{(x - 1)^2}{9} = 1$
25. $9x^2 + 4(y - 2)^2 = 36$
26. $9y^2 + 25(x + 1)^2 = 225$
27. $y^2 + 8x - 6y + 41 = 0$
28. $y^2 - 2x - 6y + 13 = 0$
29. $x^2 + y^2 - 4x + 2y = 0$

30. $x^2 + y^2 + 6x + 4y - 3 = 0$
31. $9x^2 + 4y^2 + 36x - 24y + 36 = 0$
32. $x^2 + 4y^2 - 2x + 16y + 13 = 0$
33. $4x^2 - y^2 - 32x + 4y + 64 = 0$
34. $x^2 - 4y^2 - 2x - 16y - 19 = 0$
35. $4x^2 + 9y^2 - 16x - 20 = 0$
36. $x^2 + y^2 + 2x - 4y + 1 = 0$
37. $2y^2 - 16y - x + 35 = 0$
38. $9x^2 - 16y^2 + 36x + 96y + 36 = 0$
39. $16x^2 - 9y^2 - 32x - 54y + 79 = 0$
40. $25x^2 + 16y^2 + 100x - 32y - 284 = 0$

In Problems 41–48 draw the graph of the given function by letting y equal the function value, squaring, and writing the result as the standard form for a conic. Take into account the necessary restrictions.

41. $f(x) = \sqrt{4 - x^2}$ 42. $g(x) = -\sqrt{1 - x}$

43. $F(x) = \dfrac{2}{3}\sqrt{9 - x^2} - 1$

44. $G(x) = \sqrt{x + 2} + 3$

45. $\phi(x) = \dfrac{2}{3}\sqrt{9 + x^2} + 2$

46. $\psi(x) = \sqrt{x^2 + 1} - 4$

47. $f(x) = -\sqrt{1 - 2x}$ 48. $h(x) = -\sqrt{9 - 4x^2}$

In Problems 49 and 50 find the equation of the circle satisfying the given conditions.

49. Endpoints of a diameter are $(-1, 3)$ and $(-5, 7)$.
50. A diameter is the portion of the line $3x - 2y = 12$ cut off by the coordinate axes.

B

In Problems 51–54 write the equation in standard form, identify, and sketch.

51. $3x^2 + 3y^2 - 8x + 10y - 7 = 0$
52. $3y^2 + 2x - 5y - 4 = 0$
53. $5x^2 + 3y^2 + 7x - 2y - 11 = 0$
54. $12x^2 - 3y^2 - 18x - 3y - 10 = 0$

In Problems 55–58 draw both curves on the same coordinate system and shade the area between them.

55. $y = \sqrt{3 + x^2}$, $x - 2y + 5 = 0$
56. $y = \sqrt{2x - x^2}$, $y = (x - 1)^2$
57. $y^2 = 8x$, $8x^2 + 5y = 12$
58. Inside $x^2 + y^2 + 4x - 2y + 1 = 0$ and to the right of $y^2 - 3x - 2y = 5$
59. Find the equation of the tangent line to the circle $x^2 + y^2 - 4x + 6y - 12 = 0$ at the point $(5, -7)$. **Hint.** The tangent line is perpendicular to the radius drawn to the point.
60. A circle of radius 5 has its center on the line $3x - 2y + 6 = 0$, and the circle is tangent to the y axis. Find its equation. (There are two solutions.)
61. Show that the circles $x^2 + y^2 - 8x + 2y - 19 = 0$ and $x^2 + y^2 + 4x - 14y + 37 = 0$ are tangent to each other. **Hint.** Show that the distance between their centers equals the sum of their radii.

REVIEW EXERCISE SET

A

1. Find the slopes of the lines joining the following pairs of points:
   a. $(-1, 3)$ and $(5, -4)$
   b. $(-2, -6)$ and $(-4, 2)$
2. In each of the following a line passes through the given point and has the given slope. Find a second point on the line.
   a. $(2, 5)$, slope $\frac{2}{3}$ **b.** $(-3, 2)$, slope $-\frac{5}{2}$

3. Find the unknown if the line joining the two points has the given slope.
   a. $(5, -3)$ and $(7, y)$; slope 3
   b. $(-4, 2)$ and $(x, -6)$; slope $-\frac{4}{3}$

In Problems 4–7 find the equations of the lines satisfying the given conditions.

4. **a.** Slope $-\frac{1}{2}$, passing through $(-5, 3)$
   b. Passing through $(-1, -4)$ and $(2, -6)$

5. a. Slope $\frac{3}{2}$, y intercept -2
   b. Passing through $(2, 7)$ and $(-1, 4)$
6. a. Slope $\frac{3}{4}$, passing through $(-1, -5)$
   b. Passing through $(-2, 4)$ and $(5, 7)$
7. a. y intercept 3, slope $\frac{3}{7}$
   b. Passing through $(5, -3)$ and $(-2, 6)$

In Problems 8 and 9 give the slope, x intercept, and y intercept, and draw the graphs.

8. a. $5x - 4y + 20 = 0$ **b.** $3(y - 4) = 6 - 2x$
9. a. $3(x - 4) = 2(y + 3)$
   b. $\dfrac{2x - 3}{5} = \dfrac{1 - y}{3}$

10. Find the length of the line segment joining the given points.
   a. $(-2, 1)$ and $(3, -4)$
   b. $(0, 3)$ and $(4, 0)$
   c. $(4, -2)$ and $(-1, 10)$
   d. $(6, 9)$ and $(-2, -6)$
   e. $(7, -5)$ and $(-2, 3)$
11. Find the midpoint of each line segment in Problem 10.
12. Find y so that the distance between $(-3, 2)$ and $(5, y)$ is 10. (There are two solutions.)
13. Show that the triangle with vertices $(-1, 5)$, $(3, -2)$, and $(-2, -3)$ is isosceles.

In Problems 14–16 find the equations of the lines satisfying the given conditions.

14. a. Passing through $(3, 5)$ and parallel to $2x - 4y + 3 = 0$
   b. Passing through $(-1, 2)$ and perpendicular to $x + 5y - 6 = 0$
15. a. Perpendicular to the line joining $(2, 6)$ and $(-3, 2)$ and passing through $(1, -1)$
   b. Parallel to the line joining $(1, 4)$ and $(-3, 7)$ and passing through $(2, -5)$
16. a. Parallel to the line joining $(-1, 2)$ and $(5, 4)$ and passing through the origin
   b. Perpendicular to the line $3x + 2y - 4 = 0$ and having x intercept -1
17. For the triangle with vertices at $A(3, -2)$, $B(-1, 4)$, and $C(5, 6)$, find the following:
   a. The equation of the line through C and parallel to the side AB
   b. The equation of the altitude drawn from C

18. The line through $(-4, -1)$ and $(2, k)$ is parallel to the line $5x + 2y + 4 = 0$. Find k.
19. Find the equation of the perpendicular bisector of the line segment joining $(2, 7)$ and $(-4, -3)$.
20. a. Find the equation of the family of lines parallel to $3x - 5y + 2 = 0$. Find the equation of the member of this family that has y intercept 2.
   b. Find the equation of the family of lines perpendicular to $2x + 3y - 4 = 0$. Find the equation of the member of this family that has an x intercept at the same point as that of the given line.
21. a. Find the equation of the family of lines parallel to the line $8x + 3y + 7 = 0$. Find the equation of the member of this family that passes through the point $(2, -3)$.
   b. Find the equation of the family of lines perpendicular to the line $y = 3x - 4$. Find the equation of the member of this family that passes through the point $(-1, 5)$.
22. For a certain linear function f it is given that $f(2) = -1$ and $f(5) = 3$. Find $f(x)$ and draw the graph of f.
23. If g is a polynomial function of degree 1, and $g(0) = -3$ and $g(2) = -1$, find $g(x)$. What is the slope of the graph of g?
24. For a certain linear function f it is given that $f(-1) = 2$ and $f(3) = -4$. Find $f(x)$ and draw the graph of f.
25. Let $f(x) = (3x + 4)/5$ and $g(x) = 2 - 3x$. Find $f \circ g$ and $g \circ f$, and draw their graphs.
26. Let $g(x) = 4x - 5$ and $h(x) = (x - 2)/3$. Find $g \circ h$ and $h \circ g$, and draw their graphs.
27. The temperature of the air diminishes approximately linearly with altitude h, for altitudes of up to 6 miles, dropping about $2°F$ for each 1,000 feet of altitude. If T_0 is the temperature on the ground, write the temperature T as a linear function of the altitude h(feet). If $T_0 = 72°F$, find T at an altitude of 18,000 feet.
28. A bicycle shop owner sells an average of 20 tenspeed bicycles a month when the price is \$175. When the price is reduced by 20%, sales go up by 50%. Assuming a linear relationship between price p and number of sales n, find the equation relating them. How many ten-speed bicycles should the shop owner expect to sell per month if the price is reduced by 10% from the original price?

In Problems 29 and 30 find the vertex and the axis of the parabola, and draw its graph.

29. a. $y = x^2 - 6x + 4$ **b.** $y = -2x^2 + 4x - 3$

30. a. $y = 4x - x^2$ **b.** $y = \dfrac{x^2}{2} + 2x + 3$

In Problems 31 and 32 find the maximum or minimum value of each function and state the value of x that gives this maximum or minimum value.

31. a. $f(x) = 3x^2 - 9x + 4$
　　b. $g(x) = 8 - 4x - x^2$
32. a. $F(x) = 2x^2 - 3x + 7$
　　b. $h(x) = 3x(2 - x) + 4$
33. The cost of producing x parts per month for a certain piece of equipment is given by the formula

$$C(x) = 2x^2 - 400x + 30{,}000$$

How many parts should be produced to minimize cost? What is the minimum cost?

34. The profit function $P(x)$ from producing and selling a certain type of pocket calculator is given by

$$P(x) = -0.02x^2 + 60x - 35{,}000$$

where x is the number produced and sold per week, and $P(x)$ is given in dollars. How many calculators should the manufacturer produce and sell each week to maximize profits? What is the maximum profit?

In Problems 35–37 solve the inequalities by the graphical method.

35. a. $x^2 - x - 2 < 0$ **b.** $3 - 2x - x^2 > 0$
36. a. $3x^2 - 5x - 2 \geq 0$ **b.** $2x^2 - 5x + 2 \leq 0$
37. a. $2(3x^2 + 2) \geq 11x$ **b.** $8 \geq x(2 + 3x)$

In Problems 38–45 identify the curve and draw its graph.

38. a. $x^2 + y^2 = 25$ **b.** $16x^2 + 25y^2 = 400$
39. a. $y^2 + 2x = 0$ **b.** $4x^2 - y^2 = 4$

40. a. $x - 2 = \dfrac{1}{4}(y + 1)^2$

　　b. $y^2 - x^2 = 1$
41. a. $4y^2 - 9x^2 + 54x + 8y = 41$
　　b. $y = 3 - 2(x + 1)^2$

42. a. $9x^2 + 4y^2 - 18x + 16y = 11$
　　b. $y^2 - 4x + 6y + 9 = 0$
43. a. $x^2 + 4x + 3y = 2$
　　b. $x^2 + y^2 + 2x - 8y + 1 = 0$
44. a. $y = \sqrt{2(x - 1)} - 2$
　　b. $y = \sqrt{9 - x^2} + 1$
45. a. $y = \sqrt{3 - x^2 - 2x}$
　　b. $y = 2 - \sqrt{x - 4}$

B

46. Show that the points $(-5, 2), (-3, -2), (6, 1)$, and $(4, 5)$ are the vertices of a parallelogram. Find the equations of the sides.
47. For the triangle with vertices at $(0, 5), (-4, -3)$, and $(6, 1)$ find the equations of the following:
　a. The altitudes　　**b.** The medians
　(See Problems 56 and 57, Exercise Set 2.)
48. A space shuttle is placed in a circular orbit around the earth. When its engine is fired, it will move in a straight line tangent to the orbit. If with respect to a certain coordinate system for the plane of the orbit the center of the earth is at $(-1, 2)$ and the engine is fired when the spacecraft is at $(2, 3)$, what is the equation of the line along which the space shuttle will move?
　Note. The tangent to the circle at a point is perpendicular to the radius drawn to that point.
49. By calculus it can be shown that the slope of the tangent line to the curve $y = x^4 - 2x^2 + 3x - 7$ at any point (x, y) is given by $m = 4x^3 - 4x + 3$. Find the equations of the tangent line and the normal line (the line perpendicular to the tangent) at the point for which $x = -1$.
50. In calculus it can be shown that the slope of the tangent line at any point (x, y) on the curve $y = 1/x$ is given by the formula $m = -1/x^2$. Find the equations of the tangent line and the normal line at the point for which $x = -1$. Draw the graph, showing the tangent and normal.
51. Use the graphical method to solve the inequality $x^3 + 4x^2 - x - 4 \geq 0$.

In Problems 52 and 53 identify the curve and draw its graph.

52. a. $3x^2 + 3y^2 + 2x - 4y - 5 = 0$
　　b. $8x^2 - 8x + 12y + 11 = 0$

53. a. $3x^2 - 12y^2 - 3x - 18y = 10$
  b. $2x^2 + 4y^2 + 5x + 6y - 4 = 0$

54. Prove that the line $3x - 4y + 15 = 0$ is tangent to the circle $x^2 + y^2 - 4x + 2y - 20 = 0$ at the point $(-1, 3)$.

55. At a certain movie theater the price of admission is $3.00, and the average daily attendance is 200. As an experiment the manager reduces the price by 5¢ and finds that the average attendance increases by 5 people per day. Assuming that for each further 5¢ reduction the average attendance would rise by 5, find the number of 5¢ reductions that would result in the maximum revenue.

56. A railroad company agrees to run an excursion train for a group under the following conditions: If 200 or fewer people go, the rate will be $10 per ticket. The rate for each ticket will be reduced by 2¢ for each person in excess of 200 who go. What number of people will produce the maximum revenue for the railroad?

57. Find the dimensions of the rectangle of maximum area that can be inscribed in an isosceles triangle of altitude 8 and base 6. (See sketch.)

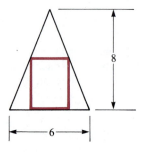

Hint. Use similar triangles to express the height of the rectangle in terms of its base.

5

HIGHER-DEGREE POLYNOMIAL AND RATIONAL FUNCTIONS

Polynomial equations of degree higher than 2 often occur in calculus, so it is important to learn ways of solving them. In the following excerpt from a calculus textbook the author first notes some of the algebraic techniques that can be used, and then he gives an example.*

The techniques of college algebra (synthetic division, deflation, the quadratic formula, rational zeros, Descartes' rules of signs) can be used to find zeros of many polynomials. . . .

Example   Sketch the graph of the function

$$p(x) = 3x^5 - 25x^4 + 80x^3 - 120x^2 + 80x - 24$$

Solution   Following the algorithm, we first compute the derivative:

$$p'(x) = 15x^4 - 100x^3 + 240x^2 - 240x + 80.$$

We find by college algebra methods that the zeros of p' are 2 (of multiplicity 3) and $\frac{2}{3}$, so these are the critical points of p. . . .

There is more to the example, including the sketch, but most of it involves calculus. The **zeros** of p' are the solutions of the equation $p'(x) = 0$, and in this chapter we will learn the algebraic techniques the author mentions for finding these zeros.

---

* Lynn E. Garner, *Calculus and Analytic Geometry* © 1988 by Dellen Publishing Company, a division of Macmillan, Inc. Reprinted by permission of Macmillan Publishing Company.

HISTORICAL NOTE

The problem of trying to find formulas analogous to the quadratic formula for the roots of polynomial equations of degree greater than 2 occupied mathematicians for centuries. For the cases $n = 3$ and $n = 4$, called the **cubic** and **quartic** equations, respectively, formulas were obtained around 1540 by the Italian mathematicians Tartaglia (cubic) and Ferrari (quartic). Attempts over the next two centuries to solve the general **quintic** (degree 5) equation all met with failure. Finally, in 1824, the brilliant Norwegian mathematician Niels Henrik Abel (1802–1829), at the age of 22, made the remarkable discovery that no such general formula for the solutions of polynomial equations of degree higher than 4 could be found. This settled the matter once and for all, but in a surprising way. It is indeed a curious fact that algebraic formulas for the solutions of degree n equations in terms of their coefficients exist for the cases $n = 1, 2, 3, 4$ but not for $n \geq 5$. It is not simply that no one has been able to find these formulas—it is *impossible* for anyone ever to find them because they do not exist. This is not to say that *no* equation of degree 5 or higher can be solved in terms of its coefficients; in some cases this can be done. The special case $ax^5 + b = 0$ can certainly be solved for one of its roots, namely, $x = \sqrt[5]{-b/a}$. Furthermore, we will show in this chapter a technique for finding the rational roots of polynomial equations (with integral coefficients) of arbitrary degree and also a way of approximating the irrational roots to any desired degree of accuracy. So the situation is not hopeless. As a matter of fact, the formulas developed by Tartaglia and Ferrari are of little practical importance. Cubic and quartic equations are typically solved not by their formulas but by the techniques to be presented here (or by numerical techniques with the aid of the computer). The quadratic formula, on the other hand, is used extensively because of its simplicity.

THE REMAINDER THEOREM AND THE FACTOR THEOREM

Let P_n denote a polynomial function of degree n:

$$P_n(x) = a_n x^n + a_{n-1}x^{n-1} + \cdots + a_1 x + a_0$$

When this is set equal to 0, it becomes a polynomial equation, and we will be seeking ways to solve such equations for $n > 2$. Recall that we use the terms **root** and **solution** interchangeably, and if r is a root of the equation, then r is also called a **zero** of the polynomial, and conversely. For example, we may say that the roots of the polynomial equation $x^2 - x - 2 = 0$ are 2 and -1, or equivalently, that the zeros of the polynomial $x^2 - x - 2$ are 2 and -1. The zeros of a polynomial are the x intercepts of its graph.

If we divide $P_n(x)$ by $x - r$, we obtain a quotient of degree $n - 1$ and a constant term as remainder. For example, let us divide $2x^3 - 3x + 4$ by $x - 2$:

$$
\begin{array}{r}
2x^2 + 4x\ + 5 \\
x - 2\overline{\smash{\big)}\,2x^3 \qquad\quad - 3x +\ \ 4} \\
\underline{2x^3 - 4x^2} \\
4x^2 - 3x \\
\underline{4x^2 - 8x} \\
5x +\ \ 4 \\
\underline{5x - 10} \\
14 \quad \text{Remainder}
\end{array}
$$

The quotient $2x^2 + 4x + 5$ is one degree less than the dividend, and the remainder 14 is a constant. We could write in this case

$$\frac{2x^3 - 3x + 4}{x - 2} = 2x^2 + 4x + 5 + \frac{14}{x - 2}$$

and obtain from multiplication by $(x - 2)$

$$2x^3 - 3x + 4 = (x - 2)(2x^2 + 4x + 5) + 14$$

This is an identity in x; that is, it is true for all values of x. In a similar way, if we denote the quotient on dividing $x - r$ into $P_n(x)$ by $Q(x)$ and the remainder by R, we obtain the identity

$$P_n(x) = (x - r)Q(x) + R \qquad\qquad (1)$$

Two results can now be obtained, and these are stated as theorems.

THE REMAINDER THEOREM   **When the polynomial $P_n(x)$ is divided by $x - r$, the remainder is equal to $P_n(r)$.**

Proof   In equation (1) substitute $x = r$. We obtain

$$
\begin{aligned}
P_n(r) &= (r - r)Q(r) + R \\
&= 0 + R \\
&= R
\end{aligned}
$$

For example, if we let $P(x) = 2x^3 - 3x + 4$, we can conclude from the division problem above that $P(2) = 14$, since when we divided by $x - 2$, we obtained 14 as the remainder. We emphasize that to get $P(2)$, we divide by $x - 2$ and read off the *remainder* as the answer.

THE FACTOR THEOREM   **The polynomial $x - r$ is a factor of the polynomial $P_n(x)$ if and only if r is a root of the polynomial equation $P_n(x) = 0$, or equivalently, if and only if r is a zero of the polynomial $P_n(x)$.**

For example, consider the polynomial $P(x) = x^3 - x^2 + x - 1$. By inspection we see that $x = 1$ is a root of the equation $P(x) = 0$. By the factor theorem it follows that $x - 1$ is a factor of $P(x)$. We can see this is true by factoring as follows:

$$x^3 - x^2 + x - 1 = x^2(x - 1) + (x - 1)$$
$$= (x^2 + 1)(x - 1)$$

Conversely, since $x - 1$ is a factor of $P(x)$, by putting $x = 1$, the polynomial equals 0.

These two theorems, although very simple, play a fundamental role in the solution of equations. The second theorem shows that the problem of finding roots of a polynomial equation and the problem of factoring a polynomial are essentially the same.

SYNTHETIC DIVISION

Since division will play a prominent role in the rest of this discussion, we will pause here to describe a shortened version of long division known as **synthetic division.** Consider a typical long division problem in which the divisor is of the special form $x - r$. For example, let us repeat the division of $2x^3 - 3x + 4$ by $x - 2$, shown in Section 2:

$$
\begin{array}{r}
2x^2 + 4x + 5 \\
x - 2 \overline{\smash{\big)}\, 2x^3 \quad\quad -3x + 4} \\
\underline{2x^3 - 4x^2} \\
4x^2 - 3x \\
\underline{4x^2 - 8x} \\
5x + 4 \\
\underline{5x - 10} \\
14
\end{array}
$$

We will go through a series of steps that simplifies this procedure. First, there really is no need to write down the terms that have been circled above. We know that by subtraction the result will be 0. Leaving these out, we have

$$
\begin{array}{r}
2x^2 + 4x + 5 \\
x - 2 \overline{\smash{\big)}\, 2x^3 \quad\quad -3x + 4} \\
\underline{- 4x^2} \\
4x^2 - 3x \\
\underline{- 8x} \\
5x + 4 \\
\underline{- 10} \\
14
\end{array}
$$

Next, note that the circled terms in the above display are the same as the coefficients in the quotient; this is because the coefficient of x in the divisor is unity. With this observation, then, we can leave out the x in the divisor. Also, we can dispense with the powers of x everywhere and let the position of the coefficient indicate the associated power of x. The display now takes the form

$$
\begin{array}{r}
2 \quad\; 4 \quad\; 5 \\
\hline
-2\overline{)2 \quad\; 0 \; -3 \quad\; 4} \\
\underline{-4 } \\
4 \;\; (-3) \\
\underline{-8 } \\
5 \quad\; (4) \\
\underline{-10} \\
14
\end{array}
$$

Note that we have written a 0 for the coefficient of the missing power. The first coefficient 2 in the quotient is the same as the first coefficient in the dividend. Each succeeding coefficient is obtained by multiplying the preceding coefficient by the -2 of the divisor and subtracting the result from the appropriate coefficient in the dividend. This can be simplified by changing the sign of the -2 in the divisor to $+2$ and *adding* rather than subtracting. Also, there is no need to repeat the circled terms above, because they occur in the dividend. This will enable us to compress the entire display as follows:

$$
\begin{array}{r}
2 \quad 4 \quad\; 5 \\
\hline
2\overline{)2 \quad 0 \; -3 \quad 4} \\
4 \quad\; 8 \quad 10 \\
\hline
4 \quad\; 5 \quad 14
\end{array}
$$

Finally, if we bring down the initial coefficient 2 of the dividend to the bottom line, we see that the answer can be read from that line, so that we can omit the top line entirely. So we have

$$
\begin{array}{r}
\underline{2|} \quad 2 \quad 0 \; -3 \quad\; 4 \\
\downarrow \quad 4 \quad\;\; 8 \quad 10 \\
\hline
2 \quad 4 \quad\;\; 5 \quad 14
\end{array}
$$

Recall that the second line is obtained by multiplying the 2 in the divisor position by the numbers 2, 4, and 5, respectively, of the bottom line. Thus, $4 = 2 \cdot 2$, $8 = 2 \cdot 4$, and $10 = 2 \cdot 5$. The entries in the third line are obtained by adding the two entries above them. Finally, the answer is read, from right to left, as the remainder, constant term, coefficient of x, and coefficient of x^2.

Remark. Although 2 appears in the divisor position, the divisor is $x - 2$.

While the procedure was illustrated with a specific example, the method is general. Note carefully that the divisor must be of the form $x - r$, and r then is the number that appears in the position of the divisor. For example, if the divisor were $x + 2$, then this must be thought of as $x - (-2)$, so that -2 is in the divisor position. Remember, too, that since the positions of the coefficients indicate the powers of x, *these coefficients must be arranged according to descending powers of* x, *and a 0 must be supplied as the coefficient of any missing power.*

EXAMPLE 1

Divide $4x^4 - 15x^3 + 8x^2 - x + 7$ by $x - 3$.

Solution

$$
\begin{array}{r|rrrrr}
3 & 4 & -15 & 8 & -1 & 7 \\
  &   & 12 & -9 & -3 & -12 \\
\hline
  & 4 & -3 & -1 & -4 & -5
\end{array}
$$

Thus, the quotient is $4x^3 - 3x^2 - x - 4$, and the remainder is -5.   ▲

EXAMPLE 2

Divide $x^5 - 2x^4 + x$ by $x + 2$.

Solution

$$
\begin{array}{r|rrrrrr}
-2 & 1 & -2 & 0 & 0 & 1 & 0 \\
   &   & -2 & 8 & -16 & 32 & -66 \\
\hline
   & 1 & -4 & 8 & -16 & 33 & -66
\end{array}
$$

Thus, the quotient is $x^4 - 4x^3 + 8x^2 - 16x + 33$, and the remainder is -66.   ▲

A Word of Caution. In synthetic division be sure to use 0 as the coefficient of each missing power of x.

Let us now consider the problem of evaluating a polynomial function $P_n(x)$ at a particular value of x, say $x = r$. There are two ways of doing this:

1. **The direct method:** Substitute r for x wherever x appears.
2. **Use the remainder theorem:** Divide $P_n(x)$ by $x - r$, using synthetic division; the answer sought is the remainder.

Consider, for example, the polynomial used previously, $P(x) = 2x^3 - 3x + 4$, and evaluate this by each method when $x = 2$. By the direct method,

$$P(2) = 2(2)^3 - 3(2) + 4 = 16 - 6 + 4 = 14$$

To apply method 2, we divide $P(x)$ by $x - 2$, using synthetic division, and read off the remainder as the answer.

$$
\begin{array}{r|rrrr}
2 & 2 & 0 & -3 & 4 \\
  &   & 4 & 8 & 10 \\
\hline
  & 2 & 4 & 5 & \boxed{14 = P(2)}
\end{array}
$$

Remark. In general, method 2 is easier to use than method 1 for polynomials of degree 3 or greater, and for this reason we will use method 2 extensively throughout the remainder of this chapter. However, you may use the direct method if you prefer to do so (except in problems explicitly calling for method 2).

A calculator can be used to evaluate a polynomial function by either method we have described. To illustrate method 2, consider the polynomial of Example 1, which we will call $Q(x)$.

$$Q(x) = 4x^4 - 15x^3 + 8x^2 - x + 7$$

We will show the keystrokes for the synthetic division in Example 1.

3 STO × 4 − 15 = × RCL + 8 = × RCL − 1 = × RCL + 7 =

The displayed answer will be -5, which is $Q(3)$. It is interesting to note that this method is equivalent to evaluating a polynomial directly when it is written in what is called **nested form.** For $Q(x)$, the nested form is

$$Q(x) = (((4x - 15)x + 8)x - 1)x + 7$$

You can verify this is correct by performing the indicated multiplications.

EXERCISE SET 3

A

In Problems 1–12 use synthetic division to divide the first polynomial by the second.

1. $x^2 - 3x - 2$ by $x - 3$
2. $x^2 - 3x + 4$ by $x - 2$
3. $x^3 - 2x^2 + x - 5$ by $x - 1$
4. $2x^3 + 3x^2 - x - 4$ by $x + 1$
5. $x^3 - 2x^2 + 7$ by $x + 2$
6. $4x^3 - 3x + 4$ by $x - 2$
7. $x^4 - 2x^2 - 3$ by $x - 3$
8. $x^4 - 3x^3 - 2x - 5$ by $x + 3$
9. $x^5 - 5x^3 - 18$ by $x + 2$
10. $4x^3 + 3x^2 - x$ by $x - 1$
11. $3x^4 - 14x^3 - 4x^2 - 15$ by $x - 5$
12. $2x^4 + 32x^2 - 3x - 7$ by $x + 4$

In Problems 13–28 make use of synthetic division and the remainder theorem to find the indicated function values.

13. $P(x) = 3x^2 - 14x + 11; P(4)$
14. $P(x) = x^3 - x^2 + 2x + 3; P(2)$
15. $P(x) = x^3 + 6x^2 + 7x - 3; P(-2)$

16. $Q(x) = 2x^3 - 3x^2 + 7; Q(-1)$
17. $f(x) = 2x^3 - 4x^2 - 3x + 7; f(3)$
18. $g(t) = 3t^4 - 5t^2 - 12; g(-2)$
19. $f(x) = 3x^4 - 10x^3 - 24x - 16; f(4)$. Also find $f(4)$ by substitution. Compare the difficulty of the two methods.
20. $P(t) = t^5 - 7t^3 - 11t + 13; P(2)$
21. $Q(t) = 3t^3 + 12t - 8; Q(-3)$
22. $F(x) = x^4 - 120; F(5)$
23. $P(s) = 3s^4 - 20s^2 - 15s + 17; P(3)$
24. $g(y) = y^4 - 3y^2 - 8; g(-3)$
25. $P(x) = x^4 + 7x^3 + 3x^2 - 4x + 9; P(-2)$
26. $Q(t) = 1 - 2t - 3t^2; Q(5)$
27. $P(x) = 4 - 5x - 3x^2 - x^3; P(2)$
28. $F(x) = 9 - x^2 + 2x^3 - x^4; F(-3)$

B

Use synthetic division to perform the divisions in Problems 29–32.

29. $(6x^4 - 5x^3 + 3x - 8) \div (x - \frac{1}{2})$
30. $(3x^4 - 4x^3 + 4x^2 - x) \div (x + \frac{2}{3})$
31. $(x^3 - 4x^2 + 2x - 3) \div (2x - 1)$

Hint. Write $2x - 1 = 2(x - \frac{1}{2})$. First divide by $x - \frac{1}{2}$, using synthetic division, and then divide the quotient by 2.

32. $(2x^4 - 4x^3 + 3x + 7) \div (3x + 2)$ (See the hint for Problem 31.)

33. By writing

$$\frac{P(x)}{ax + b} = \frac{P(x)}{a\left(x + \dfrac{b}{a}\right)} = \frac{1}{a}\left[\frac{P(x)}{x + \dfrac{b}{a}}\right]$$

justify the procedure suggested in Problem 31.

34. Show by two methods that $x = -2$ is a zero of the polynomial $P(x) = 3x^4 - 8x^3 - 15x^2 + 12x - 28$.

35. Show by synthetic division that $x^2 - 2x - 8$ is a factor of $x^4 - 2x^3 - 6x^2 - 4x - 16$.
    Hint. Show that $(x - 4)$ and $(x + 2)$ are both factors.

36. Show by synthetic division that $x^2 - 4$ is a factor of $x^5 - 2x^4 + 11x^2 - 16x - 12$. (See the hint for Problem 35.)

37. Let $P(x) = x^5 - 4x^4 - 5x^3 + 20x^2 + 4x - 16$. Use synthetic division and the remainder theorem to show that $P(4) = 0$. Show the complete factorization of $P(x)$.

38. Use synthetic division to find $f(\frac{1}{2})$ if $f(x) = 3 - x + 4x^2 - x^4$.

39. Let $P(x) = 7 + 5x - 3x^3 - 2x^4$. Find $P(-\frac{3}{2})$ by using synthetic division.

40. Let $Q(x) = 12x^4 + 5x^3 - 6x^2 - 3x + 4$. Find $Q(\frac{3}{4})$ by using synthetic division.

41. Show by synthetic division that $x = 5$ is a zero of the polynomial $2x^3 - 9x^2 - 11x + 30$. Factor this polynomial completely. What are the other zeros?

42. Show by synthetic division that $x = -4$ is a root of the equation $3x^3 + 10x^2 - 7x + 4 = 0$ and that the other roots are imaginary. Find the other two roots.

In Problems 43 and 44 find the specified function values by using synthetic division and a hand calculator.

43. If $f(x) = 3.205x^3 - 7.561x^2 - 1.235x + 8.679$, find
    a. $f(2.761)$ **b.** $f(-6.032)$

44. If $g(t) = 13.75 + 24.17t - 5.234t^2 - 10.83t^4$, find
    a. $g(-2.546)$ **b.** $g(0.3125)$

THE FUNDAMENTAL THEOREM OF ALGEBRA AND ITS COROLLARIES

As its name implies, the following theorem is fundamental to the theory of equations. Its proof is beyond the scope of this course, but two important corollaries will be proved.

THE FUNDAMENTAL THEOREM OF ALGEBRA

Every polynomial equation $P_n(x) = 0$ with $n \geq 1$ has at least one root in the field of complex numbers.

Our interest will center on polynomials with real, and usually integral, coefficients, but this theorem is valid when the coefficients are any complex numbers. The root that is asserted to exist, however, cannot be assumed to be real, even when the coefficients are. For example, the equation $x^2 + 4 = 0$ has the root $x = 2i$ (also $x = -2i$), which is not real, even though the coefficients are integers. As a corollary to the fundamental theorem, we have the following:

COROLLARY 1   **A polynomial of degree $n \geq 1$ can be factored into n (not necessarily distinct) linear factors over the complex number field.**

Proof   By the factor theorem, we know that $x - r$ is a factor of $P_n(x)$ if and only if $P_n(r) = 0$, that is, if and only if r is a root of the polynomial equation $P_n(x) = 0$. By the fundamental theorem, we know that $P_n(x)$ does have a root, say, r_1. Thus,

$$P_n(x) = (x - r_1)P_{n-1}(x)$$

where $P_{n-1}(x)$ is the quotient of $P_n(x)$ divided by $x - r_1$, and hence is of degree $n - 1$ and has the same leading coefficient, say, a_n, as $P_n(x)$. Further, $P_n(x) = 0$ if and only if

$$(x - r_1)P_{n-1}(x) = 0$$

So, applying the factor theorem and the fundamental theorem to $P_{n-1}(x)$, we know there is a root, say r_2, of $P_{n-1}(x) = 0$, so that

$$P_{n-1}(x) = (x - r_2)P_{n-2}(x)$$

and thus

$$P_n(x) = (x - r_1)(x - r_2)P_{n-2}(x)$$

where $P_{n-2}(x)$ is of degree $n - 2$ and has the same leading coefficient as $P_{n-1}(x)$ and hence as $P_n(x)$. We continue this process until arriving at the stage

$$P_n(x) = (x - r_1)(x - r_2) \cdot \cdots \cdot (x - r_{n-1})P_1(x)$$

where $P_1(x)$ is of degree 1 and has the same leading coefficient as $P_n(x)$. That is, $P_1(x) = a_n x + b = a_n[x + (b/a_n)]$. If we write $r_n = -b/a_n$, we have, finally,

$$P_n(x) = a_n(x - r_1)(x - r_2) \cdot \cdots \cdot (x - r_n) \tag{2}$$

and the proof is complete.

It must be emphasized that the factors in the above corollary may not all be distinct, which is equivalent to saying that the r_k's are not necessarily distinct. Each factor $(x - r_k)$ corresponds to a root of $P_n(x) = 0$, according to the factor theorem. We therefore have another corollary.

COROLLARY 2   **A polynomial equation of degree $n \geq 1$ has at most n roots.**

There will, in fact, be exactly n roots if all the r_k's are distinct. If, in the factorization given in equation (2), $(x - r_k)$ occurs as a factor m times, then r_k is said to be a **root of multiplicity m** of the equation $P_n(x) = 0$.

Suppose now that the coefficients in

$$P_n(x) = a_n x^n + a_{n-1} x^{n-1} + \cdots + a_1 x + a_0$$

are all real, and suppose $x = r$ is an imaginary root of the equation $P_n(x) = 0$. We will show that the conjugate $\bar{r}$ is also a root. For

$$\overline{P_n(r)} = \overline{a_n r^n + a_{n-1} r^{n-1} + \cdots + a_1 r + a_0}$$
$$= \bar{a}_n \bar{r}^n + \bar{a}_{n-1} \bar{r}^{n-1} + \cdots + \bar{a}_1 \bar{r} + \bar{a}_0$$

where we have used the properties of conjugates repeatedly. Now, since the coefficients are real, conjugation leaves them unaltered; that is,

$$\overline{P_n(r)} = a_n \bar{r}^n + a_{n-1} \bar{r}^{n-1} + \cdots + a_1 \bar{r} + a_0 = P_n(\bar{r})$$

Finally, since $P_n(r) = 0$ by hypothesis and 0 is a real number,

$$\overline{P_n(r)} = \bar{0} = 0 = P_n(\bar{r})$$

So $\bar{r}$ is a root of $P_n(x) = 0$ as asserted. This result shows that **in polynomial equations with real coefficients, imaginary roots always occur in conjugate pairs.** Thus, if $2 + 3i$ is a root of such an equation, then $2 - 3i$ is also.

EXAMPLE 3

Find a polynomial having zeros 2, -1, and $2i$.

Solution   Since $2i$ is a zero, we know that its conjugate, $-2i$, is also a zero. According to the factor theorem the polynomial

$$P(x) = (x - 2)(x + 1)(x - 2i)(x + 2i)$$
$$= (x - 2)(x + 1)(x^2 + 4)$$

will have the desired roots. When the factors are multiplied, we get (verify)

$$P(x) = x^4 - x^3 + 2x^2 - 4x - 8 \qquad \blacktriangle$$

Because imaginary roots of polynomials with real coefficients always occur in conjugate pairs, we can show that every such polynomial can be factored over the reals into linear and/or quadratic factors. Suppose, for example, that r is an imaginary zero. Then by what we have seen, so is $\bar{r}$. Hence, two of the factors are $x - r$ and $x - \bar{r}$. When these are multiplied, we get

$$(x - r)(x - \bar{r}) = x^2 - (r + \bar{r})x + r\bar{r}$$

If $r = a + bi$, then $\bar{r} = a - bi$. So $r + \bar{r} = 2a$ and $r\bar{r} = a^2 + b^2$. Thus, $x^2 - (r + \bar{r})x + r\bar{r}$ is a quadratic polynomial with real coefficients that is irreducible over the reals. This reasoning holds for each pair of factors corresponding to complex conjugate pairs of roots. The only other roots are real, and the corresponding factors are linear.

Example 3 illustrates what we have just described. The factorization of the polynomial $x^4 - x^3 + 2x^2 - 4x - 8$ over the complex members is

$$(x - 2)(x + 1)(x - 2i)(x + 2i)$$

with all factors linear, as guaranteed by Corollary 1. When the complex factors are multiplied, we get the real irreducible quadratic factor $x^2 + 4$, so the factorization over the reals is

$$(x - 2)(x + 1)(x^2 + 4)$$

EXERCISE SET 4

A

In Problems 1–5 verify the factorizations by using synthetic division.

1. $3x^3 - 5x^2 - 4x + 4 = 3(x + 1)(x - 2)(x - \frac{2}{3})$
2. $2x^3 + 5x^2 - 18x - 45 = 2(x - 3)(x + 3)(x + \frac{5}{2})$
3. $4x^3 + 13x^2 + 4x - 12 = 4(x + 2)^2(x - \frac{3}{4})$
4. $3x^4 - 5x^3 + 10x^2 - 20x - 8 =$
   $3(x + \frac{1}{3})(x - 2)(x - 2i)(x + 2i)$
5. $4x^4 - 4x^3 - 19x^2 + 24x + 45 =$
   $4(x + \frac{3}{2})^2(x - 2 - i)(x - 2 + i)$
6. Construct a polynomial with integer coefficients whose only zeros are 3, -2, and $\frac{2}{3}$.
7. Find a polynomial equation of lowest degree that has roots -3, 2, and 4.
8. Construct a polynomial with integer coefficients whose only zeros are -1, $\frac{3}{2}$, and 5.
9. Find a polynomial equation with -2 as a root of multiplicity 3 and whose only other root is 3.
10. Find a polynomial with 3 as a zero of multiplicity 2 and whose only other zero is -4.
11. Find a fourth degree polynomial equation having -2 as a double root and $-2i$ as one of its other roots.
12. If P is a third degree polynomial for which $P(1) = 0$, $P(i) = 0$, and $P(0) = 1$, find $P(x)$.
13. If Q is a third degree polynomial with the simple zero $x = 2$ and a zero of multiplicity 2 at $x = -1$, and if $Q(1) = -8$, find $Q(x)$.
14. **a.** Explain why a fifth degree polynomial equation with real coefficients cannot have more than four distinct imaginary roots.

b. Explain why a sixth degree polynomial equation with real coefficients and no repeated roots cannot have an odd number of real roots.
15. Show by direct substitution that both $2 + i$ and its conjugate are roots of the equation $3x^3 - 8x^2 - x + 20 = 0$.
16. Use synthetic division to show that both $2i$ and $-2i$ are zeros of the polynomial $3x^4 - 5x^3 + 8x^2 - 20x - 16$.
17. Give an example of a quadratic equation with
   a. No real roots   **b.** Two real roots
   c. Exactly one real root
18. Factor the polynomial $x^4 + 6x^2 + 8$ into linear factors.
19. Let $P(x) = (2x^2 - x - 1)(x^2 + 2x + 2)$. Factor P completely over **(a)** the real field and **(b)** the complex field.

B

20. Show that every polynomial of *odd* degree with real coefficients has at least one real root.
21. Verify that $x = -2$ is a root of multiplicity 3 of the equation $2x^4 + 9x^3 + 6x^2 - 20x - 24 = 0$. What is the other root?
22. Find a polynomial whose only zeros are $4 + 5i$, $4 - 5i$, and -3.
23. Find a fourth degree polynomial with integer coefficients, two of whose zeros are $4 - 3i$ and $3 + 2i$.
24. Two roots of a certain fourth degree polynomial equation are $2 - 3i$ and $3 + 4i$. Find the equation.
25. Factor $x^6 - 64$ into linear factors.

RATIONAL ROOTS

If the polynomial equation $P_n(x) = 0$ has rational numbers as coefficients, it is equivalent to one with integer coefficients. This can be seen by multiplying both sides of the equation by the LCD of the coefficients. For example, by multiplying both sides of

$$\tfrac{2}{3}x^3 - \tfrac{1}{4}x^2 + \tfrac{5}{6}x - 1 = 0$$

by 12, we obtain the equivalent equation

$$8x^3 - 3x^2 + 10x - 12 = 0$$

For this reason we will concentrate our attention on polynomial equations with integer coefficients. When the coefficients are irrational, the roots can be approximated to any desired degree of accuracy by using numerical techniques. A hand calculator is very helpful in this situation.

Even when the coefficients are integers, the roots are not necessarily rational, or even real (for example, $x^2 + 1 = 0$ has imaginary roots). However, if there *are* rational roots, there is a procedure for finding them, which we now describe.

A rational number can always be written in the form p/q, where p and q are **relatively prime** integers (that is, they have in common no integer factor other than ± 1). Suppose p and q are relatively prime and $x = p/q$ is a root of the equation

$$P_n(x) = a_n x^n + a_{n-1} x^{n-1} + \cdots + a_1 x + a_0 = 0$$

where now it is assumed that all coefficients are integers. Then we have

$$a_n \frac{p^n}{q^n} + a_{n-1} \frac{p^{n-1}}{q^{n-1}} + \cdots + a_1 \frac{p}{q} + a_0 = 0 \tag{3}$$

We multiply both sides of (3) by q^n and obtain

$$a_n p^n + a_{n-1} p^{n-1} q + a_{n-2} p^{n-2} q^2 + \cdots + a_1 p q^{n-1} + a_0 q^n = 0 \tag{4}$$

We rewrite (4) in two equivalent forms. First, subtract $a_0 q^n$ from both sides and factor the common factor p from each term remaining on the left:

$$p(a_n p^{n-1} + a_{n-1} p^{n-2} q + \cdots + a_1 q^{n-1}) = -a_0 q^n$$

From this we see that p is a factor of the left-hand side and hence also of the right-hand side. But p and q are relatively prime; hence, p is not a factor of q^n.* So p must divide a_0.

Now write equation (4) in another way—this time subtracting $a_n p^n$ from both sides and factoring q from what remains on the left:

$$q(a_{n-1} p^{n-1} + a_{n-2} p^{n-2} q + \cdots + a_0 q^{n-1}) = -a_n p^n$$

Reasoning as above, we see that q must divide a_n. These results are summarized as a theorem.

---

* Although this really requires more proof, it is intuitively clear; we will not go into more detail here.

THEOREM   **The only possible rational roots of the polynomial equation with integral coefficients**

$$a_n x^n + a_{n-1} x^{n-1} + \cdots + a_0 = 0$$

are of the form p/q, where p is a factor of the constant term a_0 and q is a factor of the leading coefficient a_n.

EXAMPLE 4 |||||||||||||||||||||||||||||||||||||||||||||||||||||||||||||||||||||||||||||||||||||||||||||||||||||||||||||||||||||||||||||||

Find all roots of the equation $3x^3 - 8x^2 + 5x - 2 = 0$.

Solution   The possible numerators of rational roots are factors of 2, namely ± 1 and ± 2. The possible denominators are factors of 3, that is, 1 or 3. We need not consider negative denominators, since we get all possible signs by allowing the numerator only to vary. Thus, a listing of all possible rational roots is

$$\pm \tfrac{1}{3} \qquad \pm \tfrac{2}{3} \qquad \pm 1 \qquad \pm 2$$

We do not know whether any of these is actually a root, but at least the list of those we have to try is finite. There are various strategies for the order in which to test the possible rational roots. One way is to order them according to size and start with the smallest positive one and work up until (1) a root is found, (2) the list of positive possibilities is exhausted, or (3) it becomes evident that there is no need to proceed further. When a root r is encountered, we usually look for further roots in the so-called **depressed equation,** which is the quotient (on dividing by $x - r$) set equal to 0. If the depressed equation is quadratic, then the quadratic formula (or simple factorization) is used. If situation (2) or (3) above is encountered before finding all roots, then we can try the negative possibilities, working in reverse order.

For this problem the procedure is as follows:

$$
\begin{array}{r|rrrr}
\tfrac{1}{3} & 3 & -8 & 5 & -2 \\
& & 1 & & \\
\hline
& 3 & -7 & &
\end{array}
$$

This is not carried further since all remaining multiplications result in non-integers, and thus there is no possibility of getting a 0 remainder.

$$
\begin{array}{r|rrrr}
\tfrac{2}{3} & 3 & -8 & 5 & -2 \\
& & 2 & -4 & \\
\hline
& 3 & -6 & 1 &
\end{array}
$$

$$
\begin{array}{r|rrrr}
1 & 3 & -8 & 5 & -2 \\
& & 3 & -5 & 0 \\
\hline
& 3 & -5 & 0 & -2
\end{array}
$$

$$
\begin{array}{r|rrrr}
2 & 3 & -8 & 5 & -2 \\
& & 6 & -4 & 2 \\
\hline
& 3 & -2 & 1 & 0
\end{array}
$$

Now we have met with success. We have found that $x = 2$ is a root of the equation, since on division by $x - 2$ we have a remainder of 0. The polynomial can be factored as $(x - 2)(3x^2 - 2x + 1)$. Thus, all other roots are those of the depressed equation

$$3x^2 - 2x + 1 = 0$$

Since this is quadratic, we solve it by the quadratic formula:

$$x = \frac{2 \pm \sqrt{4 - 12}}{6} = \frac{1 \pm i\sqrt{2}}{3}$$

Note here, as we expected, the imaginary roots are conjugates. Now we have the complete solution set:

$$\left\{ 2, \ \frac{1 + i\sqrt{2}}{3}, \ \frac{1 - i\sqrt{2}}{3} \right\}$$

Recall from equation (2) that when the zeros of a polynomial are $r_1, r_2, \ldots, r_n$ and the leading coefficient is a_n, the polynomial can be factored as

$$a_n(x - r_1)(x - r_2) \cdots (x - r_n)$$

In our case we can therefore write

$$3x^3 - 8x^2 + 5x - 2 = 3(x - 2)\left(x - \frac{1 + i\sqrt{2}}{3} \right)\left(x - \frac{1 - i\sqrt{2}}{3} \right)$$

This is the complete factorization over the complex numbers. Over the reals the factorization is

$$3x^3 - 8x^2 + 5x - 2 = (x - 2)(3x^2 - 2x + 1)$$

This illustrates again that a polynomial with real coefficients can always be factored into real factors of degree at most two, in such a way that any quadratic factors cannot be further factored over the reals (they are irreducible over R).

▲

We summarize here for emphasis how to proceed after finding a root of a polynomial equation.

If a root $x = r$ of the polynomial equation $P_n(x) = 0$ has been found, then $P_n(x)$ can be factored in the form

$$(x - r)P_{n-1}(x)$$

where $P_{n-1}(x)$ is the quotient on dividing $P_n(x)$ by $x - r$. All other roots are found by solving the depressed equation

$$P_{n-1}(x) = 0$$

In using synthetic division to evaluate a polynomial $P(x)$ for several values of x, writing the coefficients of $P(x)$ for each division becomes tedious and inefficient. We illustrate below a tabular arrangement that is more compact, using the polynomial

$$P(x) = 2x^3 - 5x^2 - 8x + 12$$

For purposes of this illustration we will evaluate $P(x)$ for a range of positive and negative integral values of x, without regard to whether these are possible rational roots.

In the table we have written the coefficients of $P(x)$ only once, and for each division we have deleted the usual second line, carrying out the calculations mentally. For example, the line in Table 1 corresponding to $x = 1$ would be obtained as follows if we were using the usual format.

$$
\underline{1 \,\big|\ \ 2 \quad -5 \quad -8 \quad\ \ 12}
$$
$$
\quad\ \ \ 2 \quad -3 \quad -11
$$
$$
\ \ 2 \quad -3 \quad -11 \quad\ \ 1
$$

The numbers $x = r$ whose function values we are seeking are placed in the first column, and the remainders on division by $x - r$, that is, the values of $P(r)$, are in the last column.

r				$P(r)$	
0	2	-5	-8	12	
1	2	-3	-11	1	**Note sign change**
2	2	-1	-10	-8	
3	2	1	-5	-3	**Note sign change**
4	2	3	4	28	
5	2	5	17	97	
-1	2	-7	-1	13	**Note sign change**
-2	2	-9	10	-8	
-3	2	-11	25	-63	

TABLE I

Observe the sign changes noted in the table. For example, $P(1) = 1$ and $P(2) = -8$. Now it can be proved, using calculus, that the graph of every polynomial function has no breaks in it (the function is **continuous**). It follows that the graph of our polynomial must cross the x axis at least once between $x = 1$ and $x = 2$, since $P(1)$ and $P(2)$ are opposite in sign. Therefore, $P(x)$ has a zero between $x = 1$ and $x = 2$, or equivalently, the equation $P(x) = 0$ has a root in this interval. Similarly, $P(x) = 0$ has a root between $x = 3$ and $x = 4$ and between $x = -1$ and $x = -2$.

Refer again to Table 1, and observe that all entries corresponding to $r = 4$ are positive, as are those in the next line. We could, in fact, have anticipated that the entries corresponding to $r = 5$ would also be positive, because they were positive when $r = 4$ was the multiplier, and so they would have to be even larger using $r = 5$. The number 4 is therefore an upper bound to the roots of $P(x) = 0$ in this case. This means that there are no roots larger than 4.

Similarly, since in Table 1 the signs in the row corresponding to $r = -2$ alternate, they will continue to alternate for $r = -3$ as well as for other negative numbers larger in absolute value than 2. Thus, -2 is a lower bound to the roots of $P(x) = 0$.

Note. If the leading coefficient of a polynomial $P(x)$ is negative, an upper bound to the roots occurs at a value $r > 0$ when all other entries in the row corresponding to $x = r$ are negative. The important thing is that all entries are of the same sign. For lower bounds, the criterion of alternating signs for $r < 0$ continues to hold regardless of the sign of the leading coefficient.

These tests for upper and lower bounds to roots are summarized below.

Upper and Lower Bounds to Roots

If $r > 0$ and all entries in the table for calculating $P(r)$ by synthetic division are of the same sign, then r is an upper bound to the roots of $P(x) = 0$. If $r < 0$ and the entries alternate in sign, then r is a lower bound to the roots. In applying this test 0 may be treated as either positive or negative.

Upper and lower bounds can often save work in testing an equation for rational roots. For example, if an upper bound is encountered and some of the untested possibilities for rational roots are larger than this upper bound, they can be discarded.

Another aid in limiting the number of trials when attempting to find roots of a polynomial equation is given by the following rule, which we state without proof.

Descartes' Rule of Signs

Let $P_n(x)$ be a polynomial with real coefficients, of degree $n \geq 1$, with terms arranged in descending (or ascending) powers of x. The number of positive real roots of the equation $P_n(x) = 0$ is either equal to the number of sign changes between successive terms of $P_n(x)$ or fewer than this by an even number. The number of negative real roots is found by applying the same rule to $P_n(-x)$.

Note. In Descartes' rule a root of multiplicity m is counted m times. Observe that the rule gives information on the number of real roots, but these roots need not be rational.

We illustrate how to apply Descartes' rule with the equation

$$P(x) = 2x^3 - 5x^2 - 8x + 11 = 0$$

sign   sign
change   change

There are two sign changes between successive terms of $P(x)$, so there are either two or zero positive real roots. Now consider $P(-x)$:

$$P(-x) = 2(-x)^3 - 5(-x)^2 - 8(-x) + 11$$
$$= -2x^3 - 5x^2 + 8x + 11$$

sign
change

Since there is only one sign change, there must be exactly one negative real root to the original equation. (We cannot have fewer than one by an even number.) All the concepts discussed in this section are employed in the next two examples.

EXAMPLE 5

Find all rational roots of $2x^4 - 7x^3 - 10x^2 + 33x + 18 = 0$.

Solution   The possible numerators are $\pm 1, \pm 2, \pm 3, \pm 6, \pm 9, \pm 18$ and possible denominators are 1, 2. Therefore, the possible rational roots are $\pm\frac{1}{2}, \pm 1, \pm\frac{3}{2}, \pm 2, \pm 3, \pm\frac{9}{2}, \pm 6, \pm 9, \pm 18$. By Descartes' rule of signs we find that there are either two or zero positive real roots and two or zero negative real roots.

From Table 2 we see that $x = 3$ is a root. Now that a root has been found, we seek the remaining roots from the depressed equation

$$2x^3 - x^2 - 13x - 6 = 0$$

r					$P(r)$
0	2	-7	-10	33	18
$\frac{1}{2}$	2	-6	-13		
1	2	-5	-15	18	36
$\frac{3}{2}$	2	-4	-16	9	
2	2	-3	-16	1	20
3	2	-1	-13	-6	0
3	2	5	2	0	

TABLE 2

First, we observe that since the original polynomial can now be factored as

$$(x - 3)(2x^3 - x^2 - 13x - 6)$$

any root of the depressed equation is also a root of the original; so we know there is no need to try possible rational roots that have already been rejected. However, it is possible that $x = 3$ is a root of multiplicity 2 or greater, and so we test $x = 3$ in the depressed equation. This procedure is general; **when a root has been found, test this same number to see if it is a root of the depressed equation.** This is important, because otherwise multiple roots may not show up. Note, too, that the coefficients of the depressed equation are the ones on the line in Table 2 with the first $r = 3$ entry, so we have used these in the synthetic division testing $r = 3$ a second time, writing the remainder of the division one column to the left. Since this remainder is 0, we conclude that 3 is a double root. The new depressed equation is quadratic, so we solve it as follows:

$$2x^2 + 5x + 2 = 0$$

$$(2x + 1)(x + 2) = 0$$

$$x = -\tfrac{1}{2} \quad \text{or} \quad x = -2$$

The complete solution set is therefore $\{3, -\tfrac{1}{2}, -2\}$, with 3 a root of multiplicity 2 (which Descartes' rule counts as two roots).   ▲

EXAMPLE 6

Find all rational roots of $x^5 - 8x^3 - 10x^2 + 12x - 20 = 0$.

Solution   The possible rational roots are $\pm 1, \pm 2, \pm 4, \pm 5, \pm 10, \pm 20$. There are either three or one positive real roots and two or zero negative real roots. The calculations are shown in Table 3.

We have encountered upper and lower bounds to the roots without finding a rational root, so there are no rational roots. We observe the sign change between

r						$P(r)$	
0	1	0	-8	-10	12	-20	
1	1	1	-7	-17	-5	-25	Note
2	1	2	-4	-18	-24	-68	sign
Upper bound → 4	1	4	8	22	100	380	change
-1	1	-1	-7	-3	15	-35	
-2	1	-2	-4	-2	16	-52	
Lower bound → -4	1	-4	8	-42	180	-740	

TABLE 3

$P(2)$ and $P(4)$, however, and conclude that $P(x) = 0$ for some value of x between 2 and 4. We find $P(3)$ in order to narrow the bounds on the root:

$$
\begin{array}{r|rrrrrr}
3 & 1 & 0 & -8 & -10 & 12 & -20 \\
  &   & 3 & 9 & 3 & -21 & -27 \\
\hline
  & 1 & 3 & 1 & -7 & -9 & -47
\end{array}
$$

So $P(3) = -47$, which shows that the root lies between 3 and 4. In the next section we will see how we can obtain the root to any desired degree of accuracy.   ▲

We now summarize the procedure we have discussed for finding the rational roots of a polynomial equation $P_n(x) = 0$ having integral coefficients.

Step 1. List in order all numbers of the form $\pm p/q$, where p is a factor of the constant term of $P_n(x)$ and q is a factor of the leading coefficient.

Step 2. Apply Descartes' rule of signs to determine the possible numbers of positive and negative real roots.

Step 3. By synthetic division or otherwise, begin trying the positive numbers obtained in Step 1. Stop if (a) a root is found, (b) a bound on the roots is found, or (c) the possibilities are exhausted.

Step 4. If a root is found and the depressed equation is of degree 2, solve for all other roots (rational, irrational, and imaginary) by use of the quadratic formula or by factoring. If the depressed equation is of degree 3 or higher, continue testing in this equation all possibilities not previously rejected, beginning with the successful root just found. Use the same stopping criteria given in Step 3.

Step 5. When stopping criterion 3(b) or 3(c) is operative, begin testing negative possibilities in the current depressed equation until 3(a), 3(b), or 3(c) occurs.

Step 6. Repeat Step 4.

This procedure will always yield all rationai roots, and in case the depressed equation at any stage is quadratic, all roots can be found. Even when it is not possible to find all roots, by observing changes in sign in the values of $P_n(x)$, we can roughly locate irrational roots.

If you have a graphing calculator, the search for rational roots can be greatly facilitated. For example, let use return to Example 5. There we determined a rather long list of possible rational roots. With a graphing calculator, we could then graph the polynomial, obtaining a graph approximately like that in Figure 1. Now while this is not accurate enough to tell exactly what the zeros are, we can consult our list and pick out the most likely candidates. Then we can test these, making use of the calculator. Note that we can tell by observing the graph the number of positive and negative roots, as well as upper and lower bounds.

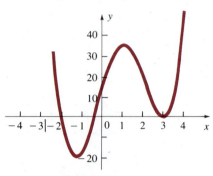

FIGURE I

EXERCISE SET 5

A

7-21

In Problems 1 and 2 list all possible rational roots. (Do not test the possibilities.)

1. a. $3x^4 - 9x^3 + 14x^2 - 8x - 4 = 0$
   b. $6x^5 - 23x^3 + 7x - 15 = 0$
2. a. $20x^6 - 13x^3 + 14x^2 - 12 = 0$
   b. $18x^4 - 7x^3 + 5x^2 - 3x - 8 = 0$

In Problems 3 and 4 find upper and lower bounds for the roots. Indicate real roots between consecutive integers.

3. $3x^3 + 5x^2 - 13x - 19 = 0$
4. $x^4 - 5x^3 - 6x^2 + 3x - 8 = 0$

In Problems 5 and 6 use Descartes' rule of signs to determine the possible numbers of positive and negative real roots.

5. a. $3x^4 - 2x^3 - 8x + 5 = 0$
   b. $2x^5 - 3x^3 - 5x^2 + 3x - 1 = 0$
6. a. $5x^5 - 8x^4 + 2x^3 - 3x - 4 = 0$
   b. $2x^4 + 3x^3 - 4x^2 - 3x - 7 = 0$

In Problems 7–24 find all rational roots. When the depressed equation is quadratic, find all roots.

7. $x^3 + 3x^2 - 4 = 0$
8. $x^3 - 12x - 16 = 0$
9. $x^3 - 8x^2 + 9x + 18 = 0$
10. $2x^3 + 7x^2 - 7x - 12 = 0$
11. $3x^3 - 4x^2 - 17x + 6 = 0$
12. $2x^3 + 3x^2 - 14x - 15 = 0$
13. $2x^3 - 3x^2 + 6x + 4 = 0$
14. $3x^3 + 10x^2 + 7x - 10 = 0$

15. $3x^3 + 7x^2 - 4 = 0$
16. $4x^3 - 8x^2 - 11x - 3 = 0$
17. $x^4 + x^3 - 8x^2 - 9x - 9 = 0$
18. $2x^4 + 2x^3 - 11x^2 - 4x + 20 = 0$
19. $x^4 - 5x^3 + 8x - 40 = 0$
20. $4x^4 - 4x^3 + x^2 - 4x - 3 = 0$
21. $6x^3 + 25x^2 - 8x - 48 = 0$
22. $6x^4 - 17x^3 + 5x^2 + 19x - 9 = 0$
23. $2x^4 - 15x^2 + 4x + 36 = 0$
24. $4x^4 - 20x^3 + 13x^2 + 30x + 9 = 0$

In Problems 25–32 factor completely over the reals.

25. $x^3 - x + 6$ 26. $2x^3 - 7x^2 + 9$
27. $x^3 - 3x - 2$ 28. $x^3 - 2x^2 - 6x - 8$
29. $2x^3 + 3x^2 - 14x - 15$
30. $x^4 - x^2 - 2x + 2$
31. $3x^3 + 22x^2 + 32x - 32$
32. $x^4 - 6x^2 - 8x - 3$

33. The product of the square of a certain even natural number and the next consecutive even number is 288. Find both numbers.

34. The cube of the reciprocal of a certain rational number exceeds four times the number by 6. Find the number.

35. The length of a box is 3 feet less than twice its width, and the height is 1 foot more than the width. If the volume is 210 cubic feet, find the dimension of the box.

36. The cost, in thousands of dollars, for manufacturing x thousand units per month of a certain component for a telephone switching system is given by

$$C(x) = -x^2 + 5x + 15$$

for $0 \le x \le 7$. The revenue (also in thousands of dollars) for selling x thousand units is

$$R(x) = x^3 - 2x^2$$

For a certain month the profit was \$60,000. Find how many units were manufactured and sold that month.

Hint. Recall that profit equals revenue minus cost.

In Problems 37–44 find all roots.

37. $12x^4 - 2x^3 - 16x^2 + 17x - 6 = 0$
38. $32x^4 - 104x^3 + 132x + 45 = 0$
39. $2x^5 + 3x^4 - 10x^3 - 15x^2 + 12x + 18 = 0$
40. $5x^5 - 28x^4 + 50x^3 - 46x^2 + 45x - 18 = 0$
41. $4x^5 - 18x^4 + 21x^3 - 14x^2 + 60x - 72 = 0$
42. $4x^5 - 35x^3 + 83x^2 - 114x + 72 = 0$
43. $4x^6 - 4x^5 - 52x^4 + 4x^3 + 49x^2 - x - 12 = 0$
44. $6x^6 - 5x^5 - 12x^4 + 5x^3 - 6x^2 + 10x + 12 = 0$
45. Show that the equation $2x^4 + 13x^3 - 40x - 24 = 0$ has only one rational root. How many real roots does it have? Justify your answer.
46. An open-top box is to be constructed from a flat rectangular sheet of cardboard 20 cm long and 16 cm wide by cutting out squares, each with side x cm, and bending up the flaps. (See the figure.) Find x if the volume of the box is 300 cm^3.

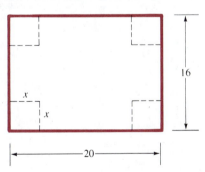

47. A right circular cone is inscribed in a sphere of diameter 3. Express the volume of the cone as a function of the distance x shown in the figure. If the volume of the cone is $4\pi/3$, find the base radius and the height of the cone.

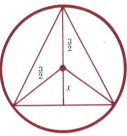

Cross section

C   (Graphing calculator)

In Problems 48–51 find all possible rational roots. Then use a graphics display to decide which of these possibilities seem most likely to actually be roots. Test these by using synthetic division. Find the complete solution set.

48. $6x^3 - 11x^2 - 14x - 24 = 0$
49. $12x^4 + x^3 - 14x^2 - 26x - 15 = 0$
50. $16x^4 - 56x^3 + 89x^2 - 66x + 18 = 0$
51. $8x^5 + 4x^4 - 46x^3 - 65x^2 - 31x - 5 = 0$

IRRATIONAL ROOTS

When a real root of a polynomial equation is not rational, the method of Section 5 does not enable us to find it unless we arrive at a depressed equation that is quadratic. We have seen, though, that we can locate such an irrational root between consecutive integers by observing a change in sign in the corresponding function values. There are various numerical techniques for finding irrational roots to any desired degree of accuracy. We will illustrate one such technique, called **linear interpolation,** in the next example. A more efficient technique, using tangent lines, is studied in calculus.

EXAMPLE 7

Approximate the real solution of the equation $x^3 - 3x^2 + 3x - 4 = 0$ to two decimal places of accuracy.

Solution   A quick check will show that there are no rational roots. By synthetic division we locate the root between consecutive integers (Table 4). If we denote the polynomial by $f(x)$, we see that $f(2) = -2$ and $f(3) = 5$, so there is a root between 2 and 3. Since 3 is an upper bound and 0 is a lower bound, there are no other real roots. On the interval [2, 3] we approximate the graph of $y = f(x)$ by the line joining the two points $(2, -2)$ and $(3, 5)$, as shown in Figure 2. The x intercept of this line is then an approximation of the zero of $f(x)$. In effect we are approximating the curve between $(2, -2)$ and $(3, 5)$ by the straight line joining them (the **secant** line). This explains why the process is called *linear* interpolation.

Lower bound → 0	1	−3	3	−4	
1	1	−2	1	−3	
2	1	−1	1	−2 ⎫	Note sign change
Upper bound → 3	1	0	3	5 ⎭	

TABLE 4

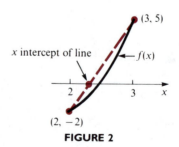

FIGURE 2

The slope of this line is 7, and its equation is

$$y + 2 = 7(x - 2)$$

or

$$y = 7x - 16$$

Setting $y = 0$, we find the x intercept to be $2\frac{2}{7}$, or approximately 2.3. So we try this as our next approximation to the root. We want to locate the root between consecutive tenths.

Because we are now dealing with decimal quantities, the arithmetic of synthetic division rapidly becomes complicated, so a calculator is almost essential from this stage onward. We find that

$$f(2.3) = -0.80$$

to two decimal places. Since $f(3)$ is positive, we conclude that the root lies between 2.3 and 3. Our goal at this stage is to locate the root between successive tenths, so we try 2.4 and get

$$f(2.4) = -0.26$$

Since this also is negative, we see from Figure 2 that the root is to the right of 2.4, so we try 2.5.

$$f(2.5) = 0.38$$

The change in sign of the function tells us that the root lies between 2.4 and 2.5. Now we use linear interpolation again, taking as our next trial the x intercept of the line segment joining $(2.4, -0.26)$ and $(2.5, 0.38)$. You should verify that this is approximately 2.44. We now calculate

$$f(2.44) = -0.0140$$

and

$$f(2.45) = 0.0486$$

Thus, the root is between 2.44 and 2.45. We will carry the process one more stage so we can see whether to round off to 2.44 or 2.45.

The x intercept of the line segment joining $(2.44, -0.0140)$ and $(2.45, 0.0486)$ is found to be 2.443, and

$$f(2.443) = 0.00469$$

Since this is positive, the root is to the left of 2.443. Testing 2.442, we find

$$f(2.442) = -0.00156$$

The root therefore lies between 2.442 and 2.443. So to two decimal places of accuracy, the answer is 2.44.

Remark.   Because the root was found to be to the left of 2.443, and we knew already it was between 2.44 and 2.45, we could have concluded without calculating $f(2.442)$ that it was closer to 2.44 than to 2.45.   ▲

We can summarize this procedure as follows. Suppose we have found that a root of $f(x) = 0$ lies between x_1 and x_2 (which may be consecutive integers, consecutive tenths, hundredths, etc.). The line joining $(x_1, f(x_1))$ and $(x_2, f(x_2))$ has the equation

$$y - f(x_1) = m(x - x_1)$$

where

$$m = \frac{f(x_2) - f(x_1)}{x_2 - x_1}$$

This line is used to approximate the actual graph of $y = f(x)$ on the interval $[x_1, x_2]$, and its x intercept provides an approximation to the root. This intercept is found by setting $y = 0$ in the equation of the line, which gives

$$x = x_1 - \frac{f(x_1)}{m} \tag{5}$$

The root is then approximated to any desired degree of accuracy by repeated application of the following two steps:

Step 1. If $f(x_1)$ and $f(x_2)$ are opposite in sign, where x_1 and x_2 are consecutive n-place decimals ($n = 0, 1, 2, \ldots$), find x by equation (5) and round off to $n + 1$ decimal places.

Step 2. Test the value of x found in Step 1 and move forward or backward, as required, by consecutive $(n + 1)$-place decimals to locate new values of x_1 and x_2 for which $f(x_1)$ and $f(x_2)$ are opposite in sign.

The process is stopped when the root is located between consecutive decimals having *one more decimal place of accuracy than is desired*. The answer is found by rounding back one place.

EXAMPLE 8

Find the real root of $x^3 - x^2 - 2x + 3 = 0$ correct to two decimal places.

Solution   By Descartes' rule of signs we determine that there are either 2 or 0 positive roots and 1 negative root. Table 5 shows that 2 is an upper bound, and that a root lies between -1 and -2.

	0	1	-1	-2	3
	1	1	0	-2	1
Upper bound →	2	1	1	0	3
	-1	1	-2	0	$3\}$ **Sign change**
	-2	1	-3	4	$-5\}$

TABLE 5

Now we use linear interpolation, following the two-step procedure given above, first taking $x_1 = -2$ and $x_2 = -1$. Thus,

$$m = \frac{f(-1) - f(-2)}{-1 - (-2)} = \frac{3 - (-5)}{1} = 8$$

So by equation (5), the x intercept is

$$x = -2 - \frac{-5}{8} = -1.4$$

We calculate the values

$$f(-1.4) = 1.096$$
$$\left.\begin{array}{l} f(-1.5) = 0.375 \\ f(-1.6) = -0.456 \end{array}\right\} \text{ sign change}$$

The root therefore lies between -1.5 and -1.6, so we now let $x_1 = -1.6$ and $x_2 = -1.5$, obtaining the new m and new x:

$$m = 8.31, \quad x = -1.6 - \frac{-0.456}{8.31} = -1.54$$

We calculate the values

$$\left.\begin{array}{l} f(-1.54) = 0.0561 \\ f(-1.55) = -0.0264 \end{array}\right\} \text{ sign change}$$

Finally, we let $x_1 = -1.55$ and $x_2 = -1.54$ and get the new m and x:

$$m = 8.25, \ x = -1.55 - \frac{-0.0264}{8.25} = -1.547$$

Since

$$f(-1.547) = -0.0015$$

and

$$f(-1.546) = 0.00676$$

we conclude that the root is between -1.546 and -1.547. The answer to two decimal places is therefore -1.55. ▲

Remark. Linear interpolation is not limited to polynomial equations. The only requirement is that in the equation $f(x) = 0$, the function f be continuous (roughly speaking, the graph has no breaks) on the original interval $[x_1, x_2]$, where $f(x_1)$ and $f(x_2)$ are opposite in sign. It should also be noted that there may be more than one root on the interval $[x_1, x_2]$. A more detailed analysis, involving calculus, is needed for cases where roots are tightly bunched together.

In solving a polynomial equation it is best to look for rational roots first. If any are found, go to the depressed equation. Only if there are no rational roots or if the depressed equation is of degree 3 or greater should the method of this section be used.

To use a graphing calculator to estimate irrational roots, first graph the function, being careful to use a sufficiently large interval on the x axis so that all of the real zeros show up. Then use the "zoom" feature repeatedly on each zero to approximate it to the desired degree of accuracy. Some calculators have a "solve" key that gives approximations to the real roots.

ACCURACY IN COMPUTATIONS

In working with approximations to true values, as we do in finding irrational roots, for example, it is important to realize that the answer to a calculation can be no more accurate than the least accurate of the numbers involved. So we frequently need to **round off** the answer. We round up or down according to whether the amount dropped is greater than or less than one-half unit in the last place retained. When it is exactly half, we follow the convention of rounding so that the last digit retained is even. Thus, to two decimals, we would round both 3.745 and 3.735 to 3.74.

Determining how many **significant digits** there are in a number is important in approximations. Any nonzero digit is always significant, and 0 is significant except when it is used solely to place the decimal, as in 0.0021. When 0 is ambiguous, we can use scientific notation. For example, we could write 1,200 as 1.2×10^3 to indicate two significant digits, as 1.20×10^3 to indicate three, and as 1.200×10^3 to indicate four significant digits. Test your understanding by counting the significant digits in each of the following: 237, 1.02, 0.045, 0.00450, 2,001, 1.2030×10^4. (Your answers should be, in order: 3, 3, 2, 3, 4, 5.)

A number A is said to be more *precise* than B if the last significant digit in A occurs farther to the right with respect to the decimal point than the last significant digit in B. So 0.0032 is more precise than 2.753 (even though the latter has more significant digits).

In **addition or subtraction** of approximate data, round the answer so that it has the same precision as the least precise of the data. In **multiplication or division** of approximate data, round the answer so that it has the same number of significant digits as the number in the data with the least number of significant digits. (Treat raising to powers and extracting roots the same as multiplication and division.)

These rules are reasonable and usually lead to reliable results, but you should be aware that error analysis is a deep subject, requiring far more study than we can go into here. An awareness of these rules, however, can prevent you from having "delusions of accuracy," especially when using a hand calculator. To illustrate, suppose the radius of a circle is measured to be 4.53 cm. On a calculator we would find the area (using $A = \pi r^2$) to be 64.46830869 cm². This, of course, is an absurd answer. Our most reasonable approximation would be 64.5 cm².

EXERCISE SET 6

A

In Problems 1–7 find the specified irrational root correct to two decimal places.

1. The real root of $x^3 - x^2 + 2x - 3 = 0$
2. The real root of $x^3 + 2x^2 + 3x - 10 = 0$
3. The real root of $x^3 - x + 4 = 0$
4. The negative root of $2x^3 - x^2 - 12x + 10 = 0$
5. The largest positive root of $2x^4 + x^3 - 13x^2 - 4x + 12 = 0$
6. The positive root of $2x^3 - 3x^2 - 12x - 5 = 0$
7. The two real roots of $3x^4 + x^3 - 3x - 27 = 0$

In Problems 8–19 find all rational and irrational roots, expressing irrational roots correct to two decimal places.

8. $x^4 + x^3 - 2x^2 - 7x - 5 = 0$
9. $x^4 - 4x^3 + 4x^2 + 8x - 16 = 0$
10. $2x^4 - 3x^3 - 3x^2 + 8x - 3 = 0$
11. $x^4 + x^3 - 7x^2 - 4x - 3 = 0$
12. $2x^4 + x^3 - 12x^2 + 27x - 27 = 0$
13. $8 + 12x + 6x^2 - x^4 = 0$
14. $4 - 4x - 3x^2 - 3x^3 - 3x^4 = 0$

B

15. $x^3 + x^2 - 7x - 5 = 0$
16. $x^3 - 2.34x^2 - 5.16x - 4.87 = 0$
17. $x^4 - 2x^2 + 5x - 8 = 0$

18. $x^5 - 6x^4 + 7x^3 + 11x^2 - 16x - 4 = 0$
19. $2x^5 - 8x^4 - 11x^3 + 21x + 20 = 0$
20. In this problem we describe another way to approximate irrational roots, called the **method of bisection.** Let f be continuous on an interval $[a, b]$ for which $f(a)$ and $f(b)$ are opposite in sign. Bisect $[a, b]$, and denote by $[a_1, b_1]$ the half for which $f(a_1)$ and $f(b_1)$ are opposite in sign. Then bisect $[a_1, b_1]$ and let $[a_2, b_2]$ be the half for which $f(a_2)$ and $f(b_2)$ are opposite in sign. Continue this process until a_n and b_n differ by less than $(10)^{-k}$, where k decimal places of accuracy are desired. As a final approximation use $(a_n + b_n)/2$. Redo Example 7, using this method.
21. Refer to Problem 20. Show that if a and b are consecutive integers, the error at the nth stage is less than $1/2^n$. If you wanted three decimal places of accuracy, how many bisections would be needed?

C (Graphing calculator)

In Problems 22–24 use the zooming feature of the graphing calculator to approximate all real roots to three decimal places.

22. $2x^3 - 3x^2 - 6x + 4 = 0$
23. $x^4 + 2x^3 - x^2 - 6x - 6 = 0$
24. $3x^4 - 6x^3 - 16x^2 + 8x + 16 = 0$

7   GRAPHING POLYNOMIAL FUNCTIONS

We can use the method of Section 5 to make a table of values to be used in graphing a polynomial function, $y = P(x)$. The zeros of the polynomial are particularly helpful, and if these are rational, we know how to find them. Real zeros that are irrational can at least be located between consecutive integers.

As we indicated earlier, the graph of a polynomial function is a smooth curve, with no breaks. The graph of a second degree polynomial is a parabola, as we know. The graph of a third degree polynomial typically has both a maximum point and a minimum point and looks like one of the curves shown in Figure 3. However, the "humps" may be straightened out, as in Figure 4.

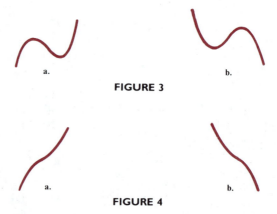

a.   b.

FIGURE 3

a.   b.

FIGURE 4

Fourth degree polynomials usually add one more maximum or minimum point, as shown in Figure 5. But again, some of the humps may not be present. The shape could be any of those shown in Figure 6.

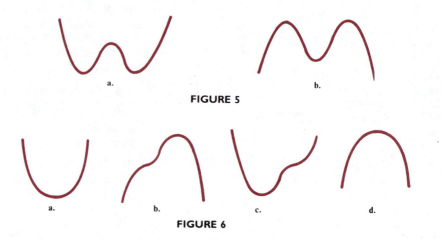

a.   b.

FIGURE 5

a.   b.   c.   d.

FIGURE 6

This pattern continues. The number of maximum points and minimum points is at most 1 less than the degree of the polynomial.

The sign of the coefficient of the highest power of x tells us the sign of the function when x is very large in absolute value. For example, in

$$y = 2x^3 - 15x^2 - 30x - 100$$

when x is very large, y will be positive, because the term $2x^3$ will dominate all other terms. Similarly, for x negative but large in absolute value, y will be negative. Thus, the general shape of the graph will be that of Figure 3(a). By locating the real zeros and a few more points we can draw the graph with reasonable accuracy.

EXAMPLE 9 ||||||||||||||||||||||||||||||||||||||||||||||||||||||||||||||||||||||||||||||||||||||||||||||||

Graph the function $y = 2x^3 - 5x^2 - 8x + 12$.

Solution   This is the function we used as an illustration in Section 5. We can summarize the function values found in Table 2 as follows:

x	0	1	2	3	4	5	-1	-2	-3
y	12	1	-8	-3	28	97	13	-8	-63

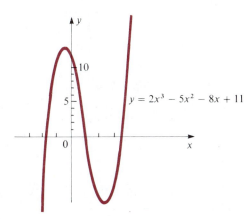

$$y = 2x^3 - 5x^2 - 8x + 11$$

FIGURE 7

The x intercepts are the zeros of the polynomial, which we found to be between 1 and 2, between 3 and 4, and between -1 and -2. By checking rational possibilities, it is easy to show that these zeros are irrational. They can be located to any desired degree of accuracy by using the method described in Section 6 or by zooming in repeatedly with a graphing calculator. The graph is shown in Figure 7. The exact location of the maximum and minimum points can be determined by calculus. They can also be approximated with a graphing calculator.

EXAMPLE 10 ||||||||||||||||||||||||||||||||||||||||||||||||||||||||||||||||||||||||||||||||||||||||||||||

Graph the function $y = x^4 - x^3 - 5x^2 + 12$.

Solution   A table of values can be obtained by direct substitution (using a calculator as needed), but we choose to use synthetic division, since this provides a quick visual way of determining upper and lower bounds on the zeros of the function and also on locating these zeros and determining their multiplicities. One important difference in using synthetic division to make a table of values as opposed to finding roots is that when we locate a rational zero (i.e., a root of the equation), we do not go immediately to the depressed equation coefficients but continue with the coefficients of the original function, since it is this function whose graph we wish to plot. After extending the table until upper and lower bounds on the zeros are reached, *then* we check to see if any zeros we have found have multiplicity greater than one. These ideas are illustrated in Table 6, and the calculation that follows it.

x					y
0	1	−1	−5	0	12
1	1	0	−5	−5	7
2	1	1	−3	−6	0
Upper bound → 3	1	2	1	3	21
−1	1	−2	−3	3	9
Lower bound → −2	1	−3	1	−2	16

TABLE 6

Since we found that $x = 2$ is a zero of the function, we can write y in the factored form

$$y = (x - 2)(x^3 + x^2 - 3x - 6)$$

Now we test to see if $x = 2$ is also a zero of the second factor.

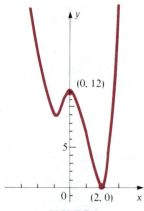

(0, 12)

5

0   (2, 0)   x

FIGURE 8

$$
\begin{array}{r|rrrr}
2 & 1 & 1 & -3 & -6 \\
  &   & 2 & 6 & 6 \\
\hline
  & 1 & 3 & 3 & 0
\end{array}
$$

This shows that 2 is a zero of the second factor and hence is a zero of multiplicity 2 of the original function. We can now factor y further as

$$y = (x - 2)^2(x^2 + 3x + 3)$$

Since the discriminant $b^2 - 4ac$ of the factor $x^2 + 3x + 3$ is $9 - 12 = -3$, it follows that there are no other real zeros.

The graph is shown in Figure 8. The significance of the fact that 2 is a zero of multiplicity 2 is that the curve is *tangent to the x axis* at $x = 2$.   ▲

In Example 10 we indicated that the significance of the factor $(x - 2)^2$ occurring in the factored form of the polynomial was that the graph was tangent to the x axis at $x = 2$. **More generally, if a factor of the form $(x - r)^k$ occurs in**

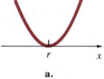

a.

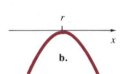

b.

FIGURE 9

the factored form of a polynomial, where $k \geq 2$, then the graph will be tangent to the x axis at $x = r$. If k is even, then in the vicinity of $x = r$, the graph will have an appearance similar to one of the forms shown in Figure 9. If k is odd, the curve will look generally like one of the forms shown in Figure 10. The larger the exponent k is, the flatter the curve will be near the point of tangency.

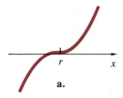

a.

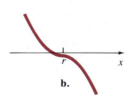

b.

FIGURE 10

EXAMPLE 11 |||||||||||||||||||||||||||||||||||||||||||||||||||||||||||||||||||||||||||||||||||||||||||||||||||||||

Find the zeros and draw the graph of the function $y = 4 + 3x^2 - x^4$.

Solution   In this case there is no need to use synthetic division, since we can factor by elementary means in the form

$$y = (4 - x^2)(1 + x^2) = (2 - x)(2 + x)(1 + x^2)$$

So the only real zeros are ± 2. The function is even and hence its graph is symmetric to the y axis. We make a table of values and show the graph in Figure 11. Again, calculus is needed to determine the exact nature of the graph.

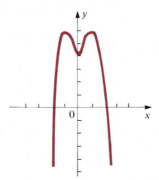

x	0	± 1	± 2	± 3
y	4	6	0	-50

FIGURE 11

EXERCISE SET 7

A

In Problems 1–21 find the rational zeros, locate other real zeros between consecutive integers, make a table of values, and graph the function.

1. $y = x^3 - 1$
2. $y = x^3 - x^2 - 6x$
3. $y = x^3 - x^2 - 9x + 9$
4. $y = 1 - x^4$
5. $y = 4x - x^3$
6. $y = (x - 4)(x + 1)(x + 3)$
7. $y = (2x - 5)(x - 6)(x + 2)$
8. $y = (x - 1)(2x + 3)(x - 3)^2$
9. $y = (x + 1)^2(2x - 1)(x - 4)$
10. $y = \frac{1}{4}(x + 2)^3(x - 1)^2$
11. $y = x^3 - 3x + 2$
12. $y = 2x^3 - 7x^2 + 9$
13. $y = x^3 + x^2 - 10x - 12$
14. $y = x^4 - 10x^2 + 9$
15. $y = 8x^2 + 2x^3 - x^4$
16. $y = 12 - 3x + 4x^2 - x^3$
17. $y = x^4 + x^3 - 13x^2 - x + 12$
18. $y = 2x^4 - 9x^3 + 7x^2 + 9x - 9$
19. $y = 2x^4 - 5x^3 + 10x - 12$
20. $y = x^5 - 4x^3$
21. $y = x^4 + 2x^3 - 12x^2 - 8x + 32$

In Problems 22–27 solve the inequalities by graphical means.

22. $(x - 3)(2x + 5)(x - 7)^3 \le 0$
23. $(x + 4)^2(3x - 4)(x - 5)(x + 6) > 0$
24. $x^3 - 3x^2 - x + 3 < 0$
25. $x^3 - 5x^2 + 2x + 8 \ge 0$
26. $x^3 - 2x^2 - 5x + 6 \le 0$
27. $x^4 - 9x^2 - 4x + 12 > 0$

Find the domain of each of the functions in Problems 28 and 29.

28. $f(x) = \sqrt{2x^3 - 9x^2 + x + 12}$

29. $g(x) = \dfrac{1}{\sqrt{x^4 - 9x^2 + 4x + 12}}$

B

35. $y = x^5 - x^4 - 15x^3 + x^2 + 38x + 24$
36. $y = 2x^5 - 5x^4 - 11x^3 + 23x^2 + 8x - 15$

In Problems 30 and 31 solve the inequalities by graphical means.

C   (Graphing Calculator)

30. $2x^4 - 7x^3 - 26x^2 + 49x + 30 < 0$
31. $x(x^3 - 4x^2 + 3) \geq 14(1 - x)$

Graph the functions in Problems 37–40. Use the zooming feature to approximate all zeros and the coordinates of all maximum and minimum points, to three decimal places.

Graph the functions in Problems 32–36.

32. $y = x^4 - 3x^3 - 8x^2 + 10x + 12$
33. $y = x^4 - 2x^3 - 5x^2 + 6x$
34. $y = 3x^4 - 5x^3 - 8x^2 + 10x + 7$

37. $y = 2x^3 - 5x^2 - 3x - 4$
38. $y = 3 + 9x^2 - 4x^3$
39. $y = 2.134x^4 + 0.473x^3 - 5.237x - 12.728$
40. $y = 7.831 - 1.245x^3 - 0.496x^5$

 # RATIONAL FUNCTIONS

A function of the form

$$f(x) = \frac{P(x)}{Q(x)}$$

where $P(x)$ and $Q(x)$ are polynomials, is called a **rational function.** We will assume to begin with that $P(x)$ and $Q(x)$ have no nonconstant factor in common. The zeros of $f(x)$ are the same as those of $P(x)$, since if the numerator of a fraction is 0 and the denominator not 0, the fraction is 0. These zeros are the x intercepts of the graph of f. The y intercept, if any, is found by putting $x = 0$.

The zeros of $Q(x)$ are not in the domain of f, but the behavior of $f(x)$ for x near these zeros is particularly useful in drawing the graph of f. The next example illustrates how to analyze this behavior.

EXAMPLE 12

Examine the behavior of the function below near $x = 2$:

$$f(x) = \frac{x + 1}{x - 2}$$

Solution   We set $y = f(x)$ and make a table of values for x close to 2.

x	1	1.5	1.9	1.99	2.01	2.1	2.5	3
y	-2	-5	-29	-299	301	31	7	4

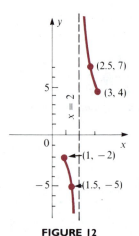

FIGURE 12

As x approaches 2 from the left, y is negative but becomes arbitrarily large in absolute value. Similarly, when x approaches 2 from the right, y becomes arbitrarily large through positive values. The graph in the neighborhood of $x = 2$ is shown in Figure 12. The line $x = 2$ is called an **asymptote** to the curve, and the curve is said to be **asymptotic to this line.** ▲

Example 12 illustrates the general result that if r is a zero of $Q(x)$ **but not a zero of $P(x)$, then the line $x = r$ is a vertical asymptote to the graph of f.**

Rational functions may also have horizontal asymptotes, and these can be determined by examining the behavior of $f(x)$ when x becomes arbitrarily large in absolute value. The next example illustrates a way of doing this.

EXAMPLE 13 ||||||||||||||||||||||||||||||||||||||||||||||||||||||||||||||||||||||||||||||||||||||||||||||||||||||||||||

Examine the behavior of the function

$$f(x) = \frac{2x^2 + x}{x^2 - 4}$$

as x becomes arbitrarily large in absolute value.

Solution   Let $y = f(x)$ and divide numerator and denominator by the highest power of x—in this case, x^2:

$$y = \frac{2x^2 + x}{x^2 - 4} = \frac{2 + (1/x)}{1 - (4/x^2)} \qquad (x \neq 0)$$

Now as x gets arbitrarily large in absolute value, the terms $1/x$ and $4/x^2$ both approach 0. So y approaches $\frac{2}{1} = 2$. For large values of $|x|$, the graph has the appearance shown in Figure 13. The line $y = 2$ is a horizontal asymptote.

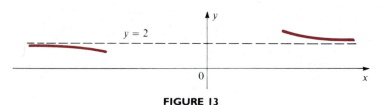

FIGURE 13 ▲

By writing

$$y = \frac{a_m x^m + a_{m+1} x^{m-1} + \cdots + a_0}{b_n x^n + b_{n-1} x^{n-1} + \cdots + b_1}$$

and following the procedure used in Example 13 of dividing numerator and denominator by the highest power of x, we can prove the following theorem. (You will be asked to supply the proof in Problem 55 of Exercise Set 8.)

HORIZONTAL ASYMPTOTE THEOREM

Let

$$f(x) = \frac{P(x)}{Q(x)}$$

where P and Q are polynomials of degree m and n, respectively, and denote their leading coefficients by a_m and b_n, respectively. Then the graph of f has a horizontal asymptote if and only if $m \leq n$. The equation of this asymptote is:

(i) $y = 0$ if $m < n$

(ii) $y = \dfrac{a_m}{b_n}$ if $m = n$

Note that in Example 13 the numerator and denominator had the same degree, so by the theorem the line $y = 2/1$, or $y = 2$, is the horizontal asymptote, in agreement with what we found. This was an example of case (ii) of the theorem, which can be stated in words as follows:

> When numerator and denominator are of the same degree, the horizontal asymptote is found by setting y equal to the quotient of the leading coefficient (i.e., coefficient of highest-degree term) of the numerator divided by the leading coefficient of the denominator.

EXAMPLE 14

In each of the following find the horizontal asymptote, or show none exists.

a. $f(x) = \dfrac{x^2 + 1}{x^3 - 8}$ **b.** $g(x) = \dfrac{3 - 4x - 2x^3}{3x^3 - 5x^2 + 7}$ **c.** $h(x) = \dfrac{x^3 + 1}{2x^2 - 3}$

Solution   **a.**   The degree of the denominator is greater than that of the numerator, so $y = 0$ is the horizontal asymptote.

b.   The numerator and denominator are of the same degree, so

$$y = -\frac{2}{3}$$

is the horizontal asymptote.

c.   There is no horizontal asymptote, since the numerator has the higher degree.

In the next two examples we make use of symmetry, intercepts, asymptotes, and also an **analysis of signs** to draw the graph.

EXAMPLE 15

Discuss completely and draw the graph of

$$y = \frac{2x}{x^2 - 9}$$

Solution   The graph is symmetric to the origin. One way to see this is to clear of fractions, getting $x^2 y - 9y = 2x$, and observe that every term is of *odd* degree, so that all terms change sign when both x and y are replaced by their negatives. The resulting equation is equivalent to the original. Replacing only x, or only y, by its negative does not yield an equivalent equation, so there is no symmetry to either axis.

The x intercept is 0, and the y intercept is 0. So the curve goes through the origin but does not cross either axis at any other point.

By factoring the denominator we find the vertical asymptotes to be $x = 3$ and $x = -3$. According to the theorem, the horizontal asymptote is $y = 0$.

If we try to draw the graph with the information we have so far, we find there is still a good bit of uncertainty. Some of this can be removed by analyzing the signs of the function in intervals between critical numbers. (Recall that a zero of either the numerator or denominator is a critical number.) The sign graph in Figure 14 shows that y is positive for $x > 3$ and negative for $0 < x < 3$. Because of symmetry to the origin, the signs to the left of the origin are determined. The general nature of the graph is now evident. Plotting one or two points gives better accuracy. The graph is shown in Figure 15.

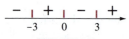

FIGURE 14

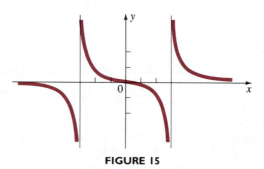

FIGURE 15

EXAMPLE 16

Analyze the function

$$f(x) = \frac{x - 2}{x^2 - 1}$$

and draw its graph.

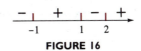

FIGURE 16

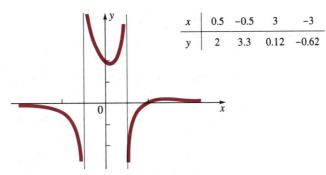

x	0.5	−0.5	3	−3
y	2	3.3	0.12	−0.62

FIGURE 17

Solution   Let $y = f(x)$. There is no symmetry. The x intercept is 2 and the y intercept also is 2. The vertical asymptotes are $x = 1$ and $x = -1$, and the horizontal asymptote is $y = 0$. This information, along with the sign graph of Figure 16, and a few additional points enable us to obtain the graph shown in Figure 17.   ▲

Remark. A common misconception is that curves never cross their asymptotes. The graph in the preceding example shows this is not true. Graphs of *functions*, however, cannot cross their vertical asymptotes.

According to the horizontal asymptote theorem, the graph of a rational function has no horizontal asymptote if the degree of the numerator exceeds that of the denominator. However, in case the numerator has degree *exactly one more than* the denominator there is an **inclined** (or **oblique**) asymptote. The procedure for finding this is illustrated in the next example. You will be asked in Problem 56 of Exercise Set 8 to consider the general case.

EXAMPLE 17 ||||||||||||||||||||||||||||||||||||||||||||||||||||||||||||||||||||||||||||||||||||||||||||||||||||

Show that

$$y = \frac{x^2 - 3x}{x - 2}$$

has an inclined asymptote. Use the result, along with other analysis, to draw the graph.

Solution   To find the inclined asymptote, first divide:

$$
\begin{array}{r}
x - 1 \\
x - 2 \overline{\smash{)}\; x^2 - 3x } \\
\underline{x^2 - 2x } \\
- \; x \\
\underline{- \; x + 2} \\
- 2
\end{array}
$$

We can therefore write the function in the form

$$
y = x - 1 - \frac{2}{x - 2}
$$

As x becomes arbitrarily large in absolute value, the term $2/(x - 2)$ approaches 0. So the graph approaches the line $y = x - 1$. That is, this line is an asymptote. It is called an inclined asymptote since it has nonzero slope.

We readily see that the graph is not symmetric to either axis or the origin. The x intercepts are 0 and 2, and the y intercept is 0. The line $x = 2$ is a vertical asymptote. This information, together with the sign graph (Figure 18) results, enables us to draw the graph in Figure 19.

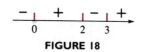

FIGURE 18

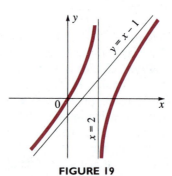

FIGURE 19

So far we have considered rational functions in which the numerator and denominator have no nonconstant factor in common. Now we consider how the graph is affected when they do have a common factor. In case this is an irreducible quadratic over the reals, so that its zeros are imaginary, it can be divided out, and the graph of the resulting function is identical to that of the original. The only other case to consider is the one in which numerator and denominator have one or more linear factors in common.

Suppose first there is only one such common linear factor, say $x - a$, and it is to the first power only. Then we can write

$$f(x) = \frac{P(x)}{Q(x)} = \frac{(x-a)P_1(x)}{(x-a)Q_1(x)} = \frac{P_1(x)}{Q_1(x)} \qquad \text{if } x \neq a$$

Note carefully that the cancellation of the factor $x - a$ is valid only if $x \neq a$. So except for this single point, the graph of f is identical to the graph of $P_1(x)/Q_1(x)$. To get the graph of f, we draw the graph of $P_1(x)/Q_1(x)$ and remove the point corresponding to $x = a$.

To illustrate, consider

$$f(x) = \frac{x^2 - 1}{x^2 - 3x + 2}$$

Factoring numerator and denominator reveals the common factor $x - 1$, which we divide out, with the restriction that $x \neq 1$:

$$f(x) = \frac{(x-1)(x+1)}{(x-1)(x-2)} = \frac{x+1}{x-2} \qquad \text{if } x \neq 1$$

We now find the x intercept -1, the y intercept $-\frac{1}{2}$, the vertical asymptote $x = 2$, and the horizontal asymptote $y = 1$, and we draw the graph, removing the point corresponding to $x = 1$. (See Figure 20.) Since it is impossible to show a gap of just one point, we indicate the missing point with an open circle. A missing point such as this is called a **removable discontinuity** (by supplying the missing point, the discontinuity is removed).

In case a common linear factor appears to some higher power, say $(x - a)^k$, the analysis goes in essentially the same way. Also, if there are other common

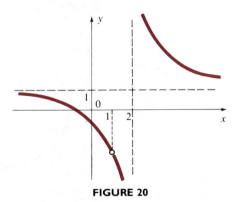

FIGURE 20

linear factors, then each one is handled as we have described. *The essential idea is to divide out all common linear factors, graph the resulting rational function, and then remove the points corresponding to each of the common linear factors.*

We conclude this section with an example of the graph of an equation of the form $y^2 = f(x)$, where f is a rational function. This does not define y as a function of x, although by taking square roots, we get the two functions $y_1 = \sqrt{f(x)}$ and $y_2 = -\sqrt{f(x)}$. Much of our analysis regarding symmetry, intercepts, and asymptotes can be adapted to equations of this type. However, we must be careful to exclude all values of x for which $f(x) < 0$, since such values would cause y to be imaginary. A sign graph for f will show these excluded values. How these affect the analysis will be brought out in the next example.

EXAMPLE 18

Analyze the equation

$$y^2 = \frac{x - 1}{x^2 - 4}$$

and draw its graph.

Solution   The graph is symmetric to the x axis but not to the y axis or the origin. The x intercept is 1. Setting $x = 0$ gives $y^2 = \frac{1}{4}$, so the y intercepts are $\frac{1}{2}$ and $-\frac{1}{2}$. A sign analysis (Figure 21) shows that the intervals $(-\infty, -2]$ and $(1, 2]$ must be excluded, since y^2 cannot be negative. The lines $x = 2$ and $x = -2$ are vertical asymptotes, since as x approaches 2 or -2 from the right, y^2 becomes infinite, and hence y approaches both $+\infty$ and $-\infty$. As x becomes arbitrarily large in the positive direction, y^2 approaches 0, and hence y approaches 0. So $y = 0$ is a horizontal asymptote on the right. We cannot allow x to approach $-\infty$, since the region to the left of -2 is excluded.

Putting all of this together, we get the graph in Figure 22.

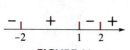

FIGURE 21

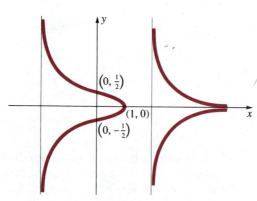

FIGURE 22

EXERCISE SET 8

A

1-11 ; 22-40

In Problems 1–6 find all vertical asymptotes.

1. a. $y = \dfrac{1}{x-1}$ **b.** $y = \dfrac{x}{x+2}$

2. a. $y = \dfrac{x+1}{x-3}$ **b.** $y = \dfrac{1}{4-x^2}$

3. a. $y = \dfrac{2x^3-5}{x^2-2x}$ **b.** $y = \dfrac{3x+7}{2x^2+5x-3}$

4. a. $y = \dfrac{x^3+1}{x^3-1}$ **b.** $y = \dfrac{x^2+4}{x^3-x^2-x+1}$

5. a. $y = \dfrac{x^2-2x+4}{x^3+4x^2-9x-36}$

   b. $y = \dfrac{1}{x^4-4x^2}$

6. a. $y = \dfrac{1-2x}{x^3+3x^2-4}$

   b. $y = \dfrac{2x^2-3x-5}{x^3-7x+6}$

In Problems 7–11 find the horizontal asymptote or show that there is none.

7. a. $y = \dfrac{2x+1}{x-3}$ **b.** $y = \dfrac{x-4}{2x-5}$

8. a. $y = \dfrac{3x}{x^2-4}$ **b.** $y = \dfrac{x^2-4}{x^2-1}$

9. a. $y = \dfrac{3x^2-4x}{x^3-x+1}$ **b.** $y = \dfrac{x^2-1}{x+2}$

10. a. $y = \dfrac{x^2-1}{x^4+2}$ **b.** $y = \dfrac{5-x^2}{2x^2+x+3}$

11. a. $y = \dfrac{10x^2+20}{1-x^3}$ **b.** $y = \dfrac{x^3+8}{100x^2+50}$

In Problems 12–21 find the inclined asymptote.

12. $y = 3x - 4 + \dfrac{2}{x}$

13. $y = 2x + 3 - \dfrac{x}{x^2-1}$

14. $y = \dfrac{x^2}{x-1}$ **15.** $y = \dfrac{2x^2+3x+1}{x-3}$

16. $y = \dfrac{x^2-4}{x}$ **17.** $y = \dfrac{x^2-5}{x+4}$

18. $y = \dfrac{x^3+1}{x^2-8}$ **19.** $y = \dfrac{2x^3+3x-1}{x^2+1}$

20. $y = \dfrac{x^4}{x^3-2}$ **21.** $y = \dfrac{3x^4+2x^2-4}{x^3-2x}$

In Problems 22–50 check for symmetry, find intercepts and asymptotes, analyze the signs, and draw the graph.

22. $y = \dfrac{2}{x+1}$ **23.** $y = \dfrac{x+1}{x^2-4}$

24. $y = \dfrac{x^2}{x^2+1}$ **25.** $y = \dfrac{x^2}{x^2-4}$

26. $y = \dfrac{x^2-4}{x^2-1}$ **27.** $y = \dfrac{x-2}{x^2-2x-3}$

28. $y = \dfrac{x^2-4}{x^2+9x}$ **29.** $y = \dfrac{x^2-1}{x+2}$

30. $y = \dfrac{x^2-4}{x}$ **31.** $y = \dfrac{4-x^2}{4+x^2}$

32. $y = \dfrac{x^2}{x-2}$ **33.** $y = \dfrac{x^2-2x}{x+2}$

34. $y^2 = \dfrac{1}{x^2-4}$ **35.** $y^2 = \dfrac{x^2}{x^2-1}$

B

36. $y = \dfrac{x^2-2x}{x+2}$ **37.** $y = \dfrac{x-2}{x^2+4x}$

38. $y = \dfrac{3-2x-x^2}{x^3-x-6}$ **39.** $y = \dfrac{2x+5}{x^2-1}$

40. $y = \dfrac{x^2-1}{x^2+2x}$ **41.** $y = \dfrac{x^2-2x-8}{x^3-3x^2+4}$

42. $y = \dfrac{x^3-1}{x^2-4}$

43. $y = \dfrac{x^3-2x^2-7x-4}{x^2+x-2}$

44. $y^2 = \dfrac{x-1}{x^3}$

45. $y^2 = \dfrac{x^2-9}{x^2-x-20}$

46. $y^2 = \dfrac{x^2(x-1)}{x-4}$

47. $x = \dfrac{1}{y^2-x^2}$

48. $4y^2 + x - xy^2 - 1 = 0$

49. $y = \sqrt{\dfrac{x-1}{x+1}}$

50. $y = \dfrac{\sqrt{x^2-2x}}{x+1}$

In Problems 51–54 draw the graph, after determining if there are any removable discontinuities.

51. $y = \dfrac{2x^2-4x}{x^3-2x^2+x-2}$

52. $y = \dfrac{x^2-2x-3}{x^3+2x^2-x-2}$

53. $y = \dfrac{x^3-2x^2-4x+8}{x^3+x^2-4x-4}$

54. $y = \dfrac{x^3-4x^2+x+6}{x^3+3x^2-6x-8}$

55. Prove the horizontal asymptote theorem.
56. Let $f(x) = P(x)/Q(x)$, where P and Q are polynomials of degrees $n+1$ and n, respectively ($n \geq 1$). Explain why $f(x)$ can be written in the

form

$$f(x) = ax + b + \dfrac{R(x)}{Q(x)}$$

where $a \neq 0$ and R is of degree less than n. How can you conclude that the line $y = ax + b$ is an inclined asymptote to the graph of f?

C (Graphing calculator)

In Problems 57–60 use the calculator to obtain the graph. Approximate intercepts and asymptotes.

57. $y = \dfrac{4x^2-3x-5}{2x^2-7x+1}$

58. $y = \dfrac{x^2-3x-5}{x^3+9}$

59. $y = \dfrac{x^3+5}{x^2-x-3}$

60. $y = \dfrac{2x^3-4x+7}{4-3x-x^3}$

In Problems 61–64 obtain the graph of $y^2 = f(x)$ by combining the graphs of $y_1 = \sqrt{f(x)}$ and $y_2 = -\sqrt{f(x)}$.

61. $f(x) = \dfrac{3x^2-7}{2x^2+5}$

62. $f(x) = \dfrac{2x^2-3x+7}{x^2+4x-3}$

63. $f(x) = \dfrac{x^3+3}{x^2-5}$

64. $f(x) = \dfrac{3x^2}{x^3+2}$

PARTIAL FRACTIONS

There are times, especially in calculus, when it is important to write a complicated rational fraction as a sum of simpler fractions. That is, we sometimes want to do just the opposite of adding fractions together. When we do this, we say that we have *resolved* (or *decomposed*) the original fraction into partial fractions. Consider the following example of adding two fractions:

$$\dfrac{2}{x-1} + \dfrac{3}{x+2} = \dfrac{2(x+2)+3(x-1)}{(x-1)(x+2)} = \dfrac{5x+1}{x^2+x-2}$$

Now suppose we are given the final fraction only and are asked to find the two original fractions that add to give this result. Of course, we know the answer in this case, but you might try covering up those original fractions and see if you can find a way of discovering what they should be.

A systematic way of doing this problem is as follows. Write

$$\dfrac{5x+1}{x^2+x-2} = \dfrac{5x+1}{(x-1)(x+2)} = \dfrac{A}{x-1} + \dfrac{B}{x+2} \tag{6}$$

where A and B are to be determined. Notice that we first factor the denominator and then indicate the sum of two fractions, each having one of these factors as its denominator, and with an unknown constant as the numerator. To find A and B, we multiply both sides of the last equation in (6) by the LCD to obtain

$$5x + 1 = A(x + 2) + B(x - 1)$$

We want to determine A and B so that this equation is an identity in x. By substituting, in turn, the particular values $x = 1$ and $x = -2$, we get

$x = 1$: $6 = A \cdot 3 + B \cdot 0$ $x = -2$: $-9 = A \cdot 0 + B(-3)$

$6 = 3A$ $-9 = -3B$

$A = 2$ $B = 3$

Thus we have the partial-fraction decomposition

$$\frac{5x + 1}{x^2 + x - 2} = \frac{2}{x - 1} + \frac{3}{x + 2}$$

This example illustrates the basic idea of partial fractions, but certain complications can arise. First, the method will work only if the original fraction is *proper*, which means that the degree of the numerator must be less than the degree of the denominator. If this is not the case, then divide until you obtain a remainder with lower degree than the denominator. Suppose, for example, that $P_m(x)$ and $P_n(x)$ are polynomials of degree m and degree n, respectively, and that $m \geq n$. Then by long division we obtain a quotient $Q(x)$ and a remainder $R(x)$ such that

$$\frac{P_m(x)}{P_n(x)} = Q(x) + \frac{R(x)}{P_n(x)}$$

where either $R(x) = 0$ or $\deg R < n$. Then we resolve $R(x)/P_n(x)$ into partial fractions. We will illustrate this procedure in Example 20.

The second complication is that the factors of the denominator may not be so simple as in our example. We know that (in theory, at least) every real polynomial can be factored into linear or quadratic factors over the reals, where one or more factors may have multiplicity greater than 1 (in which case we say the factors are *repeated*). It is useful to classify partial-fraction problems by the nature of the factors of the denominator:

Case I: All linear factors, none repeated
Case II: All linear factors, one or more repeated
Case III: At least one irreducible quadratic factor, not repeated
Case IV: At least one repeated irreducible quadratic factor

Note. In cases III and IV linear factors, repeated or not, may be present.

Our example was a case I problem, and the technique we used for determining the unknown constants A and B is called the **method of substitution.** This method works on all case I problems. In the other three cases the method of substitution will not suffice to find all unknown constants, but it is still useful whenever linear factors are present. A second method, called **comparison of coefficients,** can then be used to find the remaining constants. We will see how this method works in the next example.

EXAMPLE 19

(CASE II)

Resolve

$$\frac{x^2 - 13x + 20}{(x - 1)(x - 3)^2}$$

into partial fractions.

Solution   First note that the fraction is proper. The appropriate form of the partial-fraction decomposition is

$$\frac{x^2 - 13x + 20}{(x - 1)(x - 3)^2} = \frac{A}{x - 1} + \frac{B}{x - 3} + \frac{C}{(x - 3)^2} \tag{7}$$

Notice in particular the second term on the right. We must allow for its presence, since a fraction of this type could be involved when the common denominator is $(x - 1)(x - 3)^2$. More generally, when a linear factor of multiplicity m is present, say $(x - a)^m$, we must include fractions with denominators of the form $(x - a)^k$ for *all* positive integers $k \leq m$.

We now clear equation (7) of fractions, obtaining

$$x^2 - 13x + 20 = A(x - 3)^2 + B(x - 1)(x - 3) + C(x - 1) \tag{8}$$

We use the method of substitution, first putting $x = 1$ and then $x = 3$:

$$\begin{aligned} \textbf{x = 1:} \quad & 8 = A(-2)^2 & \textbf{x = 3:} \quad & -10 = 2C \\ & 4A = 8 & & C = -5 \\ & A = 2 \end{aligned}$$

This is as far as we can go substituting numbers that cause one or the other of the factors on the right to be zero. So we turn to the comparison of coefficients. We expand the right-hand side of equation (8) and collect like terms, obtaining

$$x^2 - 13x + 20 = (A + B)x^2 + (-6A - 4B + C)x + (9A + 3B - C)$$

Since this is to be an identity, the coefficient of each power of x on the left-hand side must be equal to the coefficient of the corresponding power of x on the right.

Comparing coefficients of x^2 gives

$$1 = A + B$$

and since $A = 2$, we conclude that $B = -1$. Although we could compare the other coefficients, there is no need to do so, since we now know A, B, and C. The partial-fraction decomposition is therefore

$$\frac{x^2 - 13x + 20}{(x - 1)(x - 3)^2} = \frac{2}{x - 1} - \frac{1}{x - 3} - \frac{5}{(x - 3)^2} \qquad \blacktriangle$$

Remarks. The method of comparison of coefficients can be used without first using the method of substitution, and it will always work. However, it is usually easier to get as much information as possible by using substitution and then turning to the comparison of coefficients.

In comparing coefficients it is not always necessary to expand the right-hand side and collect like terms, since this can sometimes be done mentally, especially for the highest power. For example, we can look at the right-hand side of equation (8) and determine that the coefficient of x^2 is $A + B$.

EXAMPLE 20

(CASE III)

Resolve

$$\frac{x^3 + 7x - 2}{x^3 - 2x - 4}$$

into partial fractions.

Solution   The fraction is improper, so we divide and get

$$\frac{x^3 + 7x - 2}{x^3 - 2x - 4} = 1 + \frac{9x + 2}{x^3 - 2x - 4} \qquad \text{(Check this.)}$$

and we concentrate on the last term. Using the methods of this chapter we factor the denominator in the form $(x - 2)(x^2 + 2x + 2)$. The factor $x^2 + 2x + 2$ is irreducible over the reals since its discriminant $b^2 - 4ac$ is negative. The proper form of the partial-fraction decomposition in this case is

$$\frac{9x + 2}{(x - 2)(x^2 + 2x + 2)} = \frac{A}{x - 2} + \frac{Bx + C}{x^2 + 2x + 2} \qquad (9)$$

This form is typical of case III problems. When the denominator is an irreducible quadratic polynomial, we must allow for an unknown linear polynomial in the numerator.

Now we clear equation (9) of fractions and then substitute $x = 2$:

$$9x + 2 = A(x^2 + 2x + 2) + (Bx + C)(x - 2) \tag{10}$$

$$x = 2: \quad 20 = A(10)$$

$$A = 2$$

Next, we expand the right-hand side of equation (10), collect like terms, and compare coefficients:

$$9x + 2 = (A + B)x^2 + (2A - 2B + C)x + (2A - 2C)$$

Coefficient of x^2: $\quad 0 = A + B$ (Observe that since x^2 does not appear on the left, its coefficient is zero.)

$$B = -A = -2$$

Coefficient of x: $\quad 9 = 2A - 2B + C$

$$9 = 4 + 4 + C$$

$$C = 1$$

The final answer is

$$\frac{x^3 + 7x - 2}{x^3 - 2x - 4} = 1 + \frac{2}{x - 2} + \frac{-2x + 1}{x^2 + 2x + 2}$$

$$= 1 + \frac{2}{x - 2} - \frac{2x - 1}{x^2 + 2x + 2} \qquad \text{(Note how the signs are handled.)}$$

▲

EXAMPLE 21 ||||||||||||||||||||||||||||||||||||||||||||||||||||||||||||||||||||||||||||||||||||||||||||||||||||||

(CASE IV)

Resolve

$$\frac{2x - 3}{x^5 + 2x^3 + x}$$

into partial fractions.

Solution   The denominator in factored form is $x(x^2 + 1)^2$, and just as with repeated linear factors we must allow for fractions with $x^2 + 1$ as well as $(x^2 + 1)^2$ in the denominator. The proper form is therefore

$$\frac{2x - 3}{x(x^2 + 1)^2} = \frac{A}{x} + \frac{Bx + C}{x^2 + 1} + \frac{Dx + E}{(x^2 + 1)^2}$$

We clear of fractions and substitute $x = 0$:

$$2x - 3 = A(x^2 + 1)^2 + (Bx + C)x(x^2 + 1) + (Dx + E)x$$

$$x = 0: \quad -3 = A$$

To compare coefficients we expand and collect terms on the right:

$$2x - 3 = (A + B)x^4 + Cx^3 + (2A + B + D)x^2 + (C + E)x + A$$

Coefficient of x^4: $0 = A + B$

$$B = -A = 3$$

Coefficient of x^3: $0 = C$

Coefficient of x^2: $0 = 2A + B + D$

$$0 = -6 + 3 + D$$

$$D = 3$$

Coefficient of x: $2 = C + E$

$$2 = 0 + E$$

$$E = 2$$

Thus

$$\frac{2x - 3}{x^5 + 2x^3 + x} = \frac{-3}{x} + \frac{3x}{x^2 + 1} + \frac{3x + 2}{(x^2 + 1)^2}$$

EXERCISE SET 9

A

Resolve each of the following into partial fractions.

1. $\dfrac{5x + 7}{x^2 + 2x - 3}$

2. $\dfrac{6x - 7}{x^2 + x - 6}$

3. $\dfrac{x - 17}{x^2 + x - 12}$

4. $\dfrac{19 - x}{x^2 - 3x - 10}$

5. $\dfrac{3x + 10}{x^2 + 5x + 6}$

6. $\dfrac{x - 4}{x^2 - 3x + 2}$

7. $\dfrac{-2}{x^2 - 9x + 20}$

8. $\dfrac{3}{x^2 - 7x + 12}$

9. $\dfrac{3x^2 + 32x + 44}{(x + 3)(x - 2)(x + 4)}$

10. $\dfrac{2x^2 + 19x - 45}{(x - 1)(x - 3)(x + 2)}$

11. $\dfrac{3x^2 - 7x - 2}{x^3 - x}$

12. $\dfrac{10x + 4}{4x - x^3}$

13. $\dfrac{x^2 + 4x - 2}{x^2 + x - 2}$

14. $\dfrac{2x^2 - x - 20}{x^2 - x - 6}$

15. $\dfrac{x^3 + x^2 - 10x - 14}{x^2 + 4x + 3}$

16. $\dfrac{2x^3 - x^2 - 4x + 5}{x^2 - 1}$

17. $\dfrac{5x + 1}{(x + 2)(x - 1)^2}$

18. $\dfrac{x^2 - 6x + 23}{(x + 3)(x - 2)^2}$

19. $\dfrac{4x^2 - 10x + 9}{(x - 1)^3}$

20. $\dfrac{x^2 + x}{(x + 2)^3}$

21. $\dfrac{4 + 4x - x^2}{x^3 + 2x^2}$

22. $\dfrac{2x^2 - 6x + 12}{x^3 - 4x^2 + 4x}$

23. $\dfrac{5x + 6}{(x - 2)(x^2 + 4)}$

24. $\dfrac{x^2 + x - 7}{(x^2 + 1)(x + 2)}$

25. $\dfrac{3 + 4x - 3x^2}{(x + 1)^2(x^2 + 3)}$

26. $\dfrac{x^3 - 4x^2 + 5x - 2}{x^4 - x^3 + x^2}$

27. $\dfrac{6 - 12x}{x^3 - 2x^2 - 5x + 6}$

28. $\dfrac{2x^2 + 17x + 5}{x^3 - 13x - 12}$

29. $\dfrac{x^2 + 3x + 6}{x^3 + x - 2}$

30. $\dfrac{6x - 3}{x^3 - 3x - 2}$

B

31. $\dfrac{ad - bc}{(ax + b)(cx + d)}$

32. $\dfrac{2rs(x + 1)}{r^2x^2 - s^2}$

33. $\dfrac{6x^2 + 20x + 28}{x^3 + 3x^2 - 4}$

34. $\dfrac{3x^4 - 3x^3 + 5x^2 - 6x + 13}{(x + 3)(x^2 + 1)^2}$

35. $\dfrac{11x^3 - 41x^2 + 72x - 32}{x(x^2 - 3x + 4)^2}$

36. $\dfrac{12x^2 + 2x - 41}{6x^2 + x - 12}$

37. $\dfrac{10x^3 - 9x^2 + 7x + 14}{10x^2 + x - 3}$

38. $\dfrac{5x^2 - 6x + 58}{6(x^3 + 4x^2 + 2x + 8)}$

39. $\dfrac{3x^2 - 4}{x^4 - 4x^3 + 8x^2 - 16x + 16}$

40. $\dfrac{3x - 5}{12x^3 - 4x^2 - 5x + 2}$

41. $\dfrac{x^3 - 3x^2 - 5x}{(x^2 + 1)(x^2 + 4)}$

REVIEW EXERCISE SET

A

In Problems 1–3 use synthetic division to perform the indicated divisions.

1. a. $2x^3 + 3x^2 - 4x + 5$ by $x - 2$
  b. $x^3 - 3x + 4$ by $x + 3$
2. a. $x^4 - 3x^3 + 5x - 7$ by $x - 3$
  b. $3x^5 - 12x^3 + 10x^2 + 22x + 4$ by $x + 2$
    (What can you conclude?)
3. a. $3x^4 - 2x^3 + x - 5$ by $x + 4$
  b. $2x^5 - 10x^3 + 16$ by $x - 2$
    (What can you conclude?)
4. a. Show by synthetic division that $x + 5$ is a factor of $x^5 + 4x^4 - 7x^3 - 9x^2 - 25$.
  b. Let $P(x) = 2x^4 - x^3 - 27x^2 + 36x$. Show that $P(3) = 0$ by synthetic division and give the complete factorization of $P(x)$.
5. Find a polynomial whose only zeros are $-2, \frac{3}{2}, 3i, -3i$.
6. Find a polynomial equation having 4 as a root of multiplicity 2 and whose only other root is -2.
7. Without testing for actual roots, list all possible rational roots and use Descartes' rule of signs to determine the possible numbers of positive and negative real roots of each of the following:
  a. $3x^5 - 2x^3 + 4x^2 - 8 = 0$
  b. $10x^4 + 7x^3 - 3x^2 - 5x + 12 = 0$

In Problems 8–13 find all rational roots, and if the depressed equation is quadratic, find all roots.

8. $2x^3 - 7x^2 + 9 = 0$
9. $x^3 - 4x^2 - 3x + 18 = 0$
10. $x^4 - 8x^2 - 24x - 32 = 0$
11. $6x^4 + 5x^3 + 20x^2 + 20x - 16 = 0$
12. $2x^4 + 6x^3 + 3x^2 + 4x + 12 = 0$
13. $4x^4 - 19x^2 + 3x + 18 = 0$

In Problems 14–16 factor completely over the reals.

14. $x^3 + 9x^2 + 15x - 25$
15. $2x^3 - 9x^2 + 18x - 20$
16. $2x^4 - 6x^3 + 7x^2 - 9x + 6$

In Problems 17–21 draw the graph.

17. $y = 2x^3 - 3x^2 - 5x$
18. $y = 25 + 21x^2 - 4x^4$
19. $y = (x - 3)^2(2x + 1)(x + 4)$
20. $y = x^3 - 12x - 16$
21. $y = 2x^4 + x^3 - 19x^2 - 9x + 9$

In Problems 22–24 solve the inequalities by graphical means.

22. $12x - x^2 - x^3 \geq 0$
23. $x(x^2 - 3) < 2$ **24.** $x^3 > 7x - 6$

25. Find the domain of the function

$$f(x) = \sqrt{2x^3 - 5x^2 + 18}$$

In Problems 26–35 discuss completely, and draw the graph.

26. $f(x) = \dfrac{2x + 3}{x - 4}$

27. $g(x) = \dfrac{4x}{1 - x^2}$

28. $y = \dfrac{x^2 - 4}{x^2 - 1}$

29. $y = \dfrac{2x^2 - 3x - 2}{x^2 - 5x}$

30. $y = \dfrac{1 - x^2}{(x - 2)^2}$

31. $y = \dfrac{x^2 - 9}{x^2 - 7x + 12}$

32. $y = \dfrac{x^2 - 1}{x}$

33. $y = \dfrac{x^2 + 2}{x - 1}$

34. $y^2 = \dfrac{x}{x - 1}$

35. $y^2 = \dfrac{x^2 + 2}{x^2 - 1}$

In Problems 36–41 resolve into partial fractions.

36. $\dfrac{x + 10}{x^2 + 2x - 8}$

37. $\dfrac{x^2 + 2x + 12}{x^2 - 4}$

38. $\dfrac{7 - 2x - x^2}{(x^2 - 1)(x - 1)}$

39. $\dfrac{2x^2 - 5x + 16}{x^3 - 2x^2 + 4x}$

40. $\dfrac{8x^2 + 6x - 8}{(x + 2)^2(x^2 + 2)}$

41. $\dfrac{x^4 - x^3 - 7x^2 + 11x}{x^3 - 3x^2 - x + 3}$

42. Find the real root of $2x^3 - 10x + 9 = 0$ correct to two decimal places.

43. Find the largest positive root of $2x^4 + x^3 - 13x^2 - 4x + 12 = 0$ correct to two decimal places.

B

44. Construct a polynomial of fourth degree with integral coefficients three of whose zeros are $-2, \frac{5}{2}$, and $3 - 2i$.

In Problems 45 and 46 find all roots.

45. $6x^4 - 13x^3 + 2x^2 - 4x + 15 = 0$

46. $6x^4 + x^3 + 2x - 24 = 0$

Draw the graphs of the functions in Problems 47 and 48.

47. $y = (2x - 1)(x + 3)^3(x - 4)^2$

48. $P(x) = x^4 + 2x^3 + 7x - 6$

Discuss the nature of the zeros of this polynomial.

49. Solve the inequality $x^4 - 9x^2 + 4x + 12 > 0$.

50. Find the domain of the function

$$f(x) = \sqrt{\dfrac{x^3 - 4x^2}{x^3 - 7x + 6}}$$

In Problems 51–55 discuss completely and draw the graph.

51. $y = \dfrac{x^2}{x^2 - 2x - 8}$

52. $y = \dfrac{x^2 + x}{x - 1}$

53. $y = \dfrac{x^3 - 27}{x^4 - 16}$

54. $y = \dfrac{x^4 - x^3 - 6x^2}{x^4 - x^3 - 5x^2 - x - 6}$

55. $y^2 = \dfrac{x(x + 2)}{x^2 - 9}$

In Problems 56 and 57 resolve into partial fractions.

56. $\dfrac{20 - 5x^2}{4x^4 + 9x^2 - 11x + 3}$

57. $\dfrac{6x^4 - 5x^3 - 3x^2 - 3}{6x^3 - 5x^2 + 6x - 5}$

58. Find all real roots of $x^3 - 3x^2 - 4x + 2 = 0$ correct to two decimal places.

59. Find the real roots of $x^4 - 2.013x^3 + 0.025x^2 + 8.976x + 3.164 = 0$ correct to three decimal places.

6

EXPONENTIAL AND LOGARITHMIC FUNCTIONS

Mathematical models of growth and decay phenomena usually involve exponential and logarithmic functions in their solution. The following example illustrates the use of both of these types of functions in finding the **half-life** of a radioactive element.*

The *half-life* of a radioactive element is the time required for half of the radioactive nuclei present in a sample to decay. It is a remarkable fact that the half-life is a constant that does not depend on the number of radioactive nuclei initially present in the sample.

To see why, let y_0 be the number of radioactive nuclei initially present in the sample. Then the number y present at any later time t will be

$$y = y_0 e^{kt}$$

We seek the value of t at which

$$y_0 e^{kt} = \frac{1}{2} y_0$$

for this will be the time when the number of radioactive nuclei present equals half the original number. The y_0's cancel in this equation to give

$$e^{kt} = \frac{1}{2}$$

$$kt = \ln \frac{1}{2} = -\ln 2$$

$$t = -\frac{\ln 2}{k}$$

This value of t is the half-life of the element . . .

* Thomas & Finney, *Calculus and Analytic Geometry*, Eighth Edition, © 1992, Addison-Wesley Publishing Co., Inc., Reading, Massachusetts. Example 5 on page 435. Reprinted with permission.

Calculus was used to derive the original equation in this example, but everything else will be accessible to you before you complete this chapter. In fact, you will be asked in some of the exercises to compute the half-life in just this way.

THE EXPONENTIAL FUNCTION

One of the very important functions that we have not yet studied is the **exponential function.** A typical example of such a function is given by

$$f(x) = 2^x$$

having as domain the set of all real numbers. We know how to evaluate this function for any integer x (positive, negative, or 0), and, in fact, for any rational number x. For example,

$$f(\tfrac{3}{2}) = 2^{3/2} = \sqrt{2^3} = \sqrt{2^2 \cdot 2} = 2\sqrt{2} \approx 2.828$$

(since $\sqrt{2} \approx 1.414$). How do we evaluate it if x is irrational? For example, what is the meaning of $f(\pi) = 2^\pi$? Unfortunately, we cannot answer these questions in an entirely satisfactory way without more knowledge of limiting procedures. Nevertheless, it is not difficult to gain an intuitive feeling for the meaning of such expressions as 2^x, where x is an irrational number.

It can be shown that every irrational number can be approximated to any desired degree of accuracy by a rational number. This is a fundamental property of the real number system. Consider the irrational number we symbolize by π, for example. We can approximate it by 3.14, 3.142, 3.1416, and so on, and these numbers are rational, since they are terminating decimals. We *define* the number 2^π, then, as the number that is *approached* by $2^{3.14}$, $2^{3.142}$, $2^{3.1416}$, and so on. The fact that a unique number is approached more and more closely as the exponent approximates π with increasing accuracy, we will have to ask the reader to accept on faith at this stage.

More generally, let a be any positive number other than 1. (We exclude 1, since $1^x = 1$ for every x.) Now we define the function

$$f(x) = a^x \tag{1}$$

When x is rational, this is a well-defined number, and when x is irrational, we define a^x to mean the number approached by a^r as r takes on rational values that approach x arbitrarily closely. The function defined by equation (1) is called the **exponential function with base a.** The domain of the function consists of all real numbers, and the range is the set of all positive real numbers. While we cannot at this stage prove the latter assertion, it is intuitively reasonable that the range is at least a subset of R^+. Since roots, powers, and reciprocals of positive numbers are positive, then for a rational and positive x, say, $x = p/q$,

$$a^x = a^{p/q} = \sqrt[q]{a^p}$$

is positive, since $a > 0$. And since $a^{-x} = 1/a^x$, this too is positive. If x is irrational, then the value of a^x is approximated arbitrarily closely by numbers of the form a^r (with r rational), which is positive. It follows, then, that a^x also is positive. Finally, if $x = 0$, we know that $a^x = 1$.

It would lead to trouble to allow the base a to be either 0 or a negative number. If $a = 0$, then $a^x = 0$ for all $x > 0$, but a^x would be undefined for $x \le 0$. If $a < 0$, then a^x would lead to imaginary numbers for infinitely many values of x (for example, for $x = \frac{1}{2}, \frac{3}{4}$, and so on).

A table of values for $f(x) = 2^x$ and the graph of f are shown in Figure 1. Observe that f is an increasing function, and hence is 1–1. It increases more and more rapidly as x increases. The curve is asymptotic to the negative x axis.

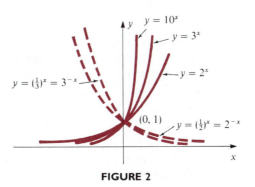

x	0	1	2	3	4	-1	-2	-3	-4	-5	-10
2^x	1	2	4	8	16	$\frac{1}{2}$	$\frac{1}{4}$	$\frac{1}{8}$	$\frac{1}{16}$	$\frac{1}{32}$	$\frac{1}{1,024}$

FIGURE 1

The most common bases are those that are greater than 1. In fact, if $0 < a < 1$, then $a^x = (1/a)^{-x}$, and $1/a > 1$; so it really is sufficient to consider bases greater than 1. Figure 2 illustrates several exponential functions on the same coordinate system. The functions all have the following features in common.

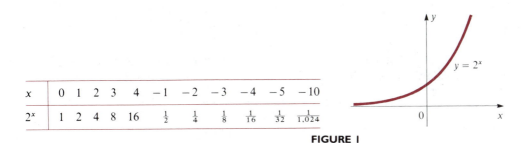

FIGURE 2

1. The y intercept is 1.
2. For $a > 1$, the negative x axis is an asymptote, and for $a < 1$, the positive x axis is an asymptote.
3. For $a > 1$, the functions are all increasing, and for $a < 1$, they are all decreasing.

Using these basic facts, we can graph more complex exponential functions, as the next example shows.

EXAMPLE 1

Draw the graph of the function defined by $y = 2^{1-|x|}$.

Solution   First we make a table of values:

x	0	± 1	± 2	± 3	± 4
y	2	1	$\frac{1}{2}$	$\frac{1}{4}$	$\frac{1}{8}$

Notice that f is an even function. Also observe that as x increases in absolute value, y approaches zero. So the x axis is an asymptote. The graph is shown in Figure 3.

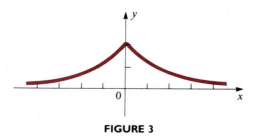

FIGURE 3

The next two examples illustrate ways in which exponential functions are used to model certain phenomena in nature. We will see further applications in Section 4 after we have studied logarithmic functions.

EXAMPLE 2

A culture of bacteria, originally numbering 1000, doubles in size every hour. Find a formula for the number $N(t)$ of bacteria present after t hours. How many bacteria are present after 8 hours?

Solution   After one hour there are

$$N(1) = 1000(2)$$

bacteria present. After two hours this number is doubled, giving

$$N(2) = 1000(2)(2) = 1000(2^2)$$

After three hours, this is again doubled, giving

$$N(3) = 1000(2^3)$$

Continuing in this way, we get the formula

$$N(t) = 1000(2^t)$$

After 8 hours, the number is

$$N(8) = 1000(2^8) = 256{,}000$$ ▲

EXAMPLE 3 ||||||||||||||||||||||||||||||||||||||||||||||||||||||||||||||||||||||||||||||||||||||||||||||||||||||||||

Suppose that when a quantity of sugar is placed in water, 10% of it dissolves each minute. Let $Q(t)$ denote the quantity present after t minutes. If there are 10 pounds initially, that is, $Q(0) = 10$, find approximately how much will be present after 15 minutes.

Solution   Each minute 10% of the sugar dissolves, so 90% remains undissolved. Thus, after 1 minute the amount remaining is

$$Q(1) = 10(0.9)$$

and after 2 minutes, 90% of $Q(1)$ is

$$Q(2) = 10(0.9)(0.9) = 10(0.9)^2$$

Continuing in this way, we get

$$Q(t) = 10(0.9)^t$$

So

$$Q(15) = 10(0.9)^{15} \approx 2.06 \text{ pounds}$$

(by calculator).   ▲

Among all possible bases one stands out as being far and away the most important in advanced courses in mathematics and in applications to real-world phenomena. It is an irrational number that is universally designated by the letter e, and its value is approximately 2.71828. The following is one way of defining the number e. Let us consider the expression

$$\left(1 + \frac{1}{n}\right)^n$$

where n designates any positive integer. We evaluate this for a few values of n (using a calculator) in Table 1. This should be fairly convincing evidence (but it is not a proof) that as n becomes larger and larger, $[1 + (1/n)]^n$ approaches more and more closely some number whose first few digits are 2.718. We say

n	$\left(1 + \dfrac{1}{n}\right)^n$
1	2.00000
2	2.25000
3	2.37037
4	2.44141
5	2.48832
10	2.59374
100	2.70481
1,000	2.71692
10,000	2.71815
1,000,000	2.71828

TABLE I

that the **limit** of $[1 + (1/n)]^n$ as n becomes infinite is e, and write

$$e = \lim_{n \to \infty} \left(1 + \frac{1}{n}\right)^n \tag{2}$$

One way in which this number e arises in a practical situation concerns money invested at compound interest. If a certain principal amount P is invested, and interest is compounded annually at an interest rate of r per year (for example, if the interest rate is 5%, then $r = 0.05$), the amount $A(t)$ of money in the account at the end of t years can be shown to be

$$A(t) = P(1 + r)^t$$

If the interest is compounded semiannually, the formula becomes

$$A(t) = P\left(1 + \frac{r}{2}\right)^{2t}$$

and if quarterly,

$$A(t) = P\left(1 + \frac{r}{4}\right)^{4t}$$

More generally, if interest is compounded k times a year, the amount after t years is

$$A(t) = P\left(1 + \frac{r}{k}\right)^{kt} \tag{3}$$

For example, some banks make a big point of the fact that their interest is compounded daily. In this case equation (3) becomes

$$A(t) = P\left(1 + \frac{r}{365}\right)^{365t}$$

Suppose that we wish to determine what the formula becomes when interest is compounded *continuously*. Some banks actually calculate their interest this way. What we must do is to let k become larger and larger in equation (3). That is, we want to find the limit of the expression on the right side of the equation as k gets arbitrarily large. In symbols, we want

$$\lim_{k \to \infty} P\left(1 + \frac{r}{k}\right)^{kt}$$

To find what this limit is, we manipulate the expression $[1 + (r/k)]^{kt}$ somewhat. First, let us make the substitution

$$n = \frac{k}{r}$$

Thus, $k = nr$. The substitution yields

$$\left(1 + \frac{r}{k}\right)^{kt} = \left(1 + \frac{1}{n}\right)^{nrt}$$

Now, as k gets larger and larger, so does n, and vice versa. So we can write the limit as*

$$\lim_{k \to \infty} P\left(1 + \frac{r}{k}\right)^{kt} = \lim_{n \to \infty} P\left(1 + \frac{1}{n}\right)^{nrt}$$

$$= P\left[\lim_{n \to \infty} \left(1 + \frac{1}{n}\right)^{n}\right]^{rt}$$

The expression in brackets is just the same as the right side of equation (2), and hence equals e. So, finally, our formula for the amount accumulated at the end of t years when the interest is compounded continuously is

$$A(t) = Pe^{rt} \qquad\qquad (4)$$

It should be emphasized that r is the annual interest rate, and t is the number of years.

EXAMPLE 4 ||||||||||||||||||||||||||||||||||||||||||||||||||||||||||||||||||||||||||||||||||||||||||||||||||||||||||||

If \$1,000 is placed in a savings account yielding interest at 6%, find the amount accumulated after 5 years if interest is compounded
a. Annually   b. Semiannually   c. Quarterly
d. Daily   e. Continuously

Note. To illustrate how problems such as this can be done on a calculator, we show keystroke displays for parts **d** and **e**.

_____

* We use certain properties of limits here that are shown to be true in calculus.

Solution **a.** $A(5) = 1,000(1 + 0.06)^5 = 1,000(1.06)^5 = 1,000(1.33822) = \$1,338.22$

b. $A(5) = 1,000\left(1 + \dfrac{0.06}{2}\right)^{5(2)} = 1,000(1.03)^{10} = 1,000(1.34392) = \$1,343.92$

c. $A(5) = 1,000\left(1 + \dfrac{0.06}{4}\right)^{5(4)} = 1,000(1.015)^{20} = 1,000(1.34686) = \$1,346.86$

d. $A(5) = 1,000\left(1 + \dfrac{0.06}{365}\right)^{5(365)} = 1,000(1.000164)^{1,825} = 1,000(1.34982)$

   $= \$1,349.82$

By calculator:

1,000 ⊠ ⦅ 1 ⊞ .06 ⊟ 365 ⦆ y^x ⦅ 5 ⊠ 365 ⦆ ⊟

The displayed answer is 1,349.8257.

e. $A(5) = 1,000e^{(0.06)5} = 1,000(e^{0.3}) = 1,000(1.34986) = \$1,349.86$
   By calculator:

1,000 ⊠ .3 e^x ⊟

The displayed answer is 1,349.8588.   ▲

Note. Some calculators have an e^x key. On others e^x is obtained either by the sequence 2nd ln x or by INV ln x. The basis for this latter sequence will be made clear in the next section.

EXAMPLE 5 ‖‖

Use a calculator to make a table of values, and draw the graph of

$$f(x) = 1 - e^{-x}$$

Solution   The graph is shown in Figure 4. The curve is asymptotic to the line $y = 1$ on the right since as x increases positively beyond bound, $e^{-x} = 1/e^x$ approaches 0,

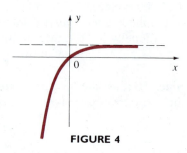

FIGURE 4

so that $1 - e^{-x}$ approaches 1. You might also observe that we could have used the ideas of Section 3 in Chapter 3 to obtain the graph of f from the basic graph of $y = e^x$, by reflections in the x and y axes and then a vertical translation of 1 unit. ▲

The familiar laws of exponents given in Chapter 1 for rational exponents also hold true for arbitrary real exponents, as long as the base is positive. A proof of this requires the use of calculus. We list the laws below for reference.

Laws of Exponents

For $a > 0$ and $b > 0$, and for all real numbers x and y,

1. $a^x \cdot a^y = a^{x+y}$ **2.** $\dfrac{a^x}{a^y} = a^{x-y}$

3. $(a^x)^y = a^{xy}$ **4.** $(ab)^x = a^x b^x$

5. $\left(\dfrac{a}{b}\right)^x = \dfrac{a^x}{b^x}$

EXERCISE SET I

A

In Problems 1 and 2 use a calculator to approximate the given numbers correct to five significant digits.

1. a. $2^{\sqrt{2}}$ **b.** 3^{π} **c.** $4^{-\sqrt{3}}$
   d. e^2 **e.** e^{-1}

2. a. $(\sqrt{5})^{\sqrt{7}}$ **b.** e^{π} **c.** π^2
   d. $2^{2/3}$ **e.** $e^{-1/2}$

In Problems 3–20 make a table of values and draw the graph. If it is needed, use a calculator.

3. $y = 4^x$ **4.** $y = 4^{-x}$
5. $y = e^x$ **6.** $y = e^{-x}$
7. $y = (\frac{2}{3})^x$ **8.** $f(x) = 2^{x/2}$
9. $F(x) = 2^{x-1}$ **10.** $g(x) = (\sqrt{3})^{-x}$
11. $h(x) = 3^{(1-x)/2}$ **12.** $y = 2^{x^2}$
13. $y = 1 - e^{-x}$ **14.** $f(x) = 1 - 2^{-x}$
15. $g(x) = 2^{1-x^2}$ **16.** $h(x) = 2^{|x|}$
17. $f(x) = (\frac{3}{4})^{x-1}$ **18.** $F(x) = (\frac{3}{2})^{x/2}$
19. $G(x) = (\frac{1}{2})^{2-x}$ **20.** $y = 3^{|x-1|}$

21. The sum of \$2,000 is placed in a savings account at $5\frac{1}{2}\%$ annual interest. Find the amount at the end of 4 years if interest is compounded
   a. Annually   **b.** Quarterly
   c. Continuously

22. One bank offers 6% interest compounded quarterly. Another offers 6% compounded continuously. How much more income would result from depositing \$1,000 for 5 years in the second account than in the first?

23. Determine which of the following is the better way to invest \$500 for 2 years:
   a. An account paying $7\frac{1}{2}\%$, compounded annually
   b. An account paying $7\frac{1}{4}\%$, compounded continuously
     If in part **a** interest were compounded quarterly, how would the amount compare with the result of part **b**?

24. A certain bacteria culture grows by 50% each hour. Let N_0 be the number present initially, and find a formula for the number $N(t)$ present after t hours. If $N_0 = 500$, find how many will be present after 7 hours. (Use a calculator.)

25. A machine that costs \$4,000 when new loses 20% of its remaining value each year in depreciation. Find its value after t years. What is its value after 5 years?

26. Assume that the world population increases by 20% each decade. If the present world population is 4 billion, find a formula for the population x decades from now. Under the given assumption, what will be the world population 3 decades from now?

B

In Problems 27–30 use a calculator to make a table of values and then draw the graph.

27. $y = xe^{-x}$

28. $f(x) = \dfrac{e^x + e^{-x}}{2}$

29. $g(x) = \dfrac{e^x - e^{-x}}{2}$

30. $h(x) = \dfrac{e^x - e^{-x}}{e^x + e^{-x}}$

Note. The functions in Problems 28, 29, and 30 are called, respectively, **the hyperbolic cosine, the hyperbolic sine, and the hyperbolic tangent, written cosh x, sinh x, and tanh x.**

In Problems 31 and 32 solve the inequalities.

31. $e^{2x} - e^x < 0$

32. $x(2^x - 2) \geq 0$

C (Graphing calculator)

33. Obtain the graph of

$$y = x^{\alpha} e^{-\beta x}$$

for several different positive integers α and β. Discuss your findings.

INVERSES OF EXPONENTIAL FUNCTIONS

As we have noted, the exponential function

$$f(x) = a^x \qquad (a > 0, \quad a \neq 1)$$

is increasing if $a > 1$ and decreasing if $0 < a < 1$. So f is 1–1, since each horizontal line intersects the curve at most once. Thus, an inverse exists for this function. For historical reasons the name given to the inverse is the **logarithm function,** and its value is symbolized by $\log_a x$. This is read "log to the base a of x." So if $f(x) = a^x$, then

$$f^{-1}(x) = \log_a x$$

It is unfortunate that the name and symbol used here do not in any way suggest the relationship of the logarithm function to the exponential; this relationship simply must be learned.

In Figure 5 we have shown the graph of $y = a^x$ for $a > 1$ and its inverse, $y = \log_a x$, on the same set of axes. Observe that the domain of $y = \log_a x$ is the set of all positive real numbers and the range is all of R.

We know from Chapter 3 that for any function f having an inverse f^{-1} the equations

$$f(f^{-1}(x)) = x \qquad \text{for all } x \text{ in the domain of } f^{-1} \tag{5}$$

and

$$f^{-1}(f(x)) = x \qquad \text{for all } x \text{ in the domain of } f \tag{6}$$

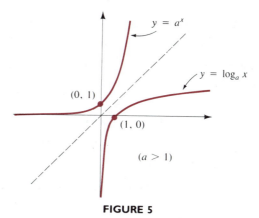

FIGURE 5

are valid. If we put $f(x) = a^x$, then $f^{-1}(x) = \log_a x$, and equations (5) and (6) give the results

$$a^{\log_a x} = x \qquad \text{for all } x > 0 \qquad (7)$$

and

$$\log_a a^x = x \qquad \text{for all real } x \qquad (8)$$

Also, we know that a point (x_0, y_0) satisfies $y = f^{-1}(x)$ if and only if the point (y_0, x_0) satisfies $y = f(x)$; that is,

$$y_0 = f^{-1}(x_0)$$

if and only if

$$x_0 = f(y_0)$$

In terms of the logarithmic and exponential functions, this becomes (omitting the subscript)

$$y = \log_a x \qquad \text{if and only if} \qquad x = a^y \qquad (9)$$

This is of fundamental importance and should be memorized. We refer to the left-hand equation in (9) as the **logarithmic form** and the right-hand equation as the **exponential form.** We can often deduce facts about logarithms by using an equivalent exponential form. For example, suppose we wish to obtain the value of $\log_2 \frac{1}{16}$. Let us denote this unknown value by y. Then

$$y = \log_2 \tfrac{1}{16}$$

and by (9) the equivalent exponential form is

$$\tfrac{1}{16} = 2^y$$

Since $\frac{1}{16} = (\frac{1}{2})^4 = 2^{-4}$, we have $2^{-4} = 2^y$, from which it follows that $y = -4$. Here are some other examples of this type.

EXAMPLE 6 ||||||||||||||||||||||||||||||||||||||||||||||||||||||||||||||||||||||||||||||||||||||||||||||||||

Find the value of x if $x = \log_3 27$.

Solution   The exponential form is $3^x = 27$, and since $27 = 3^3$, we have $3^x = 3^3$, so that $x = 3$. ▲

EXAMPLE 7 ||||||||||||||||||||||||||||||||||||||||||||||||||||||||||||||||||||||||||||||||||||||||||||||||||

Find y if $6 = \log_2 y$.

Solution   In exponential form this is $2^6 = y$. So $y = 64$. ▲

EXAMPLE 8 ||||||||||||||||||||||||||||||||||||||||||||||||||||||||||||||||||||||||||||||||||||||||||||||||||

Solve for x: $3 = \log_x 125$

Solution   We have $x^3 = 125$, so that $x = \sqrt[3]{125} = 5$. ▲

One special value of the logarithm function deserves emphasis. We know that for $a > 0$, $a^0 = 1$, so that by equations (9), taking $x = 1$ and $y = 0$, we get

$$\log_a 1 = 0$$

This is true for all admissible bases a.

It helps in changing from logarithmic to exponential form to observe that the base in each case is the same. In the logarithmic form this appears as the subscript:

$$y = \log_{\boxed{a}} x$$

$\searrow$ **Base**

and in the exponential form it is the number that is raised to a power:

$$x = \boxed{a}^y$$

$\searrow$ **Base**

Also note that in the exponential form the exponent y is the logarithm. We emphasize this fact below.

The logarithm to the base a of a number x is the exponent y for which $a^y = x$.

Remark. The base e is the most important base for logarithms, and logarithms with this base are called **natural** logarithms. They also are sometimes called **Napierian** (after John Napier, a Scot who invented logarithms in the early seventeenth century). The symbol **ln** is often used instead of the more cumbersome $\log_e$. So

$$\log_e x = \ln x$$

It is also common practice to use "log" to mean "$\log_{10}$," and this is called the **common logarithm.** Historically, common logarithms have been most used in computational work. Now, however, with the advent of hand calculators, they are seldom needed. Scientific calculators typically have both a $\boxed{\ln x}$ key and a $\boxed{\log x}$ key, for base e and base 10, respectively.

A Word of Caution. Although "log" usually is understood to mean $\log_{10}$, there are exceptions, especially in advanced mathematics books, where it may mean $\log_e$. You will need to check what meaning each author intends.

The graph of a function involving logarithms can be obtained by working from the basic logarithm function, or by changing to exponential form, as the next example shows.

EXAMPLE 9 ||||||||||||||||||||||||||||||||||||||||||||||||||||||||||||||||||||||||||||||||||||||||||||||||||||||||||||||

Draw the graph of the function defined by

$$y = \log_3(4 - x)$$

The graph of $y = \log_3 x$ is of the form shown in Figure 5. Referring to this as the basic graph, we first reflect in the y axis to get the graph of $y = \log_3(-x)$, and then shift 4 units to the right to get the graph of $y = \log_3[-(x - 4)]$, or equivalently, $y = \log_3(4 - x)$. The result is shown in Figure 6.

An alternative procedure is to write the equation in exponential form, $4 - x = 3^y$, and solve for x, getting

$$x = 4 - 3^y$$

Then points can be found by substituting for y to get values of x.

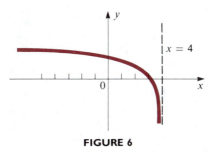

FIGURE 6

EXAMPLE 10

The formula for the amount $Q(t)$ in grams present at time t in years of a radioactive substance is given by

$$Q(t) = Q_0 e^{-0.2t}$$

where Q_0 is the amount initially. Find how long it will be until only 10% of the original quantity is left.

Solution We want to know the value of t for which $Q(t) = (0.1)Q_0$. Substituting this for $Q(t)$ yields

$$(0.1)Q_0 = Q_0 e^{-0.2t}$$
$$0.1 = e^{-0.2t}$$

To solve for t we write this in logarithmic form:

$$\ln(0.1) = -0.2t$$

So,

$$t = -\frac{\ln(0.1)}{0.2} \approx -\frac{-2.30285}{0.2} \approx 11.513 \text{ years}$$

By calculator:

$$.1 \;\boxed{\ln x}\; \boxed{+/-}\; \boxed{\div}\; .2 \;\boxed{=}$$

The displayed answer is 11.5129. ▲

EXAMPLE 11

A man deposits a sum of money in a savings account paying 5% interest, compounded continuously. How long will it take for him to double his money?

Solution Let P be the initial amount. Then, as we found in the previous section,

$$A(t) = Pe^{0.05t}$$

is the formula for the amount after t years. We want to know t when $A(t) = 2P$.

$$2P = Pe^{0.05t}$$
$$2 = e^{0.05t}$$

In logarithmic form this is

$$\ln 2 = 0.05t$$

So

$$t = \frac{\ln 2}{0.05} \approx \frac{0.6931}{0.05} \approx 13.86 \text{ years}$$ ▲

EXAMPLE 12

The human ear can accommodate an enormous range of sound wave intensities, ranging from about 10^{-12} watt per square meter, taken to be the threshold of hearing, to about 1 watt per square meter, the threshold of pain for most people. Since the range is so great, it is customary to use a logarithmic scale to describe the relative loudness of a sound. The formula

$$\beta = 10 \log_{10} \frac{I}{I_0}$$

is used, where I_0 is the intensity of the threshold of hearing (10^{-12} watt per square meter) and I is the intensity of the sound in question. The number β is said to be the number of **decibels** of the sound. Find the decibel intensity of the noise of a heavy truck passing a pedestrian at the side of a road if the sound wave intensity I of the truck is 10^{-3} watt per square meter.

Solution   Since $I_0 = 10^{-12}$ and $I = 10^{-3}$, we have

$$\beta = 10 \log_{10} \frac{I}{I_0} = 10 \log_{10} \frac{10^{-3}}{10^{-12}}$$

$$= 10 \log_{10} 10^9$$

$$= 10(9) = 90$$

So the intensity is 90 decibels.

 EXERCISE SET 2

A

Evaluate the expressions in Problems 1–11.

1. a. $\log_2 2^3$ b. $2^{\log_2 3}$
2. a. $\log_3 3^{10}$ b. $5^{\log_5 7}$
3. a. $\log_b b^x$ b. $b^{\log_b x}$
4. a. $\log_2 8$ b. $\log_3 \frac{1}{9}$
5. a. $\log_{10} 10{,}000$ b. $\log_{10} 0.001$
6. a. $\log_4 2$ b. $\log_2 0.125$
7. a. $\log_4 64$ b. $\log_2 \frac{1}{16}$
8. a. $\log_{10} 0.1$ b. $\log_3 81$
9. a. $\log_{10} 100$ b. $\ln(1/e^2)$
10. a. $e^{\ln 4}$ b. $10^{\log_{10} 7}$
11. a. $\log_9 9^5$ b. $\log_6 6^{-5}$

Find the unknown in Problems 12–24.

12. a. $\log_4 x = 2$ b. $\log_x 16 = 4$
13. a. $\log_2 8 = y$ b. $\log_y 3 = -\frac{1}{2}$
14. a. $t = \log_2 \frac{1}{8}$ b. $\log_{10} x = -2$
15. a. $\log_{10} 1{,}000 = x$ b. $\log_{10} x = -3$
16. a. $\log_u 3 = \frac{1}{2}$ b. $\log_{1/2} u = -4$
17. a. $\log_4 1 = v$ b. $\log_v \frac{4}{9} = -\frac{2}{3}$
18. a. $\log_2 y = 6$ b. $\log_x 125 = 3$
19. a. $\log_4 0.25 = t$ b. $\log_4 x = \frac{1}{2}$
20. a. $\ln 1 = y$ b. $\log_z 0.01 = -2$
21. a. $\log_9 u = -\frac{3}{2}$ b. $\log_8 4 = v$
22. a. $\log_x 16 = \frac{4}{3}$ b. $\log_{4/9} t = -\frac{3}{2}$
23. a. $\log_3(x - 1) = 2$ b. $\log_2 0.125 = y$
24. a. $w = \ln(1/e)$ b. $\log_2|w| = 3$

In Problems 25–28 write in logarithmic form.

25. a. $4^3 = 64$ b. $3^{-2} = \frac{1}{9}$
26. a. $10^4 = 10{,}000$ b. $10^{-3} = 0.001$
27. a. $2^8 = 256$ b. $2^{-3} = 0.125$
28. a. $r^s = t$ b. $z^x = m$

In Problems 29–32 write in exponential form.

29. a. $\log_2 16 = 4$ **b.** $\log_{10} 100 = 2$
30. a. $\ln 1 = 0$ **b.** $\log_3 \frac{1}{9} = -2$
31. a. $\log_5 0.2 = -1$ **b.** $\log_8 4 = \frac{2}{3}$
32. a. $\log_a x = y$ **b.** $\log_x z = t$

In Problems 33–42 draw the graph of each function by first writing an equivalent exponential form and then making a table of values.

33. $y = \log_2 x$ **34.** $y = \log_3 x$
35. $y = \log_2(x + 1)$ **36.** $y = \log_3(1 - x)$
37. $y = \log_4(2x - 3)$ **38.** $y = \log_2(1 - 3x)$
39. $y = \log_2 x^2$ **40.** $y = \ln |x|$
41. $y = \ln(-x)$ **42.** $y = 1 - \ln x$
43. If $10,000 is invested at 6% interest, compounded continuously, in approximately how many years will the amount have grown to $15,000?
44. If $5,000 is invested at $5\frac{1}{2}$%, compounded continuously, approximately how long will it take for the amount to double?
45. A woman has $10,000 invested at 6% and her brother has $9,000 invested at 7%. If in each account interest is compounded continuously, find how long it will take for the amount in the brother's account to equal the amount in his sister's.
46. Use the information given in Example 12 to find the decibel level of sound at
  a. The threshold of hearing
  b. The threshold of pain
47. At a distance of 10 meters from a rock band the intensity of sound is approximately 0.1 watt per square meter. Find the intensity in decibels.

B

In Problems 48–51 draw the graph by first writing the equivalent exponential form and solving for x.

48. $y = \ln(1 - x^2)$ **49.** $y = \log_2 |1 - x|$
50. $y = \ln\left(\dfrac{1 - x}{1 + x}\right)$ **51.** $y = \ln(x + \sqrt{x^2 + 1})$

52. In a chemical reaction a substance is converted according to the formula

$$y(t) = Ce^{-0.3t}$$

where $y(t)$ is the amount of the unconverted substance at time t minutes after the reaction began. If there are 20 grams of the substance initially, find how long it will take for only 4 grams of it to remain unconverted.

53. The pH of a solution is defined by

$$pH = -\log_{10}[H^+]$$

where $[H^+]$ is the concentration of hydrogen ions in aqueous solution, measured in moles per liter. Find the pH of a solution for which $[H^+] = 0.050$.

54. Find $[H^+]$ for a solution with pH $= 7.92$. (See Problem 54.)

C   (Graphing calculator)

55. Obtain the graph of

$$y = \frac{(\ln x)^{\alpha}}{x^{\beta}} \qquad (x > 0)$$

for several different positive integers α and β. Discuss your findings.

PROPERTIES OF LOGARITHMS

The relations (7) and (8) of Section 2 enable us to exploit the properties of the exponential function to obtain corresponding properties of the logarithm function. In what follows we will repeatedly use the fact that if $a^x = a^y$, then $x = y$, which is a consequence of the fact that the exponential function is always increasing or always decreasing and hence is one-to-one.

By equation (7) we have that for any $x > 0$ and $y > 0$,

$$x = a^{\log_a x} \qquad \text{and} \qquad y = a^{\log_a y}$$

Similarly,

$$xy = a^{\log_a xy}$$

Thus, we have the equality

$$a^{\log_a xy} = a^{\log_a x} \cdot a^{\log_a y}$$

But by the law of exponents for multiplication we can add the exponents on the right, and so obtain

$$a^{\log_a xy} = a^{\log_a x + \log_a y}$$

The exponents must therefore be equal:

$$\log_a xy = \log_a x + \log_a y \tag{10}$$

Next consider the following two representations of x^p. By equation (7),

$$x^p = a^{\log_a x^p}$$

but also,

$$x^p = (a^{\log_a x})^p = a^{p \log_a x}$$

where the last equality is a consequence of one of the laws of exponents. The two expressions for x^p must be equal:

$$a^{\log_a x^p} = a^{p \log_a x}$$

which yields

$$\log_a x^p = p \log_a x \tag{11}$$

Equations (10) and (11) can be used now to obtain a formula for the logarithm of a quotient. We have

$$\log_a \frac{x}{y} = \log_a \left(x \cdot \frac{1}{y} \right) = \log_a x + \log_a \frac{1}{y}$$

by equation (10). And by (11), with $p = -1$,

$$\log_a \frac{1}{y} = \log_a y^{-1} = -\log_a y$$

So finally,

$$\log_a \frac{x}{y} = \log_a x - \log_a y \tag{12}$$

The properties (10), (11), and (12) are basic and account for much of the utility of the logarithm function. These results are summarized in the box, where it is understood that a is any admissible base.

Properties of Logarithms

For $x > 0$ and $y > 0$,

1. $\log_a xy = \log_a x + \log_a y$

2. $\log_a \dfrac{x}{y} = \log_a x - \log_a y$

3. $\log_a x^p = p \log_a x$

In the exercises that follow we will often use "log" without explicitly indicating a base, since the results are valid regardless of what (admissible) base is being used. The properties of logarithms are used repeatedly. The following examples are illustrative. In Examples 13–16 we use the properties of logarithms to write the expression in a form that is free of logarithms of products, quotients, and powers.

EXAMPLE 13 |||||||||||||||||||||||||||||||||||||||||||||||||||||||||||||||||||||||||||||||||||||||||||||||||||||||||||||||||||||||||||

Simplify $\log x^2 \quad (x \neq 0)$

Solution   If $x > 0$, we have by Property 3,

$$\log x^2 = 2 \log x$$

If $x < 0$, we can write

$$\log x^2 = \log|x|^2 = 2 \log|x| \qquad \blacktriangle$$

EXAMPLE 14 |||||||||||||||||||||||||||||||||||||||||||||||||||||||||||||||||||||||||||||||||||||||||||||||||||||||||||||||||||||||||||

Simplify

$$\log \frac{2x}{3x^2 + 1} \qquad (x > 0)$$

Solution   Using Property 2 first and then Property 1, we have

$$\log \frac{2x}{3x^2 + 1} = \log 2x - \log(3x^2 + 1)$$

$$= \log 2 + \log x - \log(3x^2 + 1) \qquad \blacktriangle$$

A Word of Caution. The properties of logarithms enable us to simplify logarithms of products, quotients, and powers (and roots, since these can be treated as powers), but there is no way to simplify logarithms of sums or differences.

Thus, $\log(x + y)$ *cannot* be written as $\log x + \log y$, however tempting this might be. In fact, we know that $\log x + \log y$ is the same as $\log xy$, not $\log(x + y)$.

EXAMPLE 15

Simplify

$$\log \sqrt{\frac{1 - x}{1 + x}} \qquad (|x| < 1)$$

Solution   Using Properties 3 and 2, we get

$$\log \sqrt{\frac{1 - x}{1 + x}} = \log \left(\frac{1 - x}{1 + x}\right)^{1/2} = \frac{1}{2} \log \frac{1 - x}{1 + x}$$

$$= \frac{1}{2} \left[\log(1 - x) - \log(1 + x)\right] \qquad \blacktriangle$$

EXAMPLE 16

Simplify

$$\log \frac{x^3 \sqrt{1 + x^2}}{(2x + 1)^{3/2}(x^4 + 1)} \qquad (x > 0)$$

Solution $\log \dfrac{x^3 \sqrt{1 + x^2}}{(2x + 1)^{3/2}(x^4 + 1)} = \log x^3 \sqrt{1 + x^2} - \log(2x + 1)^{3/2}(x^4 + 1)$

$$= \log x^3 + \log \sqrt{1 + x^2} - [\log(2x + 1)^{3/2} + \log(x^4 + 1)]$$

$$= 3 \log x + \tfrac{1}{2} \log(1 + x^2) - \tfrac{3}{2} \log(2x + 1) - \log(x^4 + 1)$$

We first treated this as a quotient. Then we worked with the products, and finally with the powers. Notice that in the second step we bracketed the two terms preceded by a minus sign. This procedure is highly recommended, for a common pitfall is to apply the minus to one term and not the other. $\blacktriangle$

EXAMPLE 17

Solve for x: $\log x + \log(x - 2) = 3 \log 2$

Solution   Using the properties of logarithms, we rewrite this in the form

$$\log x(x - 2) = \log 2^3$$

Since the logarithm function is 1–1, it follows that $x(x - 2) = 2^3$. That is,

$$x^2 - 2x = 8$$

$$x^2 - 2x - 8 = 0$$

$$(x - 4)(x + 2) = 0$$

$$x = 4 \quad \text{or} \quad x = -2$$

When $x = -2$, neither $\log x$ nor $\log(x - 2)$ is defined, so we reject this solution. For $x = 4$ we have

$$\log 4 + \log 2 = \log 2^2 + \log 2 = 2 \log 2 + \log 2 = 3 \log 2$$

So the only solution is $x = 4$.   ▲

Remark. In problems such as the one in Example 17 it is essential that you check your answers to see that you do not have the logarithm of a negative number or zero.

EXAMPLE 18

Solve the inequality: $\log_3(x + 1) + \log_3(x + 3) \leq 1$

Solution   When both $x + 1 > 0$ and $x + 3 > 0$, this is equivalent to

$$\log_3(x + 1)(x + 3) \leq 1$$

We can take the exponential with base 3 of both sides and retain the sense of the inequality since 3^x is an increasing function. So by equation (7)

$$3^{\log_3(x^2 + 4x + 3)} \leq 3^1$$

or, equivalently,

$$x^2 + 4x + 3 \leq 3$$
$$x^2 + 4x \leq 0$$
$$x(x + 4) \leq 0$$

Solving this inequality, we find that $-4 \leq x \leq 0$. But we must have $x + 1 > 0$ and $x + 3 > 0$ in order for the original logarithm terms to be defined. All of the inequalities will be satisfied if $-1 < x \leq 0$. So the solution set is the interval $(-1, 0]$.   ▲

EXAMPLE 19

Find the value of x in each of the following in terms of natural logarithms:
a. $2^x = 3$ **b.** $4^{1-x} = 3 \cdot 5^x$

Solution   **a.**   We take the natural logarithm of both sides, and make use of the properties of logarithms:

$$\ln 2^x = \ln 3$$
$$x \ln 2 = \ln 3$$
$$x = \frac{\ln 3}{\ln 2}$$

A Word of Caution. Do not confuse $\ln 3/\ln 2$ with $\ln \frac{3}{2}$, which is equal to $\ln 3 - \ln 2$.

b. Again, we equate the natural logarithm of the left side to that of the right:

$$\ln 4^{1-x} = \ln(3 \cdot 5^x)$$
$$(1-x)\ln 4 = \ln 3 + \ln 5^x$$
$$\ln 4 - x\ln 4 = \ln 3 + x\ln 5$$
$$\ln 4 - \ln 3 = x(\ln 4 + \ln 5)$$
$$\ln \tfrac{4}{3} = x\ln(4 \cdot 5)$$
$$x = \frac{\ln \tfrac{4}{3}}{\ln 20}$$

▲

Sometimes it is important to change from one base for logarithms to another. Suppose we are given $\log_a x$ and want to find the value of this same expression involving logarithms to the base b. Let

$$y = \log_a x$$

We write this in exponential form and take the logarithm to the base b of both sides, making use of equation (11):

$$a^y = x$$
$$\log_b a^y = \log_b x$$
$$y\log_b a = \log_b x$$
$$y = \frac{\log_b x}{\log_b a}$$

Therefore, we have the change of base formula

$$\log_a x = \frac{\log_b x}{\log_b a} \tag{13}$$

If we put $x = b$, we obtain the interesting result

$$\log_a b = \frac{1}{\log_b a}$$

which can be used to give the following alternative form of equation (13):

$$\log_a x = (\log_a b)(\log_b x) \tag{14}$$

The change of base formula (13) is especially important when you wish to use a calculator to obtain a logarithm to a base other than e or base 10. Taking the base b as e, and writing $\log_e x$ as $\ln x$, formula (13) becomes

$$\log_a x = \frac{\ln x}{\ln a}$$

For example, to get $\log_2 3$ on a calculator, we have

$$\log_2 3 = \frac{\ln 3}{\ln 2} \approx 1.58496$$

Another formula, which is essentially a change of base formula for exponentials, can be obtained as follows. By equation (7),

$$a^x = b^{\log_b a^x}$$

and since $\log_b a^x = x \log_b a$, this gives

$$a^x = b^{x \log_b a} \qquad\qquad (15)$$

This formula is especially important in calculus when b is usually taken as the number e. Then we have

$$a^x = e^{x \ln a} \qquad\qquad (16)$$

This shows that every exponential can be expressed as an exponential with base e.

 EXERCISE SET 3

A

In Problems 1–26 use the properties of logarithms to simplify each expression so that the result does not contain logarithms of products, quotients, or powers. Assume appropriate restrictions have been placed on the variables so that all logarithms are defined.

1. **a.** $\log(x - 3)(x + 2)$ **b.** $\log \dfrac{x - 5}{x + 4}$

2. **a.** $\log(x - 1)^5$ **b.** $\log \sqrt{3 - 2x}$

3. **a.** $\log(x + 1)^{-2}$ **b.** $\log(5x - 3)^{2/3}$

4. **a.** $\log(x^2 - 3x)$ **b.** $\log \dfrac{1}{x + 2}$

5. **a.** $\log(x^2 - 2x - 8)$ **b.** $\log \dfrac{x - 5}{x^2 - 9}$

6. **a.** $\log \dfrac{x^2}{2x - 3}$ **b.** $\log \dfrac{x^2 - 2x}{(x + 4)^2}$

7. **a.** $\log \sqrt[3]{(1 - x)^2}$ **b.** $\log \dfrac{1}{\sqrt{3 - 4x}}$

8. **a.** $\log x \sqrt[4]{2 - 3x}$ **b.** $\log \dfrac{3x^2}{(x + 1)^3}$

9. **a.** $\log \sqrt{x^2 + 4}$ **b.** $\log \sqrt{x^2 - 4}$

10. **a.** $\log \sqrt{x^2 - 4x}$ **b.** $\log \dfrac{1}{(x - 2)^2}$

11. $\log \dfrac{x^3}{x^2 - x - 12}$

12. $\log \dfrac{(x + 3)^2}{\sqrt[3]{2x - 1}}$

13. $\log \dfrac{x\sqrt{x - 2}}{2x^2 - x - 1}$

14. $\log \dfrac{(3x - 4)(2x + 5)}{(x - 3)(x + 6)}$

15. $\log \dfrac{x^2 - 9}{x^2 - 5x - 6}$

16. $\log \sqrt{\dfrac{x + 1}{(x + 3)^3}}$

17. $\log \dfrac{3x \sqrt[4]{(x + 1)^3}}{x^2 + 1}$

18. $\log \dfrac{5x^3}{\sqrt{(4x - 3)^2}}$

19. $\log \sqrt{\dfrac{x^3}{4x - 5}}$

20. $\log \dfrac{2x - 1}{x^2 \sqrt{1 + 2x}}$

21. $\log \dfrac{x^3 - 2x^2}{\sqrt{x^2 - 1}}$

22. $\log \dfrac{(3x - 2)^{2/3}(x - 1)^4}{2x\sqrt{x + 1}}$

23. $\log \left(\dfrac{x^2 - 5x + 4}{x^2 - 4} \right)^{1/3}$

24. $\log \dfrac{2x(x^2+1)^3}{(x-3)\sqrt{3x+2}}$

25. $\log \dfrac{7x\sqrt[3]{1-x}}{(x-3)^{2/3}(x+2)^3}$

26. $\log \sqrt{\dfrac{x^2+3x+2}{3x^2+4x-8}}$

Find the values of the expressions in Problems 27–30.

27. a. $2^{3\log_2 5}$ **b.** $10^{2\log_{10} 3}$

28. a. $e^{4\ln 3}$ **b.** $e^{b\ln a}$

29. a. $\log_2 \dfrac{\sqrt[3]{16}}{8}$ **b.** $\ln \dfrac{1}{\sqrt{e^3}}$

30. a. $\ln \dfrac{e^{4/3}}{\sqrt{e}}$ **b.** $\log_3 \dfrac{\sqrt{27}}{(81)^2}$

In Problems 31–36 use the properties of logarithms to combine each expression into a single term.

31. a. $\log x - \log(x+1)$
  b. $3\log x + 2\log(x-1)$

32. a. $2\log(x-3) - 3\log(x+2)$
  b. $\log 3 + \log x - \log(x+2)$

33. a. $\frac{1}{2}\log(x-4) + \log 2 - 3\log(x+1)$
  b. $2\log(3-x) - \frac{1}{2}\log x - \frac{1}{2}\log(x+1)$

34. a. $\log 5 + \frac{3}{2}\log(x^2+1) - 3\log(x+4)$
  b. $2\log(2x+3) - [3\log(x+2) + \frac{1}{2}\log(x-1)]$

35. a. $3\log(1-x) - \frac{1}{2}[\log x + 3\log(x+2)]$
  b. $\log 3 + \frac{1}{2}\log(x+2) - 2\log x + \log C$

36. a. $\log(2x^2+1) + \frac{3}{2}\log(x+2) - \log 3 - \frac{1}{3}\log(2x-5) + \log C$
  b. $\log 7 + 3\log(5x+6) - \frac{1}{3}[2\log(3x+2) + \log(x-1)]$

In Problems 37–40 find x in terms of natural logarithms. (See Example 19.)

37. a. $3^x = 4$ **b.** $2^x = 5^{x-2}$

38. a. $3\cdot 2^{-x} = 6^{3x}$ **b.** $2^{x+2} = e^{x/3}$

39. a. $e^{-2x} = 3^{x-2}$ **b.** $4\cdot 3^{2x} = 5\cdot 2^{-x}$

40. a. $3e^{-x} = 4^{2x-1}$ **b.** $2^{x/2}\cdot 3^{1-x} = 4e^x$

In Problems 41–50 solve for x, and check your answers.

41. $\log 3 + \log(x-1) = \log(2x+5)$
42. $2\log x = \log 4 + \log(2x-3)$
43. $2\log x = \log(3x+5) - \log 2$
44. $\log x + \log(x+1) = \log 2$
45. $2\log(x+1) - \log(x-1) = \log(x-3)$

46. $\log_2(2-x) + \log_2(-x) = 3$
47. $\log_2(4-3x) + \log_2 x = 0$
48. $\log_4(x-1) + \log_4(x-2) = \frac{1}{2}$
49. $\ln(2-x) + \ln(4-x) - 2\ln x = 0$
50. $\ln x + \ln(2x-1) - \ln 3 = 0$

In Problems 51–54 solve the inequalities.

51. $\log_2 5 + \log_2(x-1) > \log_2(2x-3) + \log_2 3$
52. $\log_2 x + \log_2(x-2) \geq 3$
53. $\ln x + \ln(x-3) < \ln 4$
54. $\ln x + \ln(x-2) \leq \ln(x+4)$

In Problems 55–60 find y as a function of x.

55. $\log y = \log(x+2) - \log(x-1) + \log C$
56. $\log y - 3\log x = \log 4$
57. $\log(y+1) = 2\log x + \log(2x-3) + \log C$
58. $\ln y - \ln x = \ln(x+2)$
59. $\ln(2y-1) = 2\ln x + \ln C$
60. $\ln(y+2) = x + \ln C$
61. Write each of the following in terms of natural logarithms, and use a calculator to give the values, correct to five significant figures.
  a. $\log_3 5$ **b.** $\log_4 3$ **c.** $\log_8 12$
62. Explain the differences in meaning among the following:
  a. $\log x^2$ **b.** $(\log x)^2$ **c.** $\log(\log x)$
  Show by means of an example that these expressions yield different results.

In Problems 63–67 use a calculator to approximate the value of x to four significant figures.

63. a. $3^x = 4$ **b.** $2^{1-x} = 5$
64. a. $4^x = 3^{2/3}$ **b.** $7^{x/2} = 5^5$
65. a. $3\cdot 5^{-x} = 2^{x+3}$ **b.** $4\cdot 3^x = 7^{2x-3}$
66. a. $(3.708)^x = 13.15$ **b.** $(0.7328)^{-x} = 2.897$
67. a. $\log_7(3.025) = x$ **b.** $x\log_2 13.81 = 1$

B

In Problems 68 and 69 solve for x and check your answers.

68. $2\log_2|x| + \log_2(2-x) = 4$
69. $2(\ln x - \ln 2) + \ln(9-x^2) = \ln(x+3)$

In Problems 70 and 71 show that the given equation is true for all $x > 1$.

70. $-\log(x - \sqrt{x^2-1}) = \log(x + \sqrt{x^2-1})$
71. $\log(\sqrt{x} + \sqrt{x-1}) + \log(\sqrt{x} - \sqrt{x-1}) = 0$

Solve the in equalities in Problems 72 and 73.

72. $(x - 2)\ln(x + 1) < 0$
73. $x \ln x + 1 \geq x + \ln x$
74. Solve for x:

$$\frac{e^x - e^{-x}}{2} = 2$$

Hint. By clearing of fractions and negative exponents, obtain the equation in the form $e^{2x} - 4e^x - 1 = 0$, and observe that this is quadratic in e^x.

75. Solve for x:

$$\frac{e^x + e^{-x}}{2} = 3$$

(See the hint for Problem 74.)

76. Solve for x:

$$x^{\log_2 x} = \frac{x^4}{8}$$

Hint. Take the logarithm to base 2 of both sides.

APPLICATIONS OF EXPONENTIAL AND LOGARITHMIC FUNCTIONS

4

One of the most important applications of exponential functions is in modeling certain phenomena of nature where the rate of change of some quantity at any given instant is approximately proportional to the amount of the quantity present at that instant. For example, under certain conditions a culture of bacteria will grow at a rate proportional to the number present; the more bacteria there are, the faster the rate of growth of the culture. Even in human populations a crude model is that the rate of population increase is proportional to the size of the population at any given time. This model was used by the English demographer, Thomas Malthus, in the late 18th century to predict future world populations and is referred to as the **Malthus model.** It is reasonably accurate for relatively short time intervals, but because it does not take into account such limiting factors as war, famine, epidemics, and limits of food supply and available space, its growth predictions are much too large for longer time intervals. There are more realistic models that account for limitations on growth. We will mention one of these in the exercises.

Bacteria cultures and human (as well as other) populations normally increase with time. In the case of certain other quantities the rate of change is again proportional to the quantity present, but the growth rate is negative; that is, the quantity decreases with time. An example of this is the rate of change of a radioactive substance, called **radioactive decay.** One interesting application of this is in determining the age of certain plant or animal remains. Archeologists in particular have used this technique, called **radiocarbon dating.** (See Problem 27.)

In all such examples in which the rate of change of a quantity is proportional to the amount of the quantity present, the basic equation giving the amount present can be derived by calculus. If we designate this quantity by $Q(t)$, the result is

$$Q(t) = Q_0 e^{kt} \tag{17}$$

where Q_0 is the initial quantity; that is, $Q_0 = Q(0)$, and k is the constant of proportionality. If $k > 0$, then Q is said to **grow exponentially,** and if $k < 0$, then we say that Q **decays exponentially.**

You might observe that the equation $A(t) = Pe^{rt}$ that we derived as (4), giving the amount of money at continuously compounding interest, is one example of exponential growth. Comparing this with equation (17), P corresponds to Q_0 (the principal P is the initial amount), and r corresponds to k.

The following examples show further applications of equation (17). You will see that natural logarithms play an important role in these examples.

EXAMPLE 20 |||||||||||||||||||||||||||||||||||||||||||||||||||||||||||||||||||||||||||||||||||||||||||||||||||||||||||||||

A culture of bacteria is observed to triple in size in 2 days. Assuming it grows exponentially, how large will the culture be in 5 days?

Solution   We are not told the original size of the culture, so we leave Q_0 as unknown. Using $Q(t)$ as the quantity of bacteria present t days after the initial observation, we can say that $Q(2) = 3Q_0$, since the size triples in 2 days. We want to find $Q(5)$. First, we will use equation (17) to find $Q(t)$ in general. We know that

$$Q(t) = Q_0 e^{kt}$$

but we do not yet know the constant k. We can determine k by putting $t = 2$ on both sides.

$$Q(2) = Q_0 e^{k(2)}$$

But $Q(2) = 3Q_0$. So we have

$$3Q_0 = Q_0 e^{2k}$$

and thus

$$e^{2k} = 3$$

In logarithmic form, [see (9)] this is equivalent to

$$2k = \ln 3$$

$$k = \frac{\ln 3}{2}$$

Now we could use a calculator to get a decimal approximation for k, but there is an advantage to leaving it in exact form. We substitute for k in our basic equation to get

$$Q(t) = Q_0 e^{t(\ln 3)/2} = Q_0 (e^{\ln 3})^{t/2}$$
$$= Q_0 (3)^{t/2}$$

Note how we made use of laws of exponents and the fact that $e^{\ln 3} = 3$. (This shows the advantage of keeping k in exact form.)

Now we can find $Q(5)$:

$$Q(5) = Q_0(3)^{5/2} \approx 15.6Q_0$$

So after 5 days the culture will be about 15.6 times its original size.   ▲

A useful measure of the rate of decomposition of a radioactive substance is its **half-life,** defined to be the time required for half the original amount to decompose. This is a property inherent in the particular substance and is independent of the amount present initially. Theoretically, a radioactive substance never completely decomposes, so it is not possible to give the "whole life." The next example illustrates how the half-life is used to find k.

EXAMPLE 21

Strontium-90 has a half-life of 25 years. If 200 kg are on hand, how much will remain after 10 years? How long will it be until 90% of the original amount has decomposed?

Solution   We again use the basic formula (17):

$$Q(t) = Q_0 e^{kt} = 200e^{kt}$$

Since the half-life is 25 years, we know that after 25 years, half of the original amount will remain. So $Q(25) = 100$. Putting $t = 25$ therefore gives

$$100 = 200e^{25k}$$
$$e^{25k} = \tfrac{1}{2}$$
$$25k = \ln \tfrac{1}{2} = -\ln 2$$
$$k = -\tfrac{1}{25} \ln 2$$

Thus,

$$Q(t) = 200e^{t(-\ln 2)/25} = 200(e^{\ln 2})^{-t/25}$$
$$= 200(2)^{-t/25}$$

So after 10 years,

$$Q(10) = 200(2)^{-10/25} = 200(2)^{-2/5} \approx 151.6 \text{ kg}$$

To determine how long it will be until 90% has decomposed, we first observe that at this time 10%, or 20 kg, will remain. Thus, we want to find t for which $Q(t) = 20$; that is,

$$200(2)^{-t/25} = 20$$
$$2^{-t/25} = \tfrac{1}{10}$$

To find t, we take the natural logarithm of both sides:

$$-\frac{t}{25}\ln 2 = -\ln 10$$

$$t = \frac{25\ln 10}{\ln 2} \approx 83.05 \text{ yr} \qquad \blacktriangle$$

EXAMPLE 22

Find the half-life of a radioactive substance that is reduced by 30% in 20 hours.

Solution   Let Q_0 be the original amount and $Q(t)$ be the amount t hours later. Since 70% of the original will be left after 20 hours, we have

$$Q(20) = 0.7\, Q_0$$

Putting $t = 20$ in the basic formula $Q(t) = Q_0 e^{kt}$ therefore gives,

$$0.7Q_0 = Q_0 e^{20k}$$
$$e^{20k} = 0.7$$
$$20k = \ln 0.7$$
$$k = \tfrac{1}{20}\ln 0.7$$

So

$$Q(t) = Q_0 e^{t(\ln 0.7)/20} = Q_0(e^{\ln 0.7})^{t/20} = Q_0(0.7)^{t/20}$$

We want to find t such that $Q(t) = 0.5\, Q_0$. So t must satisfy

$$0.5\, Q_0 = Q_0(0.7)^{t/20}$$

or

$$(0.7)^{t/20} = 0.5$$

Taking the natural logarithm of both sides, we get

$$\frac{t}{20}\ln(0.7) = \ln(0.5)$$

$$t = \frac{20\ln(0.5)}{\ln(0.7)} \approx 38.87 \text{ hours} \qquad \blacktriangle$$

EXAMPLE 23

In 1930 the world population was approximately 2 billion, and in 1960 approximately 3 billion. Using the Malthus model, estimate the population in the year 2000.

Solution   Let $Q(t)$ be the population in billions with t in years, measured from 1930. Then $Q_0 = Q(0) = 2$ and $Q(30) = 3$. By the Malthus model, equation (17),

$$Q(t) = Q_0 e^{kt} = 2e^{kt}$$

Setting $t = 30$ and proceeding in the usual way, we get

$$3 = 2e^{30k}$$
$$e^{30k} = \tfrac{3}{2}$$
$$30k = \ln \tfrac{3}{2}$$
$$k = \tfrac{1}{30} \ln \tfrac{3}{2}$$

So

$$Q(t) = 2(e^{\ln 3/2})^{t/30} = 2(\tfrac{3}{2})^{t/30}$$

In the year 2000, $t = 70$. So according to this model, the population then should be

$$Q(70) = 2(\tfrac{3}{2})^{70/30} = 2(\tfrac{3}{2})^{7/3} \approx 5.15 \text{ billion}$$ ▲

Our final example is closely related to exponential decay and has to do with the rate of change of temperature of a body that is warmer than the surrounding medium. Suppose the temperature of the surrounding medium is T_m and the initial temperature of the warmer body is T_0. Let $T(t)$ be the temperature of the body t units of time later. Then, according to **Newton's law of cooling,** the rate of change of the temperature T is proportional to $T_0 - T_m$. By calculus, this leads to the result

$$T(t) = T_m + Ce^{kt} \tag{18}$$

where k is the proportionality constant and C is a constant that depends on T_0 and T_m.

EXAMPLE 24   ‖‖

A body initially at $120°C$ is placed in air at $20°C$, and it has cooled to a temperature of $80°C$ after $\tfrac{1}{2}$ hour. Find its temperature after 1 hour.

Solution   In equation (18) we know that $T_m = 20$. So $T(t) = 20 + Ce^{kt}$. Since $T(0) = T_0 = 120$, we can put $t = 0$ to get

$$120 = 20 + Ce^0 = 20 + C$$
$$C = 100$$

Thus,

$$T(t) = 20 + 100e^{kt}$$

We also know that $T(\frac{1}{2}) = 80$, so putting $t = \frac{1}{2}$ gives

$$80 = 20 + 100e^{k/2}$$
$$100e^{k/2} = 60$$
$$e^{k/2} = 0.6$$

So

$$\frac{k}{2} = \ln 0.6$$
$$k = 2\ln(0.6)$$

Thus,

$$T(t) = 20 + 100e^{2t\ln(0.6)} = 20 + 100(e^{\ln 0.6})^{2t}$$
$$= 20 + 100(0.6)^{2t}$$

We can now put $t = 1$ and get

$$T(1) = 20 + 100(0.6)^2 = 20 + 100(0.36) = 56$$

So after 1 hour the body will have cooled to approximately 56°C.

EXERCISE SET 4

A

1. A culture of bacteria originally numbers 500. After 2 hours there are 1,500 bacteria. Assuming exponential growth, find how many are present after 6 hours.

2. A radioactive substance has a half-life of 64 hours. If 200 grams of the substance are initially present, how much will remain after 4 days?

3. A culture of 100 bacteria doubles after 2 hours. How long will it take the number of bacteria to reach 3,200?

4. How long will it take 100 grams of a radioactive substance having a half-life of 40 years to be reduced to 12.5 grams?

5. One hundred kilograms of a certain radioactive substance decay to 40 kilograms after 10 years. Find how much will remain after 20 years.

6. In the initial stages a bacteria culture grows approximately exponentially. If the number doubles after 3 hours, what proportion of the original number will be present after 6 hours? After 12 hours?

7. If 100 grams of a radioactive material diminish to 80 grams in 2 years, find the half-life of the substance.

8. In the early stages a bacterial culture grows approximately exponentially. Assuming this holds true for the time periods in question, if the quantity doubles in 2 hours, how long will it take to triple?

9. A culture of 100 bacteria grows to a size of 150 after 2 hours. How long will it take to double in size? How many bacteria will be present after 5 hours? Assume exponential growth.

10. If 10% of a radioactive substance decays after 33 years, how much will remain after 200 years? What is the half-life of the substance?

11. Radium-226 has a half-life of 1,620 years. How long will it take for 80% of a given amount to decay?

12. If 75% of a quantity of the radioactive isotope uranium-232 remains after 30 years, how much will be present after 100 years? What is the half-life?

13. The half-life of thorium-228 is approximately 1.913 years. If 100 grams are on hand, how much will remain after 3 years? How long will it take for the amount left to be 10 grams?

14. A thermometer registering 20°C is placed in a

freezer in which the temperature is $-10°C$. After 10 minutes the thermometer registers 5°C. What will it register after 30 minutes?

15. A piece of metal is heated to 80°C and is then placed in the outside air at 20°C. After 15 minutes the temperature of the metal is 65°C. What will be its temperature after 15 more minutes?

16. A body heated to 120°F is brought into a room in which the temperature is 70°F. After 15 minutes the temperature of the body is 100°F. How long will it take the temperature of the body to drop to 80°F?

17. If a body at temperature T_0 is introduced into a medium of temperature T_m, with $T_m > T_0$ (the surrounding medium is warmer than the temperature of the body), the formula $T(t) = T_m + Ce^{kt}$ for Newton's law of cooling continues to apply. If an outside thermometer registering 5°C is brought into a room where the temperature is 20°C and it registers 12°C after 10 minutes, how long will it take for it to register 18°C?

18. The population of a certain city in California was 125,000 in 1960, and by 1980 it had increased to 350,000. Assume the population increases according to the Malthus model, and find what the approximate population (to the nearest thousand) will be in the year 2000.

19. The number N of items of a certain type of merchandise sold on a given day diminishes with time according to the formula

$$N = N_0 e^{-k(t-1)}$$

where N_0 is the number sold on the first day. If 20 of these items are sold on the first day and 10 on the fifth day, how many can be expected to be sold on the ninth day?

20. One model for the air pressure P at an altitude h above the ground is

$$P = P_0 e^{-kh}$$

where P_0 is the pressure at ground level. If the pressure at ground level is 15 pounds per square inch and at height 10,000 feet it is 10 pounds per square inch, find the approximate pressure at 20,000 feet.

21. The atmospheric pressure at sea level is approximately 15 pounds per square inch, and at Denver, which is 1 mile high, the pressure is approximately 12 pounds per square inch. What will be the

approximate atmospheric pressure at the top of Vail Pass, which is 2 miles high? (See Problem 20.)

22. The intensity of light passing through a layer of translucent material decreases at a rate approximately proportional to the thickness of the material. It is shown in calculus that this results in the formula

$$I = I_0 e^{-kx}$$

where I_0 is the intensity with which the light first strikes the material and I is the intensity x units inside the material. If sunlight striking water is reduced to half of its original intensity at a depth of 10 feet, what will be the intensity (as a fractional part of the original intensity) at a depth of 30 feet? At what depth will the intensity of illumination be only $\frac{1}{10}$ of the original?

B

23. A more realistic model than the Malthus model for population growth is the **Verhulst model,** which asserts that the rate of growth of the population is jointly proportional to the size of the population and the additional growth capacity. If m denotes the maximum sustainable size of the population, it is shown in calculus that the resulting formula for the size of the population at time t is given by

$$Q(t) = \frac{mQ_0}{Q_0 + (m - Q_0)e^{-mkt}}$$

The graph of Q is called a **logistic** curve. Redo Example 23, using the Verhulst model with the assumption that the maximum supportable world population is 10 billion.

24. A reasonably accurate model for the concentration $y(t)$ of a drug in the bloodstream t hours after it was injected is given by

$$y(t) = \frac{a}{b}(1 - e^{-bt}) + y_0 e^{-bt}$$

where $y_0 = y(0)$ is the initial concentration and a and b are positive constants. Suppose $b = 0.50$, $y_0 = 0.15$, and $y(1) = 0.10$. Find the approximate concentration of the drug after 2 hours.

25. The following formula for the velocity $v(t)$ of a falling body, subject to air resistance assumed to be proportional to the velocity, can be derived by

calculus:

$$v(t) = \left(v_0 - \frac{gm}{k}\right)e^{-kt/m} + \frac{gm}{k}$$

where $v_0 = v(0)$ is the initial velocity, m is the mass of the body, g is the acceleration due to gravity, and k is the constant of proportionality. Suppose a mass of 100 grams is hurled downward from a balloon high above the earth with an initial velocity $v_0 = 25$ m/sec. If $k = 20$, find the velocity 5 seconds later. What is the theoretical terminal velocity, that is, the maximum velocity assuming it continues falling indefinitely? (Take $g = 9.80$ m/sec^2.)

26. Using **Kirchhoff's second law,** it can be shown by calculus that the current $I(t)$ (in amperes) in an RL series circuit is given by

$$I(t) = \frac{V}{R}\left(1 - e^{(-R/L)t}\right)$$

where V is the voltage, R is the resistance (in ohms), L is the capacitance (in henrys), and where it is assumed that $I(0) = 0$. Suppose a 12-volt battery is connected to an RL series circuit with a resistance of 6 ohms and an inductance of 1 henry. Find the current after 2 seconds. What is the theoretical

steady state current, that is, the current approached as t increases indefinitely?

27. A radioactive isotope of carbon, called carbon-14, is found in all living organisms. During the lifetime of an organism the amount that decays is replenished through the atmosphere, so that the amount in the organism remains constant, but after the death of the organism the decayed carbon-14 is no longer replenished. Because the half-life of carbon-14 is long (approximately 5,730 yr), the remains of organisms that died thousands of years ago may still contain measurable amounts of carbon-14. By comparing the amount remaining with the amount known to be present in the particular organism when it was alive, the age of the remains can be determined. Suppose that archeologists find a human skull and determine that 10% of the original amount of carbon-14 remains (the original amount being determined by measuring the amount in a similar present-day skull). Find the approximate age of the skull.

C (Graphing calculator)

28. Obtain the graph of $Q(t)$ in Problem 23 with $m = 10$ and $k = 0.02$ for different positive values of Q_0, some smaller than m and some larger. Discuss your results.

REVIEW EXERCISE SET

A

In Problems 1 and 2 make a table of values and draw the graph of each function.

1. **a.** $y = (3/2)^x$ **b.** $y = 1 - 2^{-x}$
   c. $y = \log_2 x$ **d.** $y = \ln(1 - x)$
2. **a.** $f(x) = 2^{x/2} - 1$ **b.** $g(x) = \log_2(1 - 2x)$
   c. $h(x) = 3^{(x-1)/2}$ **d.** $F(x) = 2 - \ln(x - 2)$
3. Write in exponential form:
   a. $\log_3 9 = 2$ **b.** $\log_{10} 0.001 = -3$
   c. $\log_2 256 = 8$ **d.** $v = \log_k y$
   e. $z = \log_r s$
4. Write in logarithmic form:
   a. $4^{3/2} = 8$ **b.** $27^{-2/3} = \frac{1}{9}$
   c. $10^3 = 1,000$ **d.** $p = q^s$
   e. $n = a^t$

In Problems 5–7 find the unknown.

5. **a.** $\log_2 x = 4$ **b.** $\log_y 16 = -2$
   c. $\log_3 81 = z$
6. **a.** $y = \log_3 27$ **b.** $\log_x 0.0001 = -4$
   c. $\log_4 z = -3$
7. **a.** $x = e^{\ln 3}$ **b.** $y = \log_5 5^2$
   c. $\log_2(1 - x) = 3$

In Problems 8–10 write in a form that is free of logarithms of powers, roots, products, and quotients.

8. **a.** $\log \dfrac{2x^3}{\sqrt{x - 1}}$ **b.** $\log \dfrac{x(x - 2)}{2(x^2 + 1)^3}$

9. **a.** $\log \dfrac{x - 1}{(x + 2)^3}$ **b.** $\log \sqrt{\dfrac{2x - 1}{3x}}$

10. a. $\log \dfrac{x^2 - 9}{x^2 - 4x + 4}$ **b.** $\log \dfrac{4x^2 \sqrt{2x - 1}}{(x - 2)(x + 4)^{3/2}}$

Combine the expressions in Problems 11–13 into a single term.

11. a. $\frac{1}{2} \log(2x - 3) + \log x - 3 \log 2 + \log C$
 b. $4 \log x - \frac{1}{2} \log(x + 2) - \log(x - 1)$
12. a. $\log(3x + 4) - 2 \log x - \log(x + 2)$
 b. $3 \log(x + 2) + \frac{1}{2} \log(2x - 1) - \frac{3}{2} \log(2 - x)$
13. a. $\log 2 + 3 \log x - \frac{1}{3}[\log(x - 1) + \log(x + 2)]$
 b. $\frac{3}{2} \log(2x + 3) - \log 2 - 3 \log(x + 4)$

Evaluate the expressions in Problems 14 and 15.

14. a. $\log_{10} 10^{-12}$ **b.** $e^{\ln 5}$
  c. $2^{3 \log_2 5}$ **d.** $10^{2 \log_{10} 4}$
15. a. $\log_4 8$ **b.** $\log_2 0.125$
  c. $\log_3 \frac{1}{81}$ **d.** $\log_{10} 0.00001$
16. Solve for x in terms of natural logarithms.
  a. $3^x = 7$ **b.** $2^{x+1} = 3^{2x}$
  c. $4 \cdot 5^x = 3^{x-2}$ **d.** $3^x \cdot 2^{x+3} = 5^{2x}$
17. Solve the formula $y = C(1 - e^{-kt})$ for k.

In Problems 18 and 19 solve the inequality.

18. $e^{1/x - 1} > 1$ **19.** $2(1 + 2^x) < 3$

In Problems 20–23 solve for x and check your answer.

20. $\log(3x + 2) - \log(2x - 1) = \log 2$
21. $\log x + \log(x - 2) = \log 3$
22. $\log_2(1 - x) + \log_2(4x + 1) = 0$
23. $\log_5(3x + 1) + \log_5(x + 1) = 1$

In Problems 24 and 25 solve the inequality.

24. $\ln x + \ln(2x - 1) \le \ln 3$
25. $\log_2 x^2 - \log_2(x + 4) > 1$
26. Solve for y as a function of x.
  a. $\ln(2y - 1) = 3x + \ln C$
  b. $\log y = 2 \log(x - 1) - \log(x + 1) + \log C$
27. A radioactive substance has a half-life of 6 years. If 20 pounds are present initially, how much will remain after 2 years?
28. A culture of bacteria increases from 1,000 to 5,000 in 8 hours. Assuming the law of exponential growth applies, how many will be present after 20 hours?
29. The population of a certain city in Florida increased by 40% between 1970 and 1978. If the population in 1978 was 210,000, what will be the

population predicted by the Malthus model in the year 2000?
30. An object cools from 140°F to 120°F in 5 minutes when placed in outside air at 40°F. After how many more minutes will the temperature of the object be 70°F?
31. The sum of $5,000 is placed in a savings account paying 6% annual interest.
  a. How much money will there be after 5 years if interest is compounded quarterly? Continuously?
  b. If interest is compounded continuously, how long will it take for the amount to double?
32. At a distance of 1 meter the intensity of sound of normal conversation is about 10^{-6} watt per square meter. What is the intensity in decibels? (See Example 12.)
33. At what interest rate would an amount of money have to be invested in order to double after 10 years if interest is compounded continuously?

In Problems 34 and 35 use a calculator to approximate the unknown to four significant figures.

34. a. $5^{x-1} = 3$ **b.** $2^t = 3^{1-t}$
35. a. $4 \cdot 6^y = 7 \cdot 5^{y/2}$
 b. $(3.025)^{2x-1} = (0.7134)^{-x}$

B

In Problems 36 and 37 solve the inequalities.

36. $\dfrac{e^x - 1}{x^2 - 1} < 0$

37. $\log_2 x + \log_2(x + 2) \le 1 - \log_2(x - 1)$
38. Solve for x:

$$\frac{e^x - e^{-x}}{e^x + e^{-x}} = \frac{1}{2}$$

(See Problem 74, Exercise Set 3.)
39. If sugar is placed in water, the amount undissolved after t seconds is given by the formula $A = Ce^{-kt}$. If 30 grams of sugar are placed in water and it takes 10 seconds for half of it to dissolve, how long will it take for all but 3 grams to dissolve?
40. The so-called **learning curve** in educational psychology has an equation of the form $f(t) = C(1 - e^{-kt})$, where C and k are positive constants and t represents time. The value of $f(t)$ can be

interpreted as the amount learned at time t. This also has applications in industry.

a. Draw the graph of f for $C = 10$ and $k = 0.1$.

b. Suppose that learning vocabulary in a foreign language follows the learning curve, and that for an average first-year college student $C = 20$, with time t measured in days. Suppose also that on the tenth day of study it has been found that first-year students can be expected to learn 15 new words. Find how many words a student would be expected to learn on the fifth day.

41. In a second-order chemical reaction two substances A and B react with each other to form a third substance C. If the initial concentrations (in moles per liter) of A and B are a and b, respectively, and $x(t)$ is the concentration of C after t seconds, then it is known that the rate of change of x is jointly proportional to $a - x$ and $b - x$. This leads by calculus to the formula

$$x(t) = \frac{ab(e^{(a-b)kt} - 1)}{ae^{(a-b)kt} - b}$$

where k is the constant of proportionality. Suppose

$a = 0.32$, $b = 0.25$, and $x(2) = 0.13$. Find $x(5)$.

42. If air resistance is assumed to be proportional to the square of the velocity of a falling body, then its velocity at time t is given by

$$v(t) = \sqrt{\frac{mg}{k} \cdot \frac{e^{2t\sqrt{kg/m}} - 1}{e^{2t\sqrt{kg/m}} + 1}}$$

where m is the mass, g is the acceleration due to gravity, and k is the constant of proportionality. Suppose $m = 2$ slugs, $g = 32$ ft/sec^2, and $v(2) = 36$ ft/sec. Find $v(4)$.

43. In the **Gompertz model** for population growth, the rate of change of the population size Q is assumed to be jointly proportional to Q and $m - \ln Q$, where m is the limiting size of the population. This results in the formula

$$Q(t) = m(1 - e^{-kt}) + Q_0 e^{-kt}$$

Use this model to find the size of a population after 10 years if $Q_0 = 10$ (in millions), $m = 25$, and $Q(4) = 13$.

CUMULATIVE REVIEW EXERCISE SET II (CHAPTERS 4–6)

1. Find the domain of f.

a. $f(x) = \ln\left(\dfrac{x^2 - 4}{2x + 1}\right)$

b. $f(x) = \sqrt{3x^3 - 5x^2 - 38x + 40}$

2. Find all zeros of the polynomial $P(x) = 6x^3 - 19x^2 - 26x + 24$. Factor the polynomial with integer coefficients.

3. A piece of machinery valued at \$32,000 is estimated to have a residual value of \$2,000 after 20 years. If it is depreciated linearly, express its value after t years as a function of t. What will be its value after 16 years?

4. a. Solve for x in terms of natural logarithms:

$$2^{(3-x)/2} = 4 \cdot 3^x$$

b. Solve for x:

$$e^{3\ln(x-1)} = \ln e^{(3-x)}$$

5. Let $f(x) = 1 + e^{2x}$ and $g(x) = \ln(x - 1)$. Find $f \circ g$ and $g \circ f$, and draw their graphs. Give the domain of each.

6. Find the equation of the tangent line to the circle $x^2 + y^2 - 4x + 2y - 15 = 0$ at the point $(4, 3)$.

7. Use graphical means to solve the inequalities.

a. $x(4 - x) \geq -12$ **b.** $x^3 - 12x + 16 > 0$

8. Discuss symmetry, intercepts, and asymptotes; analyze the signs; and draw the graph.

a. $y = \dfrac{9 - x^2}{x^2 + 2x - 8}$ **b.** $y = \dfrac{x^2 - 4}{x + 4}$

9. Solve for x and check:

$$2\log_2|x - 1| = 2 - \log_2(x + 2)$$

10. Sketch the graph of the polynomial function:

$$y = (x - 2)(x + 1)^2(x - 5)^3(x + 4)$$

11. Resolve into partial fractions:

$$\frac{5x^2 - 4x - 6}{x^4 - x^2 - 2x + 2}$$

12. A certain radioactive substance has a half-life of 8 years. If 200 grams are present initially, how much will remain at the end of 12 years? How long will it be until only 10% of the original amount remains?

13. Solve the inequality:

$$2\ln x + \ln(x + 1) \geq \ln(x^2 + 3x + 2)$$

14. By calculus it can be shown that the slope of the tangent line to the curve $x^2 - y^2 = 9$ at any point (x, y) on the curve, with $y \neq 0$, is given by $m = x/y$. Find the equations of the tangent and normal lines at each point on the curve with abscissa 5. Draw the curve, showing these tangent and normal lines.

15. A body initially at 80°C is placed in outside air at 25°C, and after 15 minutes it has cooled to 60°C. What will its temperature be after 1 hour? How long will it take to reach a temperature of 30°C?

16. Sketch the graph of each of the following. Identify the graph in each case, and state whether the graph is that of a function of x.
   a. $y^2 + 2x - 4y = 0$ b. $y = -\sqrt{9 - x^2}$
   c. $y = 2x^2 - 3x - 5$ d. $4x^2 - y^2 - 2y = 17$

17. The cost in hundreds of dollars for a company to produce x tons of a certain type of fertilizer per day is given by $C = 4x^2 - 20x + 29$. Find the number of tons that should be produced each day to minimize cost. What is the minimum cost?

18. Find all roots of the equation

$$4x^4 - 16x^3 + 9x^2 + 5x + 25 = 0$$

19. Draw the graph of each of the following:
   a. $f(x) = \ln(1 + x^2)$ b. $g(x) = 1 - e^{-|x|}$
   c. $y = 2^{(x-1)/2}$ d. $y = 1 + \log_2|x - 1|$

20. a. Show that by taking logarithms of both sides of the equation $y = a^x$ and making the substitution $Y = \log y$, Y is a linear function of x.

   b. Similarly, in $y = x^a$, take logarithms, and let $X = \log x$ as well as $Y = \log y$, and show that Y is a linear function of X.
   c. For $y = 2^x$ graph log y versus x, and for $y = x^2$ graph log y versus log x.

21. Discuss completely and draw the graph.
   a. $y^2 = \dfrac{4x^2 - 9}{(x - 3)^2}$ b. $y^2 = \dfrac{x^2 + x - 2}{x^3 - 4x^2}$

22. a. Find the equations of the two circles whose diameters coincide, respectively, with the major and minor axes of the ellipse

   $$9x^2 + 4y^2 - 16y - 20 = 0$$

   Sketch all three curves on the same coordinate system.
   b. Identify, and sketch the graph of

   $$y = 2\sqrt{x^2 - 2x + 3}$$

23. A manufacturer of solar energy cells finds that the cost, in dollars, of producing x units of a certain type per month is given by $30x + 1{,}500$, and he sets the selling price per unit, depending on the number of units sold, as $120 - 0.1x$. How many units should be manufactured and sold per month to maximize profit? What is the maximum profit?

24. Prove that the points $(3, 2), (-1, -1)$, and $(6, -2)$ are vertices of an isosceles right triangle. Find the equation of the altitude drawn from the right angle vertex to the hypotenuse.

25. Solve the inequalities:
   a. $2^x - 2^{-x} < 0$ b. $\dfrac{\ln(x - 1)}{x - 3} > 0$

26. Let $f(x) = \ln(x^2 - x) - \ln(x^2 + 3) + \ln 2$.
   a. Find the domain of f.
   b. Find the zeros of f.
   c. Find all vertical asymptotes to the graph of f.
   d. Discuss the behavior of f as x gets large in absolute value. What can you conclude concerning horizontal asymptotes?
   e. Sketch the graph of f.

7

THE TRIGONOMETRIC FUNCTIONS

The *sine* function is used to model many physical phenomena that have an oscillatory nature. One such example of this is seen in the following problem, taken from a calculus textbook.*

The graph shows the average temperature in Fairbanks, Alaska, during a typical 365-day year. The equation that gives the temperature on day x is

$$y = 37 \sin\left[\frac{2\pi}{365}(x - 101)\right] + 25.$$

a. On what day is the temperature increasing the fastest?

b. **CALCULATOR** About how many degrees per day is the temperature increasing when it is increasing at its fastest?

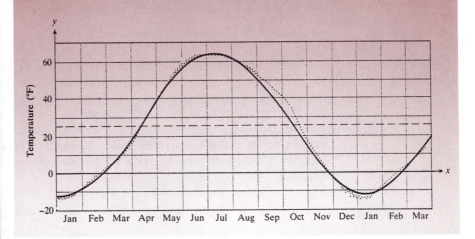

* Finney & Thomas, *Calculus*, © 1990, Addison-Wesley Publishing Company, Inc., Reading, Massachusetts. Problem 66 on page 181. Reprinted with permission.

To answer the questions posed in the problem, a knowledge of calculus is needed, but one of the things you will learn in this chapter is how to graph the sine function given in the problem.

INTRODUCTION

The introduction of trigonometry as a separate area of study is generally believed to have been due to the Greek Hipparchus in the second century B.C., but his works are lost. The first exposition of the subject still in existence comes from the work of Ptolemy, the Greek astronomer who lived in the second century A.D. Rudiments of the subject are much older, however, and can be traced to Babylon and ancient Egypt. The principal use of trigonometry up until the fifteenth century was in astronomy, because it afforded a means of making indirect measurements. The name *trigonometry* was first used by Pitiscus (a German) in 1595 as the title of his book. The name is a combination of three Greek words meaning "three-angle measurement," or triangle measurement.

As the name implies, trigonometry originally meant a study of triangles, or more generally of angles of triangles, and its primary utility lay in the fact that by means of the trigonometric functions inaccessible quantities could be measured. In more modern times, while trigonometry still is of value for this purpose, it is the analytical aspects of the subject that make it indispensable in the study of many physical phenomena. We will indicate how it is useful in indirect measurement in Chapter 9, but our emphasis here will be on the analytical properties, since these are more important in calculus.

THE MEASUREMENT OF ANGLES

When two rays emanate from the same point, as in Figure 1, we say they form an angle. The common point is called the **vertex.** In trigonometry it is useful to think of the angle as having been formed by having the rays initially coincident and then rotating one of them. We indicate the direction of rotation by a small circular arc with an arrowhead. The fixed ray is called the **initial side** and the rotated ray is called the **terminal side.** A counterclockwise rotation is called **positive;** a clockwise rotation is called **negative.** These ideas are illustrated in Figure 2.

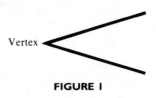

Vertex

FIGURE I

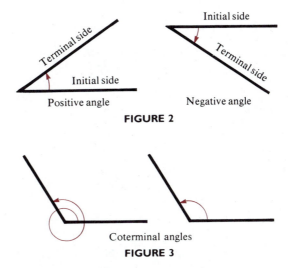

Positive angle

Negative angle

FIGURE 2

Coterminal angles

FIGURE 3

When two angles are generated by different rotations but have the same initial and terminal sides, they are said to be **coterminal.** Two coterminal angles are illustrated in Figure 3.

The most familiar way of measuring the size of angles is by means of the degree.* If a circle is divided into 360 equal parts, then an angle formed by two rays from the center passing through two consecutive points on the circle is said to have a measure of one **degree,** indicated 1°. Degrees are divided into 60 equal parts called **minutes,** and minutes are further divided into **seconds,** with 60 seconds equalling one minute. One minute is designated 1′ and one second 1″. If we write, for example, $A = 25°32′15″$, we mean that A is an angle whose measure is 25 degrees, 32 minutes, and 15 seconds. In our work we will seldom use seconds. An alternate way of indicating fractions of a degree is to use decimal notation. For example, we might have $B = 47.3°$. This form is especially useful with scientific hand calculators, since most of these require angles to be written in decimal form. It is important to be able to go from one form to the other. For example, since $1° = 60′$, we see that $0.3° = (0.3)(60′) = 18′$. So $47.3° = 47°18′$. Also, to change minutes to decimal parts of degrees, divide the number of minutes by 60. This is illustrated by

$$35°15′ = \left(35\frac{15}{60}\right)° = 35.25°$$

While the measurement of angles by degrees has the weight of history and the advantage of familiarity on its side, there is another method of measurement

---

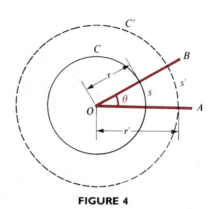

FIGURE 4

that for many purposes is more useful, especially in the study of calculus and its applications. This is the **radian.** To define it, we consider a given angle θ with initial side OA and terminal side OB as in Figure 4. Let C be any circle centered at O, and designate its radius by r. Again think of θ as having been generated by rotating OB from its initial position coinciding with OA. As it rotates, its point of intersection with the circle C sweeps out an arc on C. If θ is positive, let s be the length of this arc, and if θ is negative, let s be the negative of the length of the arc. So $s > 0$ when θ is positive, and $s < 0$ when θ is negative. We call s the **directed length of the arc on C subtended by θ. The radian measure of θ is defined to be the ratio of s to r.** So the radian measure is the ratio of two lengths: the (directed) length of the arc on C subtended by θ divided by the length of the radius of C. It is common practice to use the same symbol to designate an angle and to designate its measure. Thus, we can write

$$\theta = \frac{s}{r} \tag{1}$$

It might appear that this definition is dependent on the particular circle we use, but this is not the case. For if we take another circle, say C', with radius r', and subtended arc s', as in Figure 4, then, just as with similar triangles, we have $s/r = s'/r'$. Thus, we get the same number for the radian measure of θ regardless of which circle is used.

In practical applications it is often the case that an angle at the center of a circle is known, and the length of the subtended arc is unknown. Then we use equation (1) in the form

$$s = r\theta \tag{2}$$

When using this formula to compute the arc length, it is important to remember that θ must be in radians. So if θ is given in degrees, you must convert it to radians.

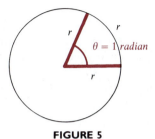

FIGURE 5

Another observation is that according to equation (1) the radian measure of θ is the ratio of two lengths, and hence the radian is dimensionless. In analyzing the dimensional properties of an expression, then, if an angle in radians appears in the expression, the dimensionality is not affected.

According to equation (1), an angle of one radian subtends, on any circle with its center at the vertex, an arc equal in length to the radius (Figure 5). This helps us to visualize the relative size of a radian. We know that the circumference of a circle is $2\pi r$. That is, we could mark off 2π arcs on the circle each equal in length to the radius (since $\pi \approx 3.1416$, we could mark off 6 full arcs equal to the radius and have a little more than a quarter of a radius left over). But for each arc on the circle of length r, there is an angle of one radian at the center. It follows then that there are 2π radians in one revolution. Since there are also $360°$ in one revolution, we have the relationships:

$$2\pi \text{ radians} = 360°$$

$$\pi \text{ radians} = 180°$$

$$1 \text{ radian} = \frac{180°}{\pi} \approx 57.3°$$

$$1 \text{ degree} = \frac{\pi}{180} \text{ radians}$$

Probably the easiest of these to remember is the second:

$$\pi \textbf{ radians} = \textbf{180°} \tag{3}$$

The others can be obtained from it. *In the future, unless the degree symbol is used in speaking of the measure of an angle, we will automatically mean radian measure.* Thus, if we write $\theta = \pi$, $\theta = \pi/6$, or $\theta = 2$, we will mean the measure of θ is π radians, $\pi/6$ radians, or 2 radians, respectively.

Relationship (3) enables us to find the radian measure of an angle whose degree measure is known, and vice versa. The following example illustrates this.

EXAMPLE 1 |||||||||||||||||||||||||||||||||||||||||||||||||||||||||||||||||||||||||||||||||||||||||

Convert the degree measures in parts **a** and **b** to radians, and convert the radian measures in parts **c** and **d** to degrees.

a. 120°   **b.** 540°   **c.** $\dfrac{7\pi}{6}$ **d.** $\dfrac{2}{3}$

Solution   **a.**   Since $1° = \pi/180$ radians, we multiply 120 by $\pi/180$.

$$120 \cdot \frac{\pi}{180} = \frac{2\pi}{3} \text{ radians}$$

b.
$$540 \cdot \frac{\pi}{180} = 3\pi \text{ radians}$$

c.   Since 1 radian $= 180/\pi$ degrees, we multiply $7\pi/6$ by $180°/\pi$.

$$\frac{7\pi}{6} \cdot \frac{180°}{\pi} = 210°$$

d.
$$\frac{2}{3} \cdot \frac{180°}{\pi} = \left(\frac{120}{\pi}\right)^{\circ} \approx 38.2°$$

Note. We will often leave the radian measure of an angle as a multiple of π, as in parts **a** and **b,** rather than give the decimal equivalent.   ▲

The next two examples illustrate uses of equations (1) and (2).

EXAMPLE 2 |||||||||||||||||||||||||||||||||||||||||||||||||||||||||||||||||||||||||||||||||||||||||

A central angle in a circle of diameter 8 centimeters subtends an arc of 3 centimeters. Find the degree measure of the angle.

Solution   Since the diameter is 8, the radius is 4. Since $s = 3$, we have by (1)

$$\theta = \frac{s}{r} = \frac{3}{4} \text{ radian}$$

To convert to degrees, we multiply by $180°/\pi$.

$$\theta = \frac{3}{4} \cdot \frac{180°}{\pi} = \left(\frac{135}{\pi}\right)^{\circ} \approx 42.97°$$ ▲

EXAMPLE 3 |||||||||||||||||||||||||||||||||||||||||||||||||||||||||||||||||||||||||||||||||||||||||

Find the arc length on a circle of radius 7 feet subtended by a central angle of 60°.

Solution   We use $s = r\theta$, but first we must find θ in radians.

$$\theta = 60 \cdot \frac{\pi}{180} = \frac{\pi}{3} \text{ radians}$$

So

$$s = 7 \cdot \frac{\pi}{3} = \frac{7\pi}{3} \text{ feet} \approx 7.33 \text{ feet} \qquad \blacktriangle$$

Suppose now that an object is traveling in a circular path at a constant rate. Its **linear speed** v is the distance along the circular arc covered per unit of time. Thus, if in time t it covers a distance of s units of arc, then

$$v = \frac{s}{t}$$

There is also an **angular speed,** which we designate by ω, associated with the moving object. This is the number of radians turned through per unit of time. Thus, if in time t a central angle of θ radians has been turned through, then

$$\omega = \frac{\theta}{t}$$

Since $s = r\theta$, we have

$$\frac{s}{t} = r \cdot \frac{\theta}{t}$$

or

$$\boldsymbol{v = r\omega} \qquad\qquad (4)$$

EXAMPLE 4 |||||||||||||||||||||||||||||||||||||||||||||||||||||||||||||||||||||||||||||||||||||||||||||||||||

A point on a flywheel of radius 2 feet is turning at an angular speed of 300 radians per second. What is the linear speed of the point?

Solution $$v = r\omega = 2 \cdot 300 = 600 \text{ feet per second} \qquad \blacktriangle$$

EXAMPLE 5 |||||||||||||||||||||||||||||||||||||||||||||||||||||||||||||||||||||||||||||||||||||||||||||||||||

A wheel of diameter 4 feet is rotating at the rate of 1,200 revolutions per minute. Find the linear speed of a point on the rim in feet per second.

Solution   In order to find the angular speed ω in radians per second, we note that each revolution contains 2π radians, and since there are 1,200 revolutions each minute, which is equivalent to 20 revolutions each second, we have

$$\omega = 2\pi(20) = 40\pi \text{ radians per second}$$

Thus,

$$v = r\omega = 2(40\pi) = 80\pi \text{ feet per second}$$

$$\approx 251.33 \text{ feet per second}$$

▲

EXERCISE SET 2   1-20

A

In Problems 1–10 find the radian measure of the angles whose degree measures are given. Leave answers in terms of π.

1. a. $30°$ b. $45°$
2. a. $60°$ b. $90°$
3. a. $120°$ b. $135°$
4. a. $150°$ b. $225°$
5. a. $240°$ b. $270°$
6. a. $300°$ b. $315°$
7. a. $330°$ b. $15°$
8. a. $-50°$ b. $72°$
9. a. $600°$ b. $-144°$
10. a. $36°$ b. $-80°$

In Problems 11–20 give the degree measure of the angles having the given radian measures.

11. a. $\dfrac{3\pi}{4}$ b. $\dfrac{5\pi}{3}$

12. a. 4π b. $\dfrac{5\pi}{2}$

13. a. $\dfrac{5\pi}{6}$ b. $\dfrac{-4\pi}{3}$

14. a. $\dfrac{-7\pi}{2}$ b. $\dfrac{3\pi}{10}$

15. a. $\dfrac{5\pi}{9}$ b. 2

16. a. $\dfrac{3\pi}{2}$ b. $\dfrac{5\pi}{4}$

17. a. 3 b. $\dfrac{\pi}{12}$

18. a. $\dfrac{7\pi}{12}$ b. $\dfrac{\pi}{5}$

19. a. -4 b. $\dfrac{11\pi}{6}$

20. a. $\dfrac{9\pi}{4}$ b. $\dfrac{13\pi}{3}$

In Problems 21–23 use a calculator to convert to decimal degrees. Express answers to the nearest hundredth of a degree.

21. a. $32°51'$ b. $102°35'$
22. a. $18°03'$ b. $321°48'$
23. a. $27°13'24''$ b. $-56°19'12''$

In Problems 24–27 use a calculator to convert to degrees and minutes, to the nearest minute.

24. a. $86.32°$ b. $29.78°$
25. a. $-13.05°$ b. $153.41°$

26. a. $\dfrac{\pi}{13}$ radians b. 2.563 radians

27. a. 3 radians b. 7.092 radians

In Problems 28–30 to convert to radians, to the nearest thousandth of a radian.

28. a. $25.426°$ b. $115.750°$
29. a. $13°22'31''$ b. $290°12'24''$
30. a. $30°$ b. $-46.023°$

In Problems 31–34, θ is a central angle on a circle of radius r, and s is the arc on the circle subtended by θ. Find the value (s, r, or θ) which is not given.

31. a. $r = 3$, $\theta = 2$ b. $s = 4$, $\theta = 3$
32. a. $r = 5$, $s = 12$ b. $r = 2$, $\theta = 30°$
33. a. $s = 8$, $\theta = 45°$ b. $s = 4$, $r = 6$

34. a. $\theta = \dfrac{\pi}{3}$, $s = 6$ b. $\theta = \dfrac{3\pi}{4}$, $r = 4$

35. Find the arc length on the equator (in miles)

subtended by an angle of 1° at the center of the earth. Assume the radius of the earth is 3,960 miles.

36. What is the length of the arc swept out by the tip of the minute hand of a clock during the time interval from 6:00 to 6:20 if the minute hand is 4 inches long?

37. The diameter of the steering wheel of a car is 15 inches, and it has 3 equally spaced spokes. Find the distance on the steering wheel between consecutive spokes.

38. A wheel is revolving at an angular speed of $5\pi/3$ radians per second. Find the number of revolutions per minute (rpm) through which the wheel is turning.

39. A flywheel is turning at 1,800 rpm. Through how many radians per second is it turning?

B

40. The earth completes one revolution about its axis in 24 hours. Find the speed in miles per hour of a point on the equator (take $r = 3,960$ miles). What is the speed in feet per second?

41. A train is traveling on a circular curve of $\frac{1}{2}$ mile radius at the rate of 30 miles per hour. Through what angle (in radians) will the train turn in 45 seconds? What is the angle in degrees?

42. A flywheel 4 feet in diameter is revolving at the rate of 50 rpm. Find the speed of a point on the rim in feet per second.

43. Assume the earth moves around the sun in a circular orbit with radius 93,000,000 miles. A complete revolution takes approximately 365 days. Find the approximate speed of the earth in its orbit in miles per hour.

44. An automobile tire is 28 inches in diameter. If the car is traveling 30 miles per hour, through how many revolutions per minute is the wheel turning?

The following two problems are taken from an engineering textbook.*

45. Two points B and C lie on a radial line of a rotating disk. The points are 2 in. apart. $V_B = 700$ ft/min and $V_C = 880$ ft/min. Find the radius of rotation for each of these points.

46. The tire of an automobile has an outside diameter of 686 mm. If the number of revolutions per minute of the wheel is 700, determine **(a)** the speed of the automobile in kilometers per hour, **(b)** the speed of the automobile in meters per second, and **(c)** the angular speed of the wheel in radians per second.

SOME SPECIAL ANGLES

An angle of 90° is called a **right angle,** and an angle of 180° is called a **straight angle.** An **acute angle** is an angle between 0° and 90°, and an **obtuse angle** is one between 90° and 180°. A triangle in which one angle is a right angle is called a **right triangle,** and the side opposite the 90° angle is called the **hypotenuse.** The other two sides are often referred to as the **legs** of the triangle. A triangle that is not a right triangle is said to be an **oblique** triangle. The following fundamental result about right triangles is proved in plane geometry. (A proof is outlined in Problem 34 at the end of this section.)

THE
PYTHAGOREAN
THEOREM

The sum of the squares of the lengths of the legs of a right triangle equals the square of the length of the hypotenuse.

---

* George H. Martin, *Kinematics and Dynamics of Machines,* 2d ed. (New York: McGraw-Hill Book Company, 1982), p. 41. Reprinted by permission.

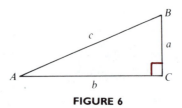

FIGURE 6

In Figure 6 we have labeled a right triangle with angles named by capital letters and sides opposite the angles named with the corresponding lowercase letters. This is a standard notation, and we will understand all right triangles to be designated this way unless specific instructions are given otherwise. The right angle is at C. We will follow the common practice of using these capital letters both to name the angles and to designate their measure. Similarly, we will use the lowercase letters as names of the sides as well as to designate their lengths. For example, we might say $A = 30°$, rather than "the degree measure of angle A is 30," and we might write $a = 2$ to mean "the length of side a is 2." This dual interpretation should not cause any difficulty. With this notation the Pythagorean theorem states that

$$a^2 + b^2 = c^2$$

The converse of this theorem is also true, and we call for a proof of this in Problem 35 at the end of this section. That is, if for a triangle with sides a, b, and c it is true that $a^2 + b^2 = c^2$, then the triangle is a right triangle, with the $90°$ angle opposite side c.

Another fundamental property of triangles proved in plane geometry is that the sum of the three angles in *any* triangle (not just a right triangle) is $180°$. In particular, if the triangle is a right triangle, so that one of the angles is $90°$, the sum of the two acute angles is $90°$. Such angles are said to be **complementary,** and each is said to be the **complement** of the other. In Figure 6, angles A and B are complementary.

There are two right triangles that we single out for special attention. The first of these is referred to as a **30°−60° right triangle** because the acute angles are $30°$ and $60°$, respectively. Let ABC be a $30°−60°$ right triangle with $A = 30°$, and $B = 60°$. If we flip the triangle over with side AC as an axis, as in Figure 7, we obtain a triangle ABB' in which all angles are $60°$. Thus, this triangle is **equiangular** (all angles equal), and hence also **equilateral** (all sides equal). So $2a = c$, or $a = c/2$. This is the result we wish to emphasize.

In a **30°−60° right triangle,** the side opposite the $30°$ angle is one-half the hypotenuse.

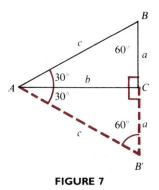

FIGURE 7

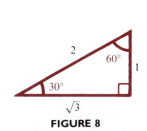

FIGURE 8

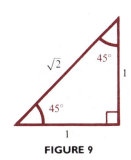

FIGURE 9

As a special case, consider a right triangle ABC with $A = 30°$, $B = 60°$, and $c = 2$. Then by what we have just shown, we know that $a = 1$. We use the Pythagorean theorem to find b.

$$b^2 = c^2 - a^2 = 4 - 1 = 3$$
$$b = \sqrt{3}$$

Figure 8 illustrates the result, and it is useful to commit this to memory. All other $30°$–$60°$ right triangles are similar to this one, and so the sides are proportional to the corresponding sides of this triangle.

The second special type of right triangle we want to consider is the **isosceles right triangle,** that is, one in which the two legs are equal. It follows that the two acute angles are also equal, and hence each is $45°$. If we let each leg be 1 unit, we find by the Pythagorean theorem that the hypotenuse is $\sqrt{1^2 + 1^2} = \sqrt{2}$. The situation is shown in Figure 9, which again should be committed to memory.

Example 6 illustrates how the concepts discussed in this section can be used to find unknown sides of certain right triangles.

EXAMPLE 6 ||||||||||||||||||||||||||||||||||||||||||||||||||||||||||||||||||||||||||||||||||||||||||||||||||||||||||

In a right triangle ABC, with right angle at C, find the unknown sides from the given information.

a. $a = 3$, $c = 5$ **b.** $A = 30°$, $c = 10$
c. $B = 45°$, $a = 8$ **d.** $A = 60°$, $b = 12$

Solution   **a.**   In Figure 10, by the Pythagorean theorem,

$$b^2 = c^2 - a^2 = 25 - 9 = 16$$
$$b = 4$$

Note. This is known as a **3–4–5 right triangle,** and it is useful to remember that *a triangle with sides 3, 4, and 5, or with sides respectively proportional to 3, 4, and 5, is a right triangle.*

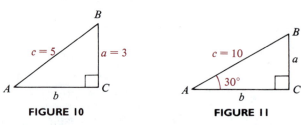

FIGURE 10   FIGURE 11

b. Since $c = 10$, we know that $a = 5$. Using proportionality and Figure 8, we see that in Figure 11 $b = 5\sqrt{3}$. Alternatively, we could use the Pythagorean theorem to find b:

$$b = \sqrt{c^2 - a^2} = \sqrt{100 - 25} = \sqrt{75} = 5\sqrt{3}$$

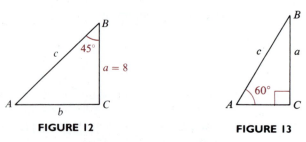

FIGURE 12   FIGURE 13

c. Since $B = 45°$, the triangle is isosceles. So $b = 8$, and from Figure 9 and proportionality we determine in Figure 12 that $c = 8\sqrt{2}$. Again, we could also find c by the Pythagorean theorem:

$$c = \sqrt{a^2 + b^2} = \sqrt{64 + 64} = \sqrt{(64) \cdot 2} = 8\sqrt{2}$$

d. In Figure 13, angle $B = 90° - A = 30°$, and since the side opposite the 30° angle is half the hypotenuse, it follows that $c = 24$. Comparison with the basic 30°–60° right triangle of Figure 8, or use of the Pythagorean theorem, gives $a = 12\sqrt{3}$. ▲

 EXERCISE SET 3

A

In Problems 1–20 *ABC* is a right triangle with right angle at *C*. In each problem find the unknown sides.

1. $A = 30°$, $c = 5$

2. $B = 60°$, $a = 3$

3. $A = 60°$, $b = 10$

4. $A = 30°$, $b = 4\sqrt{3}$

5. $A = 30°$, $b = 6$

6. $B = 30°$, $a = 12$

7. $B = 45°$, $c = 4\sqrt{2}$

8. $A = 45°$, $c = 4$

9. $A = 60°$, $c = 5$

10. $B = 60°$, $b = 3$

11. $B = 45°$, $a = 3$

12. $A = 60°$, $a = 8$

13. $a = 2$, $b = 3$

14. $a = 5$, $c = 7$

15. $a = 7$, $c = 25$

16. $b = 12$, $c = 13$

17. $a = 14$, $c = 50$

18. $a = 3$, $b = \sqrt{3}$ Also find angles A and B.

19. $a = 1, \quad b = 2\sqrt{2}$ **20.** $a = 8, \quad c = 17$

In Problems 21–26 two angles of a triangle are given. Find the third angle.

21. 23°17′, 75°34′   **22.** 43.21°, 68.35°

23. 52.76°, 103.42°   **24.** 92°47′, 21°54′

25. $\dfrac{\pi}{4}$ radians, $\dfrac{\pi}{3}$ radians

26. 1.25 radians, 0.78 radian

27. If a, b, and c are positive integers for which $a^2 + b^2 = c^2$, the numbers are said to form a **Pythagorean triple,** and we know that a triangle having sides of lengths a, b, and c, respectively, is a right triangle. Show that the following are Pythagorean triples.

a. 6, 8, 10   **b.** 5, 12, 13   **c.** 8, 15, 17
d. 7, 24, 25   **e.** 9, 40, 41

B

28. It is proved in more advanced mathematics that all Pythagorean triples (see Problem 27) are of the form $a = m^2 - n^2$, $b = 2mn$, and $c = m^2 + n^2$, where m and n are positive integers, with $m > n$.
  a. Verify that all such numbers do form a Pythagorean triple.
  b. Find the values of m and n which produce each triple in Problem 27.
  c. Determine two additional Pythagorean triples not proportional to any of those in Problem 27.

29. In the accompanying figure the 45° right triangle is constructed so that one of its legs is the hypotenuse of the 30°–60° right triangle. Find
  a. the hypotenuse of the 45° triangle.
  b. the combined areas of the two triangles.

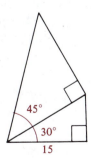

30. Find a formula for the area of an equilateral triangle with side of length a.

31. Prove that in a triangle ABC with $A = 30°$ and $C = 90°$, the median drawn from C has length a. (A **median** is the line segment from a vertex to the midpoint of the opposite side.)

32. Find a formula for the area of the inscribed circle in an equilateral triangle with side of length a.
  Hint. The angle bisectors, which in an equilateral triangle coincide with the medians and the altitudes, intersect at the center of the inscribed circle.

33. Find the area of the circumscribed circle to an equilateral triangle with side of length a. (See hint in Problem 32. The inscribed circle and circumscribed circle are concentric.)

34. By the following steps prove the Pythagorean theorem.
  a. Consider a right triangle with legs of lengths a and b and hypotenuse of length c. Construct a square of side $a + b$ as shown, and draw four replicas (I, II, III, and IV) of the given triangle inside the square as shown.
  b. Show that the quadrilateral V is a square. (Work with angles.)
  c. Write the equation that puts into mathematical form the fact that the sum of the areas I, II, III, IV, and V equals the area of the large square.
  d. By simplifying the equation in **c**, draw the desired conclusion.

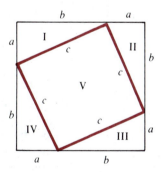

35. Follow the steps given to prove the converse of the Pythagorean theorem.
  a. Let ABC be a triangle in which $c^2 = a^2 + b^2$.

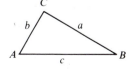

b. Construct a right triangle $A'B'C'$ having legs a and b. Let the hypotenuse be of length c'. Use the Pythagorean theorem to express c' in terms of a and b.

c. How can you conclude that $c' = c$?

d. What does this say about triangle ABC compared with triangle $A'B'C'$?

e. What do you conclude about angle C?

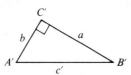

TRIGONOMETRIC FUNCTIONS OF ANGLES

In this section we introduce the six trigonometric functions, called the **sine, cosine, tangent, cotangent, secant,** and **cosecant.** We first define them with angles as domains, and in Section 5 we define them with real numbers as domains and show how the two definitions are related. If θ is an angle in the domain of each of the trigonometric functions named above, we abbreviate their values at θ by $\sin \theta$, $\cos \theta$, $\tan \theta$, $\cot \theta$, $\sec \theta$, and $\csc \theta$, respectively.

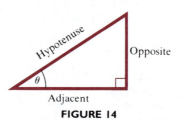

FIGURE 14

Suppose first that θ is an acute angle, and consider any right triangle having θ as one of its angles. As shown in Figure 14 we label the legs opposite to θ and adjacent to θ. In the following definition we use **opposite, adjacent,** and **hypotenuse** to mean the **lengths** of these sides. With this understanding, the six trigonometric functions of θ are defined as shown.

DEFINITION 1

$$\sin \theta = \frac{\text{opposite}}{\text{hypotenuse}} \qquad \cot \theta = \frac{\text{adjacent}}{\text{opposite}}$$

$$\cos \theta = \frac{\text{adjacent}}{\text{hypotenuse}} \qquad \sec \theta = \frac{\text{hypotenuse}}{\text{adjacent}}$$

$$\tan \theta = \frac{\text{opposite}}{\text{adjacent}} \qquad \csc \theta = \frac{\text{hypotenuse}}{\text{opposite}}$$

The domain of each function is the set of all acute angles, and the range is a subset of R. We will shortly remove the restriction that θ be acute and see how the definition can be extended to include arbitrary angles. It is important to realize that the values of the functions depend on the angle θ and not on how its measure is given. So the values are the same whether θ is measured in degrees or radians. For example, $\sin 30°$ and $\sin \pi/6$ both have the same value, namely, the sine of the angle whose degree measure is 30, or equivalently, whose radian measure is $\pi/6$.

Although the definition makes use of a right triangle, it is important to realize that the values of the functions depend only on θ and not on which right triangle having θ as one of its acute angles is used. To see this, consider any two such right triangles, as in Figure 15. Since $A = A'$ and $C = C'$ (both 90°), it follows that

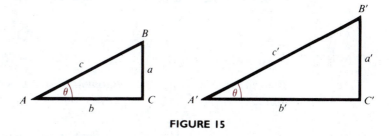

FIGURE 15

$B = B'$, so that the triangles are similar. Thus corresponding sides are proportional. In particular,

$$\frac{a}{c} = \frac{a'}{c'} \qquad \frac{b}{c} = \frac{b'}{c'} \qquad \frac{a}{b} = \frac{a'}{b'}$$

and so on. That is, we get the same values for $\sin \theta$, $\cos \theta$, $\tan \theta$, and the other functions regardless of which of the two triangles we use.

In Figure 16 we have labeled the angles and sides of a right triangle in standard notation. Observe that side a is opposite to angle A but is adjacent to angle B. Similarly, side b is adjacent to angle A but is opposite to angle B. So we have

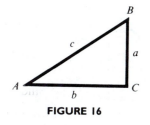

FIGURE 16

$$\sin A = \frac{a}{c} = \cos B$$

$$\cos A = \frac{b}{c} = \sin B$$

$$\tan A = \frac{a}{b} = \cot B$$

$$\sec A = \frac{c}{b} = \csc B$$

$$\csc A = \frac{c}{a} = \sec B$$

We call the sine and cosine **cofunctions** of one another. Similarly, the tangent and cotangent are cofunctions, as are the secant and cosecant. Since angles A and B are complementary, we have therefore shown the following:

Any trigonometric function of an acute angle equals the corresponding cofunction of its complement.

It is important that the trigonometric functions for acute angles be learned in terms of *opposite*, *adjacent*, and *hypotenuse*, rather than a set of letters such as those used above, because we will not always be using standard notation.

EXAMPLE 7 ιιιιιιιιιιιιιιιιιιιιιιιιιιιιιιιιιιιιιιιιιιιιιιιιιιιιιιιιιιιιιιιιιιιιιιιιιιιιιιιιιιιιιιιιιιιιιιιιιιιιιιιιιιιι

Find the values of the six trigonometric functions of

a. 30°   **b.** 45°   **c.** 60°

Solution   **a.** Using the 30°–60° right triangle shown in Figure 17, we have by Definition 1,

$$\sin 30° = \frac{1}{2} \qquad \cos 30° = \frac{\sqrt{3}}{2} \qquad \tan 30° = \frac{1}{\sqrt{3}}$$

$$\csc 30° = 2 \qquad \sec 30° = \frac{2}{\sqrt{3}} \qquad \cot 30° = \sqrt{3}$$

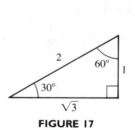

FIGURE 17

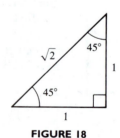

FIGURE 18

b. From the 45° right triangle of Figure 18, we obtain

$$\sin 45° = \frac{1}{\sqrt{2}} \qquad \cos 45° = \frac{1}{\sqrt{2}} \qquad \tan 45° = 1$$

$$\csc 45° = \sqrt{2} \qquad \sec 45° = \sqrt{2} \qquad \cot 45° = 1$$

c. Since 60° is the complement of 30°, we can obtain the functions of 60° as the cofunctions of $30°$, or we can use Figure 17 directly, this time taking $\sqrt{3}$ as the opposite side and 1 as the adjacent side. Either way we get

$$\sin 60° = \frac{\sqrt{3}}{2} \qquad \cos 60° = \frac{1}{2} \qquad \tan 60° = \sqrt{3}$$

$$\csc 60° = \frac{2}{\sqrt{3}} \qquad \sec 60° = 2 \qquad \cot 60° = \frac{1}{\sqrt{3}}$$

As this example shows, trigonometric functions of $30°$, $45°$, and $60°$ (or in radians, $\pi/6$, $\pi/4$, and $\pi/3$) can be obtained in exact form by using the special $30°–60°$ right triangle and the $45°$ right triangle that we have introduced. For other angles a calculator can be used. Of course, a calculator can be used for $30°$, $45°$, and $60°$ also, but it is usually advantageous to obtain these in exact form, rather than as the decimal approximations the calculator gives.

When using the calculator to obtain values of the trigonometric functions, be sure to set it in degree mode or radian mode, depending on how the angle measurement is given. Once it is set in a particular mode, it will remain in that mode (even if it is turned off) until you change it. Also, most calculators have keys only for the sine, cosine, and tangent. Observe from Definition 1 that $\cot \theta = 1/\tan \theta$, $\sec \theta = 1/\cos \theta$, and $\csc \theta = 1/\sin \theta$. Thus, for example, to find $\sec 33°$, you would set the calculator in the degree mode and use the keystroke sequence

$$33 \quad \boxed{\text{COS}} \quad \boxed{1/x}$$

The answer would show up as 1.1924 (rounded to four places).

Tables giving values of the trigonometric functions exist, and before scientific calculators became widely available, these were used extensively. Now, however, they are for all practical purposes obsolete, since the calculator is so much faster and, except for the most detailed tables, gives answers to more significant digits.

Suppose now that θ is any angle, not necessarily acute. We place the vertex of θ at the origin and its initial side along the positive x axis, as in Figure 19. An angle placed this way is said to be in **standard position.** Choose any point (x, y) other than the origin on the terminal side of θ, and denote by r the distance from $(0, 0)$ to (x, y). We define the six trigonometric functions of θ in terms of x, y, and r.

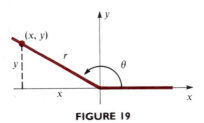

FIGURE 19

DEFINITION 2

$$\sin \theta = \frac{y}{r} \qquad\qquad \cos \theta = \frac{x}{r} \qquad\qquad \tan \theta = \frac{y}{x} \quad (x \neq 0)$$

$$\csc \theta = \frac{r}{y} \quad (y \neq 0) \qquad \sec \theta = \frac{r}{x} \quad (x \neq 0) \qquad \cot \theta = \frac{x}{y} \quad (y \neq 0)$$

The choice of the particular point (x, y) on the terminal side of θ is immaterial since, by similar triangles, the ratios would be the same if any other point were used (except the origin, which we exclude).

We note that if θ is acute, this definition is consistent with Definition 1, as can be seen by considering the right triangle having hypotenuse r, adjacent side x, and opposite side y.

EXAMPLE 8 ||||||||||||||||||||||||||||||||||||||||||||||||||||||||||||||||||||||||||||||||||||||||||||||||

The terminal side of an angle θ in standard position passes through the point $(3, -2)$. Find the values of the six trigonometric functions of θ.

Solution   We are not told whether θ is positive or negative, nor do we know if θ is a primary angle, but neither of these matters insofar as the values of the trigonometric functions are concerned. In Figure 20 we show θ as a positive angle that is between 0 and 2π (i.e., a primary angle). Using the point $(3, -2)$ as the point (x, y) in Definition 2, we find by the Pythagorean theorem that $r = \sqrt{13}$. So we have

$$\sin \theta = \frac{y}{r} = \frac{-2}{\sqrt{13}} \qquad \cos \theta = \frac{x}{r} = \frac{3}{\sqrt{13}} \qquad \tan \theta = \frac{y}{x} = -\frac{2}{3}$$

$$\csc \theta = \frac{r}{y} = -\frac{\sqrt{13}}{2} \qquad \sec \theta = \frac{r}{x} = \frac{\sqrt{13}}{3} \qquad \cot \theta = \frac{x}{y} = -\frac{3}{2} \quad \blacktriangle$$

Since r is always positive, the sine and cosine are defined for all angles θ by Definition 2. But there are certain values of θ excluded from the domain of each

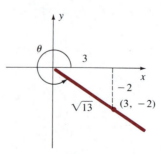

FIGURE 20

of the other functions, namely those angles that cause the denominator in the fraction defining the function to be zero. In particular, $\tan\theta$ and $\sec\theta$ are undefined when $x = 0$, and $\cot\theta$ and $\csc\theta$ are undefined when $y = 0$. Thus $\theta = 90°$, $270°$, and all angles coterminal with these are excluded from the domains of the tangent and secant. Similarly, $\theta = 0°$, $180°$, and all angles coterminal with these are excluded from the domains of the cotangent and cosecant.

We name the positive acute angle between the terminal side of θ and the x axis the **reference angle** for θ (Figure 21). For example, the reference angle of $150°$ is $30°$, and the reference angle of $315°$ is $45°$. There is a close relationship between the functions of an angle θ and of its reference angle. When θ is in standard position and we have chosen a point (x, y) on its terminal side, if we superimpose a right triangle with hypotenuse r so that the reference angle for θ is one of the angles of the triangle, then the legs of the triangle have lengths equal in absolute value to x and y, respectively (Figure 21). But whereas x and y may be either positive or negative, the lengths of the sides of a triangle are positive. We conclude that

Any trigonometric function of θ is equal in absolute value to the same function of its reference angle.

The correct sign must be affixed according to the function and the signs of x and y in the quadrant in which the terminal side lies. We illustrate this with several examples.

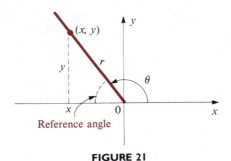

FIGURE 21

EXAMPLE 9 ||||||||||||||||||||||||||||||||||||||||||||||||||||||||||||||||||||||||||||||||||||||||||||||||||||

Find all six trigonometric functions of $225°$.

Solution   The terminal side lies in quadrant III, and the reference angle is $45°$, as shown in Figure 22. A convenient point to choose on the terminal side is $(-1, -1)$, and then $r = \sqrt{2}$. So we have

$$\sin 225° = -\frac{1}{\sqrt{2}} \qquad \cos 225° = -\frac{1}{\sqrt{2}} \qquad \tan 225° = 1$$

$$\csc 225° = -\sqrt{2} \qquad \sec 225° = -\sqrt{2} \qquad \cot 225° = 1$$

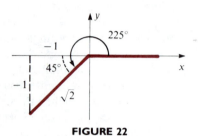

FIGURE 22

EXAMPLE 10

Find all six trigonometric functions of $5\pi/3$.

Solution   This time the angle is given in radians. With practice you will find it as convenient to work with radians as with degrees, but at this stage you may prefer to convert to degrees.

$$\frac{5\pi}{3} \text{ radians} = \frac{5\pi}{3} \cdot \frac{180°}{\pi} = 300°$$

The terminal side is in the fourth quadrant, and the reference angle is 60° (Figure 23). So we have a 30° − 60° right triangle as shown and choose $r = 2$, $x = 1$, $y = -\sqrt{3}$. Thus we have

$$\sin \frac{5\pi}{3} = -\frac{\sqrt{3}}{2} \qquad \cos \frac{5\pi}{3} = \frac{1}{2} \qquad \tan \frac{5\pi}{3} = -\sqrt{3}$$

$$\csc \frac{5\pi}{3} = -\frac{2}{\sqrt{3}} \qquad \sec \frac{5\pi}{3} = 2 \qquad \cot \frac{5\pi}{3} = -\frac{1}{\sqrt{3}}$$

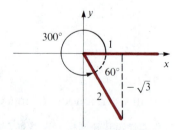

FIGURE 23

EXAMPLE 11

Find all six trigonometric functions of 510°.

Solution   The first thing to do is observe that 510° is coterminal with 150° (Figure 24). Thus, the values of the trigonometric functions of 510° are the same as those of 150°. The angle 150° lies in the second quadrant and has a reference angle of 30°. Placing our standard 30°–60° triangle as shown in the figure, we read the following function values.

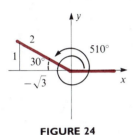

$$\sin 510° = \frac{1}{2} \qquad \cos 510° = -\frac{\sqrt{3}}{2} \qquad \tan 510° = -\frac{1}{\sqrt{3}}$$

$$\csc 510° = 2 \qquad \sec 510° = -\frac{2}{\sqrt{3}} \qquad \cot 510° = -\sqrt{3}$$

FIGURE 24

These examples illustrate how we can find exact values of all angles having as reference angles 30°, 45°, or 60°. In particular, we can find the functions of 30°, 45°, 60°, 120°, 135°, 150°, 210°, 225°, 240°, 300°, 315°, and 330°, as well as angles coterminal with any of these.

The angles 0°, 90°, 180°, 270°, and all angles coterminal with these are called **quadrantal angles.** Using Definition 2 we can find the values of the trigonometric functions of all such angles. We illustrate how to do this for 0° and 90°.

EXAMPLE 12

Find the six trigonometric functions of 0°.

Solution   In applying Definition 2 we may choose any point on the terminal side of the angle. For simplicity we choose (1, 0) in this case (Figure 25). Thus $x = 1$, $y = 0$, and $r = 1$. So we have

$$\sin 0° = \frac{0}{1} = 0 \qquad \cos 0° = \frac{1}{1} = 1 \qquad \tan 0° = \frac{0}{1} = 0$$

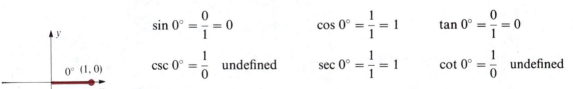

$$\csc 0° = \frac{1}{0} \text{ undefined} \qquad \sec 0° = \frac{1}{1} = 1 \qquad \cot 0° = \frac{1}{0} \text{ undefined}$$

FIGURE 25

Remark. In working with quadrantal angles it will be typical that two of the functions will be undefined. The thing to remember is that *division by 0 is not defined.*

EXAMPLE 13

Find the six trigonometric functions of 90°.

Solution   We select the point $(0, 1)$ on the terminal side, so that $x = 0$, $y = 1$, and $r = 1$ (Figure 26). Definition 2 gives the following.

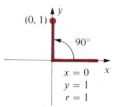

$$\sin 90° = \frac{1}{1} = 1 \qquad \cos 90° = \frac{0}{1} = 0 \qquad \tan 90° = \frac{1}{0} \quad \text{undefined}$$

$$\csc 90° = \frac{1}{1} = 1 \qquad \sec 90° = \frac{1}{0} \text{ - undefined} \qquad \cot 90° = \frac{0}{1} = 0 \qquad \blacktriangle$$

FIGURE 26

Remark. When you use a calculator to evaluate trigonometric functions, there is no need to use reference angles. For example, we can read directly

$$\tan 2500° = -0.36397 \qquad \text{and} \qquad \sin(-532 \text{ radians}) = 0.87760$$

(rounded to five significant figures).

 EXERCISE SET 4

Do not use a calculator unless the instructions call for it.

 A

In Problems 1–4 triangle ABC is a right triangle, with right angle at C and sides a, b, and c opposite the corresponding angles. Find all trigonometric functions of the specified angle.

1. $a = 2$, $b = 3$; angle A
2. $c = 5$, $b = 3$; angle B
3. $a = 5$, $c = 13$; angle B
4. $a = 7$, $b = 5$; angle A
5. For the right triangle in the accompanying figure, find all trigonometric functions of
   a. Angle R **b.** Angle T

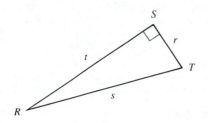

In Problems 6–11 a point on the terminal side of an angle θ in standard position is given. Find all six trigonometric functions of θ.

6. $(-4, -3)$ 7. $(-5, 12)$ 8. $(8, -15)$
9. $(-1, 2)$ 10. $(-5, -4)$ 11. $(24, -7)$

In Problems 12–24 find exact values (if they exist) of all trigonometric functions of the given angle.

12. $120°$ 13. $225°$ 14. $330°$
15. $\dfrac{4\pi}{3}$ 16. $150°$ 17. $\dfrac{3\pi}{4}$
18. $\dfrac{7\pi}{6}$ 19. $-\dfrac{\pi}{6}$ 20. π
21. $600°$ 22. $-\dfrac{5\pi}{2}$ 23. $540°$
24. $-\dfrac{7\pi}{3}$

In Problems 25–30 evaluate each function.

25. **a.** $\sin 120°$ **b.** $\cos 315°$ **c.** $\tan 210°$
    d. $\sec 240°$ **e.** $\csc 330°$

26. a. $\cos \dfrac{3\pi}{4}$ **b.** $\tan \dfrac{2\pi}{3}$ **c.** $\csc \dfrac{5\pi}{6}$

    d. $\sin \dfrac{5\pi}{4}$ **e.** $\cot \dfrac{7\pi}{6}$

27. a. $\tan 480°$ **b.** $\sec 900°$ **c.** $\sin(-270°)$

    d. $\cos \dfrac{10\pi}{3}$ **e.** $\cot \dfrac{3\pi}{2}$

28. a. $\sin \dfrac{7\pi}{6}$ **b.** $\cos 330°$ **c.** $\tan \dfrac{7\pi}{4}$

    d. $\sec 135°$ **e.** $\cot \dfrac{3\pi}{4}$

29. a. $\csc \dfrac{11\pi}{6}$ **b.** $\tan 600°$ **c.** $\sin \dfrac{4\pi}{3}$

    d. $\cos 225°$ **e.** $\sec \dfrac{7\pi}{6}$

30. a. $\cot(-300°)$ **b.** $\sec \dfrac{9\pi}{4}$

    c. $\tan\left(-\dfrac{17\pi}{6}\right)$ **d.** $\cos 660°$

    e. $\csc(-240°)$

In Problems 31–34 give an equivalent expression as a function of an acute angle.

31. a. $\sin 250°$ **b.** $\cos 132°$ **c.** $\tan 97°$
    d. $\sec 280°$ **e.** $\csc 310°$
32. a. $\cot 212°$ **b.** $\sin 562°$ **c.** $\tan 620°$
    d. $\cos(-430°)$ **e.** $\csc(-491°)$
33. a. $\cos 123°40'$ **b.** $\tan 312°15'$
    c. $\csc(-240°38')$ **d.** $\cot 263°24'$
    e. $\sin(-158°13')$

34. a. $\tan \dfrac{7\pi}{12}$ **b.** $\sec \dfrac{7\pi}{8}$ **c.** $\sin \dfrac{9\pi}{5}$

    d. $\csc \dfrac{8\pi}{9}$ **e.** $\cos \dfrac{13\pi}{12}$

35. Evaluate the expression

$$\dfrac{\sin \theta + \cos 2\theta}{\tan \theta \cot 2\theta}$$

when $\theta = 5\pi/6$.
36. Evaluate the expression

$$\left(\sec \theta - \csc \dfrac{\theta}{4}\right)(1 + \cos 3\theta)$$

when $\theta = 2\pi/3$.

Use a calculator to obtain the answers in Problems 37–40, correct to 4 significant digits.

37. a. $\sin 93.8°$ **b.** $\cos(-247°)$
    c. $\tan 115.2°$ **d.** $\sec(-23°)$
38. a. $\cot 575°$ **b.** $\csc(-327°)$
    c. $\cos 1029°$ **d.** $\tan 89.8°$

39. a. $\sec \dfrac{5\pi}{9}$ **b.** $\sin 3$

    c. $\tan 0.128$ **d.** $\cos(-10)$

40. a. $\tan \dfrac{4}{5}$ **b.** $\cot\left(-\dfrac{5\pi}{12}\right)$

    c. $\csc 5.38$ **d.** $\sin 30$

B

In Problems 41–46 draw the angle θ in standard position for which $0 \le \theta < 2\pi$ and find the other five trigonometric functions of θ.

41. $\sin \theta = -\frac{4}{5}$, θ in quadrant III
42. $\cos \theta = \frac{5}{13}$, θ in quadrant IV
43. $\sec \theta = -\frac{3}{2}$, $\tan \theta > 0$
44. $\csc \theta = 3$, $\cos \theta < 0$
45. $\tan \theta = \frac{1}{2}$, $\csc \theta < 0$
46. $\cot \theta = -\frac{1}{3}$, $\sin \theta > 0$
47. If $\sin \theta = y$ and $0 < y < 1$, find expressions for the other five trigonometric functions in terms of y. (Two solutions.)
48. If $\tan \theta = t > 0$, find expressions for the other five trigonometric functions of θ in terms of t. (Two solutions.)
49. If $\sec \theta = r > 0$, find expressions for the other five trigonometric functions in terms of r. (Two solutions).
50. For $0 < \theta < \pi/2$, express all six trigonometric functions of each of the following in terms of θ alone.

    a. $\pi - \theta$ **b.** $\pi + \theta$ **c.** $\dfrac{\pi}{2} + \theta$

    d. $\dfrac{3\pi}{2} - \theta$ **e.** $2\pi - \theta$

    Hint. Find the reference angle in each case.
51. Prove that $\cos n\pi = (-1)^n$, where n is any integer.
52. Prove that $\sin \dfrac{(2n+1)\pi}{2} = (-1)^n$, where n is any integer.

53. Prove that for all integers k,

$$\cos\frac{\pi}{3}(1+3k) = \frac{(-1)^k}{2}$$

54. Prove that for all integers k,

$$\tan\frac{\pi}{4}(1+2k) = (-1)^k$$

55. Using distances, prove that the points $A(8, 3)$, $B(7, -5)$, and $C(4, -3)$ are vertices of a right triangle. Find the six trigonometric functions of angle A.

TRIGONOMETRIC FUNCTIONS OF REAL NUMBERS

In Definition 2 the domain of each of the trigonometric functions is a set of angles. The values depend on the angles themselves and not on how they are measured. The next stage in our development is an important one, although it is somewhat subtle. We wish to give meaning to each of the trigonometric functions when its domain consists of real numbers rather than angles.

Toward this end, let us recall that in the definition of the radian measure of an angle, we used an arbitrary circle with its center at the vertex of the angle. We find it useful now to specify a circle of radius 1, called a **unit circle,** and we place its center at the origin (Figure 27). If we are given an angle θ in radians, then the directed length s of the subtended arc on this unit circle is

$$s = \theta$$

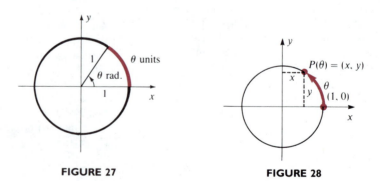

FIGURE 27   FIGURE 28

since $s = r\theta$ and $r = 1$. In other words, the linear measure of the arc length is the same as the angular measure of the central angle in radians. For example, if we are measuring distances in inches, and the central angle is 2 radians, then the arc will be of length 2 inches. The relationship is illustrated in Figure 27, and it provides us with a means of defining the trigonometric functions with real number domains that is consistent with our previous definition.

Given any real number θ (not at present to be thought of as the measure of an angle, but just a number), we mark off a distance of θ units on the unit circle, starting from $(1, 0)$ and moving counterclockwise when θ is positive and clockwise when θ is negative. Let $P(\theta)$ be the point arrived at in this manner with coordinates (x, y), as shown in Figure 28.* With this established, we define the following:

DEFINITION 3

$$\sin \theta = y \qquad\qquad \cos \theta = x \qquad\qquad \tan \theta = \frac{y}{x} \quad (x \neq 0)$$

$$\csc \theta = \frac{1}{y} \quad (y \neq 0) \qquad \sec \theta = \frac{1}{x} \quad (x \neq 0) \qquad \cot \theta = \frac{x}{y} \quad (y \neq 0)$$

The relationship between trigonometric functions of numbers and of angles should now be clear. For any given real number θ, we consider the central angle in standard position for which the directed length of the intercepted arc on the unit circle is θ. Then the radian measure of the angle is also θ, and its terminal side passes through the point $P(\theta)$, as shown in Figure 29. In Definition 2 if we use the coordinates of $P(\theta)$ as the point (x, y), then $r = 1$, and we see that the results are exactly the same as those in Definition 3.

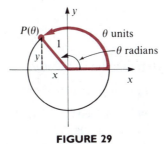

FIGURE 29

Suppose, then, that we are given the expression "sin 2." How shall we interpret it? Does it mean the sine of the number 2 or the sine of the angle whose radian measure is 2? The answer is that it does not matter, because the values are the same. Conceptually these are quite different, but sin 2, where 2 is a number, has the same value as sin 2 where 2 is the radian measure of an angle. More generally, *any trigonometric function of a real number θ has the same value as the same trigonometric function of the angle of θ radians.* Thus, insofar as values of the trigonometric functions are concerned, it makes no difference whether we think in terms of an angle measured in radians or in terms of the real number that equals the number of radians in the angle.

---

* $P(\theta)$ is thus a function of the real variable θ, whose range consists of ordered pairs (x, y) representing points on the circle.

EXAMPLE 14 ||||||||||||||||||||||||||||||||||||||||||||||||||||||||||||||||||||||||||||||||||||||||||||||||

The x coordinate of the point $P(\theta)$ on the unit circle is $-\frac{4}{5}$, and $P(\theta)$ is in the second quadrant. Find the y coordinate of $P(\theta)$ and write all six trigonometric functions of θ.

Solution   By the Pythagorean theorem we have

$$(-\tfrac{4}{5})^2 + y^2 = 1$$
$$y^2 = 1 - \tfrac{16}{25} = \tfrac{9}{25}$$
$$y = \tfrac{3}{5}$$

By Definition 3,

$$\sin \theta = y = \frac{3}{5} \qquad\qquad \csc \theta = \frac{1}{y} = \frac{5}{3}$$

$$\cos \theta = x = -\frac{4}{5} \qquad\qquad \sec \theta = \frac{1}{x} = -\frac{5}{4}$$

$$\tan \theta = \frac{y}{x} = -\frac{3}{4} \qquad\qquad \cot \theta = \frac{x}{y} = -\frac{4}{3}$$

▲

EXAMPLE 15 ||||||||||||||||||||||||||||||||||||||||||||||||||||||||||||||||||||||||||||||||||||||||||||||

Determine the value of θ, where $0 \le \theta < 2\pi$.

a. $P(\theta) = (-\sqrt{3}/2, -\frac{1}{2})$ **b.** $P(\theta) = (-1/\sqrt{2}, 1/\sqrt{2})$

Solution   **a.**   We observed from Figure 30 that the reference angle of the angle determined by the radius drawn to the given point is $30°$, or $\pi/6$ radians. So the angle itself is $7\pi/6$ radians. Therefore $\theta = 7\pi/6$.

       b.   The reference angle in this case is $45°$, or $\pi/4$ radians, as shown in Figure 31, and the angle is therefore $3\pi/4$ radians. So $\theta = 3\pi/4$.

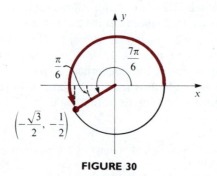

FIGURE 30

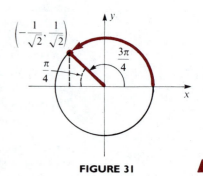

FIGURE 31   ▲

To evaluate trigonometric functions of real numbers on a calculator, treat the number as the radian measure of an angle. For example, to get sin 2, put the calculator in the radian mode (this is important), press 2, then $\boxed{\text{SIN}}$, and the answer will be given.

Remark on Notation. So far we have used θ to name the independent variable in the trigonometric functions of real numbers. Now that these functions have been defined, there is no reason not to use other letters. For example, we can write sin x or cos t. However, if the letter x (or y) is used in place of θ in figures such as Figure 27, then the axes should be renamed (say, the u axis and the v axis) so that x is not used with two different meanings. The important thing is that the point $P(x)$ on the unit circle has coordinates (cos x, sin x).

EXERCISE SET 5

A

In Problems 1–10 find all six trigonometric functions of θ.

1. $P(\theta) = (-\frac{1}{3}, -2\sqrt{2}/3)$
2. $P(\theta) = (-2/\sqrt{5}, 1/\sqrt{5})$
3. $P(\theta) = (\frac{5}{13}, -\frac{12}{13})$ 4. $P(\theta) = (-0.6, 0.8)$
5. The x coordinate of $P(\theta)$ is $\frac{1}{4}$, and $P(\theta)$ is in the fourth quadrant.
6. The y coordinate of $P(\theta)$ is $\frac{4}{5}$, and $\pi/2 \leq \theta \leq \pi$.
7. The abscissa of $P(\theta)$ is $-\frac{1}{2}$, and $\pi \leq \theta \leq 2\pi$. What is the value of θ?
8. The ordinate of $P(\theta)$ is $\sqrt{3}/2$, and the abscissa of $P(\theta)$ is negative. What is the value of θ if $0 \leq \theta < 2\pi$?
9. The abscissa of $P(\theta)$ is $\frac{1}{3}$, and the ordinate is negative.
10. The ordinate of $P(\theta)$ is $-\frac{5}{13}$, and $3\pi \leq \theta \leq 7\pi/2$.

In Problems 11 and 12 give the coordinates of the point $P(\theta)$ for the given values of θ.

11. **a.** $P(\pi/2)$ **b.** $P(\pi)$ **c.** $P(3\pi/2)$
    d. $P(5\pi/6)$ **e.** $P(4\pi/3)$
12. **a.** $P(5\pi/4)$ **b.** $P(8\pi/3)$ **c.** $P(-7\pi/4)$
    d. $P(5\pi)$ **e.** $P(31\pi/6)$

In Problems 13 and 14 indicate the approximate location of $P(\theta)$ on the unit circle for the given value of θ.

13. **a.** $\theta = 0.5$ **b.** $\theta = -2$
    c. $\theta = 4$ **d.** $\theta = -15$

14. **a.** $\theta = 8.2$ **b.** $\theta = 12.75$
    c. $\theta = -1.54$ **d.** $\theta = 4.72$

In Problems 15 and 16 give the value of θ if $0 \leq \theta < 2\pi$.

15. **a.** $P(\theta) = (\sqrt{3}/2, \frac{1}{2})$ **b.** $P(\theta) = (1/\sqrt{2}, 1/\sqrt{2})$
    c. $P(\theta) = (\frac{1}{2}, \sqrt{3}/2)$ **d.** $P(\theta) = (0, 1)$
    e. $P(\theta) = (1, 0)$
16. **a.** $P(\theta) = (-\frac{1}{2}, -\sqrt{3}/2)$
    b. $P(\theta) = (0, -1)$ **c.** $P(\theta) = (-\sqrt{3}/2, \frac{1}{2})$
    d. $P(\theta) = (\sqrt{2}/2, -\sqrt{2}/2)$
    e. $P(\theta) = (-1, 0)$

In Problems 17 and 18 use a calculator to find sin x, cos x, and tan x to 4 significant figures.

17. **a.** $x = 2.583$ **b.** $x = 0.3142$
    c. $x = -11.38$ **d.** $x = 3.872$
18. **a.** $x = -3.025$ **b.** $x = 253.9$
    c. $x = -0.2438$ **d.** $x = 1.023$
19. Find all values of θ such that $0 \leq \theta < 2\pi$ for which
    a. sin $\theta = 1$ **b.** cos $\theta = 1$
    c. tan $\theta = 1$ **d.** sin $\theta = 0$
    e. cos $\theta = 0$ **f.** tan $\theta = 0$
    g. sin $\theta = -1$ **h.** cos $\theta = -1$
    i. tan $\theta = -1$

B

20. Use the unit circle to show the following:
    a. $\sin(\pi - \theta) = \sin \theta$
    b. $\cos(\pi - \theta) = -\cos \theta$

c. $\sin(\pi + \theta) = -\sin \theta$
d. $\cos(\pi + \theta) = -\cos \theta$
e. $\tan(\pi + \theta) = \tan \theta$

21. Show that the domain of $\tan \theta$ and $\sec \theta$ is the set

$$\left\{ \theta: \ \theta \neq \frac{2n+1}{2} \pi, n = 0, \pm 1, \pm 2, \ldots \right\}$$

22. Show that the domain of $\cot \theta$ and $\csc \theta$ is the set

$$\{\theta: \ \theta \neq n\pi, n = 0, \pm 1, \pm 2, \ldots\}$$

23. Show that $|\sec \theta| \geq 1$ and $|\csc \theta| \geq 1$ for all θ in the domain of these functions.

24. Prove that the range of the tangent function is the set of all real numbers.

25. Prove that none of the trigonometric functions is 1–1.

26. Prove each of the following, where $P(\theta) = (x, y)$.
  a. $P(\theta + 2\pi) = (x, y)$
  b. $P(\theta + \pi) = (-x, -y)$

27. If $f(x + k) = f(x)$ for all x in the domain of f, then f is said to be **periodic**, of **period** k. Show that $\sin x$, $\cos x$, $\sec x$, and $\csc x$ are periodic, of period 2π, and that $\tan x$ and $\cot x$ are periodic, of period π.
Hint: Use the result of Problem 26.

 6 GRAPHS OF THE TRIGONOMETRIC FUNCTIONS

To obtain the graphs of $y = \sin x$ and $y = \cos x$ we make use of the definitions of the sine and cosine as the ordinate and abscissa, respectively, of a point on the unit circle (Figure 32). Notice that we have used x to represent arc length on the circle from $(1, 0)$ to $P(x)$ rather than the customary θ. We have used capital letters to label the axes. Table 1 indicates how $\sin x$ and $\cos x$ vary as x varies from 0 to 2π. These facts can be verified by visualizing the point $P(x)$ moving around the

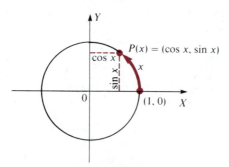

FIGURE 32

As x goes from	sin x goes from	cos x goes from
0 to $\pi/2$	0 to 1	1 to 0
$\pi/2$ to π	1 to 0	0 to -1
π to $3\pi/2$	0 to -1	-1 to 0
$3\pi/2$ to 2π	-1 to 0	0 to 1

TABLE I

circle in a counterclockwise direction, starting at $(1, 0)$. Furthermore, it should be clear from considering Figure 32 that the ordinate and abscissa of $P(x)$ (that is, $\sin x$ and $\cos x$) vary in a uniform way, with no breaks or sudden changes, as $P(x)$ moves around the circle.

Since the circumference of the unit circle is 2π, if we start at any point on the circle and move 2π units around it in either direction, we wind up at the same point. It follows that

$$\sin(x + 2\pi) = \sin x$$

and

$$\cos(x + 2\pi) = \cos x$$

These relationships can be expressed briefly by saying that the sine and cosine are **periodic,** with **period** 2π. Any function f for which $f(x + k) = f(x)$, for all x is said to be periodic of period k. The smallest positive value of k for which this is true is the **fundamental period.** The sine and cosine, and hence also the secant and cosecant, have fundamental period 2π, but the fundamental period of the tangent and cotangent is π. (See Problem 27, Exercise Set 5.)

This information, together with our knowledge of $\sin x$ and $\cos x$ for $x = \pi/6,\ \pi/4,\ \pi/3$ and related values in the other quadrants, enables us to draw the graphs with reasonable accuracy. Since $\sin x$ and $\cos x$ each has period 2π, the graphs repeat every 2π units. So once we know the graphs in the interval from 0 to 2π, we can extend them indefinitely in either direction. The graphs are shown in Figures 33 and 34. We will refer to these as the **basic sine and cosine curves.**

The maximum height attained by the sine curve and by the cosine curve is called the **amplitude.** So, for $y = \sin x$ and $y = \cos x$ the amplitude is 1.

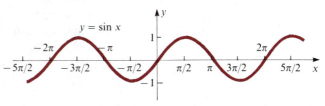

FIGURE 33

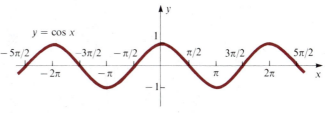

FIGURE 34

To obtain the graph of $y = \tan x$, we use the fact that, by Definition 3, $\tan x$ is the ratio of the ordinate of $P(x)$ to its abscissa. That is,

$$\tan x = \frac{\sin x}{\cos x}$$

When $x = 0$, $\tan x = 0/1 = 0$, and as x increases toward $\pi/2$, the ratio of the ordinate of $P(x)$ to its abscissa steadily increases, taking on the value 1 at $x = \pi/4$. For x near $\pi/2$ the ratio becomes very large and gets larger and larger without limit as x approaches $\pi/2$. At $x = \pi/2$ the ratio is not defined. For x slightly greater than $\pi/2$ the ratio is negative, since the ordinate is positive and the abscissa is negative, but its absolute value is large. At $x = 3\pi/4$ the ratio is -1, and at π, it is back to 0. Since $\tan x$ has period π, we can draw its graph between 0 and π and duplicate this in each succeeding interval of length π. Similarly, we can extend it to the left. With this information and our knowledge of $\tan x$ for special values of x such as $\pi/6$ and $\pi/3$, we can draw the graph, as in Figure 35. The vertical lines at odd multiples of $\pi/2$ are asymptotes.

The graph of $y = \cot x$ can be similarly obtained, and we show its graph in Figure 36.

According to Definition 3, $\sec x$ is the reciprocal of the abscissa of $P(x)$. So

$$\sec x = \frac{1}{\cos x}$$

Since $|\cos x| \leq 1$, it follows that $|\sec x| \geq 1$. The signs of $\sec x$ and $\cos x$ are the same. When $\cos x = 1$, $\sec x = 1$, and when $\cos x = 0$, $\sec x$ is undefined. This latter occurs at $x = \pi/2$, $3\pi/2$, $-\pi/2$, $5\pi/2$, and so on. As x gets close to $\pi/2$, $\cos x$ gets close to 0; so $\sec x$ gets arbitrarily large. Using this analysis, together with the values of $\sec x$ for $x = \pi/6$, $\pi/4$, $\pi/3$, and so on, we sketch the graph in Figure 37. The graph of $y = \csc x$ can be similarly obtained (Figure 38).

Both $\sec x$ and $\csc x$ have a fundamental period of 2π. No amplitude is defined for them. They each have asymptotes; for the secant they are the lines $x = \pi/2$, $-\pi/2$, $3\pi/2$, and so on, and for the cosecant the lines $x = 0$, π, $-\pi$, 2π, and so on.

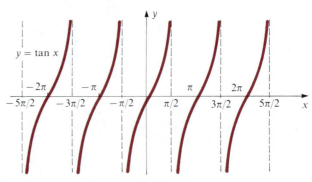

FIGURE 35

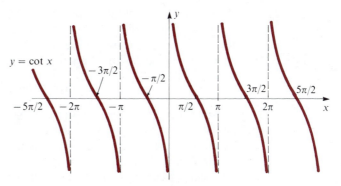

FIGURE 36

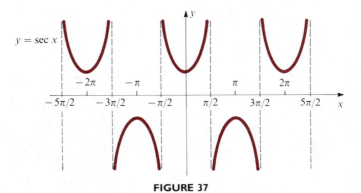

FIGURE 37

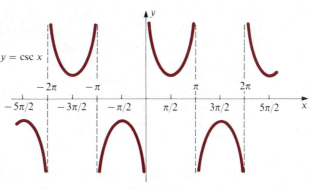

FIGURE 38

Now that we know the graphs of the basic trigonometric functions, we can use the ideas discussed in Section 3 of Chapter 3 to obtain graphs formed by vertical or horizontal translations, stretching or shrinking, and reflections. For example, let us consider the graph of

$$y = a \sin bx$$

where a and b are positive constants. The factor a causes a vertical stretching or compression, depending on whether $a > 1$ or $0 < a < 1$. In particular, when the basic sine curve attains its maximum height 1, the new curve attains the height a. So the amplitude of the new curve is a.

The factor b causes a horizontal stretching if $0 < b < 1$ and a compression if $b > 1$. By periodicity we know that $\sin(bx + 2\pi) = \sin bx$. We can write the left-hand side of this equation as $\sin b(x + 2\pi/b)$. So we have

$$\sin b\left(x + \frac{2\pi}{b} \right) = \sin bx$$

Thus, the graph of $y = a \sin bx$ repeats every $2\pi/b$ units. That is, the period is $2\pi/b$. Figure 39 illustrates a typical graph of the form $y = a \sin bx$.

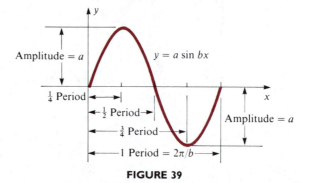

FIGURE 39

If in the equation $y = a \sin bx$ the number a is negative, then we know that the graph is reflected in the x axis. If $b < 0$, we use the fact that $\sin bx = -\sin(-bx)$, so that we again get a reflection. For example, the graph of $y = 2 \sin(-3x)$ is the same as that of $y = -2 \sin 3x$, which is the reflection of $y = 2 \sin 3x$ in the x axis. Of course, if $a < 0$ and $b < 0$, there are two reflections, and so no net change.

If we now replace x by $x - h$ and y by $y - k$, then we obtain a translation of the graph h units horizontally and k units vertically. The graph of $y - k = a \sin b(x - h)$, or equivalently,

$$y = a \sin b(x - h) + k \qquad (5)$$

for $a > 0$ and $b > 0$ is shown in Figure 40. The value h is referred to as the **phase shift.**

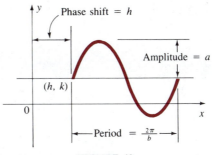

FIGURE 40

Remark. When the variable x represents time, the period is the time required to complete one cycle. The reciprocal of this, called the **frequency,** gives the number of cycles (or fractions of a cycle) completed per unit of time. Thus,

$$\text{Frequency} = \frac{1}{\text{Period}} = \frac{b}{2\pi}$$

This is a widely used concept in electronics.

A similar analysis holds for the cosine function, and a typical graph of an equation of the form $y - k = a \cos b(x - h)$ is shown in Figure 41. Because the

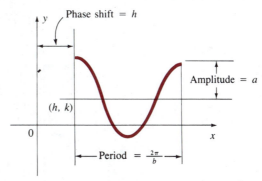

FIGURE 41

cosine is an even function, $\cos bx = \cos(-bx)$, so if b is negative, the graph is unaffected. Using the basic graphs of the other functions, we can modify them in similar ways. Note that the period of the curve $y = a \tan bx$ is π/b, and there are vertical asymptotes at odd multiples of $\pi/2b$. Amplitude is not defined for these other graphs.

EXAMPLE 16 ‖‖

Sketch the graph of $y = 3 \sin \frac{1}{2}x$.

Solution   The amplitude is 3 and the period is $2\pi \div \frac{1}{2} = 4\pi$. The graph is shown in Figure 42.

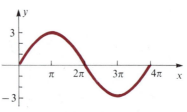

FIGURE 42   ▲

EXAMPLE 17   ııı

Sketch the graph of $y = 2 \sin(3x - \pi) + 1$.

Solution   We first factor out the 3 to put this in the form of equation (5):

$$y = 2 \sin 3\left(x - \frac{\pi}{3}\right) + 1$$

This is therefore a sine curve with amplitude 2, period $2\pi/3$, and phase shift $\pi/3$ units to the right, translated 1 unit upward. The graph is shown in Figure 43.

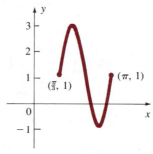

FIGURE 43   ▲

EXAMPLE 18   ııı

Sketch the graph of $y = \cos[2x + (\pi/3)]$.

Solution   We first write the equation in a form analogous to equation (23):

$$y = \cos 2\left(x + \frac{\pi}{6}\right) = \cos 2\left[x - \left(-\frac{\pi}{6}\right)\right]$$

So the phase shift h is $-\pi/6$. The curve is therefore shifted $\pi/6$ units to the left. The amplitude is 1, and the period is $2\pi/2 = \pi$ (Figure 44).

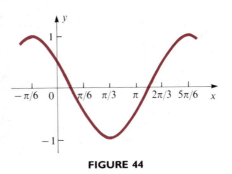

FIGURE 44

EXAMPLE 19

Sketch $y = 2 \tan \frac{1}{2}x$.

Solution   There is no amplitude defined for the tangent curve, but the effect of the coefficient 2 is to multiply all y values of the basic curve by 2; in particular, when the basic curve is 1 unit high, the new curve will be 2 units high. Since the fundamental period of the basic tangent curve is π, the period of this curve is $\pi \div \frac{1}{2} = 2\pi$. We show two complete cycles in Figure 45.

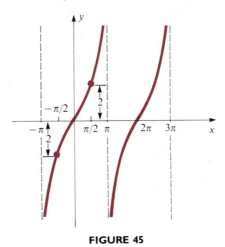

FIGURE 45

EXERCISE SET 6

A

In Problems 1–25 sketch one cycle of each of the curves. Give the period and, where appropriate, the amplitude and phase shift.

1. $y = \sin 2x$

2. $y = \cos 3x$

3. $y = 2 \sin x$
4. $y = 2 \sin 3x$
5. $y = 3 \cos(\pi x/2)$
6. $y = 4 \sin(\pi x/3)$
7. $y = 2 \sin(-3x)$
8. $y = 3 \cos(-\pi x)$
9. $y = -\sin(x/2)$
10. $y = -2 \cos x$
11. $y = \sin 2x + 1$
12. $y = 2 \cos x - 3$
13. $y = 2 \sin 3x - 1$
14. $y = 3 \cos \pi x + 2$

15. $y = \frac{1}{2} \tan \pi x$

16. $y = 2 \cot(x/2)$

17. $y = \sin\left(x - \frac{\pi}{3}\right)$

18. $y = \cos\left(x - \frac{\pi}{4}\right)$

19. $y = \sin 2\left(x + \frac{\pi}{4}\right)$

20. $y = \cos 3\left(x + \frac{\pi}{3}\right)$

B

21. $y = 2 \sin(3x + 2)$

22. $y = 3 \cos(3 - 2x)$

23. $y = 1 - 2 \sin\left(\pi x - \frac{\pi}{4}\right)$

24. $y = \frac{1}{2} \tan\left(2x - \frac{\pi}{2}\right)$

25. $y = 2 + 2 \cos\left(3x - \frac{\pi}{2}\right)$

26. Discuss the effect on the basic secant curve of introducing positive constants a and b to obtain $y = a \sec bx$.

27. Sketch $y = 2 \sec(x/3)$.

C (Graphing calculator)

28. Refer to the problem at the opening of this chapter concerning average temperatures in Fairbanks, Alaska. Obtain the graph of the equation given there on your calculator. Give the smallest positive x values that produce the coldest and hottest days (to the nearest integer). What are these coldest and hottest temperatures?

29. Graph each of the following, and discuss your results.

a. $y = \dfrac{\sin x}{x}$ $(x > 0)$

b. $y = \sin \dfrac{1}{x}$ $(x \neq 0)$

c. $y = (\sin x)^2$ **d.** $y = \sin x^2$

e. $y = x \sin \dfrac{1}{x}$ $(x \neq 0)$

 7 **THE INVERSE TRIGONOMETRIC FUNCTIONS**

Let us look again at the graph of $y = \sin x$ (Figure 46). It is immediately evident that this is not a one-to-one function, so it has no inverse. However, by a suitable restriction on the domain of the sine, an inverse can be found. The standard choice is to restrict x so that $-\pi/2 \leq x \leq \pi/2$. The sine curve with this domain will be called the **principal part** of the sine curve.

If x is so restricted, then for each y such that $-1 \leq y \leq 1$, there is exactly one

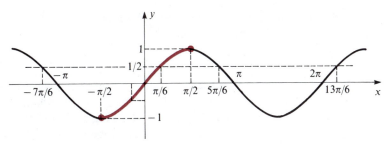

FIGURE 46

x such that $y = \sin x$. For example, if we take $y = \frac{1}{2}$, we get $x = \pi/6$. On the other hand, if $y = -\frac{1}{2}$, then $x = -\pi/6$. The equation $y = \sin x$ expresses y in terms of x; we would like to solve this equation for x in terms of y. Unfortunately, we as yet have no way of doing this. With the aid of the graph (or a calculator), we can find x for any given y, but we have no simple equation that expresses x in terms of y. What we do is invent a symbolism,

$$x = \sin^{-1} y$$

which is read "x is the inverse sine of y." Actually, we should probably read it as the "*principal* inverse sine of y," since it is the principal part of the sine curve that is used in finding x. Unless otherwise stated, we will in the future understand that $\sin^{-1} y$ means the principal value. This symbolism is consistent with the use of f^{-1} to represent the inverse of f.

There is another symbol in wide use that means the same thing as $\sin^{-1} y$; it is **arcsin y,** read "arcsine of y." Its origin probably lies in the length of arc on a unit circle used in the definition of the trigonometric functions. Thus, arcsin $\frac{1}{2}$ might be interpreted as "the length of arc on the unit circle for which the sine is $\frac{1}{2}$." We know the length in this case is $\pi/6$, so

$$\sin \frac{\pi}{6} = \frac{1}{2} \qquad \text{and} \qquad \frac{\pi}{6} = \arcsin \frac{1}{2}$$

are two ways of viewing the same fact. The first says that the sine of the number $\pi/6$ is $\frac{1}{2}$. The second says that $\pi/6$ is the number whose sine is $\frac{1}{2}$.

In what follows we will use both notations for the inverse sine (as well as inverses of the other trigonometric functions), since both are in wide use. Let us consider some examples.

EXAMPLE 20 ||||||||||||||||||||||||||||||||||||||||||||||||||||||||||||||||||||||||||||||||||||||||||||||||||||||||||||||||||

Find the value of

a. $\sin^{-1} 1$ **b.** $\sin^{-1}(-\frac{1}{2})$ **c.** $\sin^{-1} 0$

Solution   **a.** $\sin^{-1} 1 = \pi/2$, since $\pi/2$ is that value of x on the principal part of the sine curve whose sine is 1.

b.
$$\sin^{-1}(-\tfrac{1}{2}) = -\pi/6$$

It is important here to note that the answer is not $11\pi/6$, though an angle of $11\pi/6$ radians and an angle of $-\pi/6$ radians are coterminal. The distinction is most clearly seen by looking at the graph of the sine curve. On the x axis, $-\pi/6$ certainly is not the same point as $11\pi/6$, and even though the sine of each is $-\tfrac{1}{2}$, we choose $-\pi/6$ since it falls within the restricted range on x that defines the principal part of the sine curve.

c.
$$\sin^{-1} 0 = 0 \qquad\qquad \blacktriangle$$

EXAMPLE 21 ||||||||||||||||||||||||||||||||||||||||||||||||||||||||||||||||||||||||||||||||||||||||||||||||||||||||||||||||||||||||||||||||||

Evaluate

a. $\arcsin \dfrac{\sqrt{3}}{2}$ **b.** $\arcsin\left(-\dfrac{1}{\sqrt{2}}\right)$ **c.** $\arcsin(-1)$

Solution   **a.** $\arcsin \dfrac{\sqrt{3}}{2} = \dfrac{\pi}{3}$ **b.** $\arcsin\left(-\dfrac{1}{\sqrt{2}}\right) = -\dfrac{\pi}{4}$ **c.** $\arcsin(-1) = -\dfrac{\pi}{2}$

$$\blacktriangle$$

To plot the graph of the inverse of the principal part of the sine curve, we follow our usual procedure with inverses of solving for x and then interchanging x and y, giving

$$y = \sin^{-1} x$$

Now, since the roles of x and y have been switched, we must restrict x so that $-1 \le x \le 1$ and y will fall in the range $-\pi/2 \le y \le \pi/2$. The graph is just the reflection of the principal part of the sine curve in the line $y = x$ (Figure 47). Any value of x between -1 and 1 corresponds to a unique value of y between $-\pi/2$ and $\pi/2$. When x is positive, y is positive, and when x is negative, y is negative. Notice that $y = \sin^{-1} x$ if and only if $\sin y = x$ and $-\pi/2 \le y \le \pi/2$. So when we wish to evaluate $\sin^{-1} x$ for a particular x, we ask what number (or what angle in radians) between $-\pi/2$ and $\pi/2$ has x as its sine.

We consider next the inverse of the tangent function, since it has much in common with the sine. The principal part (or **principal branch**) of the graph of $y = \tan x$ is that portion between $-\pi/2$ and $\pi/2$ (Figure 48). On this portion, if any value of y is specified, a unique value of x is determined. Solving $y = \tan x$ for x leads to either

$$x = \tan^{-1} y \qquad \text{or} \qquad x = \arctan y$$

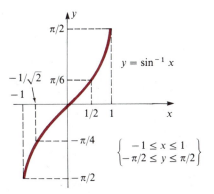

FIGURE 47

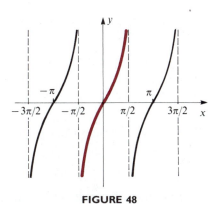

FIGURE 48

Again, we interchange the roles of x and y, and write

$$y = \tan^{-1} x \qquad \text{or} \qquad y = \arctan x$$

and restrict y such that $-\pi/2 < y < \pi/2$. The graph of this function is shown in Figure 49. With our knowledge of special angles, we conclude, for example, that

$$\tan^{-1} 1 = \frac{\pi}{4} \qquad\qquad \tan^{-1} \sqrt{3} = \frac{\pi}{3}$$

$$\tan^{-1}(-1) = -\frac{\pi}{4} \qquad\qquad \tan^{-1} 0 = 0$$

A consideration of the graph of the cosine function will make it clear that some different portion will have to be used as the principal part (Figure 50), because when x lies between $-\pi/2$ and $\pi/2$, y is always positive and hence does not assume all of its possible values. Furthermore, for a given y there is generally not a unique x determined. We choose instead the portion of the curve lying between 0 and π as the principal part. Then, for each y such that $-1 \le y \le 1$,

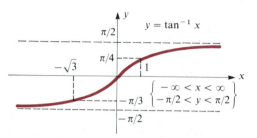

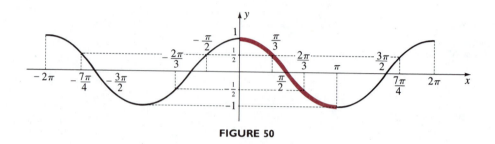

FIGURE 49

FIGURE 50

there is a unique x for which $y = \cos x$. Again, we express the dependence of x on y by writing

$$x = \cos^{-1} y \qquad \text{or} \qquad x = \operatorname{arccos} y$$

Interchanging the roles of x and y, we obtain the graph of $y = \cos^{-1} x$ (Figure 51). When x is positive, y is positive and between 0 and $\pi/2$; when x is negative, y is positive and between $\pi/2$ and π.

The inverse cotangent is seldom used, but is defined in much the same way as the inverse cosine. Its graph is shown in Figure 52.

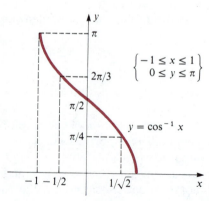

FIGURE 51

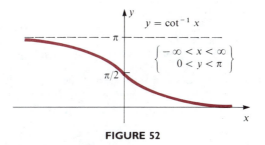

FIGURE 52

The secant and cosecant present certain problems, and there is no general agreement on what parts of the curves to invert, that is, which portions to define as the principal branches. Fortunately, this lack of agreement is not serious. The inverse cosecant is seldom used, and since it can always be circumvented, we will omit it entirely. The inverse secant is sufficiently useful to warrant consideration. For purposes of later use in calculus, there is some justification for the choice of principal parts made below. Let us consider first the graph of $y = \sec x$ (Figure 53). No single branch of the graph is suitable for defining the inverse function. While the reason is not now apparent, we select the highlighted portions in Figure 53 to define the inverse function. Thus, when $y \geq 1$, we take $0 \leq x < \pi/2$, and when $y \leq -1$, we take $\pi \leq x < 3\pi/2$. When we interchange the roles of x and y, we obtain the graph shown in Figure 54.

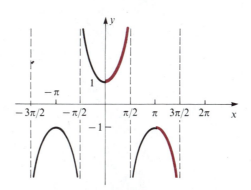

FIGURE 53

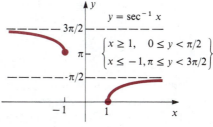

FIGURE 54

EXAMPLE 22 ||||||||||||||||||||||||||||||||||||||||||||||||||||||||||||||||||||||||||||||||||||||||||||||||||||||||

Evaluate:

a. $\tan[\sec^{-1}(-\frac{3}{2})]$ **b.** $\sin[\tan^{-1}(-\frac{1}{2})]$ **c.** $\cos^{-1}(\cos\frac{4\pi}{3})$

Solution **a.** Let $\theta = \sec^{-1}(-\frac{3}{2})$. Then θ is a third quadrant angle, since when x is nega-tive, $\pi \leq \sec^{-1} x < \frac{3\pi}{2}$. We can show θ as in Figure 55. By the Pythagorean theorem, $y = -\sqrt{5}$. Thus,

$$\tan\left[\sec^{-1}\left(-\frac{3}{2}\right)\right] = \tan\theta = \frac{y}{-2} = \frac{-\sqrt{5}}{-2} = \frac{\sqrt{5}}{2}$$

b. Let $\theta = \tan^{-1}(-\frac{1}{2})$. Then θ is a negative angle in the fourth quadrant, as shown in Figure 56. By the Pythagorean theorem, $r = \sqrt{5}$. So

$$\sin\left[\tan^{-1}\left(-\frac{1}{2}\right)\right] = \sin\theta = \frac{-1}{r} = -\frac{1}{\sqrt{5}}$$

c. At first glance it looks as though the answer should be $4\pi/3$, but this cannot be correct, since $\cos^{-1} x$ lies between 0 and π. Since $\cos 4\pi/3 = -\frac{1}{2}$, we have

$$\cos^{-1}\left(\cos\frac{4\pi}{3}\right) = \cos^{-1}\left(-\frac{1}{2}\right) = \frac{2\pi}{3}$$

since $2\pi/3$ is the only angle between 0 and π whose cosine is $-1/2$.

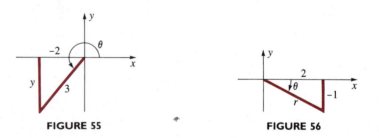

FIGURE 55   FIGURE 56   ▲

FINDING INVERSE TRIGONOMETRIC FUNCTIONS BY CALCULATOR

Calculators are programmed to give correct values of the inverses of the sine, cosine, and tangent. The appropriate key or keys to punch depends on your particular calculator. To get the inverse sine, for example, some calculators have a key labeled $\boxed{\text{SIN}^{-1}\ x}$. On most calculators, however, $\sin^{-1} x$ appears above the $\boxed{\text{SIN}\ x}$ key, and to activate it, you must first punch the $\boxed{\text{INV}}$ key, or the $\boxed{\text{2nd}}$ key (depending on the calculator). In the keystroke sequences we illustrate, we will indicate the inverse sine by $\boxed{\text{INV}}$ $\boxed{\text{SIN}\ x}$, but bear in mind that you will have to adapt this for your particular calculator. For example, to obtain $\sin^{-1}(-0.2357)$, we would set the calculator in the radian mode (unless there were a particular reason for wanting degrees) and use the keystroke sequence

$$.2357 \quad \boxed{+/-} \quad \boxed{\text{INV}} \quad \boxed{\text{SIN}}$$

The result, to four significant figures, is -0.2379. Notice that the answer satisfies the requirement that $-\pi/2 \le \sin^{-1} x \le \pi/2$. Similarly, we would obtain $\cos^{-1}(-0.2357) = 1.8087$ and $\tan^{-1}(-0.2357) = -0.2315$, and these satisfy $0 \le \cos^{-1} x \le \pi$ and $-\pi/2 < \tan^{-1} x < \pi/2$, respectively. For $x > 0$, we can obtain $\cot^{-1} x$, $\sec^{-1} x$, and $\csc^{-1} x$ by using the relationships

$$\cot^{-1} x = \tan^{-1} \frac{1}{x} \qquad \sec^{-1} x = \cos^{-1} \frac{1}{x} \qquad \csc^{-1} x = \sin^{-1} \frac{1}{x}$$

For example, to get $\sec^{-1}(1.732)$, we would use

$$1.7321 \quad \boxed{1/x} \quad \boxed{\text{INV}} \quad \boxed{\text{COS}} \quad \text{Display} \quad 0.955295875$$

So $\sec^{-1}(1.7321) = 0.9553$, to four significant figures. For $x < 0$, the following relationships must be used to get the values of $\cot^{-1} x$, $\sec^{-1} x$, and $\csc^{-1} x$ in the proper range:

$$\left. \begin{array}{l} \cot^{-1} x = \pi + \tan^{-1} \dfrac{1}{x} \\[2ex] \sec^{-1} x = 2\pi - \cos^{-1} \dfrac{1}{x} \\[2ex] \csc^{-1} x = \pi - \sin^{-1} \dfrac{1}{x} \end{array} \right\} \quad (x < 0)$$

You will be asked to verify these in the exercises (Problem 38). For example, to get $\cot^{-1}(-0.2357)$, we would use

$$.2357 \quad \boxed{+/-} \quad \boxed{1/x} \quad \boxed{\text{INV}} \quad \boxed{\text{TAN}} \quad \boxed{+} \quad \boxed{\pi} \quad \boxed{=} \quad \text{Display} \quad 1.802271549$$

So $\cot^{-1}(-0.2357)$ is approximately 1.8023, which satisfies the requirement that $0 < \cot^{-1} x < \pi$.

EXERCISE SET 7

In Problems 1–28 give the value without using a calculator.

1. **a.** $\sin^{-1}(-\sqrt{3}/2)$ **b.** $\arccos(-\sqrt{3}/2)$
2. **a.** $\tan^{-1}(-1)$ **b.** $\text{arcsec}(-2)$
3. **a.** $\text{arccot}(-1)$ **b.** $\sin^{-1} 1$
4. **a.** $\arccos 1$ **b.** $\tan^{-1} 1$

5. **a.** $\text{arcsec}\ 1$ **b.** $\sin^{-1}(-\tfrac{1}{2})$
6. **a.** $\arctan 0$ **b.** $\cos^{-1}(-1)$
7. **a.** $\arccos 0$ **b.** $\sec^{-1}(-2/\sqrt{3})$
8. **a.** $\sec^{-1}(-1)$ **b.** $\cot^{-1}(-\sqrt{3})$
9. $\sin^{-1}\left(-\dfrac{1}{2}\right) + \cos^{-1}\left(-\dfrac{1}{2}\right)$
10. $\arccos 0 - 2\arctan(-1)$

11. $2 \sin^{-1}(-1) + 3 \cos^{-1}(-1)$

12. $\cos \alpha$ if $\alpha = \sin^{-1}\left(-\dfrac{2}{3}\right)$

13. $\tan \theta$ if $\theta = \arccos\left(-\dfrac{1}{4}\right)$

14. $\sin x$ if $x = \sec^{-1}(-3)$

15. $\sec \theta$ if $\theta = \tan^{-1}\dfrac{3}{4}$

16. $\sin\left[\cos^{-1}\left(-\dfrac{1}{3}\right)\right]$
17. $\sec\left[\sin^{-1}\left(-\dfrac{1}{4}\right)\right]$

18. $\tan\left(\cos^{-1}\dfrac{3}{5}\right)$
19. $\cos\left[\arctan\left(-\dfrac{4}{3}\right)\right]$

20. $\sin\left(\tan^{-1}\dfrac{7}{12}\right)$
21. $\csc\left[\arccos\left(-\dfrac{2}{3}\right)\right]$

22. $\cos\left[\sin^{-1}\left(-\dfrac{1}{3}\right)\right]$
23. $\cot\left[\cos^{-1}\left(-\dfrac{7}{25}\right)\right]$

24. $\sin^{-1}(\sin \pi)$
25. $\tan^{-1}\left(\tan \dfrac{5\pi}{4}\right)$

26. $\sec^{-1}\left(\sec \dfrac{2\pi}{3}\right)$

27. $\cos[\arctan(-\sqrt{3}) + \text{arcsec}(-1)]$

28. $\tan[\csc^{-1}\sqrt{2} - 2\cos^{-1} 0]$

In Problems 29–34 use a calculator to evaluate each expression correct to five significant figures.

29. **a.** $\sin^{-1}\left(-\dfrac{2}{3}\right)$
b. $\tan^{-1}(-2)$

30. **a.** $\arccos\left(-\dfrac{2}{\sqrt{5}}\right)$
b. $\arcsin\left(-\dfrac{\sqrt{3}}{7}\right)$

31. **a.** $\arctan(1.2375)$
b. $\arccos(-0.38912)$

32. **a.** $\sin^{-1}(-0.87325)$
b. $\tan^{-1}(0.56732)$

B

33. **a.** $\sec^{-1}(-3)$
b. $\cot^{-1}\left(-\dfrac{4}{3}\right)$

34. **a.** $\csc^{-1}(-2.1347)$
b. $\text{arcsec}(-1.7829)$

35. Show that for $x \geq 0$,
$$\sin^{-1} x + \sin^{-1}\sqrt{1-x^2} = \dfrac{\pi}{2}.$$
Hint. Draw a figure.

36. Show that for $x < 0$,
$$\cos^{-1} x + \sin^{-1}\sqrt{1-x^2} = \pi$$

37. Show that
a. $\sin(\pi - \sin^{-1} x) = x$
b. $\cos(\pi - \cos^{-1} x) = -x$

38. Prove each of the following for $x < 0$:

a. $\cot^{-1} x = \pi + \tan^{-1}\dfrac{1}{x}$

b. $\sec^{-1} x = 2\pi - \cos^{-1}\dfrac{1}{x}$

c. $\csc^{-1} x = \pi - \sin^{-1}\dfrac{1}{x}$

In Problems 39–41 draw the graph.

39. $y = \dfrac{1}{2}\sin^{-1} 2(x+1)$

40. $y = \dfrac{\pi}{4} + \tan^{-1} 3x$

41. $y = 2\cos^{-1}(x-1) + \dfrac{\pi}{2}$

C (Graphing calculator)

42. By graphing each pair of curves and zooming in repeatedly, find the values of x correct to four decimal places for which $\sin^{-1} x = (\sin x)^{-1}$ and for which $\cos^{-1} x = (\cos x)^{-1}$. How many values of x are there for which $\tan^{-1} x = (\tan x)^{-1}$? Find the smallest such positive value correct to four decimal places.

REVIEW EXERCISE SET

A

1. Give the measure of the angle in radians:
   a. 240°   **b.** 315°   **c.** 15°
   d. 540°   **e.** 160°
2. Give the measure of the angle in degrees:
   a. $5\pi/6$ **b.** $2\pi/9$ **c.** 3π
   d. 5   **e.** $7\pi/4$
3. **a.** What is the length of an arc on a circle of radius 20 subtended by a central angle of 36°?
   b. An arc of 8 feet is subtended by a central angle θ on a circle of radius 6 feet. Find θ in degrees.
4. A wheel 18 inches in diameter is turning at the rate of 600 rpm. What is the speed in feet per second of a point on the extremity of the wheel?

In Problems 5 and 6 triangle ABC is a right triangle with right angle at C and sides a, b, c opposite the corresponding angles. Find all six trigonometric functions of the indicated angle.

5. **a.** $a = 4$, $c = 8$; A
   b. $a = 8$, $b = 15$; B
6. **a.** $a = 5$, $b = 3$; B
   b. $b = 24$, $c = 25$; A

In Problems 7 and 8 give the exact value of each function.

7. **a.** $\cos(5\pi/3)$ **b.** $\sin 210°$
   c. $\tan(-5\pi/6)$ **d.** $\cot 270°$
   e. $\sec(3\pi/4)$ **f.** $\csc 390°$
   g. $\tan(5\pi/4)$ **h.** $\sin(-90°)$
   i. $\sec 480°$ **j.** $\cos(-7\pi/6)$
8. **a.** $\sin(3\pi/2)$ **b.** $\cos(11\pi/6)$
   c. $\tan 225°$ **d.** $\sec(-3\pi)$
   e. $\cot(5\pi/3)$ **f.** $\sin 390°$
   g. $\cos(4\pi/3)$ **h.** $\tan(-315°)$
   i. $\sec(5\pi/4)$ **j.** $\csc(7\pi/6)$
9. Find all trigonometric functions of θ if
   a. $P(\theta) = \left(-\dfrac{2}{3}, \dfrac{\sqrt{5}}{3}\right)$
   b. $P(\theta) = \left(-\dfrac{12}{13}, -\dfrac{5}{13}\right)$

10. Give the coordinates $P(\theta)$ for each of the following values of θ:
    a. $\theta = -\dfrac{\pi}{2}$ **b.** $\theta = \dfrac{7\pi}{3}$
    c. $\theta = 10\pi$ **d.** $\theta = -\dfrac{7\pi}{6}$
11. Draw a unit circle and show on it the points $P\left(\dfrac{n\pi}{6}\right)$ and $P\left(\dfrac{m\pi}{4}\right)$ for $n = 0, 1, 2, \ldots, 11$ and $m = 1, 3, 5,$ and 7, along with the coordinates of each point.
12. Use a calculator to find the value of each of the trigonometric functions at x, correct to 4 decimal places:
    a. $x = \dfrac{5\pi}{7}$ **b.** $x = 31.7$
    c. $x = -0.2135$ **d.** $x = \dfrac{13}{15}$

In Problems 13–17 evaluate without using a calculator.

13. **a.** $\arccos\left(-\dfrac{1}{2}\right)$ **b.** $\tan^{-1}(-\sqrt{3})$
    c. $\arcsin(-\sqrt{3}/2)$ **d.** $\cos^{-1}(-1)$
    e. $\tan^{-1}(-1)$
14. **a.** $\sin\left[\tan^{-1}\left(-\dfrac{2}{3}\right)\right]$
    b. $\cos\left[\sin^{-1}\left(-\dfrac{2}{3}\right)\right]$
15. **a.** $\sec[\arctan(-2)]$ **b.** $\csc\left[\cos^{-1}\left(-\dfrac{1}{4}\right)\right]$
16. **a.** $\sin^{-1}\left(\sin\dfrac{5\pi}{6}\right)$ **b.** $\cos^{-1}\left[\cos\left(-\dfrac{\pi}{3}\right)\right]$
17. **a.** $3\cos^{-1}\left(-\dfrac{\sqrt{3}}{2}\right) + 2\sin^{-1}\left(-\dfrac{\sqrt{3}}{2}\right)$
    b. $2\tan^{-1}(-\sqrt{3}) - \sec^{-1}(-2)$

In Problems 18–23 draw the graph.

18. **a.** $y = 2\sin\left(\dfrac{x}{2}\right) + 1$ **b.** $y = 3\cos 2x$

19. **a.** $y = 2 \tan \dfrac{\pi x}{4}$ **b.** $y = 2 \cos \dfrac{\pi x}{2} + 3$

20. **a.** $y = \sin\left(3x + \dfrac{\pi}{2}\right)$ **b.** $y = 2 \cos\left(\pi x - \dfrac{\pi}{3}\right)$

B

21. $y = 3 - 2 \cos(3 - 2x)$

22. $y = \dfrac{1}{3} \sin^{-1} \dfrac{x}{2} + \dfrac{\pi}{4}$

23. $y = 2 \arctan(x - 1) + \dfrac{\pi}{2}$

24. If $\theta = \sin^{-1} x$, express all six trigonometric functions in terms of x, giving restrictions that must be observed. Take into consideration both positive and negative values of x.

25. Repeat Problem 24 with $\theta = \cos^{-1} x$.

26. Repeat Problem 24 with $\theta = \sec^{-1} x$.

27. Explain what restrictions must be placed on x for the equation to be true in each of the following:
    a. $\sin(\sin^{-1} x) = x$ **b.** $\sin^{-1}(\sin x) = x$
    c. $\cos(\cos^{-1} x) = x$ **d.** $\cos^{-1}(\cos x) = x$
    e. $\tan(\tan^{-1} x) = x$ **f.** $\tan^{-1}(\tan x) = x$

28. Find the domain and range of
    a. $f(x) = \tan^{-1}(\sin x)$
    b. $g(x) = \sec(\tan^{-1} x)$

29. Let $\theta = \sin^{-1}(-\sqrt{1 - x^2})$, with $|x| \le 1$. Find all six trigonometric functions of θ. Specify any restrictions.

30. Use a calculator to find the specified variables, correct to 4 significant digits.
    a. $P(\theta) = (x, -0.8764)$, $x > 0$;   find x and θ for $0 \le \theta < 2\pi$.
    b. $P(\theta) = (-0.7054, y)$, $y < 0$;   find y and θ for $-\pi < \theta \le \pi$.

8

TRIGONOMETRIC IDENTITIES AND EQUATIONS

One of the important uses of trigonometry in calculus involves what are called **trigonometric identities.** The following example from a calculus textbook illustrates a typical use of such an identity.*

Example Find the length of the arc of the parabola $y^2 = x$ from $(0, 0)$ to $(1, 1)$.

Solution Since $x = y^2$, $dx/dy = 2y$, and

$$L = \int_0^1 \sqrt{1 + \left(\frac{dx}{dy}\right)^2}\, dy = \int_0^1 \sqrt{1 + 4y^2}\, dy$$

We make the trigonometric substitution $y = \frac{1}{2}\tan\theta$, which gives $dy = \frac{1}{2}\sec^2\theta\, d\theta$ and $\sqrt{1 + 4y^2} = \sqrt{1 + \tan^2\theta} = \sec\theta$. When $y = 0$, $\tan\theta = 0$, so $\theta = 0$; when $y = 1$, $\tan\theta = 2$, so $\theta = \tan^{-1} 2 = \alpha$, say. Thus

$$L = \int_0^\alpha \sec\theta \cdot \frac{1}{2}\sec^2\theta\, d\theta = \frac{1}{2}\int_0^\alpha \sec^3\theta\, d\theta$$

$$= \frac{1}{2}\cdot\frac{1}{2}\left[\sec\theta\tan\theta + \ln|\sec\theta + \tan\theta|\right]_0^\alpha$$

$$= \frac{1}{4}(\sec\alpha\tan\alpha + \ln|\sec\alpha + \tan\alpha|)$$

Since $\tan\alpha = 2$, we have $\sec^2\alpha = 1 + \tan^2\alpha = 5$, so $\sec\alpha = \sqrt{5}$ and

$$L = \frac{\sqrt{5}}{2} + \frac{\ln(\sqrt{5} + 2)}{4}$$

The equation $\sec^2\alpha = 1 + \tan^2\alpha$ is one of the many identities we will study in this chapter. Notice, too, the use of the inverse tangent and of the natural logarithm.

---

* From *Calculus*, 2nd. ed., by James Stewart. Copyright © 1991, 1987 by Wadsworth, Inc. Reprinted by permission of Brooks/Cole Publishing Company, Pacific Grove, CA 93950.

BASIC TRIGONOMETRIC IDENTITIES

A **trigonometric identity** is an equation relating trigonometric functions that holds true for all admissible values of the variable involved. For example, we noted in the previous chapter that $\tan x = \sin x/\cos x$ and that $\sec x = 1/\cos x$. These are immediate consequences of the definition (either Definition 2 or Definition 3 of Chapter 7), and they hold true for all values of x except those for which $\cos x = 0$. That is, they hold true for all values of x in the domains of the tangent and secant, respectively. In the following table we summarize these and other identities that come directly from the definitions of the trigonometric functions. Because of customary usage, we use θ again as the independent variable.

$$\sec \theta = \frac{1}{\cos \theta} \qquad \tan \theta = \frac{\sin \theta}{\cos \theta}$$

$$\csc \theta = \frac{1}{\sin \theta} \qquad \cot \theta = \frac{\cos \theta}{\sin \theta}$$

$$\cot \theta = \frac{1}{\tan \theta}$$

We refer to the first three of these as **reciprocal identities** and the last two as **ratio identities.** In these, as in other identities, we will understand that they are valid for those values of θ for which all functions involved are defined, and only for those values. One place where the reciprocal identities are needed is on the calculator, since most calculators do not have keys for $\sec x$, $\csc x$, or $\cot x$. For example, to get $\sec x$, you would enter the value of x, then punch in succession the keys $\boxed{\cos x}$ $\boxed{1/x}$, which gives $1/\cos x$, and this is $\sec x$.

The Pythagorean theorem is the basis for another important set of identities. Using the theorem, we have, for (x, y) on the unit circle (see Figure 1),

$$x^2 + y^2 = 1$$

Since $x = \cos \theta$ and $y = \sin \theta$, we obtain the very important identity

$$\sin^2 \theta + \cos^2 \theta = 1 \tag{1}$$

where we have followed the convention of writing $\sin^2 \theta$ to mean $(\sin \theta)^2$ and similarly for $\cos^2 \theta$. More generally, for $n \geq 2$, we will write $\sin^n \theta$ instead of $(\sin \theta)^n$, and similarly for powers of the other trigonometric functions.

If we divide both sides of equation (1) by $\cos^2 \theta$, assuming $\cos \theta \neq 0$, we get

$$\left(\frac{\sin \theta}{\cos \theta}\right)^2 + 1 = \left(\frac{1}{\cos \theta}\right)^2$$

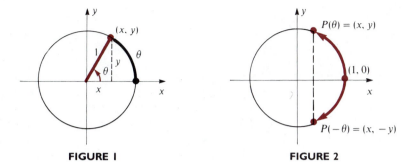

FIGURE I   **FIGURE 2**

Since $(\sin\theta)/(\cos\theta) = \tan\theta$ and $(1/\cos\theta) = \sec\theta$, we can write this in the form

$$\tan^2\theta + 1 = \sec^2\theta \qquad (2)$$

Similarly, if we divide both sides of equation (1) by $\sin^2\theta$ and use the identities $(\cos\theta)/(\sin\theta) = \cot\theta$ and $1/(\sin\theta) = \csc\theta$, we get (verify)

$$1 + \cot^2\theta = \csc^2\theta \qquad (3)$$

We collect equations 1–3 together for reference.

$$\sin^2\theta + \cos^2\theta = 1$$
$$1 + \tan^2\theta = \sec^2\theta$$
$$1 + \cot^2\theta = \csc^2\theta$$

These are known as the **Pythagorean identities** and should be learned. Notice the similarity between the last two. In applying these, it may be necessary to write them in various equivalent forms, for example, $\cos^2\theta = 1 - \sin^2\theta$ or $\sec^2\theta - 1 = \tan^2\theta$, but it is probably best to concentrate on learning them in just one form, such as the one given. You can then mentally rewrite them in various ways.

From Figure 2 we see by symmetry that if the coordinates of $P(\theta)$ are (x, y), then the coordinates of $P(-\theta) = (x, -y)$. But $P(\theta) = (\cos\theta, \sin\theta)$ and $P(-\theta) = (\cos(-\theta), \sin(-\theta))$. Thus, $\cos(-\theta) = \cos\theta$ and $\sin(-\theta) = -\sin\theta$. That is, the cosine function is even, and the sine function is odd. Now we can use the ratio and reciprocal identities to determine that the tangent, cotangent, and cosecant are odd, and that the secant is even. These results are summarized in the box. They hold for all real numbers θ for which the functions are defined.

$\sin(-\theta) = -\sin\theta$	$\csc(-\theta) = -\csc\theta$
$\cos(-\theta) = \cos\theta$	$\sec(-\theta) = \sec\theta$
$\tan(-\theta) = -\tan\theta$	$\cot(-\theta) = -\cot\theta$

EXAMPLE 1

Make use of the basic identities of this section to find the other five trigonometric functions of θ if $\tan \theta = -2$ and $\cos \theta < 0$.

Solution   First, we observe that $\cot \theta = 1/\tan \theta = -\frac{1}{2}$. Next, since $\sec^2 \theta = 1 + \tan^2 \theta = 1 + (-2)^2 = 5$, we have $\sec \theta = \pm\sqrt{5}$. But $\cos \theta$ is negative, so that $\sec \theta$, which is $1/\cos \theta$, is also negative. Thus, $\sec \theta = -\sqrt{5}$. Therefore, $\cos \theta = -1/\sqrt{5}$.

Now we use $\sin^2 \theta = 1 - \cos^2 \theta = 1 - \frac{1}{5} = \frac{4}{5}$. Since $\tan \theta$ and $\cos \theta$ are both negative, and $\tan \theta = \sin \theta/\cos \theta$, it follows that $\sin \theta$ is positive. Thus, $\sin \theta = 2/\sqrt{5}$. Finally, $\csc \theta = 1/\sin \theta = \sqrt{5}/2$.

Note. We could also have determined the signs of each of the functions by observing that for both $\tan \theta$, which is y/x, and $\cos \theta$, which is x, to be negative, we must have y positive, so that the point $P(\theta)$ is in the second quadrant, or equivalently, the angle of θ radians terminates in the second quadrant.   ▲

EXAMPLE 2

In each of the following use the basic identities to show that the first expression can be transformed into the second:

a. $\dfrac{\sin \theta}{\tan \theta}$, $\cos \theta$ **b.** $\dfrac{\cos \theta}{\tan \theta}$, $\csc \theta - \sin \theta$

Solution   **a.**   Since $\tan \theta = \sin \theta/\cos \theta$, we have

$$\frac{\sin \theta}{\tan \theta} = \frac{\sin \theta}{\frac{\sin \theta}{\cos \theta}} = \sin \theta \cdot \frac{\cos \theta}{\sin \theta} = \cos \theta$$

b.

$$\frac{\cos \theta}{\tan \theta} = \cos \theta \cdot \frac{\cos \theta}{\sin \theta} = \frac{\cos^2 \theta}{\sin \theta}$$

Now we use the identities $\cos^2 \theta = 1 - \sin^2 \theta$ and $\csc \theta = 1/\sin \theta$ to get

$$\frac{\cos^2 \theta}{\sin \theta} = \frac{1 - \sin^2 \theta}{\sin \theta} = \frac{1}{\sin \theta} - \frac{\sin^2 \theta}{\sin \theta} = \csc \theta - \sin \theta$$ ▲

By using the basic identities we can prove a multitude of others, but fortunately it is not particularly useful to try to memorize any of the results. By **proving an identity,** we mean showing that the given equation is true for all admissible values of the variable or variables involved. The general procedure is to work only on one side of the equation and, by use of the basic identities, transform it so that in the final stage it is identical to the other side. Often the more complicated side is the better to work with, since it offers more obvious possibilities for alteration, but there are times in calculus when it is better to change a simpler expression into a more complicated one. We illustrate the procedure by means of several examples.

EXAMPLE 3 |||||||||||||||||||||||||||||||||||||||||||||||||||||||||||||||||||||||||||||||||||||||||||||||||||||||||||||||||||||

Prove the identity $\tan \theta + \cot \theta = \sec \theta \csc \theta$.

Solution   Perhaps it would be better to word the instructions, "Prove that the following is an identity." For we cannot prove it is true by assuming it is true, and this is an important point of logic often missed. This precludes, for example, working on both sides of the equation, unless we verify that each step is reversible. A proper approach is to begin with the left-hand side and try to obtain the right-hand side:

$$\tan \theta + \cot \theta = \frac{\sin \theta}{\cos \theta} + \frac{\cos \theta}{\sin \theta}$$

$$= \frac{\sin^2 \theta + \cos^2 \theta}{\cos \theta \sin \theta}$$

$$= \frac{1}{\cos \theta \sin \theta}$$

$$= \frac{1}{\cos \theta} \cdot \frac{1}{\sin \theta}$$

$$= \sec \theta \csc \theta$$

and now the given equation is verified. It is true for all admissible values of θ, and in this case this means all values except 0, $\pi/2$, $3\pi/2$, and any angles coterminal with these since, at each of these values, two of the given functions are undefined. Usually we will understand that such exceptions are necessary without mentioning them explicitly.   ▲

EXAMPLE 4 |||||||||||||||||||||||||||||||||||||||||||||||||||||||||||||||||||||||||||||||||||||||||||||||||||||||||||||||||||||

Prove the identity

$$\frac{1}{\sec x + 1} = \cot x \csc x - \csc^2 x + 1$$

Solution   It would be possible to transform the right-hand side into the left, and may even be easier, but we choose to work from the left-hand side because in later applications the right-hand side will be seen to be the more desirable final form. We begin with a commonly used method, that of multiplying numerator and denominator by the same factor.

$$\frac{1}{\sec x + 1} = \frac{1}{\sec x + 1} \cdot \frac{\sec x - 1}{\sec x - 1} = \frac{\sec x - 1}{\sec^2 x - 1}$$

The object here is to bring $\sec^2 x - 1$ into the picture, because this is, by one of the Pythagorean identities, equal to $\tan^2 x$. So

$$\frac{1}{\sec x + 1} = \frac{\sec x - 1}{\tan^2 x} = \frac{\sec x}{\tan^2 x} - \frac{1}{\tan^2 x}$$

$$= \frac{1}{\cos x} \cdot \frac{\cos^2 x}{\sin^2 x} - \cot^2 x$$

$$= \frac{\cos x}{\sin^2 x} - (\csc^2 x - 1)$$

$$= \frac{\cos x}{\sin x} \cdot \frac{1}{\sin x} - \csc^2 x + 1$$

$$= \cot x \csc x - \csc^2 x + 1 \qquad \blacktriangle$$

EXAMPLE 5

Prove that

$$\frac{2 \sin^3 x}{1 - \cos x} = 2 \sin x(1 + \cos x)$$

Solution

$$\frac{2 \sin^3 x}{1 - \cos x} = \frac{2 \sin x(1 - \cos^2 x)}{1 - \cos x}$$

$$= \frac{2 \sin x(1 + \cos x)(1 - \cos x)}{1 - \cos x}$$

$$= 2 \sin x(1 + \cos x) \qquad \blacktriangle$$

EXAMPLE 6

Prove the identity

$$\frac{\csc \theta - \cot \theta}{\sec \theta - 1} = \cot \theta$$

Solution

$$\frac{\csc \theta - \cot \theta}{\sec \theta - 1} = \frac{\dfrac{1}{\sin \theta} - \dfrac{\cos \theta}{\sin \theta}}{\dfrac{1}{\cos \theta} - 1} = \frac{\dfrac{1 - \cos \theta}{\sin \theta}}{\dfrac{1 - \cos \theta}{\cos \theta}}$$

$$= \frac{1 - \cos \theta}{\sin \theta} \cdot \frac{\cos \theta}{1 - \cos \theta} = \frac{\cos \theta}{\sin \theta}$$

$$= \cot \theta \qquad \blacktriangle$$

A word is in order about the utility of proving identities. In calculus, as well as in more advanced courses, we are frequently confronted with an unwieldy expression involving trigonometric functions. Often, by use of the basic identities, these can be transformed into more manageable forms. An example of this occurs in the excerpt taken from a calculus book at the opening of this chapter, where the quantity $\sqrt{1 + \tan^2 \theta}$ was replaced by $\sec \theta$. This is a consequence of the identity $1 + \tan^2 \theta = \sec^2 \theta$, so that $\sqrt{1 + \tan^2 \theta} = \sqrt{\sec^2 \theta} = \sec \theta$ (the last equality being valid since $\sec \theta > 0$ on the interval in question).

At other times one may have to try various combinations of basic identities in order to arrive at a suitable form. It would be more accurate to describe this procedure as *deriving* an identity. The value of proving identities already provided lies in the fact that it gives a goal to shoot for in transforming an expression. Without this and with little or no experience in such activity, a student may go in circles or change to a less desirable form.

 EXERCISE SET I   1-57 odd

A

In Problems 1–8 find the remaining five trigonometric functions of θ, using the basic identities of this section.

1. $\sin \theta = -\frac{3}{5}$, $\tan \theta > 0$
2. $\sec \theta = \frac{13}{12}$, $\csc \theta < 0$
3. $\tan \theta = -\frac{4}{3}$, $\pi/2 < \theta < 3\pi/2$
4. $\cot \theta = \frac{5}{12}$, $\pi < \theta < 2\pi$
5. $\cos \theta = 2/\sqrt{5}$, $\cot \theta < 0$
6. $\csc \theta = -\frac{17}{5}$, $\tan \theta > 0$
7. $\sin \theta = \frac{1}{3}$, $5\pi/2 < \theta < 7\pi/2$
8. $\sec \theta = 3$, $3\pi < \theta < 4\pi$

In Problems 9–18 show that the first expression can be transformed into the second, using basic identities.

9. $\dfrac{\tan \theta}{\sin \theta}$, $\sec \theta$
10. $\dfrac{\cot \theta}{\csc \theta}$, $\cos \theta$

11. $\dfrac{\sin \theta}{\cot \theta}$, $\sec \theta - \cos \theta$

12. $\sec^2 \theta \sin^2 \theta$, $\sec^2 \theta - 1$
13. $\cot \theta \sec \theta$, $\csc \theta$
14. $\tan \theta \csc \theta$, $\sec \theta$

15. $\dfrac{\sec \theta}{\csc \theta}$, $\tan \theta$
16. $1 - \dfrac{\sin \theta}{\csc \theta}$, $\cos^2 \theta$

17. $\dfrac{1}{\sec^2 \theta} + \dfrac{1}{\csc^2 \theta}$, 1

18. $\sec^2 \theta - \sin^2 \theta \sec^2 \theta$, 1

In Problems 19–67 prove that the given equation is an identity by transforming the left-hand side into the right-hand side.

19. $\dfrac{1 + \sin \theta}{\tan \theta} = \cos \theta + \cot \theta$

20. $\dfrac{\sec \theta - \cos \theta}{\tan \theta} = \sin \theta$

21. $\dfrac{\sin x \cot x}{\cos x \csc x} = \sin x$

22. $\dfrac{\sin^2 x - 1}{\cos^2 x - 1} = \csc^2 x - 1$

23. $(1 + \cot^2 \theta) \tan^2 \theta = \sec^2 \theta$
24. $(\tan x + \cot x)^2 = \sec^2 x + \csc^2 x$
25. $\sec x - \sin x \tan x = \cos x$

26. $1 - \dfrac{\tan \theta \cos \theta}{\csc \theta} = \cos^2 \theta$

27. $(\sin \phi - \cos \phi)^2 = 1 - 2 \sin \phi \cos \phi$

28. $\dfrac{1 - \sec^2 \alpha}{\cos^2 \alpha - 1} = \sec^2 \alpha$

29. $\sin \theta \tan \theta = \sec \theta - \cos \theta$
30. $\sin \theta + \cos \theta \cot \theta = \csc \theta$
31. $(\tan^2 \theta + 1)(\cos^2 \theta - 1) = 1 - \sec^2 \theta$
32. $\csc \theta - \cos \theta \cot \theta = \sin \theta$
33. $\sin^4 x - \cos^4 x = 2 \sin^2 x - 1$

34. $\dfrac{(1 - \tan \theta)^2}{\sec^2 \theta} = 1 - 2 \sin \theta \cos \theta$

35. $\dfrac{\tan \theta + 1}{\sec \theta + \csc \theta} = \sin \theta$

36. $\dfrac{\cos x \cot x}{\csc x} - 1 = -\sin^2 x$

37. $\dfrac{1}{\tan \theta + \cot \theta} = \sin \theta \cos \theta$

38. $\dfrac{\cos \theta(\tan \theta - \sec \theta)}{1 - \csc \theta} = \sin \theta$

39. $\dfrac{\sec^2 x}{1 + \cot^2 x} = \tan^2 x$

40. $\dfrac{\sec x + \tan x}{\sec^2 x} = \cos x(1 + \sin x)$

41. $(1 + \csc \theta)(\sec \theta - \tan \theta) = \cot \theta$

42. $\dfrac{1 - \cos x}{\sin x} + \dfrac{\sin x}{1 - \cos x} = 2 \csc x$

43. $\tan^2 x - \sin^2 x = \tan^2 x \sin^2 x$

44. $\dfrac{1}{1 - \sin x} + \dfrac{1}{1 + \sin x} = 2 \sec^2 x$

45. $\dfrac{\tan^2 x - 1}{\sin^2 x} = \sec^2 x - \csc^2 x$

46. $\cot^4 \theta - 1 = \cot^2 \theta \csc^2 \theta - \csc^2 \theta$

47. $\sec^4 x = \tan^2 x \sec^2 x + \sec^2 x$

48. $\dfrac{\tan^2 \alpha - \sin^2 \alpha}{\sec^2 \alpha - 1} = \sin^2 \alpha$

49. $\dfrac{1 + \tan x}{1 + \cot x} = \tan x$

50. $\dfrac{\sin x}{\cot x + \csc x} = 1 - \cos x$

51. $\dfrac{1}{\sec x - \tan x} = \sec x + \tan x$

52. $\tan^4 x = \sec^2 x \tan^2 x - \sec^2 x + 1$

53. $\dfrac{\sin x}{1 - \cos x} - \dfrac{1 - \cos x}{\sin x} = 2 \cot x$

54. $\dfrac{(\tan^2 \theta - 1) \cot \theta}{\sin \theta - \cos \theta} = \sec \theta + \csc \theta$

55. $\dfrac{1 - \sec^2 x}{1 - \csc^2 x} = \tan^2 x \sec^2 x - \sec^2 x + 1$

56. $(\cot \theta - \csc \theta)^2 = \dfrac{1 - \cos \theta}{1 + \cos \theta}$

B

57. $(1 - \cos^2 \theta)(2 + \tan^2 \theta) = \sec^2 \theta - \cos^2 \theta$

58. $\tan \theta(\sin \theta + \cos \theta)^2 + (1 - \sec^2 \theta) \cot \theta = 2 \sin^2 \theta$

59. $2 \cos x - \sec x(1 - 2 \sin^2 x) = \sec x$

60. $\dfrac{\sin \theta}{1 - \cos \theta} = \csc \theta + \cot \theta$

61. $\dfrac{1}{1 - \sin \theta} = \sec^2 \theta + \tan \theta \sec \theta$

62. $\dfrac{1 - \sin x}{1 + \sin x} = 2 \sec^2 x - 2 \sec x \tan x - 1$

63. $\dfrac{\sin \theta}{1 + \sin \theta} + \sec^2 \theta = \tan \theta \sec \theta + 1$

64. $\dfrac{\sin^2 x \tan^2 x}{\tan x - \sin x} = \tan x + \sin x$

65. $\dfrac{\sec \theta - \cos \theta + \tan \theta}{\tan \theta + \sec \theta} = \sin \theta$

66. $\dfrac{\sin^3 \theta - \sin^2 \theta \cos \theta - \sin \theta \cos^2 \theta + \cos^3 \theta}{\sin \theta + \cos \theta} =$
$(\sin \theta - \cos \theta)^2$

67. $\sqrt{\dfrac{1 - \cos \theta}{1 + \cos \theta}} = \dfrac{\sin \theta}{1 + \cos \theta} \quad (0 \le \theta < \pi)$

68. By making the substitution $t = 2 \sin \theta$, where $-\pi/2 < 0 < \pi/2$, show that
$$\dfrac{t}{\sqrt{4 - t^2}} = \tan \theta$$

69. By making the substitution $t = \tan \theta$, where $-\pi/2 < \theta < \pi/2$, show that
$$\dfrac{\sqrt{1 + t^2}}{t} = \csc \theta$$

70. If $\tan \theta = t$, prove that
$$\sin \theta \cos \theta = \dfrac{t}{1 + t^2}$$

71. By making the substitution $t = \frac{3}{2} \sec \theta$, where $0 < \theta < \pi/2$ if $t > 0$ and $\pi < \theta < 3\pi/2$ if $t < 0$, show that

$$\frac{\sqrt{4t^2 - 9}}{t} = 2 \sin \theta$$

72. Make the substitution $x = \sin \theta(-\pi/2 < \theta < \pi/2)$ in the expression $x^2/\sqrt{1 - x^2}$ and show the result can be written in the form $\sec \theta - \cos \theta$.

73. Let $x = 2 \tan \theta(-\pi/2 < \theta < \pi/2)$ in $x/\sqrt{4 + x^2}$, and show the result is $\sin \theta$.

74. Substitute $x = 3 \sec \theta$, where θ lies either in the first or third quadrants, in $(x^2 - 9)^{3/2}/x$, and show the result can be written as $9(\sec \theta \tan \theta - \sin \theta)$.

75. Substitute $x = \frac{3}{2} \tan \theta(-\pi/2 < \theta < \pi/2)$ in $x/\sqrt{4x^2 + 9}$ and show that the result is $\frac{1}{2} \sin \theta$.

Remark. Problems 68–75 illustrate types of trigonometric substitutions that occur in calculus in the process known as integration.

THE ADDITION FORMULAS FOR SINE AND COSINE

2

We will develop in this section formulas for the sine and cosine of the sum and difference of two numbers (or of two angles). These are of fundamental importance in deriving other identities. We designate by α and β two numbers that we will initially restrict to be between 0 and 2π and for which $\alpha > \beta$. These restrictions will soon be removed.

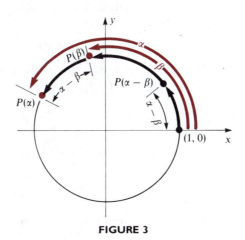

FIGURE 3

In Figure 3 we show typical positions of $P(\alpha)$ and $P(\beta)$. Now the distance along the circle from (1, 0) to $P(\alpha)$ is α, and the distance to $P(\beta)$ is β. So the arc length from $P(\beta)$ to $P(\alpha)$ is $\alpha - \beta$. If we think of sliding this latter arc along the circle until the point originally at $P(\beta)$ coincides with (1, 0), then the distance $\alpha - \beta$ is in standard position and the end point of the arc is correctly labeled $P(\alpha - \beta)$. Since the x and y coordinates of a point $P(\theta)$ are $\cos \theta$ and $\sin \theta$, respectively, we can show the coordinates of the three points $P(\alpha)$, $P(\beta)$, and $P(\alpha - \beta)$ as in Figure 4. Now, since the arc from (1, 0) to $P(\alpha - \beta)$ is equal in

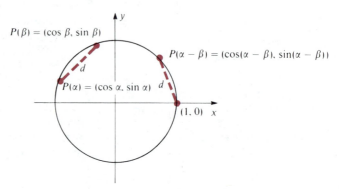

FIGURE 4

length to the arc from $P(\beta)$ to $P(\alpha)$, it follows that the chords connecting these respective pairs of points are also equal. We can get these lengths by using the distance formula:

$$\text{Distance from } (1, 0) \text{ to } P(\alpha - \beta) = \sqrt{[\cos(\alpha - \beta) - 1]^2 + [\sin(\alpha - \beta) - 0]^2}$$

$$\text{Distance from } P(\beta) \text{ to } P(\alpha) = \sqrt{(\cos \alpha - \cos \beta)^2 + (\sin \alpha - \sin \beta)^2}$$

Since these distances are the same, their squares are the same; so we obtain the equation

$$[\cos(\alpha - \beta) - 1]^2 + [\sin(\alpha - \beta)]^2 = (\cos \alpha - \cos \beta)^2 + (\sin \alpha - \sin \beta)^2$$

On squaring and collecting terms, and making use several times of the identity $\sin^2 \theta + \cos^2 \theta = 1$, this gives

$$\cos^2(\alpha - \beta) - 2 \cos(\alpha - \beta) + 1 + \sin^2(\alpha - \beta)$$

$$= \cos^2 \alpha - 2 \cos \alpha \cos \beta + \cos^2 \beta + \sin^2 \alpha - 2 \sin \alpha \sin \beta + \sin^2 \beta$$

$$2 - 2 \cos(\alpha - \beta) = 2 - 2(\cos \alpha \cos \beta + \sin \alpha \sin \beta)$$

or finally,

$$\cos(\alpha - \beta) = \cos \alpha \cos \beta + \sin \alpha \sin \beta \tag{4}$$

Now we remove the restrictions on α and β. Suppose first that $\beta > \alpha$, with α and β again in $[0, 2\pi]$. Then, since $\cos(-\theta) = \cos \theta$, we can apply (4) with α and β interchanged:

$$\cos(\alpha - \beta) = \cos(\beta - \alpha) = \cos \beta \cos \alpha + \sin \beta \sin \alpha$$

$$= \cos \alpha \cos \beta + \sin \alpha \sin \beta$$

So (4) still holds true. If $\alpha = \beta$, it again holds, since the left-hand side is cos 0, which equals 1, and the right-hand side is $\cos^2 \alpha + \sin^2 \alpha$, which also equals 1. Finally, let α and β be any two numbers whatever. Then we can write $\alpha = \alpha_1 + 2m\pi$ and $\beta = \beta_1 + 2n\pi$, where m and n are appropriately chosen integers, and α_1 and β_1 lie in the interval $[0, 2\pi]$. Then, by periodicity of the sine and cosine,

$$
\begin{aligned}
\cos(\alpha - \beta) &= \cos[(\alpha_1 + 2m\pi) - (\beta_1 + 2n\pi)] \\
&= \cos[(\alpha_1 - \beta_1) + (m - n)2\pi] \\
&= \cos(\alpha_1 - \beta_1) && \text{By periodicity} \\
&= \cos \alpha_1 \cos \beta_1 + \sin \alpha_1 \sin \beta_1 && \text{By (4)} \\
&= \cos \alpha \cos \beta + \sin \alpha \sin \beta && \text{By periodicity}
\end{aligned}
$$

Thus, equation (4) is true for all real numbers α and β.

From equation (4) we are able to get a number of other results. Consider first $\cos(\alpha + \beta)$. We write $\alpha + \beta = \alpha - (-\beta)$ and apply (4) with β replaced by $-\beta$:

$$
\cos(\alpha + \beta) = \cos[\alpha - (-\beta)] = \cos \alpha \cos(-\beta) + \sin \alpha \sin(-\beta)
$$

Since $\cos(-\beta) = \cos \beta$ and $\sin(-\beta) = -\sin \beta$, this gives

$$
\cos(\alpha + \beta) = \cos \alpha \cos \beta - \sin \alpha \sin \beta \tag{5}
$$

Next, in equation (4) let $\alpha = \pi/2$. Then, since $\cos \pi/2 = 0$ and $\sin \pi/2 = 1$, we get

$$
\cos\left(\frac{\pi}{2} - \beta\right) = \sin \beta \tag{6}
$$

In this, put $\beta = \pi/2 - \theta$ to get

$$
\cos\left[\frac{\pi}{2} - \left(\frac{\pi}{2} - \theta\right)\right] = \sin\left(\frac{\pi}{2} - \theta\right)
$$

Simplifying, we see that

$$
\sin\left(\frac{\pi}{2} - \theta\right) = \cos \theta \tag{7}
$$

Now we can get $\sin(\alpha + \beta)$ as follows:

$$
\begin{aligned}
\sin(\alpha + \beta) &= \cos\left[\frac{\pi}{2} - (\alpha + \beta)\right] && \text{By (6)} \\
&= \cos\left[\left(\frac{\pi}{2} - \alpha\right) - \beta\right] \\
&= \cos\left(\frac{\pi}{2} - \alpha\right)\cos \beta + \sin\left(\frac{\pi}{2} - \alpha\right)\sin \beta && \text{By (4)} \\
&= \sin \alpha \cos \beta + \cos \alpha \sin \beta && \text{By (6) and (7)}
\end{aligned}
$$

Notice that in applying equations (6) and (7) we have substituted different values for β and θ. Thus,

$$
\sin(\alpha + \beta) = \sin \alpha \cos \beta + \cos \alpha \sin \beta \tag{8}
$$

Writing $\alpha - \beta = \alpha + (-\beta)$ and using (8), we get

$$\sin(\alpha - \beta) = \sin[\alpha + (-\beta)] = \sin\alpha\cos(-\beta) + \cos\alpha\sin(-\beta)$$

or, since $\cos(-\beta) = \cos\beta$ and $\sin(-\beta) = -\sin\beta$,

$$\sin(\alpha - \beta) = \sin\alpha\cos\beta - \cos\alpha\sin\beta \qquad (9)$$

Formulas (4), (5), (8), and (9) are known as the **addition formulas** for the sine and cosine. They are of fundamental importance and should be memorized. We summarize them below. Notice that for the sine, the sign between the terms on the right agrees with the sign between α and β on the left, whereas for the cosine these are reversed. So there really are just two basic patterns to be learned.

The Addition Formulas for Sine and Cosine

$$\sin(\alpha + \beta) = \sin\alpha\cos\beta + \cos\alpha\sin\beta$$
$$\sin(\alpha - \beta) = \sin\alpha\cos\beta - \cos\alpha\sin\beta$$
$$\cos(\alpha + \beta) = \cos\alpha\cos\beta - \sin\alpha\sin\beta$$
$$\cos(\alpha - \beta) = \cos\alpha\cos\beta + \sin\alpha\sin\beta$$

EXAMPLE 7

Find the exact value of each of the following:
a. $\sin(5\pi/12)$ b. $\sin(\pi/12)$ c. $\cos(13\pi/12)$ d. $\cos(-\pi/12)$

Solution   a.   We can write $5\pi/12$ as $2\pi/12 + 3\pi/12 = \pi/6 + \pi/4$. So

$$\sin\frac{5\pi}{12} = \sin\left(\frac{\pi}{6} + \frac{\pi}{4}\right)$$

$$= \sin\frac{\pi}{6}\cos\frac{\pi}{4} + \cos\frac{\pi}{6}\sin\frac{\pi}{4}$$

Now we know that we may treat $\pi/6$ and $\pi/4$ as if they are radian measures of angles. Thus,

$$\sin\frac{5\pi}{12} = \left(\frac{1}{2}\right)\left(\frac{1}{\sqrt{2}}\right) + \left(\frac{\sqrt{3}}{2}\right)\left(\frac{1}{\sqrt{2}}\right) = \frac{1 + \sqrt{3}}{2\sqrt{2}}$$

b. $$\sin\frac{\pi}{12} = \sin\left(\frac{3\pi}{12} - \frac{2\pi}{12}\right) = \sin\left(\frac{\pi}{4} - \frac{\pi}{6}\right)$$

$$= \sin\frac{\pi}{4}\cos\frac{\pi}{6} - \cos\frac{\pi}{4}\sin\frac{\pi}{6}$$

$$= \left(\frac{1}{\sqrt{2}}\right)\left(\frac{\sqrt{3}}{2}\right) - \left(\frac{1}{\sqrt{2}}\right)\left(\frac{1}{2}\right) = \frac{\sqrt{3} - 1}{2\sqrt{2}}$$

c. $\cos\dfrac{13\pi}{12} = \cos\left(\dfrac{9\pi}{12} + \dfrac{4\pi}{12}\right) = \cos\left(\dfrac{3\pi}{4} + \dfrac{\pi}{3}\right)$

$$= \cos\frac{3\pi}{4}\cos\frac{\pi}{3} - \sin\frac{3\pi}{4}\sin\frac{\pi}{3}$$

$$= \left(-\frac{1}{\sqrt{2}}\right)\left(\frac{1}{2}\right) - \left(\frac{1}{\sqrt{2}}\right)\left(\frac{\sqrt{3}}{2}\right) = -\frac{1+\sqrt{3}}{2\sqrt{2}}$$

d. $\cos(-\pi/12) = \cos(\pi/12)$, since $\cos(-\theta) = \cos\theta$; so,

$$\cos\left(-\frac{\pi}{12}\right) = \cos\frac{\pi}{12} = \cos\left(\frac{3\pi}{12} - \frac{2\pi}{12}\right) = \cos\left(\frac{\pi}{4} - \frac{\pi}{6}\right)$$

$$= \cos\frac{\pi}{4}\cos\frac{\pi}{6} + \sin\frac{\pi}{4}\sin\frac{\pi}{6}$$

$$= \left(\frac{1}{\sqrt{2}}\right)\left(\frac{\sqrt{3}}{2}\right) + \left(\frac{1}{\sqrt{2}}\right)\left(\frac{1}{2}\right) = \frac{\sqrt{3}+1}{2\sqrt{2}}$$ ▲

Remark. There are other equally valid ways of writing the angles in this example. In part **a**, for example, we could have written

$$\frac{5\pi}{12} = \frac{8\pi}{12} - \frac{3\pi}{12} = \frac{2\pi}{3} - \frac{\pi}{4}$$

and then we would have used equation (9) for $\sin(\alpha - \beta)$.

EXAMPLE 8 |||||||||||||||||||||||||||||||||||||||||||||||||||||||||||||||||||||||||||||||||||||||||||||||||||||||||

If $x = \sin^{-1}\frac{3}{4}$ and $y = \cos^{-1}(-\frac{5}{6})$, find
a. $\sin(x + y)$ **b.** $\cos(x - y)$

Solution   We know by the definitions of the inverse sine and cosine that $\sin x = \frac{3}{4}$ and $\cos y = -\frac{5}{6}$, with $0 \le x \le \pi/2$ and $\pi/2 \le y \le \pi$. By the Pythagorean identity, $\sin^2 x + \cos^2 x = 1$, we get

$$\cos^2 x = 1 - \sin^2 x$$

So

$$\cos x = \pm\sqrt{1 - \sin^2 x}$$

The positive sign must be chosen, since $0 \le x \le \pi/2$. So

$$\cos x = \sqrt{1 - \left(\frac{3}{4}\right)^2} = \sqrt{1 - \frac{9}{16}} = \sqrt{\frac{7}{16}} = \frac{\sqrt{7}}{4}$$

Similarly, since $\pi/2 \le y \le \pi$, $\sin y \ge 0$, so

$$\sin y = \sqrt{1 - \cos^2 y} = \sqrt{1 - \left(-\frac{5}{6}\right)^2} = \sqrt{1 - \frac{25}{36}} = \frac{\sqrt{11}}{6}$$

We now have all the necessary ingredients for obtaining the solutions.

a. $\sin(x + y) = \sin x \cos y + \cos x \sin y = \dfrac{3}{4}\left(-\dfrac{5}{6}\right) + \left(\dfrac{\sqrt{7}}{4}\right)\left(\dfrac{\sqrt{11}}{6}\right)$

$$= \dfrac{-15 + \sqrt{77}}{24}$$

b. $\cos(x - y) = \cos x \cos y + \sin x \sin y = \dfrac{\sqrt{7}}{4}\left(-\dfrac{5}{6}\right) + \dfrac{3}{4}\left(\dfrac{\sqrt{11}}{6}\right)$

$$= \dfrac{-5\sqrt{7} + 3\sqrt{11}}{24}   ▲$$

EXAMPLE 9 |||||||||||||||||||||||||||||||||||||||||||||||||||||||||||||||||||||||||||||||||||||||

If $\cos \alpha < 0$, $\tan \alpha = \frac{4}{3}$, and $\sin \beta > 0$, $\sec \beta = -\frac{13}{5}$, find
a. $\sin(\alpha - \beta)$ **b.** $\cos(\alpha + \beta)$

Solution   From the given information we conclude that for α and β angles in standard position, α terminates in the third quadrant and β in the second quadrant. They can be shown as in Figure 5, and we can read off the functions we need.

a. $\sin(\alpha - \beta) = \sin \alpha \cos \beta - \cos \alpha \sin \beta = \left(-\dfrac{4}{5}\right)\left(-\dfrac{5}{13}\right) - \left(-\dfrac{3}{5}\right)\left(\dfrac{12}{13}\right) = \dfrac{56}{65}$

b. $\cos(\alpha + \beta) = \cos \alpha \cos \beta - \sin \alpha \sin \beta = \left(-\dfrac{3}{5}\right)\left(-\dfrac{5}{13}\right) - \left(-\dfrac{4}{5}\right)\left(\dfrac{12}{13}\right) = \dfrac{63}{65}$

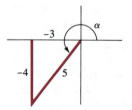   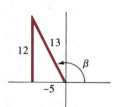

FIGURE 5   ▲

EXERCISE SET 2

A

Find the exact values in Problems 1–8.

1. $\sin(\alpha + \beta)$ and $\cos(\alpha + \beta)$, where $\alpha = \pi/4$ and $\beta = \pi/3$

2. $\sin(\alpha - \beta)$ and $\cos(\alpha - \beta)$, where $\alpha = \pi/4$ and $\beta = \pi/3$

3. a. $\sin\left(\dfrac{\pi}{3} - \dfrac{3\pi}{4}\right)$ **b.** $\cos\left(\dfrac{5\pi}{6} + \dfrac{\pi}{4}\right)$

4. a. $\sin\left(\dfrac{5\pi}{4} + \dfrac{11\pi}{6}\right)$ **b.** $\cos\left(\dfrac{4\pi}{3} - \dfrac{3\pi}{4}\right)$

5. a. $\cos\dfrac{7\pi}{12}$ **b.** $\sin\dfrac{17\pi}{12}$

6. a. $\sin\left(-\dfrac{\pi}{12}\right)$ **b.** $\cos\left(-\dfrac{5\pi}{12}\right)$

7. a. $\sin 75°$ **b.** $\cos 15°$
8. a. $\cos 255°$ **b.** $\sin 195°$
9. $\sin(\alpha - \beta)$ if $\alpha = \sin^{-1}\frac{3}{5}$ and $\beta = \cos^{-1}(-\frac{5}{13})$
10. $\cos(\alpha + \beta)$ if $\alpha = \arctan(-\frac{1}{2})$ and $\beta = \operatorname{arcsec}\frac{5}{3}$
11. $\sin(\sin^{-1}\frac{12}{13} + \cos^{-1}\frac{4}{5})$
12. $\cos[\cos^{-1}\frac{8}{17} - \sin^{-1}(-\frac{7}{25})]$
13. If $\sin \alpha = -\frac{3}{5}$, $\sin \beta = -\frac{5}{13}$, $P(\alpha)$ is in the third quadrant, and $P(\beta)$ is in the fourth quadrant, find the following.
   a. $\sin(\alpha - \beta)$ **b.** $\cos(\alpha + \beta)$
14. If $\cos \alpha = \frac{12}{13}$, $\cos \beta = -\frac{4}{5}$, $P(\alpha)$ is the fourth quadrant, and $P(\beta)$ is in the second quadrant, find:
   a. $\sin(\alpha + \beta)$ **b.** $\cos(\alpha - \beta)$
15. If $\sin \alpha < 0$, $\cos \alpha = \frac{1}{3}$, $\cos \beta < 0$, and $\sin \beta = -\frac{2}{3}$, find:
   a. $\sin(\alpha + \beta)$ **b.** $\cos(\alpha - \beta)$
16. If $\tan x = -\frac{3}{4}$, $\csc x > 0$, $\sec y = \frac{13}{5}$, and $\cot y < 0$, find:
   a. $\sin(x - y)$ **b.** $\cos(x + y)$
17. If $\sin \alpha = -2/\sqrt{5}$, $\tan \alpha > 0$, $\cos \beta = -\frac{8}{17}$, and $\csc \beta > 0$, find:
   a. $\sin(\alpha - \beta)$ **b.** $\cos(\alpha + \beta)$
18. If $\cot A = -\frac{24}{7}$, $\sec A > 0$, $\tan B = \frac{4}{3}$, and $\sin B < 0$, find:
   a. $\sin(A + B)$ **b.** $\cos(A - B)$
19. If $\sin x = -\frac{3}{5}$, $\tan x > 0$, $\sec y = -\frac{13}{5}$, and $\cot y < 0$, find:
   a. $\csc(x + y)$ **b.** $\sec(x - y)$

Establish the formulas in Problems 20–24 by using addition formulas.

20. a. $\sin(\pi + \theta) = -\sin \theta$
   b. $\cos(\pi + \theta) = -\cos \theta$

21. a. $\sin\left(\dfrac{\pi}{2} + \theta\right) = \cos \theta$

   b. $\cos\left(\dfrac{\pi}{2} + \theta\right) = -\sin \theta$

22. a. $\sin\left(\dfrac{3\pi}{2} - \theta\right) = -\cos \theta$

   b. $\cos\left(\dfrac{3\pi}{2} - \theta\right) = -\sin \theta$

23. a. $\sin(\pi - \theta) = \sin \theta$
   b. $\cos(\pi - \theta) = -\cos \theta$

24. a. $\sin\left(\dfrac{\pi}{2} - \theta\right) = \cos \theta$

   b. $\cos\left(\dfrac{\pi}{2} - \theta\right) = \sin \theta$

Prove the identities in Problems 25–33.

25. $\sin \alpha \cos \beta(\cot \alpha - \tan \beta) = \cos(\alpha + \beta)$

26. $\dfrac{\cos(\alpha + \beta)}{\cos(\alpha - \beta)} = \dfrac{1 - \tan \alpha \tan \beta}{1 + \tan \alpha \tan \beta}$

27. $\dfrac{\tan \alpha - \tan \beta}{\tan \alpha + \tan \beta} = \dfrac{\sin(\alpha - \beta)}{\sin(\alpha + \beta)}$

28. $\dfrac{\sin(\alpha + \beta) + \sin(\alpha - \beta)}{\cos(\alpha + \beta) + \cos(\alpha - \beta)} = \tan \alpha$

29. $\dfrac{\cos(\alpha - \beta) - \cos(\alpha + \beta)}{\sin(\alpha + \beta) - \sin(\alpha - \beta)} = \tan \alpha$

30. $\dfrac{\sin(\alpha + \beta)}{\sin \alpha \sin \beta} = \cot \alpha + \cot \beta$

31. $\dfrac{\cos(\alpha - \beta)}{\sin \alpha \sin \beta} = 1 + \cot \alpha \cot \beta$

32. $\sin(x + y)\cos y - \cos(x + y)\sin y = \sin x$
33. $\sin x \sin(x + y) + \cos x \cos(x + y) = \cos y$

B

34. Prove that, in general, $\sin(\alpha + \beta) \neq \sin \alpha + \sin \beta$.
35. Find all values of x for which $0 \le x < 2\pi$ and

$$\sin 5x \cos 4x = \cos 5x \sin 4x$$

36. Find all values of x for which $0 \le x < 2\pi$ and

$$2 \cos 2x \cos x = 1 - 2 \sin 2x \sin x$$

37. Derive a formula for
   a. $\sin(\alpha + \beta + \gamma)$ **b.** $\cos(\alpha + \beta + \gamma)$
   Hint. First use the associative property for addition.

Prove the identities in Problems 38–40.

38. $\sin(\alpha + \beta)\sin(\alpha - \beta) = \sin^2 \alpha - \sin^2 \beta$
39. $\cos(\alpha + \beta)\cos(\alpha - \beta) = \cos^2 \alpha - \sin^2 \beta$
40. $\sin(\alpha + \beta)\cos(\alpha - \beta) = \sin \alpha \cos \alpha + \sin \beta \cos \beta$

DOUBLE-ANGLE, HALF-ANGLE, AND REDUCTION FORMULAS

The importance of the addition formulas in Section 2 lies primarily in the fact that so many other identities can be derived from them. We will carry out the derivations for some of the most important of these.

The **double-angle formulas** are obtained from the addition formulas for sine and cosine of α and β by putting $\beta = \alpha$. If we denote the common value of α and β by θ, we obtain

$$\sin(\theta + \theta) = \sin\theta\cos\theta + \cos\theta\sin\theta$$
$$= 2\sin\theta\cos\theta$$

So

$$\sin 2\theta = 2\sin\theta\cos\theta$$

Also,

$$\cos(\theta + \theta) = \cos\theta\cos\theta - \sin\theta\sin\theta$$
$$= \cos^2\theta - \sin^2\theta$$

So

$$\cos 2\theta = \cos^2\theta - \sin^2\theta \qquad (10)$$

Two other useful forms of $\cos 2\theta$ can be obtained by replacing, in turn, $\cos^2\theta$ by $1 - \sin^2\theta$ and $\sin^2\theta$ by $1 - \cos^2\theta$. This gives

$$\cos 2\theta = (1 - \sin^2\theta) - \sin^2\theta = 1 - 2\sin^2\theta$$

and

$$\cos 2\theta = \cos^2\theta - (1 - \cos^2\theta) = 2\cos^2\theta - 1$$

So we have

$$\cos 2\theta = 1 - 2\sin^2\theta \qquad (11)$$

and

$$\cos 2\theta = 2\cos^2\theta - 1 \qquad (12)$$

Whether to use (10), (11), or (12) depends on the objective. We will shortly see some examples where a particular form is clearly preferable.

If we solve (11) for $\sin^2\theta$ and (12) for $\cos^2\theta$, we obtain

$$\sin^2\theta = \frac{1 - \cos 2\theta}{2} \qquad (13)$$

and

$$\cos^2\theta = \frac{1 + \cos 2\theta}{2} \qquad (14)$$

These forms are employed extensively in calculus.

By replacing θ by $\alpha/2$ in the last two equations and then taking square roots, we get the **half-angle formulas:**

$$\sin\frac{\alpha}{2} = \pm\sqrt{\frac{1 - \cos\alpha}{2}} \qquad (15)$$

$$\cos\frac{\alpha}{2} = \pm\sqrt{\frac{1+\cos\alpha}{2}} \qquad (16)$$

The ambiguity of sign has to be resolved in each particular instance according to the quadrant in which $\alpha/2$ lies.

EXAMPLE 10

If $\alpha = \cos^{-1}(-\frac{7}{25})$ and $\beta = \tan^{-1}(-\frac{3}{2})$, find $\cos(\frac{\alpha}{2}+2\beta)$.

Solution   We can show α and β as angles in the second and fourth quadrants, respectively, as in Figure 6. From the addition formula for the cosine we have

$$\cos\left(\frac{\alpha}{2}+2\beta\right) = \cos\frac{\alpha}{2}\cos 2\beta - \sin\frac{\alpha}{2}\sin 2\beta$$

The angle $\alpha/2$ is between 0 and $\pi/2$, since α is between $\pi/2$ and π, and so both $\sin(\alpha/2)$ and $\cos(\alpha/2)$ are positive. Thus, by the half-angle formulas

$$\sin\frac{\alpha}{2} = \sqrt{\frac{1-\cos\alpha}{2}} = \sqrt{\frac{1-\left(-\frac{7}{25}\right)}{2}} = \sqrt{\frac{16}{25}} = \frac{4}{5}$$

$$\cos\frac{\alpha}{2} = \sqrt{\frac{1+\cos\alpha}{2}} = \sqrt{\frac{1+\left(-\frac{7}{25}\right)}{2}} = \sqrt{\frac{9}{25}} = \frac{3}{5}$$

Also, by the double-angle formulas

$$\sin 2\beta = 2\sin\beta\cos\beta = 2\left(-\frac{3}{\sqrt{13}}\right)\left(\frac{2}{\sqrt{13}}\right) = -\frac{12}{13}$$

$$\cos 2\beta = \cos^2\beta - \sin^2\beta = \frac{4}{13} - \frac{9}{13} = -\frac{5}{13}$$

Making these substitutions, we get

$$\cos\left(\frac{\alpha}{2}+2\beta\right) = \left(\frac{3}{5}\right)\left(-\frac{5}{13}\right) - \left(\frac{4}{5}\right)\left(-\frac{12}{13}\right) = \frac{33}{65}$$

FIGURE 6

EXAMPLE 11

a. If $\sec 2\theta = 25/7$ and $\pi \leq 2\theta \leq 2\pi$, find $\sin \theta$ and $\cos \theta$.

b. If $\sin \alpha = 4/5$ and $5\pi/2 \leq \alpha \leq 7\pi/2$, find $\sin \alpha/2$ and $\cos \alpha/2$.

Solution   **a.** Since $\cos 2\theta = \dfrac{1}{\sec 2\theta} = \dfrac{1}{25}$, we can use formulas (13) and (14) to get

$$\sin^2 \theta = \frac{1 - \cos 2\theta}{2} = \frac{1 - \frac{7}{25}}{2} = \frac{9}{25}$$

$$\cos^2 \theta = \frac{1 + \cos 2\theta}{2} = \frac{1 + \frac{7}{25}}{2} = \frac{16}{25}$$

We are given that $\pi \leq 2\theta \leq 2\pi$, so that $\pi/2 \leq \theta \leq \pi$. Thus, $\sin \theta > 0$ and $\cos \theta < 0$. So on taking square roots of $\sin^2 \theta$ and $\cos^2 \theta$ and affixing the correct signs, we get

$$\sin \theta = \tfrac{3}{5} \qquad \text{and} \qquad \cos \theta = -\tfrac{4}{5}$$

b. Since $\sin \alpha > 0$, we conclude that α is in quadrant II; that is, $5\pi/2 \leq \alpha \leq 3\pi$. Thus, $\cos \alpha < 0$, and since $\cos^2 \alpha = 1 - \sin^2 \alpha = 9/25$, we have $\cos \alpha = -3/5$. We can find $\sin \alpha/2$ and $\cos \alpha/2$ from the half-angle formulas (15) and (16), but first we must determine the correct signs. Since $5\pi/4 \leq \alpha/2 \leq 3\pi/2$, it follows that $\alpha/2$ is in quadrant III, so that both $\sin \alpha/2$ and $\cos \alpha/2$ are negative. Thus,

$$\sin \frac{\alpha}{2} = -\sqrt{\frac{1 - \cos \alpha}{2}} = -\sqrt{\frac{1 + \frac{3}{5}}{2}} = -\frac{2}{\sqrt{5}}$$

$$\cos \frac{\alpha}{2} = -\sqrt{\frac{1 + \cos \alpha}{2}} = -\sqrt{\frac{1 - \frac{3}{5}}{2}} = -\frac{1}{\sqrt{5}} \qquad \blacktriangle$$

From the addition formulas we can obtain a class of identities sometimes called **reduction formulas,** of which (6) and (7) of Section 2 are special cases. Here are some others:

$$\sin(\pi + \theta) = \sin \pi \cos \theta + \cos \pi \sin \theta$$
$$= -\sin \theta$$

and

$$\sin(\pi - \theta) = \sin \pi \cos \theta - \cos \pi \sin \theta$$
$$= \sin \theta$$

Similarly,

$$\cos(\pi + \theta) = \cos \pi \cos \theta - \sin \pi \sin \theta$$
$$= -\cos \theta$$

and

$$\cos(\pi - \theta) = \cos \pi \cos \theta + \sin \pi \sin \theta$$
$$= -\cos \theta$$

In these, we have used the facts that $\sin \pi = 0$ and $\cos \pi = -1$. Since $\sin(3\pi/2) = -1$ and $\cos(3\pi/2) = 0$, we also have

$$\sin\left(\frac{3\pi}{2} + \theta\right) = \sin \frac{3\pi}{2} \cos \theta + \cos \frac{3\pi}{2} \sin \theta$$
$$= -\cos \theta$$

and

$$\sin\left(\frac{3\pi}{2} - \theta\right) = \sin \frac{3\pi}{2} \cos \theta - \cos \frac{3\pi}{2} \sin \theta$$
$$= -\cos \theta$$

Likewise, there are similar formulas for $\cos[(3\pi/2) \pm \theta]$.

To generalize, it appears that we should consider two cases:

Case 1. Functions of $(n\pi \pm \theta)$

Case 2. Functions of $\left[\dfrac{(2n + 1)\pi}{2} \pm \theta\right]$

where n is an arbitrary integer. Note that in Case 2, an odd multiple of $\pi/2$ is involved since $(2n + 1)$ is always odd. The following identities can now be established.

Reduction Formulas for the Sine and Cosine

Case 1. $\sin(n\pi \pm \theta) = \pm\sin \theta$
 $\cos(n\pi \pm \theta) = \pm\cos \theta$

Case 2. $\sin\left[\dfrac{(2n + 1)\pi}{2} \pm \theta\right] = \pm\cos \theta$

 $\cos\left[\dfrac{(2n + 1)\pi}{2} \pm \theta\right] = \pm\sin \theta$

To determine the correct sign on the right, it suffices to determine what the sign is when θ is acute.

These are called reduction formulas since in each case the given expression is reduced to a simpler expression. Note that in Case 1 the function on the right is the *same* as that on the left, whereas in Case 2 the function on the right is the *cofunction* of the one on the left.

We will prove the first of these formulas and leave the others as exercises. We note first that $\sin n\pi = 0$. Second, $\cos n\pi = (-1)^n$ since, when n is even, $n\pi$ is coterminal with 0, so that the cosine is $+1$; and, when n is odd, $n\pi$ is coterminal with π, so that the cosine is -1. Using the addition formula for the sine, we have

$$\sin(n\pi + \theta) = \sin n\pi \cos \theta + \cos n\pi \sin \theta$$
$$= (0) \cos \theta + (-1)^n \sin \theta$$
$$= (-1)^n \sin \theta$$

This shows that $\sin(n\pi + \theta) = \pm\sin \theta$. Furthermore, the sign on the right is $(-1)^n$, which is independent of the size of θ. So if we determine the sign when θ is acute, it will be correct for all values of θ.

EXAMPLE 12

Express each of the following in terms of $\sin \theta$ or $\cos \theta$.

a. $\sin(3\pi - \theta)$ **b.** $\cos\left(\dfrac{\pi}{2} + \theta\right)$ **c.** $\sin\left(-\dfrac{\pi}{2} + \theta\right)$

Solution   **a.** $\sin(3\pi - \theta) = \pm\sin \theta$ (Case 1)

If θ is acute, $3\pi - \theta$ is in quadrant II, so $\sin(3\pi - \theta)$ is positive. Thus,

$$\sin(3\pi - \theta) = \sin \theta$$

b. $\cos\left(\dfrac{\pi}{2} + \theta\right) = \pm\sin \theta$ (Case 2)

For θ acute, $\pi/2 + \theta$ is in quadrant II, and the cosine is negative there. So

$$\cos\left(\dfrac{\pi}{2} + \theta\right) = -\sin \theta$$

c. $\sin\left(-\dfrac{\pi}{2} + \theta\right) = \pm\cos \theta$ (Case 2)

For θ acute, $-\pi/2 + \theta$ is in quadrant IV, where the sine is negative. So

$$\sin\left(-\dfrac{\pi}{2} + \theta\right) = -\cos \theta$$ ▲

EXERCISE SET 3

A

In Problems 1–6 find $\sin 2\theta$ and $\cos 2\theta$.

1. $\sin \theta = \frac{4}{5}$ and θ terminates in quadrant II
2. $\cos \theta = -\frac{1}{3}$ and θ terminates in quadrant III

3. $\tan \theta = \frac{5}{12}$ and $\sin \theta < 0$
4. $\sec \theta = \sqrt{5}$ and $\csc \theta < 0$
5. $\cot \theta = -2$ and $0 \le \theta \le \pi$
6. $\csc \theta = \dfrac{17}{8}$ and $\dfrac{\pi}{2} \le \theta \le \dfrac{3\pi}{2}$

7. If $\sin \theta = x$ and $-\pi/2 \le \theta \le \pi/2$, find $\sin 2\theta$ and $\cos 2\theta$ in terms of x.

8. If $\cos \theta = x$ and $0 \le \theta \le \pi$, find $\sin 2\theta$ and $\cos 2\theta$ in terms of x.

In Problems 9–12 find $\sin \theta$ and $\cos \theta$.

9. $\cos 2\theta = \frac{1}{3}$ and $\pi \le 2\theta \le 2\pi$

10. $\sin 2\theta = -\frac{4}{5}$ and $\pi \le 2\theta \le \frac{3\pi}{2}$

11. $\sec 2\theta = \frac{25}{7}$ and $-\pi \le 2\theta \le 0$

12. $\tan 2\theta = -\frac{12}{5}$ and $3\pi \le 2\theta \le 4\pi$

13. Use the half-angle formulas to find

    a. $\sin \dfrac{\pi}{8}$ **b.** $\cos \dfrac{\pi}{12}$

    c. $\sin 75°$ **d.** $\cos 67.5°$

In Problems 14–17 find $\sin(\alpha/2)$ and $\cos(\alpha/2)$.

14. $\cos \alpha = \frac{7}{25}, \quad 0 \le \alpha \le \pi$

15. $\sin \alpha = -\dfrac{12}{13}, \quad \pi \le \alpha \le \dfrac{3\pi}{2}$

16. $\tan \alpha = -\frac{8}{15}, \quad 2\pi \le \alpha \le 3\pi$

17. $\sec \alpha = 3, \quad -\pi \le \alpha \le 0$

In Problems 18–29 find the exact value of the specified functions.

18. $\sin 2\theta$ if $\theta = \cos^{-1}\left(-\dfrac{3}{5}\right)$

19. $\cos 2\theta$ if $\theta = \sin^{-1}\left(-\dfrac{1}{3}\right)$

20. $\cos\left[2 \cos^{-1}\left(-\dfrac{2}{3}\right)\right]$

21. $\sin\left[2 \sin^{-1}\left(-\dfrac{3}{7}\right)\right]$ 22. $\sin\left[2 \sec^{-1}\left(-\dfrac{3}{2}\right)\right]$

23. $\cos\left[2 \tan^{-1}\left(-\dfrac{1}{2}\right)\right]$

24. $\sin \dfrac{\theta}{2}$ if $\theta = \sin^{-1} \dfrac{7}{25}$

25. $\cos \dfrac{\theta}{2}$ if $\theta = \tan^{-1}\left(-\dfrac{4}{3}\right)$

26. $\cos\left[\dfrac{1}{2} \cos^{-1}\left(-\dfrac{3}{4}\right)\right]$

27. $\sin\left[\dfrac{1}{2} \tan^{-1} \dfrac{7}{24}\right]$

28. $\sin(2\alpha - \beta)$ if $\alpha = \sin^{-1}\left(-\dfrac{1}{3}\right)$ and

    $\beta = \tan^{-1}\left(-\dfrac{2}{3}\right)$

29. $\cos\left(\alpha - \dfrac{1}{2}\beta\right)$ if $\alpha = \sec^{-1}\left(-\dfrac{5}{4}\right)$ and

    $\beta = \tan^{-1}\left(-\dfrac{\sqrt{19}}{9}\right)$

30. If $P(\theta) = \left(\dfrac{1}{\sqrt{5}}, -\dfrac{2}{\sqrt{5}}\right)$, find $P(2\theta)$.

31. If $0 \le 2\theta \le \pi$ and $P(2\theta) = (-\frac{7}{25}, \frac{24}{25})$, find $P(\theta)$.

In Problems 32–34 express the given function in terms of $\sin \theta$ or $\cos \theta$.

32. **a.** $\sin(2\pi - \theta)$ **b.** $\cos\left(\dfrac{3\pi}{2} - \theta\right)$

    c. $\sin\left(\dfrac{5\pi}{2} + \theta\right)$ **d.** $\cos(-\pi + \theta)$

    e. $\sin\left(-\dfrac{\pi}{2} - \theta\right)$

33. **a.** $\sin(\theta - \pi)$ **b.** $\cos\left(\dfrac{\pi}{2} + \theta\right)$

    c. $\sin\left(-\dfrac{3\pi}{2} - \theta\right)$ **d.** $\cos(\theta - 3\pi)$

    e. $\sin\left(\dfrac{\pi}{2} + \theta\right)$

34. **a.** $\cos\left(\dfrac{3\pi}{2} + \theta\right)$ **b.** $\sin(\pi + \theta)$

    c. $\cos(\theta - 2\pi)$ **d.** $\sin\left(-\dfrac{\pi}{2} + \theta\right)$

    e. $\cos\left(\theta - \dfrac{5\pi}{2}\right)$

Prove the identities in Problems 35–49.

35. $1 - \dfrac{\cos 2x - 1}{2\cos^2 x} = \sec^2 x$

36. $\sin^4 \theta \cot^2 \theta = \frac{1}{4} \sin^2 2\theta$

37. $\dfrac{\cos^2 \theta - \sin^2 \theta}{\sin \theta \cos \theta} = 2 \cot 2\theta$

38. $\dfrac{\sin^2 2x}{1 - \cos 2x} = 2 \cos^2 x$

39. $(\sin x - \cos x)^2 = 1 - \sin 2x$

40. $\cos^4 x - \sin^4 x = \cos 2x$

41. $\dfrac{2}{1 + \cos 2x} = \sec^2 x$

42. $\dfrac{\sin 2\theta}{\sin^2 \theta - 1} + 3 \tan \theta = \tan \theta$

43. $\tan \alpha + \cot \alpha = 2 \csc 2\alpha$

44. $\dfrac{\cot \theta + \tan \theta}{\cot \theta - \tan \theta} = \sec 2\theta$

45. $\dfrac{\tan^2 \theta}{1 + \tan^2 \theta} = \dfrac{1 - \cos 2\theta}{2}$

46. $\sin^2 \dfrac{\theta}{2} \csc^2 \theta = \dfrac{1}{2(1 + \cos \theta)}$

47. $\dfrac{2 \cos^2 \dfrac{\theta}{2}}{\sin^2 \theta} = \dfrac{1}{1 - \cos \theta}$

48. $\dfrac{2 \sin^2 \dfrac{\theta}{2} + \cos \theta}{\sec \theta} = \cos \theta$

49. $\dfrac{1 - \tan^2 x}{1 + \tan^2 x} = \cos 2x$

B

50. Prove the reduction formula for each of the following:

   a. $\sin(n\pi - \theta)$ **b.** $\cos(n\pi + \theta)$

   c. $\cos(n\pi - \theta)$ **d.** $\sin\left[\dfrac{(2n + 1)\pi}{2} + \theta\right]$

   e. $\sin\left[\dfrac{(2n + 1)\pi}{2} - \theta\right]$

   f. $\cos\left[\dfrac{(2n + 1)\pi}{2} + \theta\right]$

   g. $\cos\left[\dfrac{(2n + 1)\pi}{2} - \theta\right]$

51. If $\theta = \tan^{-1} x$ find $\sin 2\theta$ and $\cos 2\theta$ in terms of x.

52. If $\theta = \sec^{-1} x$ find:

   a. $\sin 2\theta$ **b.** $\cos 2\theta$ **c.** $\sin \tfrac{1}{2}\theta$ **d.** $\cos \tfrac{1}{2}\theta$

53. By writing $3\theta = 2\theta + \theta$, find formulas for $\sin 3\theta$ and $\cos 3\theta$ in terms of functions of θ.

54. Derive the formula

   $\sin 4\theta = 4 \sin \theta \cos \theta - 8 \sin^3 \theta \cos \theta.$

55. Derive the formula

   $\cos 4\theta = 8 \cos^4 \theta - 8 \cos^2 \theta + 1$

Prove the identities in Problems 56–62.

56. $\cos^4 x = \dfrac{3}{8} + \dfrac{\cos 2x}{2} + \dfrac{\cos 4x}{8}$

57. $\sin^6 x = \dfrac{5}{16} - \dfrac{\cos 2x}{2} + \dfrac{3 \cos 4x}{16} + \dfrac{\sin^2 2x \cos 2x}{8}$

58. $\dfrac{1 - \cos 2nx}{\sin 2nx} = \tan nx$

59. $\dfrac{\sin 3\theta}{\sin \theta} - \dfrac{\cos 3\theta}{\cos \theta} = 2$

60. $\sin 3x + \sin x = 4 \sin x \cos^2 x$

61. $\dfrac{\sin^3 x + \cos^3 x}{\sin x + \cos x} = 1 - \tfrac{1}{2} \sin 2x$

62. By calculating $\sin(\pi/12)$ in two ways, using the half-angle formulas and using the addition formulas, prove that

$$\sqrt{2 - \sqrt{3}} = \tfrac{1}{2}(\sqrt{6} - \sqrt{2})$$

FURTHER IDENTITIES

To obtain addition formulas for the tangent, it is necessary only to use the fact that $\tan \theta = \sin \theta / \cos \theta$. Thus,

$$\tan(\alpha + \beta) = \frac{\sin(\alpha + \beta)}{\cos(\alpha + \beta)} = \frac{\sin \alpha \cos \beta + \cos \alpha \sin \beta}{\cos \alpha \cos \beta - \sin \alpha \sin \beta}$$

This can be improved by dividing numerator and denominator by $\cos \alpha \cos \beta$:

$$\tan(\alpha + \beta) = \dfrac{\dfrac{\sin \alpha \cos \beta}{\cos \alpha \cos \beta} + \dfrac{\cos \alpha \sin \beta}{\cos \alpha \cos \beta}}{\dfrac{\cos \alpha \cos \beta}{\cos \alpha \cos \beta} - \dfrac{\sin \alpha \sin \beta}{\cos \alpha \cos \beta}}$$

or

$$\tan(\alpha + \beta) = \frac{\tan \alpha + \tan \beta}{1 - \tan \alpha \tan \beta} \tag{17}$$

And in a similar manner,

$$\tan(\alpha - \beta) = \frac{\tan \alpha - \tan \beta}{1 + \tan \alpha \tan \beta} \tag{18}$$

The double-angle formula for the tangent is obtained from (17) by taking $\alpha = \beta$. If we call this common value θ, we have

$$\tan(\theta + \theta) = \frac{\tan \theta + \tan \theta}{1 - \tan \theta \tan \theta}$$

So,

$$\tan 2\theta = \frac{2 \tan \theta}{1 - \tan^2 \theta}$$

We can derive half-angle formulas for the tangent as follows:

$$\tan \tfrac{1}{2}\alpha = \frac{\sin \tfrac{1}{2}\alpha}{\cos \tfrac{1}{2}\alpha} = \frac{\sin \tfrac{1}{2}\alpha}{\cos \tfrac{1}{2}\alpha} \cdot \frac{2 \cos \tfrac{1}{2}\alpha}{2 \cos \tfrac{1}{2}\alpha}$$

$$= \frac{2 \sin \tfrac{1}{2}\alpha \cos \tfrac{1}{2}\alpha}{2 \cos^2 \tfrac{1}{2}\alpha}$$

Since $2 \sin(\alpha/2) \cos(\alpha/2) = \sin 2(\alpha/2) = \sin \alpha$ and

$$2 \cos^2 \tfrac{1}{2}\alpha = 2\left(\sqrt{\frac{1 + \cos \alpha}{2}}\right)^2 = 1 + \cos \alpha$$

we obtain

$$\tan \tfrac{1}{2}\alpha = \frac{\sin \alpha}{1 + \cos \alpha}$$

If in this derivation we had multiplied numerator and denominator by $2 \sin(\alpha/2)$ instead of $2 \cos(\pi/2)$, we would have obtained the equivalent formula

$$\tan \tfrac{1}{2}\alpha = \frac{1 - \cos \alpha}{\sin \alpha}$$

(See Problem 44.)

We could obtain analogous formulas for the cotangent, secant, and cosecant in a similar way, but these are so seldom used that we will not clutter up our already formidable list with them.

We do choose to list one more group of identities known as the **sum and product formulas** for the sine and cosine. These can be obtained from the addition formulas (see Problems 41 and 42). Although these are used less frequently than the other identities we have considered, they are indispensable at times.

Sum Formulas

$$\sin A + \sin B = 2 \sin \frac{A + B}{2} \cos \frac{A - B}{2}$$

$$\sin A - \sin B = 2 \cos \frac{A + B}{2} \sin \frac{A - B}{2}$$

$$\cos A + \cos B = 2 \cos \frac{A + B}{2} \cos \frac{A - B}{2}$$

$$\cos A - \cos B = -2 \sin \frac{A + B}{2} \sin \frac{A - B}{2}$$

Product Formulas

$$\sin \alpha \cos \beta = \tfrac{1}{2}[\sin(\alpha + \beta) + \sin(\alpha - \beta)]$$

$$\cos \alpha \sin \beta = \tfrac{1}{2}[\sin(\alpha + \beta) - \sin(\alpha - \beta)]$$

$$\sin \alpha \sin \beta = \tfrac{1}{2}[\cos(\alpha - \beta) - \cos(\alpha + \beta)]$$

$$\cos \alpha \cos \beta = \tfrac{1}{2}[\cos(\alpha + \beta) + \cos(\alpha - \beta)]$$

 EXERCISE SET 4

A

1. Use (17) and (18) to find:

   a. $\tan \dfrac{5\pi}{12}$ b. $\tan \dfrac{\pi}{12}$

2. Find $\tan 2\theta$ if $P(\theta)$ is the point $(-\tfrac{1}{3}, 2\sqrt{2}/3)$.
3. Find $\tan 2\theta$ if $\sin \theta = -\tfrac{3}{5}$ and $\cos \theta = -\tfrac{4}{5}$.
4. Find $\tan(\alpha + \beta)$ if $\sin \alpha = \tfrac{4}{5}$, α is in the second quadrant, $\cos \beta = -\tfrac{5}{13}$, and β is in the third quadrant.
5. Find $\tan(\alpha - \beta)$ if $\sec \alpha = 3$, $\csc \alpha < 0$, $\csc \beta = -\sqrt{5}$, and $\cos \beta < 0$.
6. If $\sec \alpha = \tfrac{13}{5}$, $\sin \alpha < 0$, $\csc \beta = \tfrac{17}{8}$, and $\cos \beta < 0$, find $\tan(\alpha + \beta)$.
7. For α and β as in Problem 6, find $\tan 2\alpha$, $\tan 2\beta$, $\tan \tfrac{1}{2}\alpha$, and $\tan \tfrac{1}{2}\beta$.

In Problems 8–15 find the exact value of the given function.

8. $\tan(\alpha + \beta)$ if $\alpha = \sin^{-1}\left(-\dfrac{4}{5}\right)$ and

   $\beta = \cos^{-1}\left(-\dfrac{12}{13}\right)$

9. $\tan(\alpha - \beta)$ if $\alpha = \sec^{-1}\left(-\dfrac{\sqrt{13}}{3}\right)$ and

   $\beta = \cos^{-1}\left(\dfrac{1}{\sqrt{10}}\right)$

10. $\tan 2\theta$ if $\theta = \sin^{-1}\left(-\dfrac{5}{13}\right)$

11. $\tan\left[2 \tan^{-1}\left(-\dfrac{1}{2}\right)\right]$

12. $\tan \dfrac{\alpha}{2}$ if $\alpha = \sec^{-1}\left(-\dfrac{5}{4}\right)$

13. $\tan\left[\dfrac{1}{2}\tan^{-1}\left(-\dfrac{12}{5}\right)\right]$

14. $\tan\left[\tan^{-1}\dfrac{2}{3} + \tan^{-1}\dfrac{1}{4}\right]$

15. $\tan\left[2\tan^{-1}3 - \dfrac{1}{2}\tan^{-1}\left(-\dfrac{4}{3}\right)\right]$

16. Find the exact value of each of the following:

    a. $\tan 105°$ **b.** $\tan \dfrac{\pi}{12}$ **c.** $\tan \dfrac{11\pi}{12}$

    d. $\tan 195°$ **e.** $\tan \dfrac{\pi}{8}$

17. Find $\tan 2\theta$ and $\tan \frac{1}{2}\theta$ if $\csc \theta = \frac{5}{3}$ and $\sec \theta < 0$.

18. For the angles shown in the figures, find

    a. $\tan(\alpha + \beta)$ **b.** $\tan 2\alpha$
    c. $\tan(\alpha - \beta)$ **d.** $\tan 2\beta$
    e. $\tan \frac{1}{2}\alpha$ **f.** $\tan \frac{1}{2}\beta$

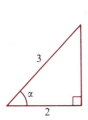

In Problems 19–22 evaluate by use of the sum and product formulas.

19. a. $\sin \dfrac{5\pi}{12} + \sin \dfrac{\pi}{12}$ **b.** $\cos \dfrac{7\pi}{12} - \cos \dfrac{\pi}{12}$

20. a. $\sin \dfrac{5\pi}{12} \cos \dfrac{7\pi}{12}$ **b.** $\cos \dfrac{3\pi}{8} \cos \dfrac{\pi}{8}$

21. a. $\sin 105° - \sin 15°$
    b. $\cos 165° + \cos 75°$

22. a. $\sin 105° \sin 15°$ **b.** $\cos 165° \sin 75°$

23. a. Write as a product: $\sin 5x + \sin 3x$
    b. Write as a sum: $\sin 5x \cos 3x$

24. a. Write as a product: $\cos 7x - \cos 5x$
    b. Write as a sum: $\sin 7x \sin 5x$

25. Derive the formula

$$\cot(\alpha + \beta) = \dfrac{\cot \alpha \cot \beta - 1}{\cot \alpha + \cot \beta}$$

26. If $\cos \theta = x$ and $\sin \theta > 0$, find $\tan 2\theta$ and $\tan \frac{1}{2}\theta$ in terms of x.

In Problems 27–37 prove the identities.

27. $\dfrac{2}{\tan 2x} = \cot x - \tan x$

28. $\csc \theta - \tan \dfrac{\theta}{2} = \cot \theta$

29. $\dfrac{1}{1 - \tan \theta} - \tan 2\theta = \dfrac{1}{1 + \tan \theta}$

30. $1 + \tan \alpha \tan \dfrac{\alpha}{2} = \sec \alpha$

31. $\cot 2\theta = \dfrac{\cot^2 \theta - 1}{2 \cot \theta}$

32. $\dfrac{2 \sin \theta - \sin 2\theta}{2 \sin \theta + \sin 2\theta} = \tan^2 \dfrac{\theta}{2}$

33. $\dfrac{\sin A - \sin B}{\cos A + \cos B} = \tan \frac{1}{2}(A - B)$

34. $\dfrac{\sin 7\theta - \sin 5\theta}{\cos 7\theta + \cos 5\theta} = \tan \theta$

35. $\dfrac{\cos 3\theta + \cos \theta}{\sin 3\theta + \sin \theta} = \cot 2\theta$

36. $\sin\left(\dfrac{3\pi}{4} + \theta\right)\sin\left(\dfrac{\pi}{4} - \theta\right) = \dfrac{1 - \sin 2\theta}{2}$

37. $2 \cos \dfrac{3\theta}{2} \sin \dfrac{\theta}{2} = \sin 2\theta - \sin \theta$

38. Find the exact value of $\tan^{-1} \frac{2}{3} - \tan^{-1}(-\frac{1}{5})$.
    Hint. Call this $\alpha - \beta$ and find $\tan(\alpha - \beta)$.

B

39. Prove that
    a. $\tan(n\pi \pm \theta) = \pm\tan \theta$

    b. $\tan\left[\dfrac{(2n + 1)\pi}{2} \pm \theta\right] = \pm\cot \theta$

    where the sign on the right-hand side can be determined by considering θ to be acute.

40. Prove the identity

$$\dfrac{\sin \theta + \sin 2\theta + \sin 3\theta}{\cos \theta + \cos 2\theta + \cos 3\theta} = \tan 2\theta$$

41. Use the addition formulas for the sine and cosine

to derive the product formulas as follows:

a. Add the formulas for $\sin(\alpha + \beta)$ and $\sin(\alpha - \beta)$ to get the formula for $\sin \alpha \cos \beta$.

b. Subtract the formula for $\sin(\alpha - \beta)$ from that for $\sin(\alpha + \beta)$ to get the formula for $\cos \alpha \sin \beta$.

c. Add the formulas for $\cos(\alpha + \beta)$ and $\cos(\alpha - \beta)$ to get the formula for $\cos \alpha \cos \beta$.

d. Subtract the formula for $\cos(\alpha + \beta)$ from that for $\cos(\alpha - \beta)$ to get the formula for $\sin \alpha \sin \beta$.

42. In the addition formulas for the sine and cosine make the following substitutions: $\alpha + \beta = A$ and $\alpha - \beta = B$. Solve these two equations simultaneously for α and β in terms of A and B, and obtain the sum formulas.

43. Prove the identity

$$\frac{\sin(x + h) - \sin x}{h} = \frac{\sin \dfrac{h}{2}}{\dfrac{h}{2}} \cos\left(x + \frac{h}{2}\right)$$

44. Derive the formula

$$\tan \frac{\alpha}{2} = \frac{1 - \cos \alpha}{\sin \alpha}$$

5 TRIGONOMETRIC EQUATIONS

The solutions of most trigonometric equations cannot be obtained exactly, and we have to settle for approximate solutions obtained through some numerical procedure, usually with the aid of a calculator. There are, however, enough such equations for which elementary techniques will yield exact answers that some time devoted to these techniques is justified.

To make clear the sorts of solutions we are looking for, let us consider the very simple equation

$$2 \sin \theta - 1 = 0$$

The problem is to discover all real numbers θ for which this equation is true. We write the equation in the equivalent form

$$\sin \theta = \frac{1}{2}$$

and then rely on our knowledge of special angles to conclude that this is satisfied if, and only if,

$$\theta = \frac{\pi}{6}, \ \frac{5\pi}{6}$$

or any other number obtained from these by adding integral multiples of 2π to each. To see this, remember that the sine of a number θ is the y coordinate of the point $P(\theta)$ on the unit circle. Thus, we are seeking those numbers θ for which the y coordinate of $P(\theta) = \frac{1}{2}$. There are only two points on the unit circle with y coordinate $\frac{1}{2}$, and our knowledge of the $30°$–$60°$ right triangle tells us that these points are $P(\pi/6)$ and $P(5\pi/6)$; see Figure 7. Since $P(\theta) = P(\theta + 2n\pi)$, we conclude that all solutions of the equation are given by $(\pi/6) + 2n\pi$ or $(5\pi/6) + 2n\pi$.

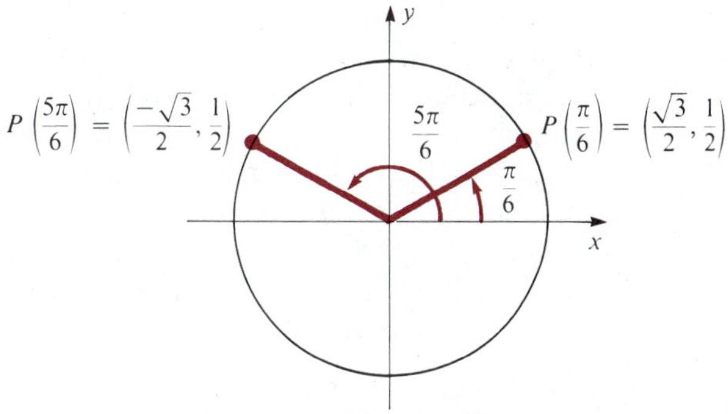

FIGURE 7

Usually, it is sufficient to give only the **primary** solutions, that is, those lying between 0 and 2π. We would know then that all other solutions are obtainable from these by adding multiples of 2π.

We can usually condense the above reasoning as follows:

1. Determine the appropriate reference angle.
2. By the sign of the function, locate all primary angles having this reference angle.
3. Write the answer as the radian measure of these angles.

In our example, since we know that $\sin 30° = \frac{1}{2}$, the reference angle is 30°. The sine is positive in quadrants I and II, so the angles are 30° and 150°. In radians these are $\pi/6$ and $5\pi/6$.

It might be useful to review at this time the values of the functions of 30°, 45°, and 60°.

$$\sin 30° = \sin \frac{\pi}{6} = \frac{1}{2} \qquad \sin 45° = \sin \frac{\pi}{4} = \frac{1}{\sqrt{2}} \qquad \sin 60° = \sin \frac{\pi}{3} = \frac{\sqrt{3}}{2}$$

$$\cos 30° = \cos \frac{\pi}{6} = \frac{\sqrt{3}}{2} \qquad \cos 45° = \cos \frac{\pi}{4} = \frac{1}{\sqrt{2}} \qquad \cos 60° = \cos \frac{\pi}{3} = \frac{1}{2}$$

$$\tan 30° = \tan \frac{\pi}{6} = \frac{1}{\sqrt{3}} \qquad \tan 45° = \tan \frac{\pi}{4} = 1 \qquad \tan 60° = \tan \frac{\pi}{3} = \sqrt{3}$$

Also,

$$\sin 0° = \sin 0 = 0 \qquad \sin 90° = \sin \frac{\pi}{2} = 1$$

$$\cos 0° = \cos 0 = 1 \qquad \cos 90° = \cos \frac{\pi}{2} = 0$$

$$\tan 0° = \tan 0 = 0 \qquad \tan 90° = \tan \frac{\pi}{2} \text{ is not defined}$$

(The cotangent, secant, and cosecant are not listed because they occur less frequently and can always be obtained by the reciprocal relations.) If you feel comfortable with radian measure by now, you can omit the degree measure entirely.

If an equation can be worked around to a form such as $\sin\theta = \pm a$, $\cos\theta = \pm b$, or $\tan\theta = \pm c$, for instance, where a and b are any of the numbers $0, 1, \frac{1}{2}, 1/\sqrt{2}$, or $\sqrt{3}/2$, and c is any of the numbers $0, 1, 1/\sqrt{3}$, or $\sqrt{3}$, then we can obtain all solutions in the manner outlined above. For other values of a, b, and c, a calculator can be used.

EXAMPLE 13

Find all values of θ, for which $0 \le \theta < 2\pi$, satisfying $2\cos^2\theta - \cos\theta - 1 = 0$.

Solution   We treat this first as a quadratic equation in $\cos\theta$ and factor:

$$(2\cos\theta + 1)(\cos\theta - 1) = 0$$

This is true if and only if $2\cos\theta + 1 = 0$ or $\cos\theta - 1 = 1$, that is, if and only if

$$\cos\theta = -\tfrac{1}{2} \quad \text{or} \quad \cos\theta = 1$$

Since $\cos 60° = \frac{1}{2}$, the reference angle in the first of these is $60°$. Since the cosine is negative in quadrants II and III, it follows that we are seeking the angles $120°$ and $240°$. In radians these are $2\pi/3$ and $4\pi/3$. The only primary angle for which $\cos\theta = 1$ is $\theta = 0°$, or 0 radians. Thus, the primary solution set for the equation is $\{0, 2\pi/3, 4\pi/3\}$.   ▲

Remark. In many equations involving trigonometric functions the variable must be treated as a real number, which is the reason for writing the above answers as the radian measures of the angles. Recall that the values of $\sin\theta$, $\cos\theta$, and the other trigonometric functions are unchanged whether we consider θ a real number or the radian measure of an angle. An example of a mixed algebraic and trigonometric equation will perhaps help to make this point clearer. The equation

$$2\sin x - x = 0$$

is clearly satisfied when $x = 0$, but by trial and error, using a calculator (or by some more sophisticated numerical procedure), we can find that another solution is

$$x \approx 1.8955$$

If we think of this x as the radian measure of an angle, then the degree measure of the angle is approximately $108.6°$. Now it would *not* be true that $x = 108.6°$; in fact, it would not even make sense, for we would have

$$2\sin 108.6° - 108.6° = 0$$

which is a totally meaningless statement.

EXAMPLE 14

Find all primary solutions of $\sin 2x + \cos x = 0$.

Solutions   This time we will omit most discussion and go only through the steps that would be expected in doing the problem.

$$\sin 2x + \cos x = 0$$

$$2 \sin x \cos x + \cos x = 0$$

$$\cos x(2 \sin x + 1) = 0$$

$$\cos x = 0 \quad \text{or} \quad 2 \sin x = -1$$

$$x = \frac{\pi}{2} \quad \text{or} \quad \frac{3\pi}{2}, \quad \sin x = -\frac{1}{2}$$

$$x = \frac{7\pi}{6} \quad \text{or} \quad \frac{11\pi}{6}$$

The solution set is $\{\pi/2, 7\pi/6, 3\pi/2, 11\pi/6\}$.   ▲

EXAMPLE 15

Find all primary solutions of $\tan \alpha - 3 \cot \alpha = 0$.

Solution
$$\tan \alpha - 3 \cot \alpha = 0$$

$$\tan \alpha - \frac{3}{\tan \alpha} = 0$$

$$\tan^2 \alpha - 3 = 0$$

$$\tan \alpha = \pm\sqrt{3}$$

The reference angle is $60°$, or $\pi/3$ radians. Therefore,

$$\alpha = \frac{\pi}{3}, \frac{2\pi}{3}, \frac{4\pi}{3}, \frac{5\pi}{3}$$

Since we multiplied by an unknown expression, namely, $\tan \alpha$, it is essential that we check to see that this was not zero for the values of α found, which is clearly the case here, since $\tan \alpha = \pm\sqrt{3} \neq 0$. So the solution set is $\{\pi/3, 2\pi/3, 4\pi/3, 5\pi/3\}$.   ▲

EXAMPLE 16

Find all primary solutions of $\sqrt{3} \cos x = 2 + \sin x$.

Solution   The fact that $\cos^2 x = 1 - \sin^2 x$ suggests that if we square both sides of the given equation, it will be easier to work with.

$$3 \cos^2 x = 4 + 4 \sin x + \sin^2 x$$

$$3(1 - \sin^2 x) = 4 + 4 \sin x + \sin^2 x$$

$$4 \sin^2 x + 4 \sin x + 1 = 0$$

$$(2 \sin x + 1)^2 = 0$$

$$\sin x = -\tfrac{1}{2}$$

$$x = \frac{7\pi}{6}, \quad \frac{11\pi}{6}$$

We again must check our answers, because squaring does not necessarily lead to an equivalent equation; it may introduce extraneous roots. So we check in the original equation.

When $x = 7\pi/6$, we get

$$\sqrt{3}\left(-\frac{\sqrt{3}}{2}\right) \overset{?}{=} 2 + \left(-\frac{1}{2}\right)$$

$$-\frac{3}{2} \neq \frac{3}{2}$$

So $7\pi/6$ is not a solution.
When $x = 11\pi/6$,

$$\sqrt{3}\left(\frac{\sqrt{3}}{2}\right) \overset{?}{=} 2 + \left(-\frac{1}{2}\right)$$

$$\frac{3}{2} = \frac{3}{2}$$

So $x = 11\pi/6$ is the only primary solution. ▲

EXAMPLE 17

Find all primary solutions of $2 \sin 3\theta - \sqrt{3} \tan 3\theta = 0$.

Solution

$$2 \sin 3\theta - \frac{\sqrt{3} \sin 3\theta}{\cos 3\theta} = 0$$

$$\sin 3\theta (2 \cos 3\theta - \sqrt{3}) = 0 \quad \text{(We multiplied by } \cos 3\theta.\text{)}$$

$$\sin 3\theta = 0 \quad \text{or} \quad \cos 3\theta = \frac{\sqrt{3}}{2} \quad \text{(So we did not multiply by zero.)}$$

Now we want all values of θ lying between 0 and 2π. We must therefore find all values of 3θ lying between 0 and 6π. In general, if $n\theta$ is involved, in order to find all values of θ between 0 and 2π, we find all values of $n\theta$ between 0 and $2n\pi$, and then divide by n. We have from $\sin 3\theta = 0$ that

$$3\theta = 0, \quad \pi, \quad 2\pi, \quad 3\pi, \quad 4\pi, \quad 5\pi$$

and

$$\theta = 0, \quad \frac{\pi}{3}, \quad \frac{2\pi}{3}, \quad \pi, \quad \frac{4\pi}{3}, \quad \frac{5\pi}{3}$$

From $\cos 3\theta = \sqrt{3}/2$, we have

$$3\theta = \frac{\pi}{6}, \quad \frac{11\pi}{6}, \quad \frac{13\pi}{6}, \quad \frac{23\pi}{6}, \quad \frac{25\pi}{6}, \quad \frac{35\pi}{6}$$

$$\theta = \frac{\pi}{18}, \quad \frac{11\pi}{18}, \quad \frac{13\pi}{18}, \quad \frac{23\pi}{18}, \quad \frac{25\pi}{18}, \quad \frac{35\pi}{18}$$

In both cases we found the first two values of 3θ, then added 2π once, and then 2π again. The complete solution set is

$$\left\{ 0, \quad \frac{\pi}{18}, \quad \frac{\pi}{3}, \quad \frac{11\pi}{18}, \quad \frac{2\pi}{3}, \quad \frac{13\pi}{18}, \quad \pi, \quad \frac{23\pi}{18}, \quad \frac{4\pi}{3}, \quad \frac{25\pi}{18}, \quad \frac{5\pi}{3}, \quad \frac{35\pi}{18} \right\} \quad \blacktriangle$$

EXAMPLE 18

Find all real solutions of $\tan x - \cot x = 2$.

Solution

$$\tan x - \cot x = 2$$

$$\frac{\sin x}{\cos x} - \frac{\cos x}{\sin x} = 2$$

$$\sin^2 x - \cos^2 x = 2 \sin x \cos x$$

$$-\cos 2x = \sin 2x$$

$$\tan 2x = -1$$

$$2x = \frac{3\pi}{4}, \quad \frac{7\pi}{4}, \quad \frac{11\pi}{4}, \quad \frac{15\pi}{4} \quad \text{(We "went around" twice.)}$$

$$x = \frac{3\pi}{8}, \quad \frac{7\pi}{8}, \quad \frac{11\pi}{8}, \quad \frac{15\pi}{8}$$

As none of these values of x makes $\sin x$, $\cos x$, or $\cos 2x$ equal to zero, we did not multiply by zero. We must also check to see that we did not lose any roots when dividing by $\cos 2x$; that is, we must check to see if any of the roots of $\cos 2x = 0$ are roots of the original equation. Since $\cos 2x = 0$ when $2x = \pi/2$, $3\pi/2$, $5\pi/2$, $7\pi/2$, and so when $x = \pi/4$, $3\pi/4$, $5\pi/4$, $7\pi/4$, we check and find that none of these is a root of the original equation. So we have found all primary solutions. To get *all* solutions, we add arbitrary multiples of 2π to each of these. So the complete solution set is

$$\left\{ \frac{3\pi}{8}, \quad \frac{7\pi}{8}, \quad \frac{11\pi}{8}, \quad \frac{15\pi}{8}, \quad \frac{19\pi}{8}, \quad \frac{23\pi}{8}, \quad \frac{27\pi}{8}, \quad \frac{31\pi}{8}, \quad \frac{35\pi}{8}, \quad \frac{39\pi}{8}, \quad \frac{43\pi}{8}, \quad \frac{47\pi}{8}, \quad \ldots \right\}$$

or we could write

$$\{3\pi/8 + 2n\pi, \ 7\pi/8 + 2n\pi, \ 11\pi/8 + 2n\pi, \ 15\pi/8 + 2n\pi: \quad n = 0, \pm 1, \pm 2, \ldots\}$$

▲

EXAMPLE 19 ||||||||||||||||||||||||||||||||||||||||||||||||||||||||||||||||||||||||||||||||||||||||||||||||||||||||||||||||||||||||

Use a calculator to find all values of x in $[0, 2\pi)$ satisfying $2 \sin^2 x - \cos^2 x = \sin x$, correct to 4 decimal places of accuracy.

Solution   Replacing $\cos^2 x$ by $1 - \sin^2 x$ and collecting terms gives

$$3 \sin^2 x - \sin x - 1 = 0$$

By the quadratic formula, we get

$$\sin x = \frac{1 \pm \sqrt{13}}{6}$$

The keystroke sequence to find x for which $\sin x = (1 + \sqrt{13})/6$ is as follows.

$$1 \ \boxed{+} \ 13 \ \boxed{\sqrt{x}} \ \boxed{=} \ \boxed{\div} \ 6 \ \boxed{=} \ \boxed{\text{INV}} \ \boxed{\text{SIN}}$$

To four decimal places we get $x = 0.8751$. Interpreted as the radian measure of an angle, this angle is in the first quadrant. But the sine is also positive in the second quadrant. So using the x value just found as a reference angle, we also have the solution

$$\pi - x = \pi - 0.8751 \approx 2.2665$$

The values of x corresponding to $\sin x = (1 - \sqrt{13})/6$ are found in a similar way. With the obvious modification in the keystroke sequence, we get $x = -0.4492$. This is a negative angle in the fourth quadrant and so is not in the interval $[0, 2\pi)$. We can get the corresponding primary solution by adding 2π. This gives $2\pi + x = 2\pi - 0.4492 \approx 5.8340$. Finally, we use 0.4492 as a reference angle to obtain the third-quadrant solution $\pi + 0.4492 \approx 3.5908$. The complete primary solution set is therefore

$$\{0.8751, 2.2665, 3.5908, 5.8340\} \quad\quad ▲$$

We could continue with examples, each possessing its own special features, but these illustrate the main techniques. It might be useful to summarize some of the points that the examples were meant to bring out:

1.   Use algebraic techniques such as factoring, multiplying by the LCD, and squaring in conjunction with basic trigonometric identities to simplify so as to obtain one or more elementary equations of the form $\sin x = \pm a$, $\cos x = \pm b$, and so on.

2. Find the radian measure of all primary angles (that is, $0 \leq x < 2\pi$) satisfying the elementary equations in 1. If *all* solutions are desired, add arbitrary multiples of 2π to each of these to obtain the complete solution set.

3. If in arriving at the elementary equations both sides of an equation were squared, or if both sides were multiplied by an expression containing a variable, it is necessary to check the answers. (In the case of multiplying by an unknown, it is sufficient to check that the multiplier was not zero.) If both sides of an equation are divided by an expression containing a variable, then the values of the variable for which this expression equals zero must be checked in the original equation to see if they are solutions. (In general, such division can be avoided by factoring.)

4. If the final elementary equation(s) is of the form $\sin nx = \pm a$, $\cos nx = \pm b$, and so on, then to get all primary values of x, find all values of nx between 0 and $2n\pi$, and divide by n. This amounts to finding the angles in the first revolution, then adding 2π a total of $n - 1$ separate times, and finally dividing everything by n.

EXERCISE SET 5

A

Find all primary solutions unless otherwise specified.

1. $2 \cos x = \sqrt{3}$
2. $\cot x + 1 = 0$
3. $\sec x = 2$
4. $\sin^2 x = 1$ (Give all solutions.)
5. $\cos x - 2 \cos^2 x = 0$
6. $\sin 2\theta - \cos \theta = 0$
7. $\cos 2x + \cos x = 0$
8. $2 \tan \psi - \tan \psi \sec \psi = 0$
9. $\sin^2 t - \cos^2 t = 1$
10. $\sec^2 \theta = 2 \tan \theta$
11. $2 \cos x - \cot x = 0$
12. $2 \cos^2 x + \sin x - 1 = 0$
13. $4 \tan^2 \alpha = 3 \sec^2 \alpha$
14. $\sin^2 x = 2(\cos x - 1)$
15. $\cos^2 x - \sin x + 5 = 0$
16. $\tan 2x - \cos 2x + \sec 2x = 0$
17. $\dfrac{1 - \cos x}{\sin x} = \sin x$
18. $2 \cos^2 2\theta - 3 \cos 2\theta + 1 = 0$
19. $\sin 6\theta + \sin 3\theta = 0$
20. $\sin^2 4\theta - \sin 4\theta - 2 = 0$
21. $\tan \dfrac{x}{2} + \cos x = 1$
22. $\sin^2 \dfrac{x}{2} - \sin^2 x = 0$
23. $2(\sin^2 2\theta - \cos^2 2\theta) + \sqrt{3} = 0$

24. $2 \cos x - \cot x \csc x = 0$
25. $2 \tan^2 \theta + \sec \theta + 2 = 0$

B

26. $\sin \theta + \cos \theta = 1$ (Give all solutions.)
27. $\cos 2x \tan 3x + \sqrt{3} \cos 2x = 0$
28. $\dfrac{\sin x}{1 - \cos x} = \dfrac{\cos x}{1 + \sin x}$
29. $\sin 2\theta - 4 \sin \theta = 3(2 - \cos \theta)$
30. $\sqrt{3} \tan \theta = 2 \sec \theta - 1$

In Problems 31–36 use a calculator to find all primary solutions, correct to four decimal places.

31. $12 \sin^2 x - \sin x - 6 = 0$
32. $\cos 2x + 2 \cos x = 0$
33. $2 \sec^2 x = 5 \tan x$ 34. $3 \sin x - \tan \tfrac{1}{2}x = 0$
35. $2 \sin x - \csc x + 3 = 0$
36. $\sec x - 2 = 10 \cos x$

In Problems 37–39 find all primary solutions without using a calculator.

37. $\tan^3 x + \tan^2 x - 3 \tan x - 3 = 0$
38. $8 \sin^4 \theta - 2 \sin^2 \theta - 3 = 0$
39. $2 \sin^3 x + 3 \sin^2 x - 1 = 0$

C (Graphing calculator)

In Problems 40–43 find all solutions to four significant figures by zooming in on the x intercepts of the graph of the function on the left-hand side.

40. $\sin x + 2x - 1 = 0$
41. $\tan x - \sin x - 2 = 0$
42. $\cos 2x + x = 0$
43. $\ln(x + 1) - 3 \cos x - 1 = 0$

 REVIEW EXERCISE SET

A

1. Evaluate each of the following without the use of a calculator:
  a. $\tan(7\pi/12)$ **b.** $\sin(23\pi/12)$
  c. $\cos(\pi/8)$ **d.** $\cos(13\pi/12)$
2. Express each of the following in terms of θ only.

  a. $\sin\left(\dfrac{3\pi}{2} - \theta\right)$ **b.** $\tan(\theta + \pi)$

  c. $\sec(4\pi - \theta)$ **d.** $\cos\left(\theta - \dfrac{\pi}{2}\right)$

  e. $\csc\left(\dfrac{9\pi}{2} + \theta\right)$

3. If $\sec \theta = \frac{5}{4}$ and $\sin \theta < 0, 0 \le \theta \le 2\pi$, find
  a. $\cos 2\theta$ **b.** $\tan 2\theta$ **c.** $\cos \frac{1}{2}\theta$
4. If $\sin \theta = \frac{4}{5}$ and $\tan \theta < 0, 0 \le \theta \le 2\pi$, find
  a. $\sin 2\theta$ **b.** $\cos 2\theta$ **c.** $\sin \frac{1}{2}\theta$
5. If $\alpha = \arcsin(-\frac{3}{5})$ and $\beta = \arccos(-\frac{5}{13})$, find
  a. $\sin(\alpha + \beta)$ **b.** $\cos(\alpha - \beta)$
  c. $\tan(\alpha + \beta)$
6. If $\alpha = \tan^{-1}(-\frac{4}{3})$ and $\beta = \cos^{-1}(-\frac{8}{17})$, find
  a. $\sin(\alpha - \beta)$ **b.** $\cos(\alpha + \beta)$
  c. $\tan(\alpha - \beta)$
7. If $\tan 2\theta = -\frac{24}{7}$ and $\pi/2 < 2\theta < \pi$, find $\sin \theta$ and $\cos \theta$.
8. If $\tan \theta = \frac{15}{8}$ and $\pi < \theta < 3\pi/2$, find $\sin(\theta/2)$, $\cos(\theta/2)$, and $\tan(\theta/2)$.
9. Evaluate
  a. $\cos[2 \arcsin(-\frac{3}{5})]$ **b.** $\sin[2 \sin^{-1}(-\frac{5}{13})]$
10. Evaluate
  a. $\tan[2 \cos^{-1}(-\frac{4}{5})]$ **b.** $\cos(\frac{1}{2} \tan^{-1} 4\sqrt{3})$

In Problems 11–24 prove that the given equations are identities.

11. $\csc \theta - \cos \theta \cot \theta = \sin \theta$

12. $\dfrac{2}{\tan \theta + \cot \theta} = \sin 2\theta$

13. $\dfrac{\cos \theta + \cot \theta}{1 + \csc \theta} = \cos \theta$

14. $\dfrac{\cos \theta}{\sec \theta - \tan \theta} = 1 + \sin \theta$

15. $\dfrac{\cot \theta - \tan \theta}{\cot \theta + \tan \theta} = \cos 2\theta$

16. $\sin \theta \tan \theta + \cos \theta = \sec \theta$

17. $\dfrac{\tan \theta}{\sec \theta + 1} + \dfrac{\sec \theta - 1}{\tan \theta} = \dfrac{2 \tan \theta}{\sec \theta + 1}$

18. $\tan \theta(\cos 2\theta + 1) = \sin 2\theta$
19. $\sin \theta + \cos \theta \cot \theta = \csc \theta$
20. $\sin 2\theta + (\cos \theta - \sin \theta)^2 = 1$

21. $1 + \dfrac{\tan^2 \theta}{\sec \theta + 1} = \sec \theta$

22. $\cot \theta - \tan \theta = 2 \cot 2\theta$

23. $\dfrac{1 + \tan^2 \theta}{1 - \tan^2 \theta} = \sec 2\theta$ **24.** $\dfrac{\sin 2\theta}{1 - \cos 2\theta} = \cot \theta$

In Problems 25–28 find all primary solutions.

25. $\sin x - 2 \sin^2 x = 0$ **26.** $\sin 2\theta + \cos \theta = 0$
27. $2 \cos^2 3t - 3 \cos 3t - 2 = 0$
28. $2 \cos^2 t = 1 - \sin t$

B

29. Evaluate
  a. $\cos[\tan^{-1} \frac{5}{12} - \sin^{-1}(-\frac{8}{17})]$
  b. $\tan^{-1} \frac{1}{3} + \tan^{-1}(-2)$

In Problems 30–35 prove the identities.

30. $\dfrac{\sin \theta + \cos 2\theta - 1}{\cos \theta - \sin 2\theta} = \tan \theta$

31. $\sin^2 \theta + \dfrac{2 - \tan^2 \theta}{\sec^2 \theta} - 1 = \cos 2\theta$

32. $\dfrac{\sec \theta - 1}{\sec \theta + 1} = 2 \csc^2 \theta - 2 \cot \theta \csc \theta - 1$

33. $\sec \theta \csc \theta = 2 \cos \theta \csc \theta = \tan \theta - \cot \theta$

34. $\dfrac{\sin 8\theta - \sin 6\theta}{\cos 8\theta + \cos 6\theta} = \tan \theta$

35. $\dfrac{\cos \theta}{1 - \sin \theta} = \tan \theta + \sec \theta$

In Problems 36–39 find all primary solutions.

36. $\tan^2 2x + 3 \sec 2x + 3 = 0$
37. $\cos 2\theta \sec \theta + \sec \theta = 1$
38. $\tan 2\theta = 2 \cos \theta$
39. $1 + \sin t = \cos t$
   Hint. Square both sides.
40. Show that if $x = \tan \theta$, where $-\pi/2 < \theta < \pi/2$, then

$$\frac{x^3}{(1 + x^2)^{3/2}} = \sin^3 \theta$$

41. By substituting $x = 2 \sec \theta$, where $0 \leq \theta < \pi/2$ if $x \geq 2$ and $\pi \leq \theta < 3\pi/2$ if $x \leq -2$, show that

$$\frac{\sqrt{x^2 - 4}}{x} = \sin \theta$$

42. Make the substitution $x = 3/2 \sin \theta$, where $-\pi/2 \leq \theta \leq \pi/2$, to show that

$$\left(\frac{3}{4}\right)\frac{\sqrt{9 - 4x^2}}{x^2} = \tfrac{4}{3} \cot \theta \csc \theta$$

C (Graphing calculator)

43. Draw the graph of $y = \sqrt{3} \cos x - \sin x$. Give the coordinates of the highest and lowest points on the curve for which $0 \leq x \leq 2\pi$.

44. Use a calculator to find all primary solutions of the equation $5 \cos \theta - 2 \tan \theta = 3 \sec \theta$, correct to four decimal places.

THE SOLUTION
OF TRIANGLES

In certain parts of calculus, students must be able to use trigonometry to solve triangles, that is, to find unknown sides or angles. The following theorem taken from a calculus book makes use of one of the important tools for doing this, called the **law of cosines;** it also makes use of **vectors.***

If θ is the angle between nonzero vectors **u** and **v**, then

$$\cos \theta = \frac{\mathbf{u} \cdot \mathbf{v}}{|\mathbf{u}||\mathbf{v}|}$$

Proof Assume first that **u** and **v** are not parallel, and choose geometric representatives of them such that each has its initial point at the origin as in [the figure] . . . . The vector $\overrightarrow{AB} = \mathbf{v} - \mathbf{u}$, and so by the law of cosines we have

$$|\mathbf{v} - \mathbf{u}|^2 = |\mathbf{u}|^2 + |\mathbf{v}|^2 - 2|\mathbf{u}||\mathbf{v}| \cos \theta$$

or, in terms of components,

$$(v_1 - u_1)^2 + (v_2 - u_2)^2 = u_1^2 + u_2^2 + v_1^2 + v_2^2 - 2|\mathbf{u}||\mathbf{v}| \cos \theta$$

so that

$$|\mathbf{u}||\mathbf{v}| \cos \theta = u_1 v_1 + u_2 v_2 = \mathbf{u} \cdot \mathbf{v}$$

---

* Leonard I. Holder, *Calculus with Analytic Geometry* (Belmont, Calif.: Wadsworth Publishing Company, © 1988, Wadsworth, Inc.), p. 548. Reprinted by permission.

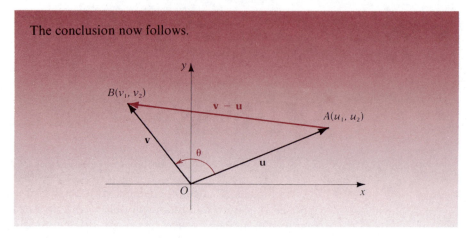

The conclusion now follows.

In this chapter we will study both vectors and the law of cosines.

RIGHT TRIANGLE TRIGONOMETRY

In many applied problems it is necessary to find one or more sides or angles of a triangle from certain known information about the triangle. For example, if we know any two sides of a right triangle, we can always find the third side, using the Pythagorean theorem. Or we may know an angle and one of the sides and want to find the other angle and sides. Finding all of the unknown angles and sides of a triangle is called **solving the triangle.** In this section we will concentrate on solving right triangles, and in the next two we will study triangles that have no right angle, called **oblique triangles.**

The Pythagorean theorem is one useful tool for solving right triangles. The primary means, however, is the use of the trigonometric functions defined in terms of right triangles (Definition 1 of Chapter 7). The following examples illustrate how this is done. Most of the calculations involve approximate values, and we round answers accordingly. For simplicity, we use equals signs throughout, even when the answers are approximate.

EXAMPLE I

Solve the right triangle ABC in which $A = 37°30'$ and $b = 17.3$ (Figure 1).

Solution   Since A and B are complementary, $B = 90° - 37°30' = 52°30'$. Using the tangent of angle A we get

$$\tan A = \frac{a}{b}$$

$$a = b \tan A = 17.3 \tan 37°30' = 13.3$$

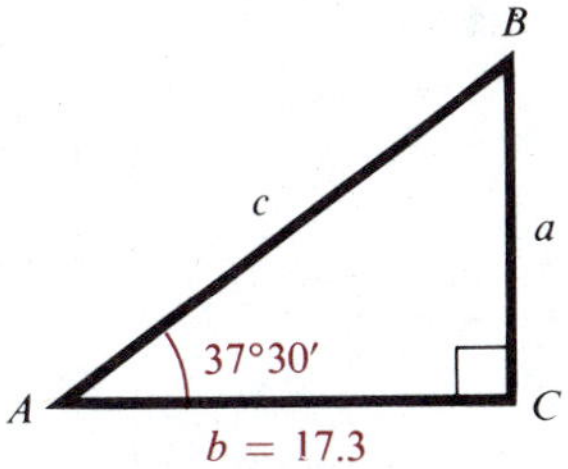

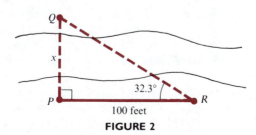

FIGURE I

By calculator, we write $37°30' = 37.5°$ and use the keystroke sequence

17.3 $\boxed{\times}$ 37.5 $\boxed{\text{TAN}}$ $\boxed{=}$ Display 13.27475

To find c, we use the cosine of A:

$$\cos A = \frac{b}{c}$$

$$c = \frac{b}{\cos A} = \frac{17.3}{\cos 37°30'} = 21.8$$

By calculator:

17.3 $\boxed{\div}$ 37.5 $\boxed{\text{COS}}$ $\boxed{=}$ Display 21.806173   ▲

One of the chief applications of right triangle trigonometry is the calculation of inaccessible distances. An illustration is given in the next example.

EXAMPLE 2 ┃┃

Figure 2 illustrates a river and two points P and Q on opposite sides. By means of a transit at P, a line of sight perpendicular to PQ is determined, and a distance of 100 feet (to the nearest foot) along this line from P to R is measured. At R the transit is again set up and the angle from RP to RQ is measured and found to be $32.3°$. It is desired to find the distance from P to Q.

FIGURE 2

Solution   Let the distance from P to Q be denoted by x. Then

$$\tan 32.3° = \frac{x}{100}$$

$$x = 100 \tan 32.3° = 63.2 \text{ feet}$$ ▲

The following terminology will be employed in certain applications.

Angle of Elevation. Angle from the horizontal upward to the line of sight of the observer (Figure 3).

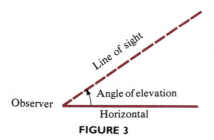

FIGURE 3

Angle of Depression. Angle from the horizontal downward to the line of sight of the observer (Figure 4).

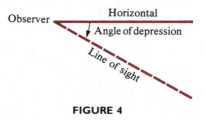

FIGURE 4

Bearing. A means of giving the direction of a path by using either North or South as a base line and then indicating an acute angle either East or West of the base line. For example, N 30° E means a path in a direction 30° to the East of North. Similarly, S 60° W means a direction which is 60° West of South. Both of these examples are illustrated in Figure 5.

Heading. An alternative means of indicating direction in which an angle between 0° and 360° measured clockwise from North is given. Figure 6 shows a heading of 225°. This means of indicating direction is widely employed in air navigation.

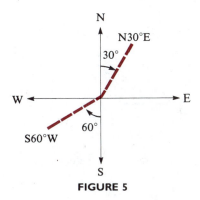

FIGURE 5

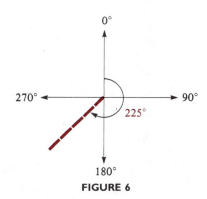

FIGURE 6

EXAMPLE 3 |||||||||||||||||||||||||||||||||||||||||||||||||||||||||||||||||||||||||||||||||||||||||||||||||||||||||

From the top of a ranger tower 132 feet high the angle of depression of an illegal campfire is 9.8°. How far is the campfire from the base of the tower? Assume the ground is level.

Solution   Refer to Figure 7. Angle B in the right triangle ABC is seen by geometry to be equal to the angle of depression. Denote the unknown distance $\overline{BC}$ by x. Then we

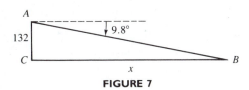

FIGURE 7

have

$$\tan B = \frac{132}{x}$$

$$x = \frac{132}{\tan B} = \frac{132}{\tan 9.8°} = 764 \text{ feet}$$ ▲

EXAMPLE 4 |||||||||||||||||||||||||||||||||||||||||||||||||||||||||||||||||||||||||||||||||||||||||||||||||||||||||

The angle of elevation of the top of a mountain from point A is 36.2°, and the angle of elevation of the top from point B, 1,000 feet (to the nearest foot) nearer the base of the mountain and on the same level as A, is 41.7° (see Figure 8). Find the height of the mountain.

Solution   We introduce the auxiliary unknown d in the figure, since it will be involved in the functions we use. There are two right triangles, each involving the unknown

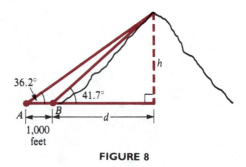

FIGURE 8

height h. In each triangle, h is the side opposite the known angle, and the adjacent side is the other about which we have at least some information. We could therefore use either the tangent or the cotangent. Either one would involve both h and d. By using both triangles we will get two equations in these two unknowns. Our objective will be to eliminate d between the two equations, thereby obtaining an equation in h alone. We choose the cotangent function, since this facilitates the elimination of d.

$$\cot 36.2° = \frac{1{,}000 + d}{h}$$

$$1{,}000 + d = h \cot 36.2°$$

$$\cot 41.7° = \frac{d}{h}$$

$$d = h \cot 41.7°$$

Now we substitute for d from the last equation into the second, and then solve for h.

$$1{,}000 + h \cot 41.7° = h \cot 36.2°$$

$$h(\cot 36.2° - \cot 41.7°) = 1{,}000$$

$$h = \frac{1{,}000}{\cot 36.2° - \cot 41.7°}$$

$$= 4{,}099$$

By calculator:

1000 ÷ (36.2 TAN 1/x − 41.7

TAN 1/x) = Display 4099.1782

So the mountain is approximately 4,100 feet above the surrounding countryside. If the elevation of points A and B above sea level is known, this can be added to 4,100 to give the height of the mountain above sea level. ▲

EXAMPLE 5

Ships A and B leave port at the same time, ship A traveling in the direction N 39°24′ E at an average speed of 25 miles per hour, and ship B traveling in the direction S 50°36′ E at an average speed of 15 miles per hour. Find how far apart the ships are after two hours and the bearing of ship B from ship A.

Solution   Since 39°24′ + 50°36′ = 90°, we see that the paths of the two ships are at right angles to each other, as shown in Figure 9. After two hours ship A has gone 50 miles and ship B 30 miles. Let d denote the distance between the ships at that

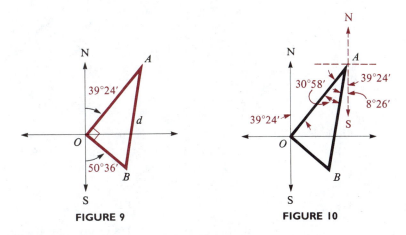

FIGURE 9   **FIGURE 10**

time. Then, by the Pythagorean theorem,

$$d^2 = (50)^2 + (30)^2 = 2{,}500 + 900 = 3{,}400$$
$$d = 58.3 \text{ miles}$$

To find the bearing of B from A, first we find the angle at A in triangle AOB. We have

$$\tan A = \tfrac{30}{50} = 0.6000$$
$$A = 30°58′$$

Now, if we draw a line from A in the south direction (see Figure 10), we see by geometry that the angle from this line to OA is 39°24′. So the angle from this line to AB is

$$39°24′ - 30°58′ = 8°26′$$

The bearing of B from A is therefore S 8°26′ W.

▲

EXAMPLE 6 ||||||||||||||||||||||||||||||||||||||||||||||||||||||||||||||||||||||||||||||||||||||||||||||

An airplane leaves an airport and travels at an average speed of 240 miles per hour on a heading of 130°. How far south and how far east of the airport is the airplane after 2 hours and 15 minutes? Assume the wind is negligible.

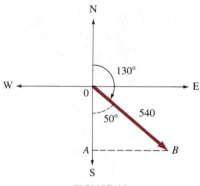

FIGURE II

Solution   In Figure 11 the origin is the airport, and B is the location of the airplane after 2 hours and 15 minutes. The angle at O in the right triangle AOB is $180° - 130° = 50°$. Since 2 hr 15 min = 2.25 hours, we use the relationship "distance = rate × time" to get

$$\overline{OB} = (240)(2.25) = 540 \text{ miles}$$

We want to know $\overline{OA}$ and $\overline{AB}$. Using right triangle AOB, we have

$$\cos 50° = \frac{\overline{OA}}{540} \quad \text{and} \quad \sin 50° = \frac{\overline{AB}}{540}$$

So

$$\overline{OA} = 540 \cos 50° = 347 \quad \text{and} \quad \overline{AB} = 540 \sin 50° = 414$$

Thus, after 2 hours and 15 minutes the airplane is approximately 347 miles south and 414 miles east of the airport.   ▲

 EXERCISE SET I

A

In Problems 1–16 triangle ABC is a right triangle with right angle at C. Solve the triangle.

1. $A = 39.2°$, $a = 16.9$
2. $a = 41.3$, $c = 57.5$

3. $B = 54.6°$, $c = 19.8$
4. $A = 43.6°$, $b = 123$
5. $A = 26°30'$, $c = 32.7$
6. $B = 55°40'$, $b = 17.9$
7. $B = 71.2°$, $a = 49.2$
8. $A = 34.3°$, $a = 1.95$

9. $a = 13.23$, $b = 18.72$
10. $B = 35°42'$, $c = 2.301$
11. $A = 42°13'$, $b = 203.7$
12. $A = 19.34°$, $c = 342.8$
13. $B = 12°34'$, $b = 3.102$
14. $a = 1.583$, $b = 2.942$
15. $b = 21.83$, $c = 38.16$
16. $B = 37°23'$, $a = 152.8$
17. From a point on the ground 162 feet from the base of a building, the angle of elevation of the top of the building is 65.7°. How high is the building?
18. A rectangle is 15.7 inches long and 9.15 inches high. Find the angle between a diagonal and the base. What is the length of the diagonal?
19. From the top of a lighthouse the angle of depression of a boat is 33.5°. If the lighthouse is 90.4 feet high, how far is the boat from the base of the lighthouse?
20. A tower 102.6 feet high is on the bank of a river. From the top of the tower the angle of depression of a point on the opposite side of the river is 28°14'. How long a cable would be required to reach from the top of the tower to the point on the opposite bank?
21. A guy wire for a telephone pole is anchored to the ground at a point 21.3 feet from the base of the pole, and it makes an angle of 64.7° with the horizontal. Find the height of the point where the guy wire is attached to the pole, and find the length of the guy wire.
22. A surveyor wants to find the distance between points A and B on opposite sides of a pond. With her transit at point A, she determines a line of sight at right angles to AB and establishes point C along this line 150 feet from A. Then she measures the angle at C from CA to CB and finds it to be 62°10'. How far is it from A to B?
23. Town A is 12.0 miles due west of town B, and town C is 15.0 miles due south of town B. How far is it from town A to town C? Assuming straight roads connect the three towns, what is the angle at A between the roads leading to B and C?
24. A bar with a regular pentagonal (five-sided) cross-section is to be milled from round stock. What diameter should the stock be if the dimension of each flat section is to be 2.50 centimeters? (See figure.)

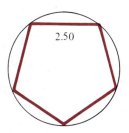

2.50

25. Points A and B are situated on the coast, with B 1,250 meters due south of A. From A the bearing of a lighthouse on an island is S 42.2° E, and from B the bearing of the lighthouse is N 47.8° E. Find the distance of the lighthouse from each of the points A and B.
26. An oil drilling platform is located in the Gulf of Mexico 3.25 miles from the nearest point, A, on shore. From a point B on the shore due east of A the bearing of the platform is S 51.2° W. How far is it from B to the platform?
27. Two forest ranger towers are 10.35 miles apart. Smoke is sighted from the first tower at a bearing of N 36°24' W, and simultaneously it is spotted from the second tower at a bearing of N 53°36' E. If the first tower is due east of the second, find the distance from each tower to the source of the smoke.
28. A jetliner travels at a heading of 312° for 45 minutes at an average speed of 480 miles per hour. Then the course is changed to a heading of 42°. After flying at this new heading for 25 minutes, maintaining the same average speed as before, what is the distance of the jetliner from the starting point? (Neglect the effect of the wind.)
29. An airplane flies in a straight path in a south-westerly direction. After one hour the plane is 205 miles west and 78 miles south of its starting point. If the velocity of the wind was negligible, determine the heading of the airplane. What was its average speed?

 B

30. The accompanying figure shows the angles of elevation of a balloon at a certain instant from

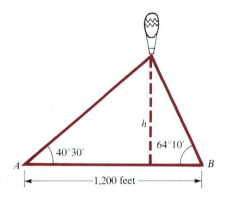

points A and B 1,200 meters apart. Find the height of the balloon at that instant.

31. From a lighthouse 120 feet high, the angles of depression of two boats directly in line with and on the same side of the lighthouse are found to be $34°30'$ and $27°40'$. How far apart are the boats?

32. At a point 250 feet from the base of a building the angle of elevation of the top of the building is $27.5°$, and the angle of elevation of the top of a statue on the top edge of the building is $30.2°$. How high is the statue?

33. Find the length of $1°$ of longitude on the circle of latitude through Chicago, $41°50'$ north. (Assume the radius of the earth is 3,960 miles.)

34. Ship A leaves port at 2:00 P.M. and travels in the direction S $34°$ E at an average speed of 15 knots (nautical miles per hour). At 2:45 P.M. ship B leaves the same port and travels in the direction N $56°$ E at an average speed of 20 knots. Find
   a. the distance between the ships at 4:00 P.M., and
   b. the bearing of ship B from ship A at 4:00 P.M.

35. From a ship, the bearings of two landmarks on shore are found to be S $37.2°$ W and S $21.7°$ W, respectively. If the landmarks are known to be 5 kilometers apart, find how far the ship is from the nearest point on shore.

36. A pilot leaves point A and flies for 35 minutes at a heading of $142°20'$ and arrives at point B. The pilot then flies at a heading of $232°20'$ for 1 hour and 12 minutes, arriving at point C. The average speed on each leg of the trip was 180 miles per hour, and wind velocity was negligible. How far is it from A to C? If the pilot wanted to fly directly from A to C, what would be the heading?

THE LAW OF SINES

As we mentioned earlier, a triangle with no right angle is called an oblique triangle. The two principal techniques used in solving oblique triangles are called the **law of sines** and the **law of cosines.** Which method to use is determined by the given information. We consider the law of sines in this section and the law of cosines in the next.

To derive the law of sines, consider an oblique triangle ABC, as shown in Figure 12. The altitude CD divides the triangle into two right triangles. From

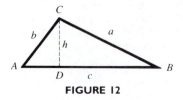

FIGURE 12

triangle ADC we obtain

$$\sin A = \frac{h}{b}$$

so that $h = b \sin A$. Similarly, from triangle BDC,

$$\sin B = \frac{h}{a}$$

so that $h = a \sin B$. The two expressions for h must be the same, giving $a \sin B = b \sin A$, or equivalently,

$$\frac{a}{\sin A} = \frac{b}{\sin B} \qquad (1)$$

Next we orient the original triangle ABC with respect to a rectangular coordinate system so that C coincides with the origin and vertex A is on the positive x axis, as in Figure 13. Let the ordinate of B be designated by y. Then,

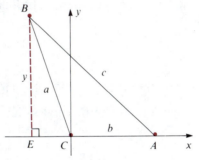

FIGURE 13

since angle C is in standard position, we have that

$$\sin C = \frac{y}{a}$$

or $y = a \sin C$. If E is the foot of the perpendicular from B to the x axis, we can use the right triangle ABE to get

$$\sin A = \frac{y}{c}$$

so that $y = c \sin A$. Equating the two values of y gives $a \sin C = c \sin A$, or equivalently,

$$\frac{a}{\sin A} = \frac{c}{\sin C} \qquad (2)$$

Finally, combining equations (1) and (2), we have the following:

The Law of Sines

$$\frac{a}{\sin A} = \frac{b}{\sin B} = \frac{c}{\sin C}$$

Although the triangle we used to derive this result had an obtuse angle (C), the derivation is valid also when all three angles are acute.

The law of sines can be employed to find the missing parts of a triangle in either of the following cases:

1. Two angles and one side are given
2. Two sides and the angle opposite one of them are given

In either case, an equation having only one unknown can be obtained from the law of sines. After solving for the unknown in the equation, all remaining unknowns can be found. The following examples illustrate typical situations.

EXAMPLE 7

In triangle ABC, $A = 22°$, $B = 110°$, and $c = 13.4$. Find sides a and b and angle C.

Solution Since the sum of the angles must be 180°, we find that $C = 180° - 132° = 48°$. From the law of sines, we select $a/\sin A = c/\sin C$. Substituting known values, we obtain

$$\frac{a}{\sin 22°} = \frac{13.4}{\sin 48°}$$

so that

$$a = \frac{(13.4)(\sin 22°)}{\sin 48°} = 6.75$$

By calculator:

$$13.4 \;\boxed{\times}\; 22 \;\boxed{\text{SIN}}\; \boxed{\div}\; 48 \;\boxed{\text{SIN}}\; \boxed{=}\; \text{Display} \quad 6.7547108$$

To find b, we employ $b/\sin B = c/\sin C$.

$$\frac{b}{\sin 110°} = \frac{13.4}{\sin 48°}$$

$$b = \frac{(13.4)(\sin 110°)}{\sin 48°} = 16.9$$

The triangle is now solved.

▲

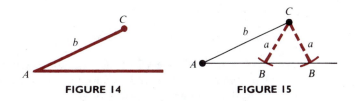

FIGURE 14   **FIGURE 15**

Before proceeding to other examples, let us analyze what the possibilities are in Case 2, in which we are given two sides and the angle opposite one of them. Suppose we are given a, b, and A. First, draw angle A with side b adjacent, thus determining vertex C (Figure 14). Now side a extends from C until it strikes the base. But it is quite evident that several possibilities exist, depending on the length of side a. A compass set at C with radius a clearly shows what can happen. If a exceeds the altitude from C but is shorter than b, then there are two distinct solutions, as Figure 15 shows. If a exceeds the altitude from C and also exceeds b, only one solution exists, as Figure 16 shows. If a equals the altitude from C, the triangle is a right triangle, and there is again just one solution, as shown in Figure 17. Finally, if a is less than the altitude from C, there is no triangle at all, as shown in Figure 18.

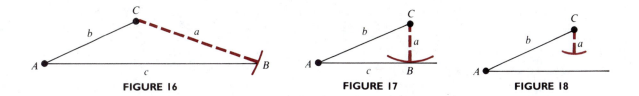

FIGURE 16   **FIGURE 17**   **FIGURE 18**

For obvious reasons, case 2 is known as the **ambiguous case.** Fortunately, it is not necessary to determine in advance which situation exists; after the law of sines is employed, the situation will become clear. It is necessary, however, to be alert and to interpret from the result whether one, two, or no solutions exist.

EXAMPLE 8 ||||||||||||||||||||||||||||||||||||||||||||||||||||||||||||||||||||||||||||||||||||||||||||||||||||||||

Solve the triangle for which $a = 125$, $b = 150$, and $A = 54.0°$.

Solution   From $a/\sin A = b/\sin B$ we get

$$\frac{125}{\sin 54°} = \frac{150}{\sin B}$$

so that

$$\sin B = \frac{(150)(\sin 54°)}{125} = 0.9708$$

By calculator:

$$150 \;\boxed{\times}\; 54 \;\boxed{\text{SIN}}\; \boxed{\div}\; 125 \;\boxed{=}\; \boxed{\text{INV}}\; \boxed{\text{SIN}}\; \text{Display} \quad 76.124805$$

So B is approximately 76.1°. But the supplement of this angle is also a possibility, since an acute angle and its supplement have the same sine. We must therefore consider an alternative value of B, $180° - 76.1° = 103.9°$, and then calculate the third angle in order to find out whether the second value of B is feasible. We have

$$C = 180° - (A + B) = 180° - (54° + 103.9°) = 22.1°$$

Thus, two distinct solutions exist. Let us designate by B_1 the value 76.1° and by B_2 the value 103.9°. Corresponding subscripts will be used for angle C and side c. Thus,

$$C_1 = 180° - (A + B_1) = 180° - (54° + 76.1°) = 49.9°$$

and $C_2 = 22.1°$, as we have already calculated.

We use $a/\sin A = c/\sin C$ to calculate side c for each case. So we have

$$\frac{125}{\sin 54°} = \frac{c_1}{\sin 49.9°} \quad \text{and} \quad \frac{125}{\sin 54°} = \frac{c_2}{\sin 22.1°}$$

These yield

$$c_1 = \frac{(125)(\sin 49.9°)}{\sin 54°} = 118$$

$$c_2 = \frac{(125)(\sin 22.1°)}{\sin 54°} = 58.1$$

The two solutions can be summarized as shown in Figure 19.

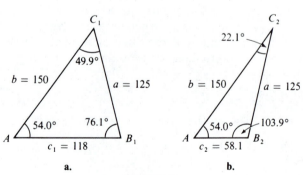

FIGURE 19

EXAMPLE 9 ||||||||||||||||||||||||||||||||||||||||||||||||||||||||||||||||||||||||||||||||||||||||||||||||||||||||||||||||||||||||||||||||||||

Solve the triangle ABC in which $b = 13.2$, $c = 10.5$, and $B = 42.0°$.

Solution   From $b/\sin B = c/\sin C$, we find that

$$\sin C = \frac{c \sin B}{b} = \frac{(10.5)(\sin 42°)}{13.2} = 0.5323$$

so that C is either $32.2°$ or its supplement, $147.8°$. But if $C = 147.8°$, then $B + C = 189.8°$, which is impossible since the sum of two angles of a triangle must be less than $180°$. Thus, only one solution exists. Using $C = 32.2°$, we get

$$A = 180° - (B + C) = 180° - (42° + 32.2°) = 105.8°$$

Finally, we use $a/\sin A = b/\sin B$ to find side a.

$$a = \frac{b \sin A}{\sin B} = \frac{(13.2)(\sin 105.8°)}{\sin 42°} = 19.0$$

The solution is shown in Figure 20.

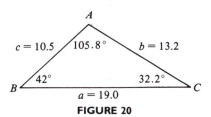

FIGURE 20

EXAMPLE 10

Solve the triangle ABC for which $a = 23.5$, $c = 12.0$, and $C = 35°$.

Solution   From $a/\sin A = c/\sin C$, we get

$$\sin A = \frac{a \sin C}{c} = \frac{(23.5)(\sin 35°)}{12.0} = 1.123$$

But this is impossible, since it is clear from the definition that the sine never exceeds 1. So there is no solution; that is, there is no triangle having a, c, and C as given.

A Word of Caution. If the sine of the angle found in the first step of solving a Case 2 problem is less than 1, you *must* consider as possible solutions both the acute angle found from a calculator *and* its supplement. The supplement will be a solution provided that it, together with the given angle, add up to an angle less than $180°$ (as in Example 8).

EXERCISE SET 2

A

Solve the triangles in Problems 1–16.

1. $a = 3.82$, $A = 43.1°$, $B = 21.7°$
2. $b = 36.8$, $A = 63.7°$, $C = 36.4°$
3. $b = 138$, $c = 87.0$, $C = 35°40'$
4. $a = 12.4$, $c = 14.2$, $A = 69.3°$
5. $A = 27.3°$, $a = 20.9$, $b = 11.3$
6. $C = 32.6°$, $b = 16.3$, $c = 7.83$
7. $B = 46.3°$, $C = 24.9°$, $c = 8.23$
8. $A = 62.7°$, $C = 71.3°$, $a = 112$
9. $A = 36°34'$, $B = 63°21'$, $b = 15.05$
10. $B = 103°21'$, $C = 22°51'$, $c = 4.312$
11. $B = 42.26°$, $b = 18.33$, $c = 11.42$
12. $C = 22.53°$, $a = 20.62$, $c = 14.38$
13. $B = 105°27'$, $b = 110.5$, $c = 72.28$
14. $A = 64.45°$, $a = 32.15$, $b = 65.82$
15. $C = 23°34'$, $a = 5.371$, $c = 3.582$
16. $A = 103°42'$, $C = 34°51'$, $b = 15.25$
17. Each of two parallel sides of a parallelogram is 27.32 inches, and one of the acute angles is 33°25'. If the longer diagonal is 132.5 inches, find the length of the other sides of the parallelogram.
18. Points A and B are on opposite sides of a river, and it is desired to find the distance between them. Point C, on the same side of the river as A, is 35.0 feet from A. In the triangle ABC, the angle at C is found to be 42°30' and the angle at A to be 103°24'. How far is it from A to B?
19. The distance x across a gorge is to be determined in order that a cable car traversing it can be constructed. One side of the gorge has only a moderate slope, so that a distance along it can be measured, as shown in the figure. The angles shown are also determined. Find the distance x.
20. Points A and B are on opposite sides of a swampy area. To find the distance between them, a point C is located 254 feet from B and 197 feet from A. The angle at A from AB to AC is found to be 47.2°. Find the distance from A to B.

B

21. A boat leaves point A and travels in the direction S 48° E at an average speed of 15 knots (nautical miles per hour). A coast guard cutter at point B, 21 nautical miles due east of A, wishes to intercept the boat. If the cutter can average 25 knots, what should its direction be (assuming the cutter leaves at the same time as the boat)? When will the interception be made?
22. Point B is on the coast 5 miles due south of A. From A the bearing of a ship is S 32.4° E, and from B the bearing of the ship is N 42.7° E. How far is it from each of the points A and B to the ship? How far is it from the ship to the nearest point on shore?
23. A pilot is planning to make a trip from A to B, 350 miles due north of A, and then back to A. There is a wind blowing from 310° at 55 miles per hour. If the average airspeed of the plane is 210 miles per hour, find the heading the pilot should take on each part of the trip. What will be the total flying time?

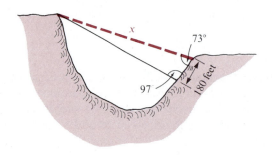

THE LAW OF COSINES

When we are given two sides of an oblique triangle and the angle between them, or when we are given all three sides, the law of sines cannot be used, because there will always be two unknowns in any of the equations. In this section we derive a formula that will handle these situations.

Let ABC be any oblique triangle, and place angle C in standard position, with vertex B on the x axis. Figure 21 illustrates this, showing C acute in **a** and obtuse in **b.** In either case, if vertex A has coordinates (x, y), we have, by

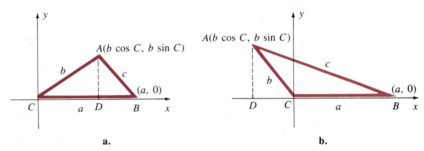

FIGURE 21

Definition 2 of Chapter 7,

$$\sin C = \frac{y}{b} \quad \text{and} \quad \cos C = \frac{x}{b}$$

So $x = b \cos C$ and $y = b \sin C$. These are shown as the coordinates of A in Figure 21. Now we apply the Pythagorean theorem to the right triangle ADB. The result is the same for parts **a** and **b.**

$$c^2 = \overline{AD}^2 + \overline{BD}^2$$

The distance $\overline{AD}$ is just the y coordinate of vertex A. Also, since the x coordinate of B is a and the x coordinate of A is $b \cos C$, we have $\overline{BD} = |a - b \cos C|$, whether C is acute or obtuse. Thus,

$$
\begin{aligned}
c^2 &= (b \sin C)^2 + (a - b \cos C)^2 \\
&= b^2 \sin^2 C + a^2 - 2\,ab \cos C + b^2 \cos^2 C \\
&= a^2 + b^2(\sin^2 C + \cos^2 C) - 2\,ab \cos C \\
&= a^2 + b^2 - 2\,ab \cos C
\end{aligned}
$$

since $\sin^2 C + \cos^2 C = 1$. This result is one form of the **law of cosines.** By placing vertices A and B, in turn, in standard position and applying the same reasoning, we get two analogous results. The three forms are

$$a^2 = b^2 + c^2 - 2bc \cos A$$

$$b^2 = a^2 + c^2 - 2ac \cos B$$

$$c^2 = a^2 + b^2 - 2ab \cos C$$

Each of these can be expressed in words as follows:

The Law of Cosines

The square of any side of a triangle equals the sum of the squares of the other two sides, minus twice the product of these two sides and the cosine of the angle between them.

If two sides and the angle between them are known, then the appropriate form of the law of cosines can be used to find the third side. For example, if you know sides b and c and angle A, you would use

$$a^2 = b^2 + c^2 - 2bc \cos A$$

to find side a. If all three sides are known, then by solving for $\cos A$, $\cos B$, or $\cos C$ in one of the forms of the law of cosines, an angle can be found. For example, A can be found from

$$\cos A = \frac{b^2 + c^2 - a^2}{2bc}$$

In either case, after one application of the law of cosines you will know all three sides and one angle. Now you need to find one of the other angles. Again, you can use a form of the law of cosines, one that you have not yet used. You may also use the law of sines, but you need to be aware of a possible pitfall. For example, suppose that, as described above, you know after one application of the law of cosines, all three sides and angle A. Then from

$$\frac{a}{\sin A} = \frac{b}{\sin B}$$

you could find $\sin B$. To use a specific example, suppose $\sin B = .9874$. From the calculator we would find that $B = 80.89°$. But remember that the sine of an acute angle and the sine of its supplement are the same. So angle B might instead be $180° - 80.89° = 99.11°$. This is not like the ambiguous case of the law of sines, where both answers may be correct. The given information here determines one and only one triangle. One way to avoid the dilemma we have is to use the law of cosines to find angle B. We would use

$$\cos B = \frac{a^2 + c^2 - b^2}{2ac}$$

Suppose this turned out to be $\cos B = -0.1583$. Then from the calculator we would find unambiguously that $B = 99.11°$.

The key fact in the above discussion is that the cosines of an acute angle and its supplement differ in sign, whereas the sines of the two are the same. It would appear, then, that in completing the solution of a triangle in which all sides and one angle are known, it would be safest to use the law of cosines again to find another angle. However, the law of sines is somewhat easier to apply, and in some cases you can determine in advance that the second angle you want to find is acute. For example, suppose again that you know (or have found) that angle A is obtuse, say $A = 120°$. Then clearly both B and C are acute, and so either could be found from the law of sines. Example 11 below illustrates this. In certain other cases a careful sketch of the triangle might show one of these remaining unknown angles as being acute.

EXAMPLE 11

In a triangle one angle is $120°$ and the two sides adjacent to this angle have lengths 3 and 5. Find all other parts of the triangle.

Solution   Let the given sides be $a = 3$, $b = 5$. Then $C = 120°$.

$$c^2 = a^2 + b^2 - 2ab \cos C$$
$$= 9 + 25 - 2(3)(5)(-\tfrac{1}{2})$$
$$= 9 + 25 + 15 = 49$$

so $c = 7$.

Since $C = 120°$, neither A nor B can be obtuse (all three angles add to give $180°$), so as discussed above, we are safe in using the law of sines. From $a/\sin A = c/\sin C$, we get

$$\sin A = \frac{a \sin C}{c} = \frac{3(\sqrt{3}/2)}{7} = \frac{3\sqrt{3}}{14} = 0.3712$$
$$A = 21.8°$$

Finally,
$$B = 180° - (A + C) = 180° - 141.8° = 38.2°$$

▲

EXAMPLE 12

Find the angles in triangle ABC for which $a = 13$, $b = 21$, $c = 15$.

Solution   We have

$$\cos A = \frac{b^2 + c^2 - a^2}{2bc} = \frac{441 + 225 - 169}{2(21)(15)} = 0.7889$$

By calculator:

21 $\boxed{x^2}$ $\boxed{+}$ 15 $\boxed{x^2}$ $\boxed{-}$ 13 $\boxed{x^2}$ $\boxed{=}$ $\boxed{\div}$ $\boxed{(}$ 2 $\boxed{\times}$ 21 $\boxed{\times}$

15 $\boxed{)}$ $\boxed{=}$ $\boxed{\text{INV}}$ $\boxed{\text{COS}}$ Display 37.918203

So $A = 37.9°$.

In this case we cannot say for sure in advance whether B or C is acute. So we choose to use the law of cosines again.

$$\cos B = \frac{a^2 + c^2 - b^2}{2ac} = \frac{169 + 225 - 441}{2(13)(15)} = -0.1205$$

and $B = 96.9°$. Thus, $C = 180° - (A + B) = 180° - 134.8° = 45.2°$. ▲

EXAMPLE 13 ||||||||||||||||||||||||||||||||||||||||||||||||||||||||||||||||||||||||||||||||||||||||||||||

A car leaves an intersection and travels at an average speed of 45 miles per hour. Ten minutes later a second car leaves the same intersection and travels on a road making an angle of 102° with the first, at an average speed of 51 miles per hour. Assuming the roads are straight, how far apart are the cars 30 minutes after the first car left?

Solution After 30 minutes the first car will have traveled $(45)(\frac{1}{2}) = 22.5$ miles. The second car travels for only 20 minutes, or $\frac{1}{3}$ hour. So it goes $(51)(\frac{1}{3}) = 17$ miles. Let x denote the distance between the cars, as in Figure 22. Then, by the law of cosines,

$$x^2 = (22.5)^2 + (17)^2 - 2(22.5)(17) \cos 102°$$

Using a calculator, we find that

$$x = 30.9 \text{ miles}$$ ▲

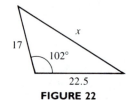

FIGURE 22

 EXERCISE SET 3

A

Solve the triangle in Problems 1–16.

1. $B = 60°$, $a = 15.1$, $c = 20.4$
2. $A = 120°$, $b = 201$, $c = 155$
3. $C = 93°50'$, $a = 5.12$, $b = 3.78$
4. $a = 3.51$, $b = 4.23$, $c = 2.08$
5. $a = 12.5$, $b = 8.72$, $c = 15.6$
6. $B = 33.8°$, $a = 13.2$, $c = 15.3$
7. $A = 110.3°$, $b = 246$, $c = 172$
8. $C = 58°40'$, $a = 42.3$, $b = 35.2$

9. $a = 2.13$, $b = 3.58$, $c = 4.79$
10. $a = 33.7$, $b = 20.7$, $c = 48.3$
11. $B = 86°35'$, $a = 33.05$, $c = 20.24$
12. $A = 98.72°$, $b = 0.245$, $c = 0.432$
13. $A = 125.4°$, $b = 3.459$, $c = 1.032$
14. $B = 17.32°$, $a = 121.3$, $c = 230.4$
15. $a = 32.05$, $b = 38.39$, $c = 21.14$
16. $a = 1.032$, $b = 0.7523$, $c = 2.154$
17. Two sides of a parallelogram are 8 centimeters and 10 centimeters, respectively, and the shorter diagonal is 12 centimeters (exact values). Find the length of the longer diagonal.

18. Two cars leave simultaneously from the same point, one going east at an average speed of 60 kilometers per hour and the other going southwest at an average speed of 80 kilometers per hour. How far apart are they after two hours?

19. A surveyor wishes to find the distance from point A to point B but cannot do so directly because there is a swamp between them. He sets up a transit at C and makes the following measurements: angle $C = 27°32'$, $\overline{CA} = 125.3$ feet, $\overline{CB} = 117.5$ feet. Find $\overline{AB}$.

20. Car A leaves an intersection and travels at an average speed of 42 miles per hour. Fifteen minutes later car B leaves the same intersection and travels on another road at an average speed of 54 miles per hour. Two hours after the first car left the intersection, the distance between the cars is 60 miles. Find the angle between the two roads. (Assume the roads are straight.)

21. A **rhombus** is a parallelogram in which all sides are equal. If each side of a rhombus has length 2 units and one of the angles is 30°, find the lengths of the two diagonals.

22. A railroad tunnel is to be cut through a mountain. To determine the length of the tunnel, a transit is set up at C, as in the accompanying figure, and the distance to points A and B, as well as the angle at C, are measured as shown. Points A and B are on the straight track at distances 20 feet and 30 feet, respectively, from the ends of the tunnel. Find the length of the tunnel.

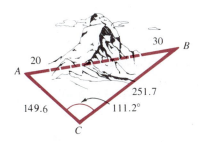

23. Airplane A leaves Chicago at 1:00 P.M. and flies at a heading of 203° at an average speed of 350 miles per hour. Airplane B leaves the same airport at 1:30 P.M. and flies at a heading of 85° at an average speed of 400 miles per hour. How far apart are the planes at 2:00 P.M.? (Neglect wind velocity.)

B

24. A regular pentagon is inscribed in a circle of radius 4 centimeters. Make use of the law of cosines to find the perimeter of the pentagon.

25. A 20-foot-high TV antenna is to be mounted at the peak of a roof that makes a 25° angle with the horizontal. A guy wire is to run from a point 5 feet below the top of the antenna to a point 12 feet down from the peak of the roof. How long a guy wire is required?

26. Let ABC be a triangle with vertex C at the origin. If A has coordinates (x_1, y_1) and B has coordinates (x_2, y_2), derive the following formula for $\cos C$:

$$\cos C = \frac{x_1 x_2 + y_1 y_2}{\sqrt{x_1^2 + y_1^2}\,\sqrt{x_2^2 + y_2^2}}$$

27. To determine the distance between two inaccessible points A and B on the opposite side of a river from an observer, two points C and D are established 50 feet apart, and angles are determined as shown. Find the distance between A and B.

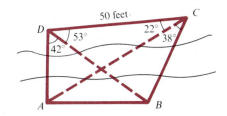

28. Derive the following formula for the distance x in the isosceles triangle shown:

$$x = \frac{2a \cos \dfrac{\alpha}{2} \sin \dfrac{\beta}{2}}{\sin \dfrac{\alpha - \beta}{2}}$$

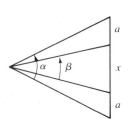

AREAS OF TRIANGLES

The area of a triangle is given by

$$\text{area} = \tfrac{1}{2}(\text{base} \times \text{altitude})$$

In a *right* triangle either leg can be taken as the base, and the other leg is then the corresponding altitude.

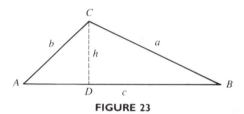

FIGURE 23

Consider now the oblique triangle in Figure 23. We can obtain the area by using the base c and the altitude h. To find h, we can use the right triangle ADC to get $\sin A = h/b$, so that $h = b \sin A$. Thus,

$$\text{area} = \tfrac{1}{2}bc \sin A \qquad (3)$$

We could also have found h from the right triangle BDC, using $\sin B = h/a$, so that $h = a \sin B$. This yields the formula

$$\text{area} = \tfrac{1}{2}ac \sin B \qquad (4)$$

By using one of the other sides as the base and finding the corresponding altitude (you can simply change the names of the vertices and sides in Figure 23), we get this result:

$$\text{area} = \tfrac{1}{2}ab \sin C \qquad (5)$$

Notice that formulas (3), (4), and (5) can all be stated in words as follows: **the area of any triangle equals one-half the product of the lengths of two of the sides, times the sine of the angle between them.**

If enough information about a triangle is known to solve it, the area can always be found by using any of the formulas (3), (4), or (5).

EXAMPLE 14 ||||||||||||||||||||||||||||||||||||||||||||||||||||||||||||||||||||||||||||||||||||||||||||||||||||||||||||

Find the area of the triangle ABC for which $a = 10$, $b = 7$, and $C = 60°$.

Solution Formula (5) can be used to give the area without having to find any other parts of the triangle. Using it, we get

$$\text{area} = \frac{1}{2}\,ab\,\sin C = \frac{1}{2}(10)(7)\!\left(\frac{\sqrt{3}}{2}\right) = \frac{35\sqrt{3}}{2} = 30.31 \qquad \blacktriangle$$

EXAMPLE 15

Find the area of the triangle for which $b = 13.5$, $A = 33.8°$, $B = 97.4°$.

Solution   This time we have to make some preliminary calculations before we can use any of the formulas (3), (4), or (5). Since we know side b, we can select either formula (3) or (5). Let us choose (5). We must find a, using the law of sines.

$$\frac{a}{\sin 33.8°} = \frac{13.5}{\sin 97.4°}$$

$$a = \frac{13.5 \sin 33.8°}{\sin 97.4°} = 7.573$$

Angle $C = 180° - (A + B) = 180° - (33.8° + 97.4°) = 48.8°$. So

$$\text{area} = \tfrac{1}{2}ab \sin C = \tfrac{1}{2}(7.573)(13.5)\sin 48.8° = 38.5 \qquad \blacktriangle$$

EXAMPLE 16

Find the area of the triangle with sides 5, 7, and 3.

Solution   Let $a = 5$, $b = 7$, and $c = 3$. We need to find any one of the angles. Suppose we choose B. By the law of cosines,

$$\cos B = \frac{a^2 + c^2 - b^2}{2ac} = \frac{25 + 9 - 49}{2(15)}$$

$$= \frac{-15}{30} = -\frac{1}{2}$$

(Now you see that B was a good choice.) So

$$B = 120°$$

From formula (4) we now have

$$\text{area} = \frac{1}{2}ac \sin B = \frac{1}{2}(5)(3)\sin 120°$$

$$= \frac{15\sqrt{3}}{4} = 6.50 \qquad \blacktriangle$$

As Examples 15 and 16 show, unless two sides and the included angle are given, some preliminary calculations must be made before the area formulas (3), (4), or (5) can be used. Of course, if you need to solve the triangle completely anyway, this presents no problem. We can, however, develop formulas for the area that can be applied directly when two angles and one side are known and when all three sides are known. We derive the first of these here and call for the second in the exercises (Problem 20). For the ambiguous case—two sides and the angle opposite one of them—a formula can be derived, but it is not very useful. It is probably best to make the necessary preliminary calculations in this case.

Suppose, then, that we are given two angles and one side. To be specific, let A, B, and b be known. Then $C = 180° - (A + B)$, and by the law of sines,

$$\frac{a}{\sin A} = \frac{b}{\sin B}$$

so $a = \dfrac{b \sin A}{\sin B}$. Now we know a, b, and the included angle C. So by (5),

$$\text{area} = \frac{1}{2} ab \sin C = \frac{1}{2}\left(\frac{b \sin A}{\sin B}\right) \cdot (b) \sin[180° - (A + B)]$$

but $\sin[180° - (A + B)] = \sin(A + B)$. So we obtain the formula

$$\text{area} = \frac{1}{2}\frac{b^2 \sin A \sin(A + B)}{\sin B} \tag{6}$$

Analogous results can be obtained by using different angles and sides. We can now redo Example 15, using formula (6), to get

$$\text{area} = \frac{1}{2}\frac{(13.5)^2 \sin 33.8° \sin(33.8° + 97.4°)}{\sin 97.4°} = 38.5$$

EXERCISE SET 4

A

In Problems 1–18 find the area of the triangle ABC.

1. $A = 30°$, $b = 24$, $c = 15$
2. $B = 150°$, $a = 54$, $c = 21$
3. $C = 105.2°$, $a = 33.1$, $b = 42.5$
4. $A = 22°30'$, $b = 15.2$, $c = 18.3$
5. $B = 62°20'$, $a = 1.42$, $c = 2.05$
6. $C = 49.7°$, $a = 12.9$, $b = 9.23$
7. $A = 34.7°$, $B = 23.9°$, $b = 17.2$
8. $B = 86.3°$, $C = 15.9°$, $b = 2.53$
9. $A = 70°20'$, $C = 32°40'$, $a = 24.5$
10. $A = 39°10'$, $B = 54°30'$, $a = 12.8$
11. $a = 1.23$, $b = 2.15$, $c = 2.87$
12. $a = 10.5$, $b = 6.23$, $c = 14.8$
13. $A = 28.5°$, $a = 21.8$, $b = 10.2$
14. $B = 47.2°$, $C = 25.8°$, $c = 7.34$
15. $B = 43.25°$, $b = 17.52$, $c = 10.27$
16. $a = 3.61$, $b = 4.72$, $c = 2.85$
17. $a = 21.5$, $b = 32.6$, $c = 50.7$
18. $A = 37°25'$, $B = 64°37'$, $b = 14.72$

B

19. Find the area of each triangle ABC for which $C = 22°31'$, $b = 6.432$, and $c = 3.871$.
20. In formula (3) replace $\sin A$ by $\sqrt{1 - \cos^2 A}$, and use the law of cosines to write $\cos A$ in terms of a, b, and c. Show that the result can be written in the form

    area $=$

    $$\frac{1}{4}\sqrt{(a + b + c)(a + b - c)(a + c - b)(b + c - a)}$$

21. Let $s = \frac{1}{2}(a + b + c)$. Show that the result of Problem 20 can be put in the form

    $$\text{area} = \sqrt{s(s - a)(s - b)(s - c)}$$

    This is known as **Heron's formula.**
22. Use Heron's formula (see Problem 21) to find the area of the triangle with vertices $(2, 1)$, $(-4, -7)$, and $(-1, 5)$.
23. **a.** Using the accompanying figure, show that the area of the triangle is rs, where $s = \frac{1}{2}(a + b + c)$. **Hint.** Add areas of six right triangles.

ction 5   Vectors   **409**

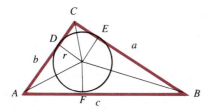

r of the inscribed circle is given by

$$r = \sqrt{\frac{(s-a)(s-b)(s-c)}{s}}$$

24. Let the lengths of the two diagonals of an arbitrary quadrilateral be d_1 and d_2, and let α be their angle of intersection (either of the two angles). Prove that the area of the quadrilateral is given by $\frac{1}{2}d_1 d_2 \sin\alpha$.

b. Use the result of part **a**, together with Heron's formula (Problem 21), to show that the radius

VECTORS

Certain physical quantities have both **magnitude** and **direction;** examples are force, velocity, acceleration, and displacement of a moving particle. A convenient way to represent such quantities is with a **directed line segment,** such as the one in Figure 24. The length of the segment, to some scale, represents the magnitude of the quantity in question, and the direction is indicated by the inclination of the segment and by the arrowhead. Such a directed line segment is called a **vector.** For example, the vector in Figure 24 might represent a wind velocity of 20 mi/hr blowing in a northeasterly direction. The length would then be taken as 20 units, to some convenient scale. Any quantity that has both magnitude and direction can be represented in this way and for this reason is called a **vector quantity.**

A vector

FIGURE 24

Vector quantities are in contrast with such things as area, mass, time, and distance, which can be adequately described by a single number. These are called **scalar quantities** (since they are measured according to some scale), and the numbers used to measure them are called **scalars.** For our purposes, then, scalars are just real numbers. It should be mentioned, however, that in certain more advanced treatments both vectors and scalars can be different from what we are discussing here.

Typically, boldface letters such as **u**, **v**, and **w** are used to designate vectors. We consider only vectors that lie in a plane, although most of the results have natural extensions to three (or more) dimensions. Two vectors **u** and **v** are said to be **equivalent** if they have the same magnitude and direction, and in this case we write **u** = **v**. Three equivalent vectors are illustrated in Figure 25. We do not

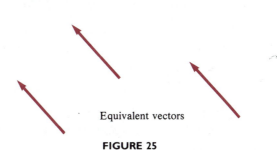

Equivalent vectors

FIGURE 25

distinguish between equivalent vectors, so that in effect we can shift a vector from one location to another so long as its original magnitude and direction are retained. Because of this freedom of movement, we say that we are working within a system of **free vectors.**

If a vector extends from a point P to a point Q, we may use the notation $\overrightarrow{PQ}$ to designate the vector. The point P is called the **initial point** and Q the **terminal point.** Sometimes we also use "tail" and "tip" instead of initial point and terminal point, respectively.

Two nonparallel vectors are added according to the **parallelogram law,** illustrated in Figure 26(a). The vectors are drawn with a common initial point, and a parallelogram is constructed with **u** and **v** as adjacent sides. The vector **u** + **v** is then defined as the vector along the diagonal from the common initial point to the opposite vertex. An alternative method is to place the initial point of **v** at the terminal point of **u**. Then **u** + **v** is the vector shown in Figure 26(b), drawn from the initial point of **u** to the terminal point of **v**. You should convince yourself that the triangle in part **b** is just the lower half of the parallelogram in part **a**. This second method is sometimes called the "tail to tip" method of adding. If **u** and **v** are parallel vectors, then the parallelogram of part **a** is degenerate. The tail to tip method still works, however.

Vector addition is consistent with observed results. For example, if **u** and v represent forces acting on an object, then the net effect is **u** + **v**; that is, the two individual forces **u** and **v** could be replaced by the force **u** + **v**, and the effect would

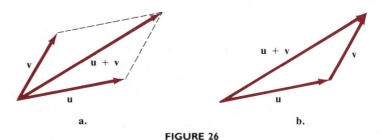

a.   b.

FIGURE 26

be the same. In this case we call **u** + **v** the **resultant** of **u** and **v**. Similarly, if **u** is a vector representing the indicated velocity of an airplane and **v** is the wind velocity vector, then the true velocity of the airplane relative to the ground is **u** + **v**.

It is convenient to introduce the notion of the **zero vector,** denoted by **0**, with a magnitude of 0 and assigned no direction. We may think of the zero vector as a single point. If **v** is a nonzero vector, then −**v** is the vector that has the same length as **v** but with a direction opposite to that of **v**. We now define subtraction by

$$\mathbf{u} - \mathbf{v} = \mathbf{u} + (-\mathbf{v})$$

Thus, **u** − **v** is the vector that, when added to **v**, gives **u**. This is illustrated in Figure 27. Notice that when **u** and **v** are drawn with the same initial point, **u** − **v**

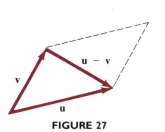

FIGURE 27

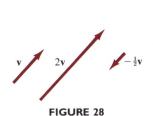

FIGURE 28

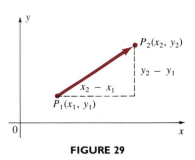

FIGURE 29

is the vector from the tip of **v** to the tip of **u**. Notice also that when we construct the parallelogram with **u** and **v** as adjacent sides, **u** − **v** is directed along the diagonal from the tip of **v** to the tip of **u**, in contrast to **u** + **v**, which is directed along the other diagonal. From this definition we see that, as we would expect,

$$\mathbf{v} - \mathbf{v} = \mathbf{0}$$

Vectors can be multiplied by scalars as follows. If $k > 0$ and **v** is a nonzero vector, then $k\mathbf{v}$ is a vector that has the same direction as **v** and magnitude k times the magnitude of **v**. If $k < 0$, then $k\mathbf{v}$ has direction opposite to that of **v** and magnitude $|k|$ times the magnitude of **v**. If $k = 0$, we define $k\mathbf{v}$ as the zero vector, and if $\mathbf{v} = \mathbf{0}$, then $k\mathbf{v} = \mathbf{0}$ for all scalars k. In Figure 28 we picture a vector **v**, along with the vectors $2\mathbf{v}$ and $-\frac{1}{2}\mathbf{v}$.

Further insight into the properties of vectors can be gained by introducing a rectangular coordinate system. Suppose that a vector $\mathbf{v} = \overrightarrow{P_1 P_2}$, where the coordinates of P_1 and P_2 are (x_1, y_1) and (x_2, y_2), respectively. As shown in Figure 29, the horizontal displacement from P_1 to P_2 is $x_2 - x_1$, and the vertical displacement is $y_2 - y_1$. We call $x_2 - x_1$ the **horizontal component** (or x component) and $y_2 - y_1$ the **vertical component** (or y component) of **v**. For example, if the coordinates of P_1 are $(3, 2)$ and of P_2 are $(7, 5)$, then the horizontal component of **v** is $7 - 3 = 4$ and the vertical component is $5 - 2 = 3$. So every vector has a unique pair of components, and equivalent vectors have the same components. Conversely, if we are given a pair of components, then these uniquely determine the collection of equivalent vectors that have these components. For example, given the x component 3 and y component 2, we can determine all vectors that have these components. The simplest of these is the one with initial point at the origin and terminal point at $(3, 2)$. Since we do not distinguish between equivalent vectors, we can in effect say that a vector is uniquely determined by its components.

This identification of a vector with its components enables us to look at vectors in a new way. We use the symbol $\langle a, b \rangle$ to indicate a vector with x component a and y component b, and we call this ordered pair of numbers itself a vector. When we wish to distinguish between vectors as directed line segments and vectors as ordered pairs, we say the former is a **geometric vector** and the latter is an **algebraic vector.** By the discussion above, given a geometric vector **v**, we can determine the corresponding algebraic vector $\langle a, b \rangle$ and conversely, and we write $\mathbf{v} = \langle a, b \rangle$. Any geometric vector corresponding to $\langle a, b \rangle$ is called a **geometric representative** of $\langle a, b \rangle$.

EXAMPLE 17

a. Express the vector $\overrightarrow{AB}$ in algebraic form, where $A = (-1, 2)$ and $B = (3, -4)$. Draw the geometric vector.

b. Draw the geometric representative of the vector $\langle -2, 3 \rangle$ whose initial point is $(4, -1)$. What is its terminal point?

Solution **a.** $\overrightarrow{AB} = \langle 3 - (-1), -4 - 2 \rangle = \langle 4, -6 \rangle$
The vector is shown geometrically in Figure 30.

b. Beginning at $(4, -1)$, we go 2 units to the left and 3 units up, giving the terminal point $(2, 2)$, as shown in Figure 31.

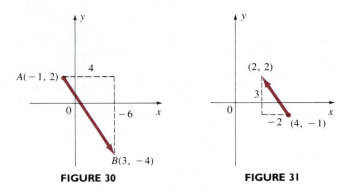

FIGURE 30   **FIGURE 31**

For vectors $\mathbf{u} = \langle u_1, u_2 \rangle$ and $\mathbf{v} = \langle v_1, v_2 \rangle$, addition, subtraction, and multiplication by a scalar are accomplished as follows:

$$\mathbf{u} + \mathbf{v} = \langle u_1 + v_1, u_2 + v_2 \rangle$$
$$\mathbf{u} - \mathbf{v} = \langle u_1 - v_1, u_2 - v_2 \rangle$$
$$k\mathbf{u} = \langle ku_1, ku_2 \rangle$$

These can be shown to be consistent with the geometric definitions given earlier.

As we have indicated, the length of a vector denotes the magnitude of the quantity it represents. When a vector is given algebraically, say $\mathbf{v} = \langle a, b \rangle$, then the length of a corresponding geometric vector is $\sqrt{a^2 + b^2}$ by the Pythagorean theorem. We call this the **magnitude** of $\mathbf{v}$ and denote it by $|\mathbf{v}|$. So

$$|\mathbf{v}| = \sqrt{a^2 + b^2}$$

where a and b are the horizontal and vertical components, respectively, of $\mathbf{v}$.

Remark. It is reasonable to use the same symbol to designate the magnitude of a vector that we use to denote the absolute value of a real number. If x is a real number, then $|x|$ can be interpreted geometrically as the distance from 0 to x on a number line. Similarly, when a vector $\mathbf{u}$ is interpreted geometrically with initial point at the origin, $|\mathbf{u}|$ is the distance from the origin to the terminal point of $\mathbf{u}$.

EXAMPLE 18 ||||||||||||||||||||||||||||||||||||||||||||||||||||||||||||||||||||||||||||||||||||||||||||||||||||||||||

Two forces, $\mathbf{F}_1$, and $\mathbf{F}_2$, of magnitudes $|\mathbf{F}_1| = 70$ lb and $|\mathbf{F}_2| = 50$ lb, act on an object. If the angle between $\mathbf{F}_1$ and $\mathbf{F}_2$ is $50°$, find the magnitude and direction of the resultant force.

Solution   Denote the resultant force by $\mathbf{R}$. Then $\mathbf{R} = \mathbf{F}_1 + \mathbf{F}_2$. From Figure 32 and the law of cosines, we have

$$|\mathbf{R}|^2 = |\mathbf{F}_1|^2 + |\mathbf{F}_2|^2 - 2|\mathbf{F}_1||\mathbf{F}_2|\cos 130°$$
$$= (70)^2 + (50)^2 - 2(70)(50)\cos 130°$$

from which we obtain, by calculator, $|\mathbf{R}| = 109$, approximately. To find the angle θ that $\mathbf{R}$ makes with $\mathbf{F}_1$, we use the law of sines.

$$\frac{50}{\sin \theta} = \frac{109}{\sin 130°}$$

$$\sin \theta = \frac{50 \sin 130°}{109}$$

By calculator, we get $\theta = 20.6°$.   ▲

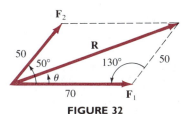

FIGURE 32

EXAMPLE 19 ||||||||||||||||||||||||||||||||||||||||||||||||||||||||||||||||||||||||||||||||||||||||||||||||||||||||||

An airplane is flying at an indicated heading of $230°$ at an indicated airspeed of 260 miles per hour. A 60 mile per hour wind is blowing from $110°$. Find the actual speed and direction of the airplane (relative to the ground).

Solution   Refer to Figure 33. The angle θ is found to be $120°$. (Verify.) So by the law of cosines, the resultant velocity $\mathbf{R}$ satisfies

$$|\mathbf{R}|^2 = (260)^2 + (60)^2 - 2(260)(60)\cos 120°$$
$$= 86{,}800$$
$$|\mathbf{R}| = 294.6$$

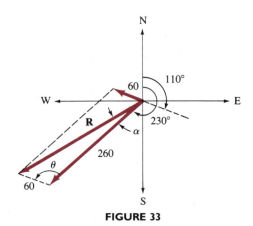

FIGURE 33

By the law of sines,

$$\frac{60}{\sin \alpha} = \frac{294.6}{\sin 120°}$$

$$\sin \alpha = \frac{60 \sin 120°}{294.6}$$

$$\alpha = 10.2°$$

So the actual heading is $230° + 10.2° = 240.2°$. ▲

We have seen how to add and subtract vectors and how to multiply vectors by scalars, but we have not yet defined the product of two vectors. There are two useful ways of doing this, one resulting in a scalar and one in a vector. We consider only the former case here and defer the vector product (also called the **cross product**) to a calculus course.

Let $\mathbf{u} = \langle u_1, u_2 \rangle$ and $\mathbf{v} = \langle v_1, v_2 \rangle$. Then the **dot product** of $\mathbf{u}$ and $\mathbf{v}$, written $\mathbf{u} \cdot \mathbf{v}$, is defined by

$$\mathbf{u} \cdot \mathbf{v} = u_1 v_1 + u_2 v_2$$

For example, if $\mathbf{u} = \langle 2, -3 \rangle$ and $\mathbf{v} = \langle 4, 2 \rangle$, then $\mathbf{u} \cdot \mathbf{v} = (2)(4) + (-3)(2) = 8 - 6 = 2$.

There is an interesting and useful relationship between the dot product and the angle between two nonzero vectors, defined to be the smallest positive angle between geometric representatives of the vectors. We state this as a theorem.

THEOREM I   **If θ is the angle between two nonzero vectors u and v, then**

$$\cos \theta = \frac{\mathbf{u} \cdot \mathbf{v}}{|\mathbf{u}||\mathbf{v}|}$$

Proof. We will assume **u** and **v** are not parallel, although it is not difficult to show that the result holds true in this case also. Take geometric vectors **u** and **v**, each with initial point at the origin. Then if $\mathbf{u} = \langle u_1, u_2 \rangle$ and $\mathbf{v} = \langle v_1, v_2 \rangle$, the terminal point A of **u** is (u_1, u_2) and the terminal point B of **v** is (v_1, v_2), as shown in Figure 34. The vector $\overrightarrow{AB} = \mathbf{v} - \mathbf{u}$. So by the law of cosines, we have

$$|\mathbf{v} - \mathbf{u}|^2 = |\mathbf{u}|^2 + |\mathbf{v}|^2 - 2|\mathbf{u}||\mathbf{v}| \cos \theta$$

or, in terms of components,

$$(v_1 - u_1)^2 + (v_2 - u_2)^2 = u_1^2 + u_2^2 + v_1^2 + v_2^2 - 2|\mathbf{u}||\mathbf{v}| \cos \theta$$
$$v_1^2 - 2u_1v_1 + u_1^2 + v_2^2 - 2u_2v_2 + u_2^2 = u_1^2 + u_2^2 + v_1^2 + v_2^2 - 2|\mathbf{u}||\mathbf{v}| \cos \theta$$
$$-2(u_1v_1 + u_2v_2) = -2|\mathbf{u}||\mathbf{v}| \cos \theta$$

Thus,

$$\cos \theta = \frac{u_1v_1 + u_2v_2}{|\mathbf{u}||\mathbf{v}|}$$

and since $\mathbf{u} \cdot \mathbf{v} = u_1v_1 + u_2v_2$, the conclusion follows.

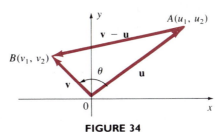

FIGURE 34

EXAMPLE 20

Find the angle between the vectors $\overrightarrow{PQ}$ and $\overrightarrow{PR}$ in Figure 35.

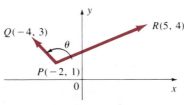

FIGURE 35

Solution   Let $\mathbf{u} = \overrightarrow{PQ}$ and $\mathbf{v} = \overrightarrow{PR}$. Then $\mathbf{u} = \langle -2, 2 \rangle$ and $\mathbf{v} = \langle 7, 3 \rangle$. So from Theorem 1,

$$\cos \theta = \frac{\mathbf{u} \cdot \mathbf{v}}{|\mathbf{v}||\mathbf{v}|} = \frac{(-2)(+7) + (2)(3)}{\sqrt{49 + 9}\sqrt{4 + 4}} = \frac{-8}{\sqrt{58}\sqrt{8}}$$

By calculator we find $\theta = 111.8°$.

EXERCISE SET 5

A

In Problems 1–12, $\mathbf{u} = \langle -2, 3 \rangle$, $\mathbf{v} = \langle 4, 2 \rangle$, and $\mathbf{w} = \langle -1, -2 \rangle$. In each case give the result as an algebraic vector, and also give a geometric construction illustrating the specified operations.

1. $\mathbf{u} + \mathbf{v}$
2. $\mathbf{v} + \mathbf{w}$
3. $\mathbf{u} + \mathbf{w}$
4. $\mathbf{u} + \mathbf{v} + \mathbf{w}$
5. $\mathbf{u} - \mathbf{v}$
6. $\mathbf{v} - \mathbf{w}$
7. $\mathbf{w} - \mathbf{u}$
8. $2\mathbf{u} + \frac{1}{2}\mathbf{v}$
9. $\mathbf{u} - 2\mathbf{w}$
10. $\mathbf{u} + \frac{3}{2}\mathbf{v} - \mathbf{w}$
11. $-2\mathbf{u} + \mathbf{v} - 3\mathbf{w}$
12. $2\mathbf{u} - \frac{1}{2}\mathbf{v} + 3\mathbf{w}$

In Problems 13–16 find the algebraic vector $\overrightarrow{P_1P_2}$.

13. $P_1 = (3, 4)$, $P_2 = (-1, 2)$
14. $P_1 = (-4, -2)$, $P_2 = (3, -1)$
15. $P_1 = (0, 4)$, $P_2 = (-3, 0)$
16. $P_1 = (7, -3)$, $P_2 = (-1, 8)$

In Problems 17–20 draw the vector $\mathbf{v} = \overrightarrow{P_1P_2}$ for the given vector $\mathbf{v}$ and the given point P_1. Determine the coordinates of P_2.

17. $\mathbf{v} = \langle 3, -2 \rangle$; $P_1 = (0, 0)$
18. $\mathbf{v} = \langle -2, 4 \rangle$; $P_1 = (1, 2)$
19. $\mathbf{v} = \langle 0, 3 \rangle$; $P_1 = (-2, -3)$
20. $\mathbf{v} = \langle -3, -4 \rangle$; $P_1 = (4, 2)$

In Problems 21–26 find $|\mathbf{v}|$.

21. $\mathbf{v} = \langle 3, 4 \rangle$
22. $\mathbf{v} = \langle -8, 6 \rangle$
23. $\mathbf{v} = \langle 8, 15 \rangle$
24. $\mathbf{v} = \langle -12, -5 \rangle$
25. $\mathbf{v} = \overrightarrow{P_1P_2}$, where $P_1 = (2, -3)$ and $P_2 = (-4, -1)$
26. $\mathbf{v} = \overrightarrow{P_1P_2}$, where $P_1 = (-1, 3)$ and $P_2 = (5, -2)$

In Problems 27–30 find $\mathbf{u} \cdot \mathbf{v}$.

27. $\mathbf{u} = \langle 4, 7 \rangle$, $\mathbf{v} = \langle -5, 2 \rangle$
28. $\mathbf{u} = \langle -3, -6 \rangle$, $\mathbf{v} = \langle 5, -2 \rangle$
29. $\mathbf{u} = \overrightarrow{P_1P_2}$, $\mathbf{v} = \overrightarrow{Q_1Q_2}$, where $P_1 = (0, 2)$, $P_2 = (-1, 4)$, $Q_1 = (-3, 1)$, $Q_2 = (2, -2)$
30. $\mathbf{u} = \overrightarrow{AB}$, $\mathbf{v} = \overrightarrow{CD}$, where $A = (3, 4)$, $B = (-1, 1)$, $C = (3, 0)$, $D = (-2, -3)$

In Problems 31–34 find the angle between $\mathbf{u}$ and $\mathbf{v}$.

31. $\mathbf{u} = \langle 4, -4 \rangle$, $\mathbf{v} = \langle 1, 7 \rangle$
32. $\mathbf{u} = \langle 1, -2 \rangle$, $\mathbf{v} = \langle -1, 1 \rangle$
33. $\mathbf{u} = \overrightarrow{P_1P_2}$, $\mathbf{v} = \overrightarrow{P_1P_3}$, where $P_1 = (3, 2)$, $P_2 = (-1, 4)$, $P_3 = (7, -6)$
34. $\mathbf{u} = \overrightarrow{PQ}$, $\mathbf{v} = \overrightarrow{PR}$, where $P = (4, 5)$, $Q = (2, -1)$, $R = (-1, 2)$
35. A force of 150 pounds being exerted on a body is acting horizontally to the right. A second force of 68 pounds exerted on the body is acting vertically upward. Find the magnitude and direction of the resultant force.
36. A pilot sets his heading at 120° and flies at an indicated airspeed of 195 miles per hour. There is a wind blowing from 210° at 62 miles per hour. What are the true course and speed of the airplane relative to the ground?
37. A pilot flew to a destination due south of her starting point. If the wind was from 300° at 80 miles per hour and her actual ground speed averaged 160 miles per hour, find her indicated heading and airspeed.
38. A 1,000-newton weight is resting on a plane that is inclined at an angle of 28° with the horizontal. Find the components of the weight parallel to and perpendicular to the plane.
39. A hot-air balloon is rising vertically at the rate of 70 feet per second. If the wind is blowing horizontally at the rate of 30 miles per hour, find the actual path of the balloon and its speed along that path.
40. The pilot of a small plane plans a trip from town A to town B, 400 miles due south of A. He will cruise at an indicated airspeed of 180 miles per hour. The wind is from the west at 40 miles per hour. What should his heading be to compensate for the wind? How long will it take to make the trip?
41. The magnitude of the resultant $\mathbf{R}$ of two forces $\mathbf{F}_1$ and $\mathbf{F}_2$ acting on an object is $|\mathbf{R}| = 223.4$ pounds, and $|\mathbf{F}_2| = 111.5$ pounds. If the angle between $\mathbf{F}_1$ and $\mathbf{R}$ is 27°34′, and the angle between $\mathbf{F}_1$ and

F_2 is acute, find $|F_1|$ and the angle between F_1 and F_2.

42. Find the magnitude and direction of the resultant of two forces of 25 pounds and 30 pounds, respectively, acting on an object, if the angle between the forces is 23°.

43. If the angle between the vectors v and w is 47.2°, with $|v| = 21.5$ and $|w| = 37.2$, find $|v + w|$ and the angle from v to $v + w$.

44. A river that flows north to south has a current of 8 miles per hour. A motorboat leaves the west bank and heads in the direction S 75° E at an indicated speed, relative to the water, of 20 miles per hour. Find the actual speed and direction of the boat.

B

45. **a.** If the angle between two vectors is $\pi/2$, we say the vectors are **orthogonal**. Show that two nonzero vectors are orthogonal if, and only if, their dot product is 0.
    b. Show that the vectors $\overrightarrow{AB}$ and $\overrightarrow{CD}$ are orthogonal, where $A = (-8, 11)$, $B = (2, 5)$, $C = (9, -1)$, $D = (12, 4)$.

46. Use vectors to show that the triangle with vertices $(2, 1)$, $(6, 9)$, and $(-2, 3)$ is a right triangle. (See Problem 45.)

47. Three forces of magnitude $|F_1| = 30$ pounds, $|F_2| = 45$ pounds, and $|F_3| = 56$ pounds are acting on an object as shown. Find the magnitude of the resultant, and give its angle from F_1.

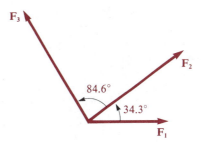

48. A boat traveling at 23 miles per hour heads in the direction N 42.5° W for 45 minutes and then changes to the direction N 36.8° E and goes on this course for 1 hour and 15 minutes. How far is the boat from its starting point?

49. The vectors $\langle 1, 0 \rangle$ and $\langle 0, 1 \rangle$ are often denoted by i and j, respectively. Show that if $u = \langle a, b \rangle$ is any vector, then u can be written in the form $u = ai + bj$.

50. The pilot of a light airplane plans to fly from town A to town B, a distance of 650 miles. The bearing of town B from town A is S 48°32′ E. The pilot will fly at an average airspeed of 175 miles per hour. If a constant wind is blowing from the northeast at 35 miles per hour, what should the pilot's heading be? How long will it take to make the trip?

51. The heading of a jetliner flying at an indicated airspeed of 465 miles per hour is 220°. If a wind of 90 miles per hour is blowing from 300°, find the ground speed of the plane and its actual course.

 REVIEW EXERCISE SET

A

Solve the right triangles in Problems 1–14. Wherever possible, solve without using a calculator. The right angle in each case is C.

1. $A = 60°$, $b = 5$
2. $B = 30°$, $c = 9$
3. $B = 60°$, $a = 4$
4. $A = 45°$, $a = 12$
5. $B = 45°$, $c = 10$
6. $A = 30°$, $c = 5$
7. $a = 7$, $c = 14$
8. $a = 8$, $b = 15$
9. $A = 33°$, $c = 13$
10. $a = 4$, $b = 7$

11. $b = 102$, $c = 151$
12. $A = 67°$, $a = 12.6$
13. $B = 26.2°$, $c = 33.5$
14. $B = 62.3°$, $a = 2.53$

Solve the triangles in Problems 15–26.

15. $A = 34.3°$, $B = 26.8°$, $a = 13.4$
16. $B = 54°30′$, $C = 48°20′$, $a = 102.5$
17. $C = 32°$, $b = 15$, $c = 8$
18. $a = 4$, $b = 7$, $C = 110°$
19. $a = 22.3$, $b = 17.9$, $c = 10.5$

20. $A = 110°$, $C = 28°$, $c = 12$
21. $A = 63.2°$, $a = 14.3$, $b = 25.2$
22. $B = 36°20'$, $b = 34.5$, $c = 12.7$
23. $a = 8$, $b = 9$, $c = 13$
24. $a = 2.05$, $c = 3.72$, $B = 61°40'$
25. $A = 25.2°$, $c = 27.4$, $a = 15.1$
26. $a = 7.214$, $b = 5.056$, $c = 9.524$

In Problems 27–30 find the area of triangle ABC.

27. $a = 32.5$, $b = 51.6$, $C = 25.8°$
28. $b = 1.22$, $c = 3.56$, $C = 102°10'$
29. $a = 13.4$, $b = 8.23$, $c = 6.48$
30. $b = 13.2$, $c = 10.5$, $B = 42.0°$
31. A 40-foot-high building is at the edge of a river. Directly opposite on the other edge an observer finds that the angle of elevation of the top of the building is $33°10'$. How wide is the river at that point?
32. The sides of a parallelogram are 6 and 10, and the longer diagonal is 14. Find the interior angles of the parallelogram.
33. A building lot is in the shape of a triangle in which two sides are 100 feet and 120 feet, and the angle between them is $55°$. Find the length of the third side and the area of the lot.
34. A portion of a modern metal sculpture consists of a large triangular plate mounted vertically, with one vertex at the top and with a horizontal base. The length of the base is 12 feet, and the base angles are $30°$ and $50°$, respectively. Find the lengths of the other two sides of the triangle. What is its area?
35. The angle between two vectors v_1 and v_2 is $27.5°$, and the angle from v_1 to $v_1 + v_2$ is $16.1°$. If $|v_1| = 10.7$, find $|v_2|$ and $|v_1 + v_2|$.
36. Two forces, of magnitudes $|F_1| = 8.67$ newtons and $|F_2| = 6.43$ newtons, act on an object. If the angle between F_1 and F_2 is $72°33'$, find the magnitude and direction of the resultant.
37. Two forces, of magnitudes $|F_1| = 36.5$ pounds and $|F_2| = 53.4$ pounds, are acting on an object. If the resultant R makes an angle of $32°15'$ with F_1, find $|R|$. What is the angle between F_1 and F_2?
38. Find the angle between u and v.
    a. $u = \langle 3, 5 \rangle$, $v = \langle -1, 2 \rangle$
    b. $u = \overrightarrow{PQ}$, $v = \overrightarrow{PR}$, where $P = (-5, 7)$, $Q = (-2, 2)$, $R = (3, 1)$

39. Use vector methods to show that the points $(3, -1)$, $(5, 4)$, $(-5, 8)$, and $(-7, 3)$ are vertices of a rectangle. What is its area?

B

40. Use vector methods to find the interior angles of the triangle with vertices $(3, -5)$, $(-1, -2)$, and $(8, 7)$. Find the area of this triangle.
41. An observer is 100 feet from the base of a building. She finds the angles of elevation to the bottom and top of a flagpole on the roof of the building to be $59.3°$ and $62.4°$, respectively. How high is the flagpole?
42. Points A and B are on opposite sides of a pond, and it is desired to find the distance between them. Point C is located so that $|\overrightarrow{AC}| = 254$ feet and $|\overrightarrow{BC}| = 198$ feet. The angle at A from $\overrightarrow{AB}$ to $\overrightarrow{AC}$ is found to be $47°21'$. Find the distance from A to B.
43. At 4:00 P.M. an airplane leaves Chicago and heads due south at a cruising speed of 400 miles per hour. At 5:00 P.M. another plane leaves from the same airport and travels at a heading of $220°$, cruising at 500 miles per hour. How far apart are the two planes at 6:00 P.M.?
44. Two guy wires are attached on opposite sides of a pole. They make angles of $36°$ and $42°$ with the horizontal, and the points where they meet the ground are 50 feet apart. Find the length of each wire.
45. A ranger in a lookout tower spots a fire on a line N $40°$ W. A second ranger at a tower 10 miles to the west of the first tower also sees the fire and finds it to be on a line N $35°$ E. How far is the fire from each lookout tower?
46. If there were no wind, a pilot would head in the direction $215°$ to go from city A to city B. The distance between the cities is 425 miles. If a 40 mile per hour wind is blowing from $290°$, and if the average airspeed of the plane is 185 miles per hour, find the heading the pilot should take to go from A to B. If she also wishes to return to A, what should her heading be? What will the total flying time be for the round trip?
47. From the top of a lighthouse 120 feet high the angle of depression of ship A is $12.8°$ and the angle of depression of ship B is $14.2°$. The bearing of ship A from the lighthouse is N $35.7°$ E and the

bearing of B is S 64.3° E. Find the distance between the ships and the bearing of ship B from ship A.

48. A hill slopes at an angle of 23°10′, and a monument is erected on top of the hill, as shown in the figure. From point A at the base of the hill the angle of elevation of the top of the monument is 34°37′, and from point B, 200 feet up the hill from A, the angle of elevation of the top of the monument is 41°06′. How high is the monument?

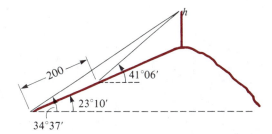

10

FURTHER APPLICATIONS OF TRIGONOMETRIC FUNCTIONS

Many calculus problems are more easily solved when **polar coordinates** are used rather than rectangular coordinates. The following example illustrates such a problem.*

Find the area of the region that lies inside the circle $r = 3 \sin \theta$ and outside the cardioid $r = 1 + \sin \theta$.

Solution   The cardioid . . . and the circle are sketched. . . . They intersect when $3 \sin \theta = 1 + \sin \theta$, which gives $\sin \theta = \frac{1}{2}$, so $\theta = \pi/6$, $5\pi/6$. The desired area can be found by subtracting the area inside the cardioid between $\theta = \pi/6$ and $\theta = 5\pi/6$ from the area inside the circle from $\pi/6$ to $5\pi/6$. Thus,

$$A = \frac{1}{2} \int_{\pi/6}^{5\pi/6} (3 \sin \theta)^2 \, d\theta - \frac{1}{2} \int_{\pi/6}^{5\pi/6} (1 + \sin \theta)^2 \, d\theta$$

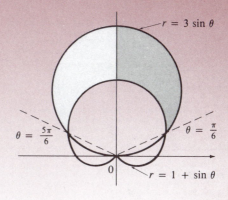

* From *Calculus*, 2nd ed. by James Stewart. Copyright © 1991, 1987 by Wadsworth, Inc. Reprinted by permission of Brooks/Cole Publishing Company, Pacific Grove, CA 93950.

Since the region is symmetric about the vertical axis $\theta = \pi/2$, we can write

$$A = 2\left[\frac{1}{2}\int_{\pi/6}^{5\pi/6}(9\sin^2\theta\,d\theta - \frac{1}{2}\int_{\pi/6}^{5\pi/6}(1 + 2\sin\theta + \sin^2\theta)\,d\theta\right]$$

$$= \int_{\pi/6}^{5\pi/6}(8\sin^2\theta - 1 - 2\sin\theta)\,d\theta = \cdots = \pi$$

We will study polar coordinates in this chapter along with some other topics that make use of trigonometric functions, and will learn about cardioids as well as polar equations of circles.

POLAR COORDINATES

In all of our graphing so far we have used the rectangular coordinate system. For certain graphing problems, however, an alternative system, called the **polar coordinate system,** has definite advantages. In this system points in the plane are located by giving radial distances from a fixed point, called the **pole,** and angles from a fixed ray called the **polar axis.** The polar axis extends horizontally to the right from the pole, and we make it into the positive half of a number line, with the pole corresponding to zero. Now let θ be any angle with the pole as its vertex and the polar axis as its initial side, as in Figure 1. For a positive number r, if we measure r units from the pole along the terminal side of θ, a unique point P is determined, and we call the ordered pair (r, θ) the **polar coordinates** of P. If r is negative, we measure $|r|$ units from the pole on the ray directed opposite to the terminal side of θ. In Figure 2 we illustrate the points $(2, \pi/6)$ and $(-2, \pi/6)$.

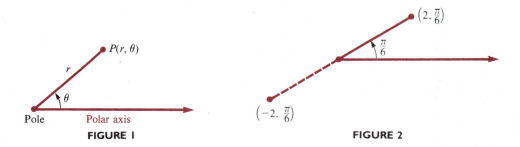

FIGURE 1   FIGURE 2

In the manner described, the ordered pair (r, θ) determines a unique point P in the plane. On the other hand, if we are given P, there are infinitely many pairs of polar coordinates for it, since there are infinitely many angles having terminal side through P. For example, $(2, \pi/6)$, $(2, 13\pi/6)$, $(2, -11\pi/6)$, and $(-2, 7\pi/6)$ represent the same point. (Verify this.)

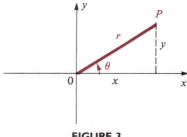

FIGURE 3

If we superimpose a rectangular coordinate system on a polar system so that the origin coincides with the pole and the positive x axis with the polar axis, as in Figure 3, we can determine relationships between rectangular and polar coordinates. Let P have coordinates (x, y) in the rectangular system and (r, θ) in the polar system. The following relationships hold true:

$$\left. \begin{aligned} \tan \theta &= \frac{y}{x} \quad (x \neq 0) \\ r^2 &= x^2 + y^2 \end{aligned} \right\} \tag{1}$$

$$\left. \begin{aligned} x &= r \cos \theta \\ y &= r \sin \theta \end{aligned} \right\} \tag{2}$$

The next two examples illustrate how we can use these equations to change from one coordinate system to the other.

EXAMPLE I ||||||||||||||||||||||||||||||||||||||||||||||||||||||||||||||||||||||||||||||||||||||||||||||||||||||||||||

a. Change $(\sqrt{3}, -1)$ to polar coordinates (r, θ), where $r > 0$ and $0 \le \theta < 2\pi$.
b. Change $(-\sqrt{2}, 3\pi/4)$ to rectangular coordinates.

Solution **a.** From equations (1) we find that

$$\tan \theta = \frac{-1}{\sqrt{3}}$$

$$r^2 = 4$$

Since $x > 0$ and $y < 0$, we know that θ is in quadrant IV, so $\theta = 11\pi/6$. The polar coordinates are therefore $(2, 11\pi/6)$.

b. From equations (2)

$$x = -\sqrt{2} \cos \frac{3\pi}{4} = -\sqrt{2}\left(-\frac{1}{\sqrt{2}}\right) = 1$$

$$y = -\sqrt{2} \sin \frac{3\pi}{4} = -\sqrt{2}\left(\frac{1}{\sqrt{2}}\right) = -1$$

So the rectangular coordinates are $(1, -1)$. ▲

EXAMPLE 2 ||||||||||||||||||||||||||||||||||||||||||||||||||||||||||||||||||||||||||||||||||||||||||||||||||

a. Find the rectangular equation of the curve whose polar equation is $r = 2 \cos \theta$.

b. Find the polar equation of the curve whose rectangular equation is $y^2 = 2x$.

Solution   a. For $r \neq 0$ we replace $\cos \theta$ by x/r to obtain

$$r = 2\frac{x}{r}$$

$$r^2 = 2x$$

Now we use the fact that $r^2 = x^2 + y^2$ to get the desired rectangular equation:

$$x^2 + y^2 = 2x$$

which we recognize as being a circle. The fact that the origin is on the circle assures us that no further points on the graph are obtained if $r = 0$, since the only point for which $r = 0$ is the pole, which coincides with the origin.

b. From equations (2) we obtain

$$r^2 \sin^2 \theta = 2r \cos \theta$$

$$r(r \sin^2 \theta - 2 \cos \theta) = 0$$

$$r = 0 \quad \text{or} \quad r = 2 \cos \theta \csc^2 \theta \quad (\theta \neq 2k\pi)$$

The solution $r = 0$ is contained in the second equation, since when $\theta = \pi/2$, the equation $r = 2 \cos \theta \csc^2 \theta$ gives $r = 0$. So $r = 2 \cos \theta \csc^2 \theta$ contains all points on the curve. Thus, the answer is

$$r = 2 \cos \theta \csc^2 \theta \qquad \blacktriangle$$

To graph a polar equation, we can make a table of values by assigning values to θ and calculating r. We typically assign to θ the radian measure of special angles we have studied, and we use our knowledge of the behavior of the functions as θ varies from 0 to 2π. As an alternative, the graph is sometimes more easily obtained by changing to the corresponding rectangular equation. However, some of the more interesting curves in polar coordinates have complicated rectangular equations. The equation $r \cos \theta = 2$ corresponds to $x = 2$ in rectangular coordinates, so we know that its graph is a vertical line. In this case it is clearly easier to work with the rectangular equation. On the other hand, the equation $r = 1 + \cos \theta$ corresponds to $x^4 - 2x^3 - 2xy^2 + 2x^2y^2 - y^2 = 0$. (Verify this.) If you attempt to draw the graph from this rectangular equation, you will soon be convinced that the polar equation is easier to work with. In the next example we show its graph.

EXAMPLE 3 ‖‖

Draw the graph of the equation $r = 1 + \cos \theta$.

Solution   A reasonably accurate graph can be drawn from the points in the following table. If more points were needed, we could use a calculator. Note that since the cosine is even, we get the same value of r for θ and for $-\theta$. The values of r are rounded to two places, using the approximations $\sqrt{3}/2 \approx 0.87$ and $\sqrt{2}/2 \approx 0.71$.

θ	0	$\pm\dfrac{\pi}{6}$	$\pm\dfrac{\pi}{4}$	$\pm\dfrac{\pi}{3}$	$\pm\dfrac{\pi}{2}$	$\pm\dfrac{2\pi}{3}$	$\pm\dfrac{3\pi}{4}$	$\pm\dfrac{5\pi}{6}$	π
r	2	1.87	1.71	1.50	1	0.50	0.29	0.13	0

The graph is shown in Figure 4. This is an example of a **cardioid,** so-called because of its heart-like shape.

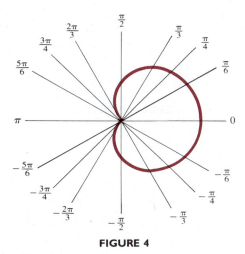

FIGURE 4

EXAMPLE 4 ‖‖

Draw the graph of the equation $r = 2 \cos 2\theta$.

Solution   Again we use the fact that the cosine is an even function. Also note that $\cos 2\theta$ has period π, so we can take θ from $-\pi/2$ to $\pi/2$. We choose θ values such that 2θ takes on the special values whose functions we know.

θ	0	$\pm\dfrac{\pi}{12}$	$\pm\dfrac{\pi}{8}$	$\pm\dfrac{\pi}{6}$	$\pm\dfrac{\pi}{4}$	$\pm\dfrac{\pi}{3}$	$\pm\dfrac{3\pi}{8}$	$\pm\dfrac{5\pi}{12}$	$\pm\dfrac{\pi}{2}$
r	2	1.73	1.41	1	0	-1	-1.41	-1.73	-2

The graph is shown in Figure 5. This is an example of a **rose curve,** and it is said to have four **petals.** In general, the graph of an equation of the form $r = a \cos n\theta$ for $n \geq 2$ will be a rose curve with $2n$ petals, if n is even. On the other hand, if n is odd, the rose curve will have just n petals.

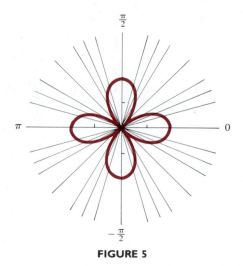

FIGURE 5

EXAMPLE 5

Draw the graph of the equation $r^2 = 4 \cos 2\theta$.

Solution   We may take θ over any interval of length π, since this is the period of $\cos 2\theta$. Let us choose $0 \leq \theta < \pi$. Note, however, that $r = \pm 2\sqrt{\cos 2\theta}$, so values of θ for which $\cos 2\theta < 0$ must be excluded, namely, $\pi/4 < \theta < 3\pi/4$.

θ	0	$\dfrac{\pi}{12}$	$\dfrac{\pi}{6}$	$\dfrac{\pi}{4}$	$\dfrac{3\pi}{4}$	$\dfrac{5\pi}{6}$	$\dfrac{11\pi}{12}$	π
r	± 2	± 1.87	± 1.41	0	0	± 1.41	± 1.86	± 2

The graph, shown in Figure 6, resembles the figure 8. It is called a **lemniscate.**

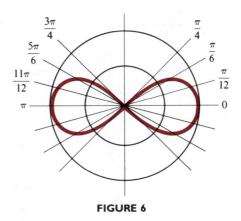

FIGURE 6

We summarize below some of the most common polar graphs, along with their equations and their names. For completeness, we include those we have illustrated in Examples 3, 4, and 5.

Straight Lines

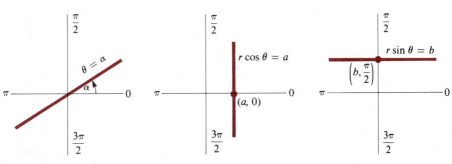

Line through pole making an angle α with polar axis

$r \cos \theta = a$
Same as $x = a$

$r \sin \theta = b$
Same as $y = b$

Circles

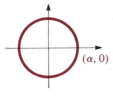

$r = a$
Circle of radius a,
center at pole

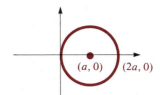

$r = 2a \cos \theta$
Circle of radius a,
center on polar axis,
passing through pole

Limaçons: $r = a + b \cos \theta$

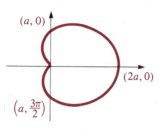

$a = b$
$r = a(1 + \cos \theta)$
Called a *cardioid* because
of heartlike shape

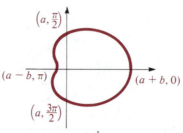

$a > b$
Limaçon without loop
(If $a \geq 2b$, the "dimple"
on the left is absent.)

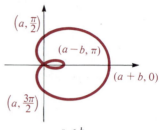

$a < b$
Limaçon with loop

Lemniscates: $r^2 = a^2 \cos 2\theta$

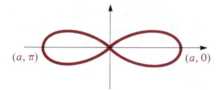

Rose Curves: $r = a \cos n\theta$

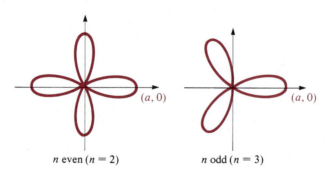

n even ($n = 2$) n odd ($n = 3$)

When n is even, there are $2n$ petals; when n is odd, there are n petals.

You may have noticed that each of the equations we have given involves the cosine function only and that the curves are all symmetric to the polar axis, since $\cos(-\theta) = \cos \theta$. It can be shown that if $\cos \theta$ is replaced by $\sin \theta$, the effect is to rotate the graph in a counterclockwise direction by $90°$. More generally, the graph of $r = f(\theta - \alpha)$ is the same as that of $r = f(\theta)$ rotated through an angle α. Since $\cos(\theta - \pi/2) = \sin \theta$, this confirms that replacing $\cos \theta$ by $\sin \theta$

rotates the graph $\pi/2$ radians, or 90°. Similarly, since $\cos(\theta - \pi) = -\cos\theta$ and $\cos(\theta - 3\pi/2) = -\sin\theta$, replacing $\cos\theta$ by $-\cos\theta$ or by $-\sin\theta$ rotates the graph 180° or 270°, respectively. In the case of $\cos n\theta$ being replaced by $\sin n\theta$, the rotation is $90°/n$, with similar results for the other cases. For example, the graph of the equation $r^2 = 2a\sin 2\theta$ is the lemniscate $r^2 = 2a\cos 2\theta$ rotated through an angle of 45°.

Polar graphs can be obtained on a graphing calculator, but to understand how to do this, you will need to know something about parametric equations, which we will study in the next section. So we will defer a discussion of this, as well as exercises on the graphing calculator, to Section 2.

We conclude this section with an example of a type of problem often encountered in calculus.

EXAMPLE 6

Draw the curves $r = 3\sin\theta$ and $r = 1 + \sin\theta$ on the same polar coordinate system, and shade the area inside the first curve and outside the second. Also find all points of intersection of the curves.

Solution   According to our summary of polar curves, the graph of $r = 2a\cos\theta$ is a circle of radius a, with center at $(a, 0)$. So $r = 3\cos\theta$ is a circle of radius $\frac{3}{2}$ centered at $(\frac{3}{2}, 0)$. Replacing $\cos\theta$ by $\sin\theta$ has the effect of rotating the curve through 90°. So $r = 3\sin\theta$ is a circle of radius $\frac{3}{2}$, centered at $(\pi/2, 3/2)$. (We could also have arrived at this result directly, as in Example 2a, by changing to rectangular coordinates.) Similarly, $r = 1 + \sin\theta$ is the cardioid of Example 3, rotated through 90°.

To find the points of intersection, we set the two r values equal to one another and solve for θ:

$$3\sin\theta = 1 + \sin\theta$$

$$2\sin\theta = 1$$

$$\sin\theta = \frac{1}{2}$$

$$\theta = \frac{\pi}{6}, \frac{5\pi}{6}$$

Substituting into either of the original equations gives $r = \frac{3}{2}$. So the curves intersect at $(\pi/6, 3/2)$ and $(5\pi/6, 3/2)$. The graphs, showing the shaded area inside the circle and outside the cardioid, are shown in Figure 7. Observe that the curves also intersect at the pole. Why did we not find this when we solved the two equations simultaneously? The answer lies in the fact that the curves pass through the pole for different values of θ. The circle goes through the pole when

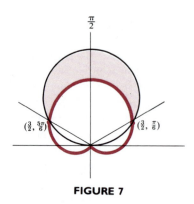

FIGURE 7

$\theta = \pi/2$, and the cardioid goes through the pole when $\theta = \pi$. This type of "nonsimultaneous" point of intersection can usually be determined from the graphs.

EXERCISE SET 1

A

In Problems 1 and 2 plot the points in a polar coordinate system.

1. a. $\left(1, \dfrac{2\pi}{3}\right)$ **b.** $\left(2, -\dfrac{\pi}{2}\right)$ **c.** $\left(-2, \dfrac{3\pi}{2}\right)$

d. $\left(3, \dfrac{\pi}{6}\right)$ **e.** $\left(-1, -\dfrac{5\pi}{4}\right)$

2. a. $(4, \pi)$ **b.** $\left(-2, \dfrac{\pi}{2}\right)$ **c.** $\left(3, -\dfrac{5\pi}{6}\right)$

d. $\left(0, \dfrac{\pi}{5}\right)$ **e.** $\left(-3, -\dfrac{5\pi}{3}\right)$

3. Give five different sets of polar coordinates for the point $(2, 2\pi/3)$. Include at least one set for which $r < 0$ and one for which $\theta < 0$.

In Problems 4 and 5 give the corresponding rectangular coordinates for the given polar coordinates.

4. a. $\left(2, \dfrac{\pi}{6}\right)$ **b.** $\left(-\sqrt{2}, \dfrac{\pi}{4}\right)$ **c.** $\left(4, \dfrac{2\pi}{3}\right)$

d. $\left(6, \dfrac{3\pi}{2}\right)$ **e.** $\left(-8, -\dfrac{5\pi}{3}\right)$

5. a. $\left(2\sqrt{2}, -\dfrac{7\pi}{4}\right)$ **b.** $\left(6, \dfrac{11\pi}{6}\right)$

c. $\left(-10, -\dfrac{9\pi}{4}\right)$ **d.** $(4, 3\pi)$ **e.** $\left(\sqrt{8}, \dfrac{9\pi}{4}\right)$

In Problems 6 and 7 give the polar coordinates (for which $r > 0$ and $0 \le \theta < 2\pi$) corresponding to the points with the given rectangular coordinates.

6. a. $(1, \sqrt{3})$ **b.** $(-2, 2)$ **c.** $(0, 5)$

d. $(-4, 0)$ **e.** $(-3, -\sqrt{3})$

7. a. $(4, -4)$ **b.** $(-2\sqrt{3}, 2)$ **c.** $(0, -2)$

d. $\left(\dfrac{1}{2}, -\dfrac{\sqrt{3}}{2}\right)$ **e.** $(-3\sqrt{2}, -\sqrt{6})$

In Problems 8–21 change to an equivalent polar equation.

8. $x^2 + y^2 = 16$ **9.** $x - y = 0$

10. $x + y = 0$ **11.** $y = \sqrt{3}x$

12. $3x - 2y = 5$ **13.** $x - 2 = 0$

14. $y + 3 = 0$ **15.** $x^2 - y^2 = 1$

16. $xy = 4$ **17.** $x^2 + y^2 = 4x$

18. $y^2 = 2x + 1$ **19.** $(x + 1)^2 + y^2 = 1$

20. $x^2 y^2 = 4(x^2 + y^2)^{3/2}$

21. $x^2 - y^2 = \dfrac{2xy}{x^2 + y^2}$

In Problems 22–35 change to an equivalent rectangular equation.

22. $r = 1$

23. $\theta = \dfrac{\pi}{4}$

24. $r \cos \theta + 1 = 0$

25. $r \sin \theta = 3$

26. $r = 2 \sec \theta$

27. $r = -4 \csc \theta$

28. $r(3 \cos \theta - 2 \sin \theta) = 4$

29. $r(2 \cos \theta + 5 \sin \theta) = 7$

30. $r = 4 \cos \theta$

31. $r = -2 \sin \theta$

32. $\theta = \tan^{-1} \frac{3}{2}$

33. $r^2 = \cos 2\theta$

34. $r = \dfrac{1}{1 - \cos \theta}$

35. $r = \dfrac{1}{1 + 2 \sin \theta}$

In Problems 36–59 identify the curve and draw its graph.

36. $r = 3$

37. $r = -2$

38. $\theta = \dfrac{3\pi}{4}$

39. $r \cos \theta = 2$

40. $r = -3 \csc \theta$

41. $r = \dfrac{6}{2 \cos \theta - 3 \sin \theta}$

42. $r = \dfrac{12}{3 \cos \theta + 4 \sin \theta}$

43. $r = 4 \cos \theta$

44. $r = 1 - \cos \theta$

45. $r = 1 + \sin \theta$

46. $r = 5 - 4 \cos \theta$

47. $r = 4 - 5 \sin \theta$

48. $r^2 = 9 \sin 2\theta$

49. $r = 2 \cos 2\theta$

50. $r = \sin 3\theta$

51. $r = -\cos 3\theta$

52. $r = 3 \sin 2\theta$

53. $r^2 = \cos^2 \theta - \sin^2 \theta$

54. $r = 2 + \sin \theta$

55. $r = 2 \cos \theta - 1$

56. $r = 2 \cos\left(\theta - \dfrac{3\pi}{4}\right)$

57. $r = 2 + 2 \cos\left(\theta + \dfrac{\pi}{6}\right)$

58. $r = 3(\sin \theta - 1)$

59. $r^2 = 8 \sin \theta \cos \theta$

B

Draw the graphs in Problems 60–66.

60. $r = 2 \sin^2 \dfrac{\theta}{2}$

61. $r = \dfrac{4}{2 + \cos \theta}$

62. $r = \cos \theta + \sin \theta$
    Hint. Write in the form $r = a \cos(\theta - \alpha)$.

63. $r = 3 \cos \theta - 4 \sin \theta$ (See the hint in Problem 62.)

64. $r = 2 \cos \dfrac{\theta}{2}$

65. $r = \sqrt{4 \sin \theta}$

66. $r = 3 \cos 5\theta$

67. By changing to rectangular coordinates, show that the graph of $r = 4 \sec^2(\theta/2)$ is a parabola. Draw the graph.

Problems 68–71 involve spiral curves. Draw them for the indicated values of a.

68. **Spiral of Archimedes,** $r = a\theta$ $(a = 1)$

69. **Hyperbolic spiral,** $r = a/\theta$ $(a = 2)$

70. **Logarithmic spiral,** $r = e^{a\theta}$ $(a = 1)$

71. **Parabolic spiral,** $r^2 = a\theta$ $(a = 1)$

72. Graph the **bifolium** $r = a \sin 2\theta \cos \theta$ for $a = 1$.

73. Graph the **cissoid** $r = a \sin \theta \tan \theta$ for $a = 2$.

74. Graph the **conchoid** $r = a \sec \theta + b$ for $a = 1$ and $b = -2$.

In Problems 75–79 show the two curves on the same coordinate system, indicate the specified area by shading, and find all points of intersection.

75. Outside $r = 2$ and inside $r = 2(1 + \cos \theta)$

76. Outside $r = 1 + \sin \theta$ and inside $r = 3 \sin \theta$

77. Inside $r = 3 \cos \theta$ and inside $r = 2 - \cos \theta$

78. Inside $r = -4 \cos \theta$ and outside $r = 4 + 4 \cos \theta$

79. Inside $r = 5 \cos \theta$ and outside $r = 2 + \cos \theta$

2 PARAMETRIC EQUATIONS

As we have seen, some curves are more conveniently represented by polar equations than by rectangular equations. There is, however, still another way of representing curves by equations that in some cases is more natural and is better suited to analysis of the curves than either rectangular or polar equations. In this

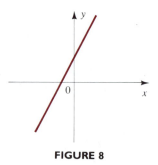

FIGURE 8

third way, a rectangular coordinate system is used, but x and y are each written separately as functions of an auxiliary variable, called a **parameter.** By assigning permissible values to this parameter, points (x, y) on the graph can be found. For example, consider the equations

$$\begin{cases} x = t - 1 \\ y = 2t \end{cases}$$

where t can be any real number. Here the letter t is the parameter, and the two equations together are called **parametric equations.** To draw the graph of the curve represented by these equations, we can construct a table as shown below and then plot the points (x, y) on a rectangular coordinate system (Figure 8).

t	0	1	2	3	-1	-2	-3
x	-1	0	1	2	-2	-3	-4
y	0	2	4	6	-2	-4	-6

The graph appears to be a straight line. Note that the parameter t does not appear on the graph. We use it to find x and y but plot the points (x, y) only.

  Sometimes it is desirable to find the rectangular equation of a curve whose parametric equations are given. In our example, we could solve for t from the first equation, getting $t = x + 1$, and substitute into the second, getting $y = 2x + 2$, which confirms the fact that the graph is a straight line. In the examples that follow, we will see some other techniques for eliminating the parameter, but it should be noted that it is not always possible to do so. It is important to be aware also that the parametric equations may impose certain restrictions on the graph that are not present in the rectangular equation. We will illustrate this in some of the examples.

  In general, parametric equations for a plane curve C are of the type

$$\begin{cases} x = f(t) \\ y = g(t) \end{cases}$$

where t varies over some prescribed interval I (which may be all of R). The totality of points (x, y), obtained by letting t vary over I, is the curve C. When C is described in this way, we say it is *represented parametrically*. The letter t is commonly used as a parameter, since in applications it often represents time. However, other letters may be used. The next three examples further illustrate these concepts.

EXAMPLE 7

Let C be the curve defined parametrically by

$$\begin{cases} x = 2t \\ y = t^2 \end{cases} \quad -2 \le t \le 2$$

Eliminate the parameter and draw the graph of C.

Solution   From the first equation we get $t = x/2$, and on substituting into the second, we have

$$x^2 = 4y$$

which is the equation of a parabola. We note, however, that with the given domain for t, the graph extends only from $(-4, 4)$ to $(4, 4)$, as shown in Figure 9.

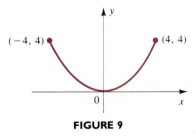

FIGURE 9

EXAMPLE 8

Show that

$$\begin{cases} x = 2 \cos t \\ y = 2 \sin t \end{cases} \quad 0 \le t \le 2\pi$$

is a parametric representation of a circle.

Solution   To eliminate the parameter, we square both sides of each equation and then add:

$$x^2 + y^2 = 4 \cos^2 t + 4 \sin^2 t = 4(\cos^2 t + \sin^2 t) = 4$$

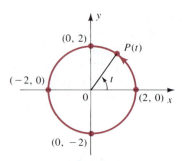

FIGURE 10

or $x^2 + y^2 = 4$, which is the equation of a circle of radius 2, centered at the origin. It is instructive to observe how the circle is traced out as t varies over $[0, 2\pi]$. We can think of the point $P(t) = (x, y)$ as a moving point whose position is determined by t. We see from the given equations that $P(0) = (2, 0)$, $P(\pi/2) = (0, 2)$, $P(\pi) = (-2, 0)$, $P(3\pi/2) = (0, -2)$, and $P(2\pi) = (2, 0)$. Thus, the circle is traced out in a counterclockwise direction, starting from $(2, 0)$.

In this case we can see a geometric interpretation of the parameter t. As shown in Figure 10, it is the radian measure of the angle in standard position with terminal side $0P$.   ▲

EXAMPLE 9 ||||||||||||||||||||||||||||||||||||||||||||||||||||||||||||||||||||||||||||||||||||||||||||||||||||||

Discuss the curve C defined parametrically by

$$\begin{cases} x = 1 + \cos 2t \\ y = \sin t \end{cases} \qquad -\frac{\pi}{2} \le t \le \frac{\pi}{2}$$

Solution   Observe first that for t in the given interval we have $0 \le x \le 2$ and $-1 \le y \le 1$. To aid in identifying the curve, we eliminate the parameter, making use of the identity $\cos 2t = 1 - 2 \sin^2 t$. Since $y = \sin t$, we get

$$x = 1 + \cos 2t = 1 + (1 - 2 \sin^2 t) = 2 - 2y^2$$

or

$$y^2 = -\tfrac{1}{2}(x - 2)$$

This is the equation of a parabola with vertex at $(2, 0)$, opening to the left. But because of the restrictions on x and y imposed by the parametric equations, the graph of C consists only of that portion of the parabola shown in Figure 11. Note how the curve is traced out as t varies from $-\pi/2$ to $\pi/2$.   ▲

As we have seen in Examples 7 and 9, the graph of a curve defined parametrically may be only a portion of the graph of the corresponding rectangular equation. Restrictions on x and y may come about because of the interval for t (as in Example 7) or by the equations themselves (as in Example 9).

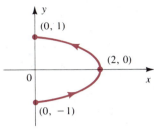

FIGURE 11

The important thing is that **you must take such restrictions into account when using the rectangular equation to draw the graph.**

The next example illustrates how parametric equations sometimes arise naturally when describing the motion of a moving object.

EXAMPLE 10

A projectile is fired at an angle of inclination α (where $0 < \alpha < \pi/2$) at an initial speed of v_0 feet per second. Find parametric equations for the path taken by the projectile. Neglect air resistance, and assume the ground is level.

Solution   Choose axes as shown in Figure 12, and let (x, y) be the coordinates of the projectile after t seconds. If there were no gravity, the projectile would simply continue along the straight line of its initial velocity vector and be at a distance $v_0 t$ on it at time t. But gravity causes the y coordinate to be diminished by $\frac{1}{2}gt^2$ after t seconds. So from the figure, we see, using right triangle methods, that

$$\begin{cases} x = v_0 t \cos \alpha \\ y = v_0 t \sin \alpha - \frac{1}{2}gt^2 \end{cases}$$

These, then, are parametric equations for the path. We must have $t \geq 0$, and t is also limited by when the projectile strikes the ground. You will be asked to find this upper limit on t in Problem 22, Exercise Set 2. Also, you will be asked to find the rectangular equation of the path (Problem 21), as well as the maximum horizontal and vertical distances for the projectile in Problem 22.

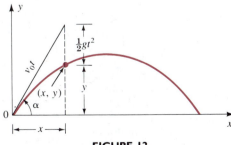

FIGURE 12

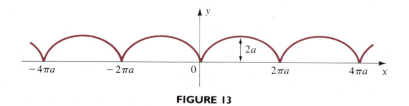

FIGURE 13

Finally, we derive parametric equations for an important curve known as the **cycloid.** You can think of a cycloid as the path traced out by a point on the circumference of a wheel as the wheel rolls in a straight line along a level path. For example, a point on a bicycle tire would trace out a cycloid as the bicycle rolls along. The resulting curve is shown in Figure 13, where a is the radius of the wheel. Note that the curve is periodic, with period $2\pi a$.

To get the equations, refer to Figure 14, where we show a typical position of the wheel, with the point P that traces out the cycloid having originally been coincident with the origin. Then, since the length of the arc $\overset{\frown}{PC}$ is equal to the length of the segment $\overline{OC}$ (why?), we have

$$x = \overline{OC} - \overline{PB} = \overset{\frown}{PC} - \overline{PB} = a\phi - a\sin\phi$$

and

$$y = \overline{BC} = \overline{AC} - \overline{AB} = a - a\cos\phi$$

So we have

$$\begin{cases} x = a(\phi - \sin\phi) \\ y = a(1 - \cos\phi) \end{cases} \qquad -\infty < \phi < \infty$$

for the parametric equations.

A fact of historical as well as practical importance is that the cycloid is the "curve of quickest descent," in the following sense. Let A and B be two points not on a vertical line. As shown in Figure 15, suppose a wire joins A and B and a bead is placed on the wire at A. Among all possible shapes into which such a wire can be bent, the one that causes the bead to reach B in the least time is part of an arch of a cycloid, with the origin at A. This is rather surprising, since one might think a straight line would be best, since this is the shortest distance. The curve of quickest descent is called a **brachistochrone.** The discovery that the solution to the

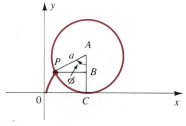

FIGURE 14

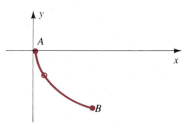

FIGURE 15

brachistochrone problem is the cycloid was made independently by a number of seventeenth-century mathematicians, including Johann Bernoulli and Blaise Pascal.

In the exercises you will be asked to find the rectangular equation of the cycloid. You will discover that graphing the curve directly from the parametric equations is easier than using the rectangular equation.

Polar equations can be converted to parametric equations as follows. Suppose $r = f(\theta)$. Then since $x = r \cos \theta$ and $y = r \sin \theta$, we can write

$$\begin{cases} x = f(\theta) \cos \theta \\ y = f(\theta) \sin \theta \end{cases}$$

and then x and y are expressed in terms of the parameter θ. Consider the cardioid $r = 1 + \cos \theta$, for example. In the parametric form above we would have

$$\begin{cases} x = (1 + \cos \theta) \cos \theta \\ y = (1 + \cos \theta) \sin \theta \end{cases}$$

This parametric form of a polar equation is particularly useful when graphing a polar curve on a graphing calculator. To do this, you set the calculator in the parametric mode and then define the two variables x and y (which may instead be called $X_1 T$ and $X_2 T$). The parameter will probably be called T instead of θ. In the C exercises that follow you will be asked to do this for certain polar curves. It is interesting to observe how the curves are traced out as the parameter increases.

 EXERCISE SET 2

A

In Problems 1–6 draw the curve represented by the parametric equations by assigning values to the parameter to determine points (x, y).

1. $\begin{cases} x = 2t - 1 \\ y = t + 2 \end{cases}$ $-\infty < t < \infty$

2. $\begin{cases} x = t^2 \\ y = t - 1 \end{cases}$ $-\infty < t < \infty$

3. $\begin{cases} x = 2 - t \\ y = 2t \end{cases}$ $0 \leq t \leq 2$

4. $\begin{cases} x = \sin t \\ y = \cos t \end{cases}$ $-\pi \leq t \leq \pi$

5. $\begin{cases} x = \sin^2 t \\ y = \cos t \end{cases}$ $0 \leq t \leq \pi$

6. $\begin{cases} x = \sqrt{t - 1} \\ y = t - 2 \end{cases}$ $1 \leq t \leq \infty$

In Problems 7–20 draw the curve represented parametrically by first noting any restrictions and then eliminating the parameter to obtain the rectangular equation.

7. $\begin{cases} x = 2t - 3 \\ y = -t - 2 \end{cases}$ $-\infty < t < \infty$

8. $\begin{cases} x = 3 - 4t \\ y = 3t - 1 \end{cases}$ $-1 \leq t \leq 1$

9. $\begin{cases} x = \cos t - 1 \\ y = \sin t + 2 \end{cases}$ $0 \leq t \leq 2\pi$

10. $\begin{cases} x = \cos 2t \\ y = 3 \sin t \end{cases}$ $0 \leq t \leq 2\pi$

11. $\begin{cases} x = \cos 2t \\ y = \cos t \end{cases}$ $0 \le t \le \pi$

12. $\begin{cases} x = \sin t - 1 \\ y = \cos 2t - 1 \end{cases}$ $-\dfrac{\pi}{2} \le t \le \dfrac{\pi}{2}$

13. $\begin{cases} x = \sec t \\ y = \tan^2 t \end{cases}$ $-\dfrac{\pi}{2} < t < \dfrac{\pi}{2}$

14. $\begin{cases} x = e^t \\ y = e^{2t} \end{cases}$ $-\infty < t < \infty$

15. $\begin{cases} x = e^t \\ y = e^{-t} \end{cases}$ $-\infty < t < \infty$

16. $\begin{cases} x = \ln t \\ y = t \end{cases}$ $0 < t < \infty$

17. $\begin{cases} x = t^2 \\ y = t^3 \end{cases}$ $-\infty < t < \infty$

18. $\begin{cases} x = 2 \tan t \\ y = \cot t \end{cases}$ $0 < t < \dfrac{\pi}{2}$

19. $\begin{cases} x = \sin \theta - 1 \\ y = \cos \theta + 2 \end{cases}$ $0 \le \theta \le 2\pi$

20. $\begin{cases} x = 2 \tan \theta - 3 \\ y = 3 \sec^2 \theta + 2 \end{cases}$ $-\dfrac{\pi}{2} < \theta < \dfrac{\pi}{2}$

B

21. Find the rectangular equation of the path of the projectile in Example 10, and identify the curve.
22. For the projectile in Example 10 find (a) the time at which the projectile strikes the ground, (b) the maximum horizontal distance covered by the projectile, and (c) the maximum height the projectile attains.
23. Find the rectangular equation of the cycloid.
24. Draw the graph of one arch of the cycloid with $a = 2$. Take ϕ in the interval $[0, 2\pi]$, and use a calculator to obtain points.

25. Draw the curve defined parametrically by $x = \cos^3 \theta$, $y = \sin^3 \theta$, for $0 \le \theta \le 2\pi$. Find the rectangular equation of this curve (the **four cusp hypocycloid**).
26. With the aid of a calculator, make a table of values for t in the interval $[0, 2\pi]$, and draw the graph of the curve defined by

$$\begin{cases} x = \cos t + t \sin t \\ y = \sin t - t \cos t \end{cases}$$

(This is called an **involute of a circle**.)

C (Graphing calculator)

In Problems 27–30 obtain the graph by using a graphing calculator set in parametric mode. Observe the manner in which the graph is traced out.

27. $\begin{cases} x = 8 \cos^3 t \\ y = 8 \sin^3 t \end{cases}$ $0 \le t \le 2\pi$

28. $\begin{cases} x = t - \sin t \\ y = 1 - \cos t \end{cases}$ $-4\pi \le t \le 4\pi$

29. $\begin{cases} x = t^2 - 3 \\ y = t^3 - 4t - 1 \end{cases}$ $-3 \le t \le 3$

30. $\begin{cases} x = 4 \cos t - \cos 4t \\ y = 4 \sin t - \sin 4t \end{cases}$ $0 \le t \le 2\pi$

In Problems 31–36 rewrite the polar equation in parametric form and obtain its graph. Observe how it is traced out. The range for θ is from 0 to 2π unless otherwise specified.

31. $r = 1 + \cos \theta$ **32.** $r = 4 + 3 \sin \theta$
33. $r = 4 - 2 \cos \theta$ **34.** $r = 6 \cos 4\theta$

35. $r = \dfrac{\theta}{2}, \quad 0 \le \theta \le 25$

3 TRIGONOMETRIC FORM OF COMPLEX NUMBERS

An interesting and useful relationship exists between complex numbers and trigonometric functions. Let $z = x + yi$ be any complex number. We associate with this the point (x, y) in a rectangular coordinate system. This provides a $1-1$ correspondence between points in the plane and complex numbers. For example, the complex number $z = 2 - 3i$ corresponds to the point $(2, -3)$, and conversely.

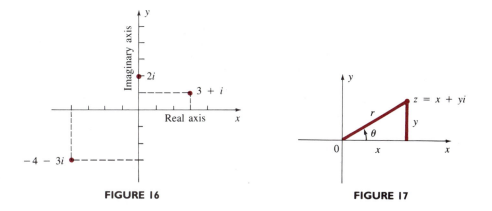

FIGURE 16   **FIGURE 17**

When we identify a point (x, y) in this way with the complex number $x + yi$, we refer to the plane as the **complex plane,** and we sometimes call the point the complex number itself. In the complex plane we sometimes refer to the x axis as the **real axis** and the y axis as the **imaginary axis,** since the real part x of $z = x + yi$ is the x coordinate of (x, y), and the imaginary part y is the y coordinate. These ideas are illustrated in Figure 16, where we have shown the numbers $3 + i$, $2i$, and $-4 - 3i$ represented geometrically by $(3, 1)$, $(0, 2)$, and $(-4, -3)$, respectively.

Now let us superimpose a polar coordinate system on the complex plane, as in Figure 17. Then, since $x = r \cos \theta$ and $y = r \sin \theta$, we can write the complex number $z = x + yi$ as

$$z = r(\cos \theta + i \sin \theta) \tag{3}$$

Equation (3) is called the **trigonometric** (or **polar**) **form** of the complex number z. The form $z = x + yi$ is called its **rectangular form.**

Because the polar coordinates of a point are not unique, a complex number has many equivalent trigonometric representations. For example, $2\left(\cos \dfrac{\pi}{3} + i \sin \dfrac{\pi}{3} \right)$, $2\left(\cos \dfrac{7\pi}{3} + i \sin \dfrac{7\pi}{3} \right)$, and $2\left[\cos\left(-\dfrac{5\pi}{3} \right) + i \sin\left(-\dfrac{5\pi}{3} \right) \right]$ all represent the same complex number, whose rectangular form is $1 + i\sqrt{3}$.

For the trigonometric form of a complex number we will always choose $r \geq 0$. Thus, for $z = x + yi$,

$$r = \sqrt{x^2 + y^2}$$

This number is called the **modulus** of z and is denoted by $|z|$. This is in keeping with the notation for the length of a vector. In fact, we may think of z as being identified with the vector $\overrightarrow{OP}$, where $P = (x, y)$. The angle θ in the trigonometric form is called an **argument** of z. Normally we use the primary argument, that is, we take θ in the interval $[0, 2\pi)$.

EXAMPLE 11

Find the trigonometric form of the complex number $z = \sqrt{3} + i$, and give its modulus and primary argument.

Solution   We represent z by the point $(\sqrt{3}, 1)$ in rectangular coordinates. As we see from Figure 18, this corresponds to $r = 2$ and $\theta = \pi/6$ in polar coordinates. So by

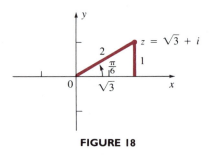

FIGURE 18

Equation (3),

$$z = 2\left(\cos\frac{\pi}{6} + i \sin\frac{\pi}{6}\right)$$

is the trigonometric form. Its modulus is 2 and its primary argument is $\pi/6$.   ▲

One of the advantages of the trigonometric form of complex numbers is that products, quotients, powers, and roots are particularly easy to calculate with numbers in this form. Consider first the product of two such numbers, say $z_1 = r_1(\cos\theta_1 + i \sin\theta_1)$ and $z_2 = r_2(\cos\theta_2 + i \sin\theta_2)$. On multiplication, making use of the fact that $i^2 = -1$ and arranging terms, we have

$$z_1 z_2 = r_1 r_2[(\cos\theta_1 \cos\theta_2 - \sin\theta_1 \sin\theta_2) + i(\sin\theta_1 \cos\theta_2 + \cos\theta_1 \sin\theta_2)]$$

By the addition formulas for the cosine and the sine, this can be written in the form

$$z_1 z_2 = r_1 r_2[\cos(\theta_1 + \theta_2) + i \sin(\theta_1 + \theta_2)] \tag{4}$$

So the modulus of the product is the product of the moduli, and the argument of the product is the sum of the arguments of the two numbers.

EXAMPLE 12

Find the product of the two complex numbers

$$z_1 = 2\left(\cos\frac{\pi}{6} + i \sin\frac{\pi}{6}\right) \qquad \text{and} \qquad z_2 = 3\left(\cos\frac{\pi}{3} + i \sin\frac{\pi}{3}\right)$$

Solution   By equation (4), the product is

$$z_1 z_2 = 2 \cdot 3 \left[\cos\left(\frac{\pi}{6} + \frac{\pi}{3}\right) + i \sin\left(\frac{\pi}{6} + \frac{\pi}{3}\right) \right]$$

$$= 6\left(\cos\frac{\pi}{2} + i \sin\frac{\pi}{2} \right)$$

This can be put in rectangular form by evaluating the trigonometric functions:

$$z_1 z_2 = 6(0 + i \cdot 1) = 0 + 6i = 6i \qquad \blacktriangle$$

Consider next the quotient

$$\frac{z_1}{z_2} = \frac{r_1(\cos\theta_1 + i\sin\theta_1)}{r_2(\cos\theta_2 + i\sin\theta_2)}$$

If we multiply numerator and denominator by the conjugate of the denominator, this can be put in the form (see Problem 45)

$$\frac{z_1}{z_2} = \frac{r_1}{r_2}\left[\cos(\theta_1 - \theta_2) + i\sin(\theta_1 - \theta_2)\right] \qquad (5)$$

EXAMPLE 13 ||||||||||||||||||||||||||||||||||||||||||||||||||||||||||||||||||||||||||||||||||||||||||||||||||||||||||||

Find the quotient z_1/z_2 if

$$z_1 = 4\left(\cos\frac{5\pi}{6} + i \sin\frac{5\pi}{6} \right) \qquad \text{and} \qquad z_2 = 2\left(\cos\frac{\pi}{2} + i \sin\frac{\pi}{2} \right)$$

Solution   By equation (5),

$$\frac{z_1}{z_2} = \frac{4}{2}\left[\cos\left(\frac{5\pi}{6} - \frac{\pi}{2}\right) + i \sin\left(\frac{5\pi}{6} - \frac{\pi}{2}\right) \right] = 2\left(\cos\frac{\pi}{3} + i \sin\frac{\pi}{3} \right)$$

In rectangular form this is

$$\frac{z_1}{z_2} = 2\left(\frac{1}{2} + i\frac{\sqrt{3}}{2} \right) = 1 + i\sqrt{3} \qquad \blacktriangle$$

It is in raising to a power that the trigonometric form has the greatest advantage. We will state without proof the following important result due to the French mathematician Abraham De Moivre (1667–1754):

DE MOIVRE'S THEOREM   **Let $z = r(\cos\theta + i\sin\theta)$ be the trigonometric form of any complex number. Then for any natural number n**

$$z^n = r^n(\cos n\theta + i\sin n\theta)$$

Thus, the modulus of z^n is the nth power of the modulus of z, and the argument is n times the argument of z. That the result is plausible follows from a consideration of a few powers of z. By equation (4),

$$z^2 = r \cdot r[\cos(\theta + \theta) + i \sin(\theta + \theta)]$$
$$= r^2(\cos 2\theta + i \sin 2\theta)$$

Similarly, applying equation (4) again,

$$z^3 = z^2 \cdot z = r^2 \cdot r[\cos(2\theta + \theta) + i \sin(2\theta + \theta)]$$
$$= r^3(\cos 3\theta + i \sin 3\theta)$$

This process could be continued, and for any given power of z, De Moivre's theorem would be confirmed. However, this does not constitute a proof of the theorem. It can be proved by using a technique called **mathematical induction,** which we study in Chapter 13.   ▲

EXAMPLE 14 ||||||||||||||||||||||||||||||||||||||||||||||||||||||||||||||||||||||||||||||||||||||||||||||||||||||

Expand $(1 + i)^6$.

Solution   With the aid of a sketch (Figure 19) we determine that $r = \sqrt{2}$ and $\theta = \pi/4$. Writing $z = 1 + i$, we have

$$z^6 = (1 + i)^6 = \left[\sqrt{2}\left(\cos \frac{\pi}{4} + i \sin \frac{\pi}{4} \right) \right]^6$$

$$= (\sqrt{2})^6\left(\cos \frac{6\pi}{4} + i \sin \frac{6\pi}{4} \right)$$

$$= 8\left(\cos \frac{3\pi}{2} + i \sin \frac{3\pi}{2} \right)$$

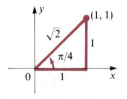

FIGURE 19

In rectangular form the answer is

$$(1 + i)^6 = 8[0 + i(-1)] = -8i$$

Remark. You might wish to compare this solution with the work involved in doing this problem by expanding by the binomial theorem and simplifying the result. You will find that the method we used is much simpler.   ▲

Finally, we consider the problem of taking roots of complex numbers. Again, let $z = r(\cos \theta + i \sin \theta)$, and suppose we wish to define $\sqrt[n]{z} = z^{1/n}$. Since adding an integral multiple of 2π to the argument θ does not change z, we can write, for any integer k,

$$z = r[\cos(\theta + 2k\pi) + i\,\sin(\theta + 2k\pi)]$$

For n a positive integer greater than 1, we define

$$z^{1/n} = r^{1/n}\left[\cos\left(\frac{\theta + 2k\pi}{n}\right) + i\,\sin\left(\frac{\theta + 2k\pi}{n}\right)\right] \qquad (6)$$

for $k = 0, 1, 2, \ldots, n - 1$.

Remarks. If we take larger values of k, we get repetitions. For example, $k = n$ gives the same value as $k = 0$. (Verify.) There are exactly n distinct nth roots. They are equally spaced on a circle of radius $r^{1/n}$. Note that since r is a real number, $r^{1/n}$ means the principal nth root of r.

The motivation for this definition lies in the fact that the nth power of the right-hand side of equation (6) is z, as we now show. By De Moivre's theorem,

$$\left[r^{1/n}\left(\cos\frac{\theta + 2k\pi}{n} + i\,\sin\frac{\theta + 2k\pi}{n}\right)\right]^n = (r^{1/n})^n\left[\cos n\left(\frac{\theta + 2k\pi}{n}\right)\right.$$
$$\left. + i\,\sin n\left(\frac{\theta + 2k\pi}{n}\right)\right]$$
$$= r[\cos(\theta + 2k\pi) + i\,\sin(\theta + 2k\pi)]$$
$$= z$$

EXAMPLE 15

Find all cube roots of -8.

Solution   Let $z = -8 = -8 + 0i$. Then $r = 8$ and $\theta = \pi$ (Figure 20). So by equation (6),

$$\sqrt[3]{z} = z^{1/3} = 8^{1/3}\left[\cos\left(\frac{\pi + 2k\pi}{3}\right) + i\,\sin\left(\frac{\pi + 2k\pi}{3}\right)\right]$$

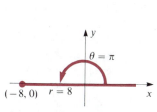

FIGURE 20

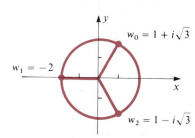

FIGURE 21

and the three distinct roots correspond to $k = 0$, $k = 1$, and $k = 2$. Let w_0, w_1, and w_2, be the roots corresponding to these values of k. Then

$$w_0 = 2\left(\cos\frac{\pi}{3} + i\sin\frac{\pi}{3}\right) = 2\left(\frac{1}{2} + i\frac{\sqrt{3}}{2}\right) = 1 + i\sqrt{3}$$

$$w_1 = 2\left(\cos\frac{\pi + 2\pi}{3} + i\sin\frac{\pi + 2\pi}{3}\right) = 2(\cos\pi + i\sin\pi) = -2$$

$$w_2 = 2\left(\cos\frac{\pi + 4\pi}{3} + i\sin\frac{\pi + 4\pi}{3}\right) = 2\left(\cos\frac{5\pi}{3} + i\sin\frac{5\pi}{3}\right)$$

$$= 2\left(\frac{1}{2} - \frac{i\sqrt{3}}{2}\right) = 1 - i\sqrt{3}$$

These roots are shown graphically in Figure 21.   ▲

EXERCISE SET 3

A

In Problems 1–10 find the trigonometric form of the complex number. Draw a sketch in each case.

1. $1 - i$

2. $2 + 2i$

3. $\sqrt{3} - i$

4. $1 + i\sqrt{3}$

5. -4

6. $-8i$

7. $-2 - 2i\sqrt{3}$

8. $4\sqrt{2} - 4i\sqrt{2}$

9. 5

10. $i - 1$

In Problems 11–20 find the rectangular form of the complex number.

11. $2\left(\cos\frac{4\pi}{3} + i\sin\frac{4\pi}{3}\right)$

12. $4\left(\cos\frac{3\pi}{4} + i\sin\frac{3\pi}{4}\right)$

13. $5\left(\cos\frac{3\pi}{2} + i\sin\frac{3\pi}{2}\right)$

14. $7(\cos\pi + i\sin\pi)$

15. $6\left(\cos\frac{11\pi}{6} + i\sin\frac{11\pi}{6}\right)$

16. $2\left(\cos\frac{5\pi}{3} + i\sin\frac{5\pi}{3}\right)$

17. $\sqrt{2}\left(\cos\frac{7\pi}{4} + i\sin\frac{7\pi}{4}\right)$

18. $5(\cos 0 + i\sin 0)$

19. $3(\cos 150° + i\sin 150°)$

20. $8(\cos 315° + i\sin 315°)$

In Problems 21–27 find $z_1 \cdot z_2$ and z_1/z_2 by using equations (4) and (5). When possible without using a calculator, write answers in rectangular form.

21. $z_1 = 2\left(\cos\frac{\pi}{4} + i\sin\frac{\pi}{4}\right)$, $z_2 = 3\left(\cos\frac{5\pi}{4} + i\sin\frac{5\pi}{4}\right)$

22. $z_1 = 8(\cos\pi + i\sin\pi)$, $z_2 = 4\left(\cos\frac{2\pi}{3} + i\sin\frac{2\pi}{3}\right)$

23. $z_1 = 4\left(\cos\frac{\pi}{2} + i\sin\frac{\pi}{2}\right)$, $z_2 = 2\left(\cos\frac{3\pi}{4} + i\sin\frac{3\pi}{4}\right)$

24. $z_1 = \sqrt{2} + i\sqrt{2}$, $z_2 = 4\sqrt{2} - 4i\sqrt{2}$

25. $z_1 = \sqrt{3} - i$, $z_2 = 2 + 2i\sqrt{3}$

26. $z_1 = 3(\cos 70° + i\sin 70°)$, $z_2 = 5(\cos 50° + i\sin 50°)$

27. $z_1 = 4(\cos 100° + i \sin 100°)$,
$z_2 = 6(\cos 70° + i \sin 70°)$

In Problems 28–34 find the indicated power by using De Moivre's theorem. Express answers in rectangular form.

28. $\left[2\left(\cos \dfrac{\pi}{6} + i \sin \dfrac{\pi}{6}\right)\right]^4$

29. $\left[3\left(\cos \dfrac{3\pi}{4} + i \sin \dfrac{3\pi}{4}\right)\right]^6$

30. $[2(\cos 240° + i \sin 240°)]^5$

31. $(\sqrt{3} - i)^8$

32. $(1 - i)^{10}$

33. $(2 - 2i\sqrt{3})^4$

34. $(-\sqrt{2} - i\sqrt{2})^5$

B

In Problems 35–40 find all the roots indicated. Express answers in rectangular form whenever this can be done without a calculator.

35. Cube roots of $8i$

36. Fourth roots of $-\dfrac{1}{2} - \dfrac{i\sqrt{3}}{2}$

37. Sixth roots of 1

38. Fourth roots of $-8 + 8i\sqrt{3}$

39. Square roots of $-16i$

40. Fifth roots of $-16\sqrt{3} + 16i$

In Problems 41–44 find the complete solution set.

41. $x^6 + 64 = 0$ **42.** $x^4 + 256 = 0$

43. $x^6 = 64$ **44.** $x^3 + 27i = 0$

45. Verify equation (5).

REVIEW EXERCISE SET

A

1. Give the rectangular coordinates corresponding to the given polar coordinates.

a. $\left(2, -\dfrac{\pi}{2}\right)$ **b.** $\left(3, \dfrac{5\pi}{4}\right)$

c. $\left(-1, -\dfrac{3\pi}{4}\right)$ **d.** $\left(4, \dfrac{10\pi}{3}\right)$

2. Give polar coordinates (r, θ) with $r \geq 0$, $0 \leq \theta < 2\pi$, for the points with the given rectangular coordinates.

a. $(-3, 0)$ **b.** $(4\sqrt{2}, -4\sqrt{2})$

c. $(-\sqrt{3}, 3)$ **d.** $(-\sqrt{3}, -1)$

In Problems 3–6 change to an equivalent polar equation.

3. **a.** $3x - 5y = 7$ **b.** $y = \dfrac{2}{x}$

4. **a.** $x^2 + y^2 - 2y = 0$ **b.** $y = x^2$

5. **a.** $(x - y)^2 = x$ **b.** $x^2 = 1 - 2y$

6. **a.** $(x^2 + y^2)^2 = 4xy$ **b.** $x^2 - y^2 = 2x$

In Problems 7–10 change to an equivalent rectangular equation.

7. **a.** $r(1 + \tan \theta) = 2 \sec \theta$ **b.** $\theta = 2 \tan^{-1} \dfrac{2}{3}$

8. **a.** $r = 2(\sin \theta + \cos \theta)$ **b.** $r(2 + \sin \theta) = 1$

9. **a.** $r^2 = 2 \sin 2\theta$ **b.** $r = 2 \cos \theta - \sin \theta$

10. **a.** $r \csc \theta = 3$ **b.** $r^2 = 1 - 2 \sin^2 \theta$

In Problems 11–15 identify the curve and draw its graph.

11. **a.** $r = \dfrac{3}{2 \cos \theta + 4 \sin \theta}$ **b.** $r = -3 \sec \theta$

12. **a.** $r \sec \theta = 3$ **b.** $r + 4 = 0$

13. **a.** $r = 1 - \sin \theta$ **b.** $r = 2 + \cos \theta$

14. **a.** $r = 2 \cos 3\theta$ **b.** $r = 3 - 2 \sin \theta$

15. **a.** $r = \cos \theta - 1$ **b.** $r^2 = 2 \cos^2 \theta - 1$

In Problems 16–20 draw the curve represented para-metrically by first noting any restrictions and then eliminating the parameter.

16. $\begin{cases} x = 2t - 3 \\ y = 3t + 4 \end{cases}$ $-\infty < t < \infty$

17. $\begin{cases} x = 1 - \sin t \\ y = \cos t \end{cases}$ $0 \le t < 2\pi$

18. $\begin{cases} x = \cos t \\ y = \sec t \end{cases}$ $-\dfrac{\pi}{2} < t < \dfrac{\pi}{2}$

19. $\begin{cases} x = 1 + \sec t \\ y = 1 - \tan^2 t \end{cases}$ $-\dfrac{\pi}{2} < t < \dfrac{\pi}{2}$

20. $\begin{cases} x = 2 \sin \theta \\ y = \cos 2\theta \end{cases}$ $-\dfrac{\pi}{2} \le \theta \le \dfrac{\pi}{2}$

In Problems 21 and 22 give the equivalent trigono-metric form for which $0 \le \theta < 2\pi$.

21. **a.** $-2 + 2i$ **b.** $-3i$

22. **a.** $\sqrt{3} + 3i$ **b.** $\dfrac{i - \sqrt{3}}{2}$

In Problems 23 and 24 give the equivalent rectangular form.

23. **a.** $3\left(\cos \dfrac{5\pi}{2} + i \sin \dfrac{5\pi}{2} \right)$

b. $4\left(\cos \dfrac{7\pi}{6} + i \sin \dfrac{7\pi}{6} \right)$

24. **a.** $2\sqrt{2}\left[\cos\left(-\dfrac{\pi}{4}\right) + i \sin\left(-\dfrac{\pi}{4}\right) \right]$

b. $2\sqrt{3}\left(\cos \dfrac{5\pi}{3} + i \sin \dfrac{5\pi}{3} \right)$

In Problems 25 and 26 find $z_1 \cdot z_2$ and z_1/z_2. Express answers in rectangular form.

25. $z_1 = 8\left(\cos \dfrac{5\pi}{3} + i \sin \dfrac{5\pi}{3} \right)$,

$z_2 = 2\left(\cos \dfrac{3\pi}{2} + i \sin \dfrac{3\pi}{2} \right)$

26. $z_1 = 10\left(\cos \dfrac{3\pi}{4} + i \sin \dfrac{3\pi}{4} \right)$,

$z_2 = 5\left(\cos \dfrac{5\pi}{12} + i \sin \dfrac{5\pi}{12} \right)$

In Problems 27 and 28 use DeMoivre's theorem to obtain the result. Express the answer in rectangular form.

27. $\left[3\left(\cos \dfrac{\pi}{4} + i \sin \dfrac{\pi}{4} \right) \right]^3$

28. $(1 - \sqrt{3}\,i)^6$

In Problems 29 and 30 find the indicated roots. Express answers in rectangular form.

29. Sixth roots of 64

30. Fourth roots of $-8 - 8i\sqrt{3}$

B

31. Change the polar equation $r = \csc^2 \dfrac{\theta}{2}$ to an equiv-alent rectangular equation, and draw its graph.

32. Find all points of intersection of the curves $r = 2 \cos \theta$ and $r^2 = 2\sqrt{3} \sin 2\theta$, and draw the graphs.

33. Draw the curves $r = 3 - 4 \cos \theta$ and $r = 2\sqrt{3}(1 - \cos \theta)$ on the same polar coordinate system, and shade the area that lies outside the first curve and inside the second. Find all points of intersection.

34. Draw the graph of the curve defined by

$$\begin{cases} x = \cos^2 \theta \\ y = \tfrac{1}{2} \sin 2\theta \end{cases} \quad 0 \le \theta < \pi$$

after first eliminating the parameter.

35. An **epicycloid** is a curve generated by a point on a circle that rolls externally around a fixed circle. If the fixed circle has radius a and the rolling circle has radius b, it can be shown that parametric equa-tions of the epicycloid are

$$\begin{cases} x = (a + b)\cos \theta - b \cos\left(\dfrac{a + b}{b}\right)\theta \\[2mm] y = (a + b)\sin \theta - b \sin\left(\dfrac{a + b}{b}\right)\theta \end{cases}$$

Draw the epicycloid for which $a = 4$ and $b = 1$. Take θ from 0 to 2π.

36. With the aid of a calculator find all solutions of the equation $x^5 + 1 = 0$. Express answers in rect-angular form. Use four significant figures.

CUMULATIVE REVIEW EXERCISE SET III (CHAPTERS 7–10)

1. In each of the following a right triangle ABC is given, with right angle at C. Find all missing sides and angles without the use of a calculator.
  a. $A = 30°$, $b = 12$ **b.** $B = 60°$, $a = 4$
  c. $A = 45°$, $c = 16$ **d.** $a = 10$, $c = 20$
  e. $b = 6\sqrt{2}$, $c = 12$

2. Without using a calculator find all six trigonometric functions of each of the following. Leave answers in exact form.
  a. $4\pi/3$ **b.** $-3\pi/4$ **c.** $17\pi/6$
  d. $75°$ **e.** $990°$ **f.** $17\pi/12$
  g. $-22.5°$ **h.** $(2n-1)\pi$, n an integer

3. Evaluate without a calculator:
  a. $\sin[2\sin^{-1}(-\frac{3}{5})]$
  b. $\cos[\cos^{-1}(-\frac{5}{13}) + \tan^{-1}(-\frac{4}{3})]$
  c. $\tan \frac{1}{2}\theta$, where $\sec\theta = \frac{3}{2}$ and $\sin\theta < 0$
  d. $\csc\left(\dfrac{\pi}{2} - \theta\right)$, where $\cot\theta = \frac{1}{2}$ and $\cos\theta < 0$

4. Prove the identities:
  a. $\dfrac{\sec\theta - \cos\theta + \tan\theta}{\tan\theta + \sec\theta} = \sin\theta$
  b. $\dfrac{(\sin\theta - \cos\theta)^2}{\cos 2\theta} = \dfrac{1 - \tan\theta}{1 + \tan\theta}$

5. Let
$$z_1 = 4\left(\cos\frac{4\pi}{3} + i\sin\frac{4\pi}{3}\right) \quad \text{and}$$
$$z_2 = 2\left(\cos\frac{\pi}{6} + i\sin\frac{\pi}{6}\right)$$
  Find
  a. $z_1 z_2$ **b.** z_1/z_2 **c.** z_2^4
  d. The square roots of z_1
  Express answers in rectangular form.

6. Points A and B are on level ground on a line with the base of a tower. From the top of the tower the angles of depression of A and B, respectively, are $24.6°$ and $31.2°$. If points A and B are known to be 20 meters apart, find the height of the tower.

7. Evaluate without using a calculator:
  a. $\cos[2\sin^{-1}(-\frac{1}{3})]$
  b. $\sin[2\arctan(-3)]$

c. $\tan\frac{1}{2}[\cos^{-1}(-\frac{7}{25})]$
  d. $\cos\frac{1}{2}\alpha$, where $\csc\alpha = -\frac{5}{4}$ and $\pi \le \alpha \le 3\pi/2$

8. Find all primary solutions:
  a. $2\sin x + \sqrt{3}\tan x = 0$
  b. $2\sin^2\frac{1}{2}x - 2\cos^2 x = 1$

9. A chord on a circle of radius 6 inches is 10 inches long. Find
  a. The central angle formed by the radii to the end points of the chord
  b. The arc length cut off by the chord
  c. The area of the sector formed by this arc and the radii
  d. The area of the triangle formed by the chord and the radii

10. In each of the following, triangle ABC is a right triangle, with right angle at C. Find all unknown sides and angles.
  a. $A = 16.3°$, $b = 4.68$
  b. $B = 52°40'$, $c = 171.5$
  c. $a = 13.7$, $b = 21.3$
  d. $b = 50$, $c = 130$

11. a. A curve on a railroad track is in the form of a circular arc, which is 1.5 kilometers long. If the central angle corresponding to this arc is $72°$, find the radius of the circle.
  b. A girl is riding a bicycle with 28-inch-diameter tires, which are rotating at the rate of 200 revolutions per minute. Approximately what is the speed of the bicycle in miles per hour?

12. Find all unknown sides and angles of the triangles ABC. Also, find their areas.
  a. $A = 32.4°$, $B = 47.8°$, $c = 25.3$
  b. $B = 28.3°$, $b = 23.5$, $c = 39.2$
  c. $A = 98°24'$, $b = 124.3$, $c = 89.76$
  d. $a = 32.5$, $b = 26.8$, $c = 18.6$

13. Prove these identities:
  a. $\dfrac{\sin\theta}{\sec\theta - 1} + \dfrac{\sin\theta}{\sec\theta + 1} = 2\cot\theta$
  b. $\dfrac{\sin 2x \cos x}{(1 + \cos 2x)\cos^2(x/2)} = 2\tan(x/2)$

14. Find all primary solutions.
  a. $\sin x + \sin 2x = 0$
  b. $\cos 4x = \cos 2x$

15. Find all solutions to the equation $x^3 + 64i = 0$.

16. Evaluate
   a. $\sin[\cos^{-1}(-\frac{4}{5}) - \tan^{-1}(-\frac{12}{5})]$
   b. $\arctan(-2) + \arctan\frac{1}{2}$
   c. $\cos(\alpha + 2\beta)$, where $\sec\alpha = \frac{13}{5}$, $\csc\alpha < 0$, $\cot\beta = 2$, $\sec\beta < 0$

17. Draw the graphs:
   a. $y = 1 + 2\sin\left(\dfrac{x}{2} - \dfrac{\pi}{3}\right)$

   b. $y = \frac{1}{2}\cos\left(\pi x + \dfrac{\pi}{4}\right)$

18. An airplane leaves an airport at 2:00 P.M. and flies at a heading of 148° at an average cruising speed of 240 miles per hour. Another plane leaves the same airport at 2:15 P.M. with a heading of 25° and cruises at an average speed of 400 miles per hour. If they fly at the same altitude, how far apart are they at 3:30 P.M.?

19. Prove these identities:
   a. $\dfrac{2\cos^3\theta}{1 + \sin\theta} + \sin 2\theta = 2\cos\theta$

   b. $\tan\left(\dfrac{\alpha}{2} - \dfrac{\pi}{4}\right) = \dfrac{\sin\alpha - 1}{\cos\alpha}$

20. Prove these identities:
   a. $\cot\theta\sin 2\theta - \cos 2\theta = 1$

   b. $\dfrac{\sin\theta}{1 - \cos\theta} - \tan\dfrac{\theta}{2} = 2\cot\theta$

21. Draw the graphs:
   a. $y = \frac{1}{3}\tan(\pi x/2)$
   b. $y = \frac{1}{2}\sin^{-1}(2x) + \pi/3$

22. Observer A is stationed at a point due west of a monument, and measures the angle of elevation of the top of the monument to be 45°. Observer B is 100 feet due south of observer A, and B measures the angle of elevation of the top of the monument to be 30°. Without using a calculator, find the height of the monument.

23. Find the complete solution set of each of the following:
   a. $\sqrt{3}\sin x = \cos x - 1$

   b. $\dfrac{\cos 3x}{\sin x} + \dfrac{\sin 3x}{\cos x} = \csc 2x$

24. An offshore oil drilling rig in the Gulf of Mexico is viewed from points A and B on the shore, with the point on the shore nearest the rig lying be-

tween A and B. In the triangle formed by the rig and the points A and B, the angle at A is 76°25′, and the angle at B is 68°54′. If A and B are 6 kilometers apart, find the distance from each of these points to the drilling rig. How far is the rig from the nearest point on shore?

25. Draw the graphs of $r = 3 + 2\cos\theta$ and $r = 1 - 2\cos\theta$ on the same polar coordinate system. Find all points of intersection of the two curves.

26. Prove that the area of the parallelogram having the vectors $\mathbf{v}$ and $\mathbf{w}$ as adjacent sides is $|\mathbf{v}||\mathbf{w}|\sin\theta$, where θ is the angle between $\mathbf{v}$ and $\mathbf{w}$.

27. In the accompanying figure, triangle ABC is in the horizontal plane and line CD is vertical. If side c and angles α, β, and ϕ are known, derive the following formula for the length h of side CD:

$$h = \dfrac{c\sin\alpha\tan\phi}{\sin(\alpha + \beta)}$$

If $c = 21.34$, $\alpha = 18°37′$, $\beta = 43°16′$, and $\phi = 56°42′$, find h. Also find angle θ.

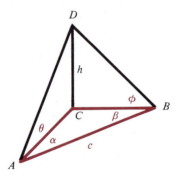

28. The pilot of a small airplane flew from town A to town B and returned the following day. The bearing of B from A is N 42.3° E, and its distance from A is 452 miles. On the trip from A to B, a 33.4 mile per hour wind was blowing from 270°; on the return trip the wind was from 305° at 29.7 miles per hour. If the pilot cruised at 182 miles per hour airspeed, find her heading on each leg of the trip. Also find her total time in the air.

29. Graph the curve with parametric representation:

$$\begin{cases} x = 2\sin^2\dfrac{\theta}{2} \\ y = 2\sin^2\theta \end{cases} \quad 0 \le \theta \le \pi$$

30. Factor the polynomial $x^6 + 64$ completely over the complex numbers.

31. Graph the curve defined parametrically by

$$\begin{cases} x = e^t + e^{-t} \\ y = e^t - e^{-t} \end{cases} \quad -\infty < t < \infty$$

32. Draw the graphs of $r = 2 \cos 3\theta$ and $r = -2 \cos \theta$ and shade the area inside the first curve that is outside the second. Find all points of intersection of the two curves.

33. Make use of the rectangular equation to graph the curve defined by

$$\begin{cases} x = \sin^2 t \\ y = \ln \sec t \end{cases} \quad 0 \le t < \frac{\pi}{2}$$

34. Let $f(x) = \sin x$. Find the slope of the line segment joining the points $(\pi/2, f(\pi/2))$ and $(5\pi/6, f(5\pi/6))$. What is the length of this line segment?

35. Let

$$f(x) = \frac{2 \sin x + 1}{2 \cos^2 x + \cos x - 1} \quad \text{for } -\pi < x \le \pi$$

Find all vertical asymptotes for the graph of f. Also, find the x and y intercepts.

SYSTEMS OF EQUATIONS AND INEQUALITIES

Some calculus problems involve the simultaneous solution of a system of equations. These may be linear or nonlinear. The following example shows that the solution of a linear system is used in finding the equation of a plane.*

Find an equation of the plane passing through the points $(2, 1, 3), (1, 3, 2),$ $(-1, 2, 4)$.
Solution   Since the three points lie in the plane, each of them satisfies Equation . . . .
We have

$$(2, 1, 3): \qquad 2A + B + 3C + D = 0$$

$$(1, 3, 2): \qquad A + 3B + 2C + D = 0$$

$$(-1, 2, 4): \quad -A + 2B + 4C + D = 0$$

Solving for A, B, C in terms of D, we obtain

$$A = -\frac{3}{25} D, \qquad B = -\frac{4}{25} D, \qquad C = -\frac{5}{25} D$$

We can now choose any value for D. It is convenient to take $D = -25$. We get the equation

$$3x + 4y + 5z - 25 = 0$$

The system of three equations in four unknowns that occurs here is an example of a **dependent system.** We will learn an efficient way of solving such a system in this chapter.

---

* Reprinted with the permission of Jones and Bartlett Publishers, Inc., Boston, MA, from *Calculus with Analytic Geometry*, 4th Edition, by Murray H. Protter and Philip E. Protter, 1988, page 546.

LINEAR SYSTEMS

We saw in Chapter 4 that the general form of the equation of a line is

$$Ax + By + C = 0$$

This is also called a **general linear equation in two variables.** The word *linear* continues to apply, however, regardless of the number of variables as long as the equation is of degree 1. Thus, $2x - 3y + 4z = 0$ and $5x_1 - 2x_2 + 3x_3 - 4x_4 + 8x_5 = 2$ are linear, whereas $x^2 - 2x + 3y = 4$, $xy - 3x + 4y - 7$, and $\sqrt{x_1} + x_2 = 4$ are not. Techniques for solving **systems of linear equations** form the basis of a large part of what is called **linear algebra.** A complete treatment of linear systems is not appropriate in this course, but a review of techniques studied in high school algebra, with an introduction to more advanced techniques, is in order.

We consider first **two linear equations in two variables.** For example,

$$\begin{cases} 2x - 3y = 4 \\ 5x + 4y = 7 \end{cases}$$

Our object is to find all pairs (x, y) that simultaneously satisfy both equations. The collection of all such pairs is the solution to the system. In this context we will refer to x and y as the unknowns, since we are seeking to determine values for them.

Since two linear equations correspond geometrically to two lines, we can reason in advance that one of three situations will occur:

1. The lines intersect in one and only one point, and its coordinates constitute the unique solution. The equations are said to be **consistent.**
2. The lines are parallel and hence do not intersect, so there is no simultaneous solution. The equations are said to be **inconsistent.**
3. The lines coincide and hence intersect at infinitely many points, so there are infinitely many solutions. The equations are said to be linearly **dependent.**

Let us consider the example given above.

EXAMPLE 1

Solve the system: $\begin{cases} 2x - 3y = 4 \\ 5x + 4y = 7 \end{cases}$

Solution   By appropriate multiplications and additions or subtractions, we seek to eliminate one of the unknowns. For example, y can be eliminated by multiplying the top equation by 4 and the bottom equation by 3 and then adding the resulting

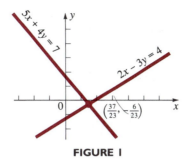

FIGURE I

equations, as follows:

$$8x - 12y = 16$$
$$15x + 12y = 21$$
$$\overline{23x \qquad\quad = 37}$$
$$x = \tfrac{37}{23}$$

This is the x coordinate of a point on both lines. To find the y coordinate, we can substitute this value of x into either equation. By substituting into the first equation, we have

$$2(\tfrac{37}{23}) - 3y = 4$$
$$-3y = \tfrac{18}{23}$$
$$y = \tfrac{-6}{23}$$

So this problem illustrates a consistent system. There is a unique solution $(\tfrac{37}{23}, \tfrac{-6}{23})$. The situation is graphed in Figure 1. ▲

The technique used to solve the system in Example 1 is known as **elimination.** We eliminated y from the two equations and obtained an equation in x alone. Then we were able to solve for x, and y was found by substituting this value of x into one of the original equations. Similarly, we could have eliminated x from the two equations and then used the resulting value of y in either equation to find x. In general, to eliminate an unknown we use one of two methods: elimination by *addition or subtraction*, or elimination by *substitution*.

To eliminate an unknown from two equations by addition or subtraction, the coefficients of that unknown must be the same in absolute value. If they are not the same, then we make them the same by multiplying the equations by appropriate numbers. In Example 1 we multiplied both sides of the first equation by 4 and both sides of the second equation by 3. When the coefficients are the same in absolute value, we *add* if the coefficients are opposite in sign (as in Example 1) and *subtract* if they are like in sign.

To eliminate an unknown from two equations by substitution, we solve one of the equations for that unknown in terms of the other (or others) and substitute this into the other equation. Consider, for example, the system

$$\begin{cases} 3x - 4y = 7 \\ 2x - y = 3 \end{cases}$$

We can solve the second equation for y to get $y = 2x - 3$ and then substitute $2x - 3$ in place of y in the first equation:

$$3x - 4(2x - 3) = 7$$
$$3x - 8x + 12 = 7$$
$$-5x = -5$$
$$x = 1$$

Having found x, we obtain y from $y = 2x - 3$. This gives

$$y = 2(1) - 3$$
$$= -1$$

So this is a consistent system with unique solution $(1, -1)$.

The next two examples illustrate inconsistent and dependent systems, respectively.

EXAMPLE 2 ||||||||||||||||||||||||||||||||||||||||||||||||||||||||||||||||||||||||||||||||||||||||||||||||||||||||

Solve the system: $\begin{cases} 2x - 3y = 4 \\ 4x - 6y = 15 \end{cases}$

Solution After multiplying both sides of the top equation by 2 and subtracting, we get $0 = -7$, which is clearly impossible. There is no simultaneous solution; that is, the lines are parallel. In fact, we could have seen in advance that the two lines had the same slope. The situation is graphed in Figure 2.

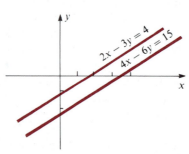

FIGURE 2

EXAMPLE 3 ||||||||||||||||||||||||||||||||||||||||||||||||||||||||||||||||||||||||||||||||||||||||||||||||

Solve the system: $\begin{cases} 2x - 3y = 4 \\ 4x - 6y = 8 \end{cases}$

Solution   This system differs from the one in Example 2 only in the fact that the entire second equation is obtainable from the first by multiplying both sides by 2. But this means that the second equation is really equivalent to the first and so has the same graph. Attempting to solve simultaneously would yield $0 = 0$. This is an example of a dependent system. We could say that the simultaneous solutions are the pairs (x, y) corresponding to all points on the line given by either equation. From the first equation we can solve for y to get

$$y = \frac{2x - 4}{3}$$

Now let x take on an arbitrary value, say, $x = c$; then $y = (2c - 4)/3$. So we could say that the solution set is the set of all pairs

$$\left(c, \frac{2c - 4}{3} \right)$$

where c may range over all real numbers.   ▲

Remark. The form of the solution set for a dependent system is not unique. We can see this from the preceding example by solving the first equation for x, rather than y, getting

$$x = \frac{3y + 4}{2}$$

Then if we let y be any real number c, we can write the solution set as the set of all ordered pairs of the form

$$\left(\frac{3c + 4}{2}, c \right)$$

where c is any real number. Even though this solution looks different from the solution found in Example 3, when we allow c to range over all real numbers, the set of ordered pairs obtained from either solution is the same as the set obtained from the other.

EXAMPLE 4 ||||||||||||||||||||||||||||||||||||||||||||||||||||||||||||||||||||||||||||||||||||||||||||||||

Solve the system: $\begin{cases} x + 2y + \ z = 3 \\ 2x - \ y - \ z = 0 \\ 3x + 4y + 2z = 8 \end{cases}$

Solution   We will again use the process of elimination. The object is to eliminate the *same* unknown from two pairs of equations, thus obtaining two equations in two unknowns. Then one further elimination enables us to find one of the unknowns, and the others are found by substitution.

It appears that z would be particularly easy to eliminate in this case. We add corresponding sides of the first two equations:

$$
\begin{array}{l}
x + 2y + z = 3 \\
\underline{2x - \ y - z = 0} \\
3x + \ y \qquad = 3
\end{array}
\qquad (1)
$$

Now returning to the original system, we must eliminate z from another pair of equations. It is essential to stay with z so that a second equation containing only x and y can be obtained. We multiply both sides of the second equation by 2 and add it to the third:

$$
\begin{array}{l}
4x - 2y - 2z = 0 \\
\underline{3x + 4y + 2z = 8} \\
7x + 2y \qquad = 8
\end{array}
\qquad (2)
$$

Now we wish to solve equations (1) and (2) simultaneously, and to this end we multiply both sides of equation (1) by -2 and add the result to equation (2):

$$
\begin{array}{l}
-6x - 2y = -6 \\
\underline{\ 7x + 2y = \quad 8} \\
\ \ x \qquad = \quad 2
\end{array}
$$

Having found x, we can substitute in either equation (1) or (2) to find y. We select (1):

$$
3(2) + y = 3
$$
$$
y = -3
$$

Finally, we substitute both x and y into any one of the original equations to find z. From the first equation we get

$$
2 + 2(-3) + z = 3
$$
$$
2 - 6 + z = 3
$$
$$
z = 7
$$

The final answer for the simultaneous solution is the triple $(2, -3, 7)$, that is, $x = 2$, $y = -3$, and $z = 7$.   ▲

Keeping track of things in the case of three equations and three unknowns is difficult enough, but with more than three the difficulty increases rapidly. For example, with four equations, three pairs are selected and from each pair the *same* unknown is eliminated. The result is a system of three equations in three unknowns, which is solved as in Example 4. When the three unknowns are found, they are substituted into any one of the original equations to find the fourth unknown.

In the next example the equations are not linear, but by expressing both sides of each of the equations to the same base, we can equate exponents, and this results in a linear system.

EXAMPLE 5

Solve the system

$$\begin{cases} 2^x \cdot 4^{-y} = 32 \\ 3^{4-x} \cdot 9^{3y-2} = 27 \end{cases}$$

Solution   We can write all terms of the first equation with base 2 and those of the second equation with base 3, as follows.

$$\begin{cases} 2^x \cdot (2^2)^{-y} = 2^5 \\ 3^{4-x}(3^2)^{3y-2} = 3^3 \end{cases} \quad \text{or} \quad \begin{cases} 2^{x-2y} = 2^5 \\ 3^{-x+6y} = 3^3 \end{cases}$$

Equating exponents of like powers gives

$$\begin{cases} x - 2y = 5 \\ -x + 6y = 3 \end{cases}$$

Adding, we get $4y = 8$, so $y = 2$. Substituting into the top equation gives $x - 4 = 5$, or $x = 9$. So the simultaneous solution is $(9, 2)$.   ▲

Equations in two or more unknowns often arise in real-world problems. Sometimes problems that could be treated with one unknown are easier to handle with two, as illustrated in the following examples.

EXAMPLE 6

A saline solution containing 5% salt is to be mixed with one containing 8% salt to obtain 5 liters of a 6% salt solution. How many liters of each solution should be used?

Solution   Let

$$x = \text{Number of liters of 5\% solution required}$$
$$y = \text{Number of liters of 8\% solution required}$$

Then

$$0.05x = \text{Amount of pure salt in first solution}$$
$$0.08y = \text{Amount of pure salt in second solution}$$
$$(0.06)5 = \text{Amount of pure salt in final solution}$$

We need two equations relating x and y. Since there are to be 5 liters in all, we must have

$$x + y = 5$$

Further, the amount of pure salt from the two components must equal the amount of pure salt in the final mixture:

$$0.05x + 0.08y = (0.06)5$$

or, after multiplying by 100 and simplifying,

$$5x + 8y = 30$$

So the system we wish to solve is

$$\begin{cases} x + y = 5 \\ 5x + 8y = 30 \end{cases}$$

By multiplying both sides of the top equation by 5 and subtracting from the bottom, we get

$$
\begin{aligned}
5x + 8y &= 30 \\
5x + 5y &= 25 \\
\hline
3y &= 5 \\
y &= \tfrac{5}{3}
\end{aligned}
$$

Finally,

$$x = 5 - y = 5 - \frac{5}{3} = \frac{15 - 5}{3} = \frac{10}{3}$$

So $3\frac{1}{3}$ liters of the 5% solution are needed, and $1\frac{2}{3}$ liters of the 8% solution are needed. ▲

EXAMPLE 7

A boy rows 6 miles upstream in 1 hour and 20 minutes, and the return trip takes 1 hour. Find how fast he can row in still water and find the rate of the current.

Solution   Let

$$x = \text{Rate in miles per hour that the boy can row in still water}$$

$$y = \text{Rate in miles per hour of the current}$$

Then

$$x - y = \text{Actual rate of progress upstream}$$

$$x + y = \text{Actual rate of progress downstream}$$

We apply the fundamental relation $d = rt$ two times, once for going upstream and once for going downstream:

$$\begin{cases} 6 = (x - y) \cdot \frac{4}{3} & (\frac{4}{3} \text{ hours} = 1 \text{ hour and } 20 \text{ minutes}) \\ 6 = (x + y) \cdot 1 \end{cases}$$

These can be simplified to the form

$$\begin{cases} 2x - 2y = 9 \\ x + y = 6 \end{cases}$$

The solution is now easily found:

$$\begin{array}{r} 2x - 2y = \ 9 \\ 2x + 2y = 12 \\ \hline 4x \quad\quad = 21 \\ x = \frac{21}{4} \end{array}$$

So

$$y = 6 - \frac{21}{4} = \frac{24 - 21}{4} = \frac{3}{4}$$

The boy can row $5\frac{1}{4}$ miles per hour in still water, and the rate of the current is $\frac{3}{4}$ mile per hour. ▲

EXAMPLE 8

Joan has a tractor mower and Charles has a hand-operated power mower. They regularly mow a certain large lawn in 4 hours by working together. On one occasion, after working together for 3 hours on the lawn, Joan's mower breaks down, and it takes Charles another 3 hours to finish the job alone. How long would it take each of them to do the entire job alone?

Solution Let

$$x = \text{Number of hours for Joan alone}$$

$$y = \text{Number of hours for Charles alone}$$

Then

$$\frac{1}{x} = \text{Fractional part of the job Joan does per hour}$$

$$\frac{1}{y} = \text{Fractional part of the job Charles does per hour}$$

If each works 4 hours, the entire job is done:

$$\frac{4}{x} + \frac{4}{y} = 1$$

When Joan works only 3 hours, Charles has to work an additional 3 hours, or 6 hours in all, to accomplish the entire job:

$$\frac{3}{x} + \frac{6}{y} = 1$$

Now we have the system

$$\begin{cases} \dfrac{4}{x} + \dfrac{4}{y} = 1 \\ \dfrac{3}{x} + \dfrac{6}{y} = 1 \end{cases}$$

These are not linear equations, but we can use the same techniques we used above to solve them. It is probably easiest in this situation not to clear the fractions at the outset. We multiply the top equation by 3 and the bottom one by 4:

$$\frac{12}{x} + \frac{12}{y} = 3$$

$$\frac{12}{x} + \frac{24}{y} = 4$$

Now we subtract the top equation from the bottom one to obtain

$$\frac{12}{y} = 1$$

Clearing the fractions now yields $y = 12$. We may substitute into either of the original equations. We choose the first:

$$\frac{4}{x} + \frac{4}{12} = 1$$

$$\frac{4}{x} = \frac{2}{3}$$

$$x = 6$$

So Joan could do the job alone in 6 hours, and Charles could do it alone in 12 hours. ▲

EXAMPLE 9

The units digit of a two-digit number is 1 less than twice the tens digit. If the digits are reversed, the new number is 27 greater than the old number. Find the original number.

Solution   Let

$$x = \text{Tens digit}$$
$$y = \text{Units digit}$$

Then

$$10x + y = \text{Original number}$$
$$10y + x = \text{Number with digits reversed}$$

Thus, from the given information,

$$y = 2x - 1$$
$$10y + x = 10x + y + 27$$

After simplifying the second equation, the system to be solved can be written in the form

$$\begin{cases} y = 2x - 1 \\ x - y = -3 \end{cases}$$

We eliminate y by substitution from the first equation into the second:

$$x - (2x - 1) = -3$$
$$-x + 1 = -3$$
$$x = 4$$

So $y = 2(4) - 1 = 7$. Thus, the original number is 47.

Remark. In Example 9 we wrote the number as $10x + y$, since in our positional system the tens digit is understood to be multiplied by 10. Thus, 47 means $(4 \times 10) + 7$. Similarly, if x were the hundreds digit, y the tens digit, and z the units digit of a number, we would indicate the number as $100x + 10y + z$.

  EXERCISE SET I

A

Solve the systems in Problems 1–22.

1. $\begin{cases} x - y = 2 \\ x + y = 6 \end{cases}$

2. $\begin{cases} x + 2y = -4 \\ x - 3y = 6 \end{cases}$

3. $\begin{cases} 3x - y = 0 \\ 5x - 2y = 1 \end{cases}$

4. $\begin{cases} 2x - 3y = 18 \\ 3x + 4y = -7 \end{cases}$

5. $\begin{cases} 3x - 2y = 6 \\ x + 2y = 10 \end{cases}$

6. $\begin{cases} 2x - y = -2 \\ 3x + 2y = 1 \end{cases}$

7. $\begin{cases} 4x - y = 6 \\ 3x + 2y = 5 \end{cases}$

8. $\begin{cases} 5s - 2t = 4 \\ 3s - 2t = 0 \end{cases}$

9. $\begin{cases} 2x + 3y = 2 \\ 3x - 4y = 20 \end{cases}$

10. $\begin{cases} 5x + 2y = 8 \\ 7x - 3y = -12 \end{cases}$

11. $\begin{cases} 5s + 4t = -1 \\ 2s - 3t = -28 \end{cases}$

12. $\begin{cases} 5m - 2n = -2 \\ 4m + 5n = 16 \end{cases}$

13. $\begin{cases} x + 3y - 2z = -1 \\ 2x - y + 3z = 5 \\ -x + 5y - 4z = 1 \end{cases}$

14. $\begin{cases} 2x - 3y - 4z = -4 \\ 3x + 4y + 2z = 1 \\ 5x - 2y - 3z = 0 \end{cases}$

15. $\begin{cases} 2x + y + z = 7 \\ x - y + 2z = 11 \\ 5x + y - 2z = 1 \end{cases}$

16. $\begin{cases} 3x + 4y - z = 4 \\ 2x - 5y + 3z = -10 \\ 4x + 3y + z = 10 \end{cases}$

17. $\begin{cases} \dfrac{2}{x} - \dfrac{3}{y} = 4 \\ \dfrac{1}{x} + \dfrac{4}{y} = -10 \end{cases}$

18. $\begin{cases} \dfrac{2}{x} + \dfrac{3}{y} = -2 \\ \dfrac{5}{x} + \dfrac{4}{y} = \dfrac{5}{6} \end{cases}$

19. $\begin{cases} ax + by = c \\ bx - ay = d \end{cases}$

20. $\begin{cases} x + y = -1 \\ y + 3z = 2 \\ 4z - w = 0 \\ 2x - w = 0 \end{cases}$

21. $\begin{cases} 3^{x-1} \cdot (27)^{2-y} = \dfrac{1}{9} \\ 4^x \cdot 8^{1-y} = 16 \end{cases}$

22. $\begin{cases} 2^{-x+1} \cdot 4^{y+2} = 8 \cdot 2^{-4x} \\ \left(\dfrac{1}{3}\right)^{2x-1} \cdot 9^y = 81 \end{cases}$

In Problems 23 and 24 show that the system is dependent, and give the solution set. Draw the graph.

23. $\begin{cases} 3x - 4y = 5 \\ 8y - 6x + 10 = 0 \end{cases}$

24. $\begin{cases} 4x - 10y = 14 \\ -6x + 15y = -21 \end{cases}$

Show that the systems in Problems 25 and 26 are inconsistent, and draw their graphs.

25. $\begin{cases} 8x - 6y = 3 \\ 3y - 4x = 2 \end{cases}$

26. $\begin{cases} 8x - 10y = 7 \\ -12x + 15y = 11 \end{cases}$

27. Show that the following system is consistent. Draw the graph.

$$\begin{cases} 2x + 5y = 6 \\ x - 2y = -7 \\ 5x + 8y = 1 \end{cases}$$

28. Show that the lines whose equations are $2x - 3y = 0$, $4x + 5y = 22$, and $3x - 2y = 5$ intersect at a common point. Draw their graphs.

29. Find the two numbers whose sum is 127 and whose difference is 63.

30. The difference between two numbers is 5. Three times the larger number minus four times the smaller number is 1. Find the numbers.

31. The perimeter of a certain rectangle is 24. If the width were doubled and the length tripled, the perimeter would be 64. Find the length and width.

32. Rowing with the current, a 10-mile trip on a river takes a boy 1 hour and 20 minutes. The return trip takes 4 hours. Find how fast the boy can row in still water and the rate of the current.

33. A girl covered a distance of $3\frac{1}{4}$ miles by alternating between jogging and walking. She walked a total of 15 minutes and jogged a total of 30 minutes. The next day she walked a total of 30 minutes and jogged 45 minutes, covering a total distance of $5\frac{1}{4}$ miles. Assuming her rates of walking and of jogging remained the same from one day to the next, find these rates.

34. The sum of the digits of a certain two-digit number is 10. By interchanging the units and tens digits, the value of the number is increased by 36. Find the number.

35. The sum of the digits of a two-digit number is 14. If the digits are reversed, the value of the number is decreased by 18. Find the number.

36. By working together for 4 hours Bill and his younger brother Tom weed one-half of a large garden. The next day they work together for 2 hours, and then Tom gets tired and quits. It takes Bill an additional 3 hours to complete the job. How long would it have taken Bill to weed the entire garden alone? How long would it have taken Tom?

37. A woman has two investments, one yielding 5% interest and one 7% interest annually. The amount at 7% lacks $1,000 of being double the amount at 5%. The total interest after 1 year is $690. Find how much she has invested in each account.

38. A customer in a coffee house wants a blend of two kinds of coffee: Colombian, selling at $4.00 per pound, and Brazilian, selling for $3.60 per pound. He gets 4 pounds of the blend, and the cost is $15.40. How many pounds of each kind did he buy?

39. A chemist wishes to mix a solution containing 10% sulfuric acid with a solution containing 25% sulfuric acid. She wants 30 cubic centimeters in all, and she wants the final mixture to be 18% sulfuric acid. How much of each concentration should she use?

40. An alloy that is 35% copper is to be combined with an alloy that is 60% copper to obtain 180 pounds of an alloy that is 50% copper. How many pounds of each kind of alloy should be used?

41. Twenty pounds of English breakfast tea is to be made by blending Indian and Ceylon teas. The Indian tea is priced at $6.25 per pound, and the Ceylon tea at $7.00 per pound. The price of the English breakfast tea is to be $6.60 per pound. How much Indian tea and how much Ceylon tea should be used?

42. The prices of tickets at a movie theater are $2.75 for adults and $1.50 for children. For a certain show there were 400 tickets sold for a total of $875. How many adults and how many children attended?

43. Four pounds of a certain brand of coffee and three dozen large eggs cost $19.70. Two pounds of the same brand of coffee and four dozen large eggs cost $12.10. Find the cost of 1 pound of the coffee and one dozen large eggs.

B

Solve the systems in Problems 44–48.

44. $\begin{cases} 3x - 2y + 4z = 5 \\ 2x + 3y - 2z = 6 \\ 4x - 5y + 3z = -5 \end{cases}$

45. $\begin{cases} 3x + 5y - 7z = 6 \\ 4x - 2y + 8z = -5 \\ 6x + 7y - 3z = 4 \end{cases}$

46. $\begin{cases} 5x - 2y + 3z = 6 \\ 6x - 3y + 4z = 10 \\ -4x + 4y - 9z = 4 \end{cases}$

47. $\begin{cases} x + 2y - 3z - 2w = 1 \\ 3x - 4y + z - w = 3 \\ 2x + 4z + 3w = 4 \\ 3y - 2z + w = 10 \end{cases}$

48. $\begin{cases} 2x + 4y - z - 3w = 6 \\ x + 3y + 2z + 4w = -4 \\ 3x - 5y + 3z + 2w = 13 \\ 4x - 6y + z - 3w = 22 \end{cases}$

49. Show that the following system is inconsistent:

$$\begin{cases} 3x - 2y + 4z = 5 \\ 4x - y + 2z = 3 \\ x + y - 2z = 7 \end{cases}$$

In Problems 50 and 51 show that the system is dependent and give the solution set.

50. $\begin{cases} x - 3y + 4z = -1 \\ 2x + y - 5z = 3 \\ 7x - 7y + 2z = 3 \end{cases}$

51. $\begin{cases} 2x - 3y + 4z - w = 3 \\ x + y - 3z + 2w = -1 \\ 3x - y - 3z + 8w = 5 \end{cases}$

52. Find a, b, and c so that the parabola $y = ax^2 + bx + c$ passes throught the points $(1, -5), (-2, 10)$, and $(3, 5)$.

53. Let $f(x) = ax^2 + bx + c$. If $f(2) = 5$, $f(-1) = 2$, and $f(4) = -3$, find a, b, and c.

54. Determine A, B, and C in the equation $x^2 + y^2 + Ax + By + C = 0$ for a circle if the circle passes through the points $(2, 2)$, $(-5, 1)$, and $(4, -2)$.

55. Find the equation of the circle that circumscribes the triangle whose sides are formed by the lines $4x - 3y = 21$, $x + 3y = 9$, and $2x + y = 3$. **Hint.** Use the idea of Problem 54.

56. A car leaves town A going toward town B, which is 50 miles away on a straight road. At the same time, two cars leave B, one going toward A and one going away from A, both traveling at the same rate of speed. The car that left from town A meets the car going from B toward A after 20 minutes, and overtakes the one going from B away from A after $2\frac{1}{2}$ hours. Find the speed of each car.

57. A commuter finds that if she leaves the office 5 minutes earlier than usual, she can average 36 miles per hour and arrive home 10 minutes earlier than usual. However, if she leaves the office 5 minutes later than usual, she can average only 25 miles per hour and arrives home 11 minutes later than usual. Find the time it normally takes her to get home and the distance from the office to her home.

58. Brine containing 40% salt is diluted by adding

pure water, resulting in a 25% salt solution. Then 20 more gallons of pure water are added, diluting the mixture to a 20% salt solution. Find how much brine there was originally and how many gallons of water were added the first time.

59. Pumps A, B, and C working simultaneously can fill a tank in 10 hours. Pumps B and C working together can fill it in 15 hours. If all three pumps work together for 4 hours and then pump C is shut off, it takes 8 more hours to fill the tank. How long would it take each pump alone to fill the tank?

REDUCTION TO TRIANGULAR FORM AND THE USE OF MATRICES

2

We will illustrate by means of the following example a technique for solving linear systems that makes the elimination of unknowns more systematic. This procedure is especially useful for systems of more than three equations and three unknowns.

EXAMPLE 10

Solve the system:
$$\begin{cases} x + 2y + 3z + 2w = 1 \\ x + 3y - z - 2w = -1 \\ 2x + 3y - 2z - w = 7 \\ 3x + 4y + z - w = -4 \end{cases}$$

Solution The first step is to eliminate x from the bottom three equations. This is accomplished by adding to each of these, in turn, certain multiples of the first equation. Thus, we add to the second equation -1 times the first, to the third -2 times the first, and to the fourth -3 times the first. The result is the new system

$$x + 2y + 3z + 2w = 1$$
$$y - 4z - 4w = -2$$
$$-y - 8z - 5w = 5$$
$$-2y - 8z - 7w = -7$$

Next, we eliminate y from the bottom two equations by adding the second equation to the third and then adding to the fourth equation 2 times the second:

$$x + 2y + 3z + 2w = 1$$
$$y - 4z - 4w = -2$$
$$-12z - 9w = 3$$
$$-16z - 15w = -11$$

The third equation can be simplified by dividing by -3:

$$x + 2y + 3z + 2w = 1$$
$$y - 4z - 4w = -2$$
$$4z + 3w = -1$$
$$-16z - 15w = -11$$

Now we eliminate z from the bottom equation by adding to it 4 times the third:

$$x + 2y + 3z + 2w = 1$$
$$y - 4z - 4w = -2$$
$$4z + 3w = -1$$
$$-3w = -15$$

The system is now in what is called **triangular form** and can be solved by working from the bottom up. We obtain from the last equation $w = 5$. Then, substituting this into the third equation, we get

$$4z + 15 = -1$$
$$4z = -16$$
$$z = -4$$

Substituting both w and z into the second equation yields

$$y + 16 - 20 = -2$$
$$y = 2$$

Finally, we get from the first equation

$$x + 4 - 12 + 10 = 1$$
$$x = -1$$

The desired solution, then, is given by the array $(-1, 2, -4, 5)$, where it is understood that the values shown are for $x, y, z,$ and $w,$ respectively.   ▲

Students should be warned that not all systems reduce to triangular form as easily as the one in Example 10. The arithmetic necessary to eliminate the desired unknowns often is more complicated. But this method always works, and it is systematic.

We will show an abbreviated version of Example 10 below, but before doing so, it will be useful to list the admissible operations for solving a system of equations. Any of the following change the system to an equivalent one, that is, to one having the same solution set:

1. Both sides of any equation may be multiplied by a nonzero number.
2. Two equations may be interchanged.
3. A multiple of any equation may be added to any other equation.

Now let us return to the original system in Example 10:

$$\begin{cases} x + 2y + 3z + 2w = 1 \\ x + 3y - z - 2w = -1 \\ 2x + 3y - 2z - w = 7 \\ 3x + 4y + z - w = -4 \end{cases}$$

We are going to represent this by means of the coefficients and the constant terms only, with the understanding that whenever it is convenient to do so, we can reinsert the letters representing the unknowns. We write this as follows:

$$\begin{bmatrix} 1 & 2 & 3 & 2 & | & 1 \\ 1 & 3 & -1 & -2 & | & -1 \\ 2 & 3 & -2 & -1 & | & 7 \\ 3 & 4 & 1 & -1 & | & -4 \end{bmatrix}$$

The vertical line is used to indicate the separation of the left-hand sides from the right. Such an array of numbers is called a **matrix,** and when used to represent a system of equations, as in the present case, it is called an **augmented matrix.** This description is used to distinguish the full matrix shown from the **coefficient matrix** only, that is,

$$\begin{bmatrix} 1 & 2 & 3 & 2 \\ 1 & 3 & -1 & -2 \\ 2 & 3 & -2 & -1 \\ 3 & 4 & 1 & -1 \end{bmatrix}$$

Now we are going to retrace the steps shown in Example 10, using the augmented matrix. We show 0 where a coefficient is eliminated. A brief indication of each step is given as a reminder of what was done to the preceding matrix to get the current one. First, we exhibit again the augmented matrix.

$$\begin{bmatrix} 1 & 2 & 3 & 2 & | & 1 \\ 1 & 3 & -1 & -2 & | & -1 \\ 2 & 3 & -2 & -1 & | & 7 \\ 3 & 4 & 1 & -1 & | & -4 \end{bmatrix}$$

$$\begin{bmatrix} 1 & 2 & 3 & 2 & | & 1 \\ 0 & 1 & -4 & -4 & | & -2 \\ 0 & -1 & -8 & -5 & | & 5 \\ 0 & -2 & -8 & -7 & | & -7 \end{bmatrix} \begin{array}{l} \text{Row } 2 + [(-1) \cdot \text{Row } 1] \\ \text{Row } 3 + [(-2) \cdot \text{Row } 1] \\ \text{Row } 4 + [(-3) \cdot \text{Row } 1] \end{array}$$

$$\begin{bmatrix} 1 & 2 & 3 & 2 & | & 1 \\ 0 & 1 & -4 & -4 & | & -2 \\ 0 & 0 & -12 & -9 & | & 3 \\ 0 & 0 & -16 & -15 & | & -11 \end{bmatrix} \begin{array}{l} \text{Row } 3 + (1 \cdot \text{Row } 2) \\ \text{Row } 4 + (2 \cdot \text{Row } 2) \end{array}$$

$$\left[\begin{array}{cccc|c} 1 & 2 & 3 & 2 & 1 \\ 0 & 1 & -4 & -4 & -2 \\ 0 & 0 & 4 & 3 & -1 \\ 0 & 0 & -16 & -15 & -11 \end{array}\right] \quad \text{Row } 3 \cdot (-\tfrac{1}{3})$$

$$\left[\begin{array}{cccc|c} 1 & 2 & 3 & 2 & 1 \\ 0 & 1 & -4 & -4 & -2 \\ 0 & 0 & 4 & 3 & -1 \\ 0 & 0 & 0 & -3 & -15 \end{array}\right] \quad \text{Row } 4 + (4 \cdot \text{Row } 3)$$

The matrix is now in triangular form, so we are ready to find the solution to the system. The last row of the matrix represents the equation

$$-3w = -15$$

so that $w = 5$. In turn we write the other equations, from the bottom up, and substitute known values as before. ▲

EXAMPLE 11

Solve the system:
$$\begin{cases} 3x + 2y - 4z = -7 \\ 7x - 8y - 5z = 5 \\ -8x + 5y + 6z = -1 \end{cases}$$

Solution   The augmented matrix is

$$\left[\begin{array}{ccc|c} 3 & 2 & -4 & -7 \\ 7 & -8 & -5 & 5 \\ -8 & 5 & 6 & -1 \end{array}\right]$$

It is an advantage to have 1 or -1 as the first element in Row 1, since it is then easy to choose appropriate multiples to eliminate the elements below it. We can always make the first element 1 by multiplying the first row by its reciprocal; in this case we would multiply by $\tfrac{1}{3}$. But this would introduce fractions, and it is generally simpler to avoid doing this. In this case we can add Row 2 to Row 3 and then interchange Rows 1 and 3:

$$\left[\begin{array}{ccc|c} 3 & 2 & -4 & -7 \\ 7 & -8 & -5 & 5 \\ -1 & -3 & 1 & 4 \end{array}\right] \quad \text{Row } 3 + \text{Row } 2$$

$$\left[\begin{array}{ccc|c} -1 & -3 & 1 & 4 \\ 7 & -8 & -5 & 5 \\ 3 & 2 & -4 & -7 \end{array}\right] \quad \begin{array}{l} \text{Row 3 interchanged with Row 1} \\ \\ \text{Row 1 interchanged with Row 3} \end{array}$$

Now we proceed to introduce zeros under the -1:

$$\begin{bmatrix} -1 & -3 & 1 & | & 4 \\ 0 & -29 & 2 & | & 33 \\ 0 & -7 & -1 & | & 5 \end{bmatrix} \quad \begin{array}{l} \\ \text{Row } 2 + (7 \cdot \text{Row } 1) \\ \text{Row } 3 + (3 \cdot \text{Row } 1) \end{array}$$

We are faced with messy arithmetic now unless we make some further simplifying moves. It appears that if we add -4 times Row 3 to Row 2, things will improve significantly:

$$\begin{bmatrix} -1 & -3 & 1 & | & 4 \\ 0 & -1 & 6 & | & 13 \\ 0 & -7 & -1 & | & 5 \end{bmatrix} \quad \begin{array}{l} \\ \text{Row } 2 + [(-4) \cdot \text{Row } 3] \\ \\ \end{array}$$

Finally, we add -7 times Row 2 to Row 3:

$$\begin{bmatrix} -1 & -3 & 1 & | & 4 \\ 0 & -1 & 6 & | & 13 \\ 0 & 0 & -43 & | & -86 \end{bmatrix} \quad \begin{array}{l} \\ \\ \text{Row } 3 + [(-7) \cdot \text{Row } 2] \end{array}$$

The third equation now reads

$$-43z = -86$$

so

$$z = 2$$

The second equation is

$$-y + 6z = 13$$

which on substituting $z = 2$ yields

$$y = -1$$

The first equation can now be solved for x:

$$-x - 3y + z = 4$$
$$-x + 3 + 2 = 4$$
$$-x = -1$$
$$x = 1$$

The desired solution is $(1, -1, 2)$.

The techniques to simplify the arithmetic illustrated in this example should be kept in mind as you do problems by this method. ▲

EXAMPLE 12

Solve the system below by reducing the augmented matrix to triangular form.

$$\begin{cases} 4x_1 + 11x_2 + 6x_3 + 22x_4 = 16 \\ x_1 + 2x_2 + x_4 = 7 \\ 2x_1 + 5x_2 + x_3 + 10x_4 = 7 \end{cases}$$

Solution   This example will show that the method we are employing will work even when the number of unknowns is different from the number of equations. To save one step, we will mentally interchange the first two equations before writing the matrix:

$$\begin{bmatrix} 1 & 2 & 0 & 1 & | & 7 \\ 4 & 11 & 6 & 22 & | & 16 \\ 2 & 5 & 1 & 10 & | & 7 \end{bmatrix}$$

$$\begin{bmatrix} 1 & 2 & 0 & 1 & | & 7 \\ 0 & 3 & 6 & 18 & | & -12 \\ 0 & 1 & 1 & 8 & | & -7 \end{bmatrix} \quad \begin{array}{l} \text{Row } 2 + [(-4) \cdot \text{Row } 1] \\ \text{Row } 3 + [(-2) \cdot \text{Row } 1] \end{array}$$

We could now interchange Rows 2 and 3, but we choose to multiply Row 2 by $\frac{1}{3}$ instead:

$$\begin{bmatrix} 1 & 2 & 0 & 1 & | & 7 \\ 0 & 1 & 2 & 6 & | & -4 \\ 0 & 1 & 1 & 8 & | & -7 \end{bmatrix} \quad \frac{1}{3} \cdot \text{Row } 2$$

Now we subtract Row 2 from Row 3:

$$\begin{bmatrix} 1 & 2 & 0 & 1 & | & 7 \\ 0 & 1 & 2 & 6 & | & -4 \\ 0 & 0 & -1 & 2 & | & -3 \end{bmatrix} \quad \text{Row } 3 + [(-1) \cdot \text{Row } 2]$$

This is as far as we can go in the triangular form. The third equation now reads

$$-x_3 + 2x_4 = -3$$

While this cannot be solved for x_3 as a specific number, it can be solved in terms of x_4:

$$x_3 = 3 + 2x_4$$

From the second equation we get

$$x_2 + 2x_3 + 6x_4 = -4$$
$$x_2 = -4 - 2(3 + 2x_4) - 6x_4$$
$$= -4 - 6 - 4x_4 - 6x_4$$
$$= -10 - 10x_4$$

and from the first,

$$x_1 + 2x_2 + x_4 = 7$$
$$x_1 = 7 - 2(-10 - 10x_4) - x_4$$
$$= 7 + 20 + 20x_4 - x_4$$
$$= 27 + 19x_4$$

Summarizing, we have found

$$\begin{cases} x_1 = 27 + 19x_4 \\ x_2 = -10 - 10x_4 \\ x_3 = 3 + 2x_4 \\ x_4 = x_4 \end{cases}$$

We may select any real value for x_4. Then x_1, x_2, and x_3 are determined. Let us designate x_4 by c. Then the solution set can be written as

$$\{(27 + 19c, -10 - 10c, 3 + 2c, c): \quad c \in R\}$$

Thus, there are infinitely many points that satisfy all three equations.   ▲

EXAMPLE 13

Determine whether the following system is consistent by reducing the augmented matrix to triangular form.

$$\begin{cases} x - y + 2z = 3 \\ 2x + y \qquad = 4 \\ x + 3y - z = 2 \\ -x + 4y + 2z = 3 \end{cases}$$

Solution   In this case there are more equations than unknowns. While we might expect three equations in three unknowns to have a common solution, it is a special case when this solution also works for a fourth equation. So it should come as no surprise if these equations turn out to be inconsistent.

$$\begin{bmatrix} 1 & -1 & 2 & | & 3 \\ 2 & 1 & 0 & | & 4 \\ 1 & 3 & -1 & | & 2 \\ -1 & 4 & 2 & | & 3 \end{bmatrix}$$

$$\begin{bmatrix} 1 & -1 & 2 & | & 3 \\ 0 & 3 & -4 & | & -2 \\ 0 & 4 & -3 & | & -1 \\ 0 & 3 & 4 & | & 6 \end{bmatrix} \quad \begin{array}{l} \\ \text{Row } 2 + [(-2) \cdot \text{Row } 1] \\ \text{Row } 3 + [(-1) \cdot \text{Row } 1] \\ \text{Row } 4 + (1 \cdot \text{Row } 1) \end{array}$$

$$\begin{bmatrix} 1 & -1 & 2 & | & 3 \\ 0 & -1 & -1 & | & -1 \\ 0 & 4 & -3 & | & -1 \\ 0 & 3 & 4 & | & 6 \end{bmatrix} \quad \begin{array}{l} \\ \text{Row } 2 + [(-1) \cdot \text{Row } 3] \end{array}$$

$$\begin{bmatrix} 1 & -1 & 2 & \bigm| & 3 \\ 0 & -1 & -1 & \bigm| & -1 \\ 0 & 0 & -7 & \bigm| & -5 \\ 0 & 0 & 1 & \bigm| & 3 \end{bmatrix} \quad \begin{array}{l} \text{Row } 3 + (4 \cdot \text{Row } 2) \\ \text{Row } 4 + (3 \cdot \text{Row } 2) \end{array}$$

$$\begin{bmatrix} 1 & -1 & 2 & \bigm| & 3 \\ 0 & -1 & -1 & \bigm| & -1 \\ 0 & 0 & 1 & \bigm| & 3 \\ 0 & 0 & -7 & \bigm| & -5 \end{bmatrix} \quad \begin{array}{l} \text{Row } 4 \text{ interchanged with Row } 3 \\ \text{Row } 3 \text{ interchanged with Row } 4 \end{array}$$

$$\begin{bmatrix} 1 & -1 & 2 & \bigm| & 3 \\ 0 & -1 & -1 & \bigm| & -1 \\ 0 & 0 & 1 & \bigm| & 3 \\ 0 & 0 & 0 & \bigm| & 16 \end{bmatrix} \quad \text{Row } 4 + (7 \cdot \text{Row } 3)$$

The last equation now reads

$$0 = 16$$

which is not possible. Therefore, the system is inconsistent.

EXERCISE SET 2

A

By reducing the augmented matrix to triangular form, solve the following systems or show that no solution exists. If the system is dependent, show the solution set as in Example 12.

1. $\begin{cases} x + 3y - 2z = 3 \\ y + z = 4 \\ x + 2y + 4z = 6 \end{cases}$

2. $\begin{cases} x - y + z = 1 \\ 2x + y + z = 4 \\ x + 5y - 2z = 1 \end{cases}$

3. $\begin{cases} x + 2y + z = 3 \\ 2x - y - z = 0 \\ 3x + 4y + 2z = 8 \end{cases}$

4. $\begin{cases} x + 2y + z = 1 \\ 2x + 5y + 4z = 2 \\ -x - 3y + 5z = 7 \end{cases}$

5. $\begin{cases} x + y - z = 3 \\ 3x + 4y + 2z = 3 \\ 4x - 2y - 8z = -4 \end{cases}$

6. $\begin{cases} x + 3y - 2z = -1 \\ x - 5y + 4z = -1 \\ 2x - y + 3z = 5 \end{cases}$

7. $\begin{cases} x - y = 1 \\ y + z = 1 \\ x + z = 2 \end{cases}$

8. $\begin{cases} x + 4y - 2z = -3 \\ 3x - 2y + z = 12 \\ 2x - 3y + z = 11 \end{cases}$

9. $\begin{cases} 2x - 3y + 2z = -2 \\ 3x - 8y - z = 2 \\ -5x + 10y - 3z = 4 \end{cases}$

10. $\begin{cases} 2x - 3y + 4z = 5 \\ x + y - 2z = 1 \\ 7x - 3y + 2z = 13 \end{cases}$

11. $\begin{cases} 2x + y + z = 7 \\ x - y + 2z = 11 \\ 5x + y - 2z = 1 \end{cases}$

12. $\begin{cases} 2x + 4y + 3z = -2 \\ x - 3y + 4z = 9 \\ 5x + 7y - z = -9 \end{cases}$

13. $\begin{cases} x + 2y + 3z = 9 \\ x + 5y + 4z = 13 \end{cases}$

14. $\begin{cases} 2x + y - 3z = 1 \\ x + 2y + z = 3 \\ 4x - y - 11z = 5 \end{cases}$

15. $\begin{cases} 2x - 3y + 4z = 0 \\ x - y - 2z = 0 \\ x - 3y + 14z = 0 \end{cases}$

16. $\begin{cases} x + y - z + w = 2 \\ 2x - 3y + 4z + 5w = -4 \\ 4x + 5y - 2z - 3w = 8 \end{cases}$

17. $\begin{cases} 4x - 2y + 3z = 5 \\ 3x + y + 2z = 3 \\ 2x - 3y + 5z = 4 \end{cases}$

18. $\begin{cases} x + 4y + z = 7 \\ 3x - 2y + 2z = 12 \\ x + 2y - z = 2 \end{cases}$

19. $\begin{cases} 3x - 2y - z = 2 \\ 5x + 3y - 4z = 6 \\ 4x - 9y + z = 3 \end{cases}$

20. $\begin{cases} 2x + 3y + 3z = 9 \\ 5x - 2y + 8z = 6 \\ 4x - y + 5z = -1 \end{cases}$

21. $\begin{cases} 5x - 2y + 3z = 6 \\ 6x - 3y + 4z = 10 \\ -4x + 4y - 9z = 4 \end{cases}$

22. $\begin{cases} 3x + 4y - z = 4 \\ 2x - 5y + 3z = -10 \\ 4x + 3y + z = 10 \end{cases}$

23. $\begin{cases} x + 2y - z + w = 1 \\ 2x - y + z - 2w = 0 \\ -x - 3y + 2z - 3w = -5 \end{cases}$

24. $\begin{cases} 2x - 3y - 2z = 1 \\ 3x - 4y - 3z = 2 \\ 4x + y + z = 10 \end{cases}$

25. $\begin{cases} 2x - 3y + 4z = 0 \\ x - y + 3z = 0 \\ 3x - 5y + 5z = 0 \end{cases}$

26. $\begin{cases} 5x - 6y + z = 1 \\ 9x - 8y + 7z = 1 \\ -3x + 4y - z = 5 \end{cases}$

27. $\begin{cases} 8x - 5y + 6z = 0 \\ 3x - 2y - 2z = -10 \\ -11x + 9y - 2z = 16 \end{cases}$

28. $\begin{cases} 2x + y + z = 1 \\ 5x + 3y - 2z = -4 \\ 3x + y + 6z = 8 \end{cases}$

29. $\begin{cases} x - y - z = -1 \\ y - z + w = 0 \\ x + 2z - 2w = 1 \\ 2x + y - w = -1 \end{cases}$

30. $\begin{cases} x + y + z + w = 1 \\ 3y - w = 0 \\ x - z = 1 \\ y - 2z - w = -2 \end{cases}$

31. $\begin{cases} x_1 + x_2 - x_3 + 2x_4 = 1 \\ 2x_1 - x_2 - 3x_3 + 2x_4 = -6 \\ -x_1 + 2x_2 + x_3 + 3x_4 = 2 \\ x_1 - x_2 - 2x_3 + x_4 = 0 \end{cases}$

B

32. $\begin{cases} x + y + 3z - 2w = -2 \\ 2x + 4y - 2z + 4w = 1 \\ -3x - 5y - 4z - w = 4 \\ 4x + 6y - 2z + 5w = 1 \end{cases}$

33. $\begin{cases} 2x_1 - x_2 + 3x_3 + x_4 = 0 \\ 4x_1 - 3x_2 + x_3 - 2x_4 = 1 \\ -3x_1 + 2x_3 - 4x_4 = -2 \\ 6x_1 - 4x_2 - 5x_3 + 3x_4 = -3 \end{cases}$

34. $\begin{cases} x - y - 2z + w = -2 \\ 3x + 2y + 3z + 2w = 1 \\ 2x - 7y - 4z - 3w = 1 \end{cases}$

35. $\begin{cases} 2x - 3y + 5z + 4w = -3 \\ 3x - 2y - 2z - w = 1 \\ 5x + 2y + 6z - 8w = 1 \\ -7x + 4y - 3z + 2w = 2 \end{cases}$

36. $\begin{cases} 3x - 2y + z - w = -1 \\ x + y - 3z + 3w = 3 \\ 2x - 3y + 5z - 4w = 1 \\ 4x + 2y + 4z - w = -4 \end{cases}$

37. $\begin{cases} 2x - 3y + 4z + 2w = 5 \\ -3x + 2y - z + w = 2 \\ 4x - 5y + 2z - 3w = 6 \\ 5x - 4y - z - 6w = 1 \end{cases}$

38. $\begin{cases} x_1 - 2x_2 + x_3 - x_4 = 2 \\ x_1 - x_2 - x_4 = 6 \\ 2x_1 - 2x_2 + x_3 = 9 \\ -x_1 + 5x_2 - 2x_3 + 5x_4 = 6 \end{cases}$

39. $\begin{cases} 2s - t + 3u + v = 0 \\ 3s + 2t + 4u - v = 0 \\ 5s - 2t - 2u - v = 0 \\ -2s - 3t + 7u + 5v = 0 \end{cases}$

40. $\begin{cases} x + y + 2z - w = 1 \\ -x + 2y + 3z - 2w = -3 \\ 2x - y - z + w = 4 \\ x + 2y + z - w = 1 \end{cases}$

41. $\begin{cases} x_1 - 2x_2 + 3x_4 + x_5 = 2 \\ -x_1 + 3x_2 + 2x_3 - 4x_4 + x_5 = -1 \\ x_1 - 2x_2 + x_3 + 3x_4 = 0 \\ x_2 + 2x_3 - 2x_4 + 2x_5 = 7 \\ 2x_1 - 4x_2 + 2x_3 + 6x_4 + x_5 = 5 \end{cases}$

DETERMINANTS

In Section 2 our approach made use of the augmented matrix of a system of equations. In Section 4 we will present an alternative procedure that uses the coefficient matrix and some other matrices associated with it. This procedure works only when the number of unknowns and the number of equations are the same, in which case the coefficient matrix is said to be **square,** since it has the same number of rows as columns. However, it will first be necessary to introduce a new concept—the **determinant** of a matrix. Various symbols are used to designate the determinant of a matrix; we will use the one that is probably the most common, namely, vertical bars on each side of the matrix. For example, the determinant of the matrix

$$\begin{bmatrix} 2 & 3 & 1 \\ 4 & -2 & 5 \\ 1 & 2 & 7 \end{bmatrix}$$

is designated by

$$\begin{vmatrix} 2 & 3 & 1 \\ 4 & -2 & 5 \\ 1 & 2 & 7 \end{vmatrix}$$

The number of rows (or columns) in a square matrix is called the **order** of the matrix. The determinant of a matrix of order 2 is defined as follows: Let

$$\begin{bmatrix} a & b \\ c & d \end{bmatrix}$$

designate an arbitrary matrix of order 2. Then

$$\begin{vmatrix} a & b \\ c & d \end{vmatrix} = ad - bc$$

The right-hand side is suggested by the arrows on the left. The diagonal elements from upper left to lower right, called the **principal diagonal elements,** are multiplied, and from their product is subtracted the product of the other diagonal elements. For example, we have

$$\begin{vmatrix} 2 & 3 \\ 4 & 5 \end{vmatrix} = 10 - 12 = -2$$

and

$$\begin{vmatrix} 3 & -1 \\ 5 & -2 \end{vmatrix} = -6 - (-5) = -6 + 5 = -1$$

The procedure for finding the determinant of a square matrix of order higher than 2 is more complicated and will require several preliminary steps. We will use a general matrix of order 3 to illustrate,

$$\begin{bmatrix} a_1 & b_1 & c_1 \\ a_2 & b_2 & c_2 \\ a_3 & b_3 & c_3 \end{bmatrix}$$

(subscripts are employed to avoid the use of too many different letters), but everything we do can be extended in a straightforward way to higher orders. We refer to the entries in the matrix as **elements** of the matrix. It is important to distinguish between rows (which go across) and columns (which go up and down). For example, consider the matrix

$$\begin{bmatrix} 2 & -1 & 3 \\ 4 & 2 & -7 \\ -1 & 0 & 5 \end{bmatrix}$$

The element in Row 2 and Column 3 is -7, and the element in Row 3 and Column 2 is 0.

If we delete the row and column of a given element in a square matrix, we are left with a submatrix of order 1 less than that of the original matrix. The determinant of this submatrix is called the **minor** of the element in question. For example, let us again consider the matrix

$$\begin{bmatrix} 2 & -1 & 3 \\ 4 & 2 & -7 \\ -1 & 0 & 5 \end{bmatrix}$$

and determine the minor of the element -7 in Row 2 and Column 3. We show the row

$$\begin{bmatrix} 2 & -1 & 3 \\ 4 & 2 & 7 \\ -1 & 0 & 5 \end{bmatrix}$$

and column containing this element as being deleted by dashed lines. The minor of -7 is therefore the determinant

$$\begin{vmatrix} 2 & -1 \\ -1 & 0 \end{vmatrix} = 2(0) - (-1)(-1) = 0 - 1 = -1$$

In a similar way we could find the minor of each of the elements.

If we multiply the minor of the element in the ith row and jth column by $(-1)^{i+j}$, we obtain what is called the **cofactor** of the element. So the cofactor of an element differs from the minor at most in sign. If $i + j$ is even, $(-1)^{i+j} = 1$, so the cofactor equals the minor. If $i + j$ is odd, $(-1)^{i+j} = -1$, so the cofactor is the negative of the minor. From this definition we can devise the following scheme for determining the sign that should be affixed to the minor to give the cofactor:

$$\begin{bmatrix} + & - & + \\ - & + & - \\ + & - & + \end{bmatrix}$$

Starting with a plus sign in the upper left, the signs alternate as we move either horizontally or vertically.

EXAMPLE 14

Find the cofactor of each element in the first row of the matrix.

$$\begin{bmatrix} 2 & 1 & -3 \\ -1 & 4 & 2 \\ 5 & 3 & 6 \end{bmatrix}$$

Solution   To get the minor of the first element, 2, we cross out (mentally or otherwise) the first row and first column:

$$\begin{bmatrix} 2 & 1 & -3 \\ -1 & 4 & 2 \\ 5 & 3 & 6 \end{bmatrix}$$

The minor is

$$\begin{vmatrix} 4 & 2 \\ 3 & 6 \end{vmatrix} = 24 - 6 = 18$$

Since by the scheme of signs given above, we multiply by $+1$ to get the cofactor, it follows that the cofactor of this element also is 18.

Similarly, we determine the minor of the second element in Row 1:

$$\begin{bmatrix} -2 & 1 & -3 \\ -1 & 4 & 2 \\ 5 & 3 & 6 \end{bmatrix}$$

$$\begin{vmatrix} -1 & 2 \\ 5 & 6 \end{vmatrix} = -6 - 10 = -16$$

To get the cofactor, we must multiply by -1, according to the scheme of signs for determining cofactors. So the cofactor of this element is 16.

For the third element in Row 1 we proceed in a similar way:

$$\begin{bmatrix} -2 & 1 & -3 \\ -1 & 4 & 2 \\ 5 & 3 & 6 \end{bmatrix}$$

$$\begin{vmatrix} -1 & 4 \\ 5 & 3 \end{vmatrix} = -3 - 20 = -23$$

The cofactor is obtained by multiplying this minor by $+1$, so the cofactor is -23. ▲

With a little practice you will probably be able to calculate cofactors of elements in a matrix of order 3 mentally. It is a good idea to write the correct factor $+1$ or -1 first, according to the position of the element, and then calculate the minor. To illustrate, consider again the matrix in Example 14. The cofactor of the element -1 in the second row and first column is $-(6 + 9) = -15$; the cofactor of the element 3 in the third row and second column is $-(4 - 3) = -1$; and so on. Note that the cofactor of an element depends on the position of the element but not on the value of the element itself.

We are now ready to state a **procedure for finding the value of the determinant of a square matrix of order 3 or greater.**

1. **Select any row or column of the matrix.**
2. **Multiply each element of that row or column by its cofactor.**
3. **Add the results.**

We will illustrate with the matrix used in Example 13.

EXAMPLE 15 ||||||||||||||||||||||||||||||||||||||||||||||||||||||||||||||||||||||||||||||||||||||||||||||||||||||||||||||||||||||||||||||||||||||||||

Find the determinant of the matrix.

$$\begin{bmatrix} 2 & 1 & -3 \\ -1 & 4 & 2 \\ 5 & 3 & 6 \end{bmatrix}$$

Solution   Let us select Row 1, since we have already found the cofactors of the elements in that row in Example 14, namely, 18, 16, and -23, respectively. Next, we multiply each element in the row by its own cofactor and then add the results. Thus,

$$\begin{vmatrix} 2 & 1 & -3 \\ -1 & 4 & 2 \\ 5 & 3 & 6 \end{vmatrix} = 2(18) + 1(16) + (-3)(-23)$$

$$= 36 + 16 + 69 = 121$$

Note. In Step 1 above we specified that *any* row or column may be selected at the outset. It can be proved (See Problem 26, Exercise Set 3) that the same result will occur regardless of this initial choice, so long as Steps 2 and 3 are carried out for the row or column selected. To illustrate, let us select Column 2 in the example above (we say "expand by Column 2"). The cofactors can be calculated mentally as we move down the column.

$$\begin{vmatrix} 2 & 1 & -3 \\ -1 & 4 & 2 \\ 5 & 3 & 6 \end{vmatrix} = 1[-(-6-10)] + 4[+(12+15)] + 3[-(4-3)]$$

$$= 16 + 4(27) + 3(-1) = 16 + 108 - 3 = 121 \quad \blacktriangle$$

EXAMPLE 16

Find the determinant of the matrix

$$\begin{bmatrix} 2 & 3 & 0 \\ -1 & 2 & -4 \\ -3 & -1 & 5 \end{bmatrix}$$

Solution   It is convenient to select either Row 1 or Column 3. (Why?) Let us expand by Column 3. We can ignore the cofactor of the first element 0, since whatever it is, the result will be 0 after multiplication. So we have

$$\begin{vmatrix} 2 & 3 & 0 \\ -1 & 2 & -4 \\ -3 & -1 & 5 \end{vmatrix} = 0 + (-4)[-(-2+9)] + 5[4+3]$$

$$= (-4)(-7) + 5(7) = 28 + 35 = 63 \quad \blacktriangle$$

While the procedure outlined for finding determinants is applicable to matrices of any order greater than 2, the work involved rapidly becomes excessive as the orders increase. For example, finding the determinant of a matrix of order 4 involves multiplying each of four elements (any row or column) by its respective cofactor, each of which involves evaluating the determinant of an order 3 matrix. It is not hard to visualize how difficult it would be to find the determinant of a matrix of order 5. In these more complicated situations there are admissible preliminary steps that can be taken to introduce zeros into a given row or column without changing the value of the determinant. The more zeros present, the easier the evaluation of the determinant. We will not pursue these methods here, since in the applications of determinants to systems of equations the method to be described is practical only for relatively small systems.

EXERCISE SET 3

A

Evaluate the determinants in Problems 1–3.

1. a. $\begin{vmatrix} 2 & 3 \\ -1 & 4 \end{vmatrix}$ **b.** $\begin{vmatrix} -3 & 5 \\ 12 & 6 \end{vmatrix}$ **c.** $\begin{vmatrix} 4 & -8 \\ -6 & -3 \end{vmatrix}$

2. a. $\begin{vmatrix} 3 & -1 \\ 4 & -2 \end{vmatrix}$ **b.** $\begin{vmatrix} 5 & 6 \\ -8 & 3 \end{vmatrix}$ **c.** $\begin{vmatrix} -10 & 3 \\ 14 & -5 \end{vmatrix}$

3. a. $\begin{vmatrix} 9 & 6 \\ 3 & 2 \end{vmatrix}$ **b.** $\begin{vmatrix} 0 & 5 \\ -1 & 4 \end{vmatrix}$ **c.** $\begin{vmatrix} 8 & 4 \\ -4 & -2 \end{vmatrix}$

Problems 4–8 refer to the matrix

$$\begin{bmatrix} 3 & -2 & 1 \\ -4 & 5 & -3 \\ -6 & -1 & 7 \end{bmatrix}$$

In each case, for the position specified, find:
a. The element in that position
b. The minor of that element
c. The cofactor of that element

4. Row 2, Column 3   **5.** Row 3, Column 2
6. Row 1, Column 3   **7.** Row 3, Column 1
8. Row 3, Column 3
9. Evaluate the determinant of the matrix used in Problems 4–8 by expanding on:
   a. The third column   **b.** The third row

Evaluate the determinants in Problems 10–18.

10. $\begin{vmatrix} 2 & -1 & -3 \\ 4 & 2 & -5 \\ -1 & 3 & 6 \end{vmatrix}$ **11.** $\begin{vmatrix} 3 & 0 & -1 \\ 2 & 5 & 0 \\ 1 & 3 & 1 \end{vmatrix}$

12. $\begin{vmatrix} 1 & 5 & 3 \\ -2 & -4 & 10 \\ 6 & 8 & -3 \end{vmatrix}$ **13.** $\begin{vmatrix} 3 & -1 & -5 \\ 0 & 2 & -4 \\ 5 & 1 & 6 \end{vmatrix}$

14. $\begin{vmatrix} 2 & 3 & -1 \\ -4 & 2 & 3 \\ 5 & -2 & 6 \end{vmatrix}$ **15.** $\begin{vmatrix} 1 & -1 & 4 \\ 3 & 2 & 0 \\ 4 & -1 & 6 \end{vmatrix}$

16. $\begin{vmatrix} 5 & 1 & 2 \\ -2 & 4 & 3 \\ 7 & 2 & -1 \end{vmatrix}$ **17.** $\begin{vmatrix} -2 & 1 & -5 \\ 0 & 3 & -4 \\ 6 & -1 & 2 \end{vmatrix}$

18. $\begin{vmatrix} 4 & -2 & -3 \\ -5 & 3 & -4 \\ 2 & -1 & -2 \end{vmatrix}$

19. Find all values of x satisfying

$$\begin{vmatrix} 1 & x & 0 \\ 2 & -4 & x-1 \\ 1 & -x & 2 \end{vmatrix} = 0$$

20. Find all values of t satisfying

$$\begin{vmatrix} -t & 2 & -1 \\ 0 & t & 4 \\ t+2 & 1 & -2 \end{vmatrix} = 0$$

21. Prove that for any number k

$$\begin{vmatrix} a & b \\ c & d \end{vmatrix} = \begin{vmatrix} a+kb & b \\ c+kd & d \end{vmatrix}$$

and

$$\begin{vmatrix} a & b \\ c & d \end{vmatrix} = \begin{vmatrix} a+kc & b+kd \\ c & d \end{vmatrix}$$

22. It can be proved that the area of a triangle with vertices (x_1, y_1), (x_2, y_2), and (x_3, y_3) is given by

$$A = \frac{1}{2} \begin{vmatrix} x_1 & y_1 & 1 \\ x_2 & y_2 & 1 \\ x_3 & y_3 & 1 \end{vmatrix}$$

Use this to find the area of the triangle with vertices $(2, -3)$, $(-4, 1)$, and $(-3, -5)$.

B

Evaluate the determinants in Problems 23–25.

23. $\begin{vmatrix} 3 & 2 & -1 & 5 \\ 1 & 0 & 2 & 4 \\ -1 & 1 & 3 & 0 \\ 2 & 0 & -1 & 1 \end{vmatrix}$
  24. $\begin{vmatrix} 1 & 0 & -2 & 3 \\ 0 & -2 & 5 & 0 \\ 2 & 3 & -1 & 1 \\ -1 & -2 & 3 & 4 \end{vmatrix}$

25. $\begin{vmatrix} 2 & -1 & 3 & 1 \\ 4 & 2 & 1 & 3 \\ -1 & 1 & 2 & 4 \\ 3 & 5 & -2 & 6 \end{vmatrix}$

26. Verify the fact that the same final result occurs regardless of which row or column is used in the calculation of the determinant of the order 3 matrix

$$\begin{bmatrix} a_1 & b_1 & c_1 \\ a_2 & b_2 & c_2 \\ a_3 & b_3 & c_3 \end{bmatrix}$$

27. Prove that if each element in any row or column

of a square matrix A is multiplied by a constant k, the determinant of the resulting matrix is $k|A|$.

28. It can be proved that if any two rows or any two columns of a square matrix A are interchanged, the determinant of the resulting matrix is $-|A|$. Use this result to show that if any two rows or columns of a square matrix A are identical, then $|A| = 0$.

29. Prove that

$$\begin{vmatrix} x & y & 1 \\ x_1 & y_1 & 1 \\ x_2 & y_2 & 1 \end{vmatrix} = 0$$

is the equation of a line passing through the points (x_1, y_1) and (x_2, y_2).

30. Prove that if all elements above or below the principal diagonal of a square matrix are 0, then the determinant of the matrix equals the product of the elements along the principal diagonal.

31. Solve for x:

$$\begin{vmatrix} x & 0 & 0 & 0 \\ 15 & x-1 & 0 & 0 \\ -5 & 24 & x+2 & 0 \\ 24 & -31 & 57 & 2x+3 \end{vmatrix} = 0$$

CRAMER'S RULE

To see how determinants can play a role in solving systems of equations, let us consider the general case of two equations in two unknowns:

$$\begin{cases} a_1x + b_1y = c_1 \\ a_2x + b_2y = c_2 \end{cases} \tag{3}$$

We eliminate y by multiplying the top equation by b_2 and the bottom by b_1 and then subtracting:

$$\begin{array}{l} a_1b_2x + b_1b_2y = c_1b_2 \\ a_2b_1x + b_1b_2y = c_2b_1 \\ \hline (a_1b_2 - a_2b_1)x = c_1b_2 - c_2b_1 \end{array} \tag{4}$$

Assume for the moment that the coefficient of x is not 0. Then we can solve for x to get

$$x = \frac{c_1b_2 - c_2b_1}{a_1b_2 - a_2b_1}$$

This can be written as a quotient of determinants:

$$x = \frac{\begin{vmatrix} c_1 & b_1 \\ c_2 & b_2 \end{vmatrix}}{\begin{vmatrix} a_1 & b_1 \\ a_2 & b_2 \end{vmatrix}} \tag{5}$$

Returning to the original system of equations (3), we multiply the top equation by a_2 and the bottom by a_1, and then subtract, in order to eliminate x:

$$a_1 a_2 x + a_2 b_1 y = a_2 c_1$$
$$a_1 a_2 x + a_1 b_2 y = a_1 c_2$$
$$\overline{(a_1 b_2 - a_2 b_1) y = a_1 c_2 - a_2 c_1} \tag{6}$$

(We subtracted the top equation from the bottom one.) Again, assuming the coefficient of y is nonzero, we get

$$y = \frac{a_1 c_2 - a_2 c_1}{a_1 b_2 - a_2 b_1}$$

which can be written as

$$y = \frac{\begin{vmatrix} a_1 & c_1 \\ a_2 & c_2 \end{vmatrix}}{\begin{vmatrix} a_1 & b_1 \\ a_2 & b_2 \end{vmatrix}} \tag{7}$$

Let us now examine the situation when the coefficient of x in equation (4) or of y in (6) is 0:

$$a_1 b_2 - a_2 b_1 = 0 \tag{8}$$

We will suppose that neither b_1 nor b_2 is 0, because if either was 0, both would have to be (why?); thus, the lines would both be vertical, and so either parallel or coincident. We can rewrite equation (8) in the form

$$\frac{a_1}{b_1} = \frac{a_2}{b_2}$$

The slopes of the lines are $-a_1/b_1$ and $-a_2/b_2$, respectively; we conclude that the lines are parallel or coincident. If they are parallel, no solution exists; the system is inconsistent. If they are coincident, the system is dependent. We conclude that a *unique* solution to the system exists if and only if $a_1 b_2 - b_2 a_1 \neq 0$, that is,

$$\begin{vmatrix} a_1 & b_1 \\ a_2 & b_2 \end{vmatrix} \neq 0$$

EXAMPLE 17 ||||||||||||||||||||||||||||||||||||||||||||||||||||||||||||||||||||||||||||||||||||||||

Solve the system below by using equations (5) and (7).

$$\begin{cases} 2x - 3y = 4 \\ 7x + 5y = 6 \end{cases}$$

Solution   We have immediately

$$x = \frac{\begin{vmatrix} 4 & -3 \\ 6 & 5 \end{vmatrix}}{\begin{vmatrix} 2 & -3 \\ 7 & 5 \end{vmatrix}} = \frac{20 + 18}{10 + 21} = \frac{38}{31}$$

$$y = \frac{\begin{vmatrix} 2 & 4 \\ 7 & 6 \end{vmatrix}}{\begin{vmatrix} 2 & -3 \\ 7 & 5 \end{vmatrix}} = \frac{12 - 28}{31} = -\frac{16}{31}$$

So the solution is $(\frac{38}{31}, -\frac{16}{31})$.   ▲

It is useful to analyze the results (5) and (7):

$$x = \frac{\begin{vmatrix} c_1 & b_1 \\ c_2 & b_2 \end{vmatrix}}{\begin{vmatrix} a_1 & b_1 \\ a_2 & b_2 \end{vmatrix}} \qquad y = \frac{\begin{vmatrix} a_1 & c_1 \\ a_2 & c_2 \end{vmatrix}}{\begin{vmatrix} a_1 & b_1 \\ a_2 & b_2 \end{vmatrix}}$$

The denominators are the same, and this common denominator is seen to be the determinant of the coefficient matrix of the system. In each case the numerator differs from the denominator only in that the column of coefficients of the unknown in question is replaced by the column of constants. It is convenient to introduce notation for the matrices involved. Let D denote the coefficient matrix of the system, and let D_x and D_y be matrices that are obtained from D by replacing the column of coefficients of x and of y, respectively, by the column of constants. Then we have

$$x = \frac{|D_x|}{|D|} \qquad y = \frac{|D_y|}{|D|}$$

This same pattern holds true for higher-order systems. For example, the solution to the system of three equations in three unknowns,

$$\begin{cases} a_1 x + b_1 y + c_1 z = d_1 \\ a_2 x + b_2 y + c_2 z = d_2 \\ a_3 x + b_3 y + c_3 z = d_3 \end{cases}$$

is given by

$$x = \frac{|D_x|}{|D|} \qquad y = \frac{|D_y|}{|D|} \qquad z = \frac{|D_z|}{|D|} \qquad (D \neq 0)$$

where D_x, D_y, and D_z are obtained from the coefficient matrix D by replacing the column of coefficients of x, of y, and of z, respectively, by the column of constants appearing on the right-hand side.

As an illustration of the meanings of the matrices D, D_x, D_y, and D_z, consider the system

$$\begin{cases} x + y + z = 4 \\ 2x + y - z = 0 \\ 3x - 4y - 3z = 1 \end{cases}$$

Here,

$$D = \begin{bmatrix} 1 & 1 & 1 \\ 2 & 1 & -1 \\ 3 & -4 & -3 \end{bmatrix}$$

and so

$$D_x = \begin{bmatrix} 4 & 1 & 1 \\ 0 & 1 & -1 \\ 1 & -4 & -3 \end{bmatrix} \qquad D_y = \begin{bmatrix} 1 & 4 & 1 \\ 2 & 0 & -1 \\ 3 & 1 & -3 \end{bmatrix} \qquad D_z = \begin{bmatrix} 1 & 1 & 4 \\ 2 & 1 & 0 \\ 3 & -4 & 1 \end{bmatrix}$$

Column of constants

The general result is known as **Cramer's rule,** which we state as follows:

Cramer's Rule

Let D denote the coefficient matrix in a system of n linear equations in the n unknowns $x_1, x_2, \ldots, x_n$, where each equation is written in the form

$$a_1 x_1 + a_2 x_2 + \cdots + a_n x_n = b$$

Let D_{x_i} be the matrix obtained from D by replacing the column of coefficients of x_i by the column of constants appearing on the right-hand side of the system. Then if $D \neq 0$,

$$x_1 = \frac{|D_{x_1}|}{|D|}, \quad x_2 = \frac{|D_{x_2}|}{|D|}, \quad \cdots \quad , \quad x_n = \frac{|D_{x_n}|}{|D|}$$

Note. In applying Cramer's rule it is important that the terms involving the unknowns in each equation be written in the same order, and that the constant terms all appear on the right. For example, before applying Cramer's rule to the system

$$\begin{cases} 2y - 3x + 4z - 7 = 0 \\ x + 2z - y = 8 \\ 3z - 4y + 2x + 5 = 0 \end{cases}$$

we would rewrite it as

$$\begin{cases} -3x + 2y + 4z = 7 \\ x - y + 2z = 8 \\ 2x - 4y + 3z = -5 \end{cases}$$

EXAMPLE 18

Solve the system below by Cramer's rule.

$$\begin{cases} x + y + z = 4 \\ 2x + y - z = 0 \\ 3x - 4y - 3z = 1 \end{cases}$$

Solution

$$x = \frac{|D_x|}{|D|} = \frac{\begin{vmatrix} 4 & 1 & 1 \\ 0 & 1 & -1 \\ 1 & -4 & -3 \end{vmatrix}}{\begin{vmatrix} 1 & 1 & 1 \\ 2 & 1 & -1 \\ 3 & -4 & -3 \end{vmatrix}} = \frac{4(-7) + 1(-2)}{1(-7) - 2(1) + 3(-2)} = \frac{-30}{-15} = 2$$

We expanded by the first column in both cases.

$$y = \frac{|D_y|}{|D|} = \frac{\begin{vmatrix} 1 & 4 & 1 \\ 2 & 0 & -1 \\ 3 & 1 & -3 \end{vmatrix}}{-15} = \frac{-4(-3) - (-3)}{-15} = \frac{15}{-15} = -1$$

We expanded the numerator by the second column.

$$z = \frac{|D_z|}{|D|} = \frac{\begin{vmatrix} 1 & 1 & 4 \\ 2 & 1 & 0 \\ 3 & -4 & 1 \end{vmatrix}}{-15} = \frac{4(-11) + 1(-1)}{-15} = \frac{-45}{-15} = 3$$

Expansion was by the third column. So the solution is $(2, -1, 3)$.   ▲

Remark. Cramer's rule works only when the number of equations is the same as the number of unknowns. Furthermore, it is usually impractical for systems with more than three equations. It works well for two equations and two unknowns, and is often the easiest method in this case. Depending on how adept one becomes in expanding order 3 determinants, it can be an efficient way to solve a system of three equations in three unknowns.

EXERCISE SET 4

A

Solve each of the following by Cramer's rule.

1. $\begin{cases} 2x - 3y = 7 \\ x + 4y = 3 \end{cases}$

2. $\begin{cases} 5x + 4y = 9 \\ 7x - 3y = 6 \end{cases}$

3. $\begin{cases} 3x + y = -2 \\ 2x + 5y = 4 \end{cases}$

4. $\begin{cases} x - 2y = 4 \\ 3x - 4y = 5 \end{cases}$

5. $\begin{cases} 6x - 5y = 4 \\ 8x + 7y = -3 \end{cases}$

6. $\begin{cases} 3y - 4x + 5 = 0 \\ 6x = 2y - 3 \end{cases}$

7. $\begin{cases} 7x - 8y = 10 \\ 5x - 4y = 8 \end{cases}$

8. $\begin{cases} 5x + 6y = -3 \\ 2x - 3y = 4 \end{cases}$

9. $\begin{cases} x + 2y - 3 = 0 \\ 3x - y + 4 = 0 \end{cases}$

10. $\begin{cases} 3x + 2y = 4 \\ 3y + 5x = 2 \end{cases}$

11. $\begin{cases} 3x - 5y = 2 \\ 2x - 3y = 4 \end{cases}$

12. $\begin{cases} 9y + 7x = 10 \\ 2x + 3y - 8 = 0 \end{cases}$

13. $\begin{cases} x - y + 2z = 0 \\ 2x - 3z = 1 \\ y + z = 3 \end{cases}$

14. $\begin{cases} x + 2y - 3z = 1 \\ 2x - y + z = 2 \\ x + y + 4z = 0 \end{cases}$

15. $\begin{cases} x + y = 1 \\ 2y - z = -1 \\ x + z = 0 \end{cases}$

16. $\begin{cases} 3x + y + z = 3 \\ x - z = -5 \\ 2y - z = 0 \end{cases}$

17. $\begin{cases} 3x + 5y = 2 \\ x + y - z = -1 \\ x - 3y - z = 1 \end{cases}$

18. $\begin{cases} x + y - z = 4 \\ 2x - y + z = 0 \\ x - 2y - 3z = 1 \end{cases}$

19. $\begin{cases} 3x - y + z = 1 \\ x - z = 2 \\ 2y + 3z = 4 \end{cases}$

20. $\begin{cases} x + y + z = 1 \\ 3x + 2y - 3z = 0 \\ y + 3z = 0 \end{cases}$

B

21. $\begin{cases} (a - 1)x + 2ay = 3 \\ (a + 1)x + (2a - 1)y = 2 \end{cases}$
Solve for x and y.

22. $\begin{cases} x^2u - 3xv = 1 \\ 2xu + 4v = 2 \end{cases}$
Solve for u and v.

23. $\begin{cases} u - 2xv = -1 \\ xu + v = x \end{cases}$
Solve for u and v.

24. $\begin{cases} ax - (a + 1)y = 2 \\ (a - 1)x - ay = 1 \end{cases}$
Solve for x and y.

25. $\begin{cases} 2m^2s + (m + 2)t = 6 \\ 4(m - 1)s + 2t = 3 \end{cases}$
Solve for s and t.

26. $\begin{cases} 3x - 4y + z = 2 \\ x + 2y - z = 3 \\ 4x - 3y - 2z = 4 \end{cases}$

27. $\begin{cases} 3x - 2y + 4z = 5 \\ 2x + 3y - 2z = 6 \\ 4x - 5y + 3z = -5 \end{cases}$

28. $\begin{cases} x + y + z + w = 1 \\ 3y - w = 0 \\ x - z = 1 \\ y - 2z - w = 2 \end{cases}$

THE ALGEBRA OF MATRICES

A matrix with **_m_ rows** and **_n_ columns** is said to be an **_m_ × _n_** (read "*m* by *n*") matrix, and the **size** (or **dimension**) of the matrix is said to be _m_ × _n_. In this section we will see how the operations of addition and multiplication are defined for matrices.

Two matrices are said to be **equal** provided they are of the same size and corresponding elements are equal. For example,

$$\begin{bmatrix} a & b \\ c & d \end{bmatrix} = \begin{bmatrix} 2 & 3 \\ 5 & 7 \end{bmatrix}$$

if and only if $a = 2$, $b = 3$, $c = 5$, and $d = 7$.

To **add** two matrices of the same size, we add their corresponding elements. For example,

$$\begin{bmatrix} 2 & -1 & 3 \\ 4 & 2 & -5 \end{bmatrix} + \begin{bmatrix} 3 & 7 & -8 \\ 5 & -2 & 4 \end{bmatrix} = \begin{bmatrix} 5 & 6 & -5 \\ 9 & 0 & -1 \end{bmatrix}$$

Addition is not defined for matrices of different sizes.

A **zero matrix,** denoted by **0,** is one in which all elements are 0. There are zero matrices of all sizes. For any matrix A, if **0** is a zero matrix of the same size as A, then $A + \mathbf{0} = A$. Thus, **0** is the **additive identity.**

The **additive inverse** of a matrix A, denoted by $-A$, is the matrix whose elements are the negatives of the corresponding elements of A. For example,

$$-\begin{bmatrix} 2 & 3 \\ 4 & -2 \\ -5 & 6 \end{bmatrix} = \begin{bmatrix} -2 & -3 \\ -4 & 2 \\ 5 & -6 \end{bmatrix}$$

If A and B are matrices of the same size, then the **difference** between A and B is defined by

$$A - B = A + (-B)$$

It follows that subtraction is carried out term-by-term, as in

$$\begin{bmatrix} 4 & 2 \\ 5 & 6 \end{bmatrix} - \begin{bmatrix} 3 & 1 \\ 7 & 2 \end{bmatrix} = \begin{bmatrix} 1 & 1 \\ -2 & 4 \end{bmatrix}$$

The following properties can now be proven:

$$A + B = B + A$$
$$A + (B + C) = (A + B) + C$$
$$A + (-A) = 0$$

In each case the matrices are understood to be of the same size. For matrices of the same size, then, the commutative and associative properties hold true for addition, and each matrix has an additive inverse.

Multiplication of matrices is more complicated. First we define multiplication of a matrix by a real number. This is referred to as **scalar multiplication,** and the real number is called a **scalar.** If A is a matrix and c is a scalar, then cA is defined as the matrix obtained by multiplying every element of A by c. For example,

$$2\begin{bmatrix} 3 & 1 & -4 \\ 2 & 3 & 6 \end{bmatrix} = \begin{bmatrix} 6 & 2 & -8 \\ 4 & 6 & 12 \end{bmatrix}$$

Now let A and B denote two matrices. In order to define the product AB, we require that the number of *columns in A* equal the number of *rows in B*. Before stating the general definition of the product AB, let us consider a specific example. Suppose

$$A = \begin{bmatrix} 2 & 1 & 4 \\ 3 & -2 & 0 \end{bmatrix} \quad \text{and} \quad B = \begin{bmatrix} 5 & 4 & 3 & -2 \\ 6 & -1 & 2 & 3 \\ 2 & 3 & -2 & 4 \end{bmatrix}$$

Then A is **2** × **3** and B is **3** × 4. So the requirement that the number of columns in A equals the number of rows in B is met. To find the elements of the product AB we describe first what is known as the **inner product** of a given row of A with a given column of B. Take, for example, the first row of A and the second column of B:

$$\boxed{\begin{array}{ccc} 2 & 1 & 4 \end{array}} \quad \boxed{\begin{array}{c} 4 \\ -1 \\ 3 \end{array}} = 2(4) + 1(-1) + 4(3) = 8 - 1 + 12 = 19$$

Observe that we multiplied corresponding elements and then added. This is what we mean by inner product. Now to obtain the element in the ith row and the jth column of AB, we find the inner product of the ith row of A and the jth column of B. In our example, we have found that 19 is the element in the first row and second column of AB. We now calculate the other elements:

$$\begin{bmatrix} 2 & 1 & 4 \\ 3 & -2 & 0 \end{bmatrix}\begin{bmatrix} 5 & 4 & 3 & -2 \\ 6 & -1 & 2 & 3 \\ 2 & 3 & -2 & 4 \end{bmatrix} = \begin{bmatrix} 24 & 19 & 0 & 15 \\ 3 & 14 & 5 & -12 \end{bmatrix}$$

We have shaded the first row of A, the second column of B, and their inner product 19, which therefore appears in the first row and second column of AB. The other elements are obtained in a similar way. (You should check these.) Notice that the size of the product matrix is 2 × 4, the number of rows being the same as the number of rows in A, and the number of columns being the same as the number of columns in B.

We now generalize these concepts.

DEFINITION 1   Let A be an $m \times n$ matrix and let B be an $n \times p$ matrix, and suppose the ith row of A and the jth column of B are as shown:

$$
\begin{array}{cc}
A \\
i\text{th row} \rightarrow
\begin{bmatrix}
& \cdots\cdots\cdots \\
a_1 & a_2 & a_3 & \cdots & a_n \\
& \cdots\cdots\cdots
\end{bmatrix}
&
\overset{\displaystyle B}{
\begin{bmatrix}
\cdot & b_1 & \cdot \\
\cdot & b_2 & \cdot \\
\cdot & b_3 & \cdot \\
\vdots & \vdots & \vdots \\
\cdot & b_n & \cdot
\end{bmatrix}}
\\
& \underset{j\text{th column}}{\uparrow}
\end{array}
$$

The **inner product** of this row and this column is the number

$$a_1 b_1 + a_2 b_2 + a_3 b_3 + \cdots + a_n b_n$$

DEFINITION 2   If A and B are $m \times n$ and $n \times p$ matrices, respectively, then the **product** AB is the $m \times p$ matrix whose element in the ith row and jth column is the inner product of the ith row of A and the jth column of B.

EXAMPLE 19

$$
\begin{bmatrix}
2 & -1 \\
3 & 4 \\
5 & -3
\end{bmatrix}
\begin{bmatrix}
-2 & 3 & 6 \\
4 & 1 & -5
\end{bmatrix}
=
\begin{bmatrix}
-8 & 5 & 17 \\
10 & 13 & -2 \\
-22 & 12 & 45
\end{bmatrix}
\qquad \blacktriangle
$$

In general, $AB \neq BA$. In fact, BA may not be defined even if AB is. For example, if

$$
A = \begin{bmatrix} 3 & 1 \\ 4 & 2 \end{bmatrix}
\qquad \text{and} \qquad
B = \begin{bmatrix} 1 & 5 & 4 \\ 4 & 3 & 6 \end{bmatrix}
$$

then the product AB is defined, but BA is not. Even when both AB and BA are defined, they may not be equal. For example, let

$$
A = \begin{bmatrix} 3 & 4 \\ -2 & 5 \end{bmatrix}
\qquad \text{and} \qquad
B = \begin{bmatrix} -1 & 2 \\ 3 & -4 \end{bmatrix}
$$

Then

$$
AB = \begin{bmatrix} 3 & 4 \\ -2 & 5 \end{bmatrix}
\begin{bmatrix} -1 & 2 \\ 3 & -4 \end{bmatrix}
= \begin{bmatrix} 9 & -10 \\ 17 & -24 \end{bmatrix}
$$

and

$$
BA = \begin{bmatrix} -1 & 2 \\ 3 & -4 \end{bmatrix}
\begin{bmatrix} 3 & 4 \\ -2 & 5 \end{bmatrix}
= \begin{bmatrix} -7 & 6 \\ 17 & -8 \end{bmatrix}
$$

So $AB \neq BA$. Matrix multiplication is therefore *noncommutative*. The following properties, however, can be shown to hold (we assume the sizes are such that all products are defined):

$$A(BC) = (AB)C \qquad \text{Associative property}$$
$$A(B + C) = AB + AC \qquad \text{Left distributive property}$$
$$(B + C)A = BA + CA \qquad \text{Right distributive property}$$

The left and right distributive properties normally yield different results because of noncommutativity.

A square matrix with 1's along the major diagonal and 0's elsewhere is called an **identity matrix** and is denoted by I. Identity matrices of sizes 2×2, 3×3, and 4×4 are shown below. Each of these is designated by the same symbol, I.

$$\begin{bmatrix} 1 & 0 \\ 0 & 1 \end{bmatrix} \qquad \begin{bmatrix} 1 & 0 & 0 \\ 0 & 1 & 0 \\ 0 & 0 & 1 \end{bmatrix} \qquad \begin{bmatrix} 1 & 0 & 0 & 0 \\ 0 & 1 & 0 & 0 \\ 0 & 0 & 1 & 0 \\ 0 & 0 & 0 & 1 \end{bmatrix}$$

The appropriate size will be understood by the operations to be performed that make use of I.

The most important property of identity matrices is that for any matrix A,

$$AI = A \quad \text{and} \quad IA = A$$

So I plays a role analogous to the multiplicative identity 1 of real numbers. For example, let

$$A = \begin{bmatrix} 2 & -1 & 4 \\ 0 & 5 & 3 \end{bmatrix}$$

Then

$$AI = \begin{bmatrix} 2 & -1 & 4 \\ 0 & 5 & 3 \end{bmatrix} \begin{bmatrix} 1 & 0 & 0 \\ 0 & 1 & 0 \\ 0 & 0 & 1 \end{bmatrix} = \begin{bmatrix} 2 & -1 & 4 \\ 0 & 5 & 3 \end{bmatrix} = A$$

and

$$IA = \begin{bmatrix} 1 & 0 \\ 0 & 1 \end{bmatrix} \begin{bmatrix} 2 & -1 & 4 \\ 0 & 5 & 3 \end{bmatrix} = \begin{bmatrix} 2 & -1 & 4 \\ 0 & 5 & 3 \end{bmatrix} = A$$

Note that since A is 2×3 in this case, we had to use the 3×3 identity for right multiplication AI, and the 2×2 identity for left multiplication IA, in order for these products to be defined.

EXERCISE SET 5

A

In Problems 1–14 perform the indicated operations.

1. $\begin{bmatrix} 3 & 4 & -2 \\ 7 & 8 & 10 \end{bmatrix} + \begin{bmatrix} -1 & 2 & 5 \\ 3 & -4 & 6 \end{bmatrix}$

2. $\begin{bmatrix} 2 & 0 \\ 7 & 3 \\ -1 & 4 \end{bmatrix} + \begin{bmatrix} 5 & -3 \\ 2 & 6 \\ 3 & -7 \end{bmatrix}$

3. $\begin{bmatrix} 1 & 4 \\ -2 & 5 \end{bmatrix} - \begin{bmatrix} 3 & 2 \\ 4 & -6 \end{bmatrix}$

4. $\begin{bmatrix} 7 & -5 & 3 \\ -6 & 4 & 2 \\ 0 & 3 & -8 \end{bmatrix} - \begin{bmatrix} 5 & -8 & 1 \\ 2 & -6 & 3 \\ 4 & 0 & -2 \end{bmatrix}$

5. $\begin{bmatrix} 2 & 5 \\ 3 & -4 \end{bmatrix} + \begin{bmatrix} -1 & 2 \\ 5 & 6 \end{bmatrix} - \begin{bmatrix} 3 & -7 \\ 8 & 2 \end{bmatrix}$

6. $\begin{bmatrix} 3 & 1 & 5 \\ 2 & -3 & -6 \end{bmatrix} - \begin{bmatrix} 8 & -4 & 3 \\ -2 & 5 & -6 \end{bmatrix}$
   $+ \begin{bmatrix} 4 & 6 & 9 \\ -5 & 7 & -2 \end{bmatrix}$

7. $2A + 3B$, where $A = \begin{bmatrix} 2 & 1 \\ -1 & 4 \\ 3 & -5 \end{bmatrix}$ and
   $B = \begin{bmatrix} -3 & 2 \\ 4 & -3 \\ 7 & 8 \end{bmatrix}$

8. $4A - 5B$, where A and B are as given in Problem 7.

9. $\begin{bmatrix} 2 & 3 \\ -1 & 4 \end{bmatrix}\begin{bmatrix} 3 & -2 \\ 5 & 6 \end{bmatrix}$

10. $[1 \quad 2 \quad -4]\begin{bmatrix} 3 & 4 \\ -2 & 6 \\ 5 & -1 \end{bmatrix}$

11. $\begin{bmatrix} 2 & 1 \\ 5 & 7 \end{bmatrix}\begin{bmatrix} 4 & -2 & 5 \\ -1 & 3 & 0 \end{bmatrix}$

12. $\begin{bmatrix} 1 & 3 & -2 \\ 0 & 4 & 5 \\ -2 & 1 & 3 \end{bmatrix}\begin{bmatrix} 4 \\ 3 \\ 2 \end{bmatrix}$

13. $\begin{bmatrix} 3 & 1 \\ 2 & -4 \\ 5 & 6 \end{bmatrix}\begin{bmatrix} 1 & 3 & -4 & 2 \\ -2 & 0 & 5 & 1 \end{bmatrix}$

14. $\begin{bmatrix} 1 & 3 & -2 \\ 4 & 0 & 1 \\ -2 & 5 & 3 \end{bmatrix}\begin{bmatrix} 1 & 2 & 0 & 5 \\ -2 & 3 & 1 & 4 \\ 3 & 0 & -5 & 6 \end{bmatrix}$

In Problems 15 and 16 solve for x and y by performing the multiplications and equating elements.

15. $\begin{bmatrix} 3 & 7 \\ 2 & -1 \end{bmatrix}\begin{bmatrix} x \\ y \end{bmatrix} = \begin{bmatrix} 1 \\ 12 \end{bmatrix}$

16. $\begin{bmatrix} 3 & 7 \\ 5 & -2 \end{bmatrix}\begin{bmatrix} x \\ y \end{bmatrix} = \begin{bmatrix} 11 \\ -9 \end{bmatrix}$

17. Perform the multiplication

$$\begin{bmatrix} 2 & -3 & 1 & -1 \\ 1 & 1 & -2 & 1 \end{bmatrix}\begin{bmatrix} 1 & 2 \\ -1 & 1 \\ 5 & 4 \\ 10 & 5 \end{bmatrix}$$

What can you conclude in general about matrices A and B if $AB = 0$?

18. Let

$$A = \begin{bmatrix} 2 & 3 \\ -1 & 4 \end{bmatrix} \qquad B = \begin{bmatrix} 1 & 0 & -2 \\ 3 & -1 & 5 \end{bmatrix}$$

$$C = \begin{bmatrix} 1 & -4 \\ -2 & 3 \\ 5 & -1 \end{bmatrix}$$

Determine which of the following are defined. For those that are defined, perform the indicated multiplications.

a. ABC b. ACB c. BAC
d. BCA e. CAB f. CBA

In Problems 19 and 20 show that $AB = I$ and $BA = I$.

19. $A = \begin{bmatrix} 2 & -2 & -3 \\ -2 & 3 & 4 \\ -3 & 4 & 6 \end{bmatrix}$, $B = \begin{bmatrix} 2 & 0 & 1 \\ 0 & 3 & -2 \\ 1 & -2 & 2 \end{bmatrix}$

20. $A = \begin{bmatrix} 0 & 1 & 2 \\ 1 & -2 & -4 \\ 2 & -2 & -5 \end{bmatrix}$, $B = \begin{bmatrix} 2 & 1 & 0 \\ -3 & -4 & 2 \\ 2 & 2 & -1 \end{bmatrix}$

21. Let

$$A = \begin{bmatrix} 5 & -2 \\ 3 & -1 \end{bmatrix} \quad \text{and} \quad B = \begin{bmatrix} a & b \\ c & d \end{bmatrix}$$

By equating entries in like positions on the left and right, solve $AB = I$ for a, b, c, and d. Then verify that $BA = I$.

22. Repeat Problem 21 with

$$A = \begin{bmatrix} -3 & 4 \\ 2 & -2 \end{bmatrix}$$

B

23. Let

$$A = \begin{bmatrix} a_1 & a_2 \\ a_3 & a_4 \end{bmatrix} \qquad B = \begin{bmatrix} b_1 & b_2 \\ b_3 & b_4 \end{bmatrix}$$

$$C = \begin{bmatrix} c_1 & c_2 \\ c_3 & c_4 \end{bmatrix}$$

Prove the following:
a. $A + B = B + A$
b. $A + (B + C) = (A + B) + C$
c. $A(BC) = (AB)C$ **d.** $A(B + C) = AB + AC$

24. When AA is defined, it is customary to denote this by A^2. Prove that A^2 exists if and only if A is a square matrix.

25. For a square matrix A, we write $A^2 = AA$, $A^3 = AAA$, and so on. If

$$A = \begin{bmatrix} 1 & 1 \\ 1 & 1 \end{bmatrix}$$

find A^2, A^3, and A^4. Generalize this to A^n.

26. Use 2×2 matrices to prove that in general $(A + B)^2 \neq A^2 + 2AB + B^2$. Find a correct formula for $(A + B)^2$.

27. Use 2×2 matrices to prove that in general $(A + B)(A - B) \neq A^2 - B^2$. Find a correct formula for $(A + B)(A - B)$.

28. Let

$$A = \begin{bmatrix} 2 & -1 & 3 \\ 4 & 2 & -2 \\ -3 & 5 & 4 \end{bmatrix} \qquad X = \begin{bmatrix} x \\ y \\ z \end{bmatrix}$$

$$B = \begin{bmatrix} 13 \\ -6 \\ -1 \end{bmatrix}$$

Solve the equation $AX = B$ for X.

29. The following represents a system of m linear equations in n unknowns:

$$\begin{cases} a_{11}x_1 + a_{12}x_2 + a_{13}x_3 + \cdots + a_{1n}x_n = b_1 \\ a_{21}x_1 + a_{22}x_2 + a_{23}x_3 + \cdots + a_{2n}x_n = b_2 \\ a_{31}x_1 + a_{32}x_2 + a_{33}x_3 + \cdots + a_{3n}x_n = b_3 \\ \hspace{2cm} -------------------- \\ a_{m1}x_1 + a_{m2}x_2 + a_{m3}x_3 + \cdots + a_{mn}x_n = b_n \end{cases}$$

Find matrices A, B, and X for which the matrix equation $AX = B$ is equivalent to this system.

30. Let

$$A = \begin{bmatrix} 0 & 1 & 1 \\ 1 & 0 & 1 \\ 1 & 1 & 0 \end{bmatrix} \qquad I = \begin{bmatrix} 1 & 0 & 0 \\ 0 & 1 & 0 \\ 0 & 0 & 1 \end{bmatrix}$$

Solve the equation $|A - \lambda I| = 0$ for λ, where λ is a scalar.

INVERSES OF MATRICES

If A is a square matrix, there may be a matrix B such that $AB = BA = I$. In this case matrix B is called the **multiplicative inverse** (or simply the inverse) of A and is denoted by A^{-1}. If B is the inverse of A, then A is also the inverse of B, so that $(A^{-1})^{-1} = A$. Matrices that have inverses are said to be **nonsingular;** those that do not are called **singular.** We will give two methods for calculating inverses of nonsingular matrices. Both methods also provide means of identifying singular matrices.

Method I. If a sequence of elementary row operations exists that transforms A into I, then A is nonsingular and A^{-1} is the matrix obtained by applying this sequence of elementary row operations to I.

We illustrate this method below, where we have written I to the right of A. Thus, as an elementary row operation is applied to A, it is also applied to I. Our goal will be reached when the left-hand matrix is I. The right-hand matrix will then be A^{-1}. If it is impossible to obtain I on the left, then A is singular.

$$
\begin{array}{c} A \qquad\qquad I \\ \left[\begin{array}{ccc|ccc} 2 & -1 & 5 & 1 & 0 & 0 \\ 0 & 2 & 4 & 0 & 1 & 0 \\ 1 & -2 & 0 & 0 & 0 & 1 \end{array}\right] \end{array}
$$

$$
\left[\begin{array}{ccc|ccc} 1 & -2 & 0 & 0 & 0 & 1 \\ 0 & 2 & 4 & 0 & 1 & 0 \\ 2 & -1 & 5 & 1 & 0 & 0 \end{array}\right]
$$
Row 3 interchanged with Row 1

Row 1 interchanged with Row 3

$$
\left[\begin{array}{ccc|ccc} 1 & -2 & 0 & 0 & 0 & 1 \\ 0 & 2 & 4 & 0 & 1 & 0 \\ 0 & 3 & 5 & 1 & 0 & -2 \end{array}\right]
$$
Row $3 + (-2) \cdot$ Row 1

$$
\left[\begin{array}{ccc|ccc} 1 & -2 & 0 & 0 & 0 & 1 \\ 0 & 1 & 2 & 0 & \frac{1}{2} & 0 \\ 0 & 3 & 5 & 1 & 0 & -2 \end{array}\right]
$$
(Row 2) $\cdot (\frac{1}{2})$

$$
\left[\begin{array}{ccc|ccc} 1 & 0 & 4 & 0 & 1 & 1 \\ 0 & 1 & 2 & 0 & \frac{1}{2} & 0 \\ 0 & 0 & -1 & 1 & -\frac{3}{2} & -2 \end{array}\right]
$$
Row $1 + 2 \cdot$ Row 2

Row $3 + (-3) \cdot$ Row 2

$$
\left[\begin{array}{ccc|ccc} 1 & 0 & 0 & 4 & -5 & -7 \\ 0 & 1 & 0 & 2 & -\frac{5}{2} & -4 \\ 0 & 0 & 1 & -1 & \frac{3}{2} & 2 \end{array}\right]
$$
Row $1 + 4 \cdot$ Row 3

Row $2 + 2 \cdot$ Row 3

(Row 3) $\cdot (-1)$

Therefore A is nonsingular and

$$
A^{-1} = \left[\begin{array}{ccc} 4 & -5 & -7 \\ 2 & -\frac{5}{2} & -4 \\ -1 & \frac{3}{2} & 2 \end{array}\right]
$$

You should check to see that $AA^{-1} = A^{-1}A = I$.

Method II. First find the matrix C obtained by replacing each element of A by its cofactor. Then interchange the rows of C with the corresponding columns. The result of this interchange of rows and columns is called the **transpose** of C and is denoted by C^T. Finally, divide C^T by the determinant of A, provided this determinant is nonzero. The result is A^{-1}. Briefly,

$$A^{-1} = \frac{C^T}{|A|}$$

If $|A| = 0$, then A is singular.

Let us apply this method to the same matrix A used in illustrating method I:

$$A = \begin{bmatrix} 2 & -1 & 5 \\ 0 & 2 & 4 \\ 1 & -2 & 0 \end{bmatrix}$$

Replace each element of A by its cofactor, obtaining C:

$$C = \begin{bmatrix} 8 & 4 & -2 \\ -10 & -5 & 3 \\ -14 & -8 & 4 \end{bmatrix}$$

Transpose rows and columns:

$$C^T = \begin{bmatrix} 8 & -10 & -14 \\ 4 & -5 & -8 \\ -2 & 3 & 4 \end{bmatrix}$$

Calculate $|A|$. (Note that this can be done by taking the inner product of any row or column of A and the corresponding row or column of C.) Expanding on the first column we get

$$|A| = 2(8) + 0(-10) + 1(-14) = 2$$

So A is nonsingular and

$$A^{-1} = \frac{1}{2} \begin{bmatrix} 8 & -10 & -14 \\ 4 & -5 & -8 \\ -2 & 3 & 4 \end{bmatrix} = \begin{bmatrix} 4 & -5 & -7 \\ 2 & -\frac{5}{2} & -4 \\ -1 & \frac{3}{2} & 2 \end{bmatrix}$$

Remark. This method shows that the condition $|A| \neq 0$ is both necessary and sufficient for A to be nonsingular.

Method II is particularly simple for 2×2 matrices. Consider, for example,

$$A = \begin{bmatrix} a & b \\ c & d \end{bmatrix}$$

We have

$$C = \begin{bmatrix} d & -c \\ -b & a \end{bmatrix} \quad \text{and} \quad C^T = \begin{bmatrix} d & -b \\ -c & a \end{bmatrix}$$

so that if $|A| = ad - bc \neq 0$, then

$$A^{-1} = \frac{\begin{bmatrix} d & -b \\ -c & a \end{bmatrix}}{|A|}$$

Note that to get C^T directly from A we simply interchange the elements a and d on the major diagonal and take the negatives of the elements on the minor diagonal. For example,

$$\begin{bmatrix} 3 & -2 \\ 5 & -4 \end{bmatrix}^{-1} = \frac{\begin{bmatrix} -4 & 2 \\ -5 & 3 \end{bmatrix}}{(-12 + 10)} = \begin{bmatrix} 2 & -1 \\ \frac{5}{2} & -\frac{3}{2} \end{bmatrix}$$

For 3×3 matrices, method II may still be easier than method I, depending on how adept one is in calculating cofactors mentally. For higher-order matrices, however, method I is generally preferable.

Using inverses of matrices provides an alternative way of solving a system $AX = B$, where the number of equations is the same as the number of unknowns (so that A is square). If A^{-1} exists, we multiply both sides by A^{-1} (left multiplication on each side) to get

$$A^{-1}(AX) = A^{-1}B$$
$$(A^{-1}A)X = A^{-1}B$$
$$IX = A^{-1}B$$
$$X = A^{-1}B$$

For example, let us solve the system

$$\begin{cases} 2x - y + 5z = 3 \\ 2y + 4z = 5 \\ x - 2y = -4 \end{cases}$$

This can be written in matrix form, $AX = B$, where

$$A = \begin{bmatrix} 2 & -1 & 5 \\ 0 & 2 & 4 \\ 1 & -2 & 0 \end{bmatrix} \qquad X = \begin{bmatrix} x \\ y \\ z \end{bmatrix} \qquad B = \begin{bmatrix} 3 \\ 5 \\ -4 \end{bmatrix}$$

We have already found the inverse of A in illustrating methods I and II, so

$$X = A^{-1}B = \begin{bmatrix} 4 & -5 & -7 \\ 2 & -\frac{5}{2} & -4 \\ -1 & \frac{3}{2} & 2 \end{bmatrix} \begin{bmatrix} 3 \\ 5 \\ -4 \end{bmatrix} = \begin{bmatrix} 15 \\ \frac{19}{2} \\ -\frac{7}{2} \end{bmatrix}$$

Thus $x = 15$, $y = \frac{19}{2}$, $z = -\frac{7}{2}$.

While using inverses to solve systems of equations appears simple, the limitations on its use are severe—especially the requirements that the system be square and that A be nonsingular. Furthermore, calculation of A^{-1} is difficult for higher-order systems. In fact, there is at least as much work involved in calculating A^{-1} by method I as in solving the system by reducing the augmented matrix to triangular form as in Section 2. It should be emphasized, however, that inverses of matrices have many other uses, even if their utility is limited in solving systems of equations.

EXERCISE SET 6

A

In Problems 1–15 use method I to find the inverse or to show that the inverse does not exist. When A is nonsingular, check your result by showing that $AA^{-1} = I$.

1. $\begin{bmatrix} 2 & -1 \\ 3 & -2 \end{bmatrix}$

2. $\begin{bmatrix} -1 & 2 \\ -3 & 5 \end{bmatrix}$

3. $\begin{bmatrix} 5 & -4 \\ 3 & -2 \end{bmatrix}$

4. $\begin{bmatrix} -2 & 6 \\ -3 & 4 \end{bmatrix}$

5. $\begin{bmatrix} 1 & 0 & 0 \\ -1 & 2 & 0 \\ 1 & 4 & 3 \end{bmatrix}$

6. $\begin{bmatrix} 1 & 2 & 3 \\ 0 & 1 & 4 \\ 0 & 0 & 1 \end{bmatrix}$

7. $\begin{bmatrix} 2 & 1 & 0 \\ -3 & -4 & 2 \\ 2 & 2 & -1 \end{bmatrix}$

8. $\begin{bmatrix} 3 & -1 & 2 \\ 0 & 1 & 6 \\ 1 & 0 & 3 \end{bmatrix}$

9. $\begin{bmatrix} -1 & 1 & 1 \\ 1 & 1 & 1 \\ -1 & 1 & -1 \end{bmatrix}$

10. $\begin{bmatrix} 2 & 0 & 1 \\ 0 & 3 & -2 \\ 1 & -2 & 2 \end{bmatrix}$

11. $\begin{bmatrix} 1 & -3 & 4 \\ -2 & 5 & 1 \\ 0 & -2 & 18 \end{bmatrix}$

12. $\begin{bmatrix} 5 & 0 & -6 \\ 3 & 2 & -3 \\ -2 & 5 & 4 \end{bmatrix}$

13. $\begin{bmatrix} 2 & 7 & -3 \\ 3 & 8 & -4 \\ -10 & 6 & 7 \end{bmatrix}$

14. $\begin{bmatrix} 2 & -3 & -3 \\ 3 & 2 & -5 \\ -4 & 3 & 6 \end{bmatrix}$

15. $\begin{bmatrix} 3 & 1 & 4 \\ -5 & -2 & 1 \\ 7 & 3 & -6 \end{bmatrix}$

16–30. Repeat Problems 1–15, using method II.

In Problems 31–36 solve the system $AX = B$, using $X = A^{-1}B$.

31. $\begin{cases} x - y - 2z = 7 \\ 2x + 3y + 4z = -3 \\ -x + 2y + 3z = -8 \end{cases}$

32. $\begin{cases} 2x - y + z = 1 \\ x + 3y - z = 3 \\ 3x - 2y + 2z = -1 \end{cases}$

33. $\begin{cases} x - y - z = 0 \\ 2x + 4y + 5z = 5 \\ 5x - 3y - 2z = 3 \end{cases}$

34. $\begin{cases} 2x + y - z = 1 \\ 3x - 2y + 5z = 3 \\ 5x + 3y - 4z = 0 \end{cases}$

35. $\begin{cases} 5x + 3y + 2z = 21 \\ 4x - 2y - 3z = 7 \\ -3x + 4y + 5z = 1 \end{cases}$

36. $\begin{cases} 3x + 4y - 5z = -1 \\ -4x - 5y + 6z = 11 \\ 7x + 5z = 2 \end{cases}$

B

37. Complete the following proof that the inverse of a nonsingular matrix is unique. Suppose A is nonsingular and B and C are inverses of A. Then $AB = AC$. (Why?) Now multiply both sides of this equation on the left by B (or by C), and use the associate property. What conclusion do you draw?

38. Prove that if A and B are nonsingular $n \times n$ matrices, then AB is nonsingular and $(AB)^{-1} = B^{-1}A^{-1}$.

39. Show that an "upper triangular" matrix of the form

$$\begin{bmatrix} a_{11} & a_{12} & a_{13} \\ 0 & a_{22} & a_{23} \\ 0 & 0 & a_{33} \end{bmatrix}$$

is nonsingular if and only if $a_{ii} \neq 0$ for $i = 1, 2, 3$, and find a formula for the inverse.

In Problems 40–43 find the inverse or show that it does not exist.

40.
$$\begin{bmatrix} 1 & -1 & 0 & 0 \\ 1 & 0 & 1 & -1 \\ 0 & 1 & 1 & 0 \\ 1 & -1 & 0 & 1 \end{bmatrix}$$

41.
$$\begin{bmatrix} 1 & -1 & 2 & 0 \\ 0 & 1 & -1 & 1 \\ -1 & 1 & -1 & 2 \\ 0 & -1 & 1 & -2 \end{bmatrix}$$

42.
$$\begin{bmatrix} 1 & 0 & -2 & 3 \\ 0 & -2 & -5 & 0 \\ 2 & 3 & -1 & 1 \\ -1 & -2 & 3 & 4 \end{bmatrix}$$

43.
$$\begin{bmatrix} 2 & -1 & 3 & 1 \\ -1 & 4 & 2 & 1 \\ 3 & -1 & 5 & 4 \\ -2 & 3 & -1 & 1 \end{bmatrix}$$

44. Solve the following system, using $X = A^{-1}B$.

$$\begin{cases} x_1 - 2x_2 + x_3 - x_4 = 2 \\ x_1 - x_2 \quad\quad - x_4 = 6 \\ 2x_1 - 2x_2 + x_3 \quad\quad = 9 \\ -x_1 + 5x_2 - 2x_3 + 6x_4 = 6 \end{cases}$$

45. The matrix

$$A = \begin{bmatrix} a_{11} & a_{12} & \cdots & a_{1n} \\ a_{21} & a_{22} & \cdots & a_{2n} \\ \vdots & \vdots & & \vdots \\ a_{n1} & a_{n2} & \cdots & a_{nn} \end{bmatrix}$$

is said to be **orthonormal** if

$$a_{i1}^2 + a_{i2}^2 + \cdots + a_{in}^2 = 1 \quad \text{and}$$

$$a_{i1}a_{k1} + a_{i2}a_{k2} + \cdots + a_{in}a_{kn} = 0$$

for $i \neq k$, $i, k = 1, 2, \ldots, n$. Prove that if A is orthonormal, $A^{-1} = A^{T}$.

46. Using the result of Problem 45, show that the following matrix is orthonormal and find its inverse.

$$\begin{bmatrix} \dfrac{1}{2} & \dfrac{1}{2} & \dfrac{1}{2} & \dfrac{1}{2} \\ \dfrac{1}{\sqrt{2}} & \dfrac{-1}{\sqrt{2}} & 0 & 0 \\ \dfrac{1}{\sqrt{6}} & \dfrac{1}{\sqrt{6}} & \dfrac{-2}{\sqrt{6}} & 0 \\ \dfrac{1}{2\sqrt{3}} & \dfrac{1}{2\sqrt{3}} & \dfrac{1}{2\sqrt{3}} & \dfrac{-\sqrt{3}}{2} \end{bmatrix}$$

47. Carry out the details of the following proof of method II for finding inverses. Let

$$A = \begin{bmatrix} a_{11} & a_{12} & \cdots & a_{1n} \\ a_{21} & a_{22} & \cdots & a_{2n} \\ \vdots & \vdots & & \vdots \\ a_{n1} & a_{n2} & \cdots & a_{nn} \end{bmatrix}$$

and let A_{ij} be the cofactor of a_{ij}. Then

$$C^{T} = \begin{bmatrix} A_{11} & A_{21} & \cdots & A_{n1} \\ A_{12} & A_{22} & \cdots & A_{n2} \\ \vdots & \vdots & & \vdots \\ A_{1n} & A_{2n} & \cdots & A_{nn} \end{bmatrix}$$

Show that both AC^{T} and $C^{T}A$ yield a matrix that can be written in the form $|A|I$. (In doing this you will need to know that when the elements of any row or column of A are multiplied by the respective cofactors of any *other* row or column, the result is zero. Can you see why?) Now draw the desired conclusion.

In Problems 48 and 49 use a calculator to find the inverse.

48.
$$\begin{bmatrix} 0.231 & 1.24 & -2.79 \\ 3.75 & -2.03 & 0.462 \\ 5.16 & 6.23 & -4.68 \end{bmatrix}$$

49.
$$\begin{bmatrix} 2.801 & -0.7254 & -3.042 \\ -4.731 & 6.287 & -0.2345 \\ 0.6498 & 3.976 & -8.504 \end{bmatrix}$$

7 NONLINEAR SYSTEMS

No general theory exists to handle the great variety of nonlinear systems that can occur. One must deal with each particular situation and decide what will work, and of course there are times when nothing will give an exact solution. Nevertheless, we can consider some fairly common situations that may be solved by elementary means. We will illustrate these by examples.

EXAMPLE 20

Solve the system: $\begin{cases} 2x^2 - y + 3 = 0 \\ x + y = 6 \end{cases}$

Solution   This is a case of a quadratic equation and a linear equation. Typically, whenever one equation is linear and the other is not, the best procedure is to substitute from the linear equation into the other. In this problem we choose to solve the second equation for y (rather than x, since the substitution is simpler):

$$y = 6 - x$$
$$2x^2 - (6 - x) + 3 = 0$$
$$2x^2 + x - 3 = 0$$
$$(2x + 3)(x - 1) = 0$$
$$x = -\tfrac{3}{2} \quad \text{or} \quad x = 1$$

When $x = -\tfrac{3}{2}$, $y = 6 - (-\tfrac{3}{2}) = \tfrac{15}{2}$, and when $x = 1$, $y = 6 - 1 = 5$. So the solution set is $\{(1, 5), (-\tfrac{3}{2}, \tfrac{15}{2})\}$. The graphs of both equations are shown in Figure 3, and the solutions are indicated.

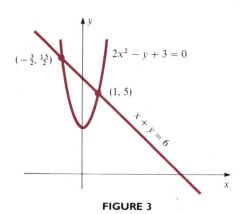

FIGURE 3

EXAMPLE 21 ||||||||||||||||||||||||||||||||||||||||||||||||||||||||||||||||||||||||||||||||||||||||||||||

Solve the system: $\begin{cases} xy + 2 = 0 \\ 2x - y = 5 \end{cases}$

Solution   This is handled in a similar way, although we could easily substitute from the first equation into the second:

$$y = 2x - 5$$
$$x(2x - 5) + 2 = 0$$
$$2x^2 - 5x + 2 = 0$$
$$(2x - 1)(x - 2) = 0$$
$$x = \tfrac{1}{2} \quad \text{or} \quad x = 2$$

When $x = \tfrac{1}{2}$, $y = 2(\tfrac{1}{2}) - 5 = -4$, and when $x = 2$, $y = 2(2) - 5 = -1$. So the solution set is $\{(\tfrac{1}{2}, -4), (2, -1)\}$. The graph of the first equation is obtained by plotting a few points (Figure 4).

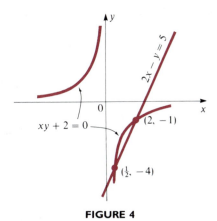

FIGURE 4

EXAMPLE 22 ||||||||||||||||||||||||||||||||||||||||||||||||||||||||||||||||||||||||||||||||||||||||||||||

Solve the system: $\begin{cases} x^2 - y^2 = 5 \\ 2x^2 + y^2 = 22 \end{cases}$

Solution   Even though these are quadratic equations, they are linear in form, and the techniques of solving linear systems can be used to find x^2 and y^2, from which x and y can be obtained. In particular, Cramer's rule could be used, but it is probably easier here just to eliminate y^2 by addition:

$$x^2 - y^2 = 5$$
$$\underline{2x^2 + y^2 = 22}$$
$$3x^2 = 27$$
$$x^2 = 9$$
$$x = \pm 3$$

Now substitute $x^2 = 9$ into the first equation (since it is simpler):

$$9 - y^2 = 5$$
$$y^2 = 4$$
$$y = \pm 2$$

The complete solution set is $\{(3, 2), (3, -2), (-3, 2), (-3, -2)\}$. The graphs are shown in Figure 5. The graph of the first equation is a hyperbola and the graph of the second is an ellipse.

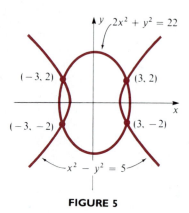

FIGURE 5

EXAMPLE 23 |||||||||||||||||||||||||||||||||||||||||||||||||||||||||||||||||||||||||||||||||||||||||||||||||||||||||||||||||||||||||||||||||||||||||||||

Solve the system: $\begin{cases} y = 3x - \dfrac{4}{x} \\ y = x^2 \end{cases}$

Solution   Here, we can eliminate y to get

$$x^2 = 3x - \frac{4}{x}$$

Multiplying by x yields

$$x^3 - 3x^2 + 4 = 0*$$

Denote the polynomial on the left-hand side by $P(x)$. We know from Chapter 5 that the only possible rational roots are $\pm 1, \pm 2, \pm 4$. We proceed by synthetic division to test these:

x				$P(x)$
0	1	-3	0	4
1	1	-2	-2	2
2	1	-1	-2	0

So $x = 2$ is a root, and the depressed equation is quadratic:

$$x^2 - x - 2 = 0$$
$$(x - 2)(x + 1) = 0$$
$$x = 2 \quad \text{or} \quad x = -1$$

Thus, $x = 2$ is a double root. When $x = 2$, $y = 4$, and when $x = -1$, $y = 1$. The solution set is $\{(2, 4), (-1, 1)\}$ with $(2, 4)$ as a double point. Graphically, the significance of the double point is that the curves are tangent to each other at that point. The graph of the first equation is somewhat more complicated this time, but again we give a sketch in Figure 6 to help visualize the situation.

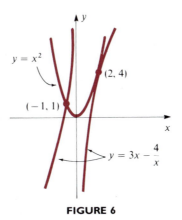

FIGURE 6

---

* This could lead to an extraneous root, since we may be unwittingly multiplying by 0. The answers should be checked to see that this is not the case.

The main device we have used in the above examples is that of elimination by substitution. This technique is especially useful whenever at least one of the unknowns appears to the first power in either of the equations. You should also be alert to the possibility of eliminating an unknown by addition or subtraction.

EXERCISE SET 7

A

Find the solution set in Problems 1–33.

1. $\begin{cases} x^2 + y^2 = 5 \\ x - y = 1 \end{cases}$

2. $\begin{cases} x^2 + y^2 = 10 \\ 2x - y = 1 \end{cases}$

3. $\begin{cases} x^2 + 2y^2 = 9 \\ x - 2y = 3 \end{cases}$

4. $\begin{cases} x^2 + y^2 = 5 \\ 3x + 4y = 5 \end{cases}$

5. $\begin{cases} y = x^2 - 4 \\ y = \dfrac{2x - 7}{3} \end{cases}$

6. $\begin{cases} y = 3 - x^2 \\ 2x - y + 4 = 0 \end{cases}$

7. $\begin{cases} x - y^2 = 4 \\ 2x - 3y = 10 \end{cases}$

8. $\begin{cases} xy = 4 \\ y - x = 3 \end{cases}$

9. $\begin{cases} 4x^2 + 3y^2 = 9 \\ y = 2x \end{cases}$

10. $\begin{cases} 3x^2 + 4y^2 = 63 \\ x - 2y + 3 = 0 \end{cases}$

11. $\begin{cases} y^2 = x \\ 2x - 3y = 5 \end{cases}$

12. $\begin{cases} y^2 = x \\ x^2 = y \end{cases}$

13. $\begin{cases} xy = 1 \\ 4x - 7y = 3 \end{cases}$

14. $\begin{cases} 2xy = -3 \\ 3x - 4y = 9 \end{cases}$

15. $\begin{cases} x^2 + y^2 = 3 \\ y^2 = 2x \end{cases}$

16. $\begin{cases} x^2 + y^2 = 6 \\ y^2 - 3x = 2 \end{cases}$

17. $\begin{cases} x^2 - y^2 = 7 \\ y = 2x - 5 \end{cases}$

18. $\begin{cases} 4x^2 - 5y^2 + 16 = 0 \\ 2x + 3y = 4 \end{cases}$

19. $\begin{cases} y^2 = 4x \\ x^2 - 2y^2 = 9 \end{cases}$

20. $\begin{cases} y = x^2 - 4 \\ 5x^2 - y^2 = 20 \end{cases}$

21. $\begin{cases} x^2 + y^2 - 2y = 24 \\ 2x - y = 4 \end{cases}$

22. $\begin{cases} 2x^2 - 3y^2 = 5 \\ x^2 - 2y^2 = 2 \end{cases}$

23. $\begin{cases} 4x^2 - y^2 = 7 \\ x^2 + 3y^2 = 31 \end{cases}$

24. $\begin{cases} xy + 6 = 0 \\ x^2 + y^2 = 13 \end{cases}$

25. $\begin{cases} 2x^2 - 3x + 4y^2 = 35 \\ x + 2y = 5 \end{cases}$

26. $\begin{cases} 3x^2 - 4y^2 = 12 \\ 7x^2 - y^2 = 8 \end{cases}$

27. $\begin{cases} 3x^2 - 7y^2 = 5 \\ 5x^2 + 3y^2 = 12 \end{cases}$

28. $\begin{cases} 3x^2 - y^2 = 1 \\ 7x^2 - 2y^2 = 5 \end{cases}$

29. $\begin{cases} y^2 = x - 3 \\ y^2 - 3y + 2x = 12 \end{cases}$

30. $\begin{cases} xy = 3 \\ 2x^2 - 3y^2 = 15 \end{cases}$

31. $\begin{cases} x^2 + y^2 + 2x - y = 8 \\ x^2 + y^2 = 5 \end{cases}$

32. $\begin{cases} x^2 + y^2 + 3x - 4y = -4 \\ x^2 + y^2 + 2x - 3y = 1 \end{cases}$

33. $\begin{cases} y = x^2 - 4 \\ x^2 + 3y^2 + 4y - 6 = 0 \end{cases}$

34. The perimeter of a rectangle is 40, and its area is 96. Find its dimensions.

35. The length of a rectangle is 4 more than twice its width, and its area is 126. Find its dimensions.

36. The difference between two numbers is 3, and their product is 270. Find the numbers.

37. The difference between two numbers is 4 and the difference between their squares is 80. Find the numbers.

38. The product of two numbers is 128, and the sum of their reciprocals is $\frac{3}{16}$. Find the numbers.

39. The perimeter of a rectangle is 34, and the length of its diagonal is 13. Find the length and width.

40. The length of a rectangle is 3 less than twice the width, and the square of the length of the diagonal is 306. Find the length and width.

41. A beam is to be sawed from a log of diameter 10 inches. If the height of the cross-section of the beam is to be twice the width, what are the maximum dimensions of the cross-section of the beam?

42. A driver travels the first 75 miles of a 95-mile trip without encountering congestion, but traffic is heavy on the remaining part, and his average speed is reduced by 10 miles per hour. The total time for the trip is 2 hours. Find his average speed on each part of the trip.

43. A man has a certain amount invested at simple interest, the annual income from which is $180. If he had invested it at another bank that pays $\frac{1}{2}\%$ more interest, he would have received an income of $198 per year. Find the amount he invested and the rate of interest.

44. A wooden box to enclose a long piece of machinery is 10 feet long, and the ends are rectangular. It costs 25¢ per square foot to construct, and the total cost is $28. If the volume is 60 cubic feet, find the dimensions.

45. A rectangle of area 12 square centimeters is inscribed in a circle of diameter 5 centimeters. Find its dimensions.

B

Find the solution set in Problems 46–49.

46. $\begin{cases} x^2 - y = 2 \\ y^2 + 2y - 3x^2 - 10x = 6 \end{cases}$

47. $\begin{cases} x^2 + y = 8 \\ x^2 - 2xy + y^2 = 16 \end{cases}$

48. $\begin{cases} xy = 1 \\ x - 4y^2 + 3 = 0 \end{cases}$ **49.** $\begin{cases} \dfrac{1}{x} + \dfrac{2}{y} = 3 \\ xy = 1 \end{cases}$

50. Find the points of intersection of the circles $x^2 + y^2 + 2x + 3y = 4$ and $x^2 + y^2 + x + 2y = 3$. Find the equation of the common chord and the equation of its perpendicular bisector.

51. Find the smaller area between the line $x + 2y = 5$ and the circle $x^2 + y^2 - 4x + 2y - 5 = 0$.

52. The height of a certain right circular cylinder is 1 inch more than twice its radius. Its volume is 9π cubic inches. Find the height and the radius.

53. A right circular cylinder of volume 72π is inscribed in a sphere of radius 5. Find the base radius and height of the cylinder.

54. The height of a right circular cone is 1 inch less than twice its base radius. If the volume is 15π cubic inches, find the base radius and the height.

55. Find the equation of the line that passes through the points of intersection of the curves whose equations are $y^2 = x$ and $x^2 + 3xy - 2y^2 = 32$.

56. Find the equation of the line passing through the points of intersection of the curves $x^2 + y^2 = 10$ and $x^2 + y^2 - 2x + 4y = 0$.

57. An isosceles triangle with base equal to its altitude is inscribed in a circle of diameter 10. Find the altitude of the triangle.

58. A boy is in a rowboat 4 miles from the point A nearest him on the shore. In order to get to the point B on the shore 10 miles from A, he rows to point C, between A and B, and jogs the rest of the way to B. If he can row 5 miles per hour and jog 7 miles per hour, and if his time rowing equals his time jogging, find how far C is from A (see sketch).

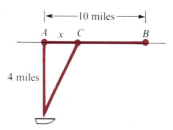

C (Graphing calculator)

In Problems 59–64 graph both curves and zoom in on points of intersection, giving their coordinates correct to three decimal places.

59. $\begin{cases} y = \sqrt{x^2 + 4} \\ x^2 - 2x - 2y + 4 = 0 \end{cases}$

60. $\begin{cases} y = 2 \sin 3x + 1 \\ y = e^x \end{cases}$ **61.** $\begin{cases} y = x^{3/2} \\ y = \ln x + 2 \end{cases}$

62. $\begin{cases} y = \dfrac{e^x - e^{-x}}{e^x + e^{-x}} \\ y = \tan^{-1}\left(\dfrac{1}{x}\right) \end{cases}$

63. $\begin{cases} y = \ln|\sec x + \tan x| \\ y = \arccos x \end{cases}$

64. $\begin{cases} y = \ln|x - 2| \\ y = 2x^2 - x^3 \end{cases}$

SYSTEMS OF INEQUALITIES

We will illustrate in this section how to determine graphic solutions to systems of inequalities. First, we must see how to solve graphically a single inequality in two variables. This is illustrated in the next two examples.

EXAMPLE 24

Show the graphic solution to the inequality $2x - y < 3$.

Solution   The inequality can be written in the equivalent form $y > 2x - 3$. We wish to describe as a region in the plane the set of all points (x, y) satisfying this. One

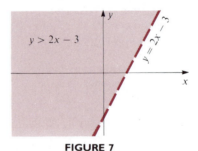

FIGURE 7

technique is to sketch first the line $y = 2x - 3$ and to observe that the solution to the given inequality consists of all points *above* this line (Figure 7). The line is dashed in the figure to indicate that it is *not* part of the solution set.   ▲

Remark. We could equally well write the inequality in the form $x < (y + 3)/2$ and conclude that the solution set consists of all points to the *left* of the line $x = (y + 3)/2$, and this is seen to agree with what we have shown.

EXAMPLE 25

Show the graphic solution of the inequality $y \leq x^2 + 4$.

Solution   The graph of $y = x^2 + 4$ is the parabola pictured in Figure 8. All points *on the parabola or below it* satisfy the given inequality. The parabola is shown as a solid curve to indicate that it *is* part of the solution set.

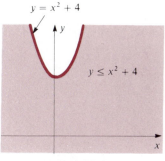

$y = x^2 + 4$

$y \le x^2 + 4$

FIGURE 8

The next two examples illustrate graphic solutions to systems of linear inequalities in two variables.

EXAMPLE 26 ꞁꞁꞁ

Find the common solution set to the system: $\begin{cases} 2x + 3y > 4 \\ x + y < 3 \end{cases}$

Solution   We rewrite the inequalities in the form

$$y > \frac{4 - 2x}{3}$$

$$y < 3 - x$$

and next draw the bounding lines, as shown in Figure 9. The simultaneous solution is the shaded region in the figure, since that is where y is greater than $(4 - 2x)/3$ and at the same time less than $3 - x$.

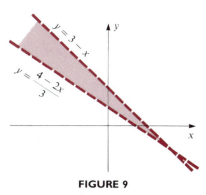

$y = 3 - x$

$y = \dfrac{4 - 2x}{3}$

FIGURE 9

EXAMPLE 27

Exhibit graphically the simultaneous solution set to the system:

$$\begin{cases} x + 3y \le 12 \\ x + y \le 6 \\ 2x + y \le 10 \\ x \ge 0 \\ y \ge 0 \end{cases}$$

Solution   The last two inequalities restrict the solution set to the first quadrant. We rewrite the other three inequalities in the form

$$y \le \frac{12 - x}{3}$$

$$y \le 6 - x$$

$$y \le 10 - 2x$$

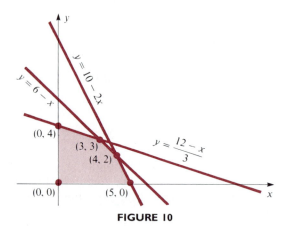

FIGURE 10

and draw the lines represented by the corresponding equalities (Figure 10). The simultaneous solution is the shaded region in the figure, including the boundary lines. In applications, it is usually important to know the vertices of the boundary. These are found by solving appropriate pairs of equations simultaneously and are shown in the figure. ▲

The next example involves absolute values and is equivalent to a system of simultaneous inequalities.

EXAMPLE 28

|||||||||||||||||||||||||||||||||||||||||||||||||||||||||||||||||||||||||||||||||||||||||||||||||||||||||||

Show the graphic solution of the inequality $|x| - |y| \leq 1$.

Solution We may rewrite the inequality in the form

$$|x| \leq |y| + 1$$

which is equivalent to

$$-|y| - 1 \leq x \leq |y| + 1$$

So if $y \geq 0$, this gives

$$-y - 1 \leq x \leq y + 1$$

whereas if $y < 0$, we obtain

$$y - 1 \leq x \leq -y + 1$$

Equivalently, we may say that when $y \geq 0$, the inequality is satisfied by all pairs (x, y) that simultaneously satisfy

$$\begin{cases} x \leq \ \ y + 1 \\ x \geq -y - 1 \end{cases}$$

and when $y < 0$, by the pairs (x, y) satisfying

$$\begin{cases} x \leq -y + 1 \\ x \geq \ \ y - 1 \end{cases}$$

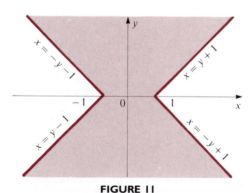

FIGURE 11

So for $y \geq 0$ we draw in the bounding lines $x = y + 1$ and $x = -y - 1$ and include all points to the *left* of the first line and at the same time to the *right* of the second; we also include points on the boundaries.

Similarly, when $y < 0$, we draw the bounding lines $x = -y + 1$ and $x = y - 1$ and include all points between as well as on these lines. The solution set is shown in Figure 11.   ▲

EXERCISE SET 8

A

Give the graphic solution to each of the following.

1. $y < 3x - 5$
2. $y > 4 - x$
3. $x + y < 2$
4. $2x - y > 3$
5. $3x + 2y \leq 4$
6. $5x - 4y \geq 3$
7. $3y - 4 < 2x - 5$
8. $5x + 2 > 3y - 4$
9. $4y - x^2 > 0$
10. $y - 1 < x^2$
11. $x > y^2$
12. $y - x^2 > 4$

13. $16y \leq x^2$

14. $\begin{cases} x - y > 0 \\ y - 2x < 4 \end{cases}$

15. $\begin{cases} x - 2y + 4 > 0 \\ x + y > -3 \end{cases}$

16. $\begin{cases} 3x - 2y > 4 \\ x + y \geq 1 \end{cases}$

17. $\begin{cases} 3x + 4y > 12 \\ 2x - y + 3 > 0 \end{cases}$

18. $\begin{cases} x \geq 2 \\ x \leq 2 + y \end{cases}$

19. $\begin{cases} y \leq 2x \\ 2x - y < 3 \end{cases}$

20. $\begin{cases} 2x - 5y < 6 \\ x + 2y \geq 4 \end{cases}$

21. $\begin{cases} y \leq 1 \\ x - y \leq 2 \end{cases}$

22. $\begin{cases} x < 1 \\ y > 1 \\ x - y + 3 > 0 \end{cases}$

23. $\begin{cases} x \geq 0 \\ x + y \leq 2 \\ 2x - 3y \leq 6 \end{cases}$

24. $|y| \leq x - 1$

25. $|2x - y| \leq 3$
26. $|y| > x - 1$
27. $|x| < 2y + 3$
28. $|x| \geq 2 - y$

29. $\begin{cases} x^2 - 4y \leq 0 \\ 2y - x < 2 \end{cases}$

30. $\begin{cases} y - \sqrt{x} < 0 \\ 4y - x > 0 \end{cases}$

31. $\begin{cases} y > x \\ y < 4 - x^2 \end{cases}$

32. $\begin{cases} x + y < 2 \\ x > y^2 \end{cases}$

33. $\begin{cases} x - y > 2 \\ x + y^2 \leq 4 \end{cases}$

34. $\begin{cases} 4y - x^2 > 0 \\ y^2 - 4x < 0 \end{cases}$

35. $\begin{cases} y \leq 9 - x^2 \\ y \geq x^2 \end{cases}$

36. $\begin{cases} x^2 + y \leq 3x \\ x^2 - 2y \leq 0 \end{cases}$

B

37. $\begin{cases} y \leq 1 + x \\ y \leq 1 - x \\ y \geq x - 1 \\ y \geq -x - 1 \end{cases}$

38. $\begin{cases} 2x + y \leq 8 \\ 2x + 3y \leq 12 \\ x \geq 0 \\ y \geq 0 \end{cases}$

39. $\begin{cases} x^2 - 2x + y \leq 0 \\ x + y > 0 \end{cases}$

40. $\begin{cases} x^2 + y < 4 \\ 2x - y < 4 \\ 2x + y + 4 > 0 \end{cases}$

41. $|x| + |y| \leq 1$
42. $|x - y| \geq 1$

LINEAR PROGRAMMING

An important application of systems of linear inequalities occurs in a technique developed in the 1940s for solving problems in which a quantity is to be maximized or minimized, subject to certain limitations called **constraints.** For example, a company might want to maximize the profit on some manufactured item or to minimize the cost. Applications of this technique, called **linear programming,** can be found in many areas but are especially important in business and industry.

In a typical linear programming problem the quantity to be maximized or minimized is a linear function of two or more variables. The method we will illustrate is applicable to functions of two variables only, but other procedures exist for dealing with larger numbers of variables. This function is known as the

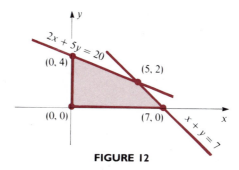

FIGURE 12

objective function, and has the form $F = ax + by + c$. The constraints on x and y are expressed in the form of linear inequalities.

For example, suppose we wish to find the maximum value of the objective function $F = 3x + 4y$, subject to the constraints given by

$$\begin{cases} x + y \le 7 \\ 2x + 5y \le 20 \\ x \ge 0 \\ y \ge 0 \end{cases}$$

The four constraints have the simultaneous solution set shown by the shaded region in Figure 12. This region is called the **set of feasible solutions** to the problem. We have shown the coordinates of each vertex of the polygon bounding the region. Finding these vertices is an essential part of the problem and involves solving two or more sets of equations simultaneously. To find the values of x and y among the feasible solutions for which the objective function F is a maximum, we make use of the following fundamental result, which we state without proof:

THEOREM   **The maximum and minimum values of the objective function in a linear programming problem, if they exist, occur at a vertex of the graph of the set of feasible solutions.***

Thus, to find the maximum value of F we test its value at each of the vertices $(0, 0), (7, 0), (5, 2),$ and $(0, 4)$. The maximum value is found to be 23, and this occurs

---

* Both maximum and minimum values do exist if the feasibility set is bounded and is **convex,** which means that for any two points of the set the line segment joining them lies entirely within the set.

when x is 5 and y is 2.

Vertex	F
(0, 0)	0
(7, 0)	21
(5, 2)	23
(0, 4)	16

The next two examples illustrate applications in economics.

EXAMPLE 29 ||||||||||||||||||||||||||||||||||||||||||||||||||||||||||||||||||||||||||||||||||||||||||||||||||||||||||||||||||||||||||||

A company manufactures two kinds of pocket calculators, one with and one without rechargeable batteries. The company can make at most 50 of the type with rechargeable batteries (type A) per day and 60 of the type without rechargeable batteries (type B) per day. Type A requires 3 work-hours to produce and type B requires 2 work-hours. The work force provides a total of 180 work-hours available per day. If the profit on each type A calculator is \$2.50, and the profit on each type B calculator is \$2.00, find how many of each type of calculator should be produced per day to give the maximum profit.

Solution   Let x be the number of type A and y the number of type B calculators to be produced. Then the profit P is given by

$$P = 2.50x + 2.00y$$

This is the objective function, which is to be maximized, subject to the constraints

$$\begin{cases} 0 \le x \le 50 \\ 0 \le y \le 60 \\ 3x + 2y \le 180 \end{cases}$$

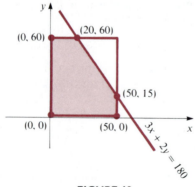

FIGURE 13

The set of feasible solutions is shown in Figure 13. Testing the value of P at each vertex, we find the maximum value to be $170, and this occurs when $x = 20$ and $y = 60$.

Vertex	P
(0, 0)	0
(50, 0)	125
(50, 15)	155
(20, 60)	170
(0, 60)	120

EXAMPLE 30

A machine tool company produces two kinds of parts, A and B. The weekly demand for each requires that at least 20 type A parts and 10 type B parts be produced. Limitations on capacity require that at most 60 type A and 40 type B parts be produced each week. In order to keep the work force fully employed, the combination of A and B parts produced must be at least 50 each week. If it costs $3 to produce each A part and $2 to produce each B part, how many parts of each type should be produced per week to minimize cost?

Solution   Let x be the number of type A parts and y the number of type B parts to be produced each week. The objective function is the cost C, given by

$$C = 3x + 2y$$

and this is to be minimized, subject to the constraints

$$20 \leq x \leq 60$$
$$10 \leq y \leq 40$$
$$x + y \geq 50$$

Figure 14 shows the set of feasible solutions. Calculating C at each vertex shows that the minimum value of $120 occurs when 20 type A parts and 30 type B parts are produced each week.

Vertex	C
(60, 10)	200
(60, 40)	260
(20, 40)	140
(20, 30)	120
(40, 10)	140

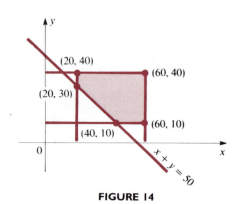

FIGURE 14

EXERCISE SET 9

A

In Problems 1–8 find the maximum or minimum value, as specified, of the objective function, subject to the given constraints. Give the values of x and y that produce this value.

1. Maximize $F = 2x + y$,

subject to $\begin{cases} 3x + 4y \leq 24 \\ x \geq 2 \\ y \geq 3 \end{cases}$

2. Minimize $C = 2x - 3y + 8$,

subject to $\begin{cases} x - y \geq 0 \\ x \leq 6 \\ y \geq 2 \end{cases}$

3. Maximize $G = 2x + 7y - 4$,

subject to $\begin{cases} x - y \leq 1 \\ 2x + 3y \leq 12 \\ x \geq 0 \\ y \geq 0 \end{cases}$

4. Minimize $F = 4x - 3y + 7$,

subject to $\begin{cases} y - 2x \leq 0 \\ x - 2y \leq 0 \\ 2 \leq y \leq 4 \end{cases}$

5. Minimize $H = 3x + 5y$,

subject to $\begin{cases} x \leq 2y + 2 \\ x \geq 6 - 2y \\ y \leq x \\ x \leq 6 \end{cases}$

6. Maximize $P = 2x + 3y$,

subject to $\begin{cases} y \leq 3x - 2 \\ 3x + 4y \leq 22 \\ y \geq 1 \end{cases}$

7. Minimize $C = 4x - y + 5$,

subject to $\begin{cases} x + y \leq 5 \\ x + 2y \geq 6 \\ x \geq 0 \\ y \geq 0 \end{cases}$

8. Maximize $P = 10x + 6y$,

subject to $\begin{cases} x + 3y \leq 18 \\ x + y \leq 8 \\ 2x + y \leq 14 \\ x \geq 0 \\ y \geq 0 \end{cases}$

9. A grocer expects to sell between 40 and 60 cases of canned applesauce during a 6-month period. He knows from experience that he can sell at least 20 cases of brand A and at most 30 cases of brand B applesauce in this time period. His profit on a case of brand A is \$3.00 and on a case of brand B is \$3.25. Assuming he sells all of the applesauce, how much should he stock for maximum profit?

10. An oil refinery needs to produce at least 100,000 gallons of unleaded gasoline and 150,000 gallons of regular leaded gasoline each month to meet demand. In order to use the plant capacity efficiently, at least 300,000 gallons of the two types combined must be produced each month, but no more than 400,000 gallons can be produced. The cost of producing unleaded gasoline is $300 per 1,000 gallons and for regular leaded gasoline it is $250 per 1,000 gallons. How many gallons of each should be produced to minimize cost?

11. A certain machine can produce two types of products, A and B. It can produce 100 type A products per hour and 150 type B products per hour. The machine can run between 6 and 12 hours per day. It is to be used at most 8 hours per day to produce product A and at least 1 but no more than 5 hours per day in the production of B. If the profit on each A item is $0.05 and on each B item $0.04, how many hours per day should the machine be used in the production of each item in order to maximize profit?

12. A woman has $20,000 that she wishes to divide between two investments, one a certificate of deposit paying $8\frac{1}{2}\%$ annual interest and requiring a minimum of $5,000, and the other a utility stock paying 10% annual interest. Because the latter involves greater risk, she wants to put no more than half as much in it as in the certificate of deposit, but she wants at least 100 shares, and it costs $20 per share. Assuming the stock price does not fluctuate, how should she invest the money for maximum income?

13. Two hardware store owners wish to buy paint of a certain kind from a wholesaler who has a stock of 1,000 gallons of the paint. The first owner will buy between 200 and 500 gallons, and the second will buy between 300 and 600 gallons. The wholesaler makes a profit of $1.10 on each gallon, but this is reduced by the cost of delivery. He estimates the cost at 5¢ per gallon for delivery to the first hardware store and 10¢ per gallon to the second. How many gallons should he sell to each owner to maximize his profits?

14. A dairy farmer plans to purchase two types of feed for his 100 cows. Type A contains 10% protein and 30% carbohydrates (by weight), and type B contains 30% protein and 40% carbohydrates. He wants to provide each cow with a minimum of 1 pound of protein and 2 pounds of carbohydrates. If type A costs 30¢ per pound and type B costs 50¢ per pound, how many pounds of each type

should he buy to minimize the cost? What is the minimum cost?

15. A machine tool company produces two types of parts that require time on a drill press and on a lathe. Part A requires 5 minutes on the drill press and 10 minutes on the lathe. Part B requires 6 minutes on the drill press and 4 minutes on the lathe. The time per day available on the drill press for producing these parts is 1 hour and 40 minutes, and the time available on the lathe is 2 hours. To meet demand, at least 5 parts of each type should be produced each day. If the profit on part A is $3 and on part B is $2, how many parts per day of each type should be made for maximum profit?

16. A baker has orders for 15 angel food cakes and 20 chocolate cakes. Each angel food cake uses 12 eggs and 1 cup of sugar, whereas each chocolate cake uses 3 eggs and 2 cups of sugar. He has available for these two items 25 dozen eggs and 36 pounds of sugar (each pound equals $2\frac{1}{4}$ cups). His profit on each angel food cake is $0.50 and on each chocolate cake is $0.75. How many cakes of each type should he make in order to meet the demand and maximize profit (assuming he can sell all the cakes he makes)?

17. A manufacturer of washing machines has two plants where three models are produced: standard, superior, and luxury. There are orders for 600 standard, 340 superior, and 200 luxury models. Plant A can produce 10 standard, 8 superior, and 3 luxury models per day, whereas plant B can produce 20 standard, 9 superior, and 8 luxury models per day. It costs $500 per day to operate plant A and $800 per day to operate plant B. How many days should each plant be used in the production of these machines to meet demand and minimize cost?

B

18. Consider the objective function $P = 5x + 2y$ and the feasibility set defined by

$$\begin{cases} x + 3y \geq 10 \\ 3x + 2y \leq 34 \\ 2x - 5y + 28 \geq 0 \\ x \geq 1 \\ y \geq 2 \end{cases}$$

Graph the feasibility set and show several members of the family obtained by setting the objective function equal to a constant, $5x + 2y = C$, for increasing values of C (if P represents profit, these are called **isoprofit** lines). Explain from what you observe why the objective function must assume both its maximum and minimum values at a vertex of the graph of the feasibility set. At which vertex do each of the extreme values of P occur, and what are the maximum and minimum values of P?

19. A dairy farmer has determined that each of his cows should get three types of food supplements, A, B, and C, each day. The minimum requirements are 10 units of A, 14 units of B, and 16 units of C. There are two kinds of feed that contain all three supplements. Feed number 1 contains 2 units each of A, B, and C per pound. Feed number 2 contains 2 units of A, 3 units of B, and 4 units of C per pound. Feed number 1 costs 10¢ per pound and feed number 2 costs 12¢ per pound. How many pounds of each type of feed should the farmer purchase to meet the daily need per cow in order to minimize cost?

20. At a furniture factory, desks and tables each go through four stations where different aspects of fabrication take place. The number of work-hours required at each station, as well as the number of work-hours available each day, are shown in the table. The profit on each item also is shown. How many desks and how many tables should be produced each day to maximize profit?

	Work-hours needed				
	Station 1	Station 2	Station 3	Station 4	Profit
Desk	1	1	2	3	$80
Table	3	1	1	2	$70
Available work-hours	9	7	10	18	

21. A woman owns two orchard supply stores. She has orders from two customers for a certain type of insecticide. Customer A wants 350 gallons and customer B wants 400 gallons. The store owner has 500 gallons of this insecticide on hand at store number 1 and 600 gallons at store number 2. The cost of delivery is estimated as shown in the table. How many gallons should be sent to each customer from store 1 and from store 2 to minimize the shipping cost?

From store	To customer	Delivery cost per gallon
1	A	$0.10
2	A	$0.12
1	B	$0.11
2	B	$0.15

REVIEW EXERCISE SET

A

In Problems 1–7 solve the systems using the method of elimination by addition or subtraction.

1. $\begin{cases} 3x - 2y = 5 \\ x + 4y = 11 \end{cases}$

2. $\begin{cases} 5x + 3y = 6 \\ 2x - 4y = 5 \end{cases}$

3. $\begin{cases} 4x + 7y = 3 \\ 5x + 8y = 0 \end{cases}$

4. $\begin{cases} x - 2y + 4z = 1 \\ 3x + y + z = 2 \\ 2x + y - z = -1 \end{cases}$

5. $\begin{cases} x - 2y + z = 2 \\ 2x - 3y - z = 0 \\ x + y - 4z = 2 \end{cases}$

6. $\begin{cases} 2x - 3y - 5z = -1 \\ x - y - 2z = 1 \\ -4x + 5y - 6z = 14 \end{cases}$

7. $\begin{cases} x - y = 1 \\ y + z = 1 \\ x - 2w = 0 \\ x + y + 3z + w = 2 \end{cases}$

Solve Problems 8–19 by reducing the augmented matrix to triangular form.

8. $\begin{cases} x - 2y + 4z = -3 \\ 3x + y - 2z = 12 \\ 2x + y - 3z = 7 \end{cases}$

9. $\begin{cases} 2x + y - z = 1 \\ x + y + z = -3 \\ -5x - 2y + 3z = -1 \end{cases}$

10. $\begin{cases} x + 2y + 3z = 1 \\ 2x - y - z = 5 \\ -x + y - 2z = 10 \end{cases}$

11. $\begin{cases} 2x + 4y - 3z = 8 \\ 3x + 2y - z = 8 \\ 5x - 2y + 4z = 0 \end{cases}$

12. $\begin{cases} 4x + 2y - z = 0 \\ 3x - 5y - 7z = 5 \\ 7x + 3y - 5z = -1 \end{cases}$

13. $\begin{cases} 2x - y + z = 3 \\ x + 3y - 2z = 4 \\ 4x - 9y + 7z = 1 \end{cases}$

14. $\begin{cases} 3x - 2y + 4z = 6 \\ 5x - 7y - 2z = 3 \\ 7x - 23y - 34z = -20 \end{cases}$

15. $\begin{cases} x - y + z = 0 \\ y - z + 3w = 1 \\ x + z + 2w = 0 \\ x + y + 3z + w = 1 \end{cases}$

16. $\begin{cases} x + y - 2z + w = 2 \\ 3x + 2y - 4z - 2w = -19 \\ 2x - y + z + 3w = 7 \\ -x - 3y + 4z + w = 4 \end{cases}$

17. $\begin{cases} x + y - z + 3w = 1 \\ 2x - y + 3z - 4w = -13 \\ 3x + 2y + z + w = 5 \end{cases}$

18. $\begin{cases} x - 2y - z - w = -4 \\ 3x + y + 2z + w = 3 \\ 2x - 3y - z + 2w = 7 \\ x - 3y - 2z - 5w = -19 \end{cases}$

19. $\begin{cases} 2x + y + 3z = 1 \\ x - 2y - z = 3 \\ 4x - 5y + 6z = 11 \\ x - y + z = 2 \end{cases}$

Evaluate the determinants in Problems 20–22.

20. a. $\begin{vmatrix} 2 & -1 & 1 \\ 1 & 3 & 4 \\ -2 & 1 & -5 \end{vmatrix}$ **b.** $\begin{vmatrix} 4 & 1 & -2 \\ 0 & 2 & -3 \\ 5 & -3 & 6 \end{vmatrix}$

21. a. $\begin{vmatrix} 3 & -5 & 4 \\ 2 & -3 & 6 \\ -1 & 2 & 3 \end{vmatrix}$ **b.** $\begin{vmatrix} -2 & 7 & 3 \\ 3 & -8 & 4 \\ 6 & -10 & -5 \end{vmatrix}$

22. a. $\begin{vmatrix} 4 & 3 & -5 \\ 7 & 8 & 2 \\ -3 & 2 & -6 \end{vmatrix}$ **b.** $\begin{vmatrix} 2 & -1 & 0 & 3 \\ 0 & 5 & -1 & 2 \\ 1 & 6 & 3 & -2 \\ -3 & 4 & 0 & -1 \end{vmatrix}$

23. Solve for x:

$$\begin{vmatrix} 1 & 0 & x \\ 2 & -x & 1 \\ 3 & x+1 & 2 \end{vmatrix} = 3$$

Find the solution to the systems in Problems 24–28 by Cramer's rule.

24. a. $\begin{cases} 5x - 11y = 6 \\ 7x - 8y = 3 \end{cases}$ **b.** $\begin{cases} 3x + 7y + 8 = 0 \\ 2x - 3y - 4 = 0 \end{cases}$

25. a. $\begin{cases} 9x - 5y = 11 \\ -6x + 3y = 7 \end{cases}$ **b.** $\begin{cases} 4x + 3y - 5 = 0 \\ 8x - 5y + 2 = 0 \end{cases}$

26. $\begin{cases} 2x + 4y + 3z = 1 \\ 3x - 5y - 2z = 5 \\ 4x - 6y - 8z = -5 \end{cases}$

27. $\begin{cases} 3x - 5y - z = 4 \\ 2x + 4y + 3z = -1 \\ 2x + y + z = 0 \end{cases}$

28. $\begin{cases} x + 3y + z = -2 \\ 4x - 2y - 5z = -1 \\ 5x + 7y + 3z = -4 \end{cases}$

In Problems 29 and 30 find the inverse by two methods.

29. $\begin{bmatrix} 3 & -2 & 0 \\ 1 & 1 & 4 \\ -2 & 3 & 5 \end{bmatrix}$ **30.** $\begin{bmatrix} 1 & -1 & -2 \\ 2 & -3 & -4 \\ 2 & 1 & -2 \end{bmatrix}$

In Problems 31 and 32 solve the system $AX = B$, using $X = A^{-1}B$.

31. The system of Problem 8.
32. The system of Problem 9.
33. Two drivers are initially 120 miles apart, and they drive toward each other. One averages 50 miles per hour and the other 40 miles per hour. Find how far each has driven when they meet.
34. A man in a motorboat makes a trip of 12 miles upstream in 1 hour. The return trip with the current takes 40 minutes. If he held the throttle wide open all the time, find how fast he would have gone in still water and the rate of the current.
35. Thirty cubic centimeters of a 25% sulfuric acid solution are obtained by mixing a 40% sulfuric acid solution with a 15% solution. How much of each was needed?
36. The total income from a community concert was $2,325. Admission for adults was $3.00 and for children $1.50. If 900 persons attended, how many were adults and how many were children?
37. The length of a certain rectangle is 1 more than twice the width, and the perimeter is 32. Find the length and the width.
38. Find how much nickel and how much zinc should be added to 200 kilograms of an alloy that is 15% nickel and 25% zinc to obtain an alloy that is 25% nickel and 30% zinc?
39. Fifty gallons of a 25% salt solution are obtained by mixing pure water with a 40% salt solution. Find how much pure water and how much of the 40% salt solution were used.
40. The sum of the digits of a two-digit number is 11. If the digits were reversed, the number would be decreased by 45. Find the number.
41. The income from an amount invested at simple interest for 1 year is $108. If the same amount were invested in an account yielding $1\frac{1}{2}\%$ more interest,

the income for the year would be $135. Find the interest rate and the amount invested.
42. Three men working together can do a certain job in 2 hours. Mr. Jones and Mr. Smith can do the job in 3 hours, and Mr. Smith and Mr. Robinson together can do it in 4 hours. How long would it take each man working alone?
43. Two pumps working simultaneously can fill a tank in $2\frac{1}{2}$ hours. Both pumps are started at the same time, but the larger one breaks down after $1\frac{1}{2}$ hours, and it takes the smaller one an additional $2\frac{1}{2}$ hours to fill the tank. Find how long it would take each pump alone to fill the tank.

Solve the systems of equations in Problems 44–52.

44. $\begin{cases} y = x^2 - 2x \\ x + 2y = 5 \end{cases}$ **45.** $\begin{cases} xy + 2 = 0 \\ 2x + 3y = 4 \end{cases}$

46. $\begin{cases} x^2 - 2y^2 = 4 \\ 3x^2 + 4y^2 = 12 \end{cases}$ **47.** $\begin{cases} y^2 - 7x^2 = 36 \\ 3x + y = 2 \end{cases}$

48. $\begin{cases} 3y^2 - x^2 = 26 \\ x + y^2 = 12 \end{cases}$ **49.** $\begin{cases} x^2 - y^2 - 2x = 0 \\ 2x - 3y = 4 \end{cases}$

50. $\begin{cases} 8xy - 6x - 5y = 3 \\ 10x + y = 1 \end{cases}$ **51.** $\begin{cases} \dfrac{1}{x} + \dfrac{2}{y} = 4 \\ 3x + 5y = 6 \end{cases}$

52. $\begin{cases} x^2 + 3xy - 2y^2 - 8 = 0 \\ 2x - y = 3 \end{cases}$

53. Find two numbers whose difference is 8 and the sum of whose squares is 274.
54. Find two numbers whose difference is 13 and the difference of whose squares is 65.
55. The diagonal of a rectangle is 25 inches and its perimeter is 62 inches. Find the length and width.
56. The length of a certain rectangle is 4 less than 3 times the width, and the area is 160. Find the length and the width.

Show the graphic solutions of the inequalities in Problems 57–60.

57. a. $3x - 5y < 8$ **b.** $y \geq x^2 - 3$

58. a. $\begin{cases} x + 2y \leq 8 \\ 2y - x > 3 \end{cases}$ **b.** $\begin{cases} y + x^2 \leq 4 \\ 2x + y \geq 0 \end{cases}$

59. a. $\begin{cases} 2y - x < 8 \\ 2x - y < 5 \\ x + y > 1 \end{cases}$ **b.** $\begin{cases} y - 2x \leq 3 \\ x + 2y \geq 2 \end{cases}$

60. **a.** $\begin{cases} -6 \le x - 3y \le 6 \\ |y - 2| \le 1 \end{cases}$ **b.** $|x + y| > 2$

In Problems 61–64 find the maximum or minimum value, as specified, subject to the given constraints.

61. Maximize $F = 10x + 5y$,

subject to $\begin{cases} 2x + 3y \le 12 \\ x + y \ge 2 \\ x \ge 1 \\ y \ge 0 \end{cases}$

62. Minimize $C = 2x + 3y + 5$,

subject to $\begin{cases} 2x + y \ge 4 \\ x \le 8 \\ 1 \le y \le 4 \end{cases}$

63. Minimize $G = 3x - y + 7$,

subject to $\begin{cases} 3x + 2y \ge 12 \\ x + 2y \ge 5 \\ x + y \le 8 \\ x \le 6 \\ y \le 4 \end{cases}$

64. Maximize $P = 60x + 100y$,

subject to $\begin{cases} x + 2y \le 14 \\ 6 \le x + y \le 10 \\ y \ge 2 \\ x \ge 0 \end{cases}$

65. A farmer has 300 acres that he will divide between corn and oats. In order to meet commitments he has made, he must plant at least 150 acres of corn and 50 acres of oats, and the demand for each suggests that he should plant at least twice as many acres in corn as in oats. If the profit from corn is $40 per acre and from oats is $30 per acre, how many acres should he plant of each for maximum profit?

66. A company manufactures two types of rotary lawn mowers, a standard model and a deluxe model. At most 100 lawn mowers can be produced each week, and no more than 40 of these can be the deluxe model. To meet demand, at least 40 standard and between 20 and 40 deluxe models should be produced each week. If the profit on the standard model is $50 and on the deluxe model $70, how many of each type should be produced each week for maximum profit?

67. The owner of an old-fashioned excursion train offers a short trip at $1.50 per person and a long trip at $5.00 per person. Each long trip he makes per day reduces by 3 the number of short trips he can make. If he makes no long trips, he can make 12 short trips, and if he makes no short trips, he can make 4 long trips. He agrees to offer at least 3 short trips and 1 long trip per day. The train holds 200 passengers, and it is 80% full on the average for short trips and 60% full for long trips. How many trips of each type should he make per day to maximize his income?

B

Solve the systems in Problems 68–73 by reducing the augmented matrix to triangular form.

68. $\begin{cases} 3x + 4y - 2z + \ w = 25 \\ x + 2y - \ z - \ w = -3 \\ 2x - \ y - 3z + 4w = 7 \\ -x - 3y + 5z - 2w = -7 \end{cases}$

69. $\begin{cases} x - \ y + 2z - \ w = 4 \\ x + 2y - \ z \quad = 5 \\ 2x - 3y + \ z - 4w = 7 \\ x + \ y - 4z - 2w = 4 \end{cases}$

70. $\begin{cases} 2x - \ y + 5z - 4w = 0 \\ 3x - 2y - 3z + 2w = 2 \\ -5x + 3y - 4z + 6w = 10 \\ 4x + \ y + 2z + 3w = -1 \end{cases}$

71. $\begin{cases} 2x - \ y + \ 4z - \ w = 0 \\ -3x + 2y + \ 3z - 2w = 0 \\ 4x - 3y - 10z + 5w = 0 \\ 5x + 2y + \ 7z + 8w = 0 \end{cases}$

72. $\begin{cases} 4x - 3y + 2z - \ w = 0 \\ 2x - \ y + \ z - 4w = 0 \\ -5x + 6y - 8z + \ w = 0 \\ 7x + 2y - 5z - 5w = 0 \end{cases}$

73. $\begin{cases} x + \ y - \ u \qquad\quad = 1 \\ 2x - \ y \qquad + 2v - \ w = 3 \\ \qquad\quad u + \ v + \ w = -1 \\ 3y - \ u + \ v - 2w = 2 \\ x \qquad + 2u \qquad + 5w = 5 \end{cases}$

In Problems 74 and 75 solve the system $AX = B$, using $X = A^{-1}B$.

74. The system of Problem 16.
75. The system of Problem 68.
76. Solve only for y by Cramer's rule:

$$\begin{cases} 2x - 3y = 3 \\ 4x + z = 5 \\ 2y - z + w = 0 \\ x + w = 4 \end{cases}$$

77. Solve for u and v by Cramer's rule:

$$\begin{cases} (a^2 - 1)u - 3av = 3 \\ au - 4v = 2 \end{cases}$$

78. Solve for x, y, and z by Cramer's rule:

$$\begin{cases} nx - my + mnz = 0 \\ x + mny - 2mz = 3n \\ 2mnx - y + 3nz = 4m \end{cases}$$

79. The points $(-1, 0), (2, -3)$, and $(3, 4)$ lie on a curve whose equation is of the form $y = ax^2 + bx + c$. Find a, b, and c.
    Hint. Each point must satisfy the equation.
80. The points $(5, 1), (-2, 0)$, and $(2, 2)$ lie on a circle whose equation is of the form $x^2 + y^2 + ax + by + c = 0$. Find a, b, and c.

In Problems 81 and 82 find the solution to the system.

81. $\begin{cases} x^2 + y^2 - 2x - 5y + 1 = 0 \\ x^2 + y^2 + 3x - 8y - 2 = 0 \end{cases}$

82. $\begin{cases} x^2 - 16y^2 = 5x - 20y \\ xy = 1 \end{cases}$

83. A 40% salt solution is mixed with a 25% solution, resulting in a 30% solution. Then 4 gallons of fresh water are added to this mixture, and the final solution is 20% salt. Find how many gallons of each original solution there were.
84. A man rows 8 miles upstream and back in 3 hours. The next day the current is twice as strong, and he makes the same trip upstream and back in 4 hours and 48 minutes. Assuming no change in his rowing effort from one day to the next, find how fast he can row in still water and the rate of the current the first day.
85. Find all points of intersection of the curves whose equations are $5x^2 - 2xy + y^2 - 12x - 16 = 0$ and $x^2 - y = 2$.
86. An isosceles triangle is inscribed in a circle of radius a. The two equal sides each have length $3a/2$. Find the length of the base and the altitude of the triangle in terms of a.
87. A girl rides a bicycle a distance of 5 miles and then has a flat tire. She has to walk 2 miles to a service station. If the total elapsed time for riding and walking was 1 hour, and if she rode at an average speed of 12 miles per hour faster than she walked, find her rate of riding and her rate of walking.
88. Two open-top boxes each have square bases. The base of the larger box is 2 inches larger on each side than that of the smaller, and the height of the larger box is 1 inch greater than that of the smaller box. If the volume of the larger box exceeds that of the smaller by 320 cubic inches, and the total surface area of the larger box is 124 square inches greater than that of the smaller, find the dimensions of the smaller box. (There are two solutions.)
89. A company produces two kinds of fertilizer. The first kind, for lawns, contains 20% nitrogen, 5% phosphorous, and 5% potash. The second kind, for gardens, contains 8% nitrogen, 10% phosphorous, and 8% potash. Each kind is packaged in 100-pound bags. The company has 1,960 pounds of nitrogen, 850 pounds of phosphorous, and 730 pounds of potash. The owner wants to have at least 30 bags of each type of fertilizer available. His profit on the lawn fertilizer is $6 per bag and on the garden fertilizer $5 per bag. How many bags of each type should he produce to maximize his profit?

THE CONIC SECTIONS

The conic sections are important in many fields, including astronomy, optics, acoustics, architecture, engineering, and art. They occur in calculus in diverse ways, often as part of a larger problem. The following theorem, taken from a calculus text, shows an interesting property of one of the conics that accounts for much of its usefulness in applications.*

Theorem   Let F be the focus of a parabola, $P_0(x_0, y_0)$ be a point on the parabola, and ℓ the tangent line to the parabola at P_0. Then ℓ makes equal angles with FP_0 and with the line through P_0 parallel to the axis of symmetry of the parabola. . . .

Proof   Without loss of generality, we can suppose that the parabola has vertex at $(0, 0)$ and equation $y^2 = 4px$. . . . Then the directrix is the line $x = -p$. We first find the equation of the tangent line ℓ to the parabola

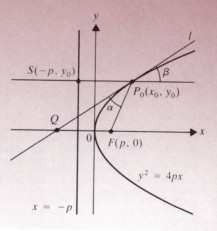

* James F. Hurley, *Calculus* (Belmont, Calif.: Wadsworth Publishing Company, © 1987, Wadsworth, Inc.), p. 487. Reprinted by permission.

at $P_0(x_0, y_0)$. Differentiating $y^2 = 4px$, we get

$$2yy' = 4p$$

so the slope of ℓ is

$$\left.\frac{dy}{dx}\right|_{(x_0, y_0)} = \frac{2p}{y_0}$$

Thus ℓ has the equation

$$y - y_0 = \frac{2p}{y_0}(x - x_0)$$

At the point $Q(x, 0)$ where ℓ meets the x-axis, we have $y = 0$. . . . Therefore

$$-y_0^2 = 2p(x - x_0) \rightarrow -4px_0 = 2px - 2px_0 \rightarrow -2px_0 = 2\,pxm$$

$$x = -x_0$$

Angle $FQP_0 = \beta$. . . . Therefore $\alpha = \beta$ if FQP_0 is an isosceles triangle. . . . From Definition . . . we have

$$d(F_1 P_0) = d(P_0, S) = x_0 + p = d(Q, 0) + d(0, F) = d(Q, F)$$

Thus the triangle is isosceles, so $\alpha = \beta$.

The point to be emphasized is not so much the result, since getting it made use of calculus, but rather the notions of the **focus, vertex, directrix,** and **equation of the parabola,** all of which we will study in this chapter.

INTRODUCTION

In Chapter 4 we introduced the curves known as conic sections and gave the standard forms of their equations when the curves were in certain positions. With the exception of the circle, whose equation we derived, our basic approach was to determine the graph, given the equation. In this chapter we will study the conic sections in more detail. Our approach will be to begin with the curves themselves and to derive their equations. Thus we begin with a *geometric* description of the curve, and we obtain an *algebraic* description (its equation).

Recall that the name *conic sections* comes from the fact that these are curves of intersection of a plane passing through a cone in different ways. (See Figure 22 in Chapter 4.) This description is primarily of historical interest. It was Apollonius of Perga, in the third century B.C., who first studied these curves extensively, writing eight books on the subject. Apollonius did not have the tools of analytical geometry at his disposal, which makes his discoveries all the more remarkable.

Present-day applications of conic sections are found in such diverse fields as art, architecture, astronomy, engineering, and physics. They play an important role in atomic and electromagnetic field theory, as well as in acoustics and optics. The orbits of planets around the sun and satellites around the earth are elliptical. Parabolic reflectors are used for radar, radio telescopes, searchlights, and solar energy devices. Certain atomic particles follow paths that are approximately hyperbolic.

Before we begin our new approach to conic sections, we introduce a notion that will be useful in formulating the definitions.

2   THE LOCUS OF A POINT

In deriving the equations of some curves, it is convenient to consider the curve as having been traced out by a moving point. A graph is really a static thing, just a collection of points, but there is no harm in supposing it to be the path of a moving point. In the final analysis, it is the equation of the path that we determine, and the notion of the moving point will disappear. To illustrate this, let us consider a circle of radius 2 with its center at the origin. One way of describing the circle is to say it is the collection of all points that are 2 units from the origin. This is the static description. It is equally valid, and useful in the derivation, to say the circle is the path traced out by a point moving so as always to be 2 units from the origin. We might call this the **kinematic description.** The path traced out by a moving point is called the **locus of the point.**

In what follows we will frequently need to use the formula for the distance between two points, which we derived in Chapter 4. It states that if the points have coordinates (x_1, y_1) and (x_2, y_2), the distance d between them is given by

$$d = \sqrt{(x_1 - x_2)^2 + (y_1 - y_2)^2}$$

EXAMPLE 1 |||||||||||||||||||||||||||||||||||||||||||||||||||||||||||||||||||||||||||||||||||||||||||||||||||||||||||||||

Find the equation of the locus of a point that moves so as to be always equidistant from $A(3, 4)$ and $B(-2, 1)$.

Solution   Let P denote the moving point, and denote its coordinates by (x, y). Our problem is to determine what restrictions must be imposed on x and y (this will be in the form of an equation) so that P will satisfy the given condition. The condition requires that $\overline{PA} = \overline{PB}$, which, translated in terms of coordinates, is equivalent to

$$\sqrt{(x - 3)^2 + (y - 4)^2} = \sqrt{(x + 2)^2 + (y - 1)^2}$$

On squaring both sides,* expanding, and then simplifying, we obtain the equivalent equations

$$x^2 - 6x + 9 + y^2 - 8y + 16 = x^2 + 4x + 4 + y^2 - 2y + 1$$

$$10x + 6y - 20 = 0$$

$$5x + 3y - 10 = 0$$

This is the equation of the locus. We recognize the answer as the equation of a straight line. It is, in fact, the equation of the perpendicular bisector of AB. ▲

EXAMPLE 2

Find the equation of the locus of a point that moves so as to be always 2 units from the origin.

Solution   This is the circle referred to above. Let $P(x, y)$ be the moving point. Then we must have

$$\overline{OP} = 2$$

or equivalently,

$$\overline{OP}^2 = 4$$

so that

$$x^2 + y^2 = 4$$

is the required equation. ▲

EXERCISE SET 2

In Problems 1–17 find the equation of the locus of a point that moves so as to satisfy the given condition.

1. Its distance from $(1, 2)$ is always 3 units; describe the locus.
2. It is always equidistant from $A(2, -1)$ and $B(-3, -4)$.
3. Its distance from $A(-1, 3)$ is always twice its distance from $B(2, -1)$.
4. It is always half as far from $A(2, 4)$ as from $B(-1, -3)$.

5. Its distance from $A(3, -4)$ is half its distance from $B(0, 2)$.
6. The ratio of its distance from $A(-2, -4)$ to its distance from $B(3, -1)$ is $\frac{3}{2}$.
7. Its distance from the point $(3, 0)$ is always equal to its distance from the line $x + 3 = 0$.
8. Its distance from $A(1, 2)$ is one-third its distance from $B(-3, 4)$.
9. Its distance from the line $y - 2 = 0$ is equal to its distance from the point $(1, -2)$.
10. Its distance from the line $x - 2 = 0$ is equal to its distance from the point $(-2, 3)$.

---

* Squaring introduces nothing extraneous in this case, because for positive numbers A and B, $\sqrt{A} = \sqrt{B}$ if and only if $A = B$, since $\sqrt{A} = -\sqrt{B}$ is impossible.

11. The sum of its distances from $P_1(2, 0)$ and $P_2(-2, 0)$ is always 6 units.
12. The difference of its distances from $P_1(0, 4)$ and $P_2(0, -4)$ is always 2 units (that is, $|\overline{P_1P} - \overline{P_2P}| = 2$).
13. Its distance from (0, 1) equals its distance from the line $y = -1$.
14. Its distance from the line $x = 6$ is half its distance from the point (0, 4).
15. Its distance from the point (0, 2) is half its distance from the line $y - 3 = 0$.

16. The sum of its distances from (1, 4) and (1, −2) is 10 units.
17. The difference of its distances from (2, 1) and (−3, 1) is always 4 units.

18. Derive the following formula for the distance d from the line $Ax + By + C = 0$ to the point (x_1, y_1):

$$d = \frac{|Ax_1 + By_1 + C|}{\sqrt{A^2 + B^2}}$$

Hint. Find the equation of the line through (x_1, y_1) and perpendicular to the given line; then get the point of intersection.

19. Find the locus of a point that moves so that its distance from the line $3x + 4y + 7 = 0$ equals its distance from the point (2, 1). (Use the result of Problem 18.)
20. Find the equation of the locus of the point whose distance from $x - 2y = 3$ is always twice its distance from the point (−1, −2).

3 THE CIRCLE

A circle can be defined, using the locus concept, as follows: *A circle with center* (h, k) *and radius* r *is the locus of a point that moves in a plane so that its distance from* (h, k) *is always equal to* r. Let $P(x, y)$ be the moving point. Then its distance from (h, k) is $\sqrt{(x - h)^2 + (y - k)^2}$. Setting this distance equal to r and then squaring both sides gives the standard form

$$(x - h)^2 + (y - k)^2 = r^2 \qquad (1)$$

that we found in Chapter 4. When the center is at the origin, this equation becomes

$$x^2 + y^2 = r^2$$

When we know the center and the radius, equation (1) is used to find the equation of the circle. When we are given an equation in the form (1), we recognize that its graph is a circle with center (h, k) and radius r. For example,

$$(x + 2)^2 + (y - 1)^2 = 5$$

is the equation of a circle with center $(-2, 1)$ and radius $\sqrt{5}$.

Recall from Chapter 4 that every equation of the form

$$x^2 + y^2 + ax + by + c = 0$$

has a graph that is a circle or a point (a degenerate circle), or there is no graph. To determine which it is, you complete the squares in x and in y to put the equation into the standard form (1). We review how to do this in the next four examples.

EXAMPLE 3

Determine the nature of the graph of the equation

$$x^2 + y^2 + 4x - 2y = 0$$

Solution   We complete the squares, taking care to obtain an equivalent equation by adding the needed quantities (the circled terms) to both sides.

$$(x^2 + 4x + ④) + (y^2 - 2y + ①) = 0 + ④ + ①$$
$$(x + 2)^2 + (y - 1)^2 = 5$$

We recognize this as the equation of the circle mentioned above, with center $(-2, 1)$ and radius $\sqrt{5}$.   ▲

EXAMPLE 4

Find the graph of the equation

$$x^2 + y^2 - 6x + 2y + 14 = 0$$

or else show there is no graph.

Solution   Proceeding as in Example 3, we have

$$(x^2 - 6x + 9) + (y^2 + 2y + 1) = -14 + 9 + 1$$
$$(x - 3)^2 + (y + 1)^2 = -4$$

Since the left-hand side cannot be negative, the equation is not satisfied by any point (x, y). So there is no graph.   ▲

EXAMPLE 5

Find the graph of

$$x^2 + y^2 + 10x + 8y + 41 = 0$$

Solution
$$(x^2 + 10x + 25) + (y^2 + 8y + 16) = -41 + 25 + 16$$
$$(x + 5)^2 + (y + 4)^2 = 0$$

The equation is satisfied by the point $(-5, -4)$ and by no other point. So the graph is the degenerate circle consisting of this point only. We could think of this as a circle with center $(-5, -4)$ and radius 0.   ▲

EXAMPLE 6

Find the graph of

$$3x^2 + 3y^2 - 2x + 5y - 4 = 0$$

Solution   The coefficient 3 of x^2 and of y^2 is a minor complication, but we remove it by dividing both sides by 3, and then complete the squares.

$$x^2 + y^2 - \frac{2}{3}x + \frac{5}{3}y - \frac{4}{3} = 0$$

$$\left(x^2 - \frac{2}{3}x + \frac{1}{9}\right) + \left(y^2 + \frac{5}{3}y + \frac{25}{36}\right) = \frac{4}{3} + \frac{1}{9} + \frac{25}{36}$$

$$\left(x - \frac{1}{3}\right)^2 + \left(y + \frac{5}{6}\right)^2 = \frac{77}{36}$$

The graph is a circle with center $(\frac{1}{3}, -\frac{5}{6})$ and radius $\sqrt{77}/6 \approx 1.46$. The only thing different about this problem was the messy arithmetic.   ▲

These examples serve to illustrate all possibilities for an equation of the form

$$x^2 + y^2 + ax + by + c = 0$$

Either it has a graph that is a circle, a degenerate circle, or else there is no graph. When you see the combination $x^2 + y^2$, or $Ax^2 + Ay^2$, in an equation that otherwise involves only linear terms, you should immediately know that if it has a graph, then that graph is a circle (or a degenerate circle).

EXERCISE SET 3

A

Find the equation of the circle having the given center and radius in Problems 1 and 2.

1. **a.** Center $(1, -3)$,   radius 2
   b. Center $(-2, -4)$,   radius 4
2. **a.** Center $(-2, 3)$,   radius 5
   b. Center $(4, 0)$,   radius 4

In Problems 3–9 write the equation in standard form, and determine whether the graph is a circle or a point, or if there is no graph. When there is a graph, draw it.

3. $x^2 + y^2 - 4x + 6y + 4 = 0$
4. $x^2 + y^2 + 4x - 12 = 0$
5. $x^2 + y^2 + 2x - 8y + 1 = 0$
6. $x^2 + y^2 - 6x + 2y + 1 = 0$
7. $x^2 + y^2 + 8x - 4y + 20 = 0$
8. $x^2 + y^2 + 4x - 6y + 14 = 0$
9. $2x^2 + 2y^2 + 3x - 5y - 3 = 0$
10. Find the equation of the circle of radius 5 whose center is in the fourth quadrant and that is tangent to both coordinate axes.
11. The line segment joining $(-2, 1)$ and $(4, 7)$ is a diameter of a certain circle. Find the equation of the circle.
12. The portion of the line $2x + 3y = 12$ that is cut off by the x and y axes is a diameter of a certain circle. Find the equation of the circle.
13. Find the equation of the tangent line to the circle $(x - 2)^2 + (y + 3)^2 = 25$ at the point $(-1, 1)$.
    Hint. The tangent line to a circle is perpendicular to the radius drawn to the point of tangency.

14. Find the equation of the tangent line to the circle $x^2 + y^2 + 4x - 10y + 19 = 0$ at the point $(-3, 2)$.
    (See the hint in Problem 13.)

15. The center of a circle of radius 6 lies on the line $x - 2y = 4$, and it is tangent to the y axis. Find its equation. (There are two solutions.)

16. Find the equations of the two circles, each having radius 8, with center on the line $y = 2x + 4$, and tangent to the x axis.

17. Show that the circles $x^2 + y^2 - 6x + 8y = 0$ and $x^2 + y^2 + 4x - 16y + 4 = 0$ are tangent to one another. Find the equation of the line joining their centers and the equation of the common tangent line.

    Hint. Find the distance between the centers and compare with the radii.

18. Show that the locus of a point that moves so that its distance from $(3, 4)$ is twice its distance from $(-1, 2)$ is a circle. Find its center and radius.

19. Find the equation of the circle passing through the three points $(2, 2)$, $(-5, 1)$, and $(4, -2)$.
    Hint. One approach is to observe that each point must satisfy $x^2 + y^2 + Ax + By + C = 0$, so A, B, and C can be determined.

20. Find the equation of the circle that circumscribes the triangle whose sides are given by $4x - 3y = 21$, $x + 3y = 9$, and $2x + y = 3$.

21. Find the points of intersection of the circles $x^2 + y^2 + 2x + 3y = 4$ and $x^2 + y^2 + x + 2y = 3$. Find the equation of the common chord and the equation of its perpendicular bisector.

22. Find the smaller area between the line $x + 2y = 5$ and the circle $x^2 + y^2 - 4x + 2y - 5 = 0$.

THE PARABOLA

A parabola is defined as the locus of a point that moves in a plane so as to be always equidistant from a fixed line and a fixed point in that plane. The fixed line is called the **directrix** of the parabola, and the fixed point is called the **focus**. The line through the focus and perpendicular to the directrix is called the **axis** of the parabola, and the point where the parabola crosses its axis is called its **vertex.** The vertex is halfway between the focus and the directrix.

If we know the equation of the directrix of a parabola and the coordinates of its focus, its equation can be obtained by requiring of the point $P(x, y)$ that $\overline{PF} = \overline{PP'}$, as shown in Figure 1. In general, this will result in a rather messy equation. If the directrix happens to be either vertical or horizontal, however, the situation is considerably simplified, and if the vertex is also at the origin, the equation is very easily obtained. We begin with this simplest position. First, suppose the directrix is vertical and is p units to the left of the y axis. Then the focus F must be p units to the right of the y axis, since the vertex is at the origin. Also, the axis of the parabola is the x axis. For any location of the point $P(x, y)$ that traces out the parabola we must have $\overline{PF} = \overline{PP'}$, where P' as shown in Figure 2 is the intersection of the directrix and a horizontal line through P. In terms of coordinates, this requirement becomes

$$\sqrt{(x - p)^2 + y^2} = x + p$$

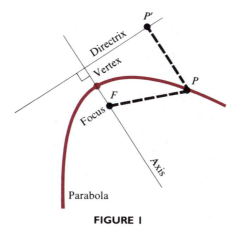

FIGURE I

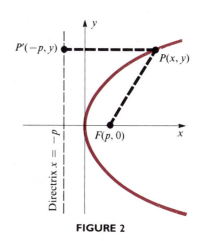

FIGURE 2

These two nonnegative numbers are equal if and only if their squares are equal:

$$x^2 - 2px + p^2 + y^2 = x^2 + 2px + p^2$$
$$y^2 = 4px$$

This, then, is the equation of the parabola described. The constant p represents the distance from the vertex to the focus, or equivalently, the distance from the vertex to the directrix. This position is sometimes referred to as the **first standard position.** The second, third, and fourth standard positions are those obtained by three successive counterclockwise rotations of $90°$ each (Figure 3). It is a relatively easy matter to show that the equations for these other three positions are $x^2 = 4py$ **(II),** $y^2 = -4px$ **(III),** and $x^2 = -4py$ **(IV).** Verification of these is left for the exercises. It is important to observe that when the axis is horizontal (positions I and III), y appears to the second power and x to the first; when the axis is vertical (positions II and IV), this situation is reversed. Also, the presence of a minus sign indicates a left-opening parabola when the axis is horizontal and a downward-opening one when the axis is vertical.

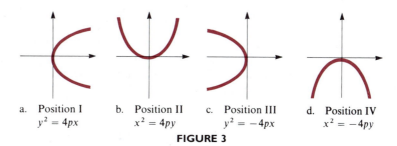

a. Position I   b. Position II   c. Position III   d. Position IV
   $y^2 = 4px$ $x^2 = 4py$ $y^2 = -4px$ $x^2 = -4py$

FIGURE 3

By the translation theorem (see Chapter 4, Section 5), the equations of parabolas in the four standard positions with vertex at (h, k) are as follows:

Position I: $(y - k)^2 = 4p(x - h)$

Position II: $(x - h)^2 = 4p(y - k)$

Position III: $(y - k)^2 = -4p(x - h)$

Position IV: $(x - h)^2 = -4p(y - k)$

Parabolas having horizontal axes have equations of positions I or III, and those with vertical axes have equations of positions II or IV. If the left-hand sides of these equations are expanded and terms are collected, we find that two basic forms emerge:

Vertical axis: $x^2 + Cx + Dy + E = 0$

Horizontal axis: $y^2 + Cx + Dy + E = 0$

We wish now to reverse the procedure; that is, if we are given an equation of one of the two types above, can we conclude that it represents a parabola? The answer is yes, and the proof consists of showing, by completing the square, that every such equation can be put in one of the forms of positions I–IV.* We illustrate this with examples.

EXAMPLE 7

Discuss the graph of $x^2 - 2x + 4y - 7 = 0$.

Solution First, complete the square in x:

$$x^2 - 2x \boxed{+ 1} = -4y + 7 \boxed{+ 1}$$
$$(x - 1)^2 = -4y + 8$$
$$(x - 1)^2 = -4(y - 2)$$

This is the form of position IV, so we know it represents a parabola having a vertical axis opening downward with its vertex at the point $(1, 2)$.

For applications, it is often sufficient to obtain a sketch of a curve, as opposed to an accurate drawing; for this purpose, locating the vertex of a parabola and recognizing its position are usually enough. If more accuracy is desired, we can get a few points on the curve by substitution. The location of the focus is seldom necessary, but it is easy to do when needed, since the distance from the vertex to the focus is p, which can be read off when the equation is in standard form. In our

---

* This assumes that in the equation $x^2 + Cx + Dy + E = 0$, $D \neq 0$ and that in $y^2 + Cx + Dy + E = 0$, $C \neq 0$. Otherwise, the graph would be two parallel or coincident lines (a degenerate parabola), or would not exist.

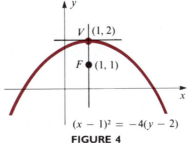

$$(x - 1)^2 = -4(y - 2)$$

FIGURE 4

example, we have $p = 1$. The focus is therefore at the point $(1, 1)$ and the graph is shown in Figure 4.   ▲

One way the focus can be useful in making a rapid sketch is by observing the length of the line segment through the focus, perpendicular to the axis, and terminating on the parabola. This segment is called the **latus rectum.** The length of the latus rectum is found easily, since by definition the distance of one of its end points from the directrix must equal its distance from the focus. But the distance from the directrix is $2p$. So the length of the latus rectum is $4p$ (Figure 5). This enables us to see how "fat" or "skinny" the parabola is.

To make a rapid sketch, we proceed as follows. From the standard form we locate the vertex and observe the position. The number p tells us the distance of the focus from the vertex. We proceed $2p$ units in either direction from the focus and perpendicular to the axis to get two more points on the curve, namely, the end points of the latus rectum. In Example 1, after having determined that $p = 1$ so that the focus is at $(1, 1)$, we know that the end points of the latus rectum are at $(-1, 1)$ and $(3, 1)$. With these points and the vertex $(1, 2)$ we are able to sketch the curve.

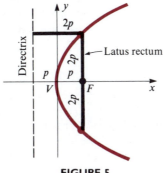

FIGURE 5

EXAMPLE 8 ||||||||||||||||||||||||||||||||||||||||||||||||||||||||||||||||||||||||||||||||||||||||||||||||||||||||||

Discuss and sketch the graph of: $y^2 - 6x + 4y + 16 = 0$

Solution
$$y^2 + 4y \,\boxed{+\ 4} = 6x - 16 \,\boxed{+\ 4}$$
$$(y + 2)^2 = 6x - 12$$
$$(y + 2)^2 = 6(x - 2)$$

This is the equation of a parabola in position I, with vertex $(2, -2)$ and $4p = 6$, so that $p = \frac{3}{2}$. The latus rectum extends 3 units to either side of the focus (Figure 6).

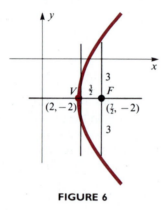

FIGURE 6

The location of the focus and the length of the latus rectum are helpful, but primary emphasis should be placed on recognition of the general nature of the curve. For example, in the equation $y^2 - 6x + 4y + 16 = 0$, we should see immediately that this represents a parabola having a horizontal axis (since y^2, rather than x^2, appears). Next in importance is the location of the vertex, which is obtained by completing the square in the squared variable and writing the equation in one of the standard forms so that (h, k) can be determined. Remember that regardless of which variable is squared, h is found with the x term and k with the y term. For example, in

$$(y - 2)^2 = -10(x + 3)$$

$h = -3$ and $k = 2$, whereas in

$$(x - 2)^2 = -10(y + 3)$$

$h = 2$ and $k = -3$.

We give one more example to show that the arithmetic does not always work out so nicely.

EXAMPLE 9

Discuss and sketch $3x^2 - 5x + 2y - 4 = 0$.

Solution   We see immediately that this is a parabola with vertical axis. To get it into standard form, first divide by 3 and then complete the square:

$$x^2 - \tfrac{5}{3}x + \tfrac{2}{3}y - \tfrac{4}{3} = 0$$

$$x^2 - \tfrac{5}{3}x \left(+ \tfrac{25}{36}\right) = -\tfrac{2}{3}y + \tfrac{4}{3}\left(+ \tfrac{25}{36}\right)$$

$$\left(x - \tfrac{5}{6}\right)^2 = -\tfrac{2}{3}y + \tfrac{73}{36}$$

$$\left(x - \tfrac{5}{6}\right)^2 = -\tfrac{2}{3}\left(y - \tfrac{73}{24}\right)$$

So the vertex is at the point $(\tfrac{5}{6}, \tfrac{73}{24})$, and it opens downward (position IV). Since $4p = \tfrac{2}{3}$, we have that $p = \tfrac{1}{6}$. The focus is therefore $\tfrac{1}{6}$ unit down from the vertex, and the latus rectum extends $\tfrac{1}{3}$ unit to either side. This time, since the focus is so close to the vertex, the latus rectum is not much help in making the sketch. The point $(0, 2)$ is seen to lie on the curve, however, by setting $x = 0$ in the original equation. Since the curve is symmetric to its axis, this effectively gives another point at the same distance on the other side of the axis, namely, $(\tfrac{5}{3}, 2)$ (Figure 7).

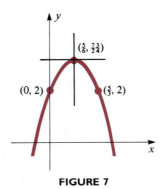

FIGURE 7

The parabola is useful in structural design and in art and architecture. Parabolas also possess an interesting and useful reflecting property. If a parabolic arc is rotated about its axis, a surface called a **paraboloid** is obtained. When sound waves or light rays strike such a surface, they are all reflected to the focus of the parabola (Figure 8). Reflecting telescopes make use of this principle, for example, as do radiotelescopes. Reversing this, if light is emitted from the focus, then all rays are reflected off the surface in rays parallel to the axis. For this reason headlights on a car are in the shape of paraboloids. Since it is impossible to have a point source of light precisely at the focus, the rays do not reflect exactly in the ideal way, but they do approximate this situation.

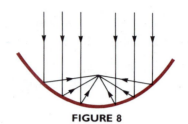

FIGURE 8

EXERCISE SET 4

A

In Problems 1–11 find the equation of the parabola described.

1. **a.** Focus $(2, 0)$; directrix $x = -2$
   b. Focus $(0, -3)$; directrix $y = 3$
2. **a.** Vertical axis; vertex at the origin; passing through $(-3, 2)$
   b. Horizontal axis; vertex at the origin; passing through $(-4, -3)$
3. **a.** Vertex $(1, -2)$; directrix $y = -4$
   b. Vertex $(-3, 1)$; focus $(-3, -1)$
4. **a.** Focus $(2, 4)$; directrix $x = 4$
   b. Focus $(-4, -2)$; vertex $(-5, -2)$
5. **a.** Vertex $(2, -3)$; end points of latus rectum $(-4, 0)$ and $(8, 0)$
   b. Directrix $x + 2 = 0$; end points of latus rectum $(1, 2)$ and $(1, -4)$
6. Vertex $(2, 3)$; focus $(2, -1)$
7. Focus $(0, 4)$; directrix $x = 6$
8. Vertex $(2, -4)$; axis vertical; passing through $(-2, 0)$
9. Vertex $(3, -2)$; axis horizontal; passing through $(-1, 2)$
10. Axis horizontal; passing through $(0, 3)$, $(-2, 1)$, and $(6, -3)$
    Hint. Substitute the given points in the equation $y^2 + Ax + By + C = 0$ to obtain three equations in the unknown constants A, B, and C.
11. Axis vertical; passing through $(1, 1)$, $(-1, 3)$, and $(3, 5)$
    Hint. Follow a procedure similar to that in Problem 10.

In Problems 12–18 write the equation in one of the standard forms I–IV and sketch the graph. Give the coordinates of the focus.

12. **a.** $y^2 = 8x$ **b.** $y^2 + 4x = 0$
    c. $x^2 = -6y$ **d.** $x^2 - 8y = 0$
13. $x^2 - 4x - 12y - 8 = 0$
14. $y^2 + 8x - 6y + 41 = 0$
15. $y = x^2 - 2x$
16. $2y^2 - 5x + 3y - 4 = 0$
17. $4y^2 + 4y + 24x - 35 = 0$
18. $x = y^2 - 3y + 4$
19. If f is a quadratic function for which $f(0) = 4$, $f(2) = 0$, and $f(-2) = 0$, find $f(x)$ and draw its graph.
20. If g is a quadratic function for which $g(1) = -1$, $g(2) = 0$, and $g(-1) = 3$, find $g(x)$ and draw its graph.

In Problems 21–24 draw the graphs and shade the areas described. Find the points of intersection of the curves.

21. Area between $y = 3x - x^2$ and $y = x - 3$
22. Area below the parabola $y = 4 - x^2$ and inside the circle $x^2 + y^2 - 8y + 14 = 0$
23. Area between $y = \sqrt{x}$ and $x - 2y = 0$
24. Area between $x^2 - 6x + 4y - 11 = 0$ and $x^2 - 6x - 8y + 1 = 0$

B

25. Find the equation of the circle that has its center at the focus of the parabola $x^2 - 2x + 8y - 23 = 0$ and that is tangent to this parabola at its vertex.

26. Verify that the equations given on page 525 for parabolas with vertices at the origin in positions II, III, and IV are correct.

In Problems 27 and 28 draw the graphs and shade the area described. Find the points of intersection of the curves.

27. Area between $y^2 = 8x$ and $8x^2 + 5y = 12$

28. Area inside the circle $x^2 + y^2 - 2x + 4y + 1 = 0$ and above the parabola $x^2 - 2x - 3y = 5$

29. Use the result of Problem 18 in Exercise Set 2 and the definition of a parabola to find the equation of a parabola having directrix $x - 2y - 4 = 0$ and focus $(-2, 0)$. Sketch the parabola.

THE ELLIPSE

An **ellipse is the locus of a point that moves in a plane so that the sum of its distances from two fixed points in the plane is constant.** The fixed points are called the **foci** of the ellipse, and the line through them is sometimes called the **focal axis.** The point on the focal axis midway between the foci is the **center,** and the points where the ellipse crosses its focal axis are called the **vertices.** The line segment joining the two vertices is called the **major axis** (so the major axis is a part of the focal axis); the line segment through the center, perpendicular to the major axis, and terminating at the ellipse is called the **minor axis.** These definitions are illustrated in Figure 9 and are intrinsic to the ellipse itself and independent of any particular orientation with respect to a coordinate system.

According to the definition, the sum of the distances of the point that traces out the ellipse from the two foci is constant. The size of this constant and the distance between the foci determine the shape of the ellipse. It is convenient in deriving equations of ellipses in standard positions with respect to the x and y axes to designate the constant referred to in the definition by $2a$ (this avoids fractions).

The simplest positions are those in which the focal axis is either horizontal or vertical and the center is at the origin. We will derive the equation for the horizontal case and leave the vertical case as an exercise. Let the coordinates of the foci be $(\pm c, 0)$, as shown in Figure 10. The point P that traces out the ellipse must satisfy

$$\overline{PF_1} + \overline{PF_2} = 2a \tag{2}$$

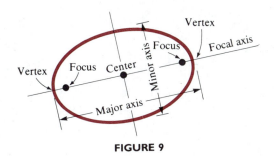

FIGURE 9

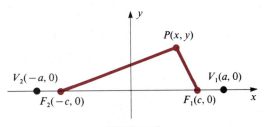

FIGURE 10

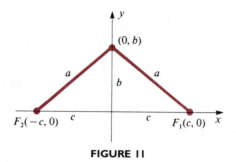

FIGURE 11

We must have $2a > 2c$, or $a > c$, since the sum of the lengths of two sides of a triangle ($\overline{PF_1} + \overline{PF_2} = 2a$) must be greater than the length of the third side ($\overline{F_1F_2} = 2c$). The coordinates of the end points of the major axis (that is, the vertices) and of the minor axis are obtained as follows. When P is at V_1, equation (2) gives

$$(x - c) + (x + c) = 2a$$

or $2x = 2a$, so that $x = a$. Thus, V_1 has coordinates $(a, 0)$. Similarly, V_2 is at $(-a, 0)$. Let the end points of the minor axis be $(0, b)$ and $(0, -b)$. Then when P coincides with $(0, b)$, we have that $\overline{PF_1} = \overline{PF_2}$; since $\overline{PF_1} + \overline{PF_2} = 2a$, it follows that $\overline{PF_1} = a$ and $\overline{PF_2} = a$. Then, by the Pythagorean theorem (see Figure 11),

$$b^2 = a^2 - c^2 \tag{3}$$

Now consider an arbitrary location of P. Condition (2) in terms of coordinates is

$$\sqrt{(x - c)^2 + y^2} + \sqrt{(x + c)^2 + y^2} = 2a$$

We seek an equation that is free of radicals and so square twice, after suitable rearrangement of terms:

$$\sqrt{(x - c)^2 + y^2} = 2a - \sqrt{(x + c)^2 + y^2}$$
$$x^2 - 2cx + c^2 + y^2 = 4a^2 - 4a\sqrt{(x + c)^2 + y^2} + x^2 + 2cx + c^2 + y^2$$
$$a\sqrt{(x + c)^2 + y^2} = a^2 + cx$$
$$a^2(x^2 + 2cx + c^2 + y^2) = a^4 + 2a^2cx + c^2x^2$$
$$(a^2 - c^2)x^2 + a^2y^2 = a^2(a^2 - c^2)$$

Or, in view of equation (3),

$$b^2x^2 + a^2y^2 = a^2b^2$$

Finally, we divide both sides by a^2b^2:

$$\frac{x^2}{a^2} + \frac{y^2}{b^2} = 1 \tag{4}$$

The fact that the squaring operations in this procedure led to equivalent equations can be demonstrated by showing that each time this was done, both sides of the equation represented positive numbers, and since two positive numbers are equal if and only if their squares are equal, the equivalence follows.

The corresponding equation for a vertical focal axis is

$$\frac{x^2}{b^2} + \frac{y^2}{a^2} = 1 \tag{5}$$

It should be emphasized that in equations (4) and (5) the number a is greater than b, since $b^2 = a^2 - c^2$. Whether the ellipse has a horizontal or vertical major axis is recognized from the equation in form (4) or (5) by observing which denominator is greater. For example,

$$\frac{x^2}{4} + \frac{y^2}{9} = 1$$

has a vertical major axis with $a = 3$, $b = 2$, whereas

$$\frac{x^2}{4} + \frac{y^2}{1} = 1$$

has a horizontal major axis with $a = 2$, $b = 1$.

If the center is at the point (h, k), then equations (4) and (5) become, by the translation theorem,

$$\frac{(x - h)^2}{a^2} + \frac{(y - k)^2}{b^2} = 1 \tag{6}$$

and

$$\frac{(x - h)^2}{b^2} + \frac{(y - k)^2}{a^2} = 1 \tag{7}$$

The shape of an ellipse, whether "fat" or "skinny," depends on the relative sizes of c and a. To get a measure of this, we define what is known as the **eccentricity** e by

$$e = \frac{c}{a}$$

Since $0 < c < a$, we have $0 < e < 1$. A small eccentricity indicates that the ellipse tends toward being circular, whereas an eccentricity close to 1 indicates that the ellipse is elongated. These facts can be seen by observing that when c/a is near 0, c is small in comparison to a, so that b, which equals $\sqrt{a^2 - c^2}$, is only slightly less than a; hence, the major and minor axes are nearly equal. On the other hand, when c/a is near 1, c and a are nearly equal, and so $b = \sqrt{a^2 - c^2}$ is close to 0. Thus, the minor axis is small in relation to the major axis.

The following examples illustrate how equations of ellipses can be found when certain properties are known.

EXAMPLE 10 ΙΙΙΙΙΙΙΙΙΙΙΙΙΙΙΙΙΙΙΙΙΙΙΙΙΙΙΙΙΙΙΙΙΙΙΙΙΙΙΙΙΙΙΙΙΙΙΙΙΙΙΙΙΙΙΙΙΙΙΙΙΙΙΙΙΙΙΙΙΙΙΙΙΙΙΙΙΙΙΙΙΙΙΙΙΙΙΙΙΙΙΙΙΙ

Find the equation of the ellipse with center at $(3, -4)$, a focus at $(0, -4)$, and a vertex at $(8, -4)$.

Solution The information given enables us to conclude the following:

1. The major axis is horizontal, so the equation is of the form

$$\frac{(x-3)^2}{a^2} + \frac{(y+4)^2}{b^2} = 1$$

2. The distance from the center to a focus is 3, so $c = 3$.
3. The distance from the center to a vertex is 5, so $a = 5$.

From this we get that $b^2 = a^2 - c^2 = 25 - 9 = 16$, so $b = 4$. The equation is therefore

$$\frac{(x-3)^2}{25} + \frac{(y+4)^2}{16} = 1$$ ▲

EXAMPLE 11 ΙΙΙΙΙΙΙΙΙΙΙΙΙΙΙΙΙΙΙΙΙΙΙΙΙΙΙΙΙΙΙΙΙΙΙΙΙΙΙΙΙΙΙΙΙΙΙΙΙΙΙΙΙΙΙΙΙΙΙΙΙΙΙΙΙΙΙΙΙΙΙΙΙΙΙΙΙΙΙΙΙΙΙΙΙΙΙΙΙΙΙΙΙΙΙΙ

Find the equation of the ellipse with vertices at $(1, 8)$ and $(1, -4)$ and with eccentricity $\frac{2}{3}$.

Solution The center is at the midpoint of the major axis, which is $(1, 2)$. Also, $a = 6$. Since $e = c/a$, we have $c/6 = \frac{2}{3}$, so that $c = 4$. Also, $b^2 = a^2 - c^2 = 36 - 16 = 20$. Since the ellipse has a vertical major axis, its equation is

$$\frac{(x-1)^2}{20} + \frac{(y-2)^2}{36} = 1$$ ▲

If equations (6) and (7) are multiplied out and terms are rearranged, each results in an equation of the type

$$Ax^2 + Cy^2 + Dx + Ey + F = 0 \tag{8}$$

where A and C are like in sign but unequal. (We are reserving the letter B for the coefficient of an xy term, which we will consider in Section 7.) Now we reverse the procedure and ask whether an equation in the form of (8) with these restrictions on A and C always represents an ellipse. To answer the question in any given case, we have only to complete the squares in x and y and write the equation in the form of equation (6) or (7), if possible. The following examples serve to illustrate the possibilities that may occur.

EXAMPLE 12

Discuss the graph of the equation: $4x^2 + 9y^2 + 16x - 18y - 11 = 0$

Solution   To complete the squares, we first factor out the coefficient of x^2 from terms involving x and the coefficient of y^2 from those involving y:

$$4(x^2 + 4x + ④) + 9(y^2 - 2y + ①) = 11 + ⑯ + ⑨$$

Notice that we added 4 and 1 inside the parentheses to complete the squares, but this had the effect of adding 16 and 9, which had to be balanced off accordingly. Simplifying, we get

$$4(x + 2)^2 + 9(y - 1)^2 = 36$$

which on dividing by 36 becomes

$$\frac{(x + 2)^2}{9} + \frac{(y - 1)^2}{4} = 1$$

This is in standard form (6) and so is the equation of an ellipse with center at $(-2, 1)$, major axis horizontal, and $a = 3$, $b = 2$. If it is desired, we could determine the coordinates of the foci and the eccentricity, since $c^2 = a^2 - b^2 = 9 - 4 = 5$. A sketch of the ellipse is given in Figure 12.

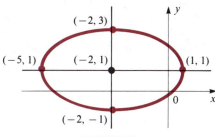

FIGURE 12

EXAMPLE 13

Discuss the graph of the equation $9x^2 + 4y^2 - 54x - 16y + 61 = 0$.

Solution   We proceed as before:

$$9(x^2 - 6x + 9) + 4(y^2 - 4y + 4) = -61 + 81 + 16$$
$$9(x - 3)^2 + 4(y - 2)^2 = 36$$
$$\frac{(x - 3)^2}{4} + \frac{(y - 2)^2}{9} = 1$$

So this is an ellipse with center at $(3, 2)$, major axis vertical, and $a = 3$, $b = 2$ (Figure 13).

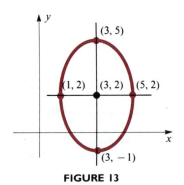

FIGURE 13

EXAMPLE 14

Discuss the graph of $4x^2 + 5y^2 - 8x + 20y + 24 = 0$

Solution   Again, as above, we have

$$4(x^2 - 2x + 1) + 5(y^2 + 4y + 4) = -24 + 4 + 20$$
$$4(x - 1)^2 + 5(y + 2)^2 = 0$$

This cannot be put in either of the standard forms (6) or (7), so it does not represent an ellipse. In fact, the left-hand side is a positive number except when $x = 1$ and $y = -2$, and at that point the equation is satisfied. The entire graph therefore consists of the single point $(1, -2)$. This is an example of a **degenerate ellipse.**

EXAMPLE 15

Discuss the graph of $4x^2 + 5y^2 - 8x + 20y + 25 = 0$.

Solution   This is just like the preceding problem except for the constant. The final form is

$$4(x - 1)^2 + 5(y + 2)^2 = -1$$

and this is not satisfied by any point. So there is no graph.

EXAMPLE 16

Discuss the graph of $3x^2 + 5y^2 - 2x + 7y - 11 = 0$.

Solution   This example is given to show that the numbers need not always work out nicely.

$$3(x^2 - \tfrac{2}{3}x + \tfrac{1}{9}) + 5(y^2 + \tfrac{7}{5}y + \tfrac{49}{100}) = 11 + \tfrac{1}{3} + \tfrac{49}{20}$$
$$3(x - \tfrac{1}{3})^2 + 5(y + \tfrac{7}{10})^2 = \tfrac{827}{60}$$

$$\frac{3(x - \frac{1}{3})^2}{\frac{827}{60}} + \frac{5(y + \frac{7}{10})^2}{\frac{827}{60}} = 1$$

$$\frac{(x - \frac{1}{3})^2}{\frac{827}{180}} + \frac{(y + \frac{7}{10})^2}{\frac{827}{300}} = 1$$

Notice that we divided the numerator and the denominator of the first term by 3 and of the second term by 5 in order to get the final form. We recognize this to be the equation of an ellipse in standard form with center at $(\frac{1}{3}, -\frac{7}{10})$, major axis horizontal, and $a = \sqrt{827/180} \approx 2.14$, $b = \sqrt{827/300} \approx 1.66$. Its graph is shown in Figure 14.

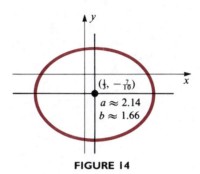

FIGURE 14

These examples serve to illustrate all possibilities for an equation of the form

$$Ax^2 + Cy^2 + Dx + Ey + F = 0$$

where A and C are either both positive or both negative, but unequal. Its graph is either an ellipse or a single point (degenerate ellipse), or there is no graph at all.

 EXERCISE SET 5

A

In Problems 1–8 find the equation of the ellipse with the given properties. Sketch the graph.

1. **a.** x intercepts ± 4; y intercepts ± 2
    b. x intercepts ± 5; y intercepts ± 8
2. Foci at $(-4, 3)$ and $(2, 3)$; a vertex at $(4, 3)$
3. Foci at $(-2, 0)$ and $(-2, 8)$; a vertex at $(-2, -1)$
4. Foci at $(2, 4)$ and $(2, -2)$; end point of minor axis $(4, 1)$
5. End points of minor axis $(4, -3)$ and $(4, -7)$; a focus at $(8, -5)$

6. Major axis vertical; a vertex at $(-3, 4)$; end point of minor axis $(-5, 1)$
7. A vertex at $(1, 3)$; corresponding focus at $(1, 0)$; eccentricity $\frac{1}{2}$
8. End points of minor axis $(-3, 3)$ and $(-3, -5)$; eccentricity $\frac{3}{5}$

In Problems 9–19 find the standard form of the equation. Give the coordinates of the center and vertices, find the eccentricity, and draw the graph.

9. **a.** $25x^2 + 4y^2 = 100$ **b.** $4x^2 + 5y^2 = 20$
10. **a.** $x^2 = 4(1 - y^2)$ **b.** $y^2 = 4(4 - x^2)$

11. a. $x^2 + 9y^2 = 4$ **b.** $5x^2 = 3(1 - 5y^2)$
12. $9x^2 + 4y^2 + 36x - 24y + 36 = 0$
13. $x^2 + 4y^2 - 2x + 16y + 13 = 0$
14. $4x^2 + 7y^2 - 48x - 70y + 319 = 0$
15. $2x^2 + 4y^2 - 5x + 6y - 4 = 0$
16. $2x^2 + 3y^2 + 4x - 12y + 8 = 0$
17. $9x^2 + 16y^2 - 54x - 63 = 0$
18. $36x^2 + 16y^2 - 36x - 48y - 99 = 0$
19. $25x^2 + 4y^2 + 150x - 20y + 225 = 0$
20. Find the equation of the parabola whose vertex and focus coincide with the upper vertex and focus of the ellipse $16x^2 + 12y^2 - 64x - 24y - 116 = 0$.
21. An arch is to be made in the shape of a semiellipse. It is to be 12 feet from end to end and 4 feet high at the center. Find how high the arch is at a point halfway from the center to one end.

22. The definition of the latus rectum for the ellipse is the same as for the parabola—a line segment through a focus, perpendicular to the major axis, and terminating on the curve. Show that the length of each latus rectum for an ellipse is $2b^2/a$.
Hint. Since this is independent of the orientation of the ellipse, you may take the ellipse with its center at the origin.
23. Find the equations of the inscribed and circumscribed circles to the ellipse $x^2 + 2y^2 - 4x + 12y - 3 = 0$. Sketch all three curves.
24. Show graphically the area bounded by the curves $y = \sqrt{25 - 4x^2}$, $3x - y + 3 = 0$, and $x = 0$. Find the point of intersection of the first two of these curves.
25. Find the equation of the ellipse with foci at $(-1, 2)$ and $(1, -2)$ and the length of major axis 10. **Hint.** Use the definition of the ellipse.

THE HYPERBOLA

The hyperbola is defined as the locus of a point that moves in a plane so that the absolute value of the difference between its distances from two fixed points in the plane is constant. Just as with the ellipse, we call the two fixed points the **foci** and the point halfway between them the **center.** We also again denote the fixed constant by $2a$. The **focal axis** is the line through the foci, and the **vertices** are the points where the hyperbola crosses the focal axis. The **transverse axis** is that portion of the focal axis between the two vertices. These definitions are illustrated in Figure 15.

When the center is at the origin and the foci are on the x axis with coordinates $(\pm c, 0)$, the equation of the hyperbola is obtained in a manner similar to that of the ellipse. We require of the point P that traces out the hyperbola $|\overline{PF_1} - \overline{PF_2}| = 2a$. The absolute values are used because we do not care which of the two distances is greater. In terms of coordinates this becomes

$$|\sqrt{(x - c)^2 + y^2} - \sqrt{(x + c)^2 + y^2}| = 2a$$

or equivalently,

$$\sqrt{(x - c)^2 + y^2} - \sqrt{(x + c)^2 + y^2} = \pm 2a$$

We isolate one radical, square, isolate the remaining radical, and square again. After simplification the result is

$$(c^2 - a^2)x^2 - a^2 y^2 = a^2(c^2 - a^2) \tag{9}$$

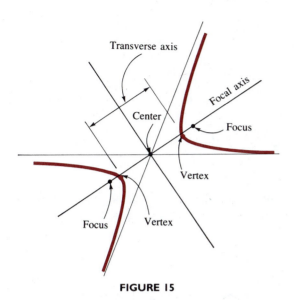

FIGURE 15

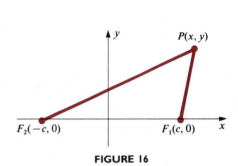

FIGURE 16

From Figure 16 we see that $\overline{F_1F_2} + \overline{PF_1} > \overline{PF_2}$, or $\overline{F_1F_2} > \overline{PF_2} - \overline{PF_1}$. Similarly, $\overline{F_1F_2} + \overline{PF_2} > \overline{PF_1}$, or $\overline{F_1F_2} > \overline{PF_1} - \overline{PF_2}$. So

$$2c = \overline{F_1F_2} > |\overline{PF_1} - \overline{PF_2}| = 2a$$

giving $c > a$. So $c^2 - a^2$ is a positive number, and we choose to designate it by b^2, that is, $b^2 = c^2 - a^2$. If we make the substitution in equation (9) and divide both sides by a^2b^2, the equation becomes

$$\frac{x^2}{a^2} - \frac{y^2}{b^2} = 1 \tag{10}$$

When $y = 0$, $x^2/a^2 = 1$, so that $x = \pm a$. The hyperbola therefore crosses the x axis at $(a, 0)$ and $(-a, 0)$. These points are the vertices. The curve does not cross the y axis, since setting $x = 0$ would lead to the impossibility $-y^2/b^2 = 1$. We name the line segment through the center, perpendicular to the tranverse axis, and b units to either side the **conjugate axis.** So, in the present case, the conjugate axis is the line segment from $(0, -b)$ to $(0, b)$. This will be helpful in drawing the graph.

If we solve equation (10) for y, we get

$$y = \pm \frac{b}{a} \sqrt{x^2 - a^2}$$

$$= \pm \frac{bx}{a} \sqrt{1 - \frac{a^2}{x^2}} \tag{11}$$

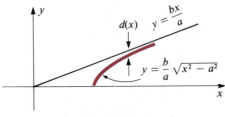

FIGURE 17

Now, when x is very large in absolute value, then a^2/x^2 is small, since a^2 is fixed. The larger $|x|$ becomes, the closer to zero a^2/x^2 becomes. So $\sqrt{1 - (a^2/x^2)} \approx 1$, and

$$y \approx \pm \frac{bx}{a}$$

The lines $y = bx/a$ and $y = -bx/a$ are asymptotes to the hyperbola.

To make matters somewhat more precise, let us consider the distance between the first-quadrant portion of the hyperbola (11) and the line $y = bx/a$. Let $d(x)$ denote this distance for any value of x (for $x > a$) (Figure 17). Then

$$d(x) = \frac{bx}{a} - \frac{b}{a}\sqrt{x^2 - a^2} = \frac{b}{a}(x - \sqrt{x^2 - a^2})$$

$$= \frac{b}{a}\left(\frac{x - \sqrt{x^2 - a^2}}{1}\right)\left(\frac{x + \sqrt{x^2 - a^2}}{x + \sqrt{x^2 - a^2}}\right)$$

$$= \frac{b}{a}\left[\frac{x^2 - (x^2 - a^2)}{x + \sqrt{x^2 - a^2}}\right]$$

$$= \frac{b}{a}\left[\frac{a^2}{x + \sqrt{x^2 - a^2}}\right]$$

$$= \frac{ab}{x + \sqrt{x^2 - a^2}}$$

Now, since the numerator is fixed, it follows that as x becomes arbitrarily large, $d(x)$ approaches 0. So the vertical distance between the curve and the line approaches 0 as the curve recedes indefinitely, and this is what is meant by the line being asymptotic to the curve. The situation is similar in the other quadrants.

The asymptotes $y = \pm(b/a)x$ are the diagonals of the rectangle built on the transverse and conjugate axes (Figure 18). We call this the **fundamental rectangle.** This provides a guide for sketching the hyperbola. Furthermore, the description of the asymptotes as the diagonals of the fundamental rectangle is independent of the location or orientation of the hyperbola. Of course, the equations of the asymptotes will be different when the hyperbola is in a different position.

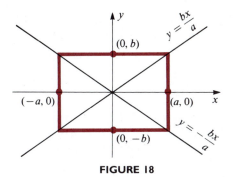

FIGURE 18

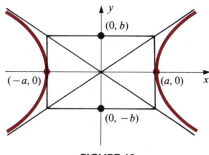

FIGURE 19

A sketch of the hyperbola described by equation (10) is given in Figure 19. It consists of two distinct branches symmetrically placed with respect to the y axis. Observe how the fundamental rectangle and its diagonals are used to aid in making the sketch.

The corresponding equation for the hyperbola with its center at the origin but with transverse axis vertical is

$$\frac{y^2}{a^2} - \frac{x^2}{b^2} = 1 \tag{12}$$

and its graph is as shown in Figure 20.

It is important to note that the position, whether horizontal or vertical, is determined by which term on the left-hand side of the equation is positive after it is in standard form (10) or (12), *not* by which of the numbers a^2 or b^2 is larger, as was true with the ellipse. For example,

$$\frac{x^2}{4} - \frac{y^2}{9} = 1$$

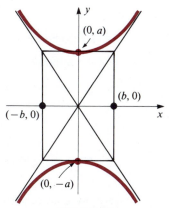

FIGURE 20

is a hyperbola with a horizontal transverse axis. Here, $a^2 = 4$ and $b^2 = 9$. Note that a^2 is always the denominator of the positive term. Contrast this with the ellipse

$$\frac{x^2}{4} + \frac{y^2}{9} = 1$$

which has a vertical major axis and $a^2 = 9$, $b^2 = 4$. In the case of the ellipse, it is the *size* of the denominator that determines the position, whereas in the case of the hyperbola, it is the *sign* of the term that is the determining factor. For the ellipse, a^2 is always larger than b^2, since

$$a^2 = b^2 + c^2$$

but for the hyperbola, either of the quantities a^2 or b^2 may be the larger, since

$$c^2 = a^2 + b^2$$

When the center is shifted to the point (h, k), we have

Horizontal axis: $\dfrac{(x - h)^2}{a^2} - \dfrac{(y - k)^2}{b^2} = 1$

Vertical axis: $\dfrac{(y - k)^2}{a^2} - \dfrac{(x - h)^2}{b^2} = 1$

If either of these equations is cleared of fractions and terms are rearranged, an equation of the form

$$Ax^2 + Cy^2 + Dx + Ey + F = 0 \tag{13}$$

is obtained, where A and C are *opposite in sign*. So every hyperbola with horizontal or vertical transverse axis has an equation of this form. In considering the converse, we again look at examples, since all possibilities are easily illustrated.

EXAMPLE 17

Discuss and sketch the graph of $4x^2 - 9y^2 - 16x - 18y - 29 = 0$.

Solution   We complete the squares, as in the case of the ellipse:

$$4(x^2 - 4x + 4) - 9(y^2 \oplus 2y + 1) = 29 + 16 - 9$$

Be careful here

$$4(x - 2)^2 - 9(y + 1)^2 = 36$$

$$\frac{(x - 2)^2}{9} - \frac{(y + 1)^2}{4} = 1$$

So this is a hyperbola in standard position with a horizontal transverse axis, center at $(2, -1)$, and $a = 3$, $b = 2$ (Figure 21). We could find the equations of

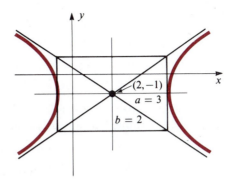

FIGURE 21

the asymptotes if desired. The slopes are $\frac{2}{3}$ and $-\frac{2}{3}$; so the equations are

$$y + 1 = \pm \tfrac{2}{3}(x - 2)$$

▲

EXAMPLE 18 ||||||||||||||||||||||||||||||||||||||||||||||||||||||||||||||||||||||||||||||||||||||||||||||||||||||||||||||||||||||||||||||||

Discuss and sketch the graph of $x^2 - 4y^2 + 4x + 24y - 28 = 0$.

Solution
$$x^2 + 4x + 4 - 4(y^2 - 6y + 9) = 28 + 4 - 36$$
$$(x + 2)^2 - 4(y - 3)^2 = -4$$
$$\frac{(y - 3)^2}{1} - \frac{(x + 2)^2}{4} = 1$$

So this is a hyperbola with a vertical transverse axis, center at $(-2, 3)$, and $a = 1$, $b = 2$ (Figure 22). The slopes of the asymptotes are $\pm \frac{1}{2}$, so their equations are

$$y - 3 = \pm \tfrac{1}{2}(x + 2)$$

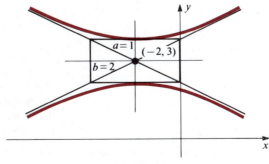

FIGURE 22

▲

EXAMPLE 19 ||||||||||||||||||||||||||||||||||||||||||||||||||||||||||||||||||||||||||||||||||||||||||||||

Discuss and sketch the graph of $16x^2 - 25y^2 - 64x + 200y - 336 = 0$.

Solution $16(x^2 - 4x + 4) - 25(y^2 - 8y + 16) = 336 + 64 - 400$

$$16(x - 2)^2 - 25(y - 4)^2 = 0$$

Since the right-hand side is 0, this is not the equation of a hyperbola. The graph can be determined, however, as follows:

$$16(x - 2)^2 = 25(y - 4)^2$$
$$4(x - 2) = \pm 5(y - 4)$$
$$y - 4 = \pm \tfrac{4}{5}(x - 2)$$

This is seen to be two lines with slopes $\tfrac{4}{5}$ and $-\tfrac{4}{5}$, and passing through the point $(2, 4)$. In fact, these lines are the asymptotes of all hyperbolas we would get if anything other than 0 had occurred on the right-hand side of the equation $16(x - 2)^2 - 25(y - 4)^2 = 0$. We might imagine the two lines as being the limiting position of a sequence of hyperbolas of the form

$$16(x - 2)^2 - 25(y - 4)^2 = K$$

where K gets smaller and smaller in absolute value. The vertices of the two branches of the hyperbolas come closer and closer as K approaches 0. The limiting position, namely, the two lines, is called a **degenerate hyperbola.**   ▲

These three examples are typical and illustrate all possibilities for equations of the form (13), where A and C are opposite in sign. So we can say that such an equation always represents a hyperbola or a degenerate of one.

The eccentricity of a hyperbola is defined just as for the ellipse, $e = c/a$, but in this case, since $c > a$, we see that $e > 1$.

 EXERCISE SET 6

 A

In Problems 1–8 find the equation of the hyperbola with the given properties. Sketch the graph.

1. **a.** Center at the origin; a vertex at $(2, 0)$; a focus at $(-3, 0)$
    b. Center at the origin; a vertex at $(0, -3)$; a focus at $(0, 4)$
2. Center at $(2, 3)$; a vertex at $(5, 3)$; a focus at $(-3, 3)$
3. Center at $(-2, -4)$; a vertex at $(-2, -7)$; a focus at $(-2, -9)$

4. Ends of transverse axis at $(-1, 2)$ and $(-1, -4)$; one end of conjugate axis at $(3, -1)$
5. Ends of conjugate axis at $(2, -7)$ and $(2, 5)$; a vertex at $(0, -1)$
6. Foci at $(0, 2)$ and $(6, 2)$; eccentricity $\tfrac{3}{2}$
7. Ends of conjugate axis at $(3, 5)$ and $(3, -1)$; eccentricity $\tfrac{5}{4}$
8. Center at $(4, -2)$; axis vertical; passing through $(5, -4)$ and $(-3, 8)$

In Problems 9–15 write the equation in standard form and sketch the graph. In each problem give the coordinates of the vertices, the eccentricity, and the equations of the asymptotes.

9. **a.** $4x^2 - 9y^2 = 36$ **b.** $4x^2 - 9y^2 + 36 = 0$
10. **a.** $x^2 = 4(y^2 + 1)$ **b.** $y^2 = x^2 + 9$
11. **a.** $9x^2 - 4y^2 = -36$ **b.** $25(x^2 - 9) = 9y^2$
12. **a.** $36x^2 - 4y^2 = 9$ **b.** $y^2 = 3x^2 + 16$
13. $4x^2 - y^2 - 32x + 4y + 56 = 0$
14. $x^2 - 4y^2 - 2x - 16y - 19 = 0$
15. $9x^2 - 16y^2 + 36x + 96y - 108 = 0$
16. $9x^2 - 16y^2 + 36x + 144y - 252 = 0$
17. $x^2 - 3y^2 - 10x - 6y + 25 = 0$
18. $3x^2 - 5y^2 - 20y - 35 = 0$
19. $36x^2 - 16y^2 + 108x + 16y + 653 = 0$
20. Find the equation of the ellipse having the same foci as the hyperbola

$$\frac{x^2}{9} - \frac{y^2}{27} = 1$$

and with eccentricity the reciprocal of that of the hyperbola.

In Problems 21–23 draw the graphs and shade the area described. Find all points of intersection of the curves involved.

21. Bounded by the curves $y = \sqrt{9 + x^2}$ and $x - 2y + 6 = 0$
22. Bounded by the curves $x^2 - y^2 = 1$ and $y = 2 - (x^2/2)$ (two areas)
23. Bounded by $y = \sqrt{1 + x^2}$ and $y = \sqrt{4 - 2x^2}$

B

24. Find the equation of the parabola whose vertex and focus coincide with the upper vertex and focus, respectively, of the hyperbola $(y^2/100) - (x^2/44) = 1$. Sketch both curves carefully and compare them. Do you think any parabola would fit this branch of the hyperbola exactly? Explain your reasoning.
25. Discuss and sketch the graph of the equation $3x^2 - 12y^2 - 3x - 18y - 10 = 0$.
26. Show the area bounded by $y = 2x - x^2$ and $11x^2 - 3y^2 + 16 = 0$, and find the points of intersection.
27. Find the equation of the path traced out by a point that moves so that the absolute value of the difference of its distances from $(1,1)$ and $(5,3)$ is always 4. Find the coordinates of the vertices and the end points of the conjugate axis, sketch the asymptotes, and draw the graph of the equation.
28. Show that the graph of the polar equation

$$r = \frac{ep}{1 + e \cos \theta}$$

is an ellipse if $e < 1$, a parabola if $e = 1$, and a hyperbola if $e > 1$. Let $p = 2$ and draw the graph for each of the values $e = \frac{1}{2}$, $e = 1$, and $e = 2$.

THE GENERAL SECOND-DEGREE EQUATION AND THE ROTATION OF AXES

The most general equation of second degree in x and y is of the form

$$Ax^2 + Bxy + Cy^2 + Dx + Ey + F = 0 \qquad (14)$$

If $B = 0$, we now know how to identify the graph of the resulting equation, because in this case we have

$$Ax^2 + Cy^2 + Dx + Ey + F = 0$$

and the identity of the graph is dependent on the coefficients of the squared terms.

1. If $A = C$, the curve is a circle or a degenerate of a circle, or there is no graph.
2. If $AC = 0$ (so that either A or C is 0, but not both, since it would not be of second degree if both were 0), the curve is a parabola in standard position or

a degenerate of a parabola. If x^2 is present, the axis is vertical, and if y^2 is present, the axis is horizontal.

3. If $AC > 0$, but $A \neq C$, then A and C are like in sign, and so the graph is an ellipse in standard position or a degenerate, or there is no graph.

4. If $AC < 0$, so that A and C are unlike in sign, the graph is a hyperbola in standard position or a degenerate (two intersecting lines).

The primary object of this section is to show that when $B \neq 0$ in equation (14), the curve is still a conic section but with a rotated axis.

Consider a cartesian coordinate system with axes x and y, and let a second cartesian coordinate system with axes x' and y' be superimposed upon the first, so that the origins coincide and the angle from the x axis to the x' axis is θ, as shown in Figure 23. It is sufficient to consider $0 < \theta < \pi/2$. Let P be a point in the plane. It has coordinates (x, y) with respect to the original axes and coordinates (x', y') with respect to the rotated axes. We wish to determine the relationships among x, y, x' and y'. Consider the construction as shown in Figure 23. It can be seen that triangle PST is similar to triangle OSR, so that the angle at P in the triangle PST is also θ. From that triangle we have

$$\sin \theta = \frac{\overline{ST}}{y'} \qquad \cos \theta = \frac{\overline{PT}}{y'}$$

so that

$$\overline{ST} = y' \sin \theta \qquad \overline{PT} = y' \cos \theta$$

Also, from triangle OSR, since $\overline{OS} = \overline{NP} = x'$,

$$\sin \theta = \frac{\overline{RS}}{\overline{OS}} = \frac{\overline{RS}}{\overline{NP}} = \frac{\overline{RS}}{x'} \qquad \cos \theta = \frac{\overline{OR}}{\overline{OS}} = \frac{\overline{OR}}{x'}$$

and thus,

$$\overline{RS} = x' \sin \theta \qquad \overline{OR} = x' \cos \theta$$

From Figure 23 we also see that

$$x = \overline{MP} = \overline{OQ} = \overline{OR} - \overline{QR} = \overline{OR} - \overline{ST} = x' \cos \theta - y' \sin \theta$$
$$y = \overline{QP} = \overline{QT} + \overline{TP} = \overline{RS} + \overline{TP} = x' \sin \theta + y' \cos \theta$$

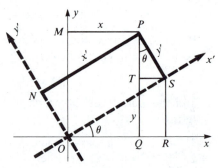

FIGURE 23

These are the desired equations relating the old coordinates to the new:

$$\begin{cases} x = x'\cos\theta - y'\sin\theta \\ y = x'\sin\theta + y'\cos\theta \end{cases} \tag{15}$$

Even though we showed P in the first quadrant with respect to both sets of axes, the equations in (15) hold true regardless of the location of P.

Now suppose we are given a second-degree equation in x and y,

$$Ax^2 + Bxy + Cy^2 + Dx + Ey + F = 0 \tag{16}$$

in which $B \neq 0$. We will show that the equation of this curve, when referred to a new set of axes x' and y' that are rotated through an angle θ from the old, will be free of the $x'y'$ term if θ is appropriately chosen. The procedure will be as follows. We consider θ as unknown, substitute from (15) into (16), and obtain an equation of the form

$$A'x'^2 + B'x'y' + C'y'^2 + D'x' + E'y' + F' = 0 \tag{17}$$

where the coefficients are functions of θ. Then we set $B' = 0$ and solve for θ. The details of the calculations are cumbersome. The result after substituting and rearranging terms is that equation (17) is obtained, with

$$\begin{cases} A' = A\cos^2\theta + B\sin\theta\cos\theta + C\sin^2\theta \\ B' = B(\cos^2\theta - \sin^2\theta) - 2(A-C)(\sin\theta\cos\theta) \\ C' = A\sin^2\theta - B\sin\theta\cos\theta + C\cos^2\theta \\ D' = D\cos\theta + E\sin\theta \\ E' = E\cos\theta - D\sin\theta \\ F' = F \end{cases} \tag{18}$$

Our goal is to determine θ so that $B' = 0$. Note first that B' can be simplified by use of the identities

$$\cos^2\theta - \sin^2\theta = \cos 2\theta \quad\text{and}\quad 2\sin\theta\cos\theta = \sin 2\theta$$

Using these, we obtain

$$B' = B\cos 2\theta - (A-C)\sin 2\theta \tag{19}$$

We set this equal to 0:

$$B\cos 2\theta - (A-C)\sin 2\theta = 0$$
$$B\cos 2\theta = (A-C)\sin 2\theta$$
$$\frac{B}{A-C} = \frac{\sin 2\theta}{\cos 2\theta}$$

or finally,

$$\tan 2\theta = \frac{B}{A-C} \tag{20}$$

provided $A \neq C$. If $A = C$, then we see from equation (19) that B' will be 0 if $\cos 2\theta = 0$, that is, if $2\theta = 90°$, or $\theta = 45°$. So if $A = C$, we take $\theta = 45°$; otherwise, find 2θ from equation (20) and from this determine $\sin \theta$ and $\cos \theta$. With θ chosen in this way, equation (17) becomes

$$A'x'^2 + C'y'^2 + D'x' + E'y' + F' = 0$$

which we know to be the equation of a conic section (or a degenerate) if it has a graph at all.

We will apply this technique to several examples.

EXAMPLE 20

By a suitable rotation of axes identify and sketch the curve having the equation $x^2 - 2xy + y^2 - 2x - 2y = 0$.

Solution   In this case, $A = C$, so we choose $\theta = 45°$. The equations (15) of rotation then become

$$x = \frac{1}{\sqrt{2}}(x' - y') \qquad \text{and} \qquad y = \frac{1}{\sqrt{2}}(x' + y')$$

Substituting, we obtain

$$\tfrac{1}{2}(x'^2 - 2x'y' + y'^2) - \tfrac{2}{2}(x'^2 - y'^2) + \tfrac{1}{2}(x'^2 + 2x'y' + y'^2)$$

$$- \frac{2}{\sqrt{2}}(x' - y') - \frac{2}{\sqrt{2}}(x' + y') = 0$$

which upon simplification becomes

$$y'^2 = \sqrt{2}x'$$

We recognize this as the equation of a parabola in standard position I with respect to the rotated axes. A sketch is shown in Figure 24. It should be emphasized that the rotated axes are a means to an end. What we really wanted was to graph the original equation with respect to the x and y axes. But this has now been done. We could erase the x' and y' axes and leave the curve intact.

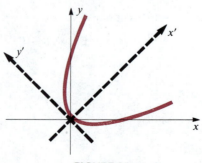

FIGURE 24

EXAMPLE 21

Identify and sketch the curve whose equation is $x^2 + 24xy - 6y^2 = 30$.

Solution   Again, we wish to rotate axes so as to eliminate the term involving $x'y'$. Using equation (19), we want to choose θ so that

$$\tan 2\theta = \frac{B}{A - C} = \frac{24}{7}$$

It is not necessary to find θ, but rather $\sin \theta$ and $\cos \theta$ in order to apply equations (15). To do this, we use the trigonometric identities

$$\sin^2 \theta = \frac{1 - \cos 2\theta}{2} \qquad \text{and} \qquad \cos^2 \theta = \frac{1 + \cos 2\theta}{2}$$

Since we are restricting θ to be between $0°$ and $90°$, we can sketch the angle 2θ as shown in Figure 25. We determine the hypotenuse, 25, by the Pythagorean theorem, and then read

$$\cos 2\theta = \tfrac{7}{25}$$

So

$$\sin^2 \theta = \frac{1 - \frac{7}{25}}{2} = \frac{9}{25} \qquad \text{and} \qquad \cos^2 \theta = \frac{1 + \frac{7}{25}}{2} = \frac{16}{25}$$

from which it follows that (since θ is acute) $\sin \theta = \tfrac{3}{5}$ and $\cos \theta = \tfrac{4}{5}$. Thus, equations (15) become

$$x = \tfrac{1}{5}(4x' - 3y') \qquad \text{and} \qquad y = \tfrac{1}{5}(3x' + 4y')$$

We substitute these into the original equation to obtain

$$\tfrac{1}{25}(16x'^2 - 24x'y' + 9y'^2) + \tfrac{24}{25}(12x'^2 + 7x'y' - 12y'^2)$$

$$-\tfrac{6}{25}(9x'^2 + 24x'y' + 16y'^2) = 30$$

which upon simplification becomes

$$10x'^2 - 15y'^2 = 30$$

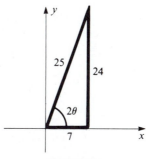

FIGURE 25

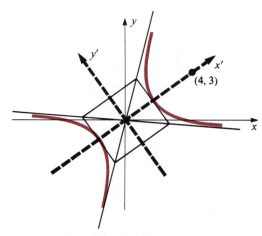

FIGURE 26

or in standard form,

$$\frac{x'^2}{3} - \frac{y'^2}{2} = 1$$

This is seen to be a hyperbola in standard position with vertices on the x' axis. To draw the rotated axes, we note that since $\sin \theta = \frac{3}{5}$ and $\cos \theta = \frac{4}{5}$, the point $(4, 3)$ lies on the x' axis (Figure 26). ▲

EXAMPLE 22

By a suitable rotation of axes, sketch the graph of $8x^2 - 4xy + 5y^2 = 36$.

Solution   The angle θ is to be such that

$$\tan 2\theta = \frac{B}{A - C} = \frac{-4}{8 - 5} = -\frac{4}{3}$$

Since we are limiting θ to be between $0°$ and $90°$, it follows that $0° \le 2\theta \le 180°$. In this case, therefore, 2θ must lie in the second quadrant. We can now determine that $\cos 2\theta = -\frac{3}{5}$ (see Figure 27). So

$$\sin^2 \theta = \frac{1 - \cos 2\theta}{2} = \frac{1 + \frac{3}{5}}{2} = \frac{4}{5}$$

and

$$\cos^2 \theta = \frac{1 + \cos 2\theta}{2} = \frac{1 - \frac{3}{5}}{2} = \frac{1}{5}$$

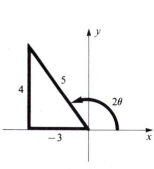

FIGURE 27

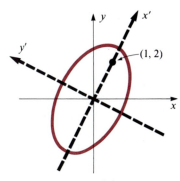

FIGURE 28

and thus,

$$\sin \theta = \frac{2}{\sqrt{5}} \qquad \cos \theta = \frac{1}{\sqrt{5}}$$

So the equations of transformation are

$$x = \frac{1}{\sqrt{5}}(x' - 2y') \qquad \text{and} \qquad y = \frac{1}{\sqrt{5}}(2x' + y')$$

On substituting these into the original equation, we obtain

$$\tfrac{8}{5}(x'^2 - 4x'y' + 4y'^2) - \tfrac{4}{5}(2x'^2 - 3x'y' - 2y'^2) + \tfrac{5}{5}(4x'^2 + 4x'y' + y'^2) = 36$$

which becomes, after simplification,

$$4x'^2 + 9y'^2 = 36$$

or finally,

$$\frac{x'^2}{9} + \frac{y'^2}{4} = 1$$

This is seen to be the equation of an ellipse having its major axis on the x' axis. As in the previous example, we sketch the new axes in relation to the old by observing that since $\cos \theta = 1/\sqrt{5}$ and $\sin \theta = 2/\sqrt{5}$, the point $(1, 2)$ lies on the x' axis (Figure 28). ▲

One particular type of equation with an xy term should be singled out. It is an equation of the form

$$xy = K \qquad\qquad\qquad (21)$$

In this case, $A = C = 0$, so we rotate through an angle of $45°$. Thus,

$$x = \frac{1}{\sqrt{2}}(x' - y') \qquad \text{and} \qquad y = \frac{1}{\sqrt{2}}(x' + y')$$

and equation (21) becomes

$$\tfrac{1}{2}(x'^2 - y'^2) = K$$
$$x'^2 - y'^2 = 2K \tag{22}$$

This is seen to be a hyperbola with axes of equal length. It is called an **equilateral hyperbola.** If $K > 0$, its transverse axis is the x' axis, and if $K < 0$, the y' axis. If the center of the hyperbola is at the point (h, k), then equation (21) is replaced by

$$(x - h)(y - k) = K \tag{23}$$

The following examples illustrate problems of this type.

EXAMPLE 23

Sketch the curve $xy = 2$.

Solution   After rotating $45°$, the equation becomes

$$x'^2 - y'^2 = 4$$

according to equation (22). The graph is sketched as shown in Figure 29. Note that the x and y axes are asymptotes.

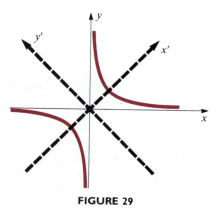

FIGURE 29

EXAMPLE 24

Discuss and sketch $xy + 2x - y = 10$.

Solution   We could substitute $x = (1/\sqrt{2})(x' - y')$ and $y = (1/\sqrt{2})(x' + y')$. Alternately, we could try to put this in the form of equation (23) by factoring. To do this, we factor x from the first two terms and then complete the factor on y:

$$x(y + 2) - (y + 2) = 10 - 2$$

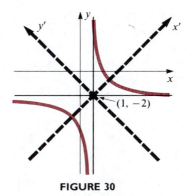

FIGURE 30

That is, we add to both sides the appropriate quantity to obtain a common binomial factor involving y, in this case $y + 2$. Now the left-hand side can be factored as

$$(x - 1)(y + 2) = 8$$

This is in the form of equation (23) and so is an equilateral hyperbola with center $(1, -2)$ and axis inclined at $45°$ (Figure 30).

In the general second degree equation (14),

$$Ax^2 + Bxy + Cy^2 + Dx + Ey + F = 0$$

the quantity $B^2 - 4AC$ is called the **discriminant.** When the axes are rotated through an angle θ, as we have seen, the result is of the form (17),

$$A'x'^2 + B'x'y' + C'y'^2 + D'x' + E'y' + F' = 0$$

where the new coefficients are given in terms of θ and the old coefficients by equations (18). The discriminant of the transformed equation is $B'^2 - 4A'C'$. While our main objective of this section was to choose θ so that $B' = 0$, the equations (18) for the transformed coefficients are valid for all values of θ.

It is an interesting and useful fact that under any rotation whatever (that is, for all values of θ), the discriminant of equation (17) is the same as the discriminant of equation (14):

$$B'^2 - 4A'C' = B^2 - 4AC \qquad (24)$$

This result may be stated in the following way:

> The discriminant of a quadratic equation in two variables is invariant under rotation of axes.

The proof of this is straightforward but tedious (so it will be left for the exercises!). The values of A', B', and C', as given by equations (18), are substituted into the left-hand side of equation (24), and by appropriate use of trigonometric identities and algebraic simplification, this eventually reduces to the right-hand side.

Now let us suppose that θ is chosen so as to cause $B' = 0$. Then equation (17) is

$$A'x'^2 + C'y'^2 + D'x' + E'y' + F' = 0$$

and we know this to be

1. **A parabola if either A' or C' is 0**
2. **An ellipse if A' and C' are like in sign**
3. **A hyperbola if A' and C' are opposite in sign**

(The graph may also be a degenerate of these, or there may be no graph.) Furthermore, the discriminant becomes

$$B'^2 - 4A'C' = -4A'C'$$

since $B' = 0$. Thus, in case 1 the discriminant is 0; in case 2 it is negative; and in case 3 it is positive. Since the new discriminant equals the old, we may ascertain the following by looking at the discriminant of the original equation: If equation (14) has a graph, then it is

1. **A parabola if $B^2 - 4AC = 0$**
2. **An ellipse if $B^2 - 4AC < 0$**
3. **A hyperbola if $B^2 - 4AC > 0$**

In each case, allowance must be made for a degenerate of the given conic.

EXAMPLE 25 |||||||||||||||||||||||||||||||||||||||||||||||||||||||||||||||||||||||||||||||||||||||||||||||||||||||||||||||||

Identify each of the following, assuming that a graph exists.
a. $2x^2 - 3xy + 5y^2 = 7$ b. $x^2 + 9xy + 3y^2 - 2x + 5y = 4$
c. $8x^2 - xy - 2y^2 + x + 3y = 0$ d. $9x^2 - 24xy + 16y^2 + 7x - 5y = 8$

Solution   We calculate the discriminant in each case.
a. $B^2 - 4AC = 9 - 40 = -31 < 0$;   ellipse
b. $B^2 - 4AC = 81 - 12 = 69 > 0$;   hyperbola
c. $B^2 - 4AC = 1 + 64 = 65 > 0$;   hyperbola
d. $B^2 - 4AC = 576 - 576 = 0$;   parabola   ▲

EXERCISE SET 7

A

In Problems 1–8 rotate the axes so as to remove the $x'y'$ term. Identify the curve and sketch.

1. $x^2 + 2xy + y^2 + 2\sqrt{2}(x - y) = 0$
2. $x^2 - 3xy + 5y^2 = 22$
3. $6x^2 - 24xy - y^2 = 150$
4. $4x^2 + 4xy + y^2 + 5x - 10y = 0$
5. $2x^2 - 5xy + 2y^2 = 18$
6. $20x^2 - 8xy + 5y^2 = 84$
7. $9x^2 - 12xy + 4y^2 + 8\sqrt{13}x + 12\sqrt{13}y = 52$
8. $x^2 + 3xy - 3y^2 = 42$

Sketch the equilateral hyperbolas in Problems 9–15.

9. $xy + 2 = 0$ **10.** $y = 4/x$
11. $(x - 3)(y + 2) = 8$
12. $(2x + 1)(3x - 4) = 12$
13. $xy - 2x + y = 0$ **14.** $xy - 3x - 2y = 2$
15. $xy - 4x + 3y = 4$

In Problems 16 and 17 identify the graph, assuming a graph exists.

16. a. $3x^2 + 2xy - y^2 + 4x - 3y = 7$
   b. $x^2 - 4xy + 4y^2 - 3x + 2y - 5 = 0$

c. $2x^2 - 3xy + 5y^2 = 4$
d. $x^2 + 3xy - 7x + 4y = 0$
e. $5x^2 + 7xy + 3y^2 - 2x = 8$
17. a. $2xy - y^2 + 4x - 5y = 7$
   b. $xy = x^2 - 2y^2 + 4$
   c. $y^2 = 3x - 10x^2 + 4xy - 9$
   d. $x^2 + 4y^2 = 4xy - 2x + 3y + 11$
   e. $x(3x - 2y) + y(x + 5y) = 3$

B

In Problems 18–20 identify and sketch the curves after a suitable rotation of axes.

18. $2x^2 - \sqrt{3}xy + y^2 + 5\sqrt{3}x - 5y = 0$
19. $9x^2 + 24xy + 16y^2 - 110x + 20y + 125 = 0$
20. $x^2 - 4xy - 2y^2 + 22x + 4y = 5$
21. Verify equations (18).
22. Prove that the discriminant is invariant under rotation, that is, that $B'^2 - 4A'C' = B^2 - 4AC$.
23. Prove that the quantity $A + C$ in the general quadratic equation is invariant under rotation, that is, that $A' + C' = A + C$.

REVIEW EXERCISE SET

A

In Problems 1–11 write the equations in standard form and draw their graphs.

1. a. $y^2 = 4 - x^2$ **b.** $y^2 = 4 + x^2$
   c. $y^2 = 4 - x$ **d.** $y = 4 + x^2$
2. a. $16x^2 + 25y^2 = 400$
   b. $y^2 = 16 - 4x^2$
3. a. $x = y^2 - 2y$ **b.** $4x^2 = y^2 + 4$
4. $x^2 + y^2 - 2x + 4y - 20 = 0$
5. $x^2 + 4y^2 + 6x - 16y + 21 = 0$
6. $y^2 - 4x - 10y + 13 = 0$
7. $4x^2 - 9y^2 - 16x - 18y + 43 = 0$

8. $25x^2 + 16y^2 - 192y + 176 = 0$
9. $x^2 + y^2 - 10x + 8y + 41 = 0$
10. $y^2 - x^2 + 4x + 6y + 9 = 0$
11. $2x^2 - 12x - 15y - 42 = 0$

In Problems 12–24 find the equations of the conics described.

12. Circle with center $(4, -3)$ and radius 3
13. Parabola with focus $(1, -2)$ and vertex $(3, -2)$
14. Ellipse with vertices $(3, 2)$ and $(-5, 2)$, and foci $(1, 2)$ and $(-3, 2)$
15. Hyperbola with vertices $(2, 5)$ and $(2, -1)$, and eccentricity $\frac{5}{3}$

16. Parabola with vertical axis, vertex at $(-2, 5)$, and passing through the point $(2, -1)$

17. Hyperbola with same vertices as the vertices of the ellipse $9x^2 + 4y^2 = 36$ and with asymptotes $y = \pm 3x/2$

18. Ellipse with eccentricity $\frac{1}{2}$, and foci $(-1, 3)$ and $(-11, 3)$

19. Parabola with directrix $y + 2 = 0$ and focus $(3, -6)$

20. Hyperbola with vertices $(4, 3)$ and $(10, 3)$, and foci $(3, 3)$ and $(11, 3)$

21. Circle having as a diameter the latus rectum of the parabola $x^2 - 6x - 16y + 9 = 0$

22. Ellipse whose foci coincide with the foci of the hyperbola $x^2 - 3y^2 - 2x - 12y + 1 = 0$ and whose eccentricity is the reciprocal of that of the hyperbola

23. Parabola with vertical axis and passing through the points $(0, 4)$, $(3, 1)$, and $(-2, -4)$

24. Circle having as diameter the transverse axis of the hyperbola $2x^2 - y^2 + 8x + 4y - 1 = 0$

25. Find the equation of the tangent line to the circle $x^2 + y^2 - 6x + 8y + 17 = 0$ at the point $(1, -2)$.

26. The underneath side of a masonry bridge is in the form of a semiellipse whose span is 80 feet and whose maximum height is 30 feet. Find the height at 10-foot intervals from one end to the other.

27. Assuming there is a graph and it is not of a degenerate conic, identify the curves whose equations are as follows:

  a. $2x^2 - 5xy - 3y^2 + 5x - 7y + 8 = 0$

  b. $4x^2 - 20xy + 2y^2 - 3x + 8y - 7 = 0$

  c. $9x^2 - 24xy + 16y^2 - 3 = 0$

  d. $x(x + 5y) = y(3 - 8y)$

In Problems 28–31 rotate the axes so as to eliminate the $x'y'$ term. Identify the curve and sketch.

28. $2x^2 + xy + 2y^2 - 30 = 0$

29. $9x^2 - 24xy + 16y^2 + 4x + 3y = 0$

30. $2x^2 - 3xy + 6y^2 - 39 = 0$

31. $x^2 - 4xy + y^2 - 3 = 0$

32. Sketch the equilateral hyperbolas.

  a. $(x - 1)(y + 3) = 4$

  b. $xy - 4x + 2y - 6 = 0$

B

33. Find the equation of the locus of a point that moves so that the difference of its distances from $(2, -1)$ and $(-1, 0)$ is equal to 2. Identify the curve.

34. Find the equation of the locus of a point that moves so that its distance from the point $(3, -1)$ is equal to its distance from the line $4x - 3y + 2 = 0$. Identify the curve. (See Problem 18, Exercise Set 2.)

35. The end points of the conjugate axes of a hyperbola are $(4, -1)$ and $(4, 3)$, and the eccentricity is $\frac{3}{2}$. Find the equation of the hyperbola and draw its graph.

36. A parabola that opens upward has its vertex at the center of the circle $x^2 + y^2 - 4x + 6y - 7 = 0$, and its latus rectum is a chord of the circle. Find the equation of the parabola.

13

SEQUENCES AND SERIES

Sequences and series play an important role in calculus. One category of series, called a **geometric series,** is of special importance. An example of such a series is illustrated below.*

Determine whether the series

$$1 + \tfrac{1}{3} + \tfrac{1}{9} + \tfrac{1}{27} + \cdots + (\tfrac{1}{3})^k + \cdots = \sum_{k=0}^{\infty} (\tfrac{1}{3})^k$$

converges, and if so, find its sum.

The sum . . . is an example of a geometric series; each term is a constant multiple $(\tfrac{1}{3})$ of the preceding term. It is convenient to form the sequence of partial sums as follows:

$$s_0 = 1, \qquad s_1 = 1 + \tfrac{1}{3}, \qquad s_2 = 1 + \tfrac{1}{3} + \tfrac{1}{9}$$

and in general

$$s_n = 1 + \tfrac{1}{3} + \tfrac{1}{9} + \cdots + (\tfrac{1}{3})^n$$

. . . We wish to investigate the possible limiting behavior of $\{s_n\}$ as $n \to \infty$. Unfortunately, this is not immediately clear. . . . However, let us rewrite s_n in a more useful form. . . .

$$s_n = \frac{1 - (\tfrac{1}{3})^{n+1}}{1 - (\tfrac{1}{3})}; \qquad n = 0, 1, 2, \ldots$$

. . . , hence it follows . . . that

$$\lim_{n \to \infty} s_n = \lim_{n \to \infty} \frac{1 - (\tfrac{1}{3})^{n+1}}{1 - \tfrac{1}{3}} = \frac{1}{2/3} = \frac{3}{2}$$

* William E. Boyce and Richard C. DiPrima, *Calculus* (New York: John Wiley & Sons, 1988), p. 591. Reprinted by permission.

We have, as usual, omitted a number of steps in order to concentrate on the ideas we want to stress. In this chapter we will study geometric series and will see how to obtain the given expression for s_n, as well as its limit.

1 SEQUENCES

A sequence is defined as follows:

DEFINITION 1   **A sequence is a function whose domain is the set N of natural numbers. If f is such a function, then for each natural number n, $f(n)$ is called the nth term of the sequence.**

We will be concerned primarily with sequences whose terms are real numbers. It is customary to use subscripted variables to designate the terms of a sequence, rather than the usual functional notation. For example, we frequently use a_n instead of $f(n)$ for the nth term. For n arbitrary, the nth term of a sequence is also called the **general term.**

As an example, consider a sequence whose general term is given by

$$a_n = \frac{n}{2n + 1}$$

The sequence is now completely defined, and we can find as many of the terms as we wish. For $n = 1, 2,$ and 3 we have

$$a_1 = \frac{1}{2 \cdot 1 + 1} = \frac{1}{3}$$

$$a_2 = \frac{2}{2 \cdot 2 + 1} = \frac{2}{5}$$

$$a_3 = \frac{3}{2 \cdot 3 + 1} = \frac{3}{7}$$

It is customary to exhibit the terms in the following way:

$$\frac{1}{3}, \frac{2}{5}, \frac{3}{7}, \cdots, \frac{n}{2n + 1}, \cdots$$

The terms are separated by commas, three dots are shown following the last specific term, and the nth term is shown, followed by three more dots to indicate indefinite continuation. Often, we refer to such a listing of the terms of a sequence as the sequence itself, although strictly speaking the sequence is the function, and the terms constitute the range of the function.

When no ambiguity is likely to result, it is satisfactory to omit the nth term in exhibiting a sequence. For instance, we might write

$$2, 4, 6, 8, \ldots$$

and it would be understood that the sequence continues in the manner which is suggested by the specific terms shown. It should be emphasized that there may be many possibilities for the continuation of this sequence but only one *reasonable* way to continue, and we will understand this to be what is intended. Thus, in this case we would infer that the general term is given by $a_n = 2n$. Contrast this with

$$1, 13, 18, 36, \ldots$$

Here there is no clear pattern established, and this would not be considered a valid representation of a sequence.

To summarize, we can define a sequence either by giving the general term or by listing the first few terms of the sequence. When listing the terms, all uncertainty is removed if the nth term is shown in the list; however, it is permissible to omit it if the pattern is clearly established by the specific terms listed.

Another way of defining a sequence makes use of what is called a **recursion formula.** Consider, for example, the sequence for which

$$a_1 = 2$$

$$a_n = \frac{a_{n-1}}{3} \qquad n \geq 2$$

The second equation, in which a_n is given as a function of a_{n-1}, is the recursion formula. We are given a_1. Then, by the recursion formula with $n = 2$, we find

$$a_2 = \frac{a_1}{3} = \frac{2}{3}$$

Now, using the recursion formula with $n = 3$, we get

$$a_3 = \frac{a_2}{3} = \frac{\frac{2}{3}}{3} = \frac{2}{9}$$

We can continue in this way to find as many terms as are desired.

It is sometimes useful to employ a **finite sequence,** by which we mean a sequence that terminates at some stage. More precisely, we have the following definition:

DEFINITION 2   **A finite sequence is a function whose domain is a subset of N of the form $\{1, 2, 3, \ldots, n\}$ for some fixed natural number n.**

In contrast to this, a sequence as defined in Definition 1 is sometimes called an **infinite sequence,** but when we say *sequence* we will always mean *infinite sequence* unless otherwise specified.

EXAMPLE 1 |||||||||||||||||||||||||||||||||||||||||||||||||||||||||||||||||||||||||||||||||||||||||||||||||||||

Write the first five terms of the sequence whose nth term is given.

a. $a_n = 2n - 1$ **b.** $a_n = \dfrac{(-1)^{n-1}}{n}$

c. $a_n = \dfrac{2^n - 1}{2^n}$ **d.** $a_n = \dfrac{(-1)^{n-1} n}{n + 1}$

Solution   In each case we substitute $n = 1, 2, 3, 4$, and 5:

a. $1, 3, 5, 7, 9, \ldots$ **b.** $1, -\frac{1}{2}, \frac{1}{3}, -\frac{1}{4}, \frac{1}{5}, \ldots$

c. $\frac{1}{2}, \frac{3}{4}, \frac{7}{8}, \frac{15}{16}, \frac{31}{32}, \ldots$ **d.** $\frac{1}{2}, -\frac{2}{3}, \frac{3}{4}, -\frac{4}{5}, \frac{5}{6}, \ldots$ ▲

Remark. Part **a** above shows that the nth odd natural number is $2n - 1$. Similarly, the nth even natural number is $2n$. Parts **b** and **d** illustrate a standard way of handling alternating signs. The factor $(-1)^{n-1}$ equals $+1$ for n odd and -1 for n even. These are useful facts in constructing the nth term of certain sequences.

EXAMPLE 2 |||||||||||||||||||||||||||||||||||||||||||||||||||||||||||||||||||||||||||||||||||||||||||||||||||||

Determine the nth term of each of the following sequences:

a. $1, \frac{1}{4}, \frac{1}{9}, \frac{1}{16}, \ldots$ **b.** $3, 6, 9, 12, \ldots$

c. $1, -\frac{1}{3}, \frac{1}{5}, -\frac{1}{7}, \ldots$ **d.** $5, 9, 13, 17, \ldots$

e. $1, -\frac{1}{2}, \frac{1}{4}, -\frac{1}{8}, \ldots$

Solution   **a.**   The denominators are perfect squares. We could rewrite the sequence as

$$\frac{1}{1^2}, \frac{1}{2^2}, \frac{1}{3^2}, \frac{1}{4^2}, \ldots$$

and conclude that the general term is

$$a_n = \frac{1}{n^2}$$

b.   Each term is a multiple of 3; in fact, the terms are

$$3 \cdot 1, \quad 3 \cdot 2, \quad 3 \cdot 3, \quad 3 \cdot 4, \ldots$$

and the general term is

$$a_n = 3n$$

c.   The denominators are the odd natural numbers in order, and the signs alternate. By the remark following Example 1, we can therefore write

$$a_n = \frac{(-1)^{n-1}}{2n - 1}$$

d. Observe that each term after the first is 4 greater than the preceding term. This suggests that the terms are related to multiples of 4. In fact, each term is exactly 1 more than a multiple of 4. We can rewrite the terms as

$$4(1) + 1, \quad 4(2) + 1, \quad 4(3) + 1, \quad 4(4) + 1, \ldots$$

and infer that the nth term is

$$a_n = 4n + 1$$

e. The denominators appear to be powers of 2, starting with 2^0:

$$\frac{1}{2^0}, \quad \frac{-1}{2^1}, \quad \frac{1}{2^2}, \quad \frac{-1}{2^3}, \ldots$$

So we have

$$a_n = \frac{(-1)^{n-1}}{2^{n-1}}$$ ▲

EXAMPLE 3

Write the first six terms of the sequence defined recursively by

a. $a_1 = -2$
$a_n = 1 - 2a_{n-1}, \quad n \geq 2$

b. $a_1 = 1$
$a_2 = 3$
$a_n = \dfrac{a_{n-2} + a_{n-1}}{2}, \quad n \geq 3$

Solution

a. $a_2 = 1 - 2a_1 = 1 - 2(-2) \quad = 5$
$a_3 = 1 - 2a_2 = 1 - 2(5) \quad\quad = -9$
$a_4 = 1 - 2a_3 = 1 - 2(-9) \quad = 19$
$a_5 = 1 - 2a_4 = 1 - 2(19) \quad\ = -37$
$a_6 = 1 - 2a_5 = 1 - 2(-37) = 75$

b. $a_3 = \dfrac{a_1 + a_2}{2} = \dfrac{1 + 3}{2} = 2$

$a_4 = \dfrac{a_2 + a_3}{2} = \dfrac{3 + 2}{2} = \dfrac{5}{2}$

$a_5 = \dfrac{a_3 + a_4}{2} = \dfrac{2 + \frac{5}{2}}{2} = \dfrac{4 + 5}{4} = \dfrac{9}{4}$

$a_6 = \dfrac{a_4 + a_5}{2} = \dfrac{\frac{5}{2} + \frac{9}{4}}{2} = \dfrac{10 + 9}{8} = \dfrac{19}{8}$ ▲

EXERCISE SET I

A

In Problems 1–10 write the first five terms of the sequence whose nth term is given by the indicated expression.

1. $a_n = \dfrac{2n}{3n+1}$

2. $a_n = \dfrac{1}{n^2+1}$

3. $a_n = \dfrac{(-1)^{n-1}}{n(n+1)}$

4. $a_n = \dfrac{(-1)^{n-1}n}{2n+1}$

5. $a_n = \left(\dfrac{2}{3}\right)^{n-1}$

6. $a_n = \dfrac{(-1)^{n-1}n^2}{1+n}$

7. $a_n = \dfrac{2^n}{3^n+1}$

8. $a_n = \dfrac{(-1)^{n-1}}{2n^2-3n+5}$

9. $a_n = \dfrac{2}{3n+4}$

10. $a_n = \dfrac{(-1)^{n-1}}{e^n}$

In Problems 11–20 write the next three terms of the sequence.

11. $1, \frac{1}{3}, \frac{1}{9}, \frac{1}{27}, \frac{1}{81}, \dots$

12. $2, 4, 6, 8, 10, \dots$

13. $1, \frac{1}{3}, \frac{1}{5}, \frac{1}{7}, \frac{1}{9}, \dots$

14. $1, -1, 1, -1, 1, \dots$

15. $1, \frac{1}{2}, \frac{1}{4}, \frac{1}{8}, \frac{1}{16}, \dots$

16. $1, -\frac{1}{4}, \frac{1}{7}, -\frac{1}{10}, \frac{1}{13}, \dots$

17. $1, \frac{4}{5}, \frac{6}{10}, \frac{8}{17}, \frac{10}{26}, \dots$

18. $\frac{1}{2}, \frac{2}{3}, \frac{3}{4}, \frac{4}{5}, \dots$

19. $\frac{1}{5}, -\frac{2}{7}, \frac{3}{9}, -\frac{4}{11}, \frac{5}{13}, \dots$

20. $\frac{1}{2}, \frac{3}{4}, \frac{9}{8}, \frac{27}{16}, \frac{81}{32}, \dots$

In Problems 21–30 determine an expression for the nth term of the sequence.

21. $1, \frac{1}{3}, \frac{1}{5}, \frac{1}{7}, \dots$

22. $1, -\frac{1}{2}, \frac{1}{4}, -\frac{1}{8}, \dots$

23. $\frac{1}{3}, \frac{1}{6}, \frac{1}{9}, \frac{1}{12}, \dots$

24. $\frac{1}{2}, -\frac{2}{3}, \frac{3}{4}, -\frac{4}{5}, \dots$

25. $\frac{1}{2}, \frac{3}{4}, \frac{7}{8}, \frac{15}{16}, \dots$

26. $\frac{1}{2}, \frac{2}{5}, \frac{3}{10}, \frac{4}{17}, \frac{5}{26}, \dots$

27. $1, \frac{1}{4}, \frac{1}{7}, \frac{1}{10}, \dots$

28. $\frac{1}{7}, -\frac{1}{11}, \frac{1}{15}, -\frac{1}{19}, \frac{1}{23}, \dots$

29. $0.5, 0.05, 0.005, 0.0005, \dots$

30. $-\frac{2}{5}, \frac{4}{7}, -\frac{6}{9}, \frac{8}{11}, \dots$

In Problems 31–35 write the first five terms of the sequence.

31. $a_1 = 2$
$a_n = \frac{2}{3}a_{n-1}, \quad n \geq 2$

32. $a_1 = 1$
$a_n = 2a_{n-1} - 3, \quad n \geq 2$

33. $a_1 = 4$
$a_n = -\dfrac{a_{n-1}}{3}, \quad n \geq 2$

34. $a_1 = 1$
$a_n = \dfrac{a_{n-1}}{n}, \quad n \geq 2$

35. $a_1 = 2$
$a_n = -na_{n-1}, \quad n \geq 2$

B

In Problems 36–40 determine an expression for the nth term of the sequence.

36. $\frac{1}{7}, \frac{3}{10}, \frac{5}{13}, \frac{7}{16}, \dots$

37. $-1, \frac{2}{5}, \frac{3}{15}, \frac{4}{29}, \frac{5}{47}, \dots$

38. $1, -\frac{1}{2}, \frac{1}{6}, -\frac{1}{24}, \frac{1}{120}, -\frac{1}{720}, \dots$

39. $\frac{1}{4}, -\frac{2}{10}, \frac{4}{28}, -\frac{8}{82}, \frac{16}{244}, \dots$

40. $\frac{1}{2}, -\frac{3}{6}, \frac{5}{12}, -\frac{7}{20}, \frac{9}{30}, \dots$

In Problems 41–43 write the first five terms of the sequence.

41. $a_1 = 1$
$a_2 = 1$
$a_n = a_{n-1} + a_{n-2}, \quad n \geq 3$
This is called the **Fibonacci sequence.**

42. $a_1 = 2$
$a_n = \sqrt{2 + a_{n-1}}, \quad n \geq 2$

43. $a_1 = 1$
$a_2 = 2$
$a_n = \dfrac{a_{n-1}}{a_{n-2}}, \quad n \geq 3$

44. Let $f(n) = (2i)^n$, where $n \in N$. Write the values of f for $n = 1, 2, 3, 4, 5$.

45. Let

$$x_n = \frac{n+4}{n} \quad \text{and} \quad y_n = \frac{2n}{n^2+1}$$

Plot the sequence of points (x_n, y_n) for $n = 1, 2, 3, 4, 5$. By taking $n = 10, 100, 1,000$, conjecture what point the sequence is approaching as a limit.

ARITHMETIC AND GEOMETRIC SEQUENCES

When each term after the first term of a sequence is formed by adding a fixed amount to the preceding term, the sequence is called **arithmetic.** A more precise definition is given below.

DEFINITION 3

The sequence $a_1, a_2, a_3, \ldots, a_n, \ldots$ is said to be an arithmetic sequence if there exists a constant d such that for all $n \geq 1$,

$$a_{n+1} - a_n = d$$

The number d is called the common difference.

Note. An arithmetic sequence is also sometimes called an **arithmetic progression.**

EXAMPLE 4

Show that the sequence $2, 5, 8, 11, \ldots, 3n - 1, \ldots$ is arithmetic, and find the common difference.

Solution From the first four terms we observe that the difference between successive terms is the constant 3. To test to see if this pattern continues, we use Definition 3. We know that $a_n = 3n - 1$. To find a_{n+1}, we replace n by $n + 1$. This gives $a_{n+1} = 3(n + 1) - 1 = 3n + 2$. So

$$a_{n+1} - a_n = (3n + 2) - (3n - 1) = 3$$

and since this is a constant, independent of n, it follows that the given sequence is arithmetic, with common difference 3. ▲

If $a_1, a_2, a_3, \ldots, a_n, \ldots$ is an arithmetic sequence with common difference d, then by Definition 3 we have

$$
\begin{aligned}
a_2 - a_1 &= d \quad &\text{or} \quad & a_2 = a_1 + d \\
a_3 - a_2 &= d \quad &\text{or} \quad & a_3 = a_2 + d = a_1 + 2d \\
a_4 - a_3 &= d \quad &\text{or} \quad & a_4 = a_3 + d = a_1 + 3d \\
&\vdots & & \vdots \\
a_n - a_{n-1} &= d \quad &\text{or} \quad & a_n = a_{n-1} + d = a_1 + (n - 1)d
\end{aligned}
$$

The last formula enables us to find the nth term when we know a_1 and d. We restate the result for emphasis.

In an arithmetic sequence with first term a_1 and common difference d, the nth term is given by

$$a_n = a_1 + (n-1)d \tag{1}$$

Note. While this result seems evident from the pattern exhibited above, we have not *proved* it. A proof can be based on mathematical induction, and you will be asked to carry out the details in Problem 20, Exercise Set 5.

EXAMPLE 5

Find the twenty-first term of the arithmetic sequence $2, 5, 8, 11, \ldots, 3n - 1, \ldots$.

Solution   As we saw in Example 4, the common difference is 3. So by equation (1),

$$a_{21} = 2 + (21 - 1) \cdot 3 = 2 + (20) \cdot 3 = 62 \quad \blacktriangle$$

EXAMPLE 6

The fifth term of an arithmetic progression is 22 and the fourteenth term is 67. Find the common difference and the first term.

Solution   By equation (1), we have $a_5 = a_1 + 4d$ and $a_{14} = a_1 + 13d$. From the given values, then,

$$a_1 + 4d = 22$$
$$a_1 + 13d = 67$$

We eliminate a_1 by subtracting the members of the top equation from those of the bottom:

$$9d = 45$$
$$d = 5$$

Substituting into the equation $a_1 + 4d = 22$, we get

$$a_1 = 22 - 4(5) = 2 \quad \blacktriangle$$

EXAMPLE 7

Insert three arithmetic means between -4 and 20.

Solution   The instructions mean to find $a_2, a_3,$ and a_4, so that

$$a_1, a_2, a_3, a_4, a_5, \ldots$$

where $a_1 = -4$ and $a_5 = 20$, will be an arithmetic sequence. By equation (1), $a_5 = a_1 + 4d$, so that

$$20 = -4 + 4d$$
$$4d = 24$$
$$d = 6$$

So

$$a_2 = a_1 + d = -4 + 6 = 2$$
$$a_3 = a_2 + d = \quad 2 + 6 = 8$$
$$a_4 = a_3 + d = \quad 8 + 6 = 14$$

The five terms in arithmetic progression are $-4, 2, 8, 14, 20$.   ▲

A **geometric sequence** (or **geometric progression**) is a sequence in which each term after the first is obtained by multiplying the preceding term by a nonzero constant. We can state this in the following equivalent form:

DEFINITION 4   **The sequence $a_1, a_2, a_3, \ldots, a_n, \ldots$ is said to be a geometric sequence if there exists a constant $r \neq 0$ such that for all $n \geq 1$,**

$$\frac{a_{n+1}}{a_n} = r$$

The number r is called the common ratio.

EXAMPLE 8 ||||||||||||||||||||||||||||||||||||||||||||||||||||||||||||||||||||||||||||||||||||||||||||||||

Show that the sequence

$$1, \frac{1}{3}, \frac{1}{9}, \frac{1}{27}, \ldots, \frac{1}{3^{n-1}}, \ldots$$

is geometric, and find the common ratio.

Solution   It appears from the first four terms that a given term after the first is found by multiplying the preceding term by $\frac{1}{3}$, that is, that the common ratio is $\frac{1}{3}$. To show this is always true, we use Definition 4. We are given that

$$a_n = \frac{1}{3^{n-1}}$$

So

$$a_{n+1} = \frac{1}{3^{(n+1)-1}} = \frac{1}{3^n}$$

Thus,

$$\frac{a_{n+1}}{a_n} = \frac{1/3^n}{1/3^{n-1}} = \frac{3^{n-1}}{3^n} = \frac{1}{3}$$

and since this is a nonzero constant, independent of n, it follows that the given sequence is geometric, with $r = \frac{1}{3}$.   ▲

If $a_1, a_2, a_3, \ldots, a_n, \ldots$ is a geometric sequence with common ratio r, then

$$\frac{a_2}{a_1} = r \qquad \text{or} \qquad a_2 = a_1 r$$

$$\frac{a_3}{a_2} = r \qquad \text{or} \qquad a_3 = a_2 r = a_1 r^2$$

$$\frac{a_4}{a_3} = r \qquad \text{or} \qquad a_4 = a_3 r = a_1 r^3$$

$$\vdots \qquad\qquad\qquad\qquad \vdots$$

$$\frac{a_n}{a_{n-1}} = r \qquad \text{or} \qquad a_n = a_{n-1} r = a_1 r^{n-1}$$

We restate the last general result:

> In a geometric sequence with first term a_1 and common ratio r, the nth term is given by
>
> $$a_n = a_1 r^{n-1} \qquad\qquad (2)$$

As with formula (1), this result can be proved formally by techniques we will study in Section 5.

EXAMPLE 9

Find the tenth term of the geometric sequence $1, -\frac{1}{2}, \frac{1}{4}, -\frac{1}{8}, \ldots$.

Solution   We see by inspection that the common ratio r is $-\frac{1}{2}$. So

$$a_{10} = a_1 r^9 = 1\left(-\frac{1}{2}\right)^9$$

$$= -\frac{1}{2^9} = -\frac{1}{512} \qquad\qquad ▲$$

EXERCISE SET 2

A

In Problems 1–5 show that the given sequence is arithmetic. Find the common difference, and write the next five terms.

1. $3, 6, 9, 12, \ldots, 3n, \ldots$

2. $1, \dfrac{3}{2}, 2, \dfrac{5}{2}, \ldots, \dfrac{n+1}{2}, \ldots$

3. $5, 2, -1, -4, \ldots, 8 - 3n, \ldots$

4. $-3, 1, 5, 9, \ldots, 4n - 7, \ldots$

5. $5, \dfrac{7}{2}, 2, \dfrac{1}{2}, \ldots, \dfrac{13 - 3n}{2}, \ldots$

In Problems 6–10 show that the given sequence is geometric. Find the common ratio, and write the next four terms.

6. $\dfrac{2}{3}, \dfrac{4}{9}, \dfrac{8}{27}, \ldots, \left(\dfrac{2}{3}\right)^n, \ldots$

7. $1, -\dfrac{1}{2}, \dfrac{1}{4}, \ldots, \left(-\dfrac{1}{2}\right)^{n-1}, \ldots$

8. $0.2, 0.02, 0.002, \ldots, 2(0.1)^n, \ldots$

9. $\dfrac{1}{9}, \dfrac{1}{3}, 1, \ldots, \dfrac{3^{n-1}}{9}, \ldots$

10. $-2, 3, -\dfrac{9}{2}, \ldots, (-1)^n\left(\dfrac{3^{n-1}}{2^{n-2}}\right), \ldots$

Each sequence in Problems 11–16 is either arithmetic or geometric. Determine the nature of each, and find the specified term.

11. $-1, 2, 5, 8, \ldots;\ \ a_{16}$

12. $\dfrac{1}{4}, \dfrac{1}{2}, 1, 2, \ldots;\ \ a_{12}$

13. $-3, 1, -\dfrac{1}{3}, \dfrac{1}{9}, \ldots;\ \ a_{10}$

14. $1, \dfrac{1}{2}, 0, -\dfrac{1}{2}, \ldots;\ \ a_{30}$

15. $-10, -6, -2, 2, \ldots;\ \ a_{21}$

16. $8, 6, \dfrac{9}{2}, \dfrac{27}{8}, \ldots;\ \ a_7$

In Problems 17–30 identify the sequence as being arithmetic, geometric, or neither. If arithmetic, give the common difference, and if geometric, give the common ratio.

17. $-12, -6, -3, -\dfrac{3}{2}, \ldots$

18. $1, \dfrac{1}{2}, \dfrac{1}{3}, \dfrac{1}{4}, \ldots$

19. $3, 7, 11, 15, \ldots$

20. $1, 2, 3, 4, 5, \ldots$

21. $1, \dfrac{1}{4}, \dfrac{1}{9}, \dfrac{1}{16}, \ldots$

22. $1, a^2, a^4, a^6, \ldots$

23. $\dfrac{1}{2}, \dfrac{3}{4}, \dfrac{7}{8}, \dfrac{15}{16}, \ldots$

24. $0.1, 0.01, 0.001, \ldots$

25. $0.1, 0.11, 0.111, 0.1111, \ldots$

26. $2, \dfrac{1}{2}, -1, -\dfrac{5}{2}, \ldots$

27. $36, 24, 16, \dfrac{32}{3}, \ldots$

28. $-3, -\dfrac{19}{4}, -\dfrac{13}{2}, -\dfrac{33}{4}, \ldots$

29. $\dfrac{1}{2}, \dfrac{2}{3}, \dfrac{3}{4}, \dfrac{4}{5}, \ldots$

30. $x, x + 2, x + 4, x + 6, \ldots$

31. In a certain arithmetic sequence, $a_{20} = 32$ and $d = 3$. Find a_1.

32. In a certain geometric sequence, $a_8 = \dfrac{729}{512}$, and $r = \dfrac{3}{2}$. Find a_1.

33. For a certain arithmetic sequence, $a_{32} = 48$ and $a_{17} = 18$. Find a_1 and d.

34. If $x_1, x_2, x_3, \ldots$ is an arithmetic progression with $x_{15} = 19$ and $x_{28} = -\dfrac{1}{2}$, find the first five terms of the progression.

35. In a certain geometric sequence, $a_3 = -\dfrac{4}{9}$ and $a_6 = \dfrac{32}{243}$. Find a_1 and r.

36. If $t_1, t_2, t_3, \ldots$ is a geometric progression with $t_6 = 0.32$ and $t_{11} = 0.0001024$, find the first five terms of the progression.

37. Find x in the sequence $5, x, 20$ so that the sequence will be
a. Arithmetic   **b.** Geometric (two solutions)

38. Find x and y in the sequence $1, x, y, -64$ so that the sequence will be
a. Arithmetic   **b.** Geometric

39. Insert six arithmetic means between 11 and 32.

40. Insert four geometric means between 27 and $-\dfrac{32}{9}$.

B

41. If $a_1, a_2, a_3, \ldots, a_n, \ldots$ is an arithmetic sequence, and for each $n \in N$, $x_n = a_n + 5$, prove that $x_1, x_2, x_3, \ldots, x_n, \ldots$ is also an arithmetic sequence.

42. If $a_1, a_2, a_3, \ldots, a_n, \ldots$ is a geometric sequence, and for each $n \in N$, $y_n = 8a_n$, prove that $y_1, y_2, y_3, \ldots, y_n, \ldots$ is also a geometric sequence.

43. Determine the nature of the sequence ln 2, ln 4, ln 8, ln 16, ...

44. Determine the nature of the sequence for which

$$a_n = \log_{10} \frac{2}{3^n}$$

45. Prove that if $a_1, a_2, a_3, \ldots$ is a geometric sequence, then $\log a_1, \log a_2, \log a_3, \ldots$ is an arithmetic sequence. How are the common ratio of the first sequence and the common difference of the second related?

46. A ball is dropped to the ground from a height of 60 feet, and each time it bounces, it goes two-thirds as high as it was previously. How high does it go on the eighth bounce?

47. Every person has 2 parents, 4 grandparents, 8 great-grandparents, and so on. How many ancestors does a person have in the tenth preceding generation?

48. A woman wishes to plan a program of jogging in which she will jog 10 minutes the first day and a fixed number of additional minutes each day until the twenty-first day, when she wishes to be jogging for 1 hour. How many minutes of jogging should she add each day?

49. Show that if $P are invested at $r\%$, compounded annually, the amount present after 1 year, 2 years, 3 years, . . . , forms a geometric sequence. What is the common ratio? What is the nth term?

50. A woman's starting salary at a new job 5 years ago was $10,000 annually, and she has received raises of 6% each year since then. What is her salary now?

51. A piece of machinery that costs $30,000 depreciates by 20% each year. What will be its value in 6 years?

SERIES

A **series** is an indicated sum of a sequence, such as the following examples:

a. $2 + 4 + 6 + 8$
b. $1 + \frac{1}{2} + \frac{1}{4} + \frac{1}{8} + \cdots$
c. $1 + 2 + 3 + 4 + 5 + \cdots$

The first is a **finite series,** and the other two are **infinite,** as indicated by the three dots. It is clear in series **a** how to find the sum, but in series **b** and **c** the question of whether a meaningful interpretation can be given to the sum is less clear. By adding up more and more terms in series **b** you might guess that the infinite sum is 2 (and you would be right), but this requires further explanation. It is probably also not too difficult to conclude that series **c** does not add up to any finite number. So we see that for some infinite series a number can be reasonably assigned as a sum, whereas for others this is not possible. Infinite series are studied in detail in calculus. Here, we will limit ourselves primarily to **infinite geometric series,** which are series formed by adding the terms of geometric sequences. First, however, we consider *finite* arithmetic and geometric series.

It is convenient to introduce a shorthand symbol for a series, namely the **summation symbol** $\sum$. When we write, for example,

$$\sum_{k=1}^{n} a_k$$

this means the sum of the numbers a_k, where k takes on the values from 1 to n.

That is,

$$\sum_{k=1}^{n} a_k = a_1 + a_2 + a_3 + \cdots + a_n$$

The letter k as used here is called the **index of summation.** It is a "dummy variable," so-called because it does not appear in the final result, and this result would be unchanged if some other letter were used. For example,

$$\sum_{i=1}^{n} a_i = a_1 + a_2 + a_3 + \cdots + a_n$$

So

$$\sum_{k=1}^{n} a_k = \sum_{i=1}^{n} a_i$$

and, in fact, any other letter could be used as the index of summation.

EXAMPLE 10

Give the expanded form of the sums indicated.

a. $\displaystyle\sum_{k=1}^{4} \frac{1}{2k}$ **b.** $\displaystyle\sum_{i=1}^{6} 2^{i-1}$ **c.** $\displaystyle\sum_{n=1}^{5} (3n - 1)$

Solution   **a.** $\displaystyle\sum_{k=1}^{4} \frac{1}{2k} = \frac{1}{2 \cdot 1} + \frac{1}{2 \cdot 2} + \frac{1}{2 \cdot 3} + \frac{1}{2 \cdot 4}$

$$= \frac{1}{2} + \frac{1}{4} + \frac{1}{6} + \frac{1}{8}$$

b. $\displaystyle\sum_{i=1}^{6} 2^{i-1} = 2^0 + 2^1 + 2^2 + 2^3 + 2^4 + 2^5$

$$= 1 + 2 + 4 + 8 + 16 + 32$$

c. $\displaystyle\sum_{n=1}^{5} (3n - 1) = (3 \cdot 1 - 1) + (3 \cdot 2 - 1) + (3 \cdot 3 - 1) + (3 \cdot 4 - 1) + (3 \cdot 5 - 1)$

$$= (3 - 1) + (6 - 1) + (9 - 1) + (12 - 1) + (15 - 1)$$

$$= 2 + 5 + 8 + 11 + 14$$ ▲

Consider now a finite arithmetic sequence,

$$a_1, a_2, a_3, \ldots, a_n$$

The corresponding arithmetic series is

$$\sum_{k=1}^{n} a_k = a_1 + a_2 + a_3 + \cdots + a_n \tag{3}$$

Let us represent the sum by S_n. We wish to develop a formula for finding S_n. If d is the common difference, we can write series (3) in the form

$$S_n = a_1 + (a_1 + d) + (a_1 + 2d) + (a_1 + 3d) + \cdots + [a_1 + (n-1)d] \quad (4)$$

Now we are going to rewrite series (3) once again—this time in reverse order, observing that if we begin with a_n and go in reverse, we must subtract d each time to get the next term. So we obtain

$$S_n = a_n + (a_n - d) + (a_n - 2d) + \cdots + [a_n - (n-1)d] \quad (5)$$

Finally, we add (4) and (5), noting the terms that add to 0:

$$S_n + S_n = (a_1 + a_n) + (a_1 + a_n) + (a_1 + a_n) + \cdots + (a_1 + a_n)$$

Since there are n terms altogether, this can be written

$$2S_n = n(a_1 + a_n)$$

Therefore, we have the following:

The sum S_n of the terms of the finite arithmetic sequence $a_1, a_2, a_3, \ldots, a_n$ is

$$S_n = n\left(\frac{a_1 + a_n}{2}\right) \quad (6)$$

An easy way to remember this is to think of multiplying the number of terms, n, by the average of the first and last term, $(a_1 + a_n)/2$.

Remark. Although formula (6) was developed for the sum of a *finite* arithmetic sequence, we may also interpret it as the *sum of the first n terms* of an infinite arithmetic sequence.

An alternative formula for the sum S_n can be found by substituting for a_n in (6) its value from equation (1). This gives

$$S_n = n\left[\frac{a_1 + a_1 + (n-1)d}{2}\right]$$

So we have:

$$S_n = \frac{n}{2}[2a_1 + (n-1)d] \quad (7)$$

Whether to use (6) or (7) depends on the given information.

EXAMPLE 11 ||||||||||||||||||||||||||||||||||||||||||||||||||||||||||||||||||||||||||||||||||||||||||||||||||||

Find the sum of the first ten terms of the arithmetic sequence 2, 5, 8, ...

Solution   Since $a_1 = 2$ and $d = 3$, we can use equation (7) to get

$$S_{10} = \tfrac{10}{2}[2(2) + 9(3)] = 5(4 + 27) = 5(31) = 155$$ ▲

EXAMPLE 12 ||||||||||||||||||||||||||||||||||||||||||||||||||||||||||||||||||||||||||||||||||||||||||||||||||||

The first term of an arithmetic sequence is 3 and the fifteenth term is 45. Find the sum of the first fifteen terms.

Solution   This time we use equation (6):

$$S_{15} = 15\left(\frac{3 + 45}{2}\right) = 15(24) = 360$$ ▲

EXAMPLE 13 ||||||||||||||||||||||||||||||||||||||||||||||||||||||||||||||||||||||||||||||||||||||||||||||||||||

The first term of an arithmetic progression is -2, and the sum of the first ten terms is 20. Find the common difference.

Solution   By equation (7), we have

$$20 = \tfrac{10}{2}[2(-2) + 9d]$$
$$20 = 5(-4 + 9d)$$
$$4 = -4 + 9d$$
$$9d = 8$$
$$d = \tfrac{8}{9}$$ ▲

For a finite geometric sequence

$$a_1, a_2, a_3, \ldots, a_n$$

we wish also to develop a formula for S_n, where S_n represents the sum

$$S_n = \sum_{k=1}^{n} a_k = a_1 + a_2 + a_3 + \cdots + a_n$$

If r is the common ratio, we know that each term after the first is formed by multiplying the preceding one by r. So

$$S_n = a_1 + a_1 r + a_1 r^2 + a_1 r^3 + \cdots + a_1 r^{n-1} \tag{8}$$

Now if we multiply both sides of equation (8) by r,

$$rS_n = a_1r + a_1r^2 + a_1r^3 + \cdots + a_1r^{n-1} + a_1r^n$$

and then subtract this from equation (8), we get

$$S_n - rS_n = a_1 - a_1r^n$$

or

$$S_n(1 - r) = a_1(1 - r^n)$$

If $r \neq 1$, we can solve this for S_n:

$$S_n = \frac{a_1(1 - r^n)}{1 - r} \qquad r \neq 1$$

If $r = 1$, we can find S_n from equation (8), because then it reads

$$S_n = a_1 + a_1 + a_1 + \cdots + a_1$$

and since there are n terms, this gives

$$S_n = na_1 \qquad r = 1$$

We summarize these results in the box.

If $a_1, a_2, a_3, \ldots, a_n$ is a finite geometric sequence with common ratio r, then

$$S_n = \frac{a_1(1 - r^n)}{1 - r} \qquad \text{if} \quad r \neq 1 \tag{9}$$

$$= na_1 \qquad \text{if} \quad r = 1$$

where $S_n = a_1 + a_2 + \cdots + a_n$.

Again we may interpret equation (9) as being the sum of the first n terms of an infinite geometric sequence.

EXAMPLE 14

Find the sum of the first eight terms of the geometric sequence $1, -2, 4, -8, \ldots$.

Solution   We see that $r = -2$. So by equation (9),

$$S_8 = \frac{1[1 - (-2)^8]}{1 - (-2)} = \frac{1 - 256}{3}$$

$$= -\frac{255}{3} = -85$$

▲

EXAMPLE 15 ꠰꠰꠰

Find the sum

$$\sum_{k=1}^{10} \frac{1}{2^k}$$

Solution   The first thing to note is that this is a geometric series. The expanded form of the sum is

$$S_{10} = \frac{1}{2} + \frac{1}{2^2} + \frac{1}{2^3} + \cdots + \frac{1}{2^{10}}$$

The common ratio is $\frac{1}{2}$, so by equation (9) we have

$$S_{10} = \frac{\frac{1}{2}[1 - (\frac{1}{2})^{10}]}{1 - \frac{1}{2}} = 1 - \frac{1}{2^{10}}$$

$$= 1 - \frac{1}{1,024} = \frac{1,023}{1,024}$$

▲

We conclude this section with a summary of the formulas we have developed for arithmetic and geometric sequences and series.

	Arithmetic Sequence	**Geometric Sequence**
For all $n \geq 1$:	$d = a_{n+1} - a_n$	$r = \dfrac{a_{n+1}}{a_n}$
nth term:	$a_n = a_1 + (n-1)d$	$a_n = a_1 r^{n-1}$
Sum of first n terms:	$S_n = n\left(\dfrac{a_1 + a_n}{2}\right)$	$S_n = \dfrac{a_1(1 - r^n)}{1-r}$ if $r \neq 1$
	$= \dfrac{n}{2}[2a_1 + (n-1)d]$	$= na_1$ if $r = 1$

EXERCISE SET 3

A

Write the expanded forms of the series in Problems 1–10.

1. $\displaystyle\sum_{k=1}^{4} \frac{3k+1}{2k-1}$

2. $\displaystyle\sum_{i=1}^{5} \frac{i}{i^2+1}$

3. $\displaystyle\sum_{n=1}^{6} \frac{(-1)^{n-1}}{n(n+1)}$

4. $\displaystyle\sum_{j=1}^{5} \frac{2^{j-1}}{3^j}$

5. $\displaystyle\sum_{k=1}^{8} \frac{(-1)^{k-1}}{k^2+1}$

6. $\displaystyle\sum_{m=1}^{6} \frac{2^m}{m+1}$

7. $\displaystyle\sum_{k=1}^{10} (5k-3)$

8. $\displaystyle\sum_{n=1}^{5} \frac{\ln n}{n}$

9. $\displaystyle\sum_{n=1}^{6} ne^{-n}$

10. $\displaystyle\sum_{k=1}^{5} \frac{k^k}{2k-1}$

In Problems 11–18 find the sum of the first n terms of the given sequence for the specified value of n.

11. $3, 8, 13, 18 \ldots;\quad n = 20$
12. $-2, 4, -8, 16, \ldots;\quad n = 10$
13. $-6, -4, -2, 0, \ldots;\quad n = 100$
14. $1, \frac{1}{3}, \frac{1}{9}, \ldots;\quad n = 8$
15. $0.2, 0.02, 0.002, \ldots;\quad n = 12$
16. $5, \frac{7}{2}, 2, \frac{1}{2}, \ldots;\quad n = 30$
17. $0.2, 0.22, 0.24, 0.26, \ldots;\quad n = 50$
18. $-27, 18, -12, 8, \ldots;\quad n = 8$

In Problems 19–26 evaluate the indicated sums.

19. $\displaystyle\sum_{k=1}^{200} (2 + 3k)$

20. $\displaystyle\sum_{k=1}^{10} (-3)^{k-1}$

21. $\displaystyle\sum_{n=1}^{20} (2n - 1)$

22. $\displaystyle\sum_{k=1}^{12} \frac{1}{2^{k+1}}$

23. $\displaystyle\sum_{i=1}^{8} \frac{2^i}{3^{i-1}}$

24. $\displaystyle\sum_{m=1}^{50} \left(\frac{m+3}{2}\right)$

25. $\displaystyle\sum_{j=1}^{6} \frac{(-1)^{j-1} 3^j}{4^{j-1}}$

26. $\displaystyle\sum_{n=1}^{15} \left(\frac{n}{2} - 3\right)$

27. Find the sum of the first 100 natural numbers.

 B

Find the sums in Problems 28 and 29.

28. $\displaystyle\sum_{k=1}^{6} (2^k + 3k - 1)$

29. $\displaystyle\sum_{n=0}^{5} [3^n - 2(n + 1)]$

30. Find a formula for the sum of the first n odd positive integers.

31. Some pipes are stacked so that the bottom layer has 15 pipes, the next layer 14 pipes, the next 13, and so on, until there is only one pipe at the top. How many pipes are there in all?

32. A ball is dropped from a height of 40 feet, and each time it bounces, it rises to a height three-fourths as high as previously. Find the total distance the ball has covered when it hits the ground for the fifth time.

33. A man is offered two jobs, one starting at $12,000 with constant annual raises of $700, and the other starting at $10,000 with annual raises of 8%. What would be his salary at each job during the sixth year? What would be his accumulated earnings at each job through the end of the sixth year? How many years would it be before the salary of the second job exceeded the first? How many years would it be before the accumulated earnings of the second job exceeded the first?

34. A pendulum swings a distance of 20 inches initially from one side to the other, and on each subsequent swing it goes 0.8 of the distance on the previous swing. Find the total distance covered by the pendulum after ten swings.

35. If a person were given a choice of working 30 days for 1¢ the first day, 2¢ the next, 4¢ the next, and so on, each day doubling the amount earned on the previous day, or of being paid $1,000 the first day, $2,000 the next, $3,000 the next, and so on, each day adding $1,000 to the amount on the previous day, which offer should he take? How much would he make under each arrangement?

 4   INFINITE GEOMETRIC SERIES

Consider an infinite sequence

$$a_1, a_2, a_3, \ldots, a_n, \ldots$$

We designate the corresponding **infinite series** by

$$\sum_{k=1}^{\infty} a_k = a_1 + a_2 + a_3 + \cdots + a_n + \cdots$$

The symbol ∞ is read "infinity." To arrive at a reasonable definition of what is meant by such an infinite sum, we consider the finite sums

$$S_1 = a_1$$
$$S_2 = a_1 + a_2$$
$$S_3 = a_1 + a_2 + a_3$$
$$\cdots\cdots\cdots\cdots\cdots$$
$$S_n = a_1 + a_2 + \cdots + a_n$$

We call S_n the **nth partial sum** of the series $\sum\limits_{n=1}^{\infty} a_n$, and we call the sequence $S_1, S_2, S_3, \ldots, S_n, \ldots$ **the sequence of partial sums.** If it happens that for larger and larger values of n, the partial sums approach as a limit some number S, then we define S to be the sum of the infinite series. This is written symbolically as

$$S = \sum_{n=1}^{\infty} a_n = \lim_{n \to \infty} S_n$$

The symbol "$\lim\limits_{n \to \infty} S_n$" is read "the limit of S_n as n goes to infinity." Limits of sequences are studied in detail in calculus, and it is not appropriate in this course to give a precise definition. It is sufficient to say that

$$\lim_{n \to \infty} S_n = S$$

means that the partial sums S_n are arbitrarily close to S for all sufficiently large values of n.

To make these concepts more concrete, let us consider the infinite geometric series

$$1 + \frac{1}{2} + \frac{1}{4} + \frac{1}{8} + \cdots + \frac{1}{2^{n-1}} + \cdots \tag{10}$$

The partial sums are

$$S_1 = 1$$
$$S_2 = 1 + \tfrac{1}{2} = \tfrac{3}{2}$$
$$S_3 = 1 + \tfrac{1}{2} + \tfrac{1}{4} = \tfrac{7}{4}$$
$$S_4 = 1 + \tfrac{1}{2} + \tfrac{1}{4} + \tfrac{1}{8} = \tfrac{15}{8}$$
$$\cdots\cdots\cdots\cdots\cdots\cdots$$

By equation (9) we can write the nth partial sum as

$$S_n = 1 + \frac{1}{2} + \frac{1}{4} + \frac{1}{8} + \cdots + \frac{1}{2^{n-1}}$$
$$= \frac{1[1 - (\tfrac{1}{2})^n]}{1 - \tfrac{1}{2}} = 2\left[1 - \left(\frac{1}{2}\right)^n\right]$$

The sequence of partial sums,

$$1, \frac{3}{2}, \frac{7}{4}, \frac{15}{8}, \ldots, 2\left[1 - \left(\frac{1}{2}\right)^n\right], \ldots$$

appears to be approaching 2 as a limit. In fact,

$$\lim_{n \to \infty} S_n = \lim_{n \to \infty} 2\left[1 - \left(\frac{1}{2}\right)^n\right]$$

and since $(\frac{1}{2})^n$ can be made arbitrarily close to 0 by choosing n sufficiently large, it follows that $\lim_{n \to \infty} S_n = 2$.

It should be emphasized that the partial sums of an infinite series do not always approach a finite limit. If a limit is approached, we say the series **converges;** otherwise, it **diverges.** The series

$$\sum_{n=1}^{\infty} 2n = 2 + 4 + 6 + 8 + \cdots$$

clearly diverges. In fact, this is the sum of an arithmetic sequence, and by equation (6)

$$S_n = 2 + 4 + 6 + 8 + \cdots + 2n = n\left[\frac{2 + (2n)}{2}\right] = n(1 + n)$$

As n becomes arbitrarily large, S_n is unbounded.

Now we want to determine under what conditions the infinite geometric series

$$\sum_{n=1}^{\infty} a_1 r^{n-1} = a_1 + a_1 r + a_1 r^2 + a_1 r^3 + \cdots \tag{11}$$

converges. By equation (9) we know that if $r \neq 1$, the nth partial sum is given by

$$S_n = \frac{a_1(1 - r^n)}{1 - r}$$

We want to examine what happens as n gets arbitrarily large. The key lies in the term r^n. When r is less than 1 in absolute value (for example, when $r = \frac{1}{2}$), it is shown by calculus that r^n approaches 0. So we conclude that

$$\lim_{n \to \infty} S_n = \frac{a_1}{1 - r} \quad \text{if} \quad |r| < 1$$

On the other hand, if $|r| \geq 1$, it can be proved that no finite limit exists. For example, if $r = 2$, we see that 2^n gets arbitrarily large, so that S_n cannot approach a finite limit. Similarly, if $r = -2$, then $(-2)^n$ gets larger and larger numerically, but with alternating signs. So again, S_n does not approach a finite limit.

The following general result can be proved:

The infinite geometric series (11) converges if and only if $|r| < 1$. Furthermore, when $|r| < 1$, the sum S is

$$S = \frac{a_1}{1 - r} \qquad (12)$$

EXAMPLE 16

Show that the series below converges, and find its sum.

$$\sum_{n=1}^{\infty} \left(-\frac{2}{3} \right)^{n-1}$$

Solution   This is a geometric series with ratio $-\frac{2}{3}$, and since this is less than 1 in absolute value, the series converges. In expanded form the series is

$$1 - \tfrac{2}{3} + \tfrac{4}{9} - \tfrac{8}{27} + \cdots$$

By equation (12) the sum is

$$S = \frac{a_1}{1 - r} = \frac{1}{1 - (-\frac{2}{3})} = \frac{1}{1 + \frac{2}{3}} = \frac{3}{5}$$

It should be emphasized that we can never reach $\frac{3}{5}$ exactly by adding up any finite number of terms, but by definition, $\frac{3}{5}$ is the sum of the infinite series.   ▲

EXAMPLE 17

Show that each of the following series is divergent:

a. $1 + \dfrac{3}{2} + \dfrac{9}{4} + \dfrac{27}{8} + \cdots$ b. $\displaystyle\sum_{k=1}^{\infty} (-1)^{k-1}$

Solution a.   The common ratio is $\frac{3}{2}$, which is greater than 1. So the series diverges.

b.   The expanded series is $1 - 1 + 1 - 1 + \cdots$, which is a geometric series with ratio -1. Since $|-1| = 1$, the series diverges. Notice the behavior of the partial sums in this example:

$$S_1 = 1, \qquad S_2 = 0, \qquad S_3 = 1, \qquad S_4 = 0, \ldots$$

Although these do not get arbitrarily large, no specific limit is approached.   ▲

The formula (12) for the sum of an infinite geometric series can be used to find the rational number represented by a repeating decimal. This is an alternative to the procedure given in Chapter 1. The next two examples illustrate this technique.

EXAMPLE 18

Find the rational number represented by the repeating decimal 0.242424....

Solution   We treat this as the sum of the infinite series

$$0.24 + 0.0024 + 0.000024 + \cdots$$

which is geometric, with $a_1 = 0.24$ and $r = 0.01$. Thus, the sum is

$$S_n = \frac{a_1}{1 - r} = \frac{0.24}{1 - 0.01} = \frac{0.24}{0.99} = \frac{24}{99} = \frac{8}{33}$$ ▲

EXAMPLE 19

Express 2.135135135 ... as a rational number.

Solution   We write this as

$$2 + [0.135 + 0.000135 + 0.000000135 + \cdots]$$

The portion in brackets is a geometric series with $a_1 = 0.135$ and $r = 0.001$. So its sum is

$$\frac{a_1}{1 - r} = \frac{0.135}{1 - 0.001} = \frac{0.135}{0.999} = \frac{135}{999} = \frac{5}{37}$$

and the answer to the problem is

$$2 + \tfrac{5}{37} = \tfrac{79}{37}$$ ▲

We conclude this section with a glimpse of the larger problem of convergence of infinite series in general (not just geometric series). At this stage we cannot go into this subject in depth, since to do so would require a knowledge of calculus. It is a fascinating part of mathematics and presents some intriguing questions. There are, in fact, some famous unsolved problems in this field.

To get a hint of one aspect of the problem, consider the following infinite series:

a. $1 + \dfrac{1}{2} + \dfrac{1}{3} + \dfrac{1}{4} + \cdots$ **b.** $1 - \dfrac{1}{2} + \dfrac{1}{3} - \dfrac{1}{4} + \cdots$

c. $1 + \dfrac{1}{2^2} + \dfrac{1}{3^2} + \dfrac{1}{4^2} + \cdots$ **d.** $1 + \dfrac{1}{2^3} + \dfrac{1}{3^3} + \dfrac{1}{4^3} + \cdots$

Without any attempt at a proof, we state the following facts:

a. The series diverges. By choosing n large enough we can make S_n arbitrarily large, even though the sum of the first billion terms is only about 21. So the divergence is quite slow.

b. This series converges, and its sum is ln 2.

c. This series converges, and its sum is $\pi^2/6$. How this result was first arrived at by the famous seventeenth century Swiss mathematician Leonhard Euler is a fascinating chapter in the history of mathematics.

d. This series converges, but no one knows exactly what the sum is. This is one of the unsolved problems.

 EXERCISE SET 4

A

Determine which of the infinite geometric series in Problems 1–19 converge and which diverge. For those that converge, find the sum.

1. $1 + \frac{1}{3} + \frac{1}{9} + \frac{1}{27} + \cdots$

2. $1 - \frac{3}{4} + \frac{9}{16} - \frac{27}{64} + \cdots$

3. $1 + \frac{4}{3} + \frac{16}{9} + \frac{64}{27} + \cdots$

4. $\displaystyle\sum_{k=1}^{\infty} (-1.1)^{k-1}$

5. $\displaystyle\sum_{n=1}^{\infty} (-0.9)^{n-1}$

6. $\displaystyle\sum_{i=1}^{\infty} \frac{2^i}{3^{i+1}}$

7. $\displaystyle\sum_{n=1}^{\infty} \frac{(-1)^{n-1} 5^n}{6^{n-1}}$

8. $\displaystyle\sum_{n=1}^{\infty} 2^{-n}$

9. $\displaystyle\sum_{n=0}^{\infty} \left(\frac{3}{2}\right)^n$

10. $\displaystyle\sum_{n=0}^{\infty} \left(\frac{3}{2}\right)^{-n}$

11. $\displaystyle\sum_{n=1}^{\infty} 2\left(\frac{3}{5}\right)^{n-1}$

12. $\displaystyle\sum_{n=1}^{\infty} 3\left(-\frac{2}{3}\right)^{n-1}$

13. $36 - 12 + 4 + \cdots$

14. $0.01 + 0.02 + 0.04 + \cdots$

15. $\displaystyle\sum_{n=1}^{\infty} (\sqrt{2})^n$

16. $\displaystyle\sum_{k=1}^{\infty} (\sqrt{3})^{-k+1}$

17. $\displaystyle\sum_{n=1}^{\infty} 2^{-n/2}$

18. $1 + 0.1 + 0.01 + 0.001 + \cdots$

19. $1 - 0.1 + 0.01 - 0.001 + \cdots$

In Problems 20–25 find the rational number corresponding to the given decimal number using infinite series.

20. $0.333\ldots$

21. $0.151515\ldots$

22. $0.272727\ldots$

23. $1.545454\ldots$

24. $0.243243243\ldots$

25. $3.162162162\ldots$

26. Find the sum of the series $\displaystyle\sum_{n=1}^{\infty} x^n$ as a function of x. For what values of x is this valid?

27. Find the sum of the series $\displaystyle\sum_{n=1}^{\infty} (x^n/2^n)$. For what values of x is this valid?

28. If S_n is the nth partial sum of the series $\displaystyle\sum_{n=1}^{\infty} a_n$, show that $a_1 = S_1$, and for all $n \geq 2$, show that $a_n = S_n - S_{n-1}$.

29. Use the result of Problem 28 to construct an infinite series for which $S_n = (n+1)/n$. Show that the series converges and that its sum is 1.
    Hint. Write S_n in the form $1 + (1/n)$ to show convergence.

30. A ball is dropped from a height of 50 feet, and on each bounce it goes three-fourths as high as before. Approximate the total distance traveled by the ball in coming to rest.

31. The bob of a pendulum initially swings through an arc 50 cm long, and on each succeeding swing it goes 0.95 as far as on the previous swing. Find the theoretical total distance through which it swings if it continues indefinitely.

B

32. Show that

$$\frac{1}{n(n+1)} = \frac{1}{n} - \frac{1}{n+1}$$

Use this result to find the sum of the series

$$\sum_{n=1}^{\infty} \frac{1}{n(n+1)}$$

Hint. Use the first result, and write the expanded form of S_n. Observe the terms that add to 0.

33. Find the sum of the series

$$\sum_{n=1}^{\infty} \frac{(-1)^{n-1} x^{2n-1}}{3^n}$$

Show that this is valid only when $|x| < \sqrt{3}$.

34. A square is 12 cm on a side. By joining the consecutive midpoints of its sides with line segments, a second square is formed. Then the midpoints of the sides of the second square are similarly joined to form a third square, and so on, indefinitely. Find the sum of the areas of all infinitely many of the squares.

35. The theory of infinite geometric series continues to hold true for complex values of r, and formula (12) is valid as long as $|r| < 1$. Using this fact, find the sum of the series

$$1 + \frac{i}{2} - \frac{1}{4} - \frac{i}{8} + \frac{1}{16} + \frac{i}{32} - \cdots$$

36. Find the sum of the series $\sum_{n=1}^{\infty} (2x-1)^n$. Find the range of values of x for which this result is valid.

In Problems 37–40 make use of formula (12) to find the infinite geometric series having the specified sum, and state the domain of validity.

37. $\dfrac{1}{1-2x}$ **38.** $\dfrac{1}{1+(x/3)}$

39. $\dfrac{1}{2-x}$ **Hint.** Divide the numerator and denominator by 2.

40. $\dfrac{3}{4-x^2}$

41. For a certain series $\sum_{n=1}^{\infty} a_n$, the nth partial sum is $4 - (1/3^n)$. What is the sum of the series? Show that the series is geometric.
Hint. Use Problem 28 and Definition 4.

42. In calculus it is shown that

$$e^x = 1 + x + \frac{x^2}{2!} + \frac{x^3}{3!} + \cdots + \frac{x^n}{n!} + \cdots,$$

$$\text{when } n! = 1 \cdot 2 \cdot 3 \cdots n$$

With the aid of a calculator use the first ten terms of this series to estimate the value of e by taking $x = 1$. The error in this estimate can be shown to be no greater than $3/10!$. What can you conclude is the accuracy of your estimate?

MATHEMATICAL INDUCTION

In this section we present a method of proving certain statements, or formulas, about natural numbers. To illustrate how we might arrive at one such formula, let us consider the sum of the first n odd natural numbers. We observe the values for $n = 1, 2, 3,$ and 4:

$$n = 1:\quad 1 = 1$$
$$n = 2:\quad 1 + 3 = 4$$
$$n = 3:\quad 1 + 3 + 5 = 9$$
$$n = 4:\quad 1 + 3 + 5 + 7 = 16$$

Since the sum in each case is the square of the number of terms added, this suggests the following general statement:

The sum of the first *n* odd natural numbers is n^2.

Equivalently, we could write this as the formula

$$1 + 3 + 5 + \cdots + (2n - 1) = n^2$$

But have we *proved* this formula? The answer is no. We have shown it to be true for $n = 1, 2, 3$, and 4, but that is all. Even if we continued for many more values of *n* and found the result true in every case, we still would not know for sure whether it was true for *all* natural numbers. It is just such situations as this where the method of proof called **mathematical induction** comes to our rescue. (We might note that the example we have used happens to be an arithmetic series, and we know a formula for the sum; however, the method of mathematical induction applies to a much wider class of problems.)

Principle of Mathematical Induction

If a statement involving natural numbers is true for $n = 1$, and if its truth for an arbitrary natural number *k* implies its truth for $k + 1$, then the statement is true for all natural numbers.

We will take this as an axiom, although it is possible to prove it based on other fundamental properties of the natural numbers. We will, however, show its plausibility. If we have shown that the statement in question is true for $n = 1$, and if we have shown that its truth for $n = k$ implies its truth for $n = k + 1$, then we can let $k = 1$ and conclude that it is true also for $k + 1 = 2$. Then we can let $k = 2$ and conclude that it is true for $k + 1 = 3$. Next, letting $k = 3$, we see that it is true for $k + 1 = 4$, and so on and on. It can be seen, then, to be true for any given natural number, and hence for every natural number.

In order to state the principle of mathematical induction using fewer words, we introduce the symbol $P(n)$ for the statement to be proved (this can be thought of as a function with the natural numbers as domain and a set of statements as the range). The principle of mathematical induction can then be restated as follows:

If (i) $P(1)$ is true, and (ii) $P(k)$ implies $P(k + 1)$ for every natural number *k*, then $P(n)$ is true for all natural numbers.

Both conditions (i) and (ii) are essential. To prove (i), we simply substitute $n = 1$ into the statement and see if it is true. For (ii), we *assume* the statement is true for an unspecified natural number *k* and show that it follows from

FIGURE 1

this that the statement is true for $k + 1$. In general, it is more difficult to prove (ii) than (i).

Mathematical induction is often compared to lining up dominoes in such a way that if the first one is knocked down, all succeeding ones will fall (Figure 1). This is a pretty good analogy (provided we imagine infinitely many dominoes!). To say that the first one is knocked down is analogous to saying $P(1)$ is true. Having the dominoes spaced so that whenever any one of them (say the kth one) falls, the next one (the $k + 1$st one) will also fall, is analogous to saying that $P(k)$ implies $P(k + 1)$.

We illustrate the technique with several examples, the first of which is the one introduced at the beginning of this section.

EXAMPLE 20 ||||||||||||||||||||||||||||||||||||||||||||||||||||||||||||||||||||||||||||||||||||||||||||||||||||||||||||||||||||||||||||||||||||

Use mathematical induction to prove the formula

$$1 + 3 + 5 + \cdots + (2n - 1) = n^2 \tag{13}$$

Solution   Let $P(n)$ denote the given statement, that is, formula (13). To test (i), we look at $P(1)$. The left-hand side of the formula says "add all odd natural numbers up through the nth one" (which equals $2n - 1$). But for $n = 1$ the nth term is the first term, so the left-hand side consists of the single number 1. The right-hand side is 1^2, and since $1 = 1^2$, it follows that $P(1)$ is true.

Now we assume that $P(k)$ is true and try to show that $P(k + 1)$ follows, where k represents an arbitrary natural number. So we assume that

$$1 + 3 + 5 + \cdots + (2k - 1) = k^2 \tag{14}$$

We wish to work from this and by means of valid mathematical operations obtain $P(k + 1)$. Now $P(k + 1)$ is formula (13) with n replaced by $k + 1$. The left-hand side is the sum of the first $k + 1$ odd numbers and so contains all the terms on the left-hand side of $P(k)$ plus one more term, namely $2k + 1$. So we are going to add $2k + 1$ to *both* sides of formula (14)—adding it to the left to make it what we want and adding it to the right to balance this off. This yields

$$1 + 3 + 5 + \cdots + (2k - 1) + (2k + 1) = k^2 + 2k + 1$$

or on factoring the right-hand side,

$$1 + 3 + 5 + \cdots + (2k + 1) = (k + 1)^2$$

This is exactly what formula (13) gives for $n = k + 1$; in other words, it is $P(k + 1)$. So we have shown that the truth of $P(k)$ implies that of $P(k + 1)$, and both parts (i) and (ii) of the principle of mathematical induction have been shown to be true. Therefore, formula (13) is true for all natural numbers n.   ▲

EXAMPLE 21

Prove that for all natural numbers n, $2^n \geq 2n$.

Solution   Let $P(n)$ denote the given statement. We test $P(1)$, which says

$$2^1 \geq 2 \cdot 1$$

Since both sides equal 2, and $2 \geq 2$, we see that $P(1)$ is true.
Now assume $P(k)$ is true for an arbitrary natural number k:

$$2^k \geq 2k$$

To work toward $P(k + 1)$, we can multiply both sides by 2 to make the left-hand side read 2^{k+1}. This gives

$$2 \cdot 2^k \geq 2 \cdot 2k$$

or

$$2^{k+1} \geq 2k + 2k$$

Since $k \geq 1$, it follows that $2k \geq 2$. Thus,

$$2^{k+1} \geq 2k + 2 = 2(k + 1)$$

Hence,

$$2^{k+1} \geq 2(k + 1)$$

which is $P(k + 1)$. Therefore the proof by mathematical induction is complete, and we conclude that $2^n \geq 2n$ for all natural numbers n.   ▲

EXAMPLE 22

Prove by mathematical induction that for all natural numbers n, the quantity $n^2 + n$ is an even number.

Solution   Let $P(n)$ represent the statement "$n^2 + n$ is an even number." Then $P(1)$ says that $1^2 + 1$ is an even number, which is true, since $1^2 + 1 = 2$. Now assume that $P(k)$ is true, that is, $k^2 + k$ is an even number. Since by definition an even number is a multiple of 2, we can say that

$$k^2 + k = 2m$$

for some integer m. For $P(k + 1)$ we want to examine $(k + 1)^2 + (k + 1)$. This can be written

$$k^2 + 2k + 1 + k + 1 = k^2 + k + 2k + 2$$
$$= (k^2 + k) + 2(k + 1)$$

By our assumption $k^2 + k = 2m$, so

$$(k + 1)^2 + (k + 1) = 2m + 2(k + 1) = 2(m + k + 1)$$

The right-hand side is a multiple of 2 and so is even. Thus, we have shown that whenever $k^2 + k$ is even, $(k + 1)^2 + (k + 1)$ is also even. So both parts of the proof by mathematical induction are complete, and we conclude that $n^2 + n$ is even for all natural numbers n.   ▲

EXAMPLE 23 |||||||||||||||||||||||||||||||||||||||||||||||||||||||||||||||||||||||||||||||||||||||||||||||||||||||||||||||||||||||||

Prove that

$$1^2 + 2^2 + 3^2 + \cdots + n^2 = \frac{n(n + 1)(2n + 1)}{6}$$

for all natural numbers n.

Solution   Again let $P(n)$ denote the formula to be proved. For $n = 1$ this reads

$$1^2 = \frac{1(1 + 1)(2 + 1)}{6}$$

which is seen to be true.

Now assume that $P(k)$ is true:

$$1^2 + 2^2 + 3^2 + \cdots + k^2 = \frac{k(k + 1)(2k + 1)}{6}$$

Add $(k + 1)^2$ to both sides:

$$1^2 + 2^2 + 3^2 + \cdots + k^2 + (k + 1)^2 = \frac{k(k + 1)(2k + 1)}{6} + (k + 1)^2$$

This brings the left-hand side to the proper form for $P(k + 1)$. We want to see if the right-hand side also is in the proper form.

$$\frac{k(k + 1)(2k + 1)}{6} + (k + 1)^2 = \frac{k(k + 1)(2k + 1) + 6(k + 1)^2}{6}$$

$$= \frac{(k + 1)[k(2k + 1) + 6(k + 1)]}{6}$$

$$= \frac{(k+1)(2k^2 + 7k + 6)}{6}$$

$$= \frac{(k+1)(k+2)(2k+3)}{6}$$

$$= \frac{(k+1)[(k+1)+1][2(k+1)+1]}{6}$$

A reexamination of the right-hand side of the original formula $P(n)$ shows that the last expression arrived at is exactly what is obtained if n is replaced by $k+1$. So by assuming $P(k)$ to be true, we have shown that $P(k+1)$ also is true. Therefore, the proof by induction is complete.   ▲

EXERCISE SET 5

Using mathematical induction, prove that each of the following is true for all natural numbers n.

A

1. $2 + 4 + 6 + \cdots + (2n) = n(n+1)$

2. $\dfrac{1}{1\cdot2} + \dfrac{1}{2\cdot3} + \dfrac{1}{3\cdot4} + \cdots + \dfrac{1}{n(n+1)} = \dfrac{n}{n+1}$

3. $1 + 5 + 9 + \cdots + (4n-3) = n(2n-1)$

4. $n^2 + 1 \ge 2n$

5. $3 + 7 + 11 + \cdots + (4n-1) = n(2n+1)$

6. $1 + 2 + 3 + \cdots + n = \dfrac{n(n+1)}{2}$

7. $1 + 2^1 + 2^2 + 2^3 + \cdots + 2^{n-1} = 2^n - 1$

8. 3 is a factor of $n^3 + 2n$

9. If $a > 1$, then $a^n > 1$.

10. If $0 < a < 1$, then $0 < a^n < 1$.

11. $\dfrac{1}{1\cdot3} + \dfrac{1}{3\cdot5} + \dfrac{1}{5\cdot7} + \cdots + \dfrac{1}{(2n-1)(2n+1)} = \dfrac{n}{2n+1}$

12. 6 is a factor of $n(n+1)(n+2)$

13. $1\cdot2 + 2\cdot3 + 3\cdot4 + \cdots + n(n+1) = \dfrac{n(n+1)(n+2)}{3}$

14. $\displaystyle\sum_{i=1}^{n}(3i-2) = \dfrac{n(3n+1)}{2}$

15. $(ab)^n = a^n p^n$

16. 2 is a factor of $n^2 - n + 2$

17. $\displaystyle\sum_{i=1}^{n}(6i-5) = 3n^2 - 2n$

18. 3 is a factor of $4^n - 1$

19. $\displaystyle\sum_{k=1}^{n}(-1)^{k-1} = \dfrac{1-(-1)^n}{2}$

20. The nth term of an arithmetic sequence having first term a_1 and common difference d is $a_1 + (n-1)d$.

21. The nth term of a geometric sequence having first term a_1 and common ratio r is $a_1 r^{n-1}$.

B

22. $1^2 + 3^2 + 5^2 + \cdots + (2n-1)^2 = \dfrac{n(4n^2-1)}{3}$

23. If $x > -1$, then $(1+x)^n \ge 1 + nx$.

24. $1^3 + 3^3 + 5^3 + \cdots + (2n-1)^3 = n^2(2n^2-1)$

25. $1^3 + 2^3 + 3^3 + \cdots + n^3 = (1+2+3+\cdots+n)^2$
    Hint. See Problem 6.

26. $a - b$ is a factor of $a^n - b^n$
    Hint. For the second part of the proof write $a^{k+1} - b^{k+1} = a(a^k - b^k) + b^k(a-b)$.

27. $a + b$ is a factor of $a^{2n-1} + b^{2n-1}$
    Hint. For the second part of the proof write $a^{2k+1} + b^{2k+1} = a^2(a^{2k-1} + b^{2k-1}) - b^{2k-1}(a^2 - b^2)$.

PROOF OF THE BINOMIAL THEOREM

In Chapter 1 we stated the binomial theorem without proving it. Mathematical induction provides one means of proving this important result. The formula to be proved can be written as follows:

$$(a + b)^n = a^n + na^{n-1}b + \frac{n(n-1)}{1 \cdot 2} a^{n-2}b^2$$
$$+ \frac{n(n-1)(n-2)}{1 \cdot 2 \cdot 3} a^{n-3}b^3 + \cdots + b^n \tag{15}$$

We wish to show that this is true for all natural numbers n.

First we introduce a convenient symbol called the **binomial coefficient symbol**:

For $k = 1, 2, 3, \ldots, n$ we define

$$\binom{n}{k} = \frac{n(n-1)(n-2) \cdots (n-k+1)}{1 \cdot 2 \cdot 3 \cdots k} \tag{16}$$

and for $k = 0$ we define $\binom{n}{0} = 1$.

The binomial formula can now be written as

$$(a + b)^n = \binom{n}{0} a^n + \binom{n}{1} a^{n-1}b + \binom{n}{2} a^{n-2}b^2 + \cdots + \binom{n}{n} b^n$$
$$= \sum_{r=0}^{n} \binom{n}{r} a^{n-r}b^r \tag{17}$$

Note that the $(k + 1)$st term of the expansion is

$$\binom{n}{k} a^{n-k}b^k \tag{18}$$

Factorial notation enables us to give an alternate form of the binomial coefficient that is sometimes convenient. If r is any natural number, then the symbol $r!$, read "r factorial," means $1 \cdot 2 \cdot 3 \cdots r$. For example, $5! = 1 \cdot 2 \cdot 3 \cdot 4 \cdot 5 = 120$. We define $0! = 1$. We may therefore write

$$\binom{n}{k} = \frac{n(n-1)(n-2) \cdots (n-k+1)}{1 \cdot 2 \cdot 3 \cdots k}$$
$$= \frac{n(n-1)(n-2) \cdots (n-k+1)}{k!} \cdot \frac{(n-k)(n-k-1) \cdots 2 \cdot 1}{(n-k)(n-k-1) \cdots 2 \cdot 1}$$

and thus

$$\binom{n}{k} = \frac{n!}{k!(n-k)!} \qquad k = 0, 1, 2, \ldots, n \tag{19}$$

EXAMPLE 24

Evaluate each of the following in two ways, first by equation (16) and second by (19).

a. $\binom{4}{2}$ b. $\binom{5}{1}$ c. $\binom{6}{4}$ d. $\binom{8}{8}$ e. $\binom{11}{0}$

Solution a. $\binom{4}{2} = \frac{4 \cdot 3}{1 \cdot 2} = 6$ $\binom{4}{2} = \frac{4!}{2!2!} = \frac{4 \cdot 3 \cdot 2 \cdot 1}{2 \cdot 1 \cdot 2 \cdot 1} = 6$

b. $\binom{5}{1} = \frac{5}{1} = 5$ $\binom{5}{1} = \frac{5!}{1!4!} = \frac{5 \cdot 4 \cdot 3 \cdot 2 \cdot 1}{1 \cdot 4 \cdot 3 \cdot 2 \cdot 1} = 5$

c. $\binom{6}{4} = \frac{6 \cdot 5 \cdot 4 \cdot 3}{1 \cdot 2 \cdot 3 \cdot 4} = 15$ $\binom{6}{4} = \frac{6!}{4!2!} = \frac{6 \cdot 5 \cdot 4 \cdot 3 \cdot 2 \cdot 1}{4 \cdot 3 \cdot 2 \cdot 1 \cdot 2 \cdot 1} = 15$

d. $\binom{8}{8} = \frac{8 \cdot 7 \cdot 6 \cdot 5 \cdot 4 \cdot 3 \cdot 2 \cdot 1}{1 \cdot 2 \cdot 3 \cdot 4 \cdot 5 \cdot 6 \cdot 7 \cdot 8} = 1$ $\binom{8}{8} = \frac{8!}{8!0!} = \frac{8!}{(8!)1} = 1$

e. $\binom{11}{0}$ cannot be evaluated by equation (16), but it is defined to be 1; by (19) we have

$$\binom{11}{0} = \frac{11!}{0!11!} = 1$$

▲

EXAMPLE 25

Expand $(3x - y)^4$, using equation (17).

Solution $(3x - y)^4 = [3x + (-y)]^4 = \sum_{r=0}^{4} \binom{4}{r}(3x)^{4-r}(-y)^r$

$$= \binom{4}{0}(3x)^4 + \binom{4}{1}(3x)^3(-y) + \binom{4}{2}(3x)^2(-y)^2$$

$$+ \binom{4}{3}(3x)(-y)^3 + \binom{4}{4}(-y)^4$$

$$= 81x^4 + 4(27x^3)(-y) + \frac{4 \cdot 3}{1 \cdot 2}(9x^2)(y^2) + \frac{4 \cdot 3 \cdot 2}{1 \cdot 2 \cdot 3}(3x)(-y^3)$$

$$+ \frac{4 \cdot 3 \cdot 2 \cdot 1}{1 \cdot 2 \cdot 3 \cdot 4}(y^4)$$

$$= 81x^4 - 108x^3y + 54x^2y^2 - 12xy^3 + y^4$$

▲

EXAMPLE 26 ||||||||||||||||||||||||||||||||||||||||||||||||||||||||||||||||||||||||||||||||||||||||||||||||||||||

Find the seventh term in the expansion of $(x - 2y)^{10}$.

Solution   We write this as $[x + (-2y)]^{10}$ and then use (18) with $k + 1 = 7$, so that $k = 6$. Thus, the seventh term is

$$\binom{10}{6} x^{10-6}(-2y)^6 = \frac{10 \cdot 9 \cdot 8 \cdot 7 \cdot 6 \cdot 5}{1 \cdot 2 \cdot 3 \cdot 4 \cdot 5 \cdot 6} x^4 (64y^6)$$

$$= 210x^4(64y^6)$$

$$= 13{,}440x^4 y^6 \qquad \blacktriangle$$

In the proof of the binomial theorem we will need the following result, which you will be asked to prove in Problem 24, Exercise Set 6:

$$\binom{k}{r} + \binom{k}{r-1} = \binom{k+1}{r} \qquad (20)$$

Proof of the Binomial Theorem

We wish to show that

$$(a + b)^n = \binom{n}{0}a^n + \binom{n}{1}a^{n-1}b + \binom{n}{2}a^{n-2}b^2$$

$$+ \cdots + \binom{n}{n-1}ab^{n-1} + \binom{n}{n}b^n$$

is true for all natural numbers n. Let this formula be designated by $P(n)$. Then $P(1)$ reads

$$(a + b)^1 = \binom{1}{0}a + \binom{1}{1}b$$

and since $\binom{1}{0} = 1$ and $\binom{1}{1} = 1$, we see that this is true.

Now assume $P(k)$ is true:

$$(a + b)^k = \binom{k}{0}a^k + \binom{k}{1}a^{k-1}b + \binom{k}{2}a^{k-2}b^2$$

$$+ \cdots + \binom{k}{k-1}ab^{k-1} + \binom{k}{k}b^k$$

In order to make the left-hand side what we want for $P(k + 1)$, we multiply both sides by $a + b$. This gives

$$(a + b)^{k+1} = \binom{k}{0}a^{k+1} + \binom{k}{1}a^k b + \binom{k}{2}a^{k-1}b^2 + \cdots + \binom{k}{k}ab^k + \binom{k}{0}a^k b$$

$$+ \binom{k}{1}a^{k-1}b^2 + \cdots + \binom{k}{k-1}ab^k + \binom{k}{k}b^{k+1}$$

$$= \binom{k}{0}a^{k+1} + \left[\binom{k}{1} + \binom{k}{0}\right]a^k b + \left[\binom{k}{2} + \binom{k}{1}\right]a^{k-1}b^2$$

$$+ \cdots + \left[\binom{k}{k} + \binom{k}{k-1}\right]ab^k + \binom{k}{k}b^{k+1}$$

We can replace $\binom{k}{0}$ by $\binom{k+1}{0}$, since the value is 1 in either case. Also, $\binom{k}{k} = \binom{k+1}{k+1}$, since again each symbol has the value 1. Using these facts, together with equation (20), we get

$$(a + b)^{k+1} = \binom{k+1}{0}a^{k+1} + \binom{k+1}{1}a^k b + \binom{k+1}{2}a^{k-1}b^2$$

$$+ \cdots + \binom{k+1}{k}ab^k + \binom{k+1}{k+1}b^{k+1}$$

which is exactly $P(k + 1)$. Thus, $P(k)$ implies $P(k + 1)$. The proof by induction is therefore complete.

EXERCISE SET 6

A

1. Find the value of the given binomial coefficients using equation (16).

   a. $\binom{3}{2}$ b. $\binom{6}{3}$ c. $\binom{8}{2}$ d. $\binom{10}{3}$

   e. $\binom{9}{9}$ f. $\binom{10}{5}$ g. $\binom{30}{2}$ h. $\binom{16}{5}$

   i. $\binom{7}{6}$ j. $\binom{8}{6}$

2. Use equation (19) to evaluate each of the binomial coefficients in Problem 1.

Simplify the expressions in Problems 3 and 4.

3. a. $(k + 1)k!$ b. $\dfrac{(k + 1)!}{k!}$ c. $\dfrac{n!}{(n - 2)!}$

4. a. $\dfrac{1}{k!} + \dfrac{1}{(k - 1)!}$ b. $\binom{k}{2} + \binom{k}{1}$

5. Show that $\binom{r}{r} = 1$ for any natural number r.

6. Show that $\binom{k}{k} = \binom{k+1}{k+1}$

7. Show that $\binom{n}{k} = \binom{n}{n-k}$

In Problems 8–15 expand, using equation (17), and simplify.

8. $(x + y)^6$
9. $(x - y)^5$
   Hint. Write $x - y = [x + (-y)]$.
10. $(2a + b)^7$ 11. $(a - 3b)^4$ 12. $(3x + 4y)^5$
13. $(s - 2t)^6$ 14. $(3r + s^2)^4$ 15. $(x^2 - 2y)^6$

In Problems 16–20 find the specified term of the expansion.

16. $(x + y)^{10}$;   eighth term
17. $(x - y)^{12}$;   fifth term
18. $(2x + 3y)^{12}$;   ninth term
19. $(3a - b)^7$;   fourth term
20. $(a - 2b)^{13}$;   third term
21. Evaluate $(1.02)^6$ to three decimal places of accuracy by using the binomial theorem.
    Hint. Write $(1.02)^6$ as $(1 + 0.02)^6$.
22. Evaluate $(0.99)^8$ to the nearest thousandth by using the binomial theorem.

23. Show that $\binom{n}{0} + \binom{n}{1} + \binom{n}{2} + \cdots + \binom{n}{n} = 2^n$.

    Hint. Consider $(1 + 1)^n$.

B

24. Prove the identity (20).

25. Prove the identity

$$\frac{n-r+1}{r} \cdot \binom{n}{r-1} = \binom{n}{r}$$

26. It can be shown that if $|x| < 1$ and α is not a positive integer, then

$$(1+x)^\alpha = \sum_{r=0}^{\infty} \binom{\alpha}{r} x^r$$

where $\binom{\alpha}{r}$ is given by equation (16). Use this to find the first five terms of $(1+x)^{-2}$, where $|x| < 1$.

27. Use Problem 26 to write the first five terms of the expansion of $\sqrt{1+x}$, assuming $|x| < 1$.

In Problems 28 and 29 use the expansion of Problem 27 to obtain the answer to five decimal places of accuracy.

28. $\sqrt{1.02}$

29. $\sqrt{0.99}$

30. Use Problem 26 to evaluate $1/(1.01)^2$ to six decimal places of accuracy.

REVIEW EXERCISE SET

A

Write the first four terms of the sequence whose nth term is given in Problems 1–3.

1. a. $a_n = \dfrac{n}{n+1}$ **b.** $a_n = 4n - 3$

2. a. $a_n = \dfrac{(-1)^{n-1}}{\sqrt{2n-1}}$ **b.** $a_n = \dfrac{(-1)^{n-1}n}{n^2+1}$

3. a. $a_n = \dfrac{(-3)^{n-1}}{4^n}$ **b.** $a_n = \dfrac{(-1)^{n-1}}{n \log(n+1)}$

In Problems 4–6 find an expression for the nth term of each sequence.

4. a. $1, -\frac{1}{3}, \frac{1}{5}, -\frac{1}{7}, \ldots$ **b.** $1, \frac{3}{4}, \frac{9}{16}, \frac{27}{64}, \ldots$

5. a. $\dfrac{1}{1 \cdot 2}, \dfrac{1}{2 \cdot 3}, \dfrac{1}{3 \cdot 4}, \dfrac{1}{4 \cdot 5}, \ldots$

     b. $\frac{1}{2}, \frac{2}{3}, \frac{3}{4}, \frac{4}{5}, \ldots$

6. a. $1, -\frac{1}{4}, \frac{1}{16}, -\frac{1}{64}, \ldots$

     b. $1, 4, 7, 10, \ldots$

In Problems 7–10 identify the given sequence as being arithmetic or geometric, and find the term specified.

7. $1, \frac{3}{2}, 2, \frac{5}{2}, \ldots; a_{20}$ **8.** $1, \frac{3}{2}, \frac{9}{4}, \frac{27}{8}, \ldots; a_8$

9. $-2, \frac{1}{2}, -\frac{1}{8}, \frac{1}{32}, \ldots; a_7$

10. $-1, 3, 7, 11, \ldots; a_{15}$

11. Write the first five terms of the sequence defined by

     a. $a_1 = 2; a_n = 2a_{n-1} + 1, n \geq 2$

     b. $a_1 = 1; a_n = na_{n-1}, n \geq 2$

12. In a certain arithmetic sequence, $a_{12} = 7$ and $a_{18} = -5$. Find a_{30}.

13. Find y so that the sequence $3, y, 15$ will be

     a. Arithmetic   **b.** Geometric

14. Insert five arithmetic means between 3 and 11.

15. On each swing after the first a certain pendulum swings three-fourths as far as on the preceding swing. If it swings 16 inches initially, find

     a. How far it swings on the sixth swing

     b. The total distance swung during the first six swings

     c. The approximate distance covered before coming to rest

16. Write the expanded form of each of the following sums:

     a. $\displaystyle\sum_{k=1}^{5} \frac{k^2}{2k+3}$ **b.** $\displaystyle\sum_{n=1}^{6} \frac{(-1)^{n-1}n}{n^2+1}$

     c. $\displaystyle\sum_{m=0}^{5} \frac{(-3)^m}{(1+m)^2}$ **d.** $\displaystyle\sum_{i=2}^{6} \frac{2i-3}{3i+4}$

     e. $\displaystyle\sum_{n=1}^{10} e^{-n}\ln(n+1)$

Find the sums in Problems 17–26.

17. The sum of the first thirty terms of the sequence $-5, -1, 3, 7, \ldots$.

18. The sum of the first ten terms of the sequence $-4, 2, -1, \frac{1}{2}, \ldots$.

19. $\displaystyle\sum_{n=1}^{50} (3 - 2n)$ **20.** $\displaystyle\sum_{k=1}^{6} \frac{(-1)^{k-1}}{3^k}$

21. $\displaystyle\sum_{j=1}^{60}\left(\frac{1-2j}{3}\right)$

22. $\displaystyle\sum_{n=1}^{8}(-2)^{n-3}$

23. $\displaystyle\sum_{k=1}^{12}\left(3-\frac{k}{4}\right)$

24. $\displaystyle\sum_{n=1}^{\infty}\left(\frac{5}{6}\right)^{n-1}$

25. $\displaystyle\sum_{k=1}^{\infty}(-1)^{k-1}\left(\frac{3}{5}\right)^{k}$

26. $\displaystyle\sum_{n=1}^{\infty}(0.1)^{n-1}$

27. Use a geometric series to find the rational number represented by each of the following:
   a. $0.545454\ldots$ **b.** $0.148148148\ldots$
   c. $2.181818\ldots$ **d.** $27.135135135\ldots$
   e. $1.020202\ldots$

28. A manufacturer estimates that a piece of equipment costing $60,000 depreciates by 25% each year. Find its value after 5 years.

29. Logs are stacked so that each layer after the bottom one has one log fewer than the layer below it. If there are 20 logs on the bottom layer and 8 logs on the top, how many logs are in the pile?

30. A culture of bacteria doubles in size every 2 hours. If initially it contains 200 bacteria, how many will it contain 10 hours later?

In Problems 31–38 use mathematical induction to prove that the given statement is true for all natural numbers.

31. $4+8+12+\cdots+4n=2n(n+1)$

32. $3+5+7+\cdots+(2n+1)=n^2+2n$

33. $\displaystyle\sum_{r=1}^{n}\frac{3}{2r(r+1)}=\frac{3n}{2n+2}$

34. $n^2\geq 4(n-1)$ **35.** $n<a^n$ if $a\geq 2$

36. 2 is a factor of n^2+3n

37. 4 is a factor of $2n^2+6n$

38. $n^2+1<(n+1)^2$

39. Evaluate each of the following binomial coefficients:

   a. $\dbinom{25}{3}$ **b.** $\dbinom{9}{6}$ **c.** $\dbinom{4}{3}$

   d. $\dbinom{3}{0}$ **e.** $\dbinom{12}{10}$

In Problems 40 and 41 use the binomial theorem to expand, and simplify the result.

40. a. $(x+2y)^8$ **b.** $(3a-4b)^5$

41. a. $\left(x-\dfrac{y}{2}\right)^7$ **b.** $(1+x^2)^{10}$

42. Find the value of each of the following, correct to five decimal places, using the binomial theorem:
   a. $(1.01)^{10}$ **b.** $(0.98)^6$

43. Find the specified term of the given expansion:
   a. $(2x-y)^{12}$;   eighth term
   b. $(x-2)^{20}$;   tenth term

B

Find the sums in Problems 44 and 45.

44. $\displaystyle\sum_{k=1}^{8}(100k-2^k)$

45. $\displaystyle\sum_{n=1}^{6}\frac{2-n\cdot 3^n}{3^n}$

46. Find a formula for the sum of the first n positive integers. Use mathematical induction to prove your result.

47. In a certain city, housing costs have risen at an annual rate of 10% for the past 4 years. If, 4 years ago, a particular house cost $50,000, what would be its fair market value now?

48. A man started to work 10 years ago at a salary of $10,000 per year and has received annual raises of 6%. What is his salary now, and what is the total amount of money he has received during the past 10 years?

49. A ladder with 13 rungs is to be constructed so that each rung is shorter than the preceding one by a constant amount. If the initial rung is to be 24 inches long and the final one 15 inches long, what are the lengths of the intermediate rungs?

50. After college a boy agreed to pay off his debt of $2,550 to his father by paying $100 the first month, $110 the second month, $120 the third month, and so on. How many months were required to pay off the debt?

51. Prove the identity

$$\binom{n}{k}-\binom{n-1}{k}=\binom{n-1}{k-1}$$

52. Find the sum of the given infinite geometric series, and state the domain of validity:

   a. $1-\dfrac{x}{2}+\dfrac{x^2}{4}-\dfrac{x^3}{8}+\cdots$ **b.** $\displaystyle\sum_{n=1}^{\infty}\frac{x^{2n-1}}{3^{n-1}}$

53. Find the infinite geometric series having the given sum, and state the domain of validity:

a. $\dfrac{1}{1+x}$ **b.** $\dfrac{3}{2-x^2}$

In Problems 54–56 prove the given statement by mathematical induction.

54. $1 + 2 \cdot 2 + 3 \cdot 2^2 + 4 \cdot 2^3 + \cdots + n \cdot 2^{n-1} =$
$1 + (n-1) \cdot 2^n$

55. $\dfrac{1}{1 \cdot 2 \cdot 3} + \dfrac{1}{2 \cdot 3 \cdot 4} + \dfrac{1}{3 \cdot 4 \cdot 5} + \cdots +$
$\dfrac{1}{n(n+1)(n+2)} = \dfrac{n(n+3)}{4(n+1)(n+2)}$

56. $|a_1 + a_2 + a_3 + \cdots + a_n| \le$
$|a_1| + |a_2| + |a_3| + \cdots + |a_n|$

CUMULATIVE REVIEW EXERCISE SET IV (CHAPTERS 10–13)

1. Solve the following system in three ways:
  a. By elimination using addition, subtraction, or substitution
  b. By Cramer's rule
  c. By reducing the augmented matrix to triangular form

$$\begin{cases} 2x - 6y - 3z = 8 \\ x + 2y - 4z = -1 \\ 3x + 4y + z = 5 \end{cases}$$

2. a. Multiply:

$$\begin{bmatrix} 2 & -1 & 4 & 1 \\ 3 & 2 & -5 & -2 \\ -6 & 0 & 2 & 3 \end{bmatrix} \begin{bmatrix} 1 & 4 & -3 \\ 2 & 0 & 5 \\ -1 & 3 & 2 \\ -4 & 6 & -3 \end{bmatrix}$$

  b. Use inverses to find X such that $AX = B$, where

$$A = \begin{bmatrix} 2 & 3 & 1 \\ 1 & -2 & -2 \\ 4 & 9 & 3 \end{bmatrix} \quad X = \begin{bmatrix} x \\ y \\ z \end{bmatrix} \quad B = \begin{bmatrix} 4 \\ -1 \\ 6 \end{bmatrix}$$

3. Find the equation of the parabola having horizontal axis and passing through the vertices of the triangle whose sides have equations $2x - y = 3$, $2x + y + 7 = 0$, and $x + 2y = 4$. Show that this is a right triangle.

4. a. If the average annual inflation rate was 12% for the 5-year period beginning in 1980, what was the purchasing power of \$1 in 1985 as compared to its value in 1980?

b. The area bounded by the curve $y = 2^x$, the x axis, and the lines $x = 0$ and $x = 2$ is to be approximated by inscribing rectangles of width 0.1 (see sketch). Find the sum of the areas of the rectangles as a finite series and give an expression for the sum of the series. Use a calculator to approximate the result.

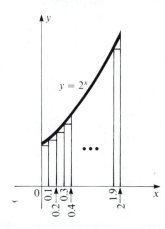

5. Find the equation of the circle whose center is the upper focus of the hyperbola $4x^2 - 6y^2 - 24x + 24y + 36 = 0$, and that passes through the center of the hyperbola. Draw both curves.

6. The proprietor of a coffee house prepared 100 pounds of a special blend of coffee using three different varieties: Colombian, Sumatran, and Brazilian. Colombian sells for \$5.30 per pound, Sumatran for \$5.00 per pound, and Brazilian for \$4.80 per pound. He calculates that the price of the blend should be $5.04 per pound. If he used 50% more Colombian than Brazilian coffee, how many pounds of each did he use?

7. The owner of a sporting goods store sells two types of roller skates: the standard skate on which she makes a profit of $5.00 per pair, and the deluxe model on which the profit is $8.00 per pair. From experience she expects to sell at least 50 but not more than 100 standard pairs and between 30 and 60 deluxe pairs each month. Because of limitations on storage capacity she can handle at most 140 pairs of skates at a time. Assuming she sells all she has in stock, how many should she stock per month to maximize profit?

8. Solve the following system by matrix methods:

$$\begin{cases} 2x + y - 3z + 2w = 4 \\ 5x - 8y - 6z + 3w = 7 \\ 3x - 2y - 4z + w = 5 \end{cases}$$

9. Solve graphically:

a. $\begin{cases} y \le 2x + 5 \\ y \le 6 - x \\ y \ge x^2 - 2x \end{cases}$

Find the points of intersection of the boundary curves.

b. $|x| + |y - 1| < 2$

10. An amusement park charges $1.20 admission for children under 18, $3.00 for adults between 18 and 65, and $1.50 for senior citizens over 65. On a given Saturday 3,950 tickets were sold, and the receipts totaled $7,110. If the number of children exceeded the total number of adults and senior citizens by 550, how many tickets of each type were sold?

11. Find the equation of the parabola whose vertex and focus, respectively, coincide with the upper vertex and focus of the ellipse

$$4x^2 + 3y^2 + 16x + 6y - 29 = 0$$

Draw both curves.

12. Let

$$A = \begin{bmatrix} 3 & 2 & 2 \\ 1 & 4 & 1 \\ -2 & -4 & -1 \end{bmatrix} \qquad I = \begin{bmatrix} 1 & 0 & 0 \\ 0 & 1 & 0 \\ 0 & 0 & 1 \end{bmatrix}$$

Solve the equation $|A - \lambda I| = 0$ for λ.

13. Identify and draw the graph of the equation $x^2 - 3xy - 3y^2 - 21 = 0$ after a suitable rotation of axes.

14. Find the indicated sums.

a. $\sum_{k=1}^{20} \left(\dfrac{3k - 5}{6} \right)$ b. $\sum_{n=0}^{\infty} e^{-2n}$

15. In a small machine shop two types of parts are produced, each of which requires time on the lathe. The first type requires 10 minutes and the second type 6 minutes. Material for the first type costs $0.50 per part and material for the second type costs $0.75 per part. Labor and equipment charges for the lathe amount to $15.00 per hour. The lathe is to be used between 6 and 8 hours per day in the production of these parts. If it is necessary to produce at least 21 parts of the first type and 15 parts of the second type each day, how many parts of each type should be produced to minimize cost?

16. Find an infinite geometric series whose sum is $1/(4 + x^2)$. For what values of x is the result valid?

17. Solve the following system and draw the graph:

$$\begin{cases} x^2 + 2y^2 = 6 \\ 2x - y = 3 \end{cases}$$

18. Find the equation of the hyperbola with eccentricity $3/\sqrt{5}$ and with transverse axis that is the line segment joining $(-4, 3)$ and $(-4, -1)$. Find the equations of the asymptotes and draw the curve.

19. Prove the formula below in two ways.

$$\sum_{k=1}^{n} \log 2^k = \left(\dfrac{n^2 + n}{2} \right) \log 2$$

a. By mathematical induction
b. By showing that the sum is an arithmetic progression

20. Rotate the axes so as to eliminate the $x'y'$ term and draw the graph of the equation $16x^2 - 24xy + 9y^2 - 25x + 50y = 75$.

21. A sequence is defined by

$$\begin{cases} a_1 = \sqrt{2} \\ a_n = \sqrt{2 + a_{n-1}} \end{cases} \qquad (n \ge 2)$$

Write the first five terms of the sequence. Prove by mathematical induction that $a_n \le 2$ for all n.

22. Nineteen pieces of pipe are to be cut so that the first piece is 2 feet long and the last piece is 14 feet

long. Each piece after the first is longer than the preceding one by a fixed amount. Find this amount. Find the sum of the lengths of all the pieces of pipe.

23. **a.** Write the sixth term of the expansion of $(x^2 - 2y)^9$ and simplify your result.

   b. Expand $(1 + x)^{-1}$ to five terms, using the binomial formula. If $|x| < 1$, it can be shown that the infinite series generated by the continuation of this expansion converges to the given function. Find the nth term of the expansion. What kind of series is this? Find the sum of the series and show that this is in agreement with the original function.

24. The ends of a barn are in the shape of a square surmounted by an isosceles trapezoid (see sketch). The top of the trapezoid is half as long as the bottom, and the overall height of the barn is 32 feet. If the total area of the square and trapezoid is 580 square feet, find the dimensions of the side

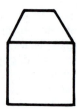

of the square and the height of the trapezoid. What is the overall perimeter?

25. Prove by mathematical induction:

$$\sin x + \sin 2x + \cdots + \sin nx$$
$$= \frac{\cos \frac{1}{2}x - \cos(n + \frac{1}{2})x}{2 \sin \frac{1}{2}x}$$
$$(x \neq 2k\pi, k = 0, \pm 1, \pm 2, \ldots)$$

Hint. Use the sum and product formulas for the sine and cosine.

ANSWERS TO SELECTED PROBLEMS

The solutions manual that accompanies this text is available from your local bookstore.

Note. Answers to certain problems can be given in different forms. Thus your answer may be correct even if it appears different from the one given.

Chapter 1

Exercise Set 1, page 7

1. a. 0.375   **b.** 1.333 ...   **c.** -0.27272 ...   **d.** 2.6   **e.** 0.592592 ...   **3.** Correct to two decimal places

5. a. Rational   **b.** Rational   **c.** Irrational   **d.** Rational   **e.** Rational

7. a. $\frac{5}{3}$ **b.** $\frac{1}{9}$ **9. a.** $\frac{44}{333}$ **b.** $\frac{105}{37}$

11. a. Distributive property   **b.** Definition of additive inverse   **c.** Definition of multiplicative inverse   **d.** Definition of additive identity   **e.** Definition of multiplicative identity   **f.** Associative property for addition   **g.** Associative property for multiplication   **h.** Commutative property for addition   **i.** Commutative property for multiplication   **j.** Property of additive identity

13. a. -12 **b.** 10   **c.** 3   **d.** $\frac{1}{5}$ **e.** 14   **15. a.** 0   **b.** 1   **c.** -1 **d.** No

17. a. No. For example, $7 - (5 - 3) = 7 - 2 = 5$, but $(7 - 5) - 3 = 2 - 3 = -1$.
   b. No. For example, $12 \div (4 \div 2) = 12 \div 2 = 6$, but $(12 \div 4) \div 2 = 3 \div 2 = \frac{3}{2}$.

19. a. 5   **b.** 0   **c.** $\frac{1}{3}$ **d.** 7   **e.** 1

21. Assume $ac = bc$ and $c \neq 0$. Then

$(ac)c^{-1} = (bc)c^{-1}$ Multiplication property of equality

$a(cc^{-1}) = b(cc^{-1})$ Associative property for multiplication

$a \cdot 1 = b \cdot 1$ Property of multiplicative inverse

$a = b$ Property of multiplicative identity

23. Taken in order, the properties used are equality property 4, property of additive identity, associative property of addition, property of additive inverse, property of additive identity.

25. a. Assume $a + z = a$ for every a in R. Then $a + z = a + 0$. So

$-a + (a + z) = -a + (a + 0)$ Addition property of equality

$(-a + a) + z = (-a + a) + 0$ Associative property for addition

$0 + z = 0 + 0$ Property of additive inverse

$z = 0$ Property of additive identity

b. Assume that for $a \neq 0$, $a \cdot z = a$. Then $az = a \cdot 1$. So

$a^{-1}(a \cdot z) = a^{-1}(a \cdot 1)$	Multiplicative property of equality
$(a^{-1} \cdot a) \cdot z = (a^{-1} \cdot a) \cdot 1$	Associative property of multiplication
$1 \cdot z = 1 \cdot 1$	Property of multiplicative inverse
$z = 1$	Property of multiplicative identity

27. Let $a \in R$, $a \neq 0$. Suppose $a \cdot b = 1$. Then

$a^{-1} \cdot (a \cdot b) = a^{-1} \cdot 1$	Multiplication property of equality
$a^{-1} \cdot (a \cdot b) = a^{-1}$	Property of multiplicative identity
$(a^{-1} \cdot a) \cdot b = a^{-1}$	Associative property for multiplication
$1 \cdot b = a^{-1}$	Property of multiplicative inverse
$b = a^{-1}$	Property of multiplicative identity

29. $a(b - c) = a[b + (-c)] = ab + a(-c) = ab + [-(ac)] = ab - ac$ **31.** $1/b = 1 \cdot b^{-1} = b^{-1}$

33. $(a^{-1} \cdot b^{-1}) \cdot (ab) = (b^{-1} \cdot a^{-1}) \cdot (ab) = b^{-1}(a^{-1} \cdot a) \cdot b = b^{-1} \cdot 1 \cdot b = b^{-1} \cdot b = 1$, so $a^{-1}b^{-1}$ is the inverse of ab

35. $(a - b)^2 = [a + (-b)]^2 = a^2 + 2a(-b) + (-b)^2$ by Problem 34

$$= a^2 - 2ab + b^2 \text{ by Further Properties 5 and 6}$$

37. **a.** -5 **b.** 0 **c.** 7 **d.** 1 **e.** 2

Exercise Set 2, page 13

1. **a.** $>$ **b.** $<$ **c.** $>$ **d.** $>$ **e.** $<$

3. **a.** $5 - 2 = 3$ is positive.   **b.** $-1 - (-3) = 2$ is positive.   **c.** $0 - (-4) = 4$ is positive.
   d. $7 - 3 = 4$ is positive.   **e.** $-2 - (-4) = 2$ is positive.

5. **a.** $1 < x < 3$ **b.** $-1 < x \leq 2$ **c.** $2 \leq x \leq 6$ **d.** $0 \leq x < 5$ **e.** $-6 \leq x < 0$

7. **a.** $\dfrac{1}{a} < \dfrac{1}{b}$ **b.** $\dfrac{1}{a} > \dfrac{1}{b}$ **c.** $\dfrac{1}{a} > \dfrac{1}{b}$ **d.** $\dfrac{1}{a} < \dfrac{1}{b}$

9. **a.** $2 \leq x \leq 5$ **b.** $-2 < x < 3$ **c.** $3 < x < 5$ **d.** $x < 2$ **e.** $x \geq 1$

11. **a.** $-3 < x < 3$ **b.** $y > 2$ or $y < -2$ **c.** $-1 \leq t \leq 1$ **d.** $w \geq 3$ or $w \leq -3$

13. **a.** $\{x: \ |x| < 3\}$ **b.** $\{x: \ |x| \leq 4\}$ **c.** $\{x: \ |x| > 2\}$ **d.** $\{x: \ |x| \geq 3\}$

15. If $a > 0$, then $a^2 = a \cdot a$ is positive since R^+ is closed under multiplication. If $a < 0$, then $(-a) > 0$, and $(-a)(-a) = a^2 > 0$.

17. If $a < b$, then $b - a$ is positive. If c is positive, then by closure of R^+ under multiplication, $(b - a) \cdot c$ is positive. But $(b - a)c = bc - ac$. So by definition, $ac < bc$. If c is negative, then $-c$ is positive. So $(b - a)(-c)$ is positive. But $(b - a)(-c) = -bc + ac = ac - bc$. So since this is positive, $bc < ac$, or equivalently, $ac > bc$.

19. If $a < b$ and $c < d$, then by property 3, $a + c < b + c$ and $b + c < b + d$. Thus, by property 6, $a + c < b + d$.

21. If $a < b$, then $\begin{cases} a/c < b/c & \text{if} \quad c > 0 \\ a/c > b/c & \text{if} \quad c < 0 \end{cases}$

If $c > 0$, by Problem 20, $c^{-1} > 0$. So by property 4, $ac^{-1} < bc^{-1}$. But $ac^{-1} = a/c$ and $bc^{-1} = b/c$. If $c < 0$, then $c^{-1} < 0$, since otherwise the product of c and c^{-1} would not be positive. Thus, by property 4, $ac^{-1} > bc^{-1}$, and so $a/c > b/c$.

23. False. For example, $6 < 7$ and $3 < 5$. But $(6 - 3) > (7 - 5)$.

25. If $a \geq 0$ and $b \geq 0$, $|a| = a$, $|b| = b$, and $|ab| = ab$, and the result is trivial. If $a < 0$ and $b \geq 0$, then $|a| = -a$, $|b| = b$ and $|ab| = -ab$, since $ab \leq 0$. So $|ab| = -ab = (-a)b = |a||b|$. If $a < 0$ and $b < 0$, $|ab| = ab = (-a)(-b) = |a||b|$.

Exercise Set 3, page 21

1. **a.** $\frac{7}{6}$ **b.** $\frac{13}{40}$ **3.** **a.** $\frac{19}{24}$ **b.** $-\frac{41}{36}$ **5.** **a.** $\frac{7}{12}$ **b.** $\frac{1}{12}$ **7.** **a.** $-\frac{2}{3}$ **b.** $\frac{5}{6}$

9. **a.** $-\frac{1}{20}$ **b.** $-\frac{1}{8}$ **11.** **a.** $\frac{1}{540}$ **b.** $\frac{41}{3360}$ **13.** 9 **15.** $-\frac{33}{40}$ **17.** $-\frac{1}{11}$ **19.** $\frac{8}{11}$ **21.** 0

23. **a.** 14 **b.** 26 **c.** 36 **d.** 32 **e.** 198

25. a. $2 \in N$ and $3 \in N$, but $2 - 3 = -1 \notin N$. If $m \in J$ and $n \in J$, then $-n \in J$, and so $m - n = m + (-n) \in J$.

   b. $2 \in J$ and $3 \in J$, but $2 \div 3 = \dfrac{2}{3} \notin J$. If $\dfrac{a}{b} \in Q$ and $\dfrac{c}{d} \in Q$, with $\dfrac{c}{d} \neq 0$, then $\dfrac{a}{b} \div \dfrac{c}{d} = \dfrac{a}{b} \cdot \dfrac{d}{c} = \dfrac{ad}{bc} \in Q$.

27. a. If $\dfrac{a}{0} = c$, then $c \cdot 0 = a$, but $c \cdot 0 = 0 \neq a$.

   b. If $\dfrac{0}{0} = c$, then $c \cdot 0 = 0$, which is true for every real c.

29. a. $\frac{155}{96}$ **b.** 0 **31.** $\frac{35}{13}$

33. $\dfrac{a}{b} \cdot \dfrac{c}{d} = (a \cdot b^{-1})(c \cdot d^{-1})$ Definition of division

   $= a \cdot (b^{-1} \cdot c)d^{-1}$ Extended associative property

   $= a \cdot (c \cdot b^{-1})d^{-1}$ Commutative property

   $= (ac)(b^{-1}d^{-1})$ Extended associative property

   $= (ac)(bd)^{-1}$ Definition of $(bd)^{-1}$, since

   $(b^{-1}d^{-1})(bd) = d^{-1}(b^{-1}b)d = d^{-1}(1)d = d^{-1}d = 1$

   $= \dfrac{ac}{bd}$ Definition of division

Exercise Set 4, page 28

1. a. 32 **b.** $\frac{1}{3}$ **c.** -27 **d.** 1 **e.** $\frac{9}{4}$ **3. a.** x^9 **b.** x^3 **c.** x^{10} **d.** $x^9 y^6$ **e.** 1

5. a. $-8a^{12}$ **b.** $\dfrac{12}{a^5}$ **c.** $\dfrac{a^2}{9}$ **d.** $\dfrac{9a^4 b^6}{4}$ **e.** $\dfrac{3}{4a^5}$ **7. a.** $\dfrac{y^8}{x^7}$ **b.** $\dfrac{r^3 t^3}{s}$ **c.** $\dfrac{3b^8}{5a^3}$ **d.** $\dfrac{2u^{2s-3t}}{5}$

9. a. $\dfrac{b^2 c^9 d^6}{a^4}$ **b.** $\dfrac{x^{13}}{y^8}$ **c.** $\dfrac{a^{8nk}}{b^{3nk}}$ **11.** -7

13. a. 3×10^9 **b.** 5×10^{-6} **c.** 2.5×10^5 **d.** 2.34×10^{-1} **e.** 3.568×10^6

15. a. 9.2×10^2 **b.** 4.0×10^3 **17. a.** 7.03×10^5 **b.** -8.6×10^{-3} **19.** 2.35×10^{13}

21. a. Let $n = -p$, where p is a positive integer. Then $a^m \cdot a^n = a^m \cdot a^{-p} = \dfrac{a^m}{a^p} = a^{m-p} = a^{m+(-p)} = a^{m+n}$

   b. Let $m = -p$, $n = -q$, where p and q are positive. Then, $a^m \cdot a^n = a^{-p} \cdot a^{-q} = \dfrac{1}{a^p} \cdot \dfrac{1}{a^q} = \dfrac{1}{a^{p+q}} = a^{-(p+q)} =$

   $a^{-p+(-q)} = a^{m+n}$

   c. $a^m \cdot a^0 = a^m \cdot 1 = a^m = a^{m+0}$

23. $\left(\dfrac{a}{b}\right)^{-m/n} = \left(\dfrac{a}{b}\right)^{(-1)\cdot(m/n)} = \left[\left(\dfrac{a}{b}\right)^{-1}\right]^{m/n} = \left(\dfrac{b}{a}\right)^{m/n}$

25. No. For example, with $a = 2$, $b = 2$, and $n = 1$, $\dfrac{1}{a^{-n} + b^{-n}} = \dfrac{1}{2^{-1} + 2^{-1}} = \dfrac{1}{\frac{1}{2} + \frac{1}{2}} = \dfrac{1}{1} = 1$, but $a^n + b^n = 2^1 + 2^1 = 4$

Exercise Set 5, page 33

1. a. $4\sqrt{19}$ **b.** $-4\sqrt[3]{7}$ **3. a.** $2\sqrt[3]{3}$ **b.** $-\sqrt{5}$ **5. a.** $11\sqrt[3]{3}$ **b.** $-2\sqrt[5]{2}$ **7.** $\dfrac{10ac^3 \sqrt{10}}{b^2}$

9. $\dfrac{6b^2 c^4}{a^5}$ **11.** $7x^4 y^2 \sqrt{2}$ **13. a.** $\dfrac{\sqrt{15}}{3}$ **b.** $\dfrac{\sqrt{10}}{5}$ **c.** $\dfrac{3\sqrt{7}}{7}$ **d.** $\dfrac{\sqrt[3]{6}}{2}$ **e.** $\dfrac{3\sqrt[3]{6}}{8}$

15. a. $\dfrac{a^2 \sqrt{6ab}}{3b}$ **b.** $\dfrac{2\sqrt{6ab}}{9b^2}$ **17. a.** $3^{7/6}$ **b.** $2^{5/6}$ **c.** $\frac{9}{32}$ **d.** $\frac{1}{64}$ **e.** $\dfrac{x^{2a}}{y^{4b}}$

19. a. $6x^{5/6}$ **b.** $x^{1/12} y^{1/6}$ **c.** $4x^{5/6}$ **21. a.** a^2 **b.** $\dfrac{8}{a}$ **c.** $\dfrac{4a}{b^4}$

23. **a.** $\dfrac{y^3}{4x^{1/2}}$ **b.** $\dfrac{b}{64a^2c^{2/3}}$ **c.** $\dfrac{x^{1/2}}{4y^{4/3}}$ **25.** **a.** $\dfrac{27x^2y}{8}$ **b.** $\dfrac{16bc^8}{81a^6}$

27. **a.** $3x^{3/2}y^{5/2}$ **b.** $3^{3/2}x^{3/2}y^6$ **c.** $4^{2/3}x^{4/3}y^{10/3}$ **29.** **a.** $3|x|y^2$ **b.** $-2x^2y^3$

31. $\sqrt[kn]{a^{km}} = a^{km/kn} = a^{m/n} = \sqrt[n]{a^m}$ **33.** $\sqrt[m]{\sqrt[n]{a}} = (a^{1/n})^{1/m} = a^{(1/n)\cdot(1/m)} = a^{1/mn} = \sqrt[mn]{a}$

35. $\left(\dfrac{a}{b}\right)^{-m/n} = \left(\dfrac{a}{b}\right)^{(-1)\cdot(m/n)} = \left[\left(\dfrac{a}{b}\right)^{-1}\right]^{m/n} = \left(\dfrac{b}{a}\right)^{m/n}$ **37.** $\dfrac{8\sqrt{2}}{3}$ **39.** $\dfrac{3\sqrt[3]{2}}{4}$ **41.** $\dfrac{11x^5\sqrt{6}}{27yz^2}$ **43.** 1.2×10^{-5}

Exercise Set 6, page 41

1. $5x^2 + 7x + 7$ **3.** $6x^2 - 5x + 10$ **5.** $t^2 - 2t + 13$ **7.** $12x^2 + 7xy + y^2$ **9.** $-7x^2 + 13x - 21$

11. $x^2 + 5x + 6$ **13.** $x^2 + 4xy + 3y^2$ **15.** $6 - x - 12x^2$ **17.** $15y^2 + 11y - 12$ **19.** $2x^4 - 5x^2 + 3$

21. $x^2 - 4$ **23.** $4a^2 - 9b^2$ **25.** $x - 4$ **27.** $x^{2n} - y^{2n}$ **29.** $x^k - y^k$ **31.** h **33.** h

35. $4x^2 + 12xy + 9y^2$ **37.** $a^4 - 4a^2b^2 + 4b^4$ **39.** $x^{2n} + 2x^ny^n + y^{2n}$ **41.** $1 - x^3$ **43.** $8 - 5x$

45. $x^2 - x - 2$ **47.** $x - 4; R = 10x - 13$ **49.** $x^6 + 6x^5y + 15x^4y^2 + 20x^3y^3 + 15x^2y^4 + 6xy^5 + y^6$

51. $8x^3 - 36x^2y + 54xy^2 - 27y^3$ **53.** $x^{16} - 8x^{14}y + 28x^{12}y^2 - 56x^{10}y^3 + 70x^8y^4 - 56x^6y^5 + 28x^4y^6 - 8x^2y^7 + y^8$

55. $x^5 + 5x^3 + 10x + \dfrac{10}{x} + \dfrac{5}{x^3} + \dfrac{1}{x^5}$ **57.** $x^{3n} + 3x^{2n} + 3x^n + 1$ **59.** $2|x|$ **61.** $x^3 + x^2 + x + 6$

63. $2x^4 + x^3 - 4x^2 + 3x - 2$ **65.** $x^2 + 2xy + y^2 - 4$ **67.** $x^2 - y^2 + 2y - 1$ **69.** $2h$

71. $x^2 + y^2 + z^2 + 2xy + 2xz + 2yz$ **73.** $x^3 - 3x^2 + 4$ **75.** $a^{3k} + 1$ **77.** $\dfrac{4}{3}x^3 - \dfrac{8}{9}x^2 + \dfrac{16}{27}x - \dfrac{32}{81}; R = -\dfrac{179}{81}$

79. $x^4 + 8x^3 + 28x^2 + 56x + 70 + \dfrac{56}{x} + \dfrac{28}{x^2} + \dfrac{8}{x^3} + \dfrac{1}{x^4}$ **81.** $8x^4 - 20x^3 + 16x^2 - 4x$

Exercise Set 7, page 48

1. **a.** $3ab(a + 2b)$ **b.** $2a^2b^2(2b - 3a^2)$ **3.** **a.** $6a^2b^2(4b - 5ab^3c + 8a^2c^2)$ **b.** $42x^3y(3y - 4x)$

5. **a.** $3x^{1/3}(x + 2)$ **b.** $8\sqrt[3]{x}(4 + 5x)$ **7.** **a.** $(x + 3)(y - 1)$ **b.** $(a - 2)(1 - b)$

9. **a.** $(x - 1)(x - y)$ **b.** $(xy + 2)(y - x)$ **11.** **a.** $(x - 3)^2$ **b.** $(a + 2)^2$

13. **a.** $(6a + 7b)^2$ **b.** $(9s - 4t)^2$ **15.** **a.** $(x - 1)(x - 4)$ **b.** $(a + 2)(a + 6)$

17. **a.** $(2x - 1)(x + 4)$ **b.** $(3x + 2)(x - 3)$ **19.** **a.** $(5x - 3)(4x + 3)$ **b.** $(8x - 3)(3x - 5)$

21. **a.** $(2x^2 + 1)(x^2 + 2)$ **b.** $(3x^2 + 4)(2x^2 + 3)$ **23.** **a.** $(3x + 2)(3x - 2)$ **b.** $(5a + 4b)(5a - 4b)$

25. **a.** $(2a + 3b)(2a - 3b)(4a^2 + 9b^2)$ **b.** $x(x + 4)(x - 4)(x^2 + 16)$

27. **a.** $(x + 3 + y)(x + 3 - y)$ **b.** $(x + y + 1)(x - y - 1)$

29. **a.** $(x - 1 - 2y)(x - 1 + 2y)$ **b.** $(3x + 2y - 3)(3x - 2y + 3)$

31. **a.** $(x^2 - x - 1)(x^2 + x - 1)$ **b.** $(a^2 - 2a - 2)(a^2 + 2a - 2)$

33. **a.** $(x - 3)(x^2 + 3x + 9)$ **b.** $(2a + 1)(4a^2 - 2a + 1)$ **35.** $(x - y)^3$ **37.** $(a - 2)^4$ **39.** $(1 + x)^5$

41. $x(3x + 7)(3x - 7)$ **43.** $(x + 2)(x - 2)(x + 3)(x - 3)$ **45.** $(a + b)^2(a - b)$ **47.** $(x + y + 2)(x - y - 2)$

49. $(4x + 3y)(9x - 16y)$ **51.** $(a + 2)(a - 2)(a^2 + 2a + 4)(a^2 - 2a + 4)$ **55.** $(x - 1)(y + 1)$ **57.** $(4a + 9b)(5a - 2b)$

59. $(x - 3)(2x - y)$ **61.** $(t - 3)^3$ **63.** $2xy(7x - 8y)(3x + 2y)$ **65.** $(2a - 3b^2)^3$ **67.** $(x^2 + 1)(x^4 - x^2 + 1)$

69. $2x(9x + 8)(2x - 5)$ **71.** $3xy^2(6x + 5y)(2x - 5y)$ **73.** $6bc(5a + 4b)(a - 2b)$ **75.** $(x + 2y + 3z)^2$

77. $(6x^2 - 2x - 5)(6x^2 + 2x - 5)$ **79.** $(x - y)(x^2 + xy + y^2 - 1)$ **81.** $(x - y - a - 2)(x - y + a + 2)$

83. $(x - y)(x + y)(x^2 + y^2)(x^2 + xy + y^2)(x^2 - xy + y^2)(x^4 - x^2y^2 + y^4)$ **85.** $(x^2 + 2)^{3/2}(x + 2)(x - 2)$

87. $(3x^2 - 2\sqrt{6}x + 4)(3x^2 + 2\sqrt{6}x + 4)$ **89.** $x(3x^2 + 1)^{1/3}(1 + x)(1 - x)$

Exercise Set 8, page 55

1. **a.** $\dfrac{x + 2}{x + 1}$ **b.** $\dfrac{2x - 1}{x + 3}$ **3.** $\dfrac{x + 2}{x - 1}$ **5.** $\dfrac{x - y}{x + y}$ **7.** $\dfrac{3x^2 + 4x}{x^2 - 4}$ **9.** $\dfrac{5x - 2}{x^2 - 2x - 8}$ **11.** $\dfrac{35 - x}{x^2 - 25}$

13. $\dfrac{-2x^2 + 10x - 9}{3x^2 - 7x + 4}$ **15.** $\dfrac{1}{x + 3}$ **17.** $\dfrac{2}{x + 2}$ **19.** $\dfrac{x^2 - 3x - 2}{x^2 - 2x}$ **21.** $\dfrac{7x - 10}{x - 3}$ **23.** $\dfrac{2x - 4}{x^2 + 3x + 2}$

25. $\dfrac{-1}{2x^2 - 3x + 1}$ **27.** xy **29.** $\dfrac{7 - x}{3 - x}$ **31.** $\dfrac{a - 2}{a + 7}$ **33.** $\dfrac{-1}{2 + h}$ **35.** $\dfrac{3}{4(x + 1)}$ **37.** $-\dfrac{a + x + 2}{(x + 1)^2(a + 1)^2}$

39. $-\dfrac{3h + 4}{(h + 1)^2}$ **41.** $\dfrac{3x + 5}{(2x - 3)(x - 2)(x + 1)}$ **43.** $\dfrac{4 - 2x}{(2x - 1)^3}$ **45.** $\dfrac{2x + 2a - 3ax}{(2 - 3x)(2 - 3a)}$ **47.** $1 + x$

Exercise Set 9, page 58

1. $\dfrac{y + x}{y - x}$ **3.** $-\dfrac{ab}{a + b}$ **5.** $\dfrac{x}{4x + 1}$ **7.** $-\dfrac{4}{x^2}$ **9.** $\dfrac{(x - 2)(x + 6)}{(x + 2)^2}$ **11.** $\dfrac{3x}{2\sqrt{x - 1}}$ **13.** $\dfrac{1 - x}{(1 - 2x)^{3/2}}$

15. $\dfrac{1}{(x^2 + 1)^{3/2}}$ **17.** $\dfrac{9 - 4x}{2(2x - 1)^{1/2}(4 - x)^{1/2}}$ **19.** $\dfrac{3x^2 + 6x}{(2x + 3)^{3/2}}$ **21.** $2(\sqrt{5} + \sqrt{3})$ **23.** $4 + 3\sqrt{2}$

25. $\sqrt{x + 3} + \sqrt{x}$ **27.** $\dfrac{1}{\sqrt{x + h} + \sqrt{x}}$ **29.** $\dfrac{3}{\sqrt{3x - 5} + 1}$ **31.** $\dfrac{3}{\sqrt{16 + 3h} + 4}$ **33.** $\dfrac{y - x}{x^2 y^2}$

35. $\dfrac{-1}{\sqrt{x}\sqrt{x + h}(\sqrt{x} + \sqrt{x + h})}$ **37.** $\dfrac{5}{2(3 - x)^{3/2}(2 + x)^{1/2}}$ **39.** $\dfrac{x^2 - 3}{3(x^2 - 1)^{4/3}}$ **41.** $\dfrac{-x(x^2 + 5)}{3(x^2 - 4)^{3/2}(x^2 - 1)^{2/3}}$

Review Exercise Set, page 59

1. **a.** 0.6875 **b.** 3.125 **c.** $0.428571428571\ldots$ **d.** $0.185185185\ldots$ **e.** $-2.1666\ldots$

3. **a.** Rational   **b.** Rational   **c.** Irrational   **d.** Rational   **e.** Rational   **f.** Irrational   **g.** Rational   **h.** Rational   **i.** Irrational   **j.** Irrational

5. **a.** $(-1, \infty)$ **b.** $(0, 2]$

c. $(-3, 3)$ **d.** $(-\infty, -2] \cup [2, \infty)$

e. $(-\infty, -2) \cup [1, \infty)$

7. **a.** $\dfrac{59}{1,680}$ **b.** $-\dfrac{3}{4}$ **9.** **a.** $\dfrac{39}{2}$ **b.** $\dfrac{5}{14}$ **11.** **a.** 3.94×10^5 **b.** -3.585×10^{-5} **c.** 5×10^{-4}

13. **a.** $\dfrac{36z^2}{x^2 y^3}$ **b.** $4a^2 b^2$ **15.** **a.** $5|x|y^2\sqrt{2x}$ **b.** $\dfrac{-2a^2}{b}\sqrt[3]{4}$

17. **a.** $\dfrac{\sqrt{6}}{3}$ **b.** $\dfrac{2\sqrt{5}}{5}$ **c.** $\dfrac{2\sqrt[3]{6}}{3}$ **d.** $\dfrac{\sqrt{6xy}}{4|y|}$ **19.** **a.** $10x^2 - 7xy - 12y^2$ **b.** $x^2 - 4y^2$

21. **a.** $x^2 + 4; R = 7$ **b.** $2x^2 - 6x + 17; R = -65x + 41$ **23.** **a.** $(3x + 1)(x - 2)$ **b.** $xy(x - 6y)^2$

25. **a.** $(2x^2 + 3)(x + 2)(x - 2)$ **b.** $(x + 2 + y)(x + 2 - y)$ **27.** **a.** $x(8x + 3y)(3x - 4y)$ **b.** $(a - 3)(a + 3)^2$

29. **a.** $(2 - x + y)(2 + x - y)$ **b.** $(3x^2 + 2y)(2x^2 - 3y)$

31. **a.** $\left(x^2 - \dfrac{y}{2}\right)\left(x^4 + \dfrac{x^2 y}{2} + \dfrac{y^2}{4}\right)$ **b.** $(x^4 + 1)(x^2 + 1)(x + 1)(x - 1)$

33. **a.** $3(2a - 3b)^2$ **b.** $(s + 4t^2)(s^2 - 4st^2 + 16t^4)$

35. **a.** $(a^2 - ab + b^2)(a^2 + ab + b^2)$ **b.** $(3x^2 - 2x - 2)(3x^2 + 2x - 2)$ **37.** $\dfrac{2}{2 - x}$ **39.** $\dfrac{6}{x - 3}$

41. $-\dfrac{3x + 8}{(x^2 - 4)^{3/2}}$ **43.** $\dfrac{1}{(1 - x^2)^{3/2}}$ **45.** **a.** $\dfrac{\sqrt{x + 3} + 2}{x - 1}$ **b.** $\dfrac{a - \sqrt{a^2 - 16}}{4}$

47. If $a > b$, then $a - b$ is positive. Therefore $(a - c) - (b - c) = a - c - b + c = a - b$ is positive, so that $a - c > b - c$.

49. If $a \neq 0$, then a^{-1} exists, and from $ab = 0$, we get

$$a^{-1}(ab) = a^{-1} \cdot 0$$
$$(a^{-1}a)b = 0$$
$$1 \cdot b = 0$$
$$b = 0$$

Thus, if $a \neq 0$, then b must be 0. So either $a = 0$ or $b = 0$.

51. 4 **53.** **a.** $(x - 2y + 2)(x - 2y - 2)$ **b.** $(x - 1)(x - 2)(x^2 + x + 1)$ **55.** $(2x - 3y + z + 3)(2x - 3y - z - 3)$

57. $\dfrac{-5}{(2x-3)^{4/3}(x+1)^{2/3}}$ **59.** $\dfrac{(3x+2)^2(6x-53)}{(2x-5)^2}$

Chapter 2

Exercise Set 2, page 68

1. $\{\frac{7}{3}\}$ **3.** $\{11\}$ **5.** $\{-1\}$ **7.** $\{-13\}$ **9.** $\{\frac{13}{6}\}$ **11.** $\{\frac{3}{2}\}$ **13.** $\{\frac{19}{8}\}$ **15.** $\{\frac{2}{5}\}$ **17.** $\{-\frac{34}{13}\}$
19. $\{\frac{26}{29}\}$ **21.** $\{\frac{4}{5}\}$ **23.** $\{\frac{9}{8}\}$ **25.** $\{\frac{6}{17}\}$ **27.** $\{4\}$ **29.** $w \approx 167.74$ **31.** $t \approx 6.289$ **33.** $t = 2(\sqrt{2}-1)$
35. $\dfrac{b+d}{c-a}$ **37.** $\dfrac{2(S-a_1n)}{n(n-1)}$ **39.** $\dfrac{A\rho}{s(\rho-r)}$ **41.** $\dfrac{EC-q}{RC}$ **43.** $k=\frac{54}{7}$ **45.** $\{\frac{7}{13}\}$

Exercise Set 3, page 71

1. $\{1,3\}$ **3.** $\{2,-5\}$ **5.** $\{\frac{1}{2},-3\}$ **7.** $\{-1,\frac{3}{2}\}$ **9.** $\{-1,6\}$ **11.** $\{-\frac{5}{3},\frac{5}{3}\}$ **13.** $\{-\frac{3}{2},\frac{4}{3}\}$ **15.** $\{2\}$
17. $\{\frac{2}{3},-\frac{3}{2}\}$ **19.** $\{-2,\frac{3}{4}\}$ **21.** $\{-2,7\}$ **23.** $\{-1,4\}$ **25.** $\{\frac{3}{2},\frac{4}{9}\}$ **27.** $\{2,-6\}$ **29.** $\{-1,1\}$
31. $\left\{\sqrt{3},\dfrac{2}{\sqrt{3}}\right\}$ **33.** $\{3,-1\}$ **35.** $\left\{\dfrac{9b}{4a},\dfrac{-3b}{2a}\right\}$ **37.** $\left\{\dfrac{m}{2},2m\right\}$ **39.** $k=\dfrac{2}{5}$ or $-\dfrac{3}{2}$

Exercise Set 4, page 75

1. $\{-1\pm\sqrt{5}\}$ **3.** $\{1,-5\}$ **5.** $\left\{\dfrac{3\pm\sqrt{29}}{2}\right\}$ **7.** $\left\{\dfrac{-5\pm\sqrt{85}}{2}\right\}$ **9.** $\left\{2,-\dfrac{1}{2}\right\}$ **11.** $\left\{\dfrac{4\pm\sqrt{34}}{3}\right\}$
13. $\left\{\dfrac{2\pm\sqrt{10}}{2}\right\}$ **15.** $\left\{\dfrac{3\pm\sqrt{105}}{8}\right\}$ **17.** $\{-1\pm\sqrt{5}\}$ **19.** $\left\{\dfrac{9\pm\sqrt{61}}{2}\right\}$ **21.** $\left\{\dfrac{-3\pm\sqrt{41}}{4}\right\}$ **23.** $\left\{\dfrac{1\pm\sqrt{13}}{6}\right\}$
25. $\left\{\dfrac{-5\pm\sqrt{15}}{5}\right\}$ **27.** $\left\{\dfrac{3\pm\sqrt{41}}{8}\right\}$ **29.** $\left\{\dfrac{5\pm\sqrt{17}}{4}\right\}$ **31.** $\left\{\dfrac{-1\pm\sqrt{17}}{4}\right\}$ **33.** $\{\sqrt{2}\}$
35. $r=\dfrac{-\pi l+\sqrt{\pi^2 l^2+\pi S}}{\pi}$ **37.** $x=\dfrac{a+b\pm\sqrt{(a-b)^2+4k}}{2}$ **39.** $r=\dfrac{-RC+\sqrt{R^2C^2-4LC}}{2LC}$
41. $\{k\sqrt{3}\}$ **43.** $\{1.47,0.642\}$

Exercise Set 5, page 81

1. **a.** $8+2i$ **b.** $-1+3i$ **c.** $-1+8i$ **d.** $3-2i$
3. **a.** $14+8i$ **b.** 10 **c.** $-76-3i$ **d.** $10-5i$ **5.** **a.** $\dfrac{2+3i}{13}$ **b.** $\dfrac{2-i}{5}$ **c.** $-\dfrac{3+2i}{13}$ **d.** $\dfrac{7+6i}{85}$
7. **a.** $\dfrac{5-12i}{13}$ **b.** $\dfrac{-10+49i}{61}$ **c.** i **d.** $\dfrac{4+7i}{13}$ **9.** **a.** $3i$ **b.** $6i\sqrt{3}$ **c.** -15 **d.** $\dfrac{-i\sqrt{2}}{4}$
11. **a.** -1 **b.** $-i$ **c.** i **d.** i **e.** -1 **13.** **a.** $\{\pm 3i\}$ **b.** $\left\{\pm\dfrac{i\sqrt{6}}{2}\right\}$
15. **a.** $\{1\pm i\}$ **b.** $\{-1\pm i\sqrt{3}\}$ **17.** **a.** $\{2\pm i\}$ **b.** $\left\{\dfrac{2\pm i\sqrt{2}}{2}\right\}$ **19.** **a.** $\left\{\dfrac{1\pm i\sqrt{5}}{3}\right\}$ **b.** $\left\{\dfrac{2\pm i\sqrt{3}}{2}\right\}$
21. **a.** Imaginary   **b.** Real and unequal   **23.** **a.** Real and unequal   **b.** Real and equal
25. **a.** $8i$ **b.** $-38+41i$ **27.** $-6+i$ **29.** **a.** $\left\{\dfrac{-3i}{2},\dfrac{-i}{2}\right\}$ **b.** $\{i,-3i\}$ **31.** $k=\dfrac{1}{5}$ or $-\dfrac{1}{3}$

Exercise Set 6, page 88

1. $\{\pm 1,\pm 2\}$ **3.** $\{\pm 3,\pm 2i\}$ **5.** $\{\frac{1}{2},-1\}$ **7.** $\{\frac{3}{5},-\frac{2}{3}\}$ **9.** $\{1,8\}$ **11.** $\{8\}$ **13.** $\{1,4\}$ **15.** $\{7\}$
17. $\{-6\}$ **19.** $\{\frac{4}{7}\}$ **21.** $\{-3\}$ **23.** $\{4,-2\}$ **25.** $\{\frac{2}{3},\frac{1}{2}\}$ **27.** No solution   **29.** $\{-7\}$ **31.** $\{-\frac{1}{3}\}$
33. No solution   **35.** $\{-1\}$ **37.** $\{\frac{3}{2}\}$ **39.** $\{\pm 1\}$ **41.** No solution   **43.** $\{0,-1\}$

45. $\left\{-\dfrac{1}{2}, 2, -1 \pm i\sqrt{3}, \dfrac{1 \pm i\sqrt{3}}{4}\right\}$ **47.** $\{\pm\sqrt{2 \pm \sqrt{2}}\}$ **49.** $\{0, 7\}$ **51.** $\{0, -\frac{2}{3}, -1\}$ **53.** $\{-\frac{2}{3}, \frac{3}{2}\}$

55. $\{-2\}$ **57.** $\{-3\}$ **59.** $\{-1, 3\}$ **61.** $\{1\}$ **63.** $\{-\frac{1}{3}, 1\}$

Exercise Set 7, page 97

1. a. Let r = average speed.   **b.** From $d = rt$ get $(50)(12.2) = r(\frac{13}{3})$.   **c.** Approximately 140.77 kilometers per hour

3. a. Let t = time for passenger train, $t - \frac{1}{2}$ = time for freight train.   **b.** $80t = 50(t + \frac{1}{2})$ **c.** 50 min

5. a. Let x = speed of jogger, $2x$ = speed of cyclist, $4/x$ = time of jogger, $4/2x$ = time of cyclist.
b. $\dfrac{4}{2x} = \dfrac{4}{x} - \dfrac{1}{4}$ **c.** 8 kilometers per hour

7. a. Let x = width of field, $3x$ = length, cost of outer fence = $3[2(3x) + 2x]$, cost of inner fences = $1.5(2x)$.
b. $24x + 3x = 1080$ **c.** 40 ft by 120 ft

9. a. Let x = width of patio, $x + 6$ = length, total width = $x + 3$, total length = $x + 12$.
b. $(x + 3)(x + 12) = 360$ **c.** 12 ft by 18 ft

11. a. Let x = the smallest number, $x + 2$ the next, and $x + 4$ the next.   **b.** $x + (x + 2) + (x + 4) = 444$
c. 146, 148, 150

13. a. Let x = grade on final, average grade = $[(3)(76) + 2x]/5$.   **b.** $\dfrac{228 + 2x}{5} = 80$ **c.** 86

15. a. Let x = number of adults, $3x$ = number of children, $(x + 3x)/2$ = number buying popcorn.
b. $4x + 1.5(3x) + 2x = 966$ **c.** 276 children, 184 bags of popcorn

17. a. Let x = rate of walking, $2/x$ = time walking, $30/40$ = time riding.   **b.** $\dfrac{3}{4} + \dfrac{2}{x} = \dfrac{5}{4}$ **c.** 4 miles per hour

19. a. Let x = amount of \$6.25 coffee, $45 - x$ = amount of \$8.50 coffee.   **b.** $6.25x + 8.50(45 - x) = 7.50(45)$
c. 20 pounds of \$6.25 coffee, 25 pounds of \$8.50 coffee

21. a. Let x = number of gallons to be drained, $10 - x$ = amount of 20% solution remaining, $0.20(10 - x)$ = amount of salt.
b. $0.20(10 - x) = 0.15(10)$ **c.** $2\frac{1}{2}$ gallons

23. a. Let x = original price, new price = $x - 0.25x = 0.75x$.   **b.** $0.75x = 487.50$ **c.** \$650

25. a. Let x = side of each square removed; dimensions of box are $12 - 2x$ by $10 - 2x$ by x.
b. $(12 - 2x)(10 - 2x) + 2x(12 - 2x) + 2x(10 - 2x) = 95$ **c.** 2.5 inches

27. a. Let x = number of pounds to be added, $0.42x$ = cost of fattening each calf, $0.62(530 + x)$ = revenue.
b. $0.62(530 + x) - (280 + 0.42x) = 225$ **c.** 882 pounds

29. a. Let x = first integer, $x + 2$ the second.   **b.** $\dfrac{1}{x} + \dfrac{1}{x + 2} = \dfrac{9}{40}$ **c.** $x = 8$, $x + 2 = 10$

31. a. Let x = number of hours for older machine, $x + 2$ = number of hours for newer machine.
b. $2000x + 3000(x + 2) = 30,000$ **c.** $x = 4$ hours 48 minutes, $x + 2 = 6$ hours 48 minutes

33. a. Let x = number purchased, $x - 5$ = number sold, $1950/x$ = cost per dress, $(1950/x) + 25$ = selling price.
b. $\left(\dfrac{1950}{x} + 25\right)(x - 5) - 1950 = 300$ **c.** $x = 30$

35. a. Let x = normal rental price.   **b.** $(x - 0.50)(300) = 200x + 48$ **c.** $x = \$1.98$

37. a. Let x = width, $x + 6$ = height, $x + 4$ = overall width, $x + 11$ = overall height.
b. $(x + 4)(x + 11) = 294$ **c.** 10 by 16

39. a. Let x = time working together, $2x/7$ = fractional part Bill does in x hours, $x/6$ = fractional part Joanne does in x hours.
b. $\dfrac{2x}{7} + \dfrac{x}{6} = 1$ **c.** $\dfrac{42}{19}$ hours ≈ 2.21 hours or 2 hours 13 minutes

41. a. Let x = number in excess of 40, $x + 40$ = total number, $15 - 0.25x$ = price of each ticket.
b. $(x + 40)(15 - 0.25x) = 625$ **c.** $x = 10$, so 50 people go.

43. a. Let x = time for large pump alone, $x + 4$ = time for small pump alone, $8/(3x)$ = fractional part large pump does in $2\frac{2}{3}$ hours, $8/[3(x + 4)]$ = fractional part small pump does in $2\frac{2}{3}$ hours.
b. $\dfrac{8}{3x} + \dfrac{8}{3(x + 4)} = 1$ **c.** 4 hours for large pump, 8 hours for small pump

45. a. Let x = number of gallons of 96 octane gasoline, $200 - x$ = number of gallons of 87 octane gasoline.
   b. $0.96x + 0.87(200 - x) = 0.93(200)$ **c.** $133\frac{1}{3}$ gallons 96 octane gasoline, $66\frac{2}{3}$ gallons 87 octane gasoline
47. a. Let x = number of additional days to fill reservoir, $\frac{1}{8}$ = fractional part first pump does per day, $\frac{1}{6}$ = fractional part second pump does per day.
   b. $\dfrac{3 + x}{8} + \dfrac{3}{6} = 1$ **c.** 1 additional day, 4 days total
49. a. Let x = distance from B to D, $150 - x$ = distance from D to C, $\sqrt{x^2 + (120)^2}$ = distance from A to D.
   b. $4\sqrt{x^2 + (120)^2} + 2(150 - x) = 720$ **c.** Two solutions: $x = 50$ feet or 90 feet
51. a. $\dfrac{0.0024t}{1990}$ **b.** 99.5 minutes

Exercise Set 8, page 103

1. $(-\infty, \frac{5}{3})$

3. $(-\infty, 4)$

5. $(-\infty, -1)$

7. $(-\infty, -1]$

9. $(-\infty, -\frac{5}{2}]$

11. $(-15, \infty)$

13. $(\frac{7}{6}, \infty)$

15. $(\frac{7}{2}, \infty)$

17. $[\frac{8}{5}, \infty)$

19. $(-\infty, \frac{14}{3}]$

21. $[1, 2]$

23. $(-\frac{11}{2}, \frac{5}{2}]$

25. $[-\frac{7}{2}, \frac{1}{2}]$

27. $[\frac{5}{6}, \frac{4}{3})$

29. $[0.49, 0.51]$

31. $x \geq 86$ **33.** $32 < °F \leq 41$ **35.** 24 minutes $\leq t \leq$ 30 minutes
37. a. False since $-2 \not< -4$ **b.** False since $3 \not< 0$ **c.** False since $-2 \not> -1$ **39.** $[\frac{13}{50}, \frac{13}{5})$
41. $(-\infty, 2] \cup (4, \infty)$ **43.** $(-\frac{5}{2}, \frac{7}{3})$

Exercise Set 9, page 107

3. $|a|^2 = |a||a| = \begin{cases} a \cdot a = a^2 & \text{if } a \geq 0 \\ (-a)(-a) = a^2 & \text{if } a < 0 \end{cases}$

$|a|^3 \neq a^3$, since $|a|^3 = |a||a||a| = \begin{cases} a \cdot a \cdot a = a^3 & \text{if } a \geq 0 \\ (-a)(-a)(-a) = -a^3 & \text{if } a < 0 \end{cases}$

$|a|^n = a^n$ if n is even, but $|a|^n = -a^n$ if n is odd
5. $(2, 8)$ **7.** $[-2, -1]$ **9.** $(-\infty, -2) \cup (-\frac{1}{2}, \infty)$ **11.** $[\frac{6}{5}, 2]$ **13.** $(-\infty, b - 2] \cup [b + 2, \infty)$ **15.** $[-1, 2]$
17. $(-3, 13)$ **19.** $(2.95, 3.05)$ **21.** $(-\infty, 2) \cup (6, \infty)$ **23.** $(-\infty, \frac{3}{2}] \cup [\frac{5}{2}, \infty)$ **25.** $(1, 3)$ **27.** $(-5, 7)$
29. $|a^{-1}||a| = |a^{-1}a| = |1| = 1$. So $|a^{-1}|$ is the inverse of $|a|$.   **35.** $\{x: \ -\frac{1}{3} \leq x \leq 3, x \neq \frac{4}{3}\}$ **37.** $\left(\dfrac{ak - \varepsilon}{bk}, \dfrac{ak + \varepsilon}{bk}\right)$

Exercise Set 10, page 112

1. $[-2, 4]$ **3.** $(-\infty, 0] \cup [5, \infty)$ **5.** $[-2, \frac{1}{2}]$ **7.** $(-\infty, -1) \cup (\frac{3}{2}, \infty)$ **9.** $(-\infty, \frac{4}{3}) \cup (\frac{4}{3}, \infty)$ **11.** $[-\frac{4}{3}, \frac{5}{2}]$
13. $[-\frac{3}{2}, -\frac{4}{3}]$ **15.** $(-3, 4)$ **17.** $(-2, 2)$ **19.** $[-4, 2]$ **21.** $(-\infty, -3) \cup [-2, 2] \cup (3, \infty)$
23. $(-2, 1) \cup (1, 2)$ **25.** $[-1, 1) \cup [2, 3)$ **27.** $(-\infty, -2) \cup (1, \infty)$ **29.** $(-1, -\frac{1}{3}) \cup (3, \infty)$
31. $[-2, -1) \cup (0, 1]$ **33.** $(-2, -\frac{3}{2}) \cup (1, 2)$ **35.** $(-1, 2] \cup (6, \infty)$ **37.** $(-\infty, -2] \cup [6, \infty)$ **39.** $(-\frac{3}{2}, 1)$
41. $0 < k \leq \frac{9}{2}$
43. Let $f(x) = (k^2 + 1)x^2 - kx + 1$. The discriminant of the equation $f(x) = 0$ is $-3k^2 - 4$, which is always negative, so there are no real roots, and since $f(0) > 0$, it follows that $f(x) > 0$ for all x.
45. $(-\infty, -2) \cup (-1, \infty)$ **47.** $(0, 1) \cup (4, \infty)$ **49.** $[-1, 0) \cup (0, 2) \cup [3, \infty)$ **51.** $\frac{4}{3} \leq d \leq \frac{8}{3}$
53. $(-\infty, -7.77) \cup (-3.24, 0.77) \cup (1.24, \infty)$

Review Exercise Set, page 113

1. a. $\{\frac{14}{13}\}$ **b.** $\{\frac{29}{36}\}$ **3. a.** $\{\frac{3}{2}, -\frac{5}{3}\}$ **b.** $\{\frac{3}{4}, \frac{4}{3}\}$ **5. a.** $\{1, 3\}$ **b.** $\left\{\dfrac{1 \pm \sqrt{13}}{3}\right\}$

7. a. $\left\{\dfrac{5 \pm i\sqrt{7}}{4}\right\}$ **b.** $\left\{\dfrac{2 \pm i\sqrt{2}}{3}\right\}$ **9. a.** $\{\frac{4}{5}, -\frac{2}{3}\}$ **b.** $\left\{\dfrac{2 \pm \sqrt{13}}{3}\right\}$ **11. a.** $\{\frac{31}{14}\}$ **b.** $\{-\frac{5}{3}\}$

13. a. $\{4, -2\}$ **b.** $\{1, 2\}$ **15. a.** $\{6\}$ **b.** $\{1\}$ **17. a.** $\{\pm 2i, \pm \frac{2}{3}\}$ **b.** $\{\frac{3}{2}, -\frac{4}{3}\}$

19. a. $23 + 2i$ **b.** $\dfrac{6 + 17i}{25}$ **c.** $-236 - 115i$ **21. a.** $\left\{\dfrac{5 \pm i\sqrt{7}}{4}\right\}$ **b.** $\left\{\dfrac{-1 \pm \sqrt{19}}{3}\right\}$

23. a. Imaginary   **b.** Real and unequal   **c.** Real and equal   **d.** Imaginary
25. Width = 13 feet, Length = 19 feet   **27.** 117 kilometers   **29.** $4\frac{2}{3}$ gallons
31. M's speed = 55 miles per hour, N's speed = 60 miles per hour   **33.** $2\frac{1}{2}$ inches   **35.** 20 kilograms
37. a. $[-1, 5)$ **b.** $(\frac{3}{8}, \frac{27}{8}]$ **39. a.** $(-3, -2] \cup (3, \infty)$ **b.** $(-\infty, -4) \cup (1, 2)$
41. a. $(-\infty, -4] \cup [2, \frac{10}{3}]$ **b.** $(-1, 3) \cup (5, \infty)$ **43. a.** $[1, \frac{5}{3}]$ **b.** $(-\infty, -1) \cup (5, \infty)$
45. a. $(-\infty, \frac{1}{6}) \cup (\frac{5}{6}, \infty)$ **b.** $[\frac{1}{3}, \frac{11}{3}]$ **47. a.** $|x - 5| < 2$ **b.** $0 < |x + 3| < 1$
49. $15 \le {}^\circ C \le 30$ **51. a.** $\{k \pm i\sqrt{k^2 - 1}\}$ **b.** $\dfrac{2ab}{a + b}$ **53.** $\{k: \ k \le \frac{2}{3}\} \cup \{k: \ k \ge 2\}$
55. $\{-2\}$ **57.** $\{\frac{1}{4}, -1\}$ **59. a.** $(-1, 0] \cup (1, 3) \cup [7, \infty)$ **b.** $(-\infty, 0) \cup (\frac{3}{2}, 2)$
61. 37, \$6.80   **63.** 60 kph, 80 kph

Chapter 3

Exercise Set 1, page 128

1. a. $2\sqrt{10}$ **b.** $(3, 2)$ **3. a.** $\sqrt{173}$ **b.** $(-\frac{3}{2}, 3)$ **5.** $x = 2$ or 8 **7.** $(4, 5)$
9. x intercept 2; y intercept -2 **11.** x intercepts $\frac{5}{2}, -1$; y intercept -5 **13.** x intercept 1; y intercept $-\frac{1}{4}$
15. x intercepts $4, -3$; y intercepts $2, -2$ **17. a.** y axis   **b.** Origin   **19. a.** y axis   **b.** y axis
21. a. x axis, y axis, Origin   **b.** y axis   **23. a.** None   **b.** Origin
25. a. $y = x + 2$ **b.** $y = x - 1$

x	0	1	-1	-2
y	2	3	1	0

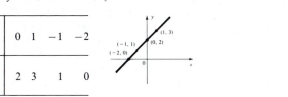

x	0	1	2	-1
y	-1	0	1	-2

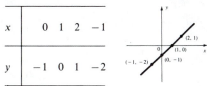

27. a. $y = 2x + 3$ **b.** $y = 4 - 3x$

x	0	1	-1	-2
y	3	5	1	-1

x	0	1	2	-1
y	4	1	-2	7

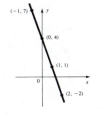

29. **a.** $y = 2 - x$

x	0	1	2	3	-1
y	2	1	0	-1	3

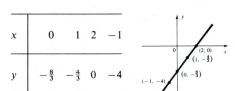

b. $y = x - 1$

x	0	1	2	-1
y	-1	0	1	-2

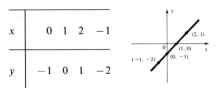

31. **a.** $y = \dfrac{4x - 8}{3}$

x	0	1	2	-1
y	$-\frac{8}{3}$	$-\frac{4}{3}$	0	-4

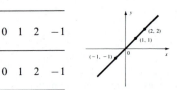

b. $y = -\dfrac{10 + 3x}{5}$

x	0	1	2	5	-1	-5
y	-2	$-\frac{13}{5}$	$-\frac{16}{5}$	-5	$-\frac{7}{5}$	1

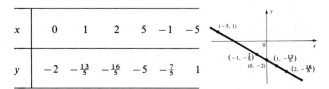

33. **a.** $y = x$

x	0	1	2	-1
y	0	1	2	-1

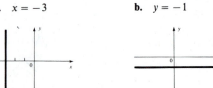

b. $y = -x$

x	0	1	2	-1
y	0	-1	-2	1

35. **a.** $x = -3$ **b.** $y = -1$

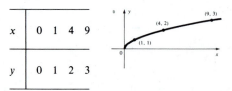

37. **a.** $y = \sqrt{x}$

x	0	1	4	9
y	0	1	2	3

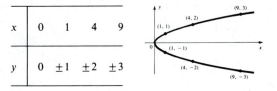

37. **b.** $y = \sqrt{x - 1}$

x	1	2	5	10
y	0	1	2	3

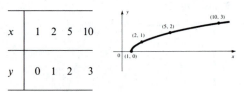

39. **a.** $x = y^2$

x	0	1	4	9
y	0	± 1	± 2	± 3

39. b. $x = y^2 + 1$

x	1	2	5	10
y	0	± 1	± 2	± 3

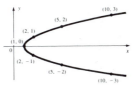

41. a. $y = |1 - x|$

x	0	1	2	3	-1
y	1	0	1	2	2

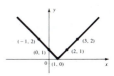

41. b. $y = 2|x| + 1$

x	0	± 1	± 2
y	1	3	5

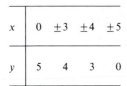

43. a. $y = \pm x$

x	0	± 1	± 2
y	0	± 1	± 2

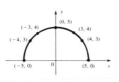

43. b. $y = \sqrt{25 - x^2}$

x	0	± 3	± 4	± 5
y	5	4	3	0

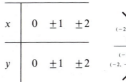

45. a. $y = \dfrac{1}{1 + x^2}$

x	0	± 1	± 2	± 3
y	1	$\frac{1}{2}$	$\frac{1}{5}$	$\frac{1}{10}$

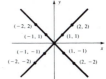

45. b. $y = \pm\sqrt{4 - x^2}$

x	0	± 1	± 2
y	± 2	$\pm\sqrt{3}$	0

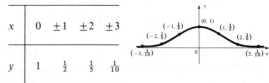

47. a. $y = x - x^2$

x	0	1	2	3	$\frac{1}{2}$	-1
y	0	0	-2	-6	$\frac{1}{4}$	-2

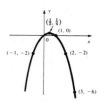

47. b. $y = (x - 1)(x + 2)$

x	0	1	2	-1	-2	-3	$-\frac{1}{2}$
y	-2	0	4	-2	0	4	$-\frac{9}{4}$

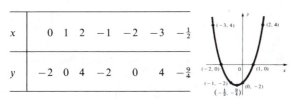

49. a. $x + y = \pm 2$
$y = 2 - x$ or $y = -2 - x$

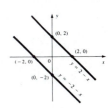

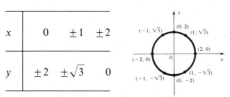

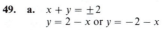

49. b. $|x| + |y| = 2$
If $x \geq 0,\ y \geq 0,\ y = 2 - x$
If $x \geq 0,\ y < 0,\ y = -2 + x$
If $x < 0,\ y \geq 0,\ y = 2 + x$
If $x < 0,\ y < 0,\ y = -2 - x$

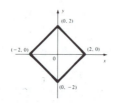

51.

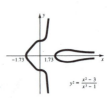

$y = x^3 - 2x^2 - 3$

53.

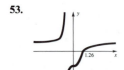

$y = \dfrac{x^3 - 2}{x^3 + 1}$

55.

$y = \sqrt[3]{4 - 2x - x^3}$

57.

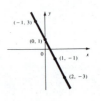

$y^2 = \dfrac{x^2 - 3}{x^3 - 1}$

Exercise Set 2, page 138

1. a. -3 **b.** -1 **c.** -7 **d.** 5 **3. a.** 0 **b.** 0 **c.** $-\frac{1}{2}$ **d.** -8

5. a. 0 **b.** $\frac{1}{11}$ **c.** -5 **d.** $\dfrac{2a - 1}{a + 3}$ **7. a.** 1 **b.** 2 **c.** $|t|$ **d.** x^2

9. a. $4x^2$ **b.** $(t + 1)^2$ **c.** $(2 + h)^2$ **d.** $(t + h)^2$ **11.** $-\dfrac{1}{2x}$ **13.** $\dfrac{-3}{1 + h}$

15. a. Domain $= R$, range $= R$ **17. a.** Domain $= R$, range $= [0, \infty)$
 b. Domain $= R$, range $= [0, \infty)$ **b.** Domain $= [1, \infty)$, range $= [0, \infty)$

19. a. Domain $= [-2, 2]$, range $= [0, 2]$ **21.** 3 **23.** $\dfrac{-2}{x(x + h)}$ **25.** $3x^2 + 3xh + h^2$
 b. Domain $= R$, range $= [0, \infty)$

27. a. -3 **b.** 2 **c.** -1 **d.** 5 **e.** $2|x| - 3$ **29.** Domain $= \{x:\ x \neq -\frac{1}{2}, 3\}$ **31.** Domain $= [0, 3]$

33. Domain $= (-\infty, -1) \cup [3, \infty)$ **35.** $A(b) = 4b$ **37. a.** $C(d) = \pi d$ **b.** $A(d) = \dfrac{\pi d^2}{4}$

39. $A(t) = 12{,}000(1.08)^t$ **41.** $F = \dfrac{9}{5}C + 32$ **45. a.** Odd   **b.** Even   **c.** Even   **d.** Neither   **e.** Odd

47. a. $R(x) = 1500x - 25x^2$ **b.** $[0, 60]$ **49.** Domain $= (-3, 1] \cup (3, \infty)$

51. **Solution 1.** $A(r) = r - \sqrt{r^2 - 1}$
 Solution 2. $A(r) = r + \sqrt{r^2 - 1}$

53. $A(b) = \dfrac{3b}{2}(4 - b)$ **55. a.** $V = \dfrac{\pi y^2}{12}(3x + 2y)$ **b.** $S = \pi y(x + y)$ **c.** $V = \dfrac{7\pi x^3}{384},\ S = \dfrac{5\pi x^2}{16}$

57. $f(x) = \begin{cases} 0.02x & \text{if } x \leq 20{,}000 \\ 400 & \text{if } x > 20{,}000 \end{cases}$

61. Set $y = f(x)$. Then if $y = 0,\ x = 0$, and if $y \neq 0,\ x = 1 \pm \sqrt{1 - y^2}$. So for $|y| \leq 1$, there is an x in R such that $f(x) = y$, and if $|y| > 1$, no such x exists.

Exercise Set 3, page 147

1.

x	0	1	2	-1
y	1	2	3	0

3.

x	0	1	2	-1
y	1	-1	-3	3

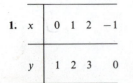

5.

x	1	4	−2
y	1	2	0

7.

x	0	1	2	3	−1
y	0	1	0	−3	−3

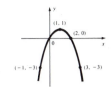

9.

x	−1	0	3	8
y	0	1	2	3

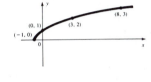

11.

x	1	2	4	$\frac{1}{2}$	−1	−2	−4	$-\frac{1}{2}$
y	2	1	$\frac{1}{2}$	4	−2	−1	$-\frac{1}{2}$	−4

13.

x	0	±1	±2	±3	±4
y	4	$\sqrt{15}$	$2\sqrt{3}$	$\sqrt{7}$	0

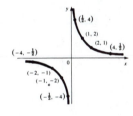

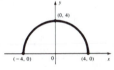

15.

x	0	1	3	−1	−2
y	1	2	4	1	1

17.

x	0	$\frac{1}{2}$	1	2	−1	−2	−3
y	1	$\frac{1}{2}$	3	3	2	−1	−2

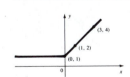

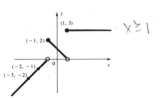

19. **a.** Function   **b.** Not a function   **c.** Function   **d.** Function   **e.** Function   **f.** Not a function

21. **a.** Domain = R, range = $(-\infty, 1]$ **b.** Domain = $[-2, 2]$, range = $[0, 2]$ **c.** Domain = R, range = $(0, \infty)$
d. Domain = R, range = $[0, 1)$

23. Increasing on $(-\infty, -10) \cup (-6, 0) \cup (4, 8)$, decreasing on $(-10, -6) \cup (0, 4) \cup (8, \infty)$

25. a.    **b.**   **27. a.**   **b.**

27. c.   **d.**   **e.**   **29. a.**

29. b.   **c.**   **d.**   **e.**

31. a.   **b.** 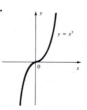   **c.** 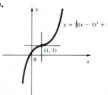   **d.**

e. 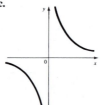   **33. a.** 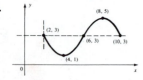   **b.**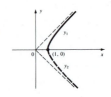

35. $y_1 = \sqrt{4 - x^2}, y_2 = -\sqrt{4 - x^2}$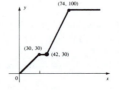

37. $y_1 = \sqrt{x^2 - 1}, x \geq 1$
$y_2 = -\sqrt{x^2 - 1}, x \geq 1$

39.

41.

Exercise Set 4, page 153

1. $(f + g)(x) = 5x - 2$
    $(f - g)(x) = x - 8$
    $(f \cdot g)(x) = 6x^2 - x - 15$
    $(f/g)(x) = \dfrac{3x - 5}{2x + 3}$

3. $(f + g)(x) = \dfrac{3x - 1}{4}$
    $(f - g)(x) = \dfrac{x - 3}{4}$
    $(f \cdot g)(x) = \dfrac{x^2 - 1}{8}$
    $(f/g)(x) = \dfrac{2(x - 1)}{x + 1}$

5. $(f + g)(x) = \sqrt{x + 4} + \sqrt{4 - x}$
    $(f - g)(x) = \sqrt{x + 4} - \sqrt{4 - x}$
    $(f \cdot g)(x) = \sqrt{16 - x^2}$
    $(f/g)(x) = \sqrt{\dfrac{4 + x}{4 - x}}$

7. $(f + g)(x) = \dfrac{x^2 + 2}{x^2 + x - 2}$
    $(f - g)(x) = \dfrac{2 + 2x - x^2}{x^2 + x - 2}$
    $(f \cdot g)(x) = \dfrac{x}{x^2 + x - 2}$
    $(f/g)(x) = \dfrac{x + 2}{x^2 - x}$

9. $(f + g)(x) = 0$
    $(f - g)(x) = \begin{cases} -2 & \text{if } x < 0 \\ 2 & \text{if } x \geq 0 \end{cases}$
    $(f \cdot g)(x) = -1$
    $(f/g)(x) = -1$

11. $(f \circ g)(x) = 2x + 3$, domain $= R$
    $(g \circ f)(x) = 2x + 1$, domain $= R$

13. $(f \circ g)(x) = 4x^2 - 12x + 10$, domain $= R$
    $(g \circ f)(x) = 2x^2 - 1$, domain $= R$

15. $(f \circ g)(x) = -\dfrac{x + 3}{2x + 5}$, domain $= \{x: \ x \neq -3, -\frac{5}{2}\}$
    $(g \circ f)(x) = \dfrac{x - 2}{3x - 5}$, domain $= \{x: \ x \neq 2, \frac{5}{3}\}$

17. $(f \circ g)(x) = \sqrt{x^2 - 4}$, domain $= \{x: \ |x| \geq 2\}$
    $(g \circ f)(x) = x - 4$, domain $= [1, \infty)$

19. $(f \circ g)(x) = \sqrt{4 - x^2}$, domain $= \{x: \ |x| \leq 2\}$
    $(g \circ f)(x) = 4 - x$, domain $= (-\infty, 4]$

21. $f(x) = 4 - 3x^2, g(x) = x^6$

23. $f(x) = 1 - x^3, g(x) = \sqrt{x}$

25. $f(x) = \dfrac{x}{x + 2}, g(x) = x^{2/3}$

27. **a.** $f + g, f - g, f \cdot g$, and f/g all even
    b. $f + g$ and $f - g$ odd, $f \cdot g$ and f/g even $(f \neq \pm g)$
    c. $f + g$ and $f - g$ neither even nor odd, $f \cdot g$ and f/g both odd $(f, g$ nonzero)

29. $[f \circ (g \circ h)](x) = [(f \circ g) \circ h](x) = \left(\dfrac{3x + 5}{x + 1}\right)^2 - 1$. In general $f \circ (g \circ h) = (f \circ g) \circ h$.

31. **a.** $F(t) = \dfrac{gR^2m}{(-\frac{1}{2}gt^2 + v_0 t + s_0)^2}$ **b.** Approximately 19,360 newtons

33. $\left[f \circ \left(\dfrac{g}{h}\right)\right](x) = \dfrac{1}{\dfrac{g(x)}{h(x)}} = \dfrac{\dfrac{1}{g(x)}}{\dfrac{1}{h(x)}} = \dfrac{(f \circ g)(x)}{(f \circ h)(x)} = \left(\dfrac{f \circ g}{f \circ h}\right)(x)$ provided x is restricted so that $g(x) \neq 0$ and $h(x) \neq 0$.

35. $(f \circ f)(x) = \begin{cases} 4x & \text{if } 0 \leq x < \frac{3}{2} \\ 4x^2 + 1 & \text{if } \frac{3}{2} \leq x < 3 \\ x^4 + 2x^2 + 2 & \text{if } x \geq 3 \end{cases}$

Exercise Set 5, page 161

1. c 3. If $f(x_1) = f(x_2), 3x_1 = 3x_2$, so $x_1 = x_2$. So f is 1–1 on R.

5. If $g(x_1) = g(x_2), 2x_1 - 3 = 2x_2 - 3$, so $2x_1 = 2x_2$, and $x_1 = x_2$. So g is 1–1 on R.

7. If $f(t_1) = f(t_2)$, $7t_1 - 9 = 7t_2 - 9$, so $7t_1 = 7t_2$, and $t_1 = t_2$. So f is 1–1 on R.

9. If $\phi(x_1) = \phi(x_2)$, $\dfrac{2x_1}{3} - \dfrac{3}{4} = \dfrac{2x_2}{3} - \dfrac{3}{4}$, so $\dfrac{2x_1}{3} = \dfrac{2x_2}{3}$, and $x_1 = x_2$. So ϕ is 1–1 on R.

11. If $f(x_1) = f(x_2)$, $\frac{3}{5}(2 - 3x_1) = \frac{3}{5}(2 - 3x_2)$, so $2 - 3x_1 = 2 - 3x_2$, and $-3x = -3x_2$, hence $x_1 = x_2$. So f is 1–1 on R.

13. If $g(t_1) = g(t_2) = 0$, then both $t_1 = 0$ and $t_2 = 0$, so $t_1 = t_2$. If $g(t_1) = g(t_2) \neq 0$, then $\dfrac{1}{t_1} = \dfrac{1}{t_2}$, and hence $t_1 = t_2$.
    So g is 1–1 on R.

15. Domain = $\{x: \ x \neq 1\}$, range = $\{y: \ y \neq 0\}$, f is 1–1.   17. Domain = R, range = $\{2\}$, f is not 1–1.

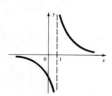

19. Domain = $\{x: \ x \neq 0\}$, range = $\{1, -1\}$, f is not 1–1.   21. $f(x_1) = f(x_2)$ implies $x_1 = x_2$, so f^{-1} exists; $f^{-1}(x) = x - 2$.

23. $g(x_1) = g(x_2)$ implies $x_1 = x_2$, so g^{-1} exists; $g^{-1}(x) = \dfrac{3x + 2}{5}$.

25. $h(x_1) = h(x_2)$ implies $x_1 = x_2$, so h^{-1} exists; $h^{-1}(x) = \dfrac{2x + 1}{x}$, $x \neq 0$.

27. $G(x_1) = G(x_2)$ implies $x_1 = x_2$, so G^{-1} exists; $G^{-1}(x) = x^2 + 4$, $x \geq 0$.

29. $F(t_1) = F(t_2)$ implies $t_1^2 - 1 = t_2^2 - 1$, so $t_1^2 = t_2^2$, and for $t_1, t_2 \geq 0$, this implies $t_1 = t_2$. So F^{-1} exists; $F^{-1}(t) = \sqrt{t + 1}$, $t \geq -1$.

31. Domain $f = \left\{t: \ t \neq -\dfrac{3}{2}\right\}$. If $f(t_1) = f(t_2)$, $\dfrac{3t_1 - 2}{2t_1 + 3} = \dfrac{3t_2 - 2}{2t_2 + 3}$, so $6t_1t_2 + 9t_1 - 4t_2 - 6 = 6t_1t_2 + 9t_2 - 4t_1 - 6$, and
    $13t_1 = 13t_2$, or $t_1 = t_2$. So f is 1–1 on its domain; $f^{-1}(t) = \dfrac{2 + 3t}{3 - 2t}$. Domain $f^{-1} = \left\{t: \ t \neq \dfrac{3}{2}\right\}$

33. $g^{-1}(x) = \dfrac{x}{2x - 1}$, $x \neq \dfrac{1}{2}$ 35.   37.

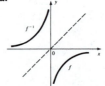

39.   41. a.   b.

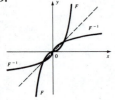

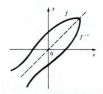

43. For $x = 2$, $x = \frac{3}{2}$, and $x = -\frac{3}{2}$, $f(x) = 0$. So f is not 1–1 and hence does not have an inverse.

45. If f is even and $x \neq 0$ is in its domain, $f(-x) = f(x)$; but $-x \neq x$. So f is not 1–1. Let $f(x) = x^3$ and $g(x) = x^3 - x$. Then f and g are odd, but f is 1–1 and g is not 1–1.

47. Since $f(x + k) = f(x)$, but $x + k \neq x$ since $k \neq 0$. Thus f is not 1–1.

49. Domain $g = R$. If $g(x_1) = g(x_2)$, $\dfrac{3 - 2x_1}{4} = \dfrac{3 - 2x_2}{4}$, so $x_1 = x_2$, and g is 1–1. If y is any real number, $\dfrac{3 - 2x}{4} = y$

provided $x = \dfrac{3 - 4y}{2}$. So the range of g is R. Thus g is 1–1 from R onto R.

51. Since $[2] = 2$ and $[2.5] = 2$, for example, $[x]$ is not 1–1. The domain of $[x]$ is R, but the range is the set J of all integers and is therefore not all of R.

53. $g(1) = g(-1)$, so g is not 1–1. Restrict the domain to be $[0, 2)$. Then $g(x_1) = g(x_2)$ implies $\dfrac{1}{\sqrt{4 - x_1^2}} = \dfrac{1}{\sqrt{4 - x_2^2}}$, so

$4 - x_1^2 = 4 - x_2^2$, and $x_1^2 = x_2^2$. Since $x \geq 0$, this implies $x_1 = x_2$. So g is 1–1 on $[0, 2)$; $g^{-1}(x) = \dfrac{\sqrt{4x^2 - 1}}{x}$.

Domain $g^{-1} = [\frac{1}{2}, \infty)$.

55. $F(1) = F(2)$, so F is not 1–1 on R. Restrict x to lie in $(-\infty, \frac{3}{2}]$. Then $F(x) = 3 - 2x$, which is 1–1 on this domain;

$F^{-1}(x) = \dfrac{3 - x}{2}$. Domain $F^{-1} = [0, \infty)$.

57. Suppose g and h are both inverses of f. Then $(g \circ f)(x) = x$ and $(h \circ f) = x$ for all x in the domain of f, so $g \circ f = h \circ f$. Thus $g \circ [(f \circ g)(x)] = h \circ [(f \circ g)(x)]$, and since $(f \circ g)(x) = x$, $g(x) = h(x)$. So $g = h$.

Exercise Set 6, page 167

1. $u = kv$ **3.** $z = kxy$ **5.** $s = k/\sqrt{t}$ **7.** $F = km_1m_2/r^2$ **9.** $s = kt^2$ **11.** $\frac{20}{3}$ **13.** 12   **15.** 336

17. 32 pounds   **19.** 22 amperes   **21.** $k = \pi/3$; $V = (1/3)\pi r^2 h$; $V = 120\pi$ **23.** 6°F per minute

25. 7 cases per day   **27.** Approximately 671 days   **29.** 375 pounds   **31.** 302.5:1

Review Exercise Set, page 169

1. a.

x	0	1	2	−1	−2
y	−1	1	3	−3	−5

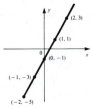

b.

x	0	1	2	3	−1	−2
y	$\frac{4}{3}$	$\frac{2}{3}$	0	$-\frac{2}{3}$	2	$\frac{8}{3}$

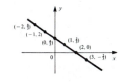

3. a.

x	0	2	4	−2
y	−2	3	8	−7

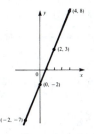

b.

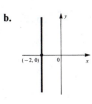

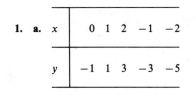

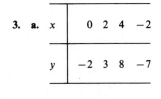

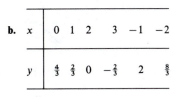

5. a.

x	0	± 1	± 2	± 3
$f(x)$	1	0	-3	-8

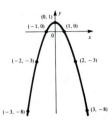

b.

x	0	1	-3	-8
$g(x)$	0	1	-1	-2

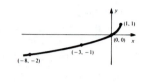

7. Domain $\{t:\ t \neq 3\}$ **a.** -1 **b.** $\frac{4}{5}$ **c.** $\dfrac{2}{1 - 3t}$ **d.** $\dfrac{2(t + \Delta t)}{t + \Delta t - 3}$

9. a. 49   **b.** 0   **c.** 4   **d.** 7   **e.** 3   **11.** $-\dfrac{x + 2}{4x^2}$ **15.** $d = s\sqrt{2}$ **17.** $A = 5{,}000(1.06)^t$

19. $C(x) = 0.30x + 30.50$ **23. a.** Even   **b.** Odd   **c.** Neither   **d.** Odd   **e.** Even

25. $(\phi \circ \psi)(x) = -\dfrac{4x + 13}{35}; (\psi \circ \phi)(x) = \dfrac{26 - 4x}{35}$ **27.** $f(x) = x^{-3/2}, g(x) = 3 - 4x$ **29.** $(g \circ f)(x) = \dfrac{T}{Ak(1 + ax)}$

31. $f^{-1}(x) = \dfrac{3x - 7}{5}$ **33.** $h^{-1}(t) = \dfrac{2 - 5t}{3}$ **35.** $g^{-1}(x) = \dfrac{2x^2 + 1}{x^2}, x > 0$ **37.** $h^{-1}(x) = \dfrac{3x + 2}{x - 1}, x \neq 1$

39. a. Not a function   **b.** Function; no inverse   **c.** Function; no inverse   **d.** Function; has inverse
e. Not a function   **f.** Function; no inverse

41. $f^{-1}(x) = x^2 - 1, x \geq 0$

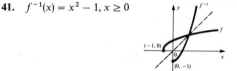

43. a.

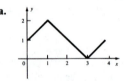

43. b.

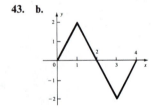

c.

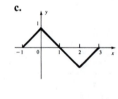

d.

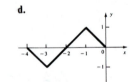

e.

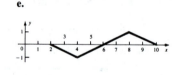

45. $(f + g)(x) = \dfrac{2x - 1}{x(x^2 - 1)}, (f - g)(x) = \dfrac{-1}{x(x^2 - 1)}$

$(f \cdot g)(x) = \dfrac{1}{x(x + 1)^2(x - 1)}, (f/g)(x) = \dfrac{x - 1}{x}$; domain $= \{x:\ x \neq 0, \pm 1\}$ in all cases.

47. 1,600 feet   **49.** $\frac{250}{3}$ cubic feet   **51. a.** $\{x:\ x \neq 2, -2, 3\}$ **b.** $(-\infty, -4) \cup [-2, 0) \cup (0, 2]$

53. $h = 10 - \dfrac{5b}{3}$

55. For $x \geq 2$, $f(x_1) = f(x_2)$ implies $4x_1 - x_1^2 = 4x_2 - x_2^2$, or $4(x_1 - x_2) - (x_1^2 - x_2^2) = 0$. Thus $(4 - x_1 - x_2)(x_1 - x_2) = 0$, so either $x_1 + x_2 = 4$ or $x_1 = x_2$. But for $x_1 \geq 2, x_2 \geq 2, x_1 + x_2 = 4$ implies $x_1 = 2$ and $x_2 = 2$. So f is 1–1; $f^{-1}(x) = 2 + \sqrt{4 - x}, x \leq 4$.

57. Since $g(-1) = g(5) = 0$, g is not 1–1 on R. For $x \leq 2$, $g(x_1) = g(x_2)$ implies $x_1^2 - 4x_1 - 5 = x_2^2 - 4x_2 - 5$, or $(x_1^2 - x_2^2) - 4(x_1 - x_2) = 0$. Thus $(x_1 + x_2 - 4)(x_1 - x_2) = 0$. If $x_1 + x_2 - 4 = 0$ and $x_1 \leq 2, x_2 \leq 2$, then $x_1 = x_2 = 2$. So either factor being 0 implies $x_1 = x_2$. So g^{-1} exists.

Cumulative Review Exercise Set I (Chapters 1–3), page 172

1. If $a < b$ and $c > 0$, then $ac < bc$. Also, if $c < d$ and $b > 0$, $bc < bd$. By transitivity, $ac < bd$. Suppose $a = -2, b = 4$, $c = -3, d = -2$. Then $-2 < 4$ and $-3 < -2$, but $(-2)(-3) = 6 \not< (-2)(4) = -8$.

3. $\dfrac{-1}{\sqrt{x}\sqrt{x+h}(\sqrt{x}+\sqrt{x+h})}$ $(h \neq 0)$

5. **a.** $\frac{79}{370}$

   b. Let $r_1 = \dfrac{-b + \sqrt{b^2 - 4ac}}{2a}, r_2 = \dfrac{-b - \sqrt{b^2 - 4ac}}{2a}$

   $r_1 + r_2 = -b/a$, which is rational
   $r_1 \cdot r_2 = c/a$, which is rational

7. $\dfrac{25(x^2 + 1)}{(3x + 4)(4x - 3)}$ 9. **a.** $\dfrac{2}{\sqrt{2x + 2h - 3} + \sqrt{2x - 3}}$ **b.** $\dfrac{17\sqrt{6}}{12}$ 11. $2\frac{1}{2}$ feet

13.

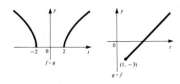

   a. 2   **b.** -1 **c.** 1   **d.** 9   **e.** -1

15. **a.** $(x + 2)(x - 2)(3 - xy)$ **b.** $2x(9x + 8y)(4x - 3y)$ 17. **a.** $s = \dfrac{Fr + 1}{r - 2F}$ **b.** $t = \dfrac{-v_0 + \sqrt{v_0^2 + 2gs}}{g}$

19. **a.** $\dfrac{4a^7 b}{125}$ **b.** 6×10^2 21. $-\dfrac{x^2 + 3}{3(x^2 - 1)^{5/3}}$ 23. **a.** $(-3, 1)$ **b.** $(-11, -5] \cup (1, 5]$

25. $(f \circ g)(t) = \sqrt{t^2 - 4}$, domain $= \{t: \ |t| \geq 2\}$, range $= \{y: \ y \geq 0\}$; $(g \circ f)(t) = t - 4$, domain $= \{t: \ t \geq 1\}$, range $= \{y: \ y \geq -3\}$

27. 90 suits sold during the sale; regular price $=$ \$210

29. $(f + g)(x) = \begin{cases} 3x & \text{if} & 0 \leq x \leq 1 \\ 3 & \text{if} & x > 1 \\ x & \text{if} & -1 \leq x < 0 \\ -1 & \text{if} & x < -1 \end{cases}$ $(f - g)(x) = \begin{cases} -x & \text{if} & 0 \leq x \leq 1 \\ -1 & \text{if} & x > 1 \\ -3x & \text{if} & -1 \leq x < 0 \\ 3 & \text{if} & x < -1 \end{cases}$

   $(f \cdot g)(x) = \begin{cases} 2x^2 & \text{if} & 0 \leq x \leq 1 \\ 2 & \text{if} & x > 1 \\ -2x^2 & \text{if} & -1 \leq x < 0 \\ -2 & \text{if} & x < -1 \end{cases}$ $(f/g)(x) = \begin{cases} \frac{1}{2} & \text{if } x > 0 \\ -\frac{1}{2} & \text{if } x < 0 \end{cases}$

   Domain of $f \pm g$ and $f \cdot g = R$; domain of $f/g = \{x: \ x \neq 0\}$.

Chapter 4

Exercise Set 2, page 186

1. a. $\frac{1}{2}$ **b.** $\frac{2}{3}$ **c.** $-\frac{5}{6}$ **d.** -1

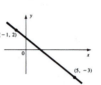

3. $m_{AB} = \frac{5}{6}$ **5.** Horizontal displacement $= 8$ **7.** $y = \frac{15}{2}$
$m_{BC} = 7$ Vertical displacement $= 1$
$m_{AC} = -\frac{2}{5}$

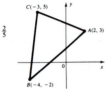

9. $m_{AB} = \frac{3}{4}$, $m_{BC} = \frac{1}{4}$, $m_{CD} = \frac{3}{4}$, $m_{AD} = \frac{1}{4}$; parallelogram, since opposite sides have same slope.

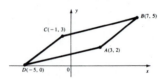

11. $3x + y + 4 = 0$ **13.** $x - y - 1 = 0$ **15.** $5x - 3y - 12 = 0$ **17.** $3x - y - 1 = 0$ **19.** $y + 2 = 0$
21. $m = \frac{2}{3}, b = -\frac{4}{3}$ **23.** $m = \frac{2}{3}, b = -\frac{5}{3}$ **25.** $m = 2, b = \frac{3}{2}$

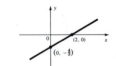
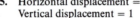

27. a. $y = 0$ **b.** $x = 0$ **c.** $y - 4 = 0$ **d.** $x + 3 = 0$ **29. a.** $2x - 5y - 10 = 0$ **b.** $6x - 5y + 4 = 0$
31. $4x - 3y - 17 = 0$ **33.** $3x + 8y - 6 = 0$ **35.** $2x + y = 0$ **37.** $y + 1 = 0$ **39.** $4x - 5y + 8 = 0$
41. Let $A = (2, 1), B = (6, 9), C = (-2, 3)$. Then $\overline{AB}^2 = 80, \overline{BC}^2 = 100, \overline{AC}^2 = 20$. So $\overline{BC}^2 = \overline{AB}^2 + \overline{AC}^2$.
43. Each side $= 10$ units   **45.** $2x - y - 7 = 0$
47. Let $A = (-2, 4), B = (3, -6), C = (6, -2)$. Then $m_{AC} = -\frac{3}{4}, m_{BC} = \frac{4}{3}$, so C is a right angle.
49. Let $A = (2, 2), B = (0, -1), C = (-4, 1), D = (-2, 4)$. Then $m_{AB} = \frac{3}{2}, m_{BC} = -\frac{1}{2}, m_{CD} = \frac{3}{2}, m_{AD} = -\frac{1}{2}$, so opposite sides are parallel.
51. $3x - 2y + C = 0; 9x - 6y + 8 = 0$ **53.** $x + 2y - 2 = 0$ **55.** $32x + 2y - 101 = 0$
57. $x = 2, 5x + 3y - 21 = 0, 2x - 3y + 7 = 0$
59. Let $A = (0, 2), B = (2, -1), C = (-4, 8)$.
   Method 1. $m_{AB} = -\frac{3}{2}, m_{BC} = -\frac{3}{2}$, so they are collinear.
   Method 2. Equation of line AB is $3x + 2y - 4 = 0$, and C satisfies this equation.
61. $(-1, 6), (-3, -8), (5, 2)$ **63.** $x^2 + y^2 - 10x + 4y + 13 = 0$
                                   circle of radius 4, center $(5, -2)$
65. $2x + 2y - 3 = 0$ **67.** $\sqrt{5x^2 - 18x + 18}$

69. **a.** $2x + y - 5 = 0$
b. $x - 2y + 5 = 0$

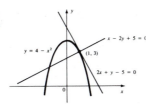

71. $m_{sec} = x + 2; m_{tan} = 4$

73. $m_{sec} = -\dfrac{1}{2x}; m_{tan} = -\dfrac{1}{4}; x + 4y = 4$

Exercise Set 3, page 191

1. Line with slope 5 and y intercept -3

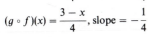

3. Slope $= \frac{3}{2}$, y intercept $= -\frac{7}{2}$

5. $f(x) = \dfrac{2x}{3} + \dfrac{11}{3}$ **7.** $g(x) = -\frac{1}{2}x + 2$ **9.** -9 **11.** $f(x) = 2x - 4$

13. $C(x) = \frac{7}{3}x + 80$; marginal cost $= \frac{7}{3}$, start-up cost $= 80$ **15.** $A(t) = P + Prt$; slope $= Pr$; $A(8) = \$4,440$

17. $R(x) = 65x + 1,500$; $R(50) = \$4,750$

19. $(f \circ g)(x) = \dfrac{1 - x}{4}$, slope $= -\dfrac{1}{4}$

$(g \circ f)(x) = \dfrac{3 - x}{4}$, slope $= -\dfrac{1}{4}$

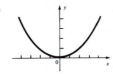

21. If slope of f is m, slope of f^{-1} is $1/m$.   **23.** $a = -C/N; b = C; V(8) = 6,000$ **25.** $L(3.546) \approx 0.5497$

Exercise Set 4, page 203

1. Vertex $(0, 0)$, axis $x = 0$

3. Vertex $(0, 3)$, axis $x = 0$

5. Vertex $(-3, 2)$, axis $x = -3$

7. Vertex $(2, -1)$, axis $x = 2$

9. Vertex $(-1, 4)$, axis $x = -1$
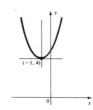

11. Vertex $(\frac{3}{2}, \frac{5}{2})$, axis $x = \frac{3}{2}$
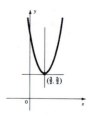

13. Vertex $(1, 1)$, axis $x = 1$

15. Vertex $(-\frac{3}{4}, \frac{41}{8})$, axis $x = -\frac{3}{4}$

17. Vertex $(1, -\frac{7}{2})$, axis $x = 1$

19. Vertex $(\frac{3}{4}, -\frac{23}{8})$, axis $x = \frac{3}{4}$
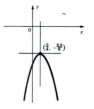

21. Maximum value $\frac{9}{4}$ when $x = \frac{3}{2}$ 23. Minimum value -1 when $x = 2$ 25. Maximum value $\frac{19}{5}$ when $x = -\frac{2}{5}$

27. Minimum value $-\frac{49}{3}$ when $x = \frac{7}{3}$ 29. 70 items; $600   31. 576 feet   33. $(-\infty, 0) \cup (4, \infty)$ 35. $[-4, 1]$

37. $[-2, \frac{4}{3}]$ 39. $(-\frac{1}{2}, 3)$ 41. $(-4, \frac{1}{2})$ 43. $(-\infty, \frac{4}{5}] \cup [\frac{3}{2}, \infty)$ 45. $[1 - \sqrt{5}, 1 + \sqrt{5}]$

47. 12.5 ft $\times$ 25 ft   49. $14.00; $1,960   51. $(-\infty, -\frac{4}{3}) \cup (\frac{5}{4}, \infty)$

53. Vertex of parabola is $(2, 1)$ and it opens upward.

55. $5 < x < 21$; maximum profit $= 64$ when $x = 13$; $p(x) = x^2 - 26x - 105$

57. $f(x) = \begin{cases} x^2 - 4 & \text{if} \quad x \geq 2 \text{ or } x \leq -2 \\ 4 - x^2 & \text{if} \quad -2 < x < 2 \end{cases}$

59. $f(x) = \begin{cases} 4 - (x + 1)^2 & \text{if} \quad -3 \leq x \leq 1 \\ (x + 1)^2 - 4 & \text{if} \quad x < -3 \text{ or } x > 1 \end{cases}$

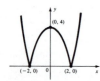

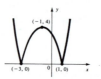

Exercise Set 5, page 214

1. Parabola, vertex $(0, 0)$, axis $y = 0$

3. Circle, center $(0, 0)$, radius 4
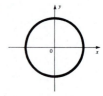

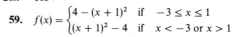

5. Ellipse, center $(0, 0)$, $a = 3$, $b = 2$

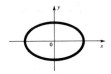

7. Hyperbola, center $(0, 0)$, $a = 3$, $b = 2$

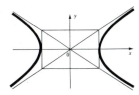

9. Ellipse, center $(0, 0)$, $a = 2$, $b = 1$

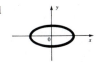

11. Parabola, vertex $(0, 0)$, axis $y = 0$

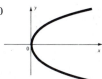

13. Hyperbola, center $(0, 0)$, $a = 2$, $b = 1$

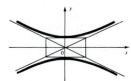

15. Circle, center $(0, 0)$, radius $\dfrac{4}{\sqrt{3}} \approx 2.31$

17. Hyperbola, center $(0, 0)$, $a = 5$, $b = 4$

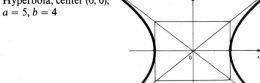

19. Parabola, vertex $(3, -2)$, axis $y = -2$

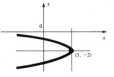

21. Ellipse, center $(-3, 2)$ $a = 4$, $b = 2$

23. Hyperbola, center $(0, 1)$, $a = 2$, $b = 3$

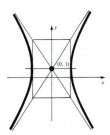

25. Ellipse, center $(0, 2)$, $a = 2$, $b = 3$

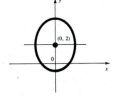

27. Parabola, vertex $(-4, 3)$, axis $y = 3$

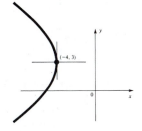

29. Circle, center $(2, -1)$, radius $\sqrt{5}$

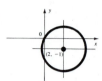

31. Ellipse, center $(-2, 3)$, $a = 2, b = 3$

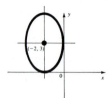

33. Hyperbola, center $(4, 2)$, $a = 1, b = 2$

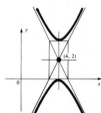

35. Ellipse, center $(2, 0)$, $a = 3, b = 2$

37. Parabola, vertex $(3, 4)$ axis $y = 4$

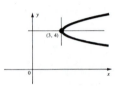

39. Hyperbola, center $(1, -3)$, $a = 3, b = 4$

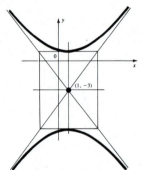

41. Semicircle, center $(0, 0)$, radius 2, $y \geq 0$

43. Semiellipse, center $(0, -1)$, $a = 3, b = 2, y \geq -1$

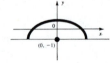

45. Upper half of hyperbola, center $(0, 2)$, $a = 3$, $b = 2, y \geq 4$

47. Lower half of parabola, vertex $(\frac{1}{2}, 0)$, axis $y = 0$, $y \leq 0$

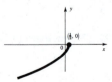

49. $x^2 + y^2 + 6x - 10y + 26 = 0$

51. $\left(x - \dfrac{4}{3}\right)^2 + \left(y + \dfrac{5}{3}\right)^2 = \dfrac{62}{9}$

Circle, center $\left(\dfrac{4}{3}, -\dfrac{5}{3}\right)$,

radius $\dfrac{\sqrt{62}}{3} \approx 2.6$

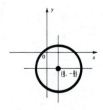

53. $\dfrac{\left(x+\dfrac{7}{10}\right)^2}{\dfrac{827}{300}} + \dfrac{\left(y-\dfrac{1}{3}\right)^2}{\dfrac{827}{180}}$

Ellipse, center $\left(-\dfrac{7}{10}, \dfrac{1}{3}\right)$,

$a = \sqrt{\dfrac{827}{300}} \approx 1.66$, $b = \sqrt{\dfrac{827}{180}} \approx 2.14$

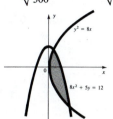

55.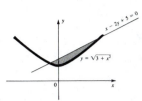

57.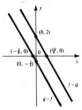

59. $3x - 4y = 43$

61. Centers are $(4, -1)$ and $(-2, 7)$, respectively; radii are 6 and 4, respectively. Distance between centers is 10 and sum of radii is 10.

Review Exercise Set, page 215

1. a. $-\dfrac{7}{6}$ **b.** -4 **3. a.** $y = 3$ **b.** $x = 2$ **5. a.** $3x - 2y - 4 = 0$ **b.** $x - y + 5 = 0$

7. a. $3x - 7y + 21 = 0$ **b.** $9x + 7y - 24 = 0$ **9. a.** $m = \dfrac{3}{2}, a = 6, b = -9$ **b.** $m = -\dfrac{6}{5}, a = \dfrac{7}{3}, b = \dfrac{14}{5}$

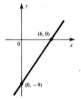

11. a. $(\tfrac{1}{2}, -\tfrac{3}{2})$ **b.** $(2, \tfrac{3}{2})$ **c.** $(\tfrac{3}{2}, 4)$ **d.** $(2, \tfrac{3}{2})$ **e.** $(\tfrac{5}{2}, -1)$

13. Let $A = (-1, 5)$, $B = (3, -2)$, $C = (-2, -3)$. Then $\overline{AB} = \overline{AC} = \sqrt{65}$.

15. a. $5x + 4y - 1 = 0$ **b.** $3x + 4y + 14 = 0$ **17. a.** $3x + 2y - 27 = 0$ **b.** $2x - 3y + 8 = 0$

19. $3x + 5y - 7 = 0$ **21. a.** $8x + 3y + C = 0, 8x + 3y - 7 = 0$ **b.** $x + 3y + C = 0, x + 3y - 14 = 0$

23. $g(x) = x - 3, m = 1$ **25.** $(f \circ g)(x) = \dfrac{10 - 9x}{5}, (g \circ f)(x) = -\dfrac{2 + 9x}{5}$

27. $T = -\dfrac{h}{500} + T_0; 36°F$

29. **a.** $y = (x - 3)^2 - 5$; vertex $(3, -5)$, axis $x = 3$

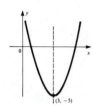

b. $y = -2(x - 1)^2 - 1$; vertex $(1, -1)$, axis $x = 1$

31. **a.** Minimum is $-\frac{11}{4}$ when $x = \frac{3}{2}$.   **b.** Maximum is 12 when $x = -2$.   **33.** 100 parts; minimum cost is $10,000.

35. **a.** $(-1, 2)$ **b.** $(-3, 1)$ **37.** **a.** $(-\infty, \frac{1}{2}] \cup [\frac{4}{3}, \infty)$ **b.** $[-2, \frac{4}{3}]$

39. **a.** Parabola, vertex $(0, 0)$, axis $y = 0$

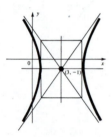

b. Hyperbola, center $(0, 0)$, $a = 1, b = 2$

41. **a.** Hyperbola, center $(3, -1)$, $a = 2, b = 3$

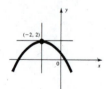

b. Parabola, vertex $(-1, 3)$, axis $x = -1$

43. **a.** Parabola, vertex $(-2, 2)$, axis $x = -2$

b. Circle, center $(-1, 4)$, radius 4

45. **a.** Upper half of circle, center $(-1, 0)$, radius 2

b. Lower half of parabola, vertex $(4, 2)$, axis $y = 2$

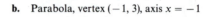

47. a. $5x + 2y - 10 = 0, 3x - 2y + 6 = 0, x + 2y - 8 = 0$ **b.** $6x + y - 5 = 0, 6x - 7y + 3 = 0, y = 1$

49. Tangent: $3x - y - 8 = 0$;   normal: $x + 3y + 34 = 0$ **51.** $[-4, -1] \cup [1, \infty)$

53. a. Hyperbola, center $\left(\dfrac{1}{2}, -\dfrac{3}{4}\right)$,   **b.** Ellipse, center $\left(-\dfrac{5}{4}, -\dfrac{3}{4}\right)$,

$a = \dfrac{2}{\sqrt{3}}, b = \dfrac{1}{\sqrt{3}}$ $a = \sqrt{\dfrac{75}{16}} \approx 2.16, b = \sqrt{\dfrac{75}{32}} \approx 1.5$

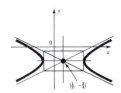

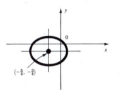

55. 10   **57.** Base 3, altitude 4

Chapter 5

Exercise Set 3, page 225

1. $x, R = -2$ **3.** $x^2 - x, R = -5$ **5.** $x^2 - 4x + 8, R = -9$ **7.** $x^3 + 3x^2 + 7x + 21, R = 60$

9. $x^4 - 2x^3 - x^2 + 2x - 4, R = -10$ **11.** $3x^3 + x^2 + x + 5, R = 10$ **13.** 3   **15.** -1 **17.** 16   **19.** 16

21. -125 **23.** 35   **25.** -11 **27.** -26 **29.** $6x^3 - 2x^2 - x + \frac{5}{2}, R = -\frac{27}{4}$ **31.** $\dfrac{x^2}{2} - \dfrac{7x}{4} + \dfrac{1}{8}, R = -\dfrac{23}{8}$

33. If $\dfrac{P(x)}{x + b/a} = Q(x) + \dfrac{R}{x + b/a}$, then

$\dfrac{P(x)}{ax + b} = \dfrac{1}{a}\left[\dfrac{P(x)}{x + b/a}\right] = \dfrac{1}{a}Q(x) + \dfrac{R}{ax + b}$.

35. Remainder 0 on division by $x - 4$ and $x + 2$.

37. Remainder 0 on division by $x - 4$. $P(x) = (x - 4)(x + 1)(x - 1)(x + 2)(x - 2)$ **39.** $-\frac{1}{2}$

41. Other zeros are 2 and $-\frac{3}{2}$.   **43. a.** 15.09   **b.** -962.4

Exercise Set 4, page 229

7. $x^3 - 3x^2 - 10x + 24 = 0$ **9.** $x^4 + 3x^3 - 6x^2 - 28x - 24 = 0$ **11.** $x^4 + 4x^3 + 8x^2 + 16x + 16$

13. $2x^3 - 6x - 4$

17. a. $x^2 + 2x + 2 = 0, b^2 - 4ac = 4 - 8 = -4 < 0$ **b.** $x^2 - x - 2 = 0, b^2 - 4ac = 1 + 8 = 9 > 0$

 c. $x^2 - 2x + 1 = 0, b^2 - 4ac = 4 - 4 = 0$ (There are many other correct answers.)

19. a. $(2x + 1)(x - 1)(x^2 + 2x + 2)$ **b.** $(2x + 1)(x - 1)(x + 1 - i)(x + 1 + i)$ **21.** The other root is $\frac{3}{2}$.

23. $x^4 - 14x^3 + 86x^2 - 254x + 325$ **25.** $(x - 2)(x + 2)(x + 1 + \sqrt{3}i)(x + 1 - \sqrt{3}i)(x - 1 + \sqrt{3}i)(x - 1 - \sqrt{3}i)$

Exercise Set 5, page 238

1. a. $\pm\frac{1}{3}, \pm\frac{2}{3}, \pm 1, \pm\frac{4}{3}, \pm 2, \pm 4$ **b.** $\pm\frac{1}{6}, \pm\frac{1}{3}, \pm\frac{1}{2}, \pm\frac{5}{6}, \pm 1, \pm\frac{3}{2}, \pm\frac{5}{3}, \pm\frac{5}{2}, \pm 3, \pm 5, \pm\frac{15}{2}, \pm 15$

3. Upper bound 3, lower bound -4; roots between 2 and 3, between -1 and -2, and between -2 and -3

5. a. 2 or 0 positive roots, 0 negative roots   **b.** 3 or 1 positive roots, 2 or 0 negative roots

7. $x = 1, -2$ (double root)   **9.** $x = -1, 3, 6$ **11.** $x = -2, 3, \frac{1}{3}$ **13.** $x = -\frac{1}{2}, 1 \pm i\sqrt{3}$ **15.** $x = -1, -2, \frac{2}{3}$

17. $x = 3, -3, \dfrac{-1 \pm i\sqrt{3}}{2}$ **19.** $x = -2, 5, 1 \pm i\sqrt{3}$ **21.** $x = \frac{4}{3}, -\frac{3}{2}, -4$ **23.** $x = -2$ (double root), $\dfrac{4 \pm i\sqrt{2}}{2}$

25. $(x + 2)(x^2 - 2x + 3)$ **27.** $(x - 2)(x + 1)^2$ **29.** $(x + 1)(x + 3)(2x - 5)$ **31.** $(x + 4)^2(3x - 2)$ **33.** 6 and 8

35. $7' \times 5' \times 6'$ **37.** $x = \dfrac{2}{3}, -\dfrac{3}{2}, \dfrac{1 \pm i}{2}$ **39.** $x = -\frac{3}{2}, \pm\sqrt{2}, \pm\sqrt{3}$ **41.** $x = 2$ (multiplicity 3), $\dfrac{-3 \pm 3i\sqrt{3}}{4}$

43. $x = 4, -3, \pm \dfrac{1}{\sqrt{2}}$ (double roots)

45. Rational root $x = -6$; irrational roots in each of the intervals $(-2, -1), (-1, 0),$ and $(1, 2)$

47. $V = \frac{1}{24}\pi(9 - 4x^2)(2x + 3)$; radius $= \sqrt{2}$, height $= 2$ **49.** $x = \dfrac{5}{3}, -\dfrac{3}{4}, \dfrac{-1 \pm i\sqrt{3}}{2}$

51. $x = -\dfrac{1}{2}$ (triple root), $\dfrac{1 \pm \sqrt{21}}{2}$

Exercise Set 6, page 245

1. 1.28   **3.** -1.80 **5.** 2.25   **7.** ± 1.73 **9.** $2, -1.51$ **11.** $-3, 2.55$ **13.** $3.24, -1.24$
15. $2.54, -0.69, -2.84$ **17.** $1.49, -2.36$ **19.** $5, -1, 1.39$ **21.** $n = 11$ **23.** $1.732, -1.732$

Exercise Set 7, page 250

1. $x = 1$ is the only real zero.

x	0	1	2	3	-1	-2
y	-1	0	7	26	-2	-9

3. Zeros are $1, 3, -3$.

x	0	1	2	3	4	-1	-2	-3	-4
y	9	0	-5	0	21	16	15	0	-35

5. Zeros are $0, 2, -2$.

x	0	1	2	3	-1	-2	-3
y	0	3	0	-15	-3	0	15

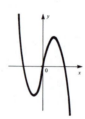

7. Zeros are $6, -2, \frac{5}{2}$

x	0	1	2	3	4	5	6	7	-1	-2	-3
y	60	45	16	-15	-36	-35	0	81	49	0	-99

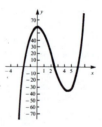

9. Zeros are -1 (double), $\frac{1}{2}, 4$.

x	0	1	2	3	4	5	-1	-2	-3
y	4	-12	-54	-80	0	324	0	30	196

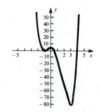

11. Zeros are 1 (double), −2.

x	0	1	2	3	−1	−2	−3
y	2	0	4	20	4	0	−16

13. Rational zero −3, irrational zeros between 3 and 4 and between −1 and −2; these irrational zeros are $1 + \sqrt{5}$ and $1 - \sqrt{5}$.

x	0	1	2	3	4	−1	−2	−3	−4
y	−12	−20	−20	−6	28	−2	4	0	−20

15. Zeros are 0 (double), 4, −2.

x	0	1	2	3	4	5	−1	−2	−3
y	0	9	32	45	0	−175	5	0	−63

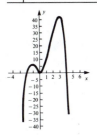

17. Zeros are 1, 3, −1, −4.

x	0	1	2	3	4	−1	−2	−3	−4	−5
y	12	0	−18	0	120	0	−30	−48	0	192

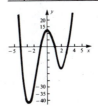

19. Real zeros are 2, $-\frac{3}{2}$.

x	0	1	2	3	−1	−2
y	−12	−5	0	45	−15	40

21. Zeros are 2 (double), −2, −4.

x	0	1	2	3	−1	−2	−3	−4	−5
y	32	15	0	35	27	0	−25	0	147

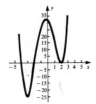

23. $(-6, \frac{4}{3}) \cup (5, \infty)$, $x \neq -4$ **25.** $[-1, 2] \cup [4, \infty)$ **27.** $(-\infty, -2) \cup (-2, 1) \cup (3, \infty)$
29. $(-\infty, -3) \cup (-1, 2) \cup (2, \infty)$ **31.** $(-\infty, -2] \cup [1, \infty)$
33. Zeros are 0, 1, 3, −2. **35.** Zeros are 2, 4, −1 (double), −3.

x	0	1	2	3	4	−1	−2	−3
y	0	0	−8	0	72	−8	0	72

x	0	1	2	3	4	5	−1	−2	−3	−4
y	24	48	0	−96	0	864	0	24	0	−432

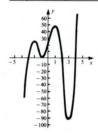

37. Zero 3.172; maximum point $(-0.260, -3.593)$; minimum point $(1.926, -14.036)$
39. Zeros 1.735, −1.339; minimum point $(0.798, -15.801)$

Exercise Set 8, page 259
1. **a.** $x = 1$ **b.** $x = -2$ **3.** **a.** $x = 0, x = 2$ **b.** $x = -3, x = \frac{1}{2}$
5. **a.** $x = 3, x = -3, x = -4$ **b.** $x = 0, x = 2, x = -2$ **7.** **a.** $y = 2$ **b.** $y = \frac{1}{2}$
9. **a.** $y = 0$ **b.** None **11.** **a.** $y = 0$ **b.** None **13.** $y = 2x + 3$ **15.** $y = 2x + 9$ **17.** $y = x - 4$
19. $y = 2x$ **21.** $y = 3x$
23. *Symmetry*: Not symmetric to x axis, y axis, or origin
Intercepts: x intercept −1, y intercept $-\frac{1}{4}$
Asymptotes: $x = \pm 2$, $y = 0$
Signs: $y > 0$ on $(-2, -1) \cup (2, \infty)$, $y < 0$ on $(-\infty, -2) \cup (-1, 2)$

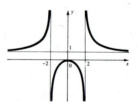

25. *Symmetry*: y axis
Intercepts: both intercepts 0
Asymptotes: $x = \pm 2$, $y = 1$
Signs: $y > 0$ on $(-\infty, -2) \cup (2, \infty)$, $y \leq 0$ on $(-2, 2)$

27. *Symmetry*: None
Intercepts: x intercept 2, y intercept $\frac{2}{3}$
Asymptotes: $x = 3$, $x = -1$, $y = 0$
Signs: $y > 0$ on $(-1, 2) \cup (3, \infty)$, $y < 0$ on $(-\infty, -1) \cup (2, 3)$

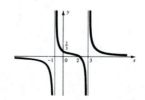

29. *Symmetry*: None
Intercepts: x intercepts ± 1, y intercept $-\frac{1}{2}$
Asymptotes: $x = -2$, $y = x - 2$
Signs: $y > 0$ on $(-2, -1) \cup (1, \infty)$, $y < 0$ on $(-\infty, -2) \cup (-1, 1)$

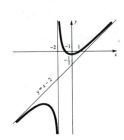

31. *Symmetry*: y axis
Intercepts: x intercepts ± 2, y intercept 1
Asymptotes: $y = -1$
Signs: $y > 0$ on $(-2, 2)$, $y < 0$ on $(-\infty, -2) \cup (2, \infty)$

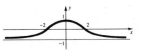

33. *Symmetry*: None
Intercepts: x intercepts 0, 2, y intercept 0
Asymptotes: $x = -2$, $y = x - 4$
Signs: $y > 0$ on $(-2, 0) \cup (2, \infty)$, $y < 0$ on $(-\infty, -2) \cup (0, 2)$

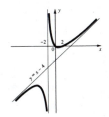

35. *Symmetry*: x axis, y axis and origin
Intercepts: x intercept 0, y intercept 0
Asymptotes: $x = \pm 1$, $y = 0$
Signs: x excluded from $(-1, 0)$ and $(0, 1)$
The origin is an isolated point.

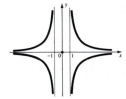

37. *Symmetry*: None
Intercepts: x intercept 2, no y intercept
Asymptotes: $x = 0$, $x = -4$, $y = 0$
Signs: $y > 0$ on $(-4, 0) \cup (2, \infty)$, $y < 0$ on $(-\infty, -4)$

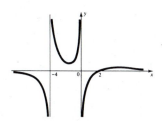

39. *Symmetry*: None
Intercepts: x intercept $-\frac{5}{2}$, y intercept -5
Asymptotes: $x = \pm 1$, $y = 0$
Signs: $y > 0$ on $(-\frac{5}{2}, -1) \cup (1, \infty)$, $y < 0$ on $(-\infty, -\frac{5}{2}) \cup (-1, 1)$

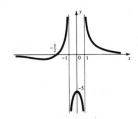

41. *Symmetry:* None
Intercepts: x intercepts 4, -2, y intercept -2
Asymptotes: $x = 2$, $x = -1$, $y = 0$
Signs: $y > 0$ on $(-2, -1) \cup (4, \infty)$, $y < 0$ on $(-\infty, -2) \cup (2, 4)$

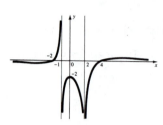

43. *Symmetry:* None
Intercepts: x intercepts 4, -1, y intercept 2
Asymptotes: $x = 1$, $x = -2$, $y = x - 3$
Signs: $y \geq 0$ on $(-2, 1) \cup (4, \infty)$, $y < 0$ on $(-\infty, -2) \cup (1, 4)$

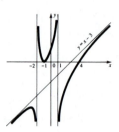

45. *Symmetry:* x axis

Intercepts: x intercepts ± 3, y intercepts: $\pm \dfrac{3}{2\sqrt{5}}$

Asymptotes: $x = 5$, $x = -4$, $y = 1$, $y = -1$
Signs: x excluded from $(-4, -3)$ and $(3, 5)$

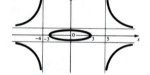

47. *Symmetry:* x axis
Intercepts: x intercept -1, no y intercept
Asymptotes: $x = 0$, $y = x$, $y = -x$
Signs: x excluded from $(-1, 0)$

49. *Symmetry:* None
Intercepts: x intercept 1, no y intercept
Asymptotes: $x = -1$, $y = 1$
Signs: x excluded from $(-1, 1)$, $y \geq 0$ on $(-\infty, -1) \cup [1, \infty)$

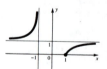

51. Removable discontinuity at $x = 2$; for $x \neq 2$, $y = \dfrac{2x}{x^2 + 1}$; x intercept: 0; y intercept: 0; vertical asymptotes: none;
horizontal asymptote: $y = 0$

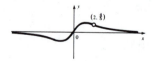

53. Removable discontinuities at $x = \pm 2$; for $x \neq \pm 2$, $y = \dfrac{x-2}{x+1}$; x intercept: none; y intercept: -2; vertical asymptote: $x = -1$, horizontal asymptote: $y = 1$

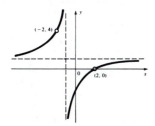

57. *Intercepts*: x intercepts 1.55, -0.80, y intercept -5
Asymptotes: $x = 3.35$, $x = 0.15$, $y = 2$

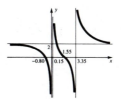

59. *Intercepts*: x intercept -1.71, y intercept -1.67
Asymptotes: $x = 2.30$, $x = -1.30$, $y = x + 1$

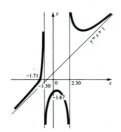

61. *Intercepts*: x intercepts ± 1.53, no y intercept
Asymptotes: $y = \pm 1.5$

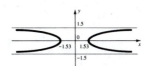

63. *Intercepts*: x intercept -1.44, no y intercept
Asymptotes: $x = \pm 2.24$

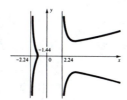

Exercise Set 9, page 265

1. $\dfrac{3}{x-1} + \dfrac{2}{x+3}$ **3.** $\dfrac{3}{x+4} - \dfrac{2}{x-3}$ **5.** $\dfrac{4}{x+2} - \dfrac{1}{x+3}$ **7.** $\dfrac{2}{x-4} - \dfrac{2}{x-5}$ **9.** $\dfrac{5}{x+3} + \dfrac{4}{x-2} - \dfrac{6}{x+4}$

11. $\dfrac{2}{x} + \dfrac{4}{x+1} - \dfrac{3}{x-1}$ **13.** $1 + \dfrac{2}{x+2} + \dfrac{1}{x-1}$ **15.** $x - 3 + \dfrac{1}{x+3} - \dfrac{2}{x+1}$ **17.** $\dfrac{1}{x-1} - \dfrac{1}{x+2} + \dfrac{2}{(x-1)^2}$

19. $\dfrac{4}{x-1} - \dfrac{2}{(x-1)^2} + \dfrac{3}{(x-1)^3}$ **21.** $\dfrac{1}{x} + \dfrac{2}{x^2} - \dfrac{2}{x+2}$ **23.** $\dfrac{2}{x-2} - \dfrac{2x-1}{x^2+4}$ **25.** $\dfrac{2}{x+1} - \dfrac{1}{(x+1)^2} - \dfrac{2x}{x^2+3}$

27. $\dfrac{1}{x-1} - \dfrac{3}{x-3} + \dfrac{2}{x+2}$ **29.** $\dfrac{5}{2(x-1)} - \dfrac{3x+2}{2(x^2+x+2)}$ **31.** $\dfrac{a}{ax+b} - \dfrac{c}{cx+d}$ **33.** $\dfrac{6}{x-1} - \dfrac{4}{(x+2)^2}$

35. $\dfrac{-2}{x} + \dfrac{2x+5}{x^2-3x+4} + \dfrac{4}{(x^2-3x+4)^2}$ **37.** $x-1+\dfrac{3}{2x-1} - \dfrac{2}{5x+3}$ **39.** $\dfrac{1}{x-2} + \dfrac{1}{(x-2)^2} - \dfrac{x}{x^2+4}$

41. $\dfrac{3x-4}{x^2+4} - \dfrac{2x-1}{x^2+1}$

Review Exercise Set, page 266

1. a. $2x^2 + 7x + 10, R = 25$ **b.** $x^2 - 3x + 6, R = -14$
3. a. $3x^3 - 14x^2 + 56x - 223, R = 887$ **b.** $2x^4 + 4x^3 - 2x^2 - 4x - 8, R = 0$; so $x - 2$ is a factor of the polynomial
5. $2x^4 + x^3 + 12x^2 + 9x - 54$
7. a. $\pm\frac{1}{3}, \pm\frac{2}{3}, \pm1, \pm\frac{4}{3}, \pm2, \pm\frac{8}{3}, \pm4, \pm8$; 3 or 1 positive roots, 2 or 0 negative roots
   b. $\pm\frac{1}{10}, \pm\frac{1}{5}, \pm\frac{3}{10}, \pm\frac{2}{5}, \pm\frac{1}{2}, \pm\frac{3}{5}, \pm\frac{4}{5}, \pm1, \pm\frac{6}{5}, \pm\frac{3}{2}, \pm2, \pm\frac{12}{5}, \pm3, \pm4, \pm6, \pm12$; 2 or 0 positive roots, 2 or 0 negative roots
9. $\{3, -2\}$; 3 is a double root   **11.** $\{\frac{1}{2}, -\frac{4}{3}, \pm2i\}$ **13.** $\{-1, -2, \frac{3}{2}\}$; $\frac{3}{2}$ is a double root
15. $(2x - 5)(x^2 - 2x + 4)$
17. Zeros are $0, -1, \frac{5}{2}$.

x	0	1	2	3	-1	-2
y	0	-6	-6	12	0	-18

19. Zeros are 3 (double), $-\frac{1}{2}, -4$.

x	0	1	2	3	4	-1	-2	-3	-4	-5
y	36	60	30	0	72	-48	-150	-180	0	576

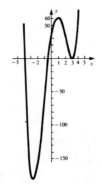

21. Zeros are $\frac{1}{2}, 3, -1, -3$.

x	0	1	2	3	4	-1	-2	-3	-4
y	9	-16	-45	0	245	0	-25	0	189

23. $(-\infty, -1) \cup (-1, 2)$ **25.** $[-\frac{3}{2}, \infty)$

27. *Symmetry*: Origin
Intercepts: Both 0
Asymptotes: $x = \pm 1$, $y = 0$
Signs: $y > 0$ on $(-\infty, -1) \cup (0, 1)$, $y < 0$ on $(-1, 0) \cup (1, \infty)$

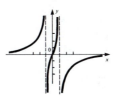

29. *Symmetry*: None
Intercepts: x intercepts 2, $-\frac{1}{2}$, no y intercept
Asymptotes: $x = 0$, $x = 5$, $y = 2$
Signs: $y > 0$ on $(-\infty, -\frac{1}{2}) \cup (0, 2) \cup (5, \infty)$, $y < 0$ on $(-\frac{1}{2}, 0) \cup (2, 5)$

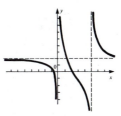

31. Removable discontinuity at $x = 3$; for $x \neq 3$, $y = \dfrac{x + 3}{x - 4}$

Symmetry: None
Intercepts: x intercept -3, y intercept $-\frac{3}{4}$
Asymptotes: $x = 4$, $y = 1$
Signs: $y > 0$ on $(-\infty, -3) \cup (4, \infty)$, $y < 0$ on $(-3, 3) \cup (3, 4)$

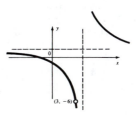

33. *Symmetry*: None
Intercepts: No x intercept, y intercept -2
Asymptotes: $x = 1$, $y = x + 1$
Signs: $y > 0$ on $(1, \infty)$, $y < 0$ on $(-\infty, 1)$

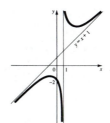

35. *Symmetry*: x axis, y axis, and origin
Intercepts: None
Asymptotes: $x = \pm 1$, $y = \pm 1$
Signs: x excluded from $(-1, 1)$

37. $1 - \dfrac{3}{x + 2} + \dfrac{5}{x - 2}$ **39.** $\dfrac{4}{x} - \dfrac{2x - 3}{x^2 - 2x + 4}$ **41.** $x + 2 - \dfrac{1}{x - 1} - \dfrac{2}{x + 1} + \dfrac{3}{x - 3}$ **43.** 2.25

45. $x = \dfrac{3}{2}, \dfrac{5}{3}, \dfrac{-1 \pm i\sqrt{3}}{2}$ **47.** Zeros are $x = \frac{1}{2}$, -3 (multiplicity 3), 4 (multiplicity 2).   **49.** $(-\infty, -3) \cup (-1, 2) \cup (2, \infty)$

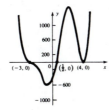

51. *Symmetry*: None
Intercepts: Both 0
Asymptotes: $x = 4$, $x = -2$, $y = 1$
Signs: $y > 0$ on $(-\infty, -2) \cup (4, \infty)$, $y \le 0$ on $(-2, 4)$

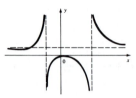

53. *Symmetry*: None
Intercepts: x intercept 3, y intercept $\frac{27}{16}$
Asymptotes: $x = \pm 2$, $y = 0$
Signs: $y > 0$ on $(-2, 2) \cup (2, 3)$, $y < 0$ on $(-\infty, -2) \cup (2, 3)$

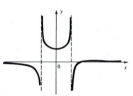

55. *Symmetry*: x axis
Intercepts: x intercepts 0, -2, y intercept 0
Asymptotes: $x = \pm 3$, $y = \pm 1$
Signs: x excluded from $(-3, -2)$ and $(0, 3)$

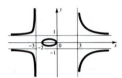

57. $x - \dfrac{3}{6x - 5} - \dfrac{x}{x^2 + 1}$ **59.** $-0.366, -1.399$

Chapter 6

Exercise Set 1, page 277

1.   **a.**   2.6651   **b.**   31.544   **c.**   0.090615   **d.**   7.3891   **e.**   0.36788

3.

x	0	1	2	3	-1	-2
y	1	4	16	64	0.25	0.0625

5.

x	0	1	2	-1	-2
y	1	2.7	7.39	0.37	0.14

7.

x	0	1	2	3	-1	-2	-3
y	1	$\frac{2}{3}$	$\frac{4}{9}$	$\frac{8}{27}$	$\frac{3}{2}$	$\frac{9}{4}$	$\frac{27}{8}$

9.

x	0	1	2	3	4	-1	-2
y	$\frac{1}{2}$	1	2	4	8	$\frac{1}{4}$	$\frac{1}{8}$

11.

x	0	1	2	3	-1	-2	-3
y	1.7	1	0.58	0.33	3	5.20	9

13.

x	0	1	2	3	−1	−2
y	0	0.63	0.86	0.95	−1.72	−6.39

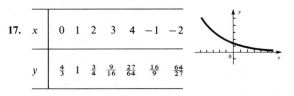

15.

x	0	±1	±2	±3
y	2	1	$\frac{1}{8}$	$\frac{1}{256}$

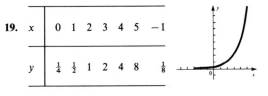

17.

x	0	1	2	3	4	−1	−2
y	$\frac{4}{3}$	1	$\frac{3}{4}$	$\frac{9}{16}$	$\frac{27}{64}$	$\frac{16}{9}$	$\frac{64}{27}$

19.

x	0	1	2	3	4	5	−1
y	$\frac{1}{4}$	$\frac{1}{2}$	1	2	4	8	$\frac{1}{8}$

21. a. $2,477.65   **b.** $2,488.42   **c.** $2,492.15

23. Total in part **a** is $577.81 and in part **b**, $578.02. So $7\frac{1}{4}\%$ continuously compounded interest is slightly better. For quarterly compounding at $7\frac{1}{2}\%$, the result is $580.11, which is better than $7\frac{1}{4}\%$ continuous interest.

25. $1,310.72

27.

x	0	1	2	3	4	−1	−2
y	0	0.37	0.27	0.15	0.07	−2.72	−14.78

29.

x	0	1	2	3
y	0	1.18	3.63	10.02

g is an odd function, so graph is symmetric to the origin.

31. $(-\infty, 0)$
33. Typical graphs are shown.

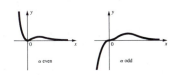

α even   α odd

Exercise Set 2, page 283

1. a. 3   **b.** 3   **3. a.** x **b.** x **5. a.** 4   **b.** −3   **7. a.** 3   **b.** −4   **9. a.** 2   **b.** −2
11. a. 5   **b.** −5   **13. a.** 3   **b.** $\frac{1}{9}$ **15. a.** 3   **b.** 0.001   **17. a.** 0   **b.** $\frac{27}{8}$
19. a. −1   **b.** 2   **21. a.** $\frac{1}{27}$ **b.** $\frac{2}{3}$ **23. a.** 10   **b.** −3   **25. a.** $\log_4 64 = 3$ **b.** $\log_3 \frac{1}{9} = -2$
27. a. $\log_2 256 = 8$ **b.** $\log_2 0.125 = -3$ **29. a.** $2^4 = 16$ **b.** $10^2 = 100$
31. a. $5^{-1} = 0.2$ **b.** $8^{2/3} = 4$

33. $x = 2^y$

x	1	2	4	8	$\frac{1}{2}$	$\frac{1}{4}$	$\frac{1}{8}$
y	0	1	2	3	-1	-2	-3

35. $x = 2^y - 1$

x	0	1	3	7	$-\frac{1}{2}$	$-\frac{3}{4}$	$-\frac{7}{8}$
y	0	1	2	3	-1	-2	-3

37. $x = \frac{1}{2}(4^y + 3)$

x	2	$\frac{7}{2}$	$\frac{19}{2}$	$\frac{13}{8}$	$\frac{49}{32}$	$\frac{5}{2}$	$\frac{7}{4}$
y	0	1	2	-1	-2	$\frac{1}{2}$	$-\frac{1}{2}$

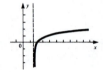

39. $x = \pm 2^{y/2}$

x	± 1	± 2	± 4	± 8	$\pm\frac{1}{2}$	$\pm\frac{1}{4}$	$\pm\frac{1}{8}$
y	0	2	4	6	-2	-4	-6

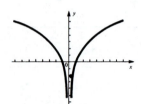

41. $x = -e^y$

x	-1	-2.7	-7.39	-20.09	-0.37	-0.14	-0.05
y	0	1	2	3	-1	-2	-3

43. Approximately 6.76 years   **45.** 10.54 years   **47.** 110 decibels
49. If $x < 1$, $x = 1 - 2^y$; if $x > 1$, $x = 1 + 2^y$

x	0	-1	-3	-7	0.5	0.75	2	3	5	1.5	1.25
y	0	1	2	3	-1	-2	0	1	2	-1	-2

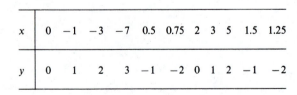

51. $x = \dfrac{e^y - e^{-y}}{2}$

x	0	1.17	3.62	10.01	-1.17	-3.62	-10.01
y	0	1	2	3	-1	-2	-3

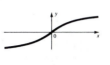

53. 1.3010   **55.** Typical graphs are shown.

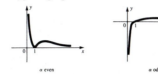

a even   a odd

Exercise Set 3, page 290

1. a. $\log(x - 3) + \log(x + 2)$ **b.** $\log(x - 5) - \log(x + 4)$ **3. a.** $-2\log(x + 1)$ **b.** $\frac{2}{3}\log(5x - 3)$
5. a. $\log(x + 2) + \log(x - 4)$ **b.** $\log(x - 5) - [\log(x + 3) + \log(x - 3)]$
7. a. $\frac{2}{3}\log(1 - x)$ **b.** $\pm\frac{1}{2}\log(3 - 4x)$ **9. a.** $\frac{1}{2}\log(x^2 + 4)$ **b.** $\frac{1}{2}[\log(x + 2) + \log(x - 2)]$
11. $3\log x - [\log(x - 4) + \log(x + 3)]$ **13.** $\log x + \frac{1}{2}\log(x - 2) - [\log(2x + 1) + \log(x - 1)]$
15. $\log(x + 3) + \log(x - 3) - [\log(x - 6) + \log(x + 1)]$ **17.** $\log 3 + \log x + \frac{3}{4}\log(x + 1) - \log(x^2 + 1)$
19. $\frac{1}{2}[3\log x - \log(4x - 5)]$ **21.** $2\log x + \log(x - 2) - \frac{1}{2}[\log(x + 1) + \log(x - 1)]$
23. $\frac{1}{3}[\log(x - 1) + \log(x - 4)] - \frac{1}{3}[\log(x + 2) + \log(x - 2)]$
25. $\log 7 + \log x + \frac{1}{3}\log(1 - x) - [\frac{2}{3}\log(x - 3) + 3\log(x + 2)]$ **27. a.** 125   **b.** 9   **29. a.** $-\frac{5}{3}$ **b.** $-\frac{3}{2}$
31. a. $\log \dfrac{x}{x + 1}$ **b.** $\log[x^3(x - 1)^2]$ **33. a.** $\log \dfrac{2\sqrt{x - 4}}{(x + 1)^3}$ **b.** $\log \dfrac{(3 - x)^2}{\sqrt{x^2 + x}}$
35. a. $\log \dfrac{(1 - x)^3}{\sqrt{x(x + 2)^3}}$ **b.** $\log \dfrac{3C\sqrt{x + 2}}{x^2}$ **37. a.** $\dfrac{\ln 4}{\ln 3}$ **b.** $\dfrac{\ln 25}{\ln 5 - \ln 2}$ **39. a.** $\dfrac{\ln 9}{\ln 3 + 2}$ **b.** $\dfrac{\ln 5 - \ln 4}{\ln 18}$
41. $x = 8$ **43.** $x = \frac{5}{2}$ **45.** No solution   **47.** $x = \frac{1}{3}$ or 1   **49.** $x = \frac{4}{3}$ **51.** $(\frac{3}{2}, 4)$ **53.** $(3, 4)$
55. $y = \dfrac{C(x + 2)}{x - 1}$ **57.** $y = Cx^2(2x - 3) - 1$ **59.** $y = \frac{1}{2}(Cx^2 + 1)$
61. a. $\dfrac{\ln 5}{\ln 3} \approx 1.4650$ **b.** $\dfrac{\ln 3}{\ln 4} \approx 0.79248$ **c.** $\dfrac{\ln 12}{\ln 8} \approx 1.1950$ **63. a.** 1.262   **b.** -1.322
65. a. -0.4260 **b.** 2.586   **67. a.** 0.5688   **b.** 0.2640   **69.** $x = 2$
71. $\log[(\sqrt{x} + \sqrt{x + 1})(\sqrt{x} - \sqrt{x - 1})] = \log(x - x + 1) = \log 1 = 0$ **73.** $(0, 1] \cup [e, \infty)$ **75.** $x = \ln(3 \pm 2\sqrt{2})$

Exercise Set 4, page 297

1. 13,500   **3.** 10 hours   **5.** 16 kg   **7.** 6.21 years   **9.** 3.42 hours; approximately 276   **11.** 3761.5 years
13. 33.7 grams; 6.35 years   **15.** 53.7°C   **17.** 32.1 min   **19.** 5   **21.** 9.6 psi   **23.** 4.68 billion
25. 40.2 m/sec; 49 m/sec   **27.** 19,035 years

Review Exercise Set, page 299

1. a.

x	0	1	2	3	-1	-2	-3
y	1	$\frac{3}{2}$	$\frac{9}{4}$	$\frac{27}{8}$	$\frac{2}{3}$	$\frac{4}{9}$	$\frac{8}{27}$

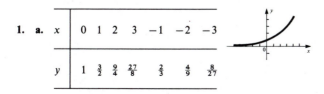

1. b.

x	0	1	2	3	-1	-2	-3
y	0	$\frac{1}{2}$	$\frac{3}{4}$	$\frac{7}{8}$	-1	-3	-7

c.

x	1	2	4	8	$\frac{1}{2}$	$\frac{1}{4}$	$\frac{1}{8}$
y	0	1	2	3	-1	-2	-3

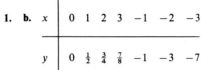

d.

x	0	-1.72	-6.39	-19.09	0.63	0.86	0.95
y	0	1	2	3	-1	-2	-3

3. a. $3^2 = 9$ **b.** $10^{-3} = 0.001$ **c.** $2^8 = 256$ **d.** $y = k^v$ **e.** $s = r^z$ **5. a.** 16 **b.** $\frac{1}{4}$ **c.** 4

7. a. 3 **b.** 2 **c.** -7 **9. a.** $\log(x - 1) - 3\log(x + 2)$ **b.** $\frac{1}{2}[\log(2x - 1) - (\log 3 + \log x)]$

11. a. $\log \dfrac{Cx\sqrt{2x - 3}}{8}$ **b.** $\log \dfrac{x^4}{(x - 1)\sqrt{x + 2}}$ **13. a.** $\log \dfrac{2x^3}{\sqrt[3]{(x - 1)(x + 2)}}$ **b.** $\log \dfrac{(2x + 3)^{3/2}}{2(x + 4)^3}$

15. a. $\frac{3}{2}$ **b.** -3 **c.** -4 **d.** -5 **17.** $k = -\dfrac{1}{t}\ln\left(1 - \dfrac{y}{C}\right)$ **19.** $x < -1$ **21.** $x = 3$ **23.** $x = \frac{2}{3}$

25. $(-4, -2) \cup (4, \infty)$ **27.** Approximately 15.87 pounds **29.** Approximately 529,750

31. a. Quarterly, \$6,734.28; continuously, \$6,749.29 **b.** Approximately 11.55 years **33.** Approximately 6.93%

35. a. 0.5670 **b.** 0.5900 **37.** $(1, \sqrt{2}]$ **39.** 33.22 seconds **41.** 0.19 **43.** 16.4 million

Cumulative Review Exercise Set II (Chapters 4–6), page 301

1. a. $(-2, -\frac{1}{2}) \cup (2, \infty)$ **b.** $[-\frac{10}{3}, 1] \cup [4, \infty)$

3. \$8,000 **5.** $(f \circ g)(x) = (x - 1)^2 + 1$, domain $= \{x: \ x > 1\}$; $(g \circ f)(x) = 2x$, domain $= R$

7. a. $y \geq 0$ on the interval $[-2, 6]$

b. $y > 0$ on $(-4, 2) \cup (2, \infty)$

9. $\{-1, 2\}$ **11.** $\dfrac{2}{x-1} - \dfrac{1}{(x-1)^2} - \dfrac{2x}{x^2 + 2x + 2}$ **13.** $[2, \infty)$ **15.** 34.02°C, 1 hour 19.6 minutes

17. $2\frac{1}{2}$ tons; $400

19. a.

x	0	±1.31	±2.53	±4.37	±7.32
y	0	1	2	3	4

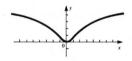

b.

x	0	±1	±2	±3	±4
y	0	0.63	0.86	0.95	0.98

c.

x	1	3	5	7	9	−1	−3	−5
y	1	2	4	8	16	$\frac{1}{2}$	$\frac{1}{4}$	$\frac{1}{8}$

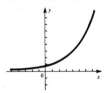

d.

x	0	$\frac{1}{2}$	$\frac{3}{4}$	$\frac{5}{4}$	$\frac{3}{2}$	2	3	5	−1	−3
y	1	0	−1	−1	0	1	2	3	2	3

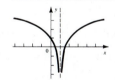

21. a. *Symmetry*: x axis
  Intercepts: x intercepts $\pm\frac{3}{2}$, no y intercepts
  Asymptotes: $x = 3$, $y = \pm 2$
  Signs: x excluded from $(-\frac{3}{2}, \frac{3}{2})$

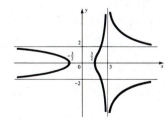

b. *Symmetry*: x axis
  Intercepts: x intercepts -2, 1, no y intercept
  Asymptotes: $x = 0$, $x = 4$, $y = 0$
  Signs: x excluded from $(-\infty, -2) \cup (1, 4]$

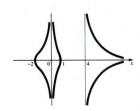

23. 450 units, maximum profit = $18,750   **25. a.** $(-\infty, 0)$ **b.** $(1, 2) \cup (3, \infty)$

Chapter 7

Exercise Set 2, page 310

1. a. $\pi/6$ **b.** $\pi/4$ **3. a.** $2\pi/3$ **b.** $3\pi/4$ **5. a.** $4\pi/3$ **b.** $3\pi/2$ **7. a.** $11\pi/6$ **b.** $\pi/12$
9. a. $10\pi/3$ **b.** $-4\pi/5$ **11. a.** $135°$ **b.** $300°$ **13. a.** $150°$ **b.** $-240°$
15. a. $100°$ **b.** $(360/\pi)° \approx 114.59°$ **17. a.** $(540/\pi)° \approx 171.89°$ **b.** $15°$
19. a. $-(720/\pi)° \approx -229.18°$ **b.** $330°$ **21. a.** $32.85°$ **b.** $102.58°$ **23. a.** $27.22°$ **b.** $-56.32°$
25. a. $-13°03'$ **b.** $153°25'$ **27. a.** $171°53'$ **b.** $406°21'$ **29. a.** 0.233 **b.** 5.065
31. a. $s = 6$ **b.** $r = \frac{4}{3}$ **33. a.** $r = 32/\pi \approx 10.19$ **b.** $\theta = \frac{2}{3}$ **35.** $s = 22\pi$ mi ≈ 69.12 mi
37. $s = 5\pi$ in. ≈ 15.7 in.   **39.** 60π rad/sec   **41.** $\frac{3}{4}$ rad, $(135/\pi)° \approx 43.0°$ **43.** 66,705 mph
45. $r_B = 7.78$ in., $r_C = 9.78$ in.

Exercise Set 3, page 314

1. $a = \frac{5}{2}, b = 5\sqrt{3}/2$ **3.** $a = 10\sqrt{3}, c = 20$ **5.** $a = 2\sqrt{3}, c = 4\sqrt{3}$ **7.** $a = 4, b = 4$ **9.** $a = 5\sqrt{3}/2, b = \frac{5}{2}$
11. $b = 3, c = 3\sqrt{2}$ **13.** $c = \sqrt{13}$ **15.** $b = 24$ **17.** $b = 48$ **19.** $c = 3$ **21.** $81°09'$ **23.** $23.82°$
25. $5\pi/12$ **29. a.** $10\sqrt{6}$ **b.** $\frac{75}{2}(\sqrt{3} + 4)$
31. Length $AB = 2a$, length $BD = a$ $\therefore \triangle CBD$ is isosceles. Since $B = 60°$ and $\angle BDC = \angle BCD$, it follows that these two
angles are both $60°$. So $\triangle CBD$ is equilateral. $\therefore CD = a$.

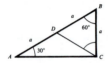

33. $\pi a^2/3$

Exercise Set 4, page 324

1. $\sin A = \dfrac{2}{\sqrt{13}}, \cos A = \dfrac{3}{\sqrt{13}}, \tan A = \dfrac{2}{3}, \cot A = \dfrac{3}{2}, \sec A = \dfrac{\sqrt{13}}{3}, \csc A = \dfrac{\sqrt{13}}{2}$

3. $\sin B = \frac{12}{13}, \cos B = \frac{5}{13}, \tan B = \frac{12}{5}, \cot B = \frac{5}{12}, \sec B = \frac{13}{5}, \csc B = \frac{13}{12}$

5. a. $\sin R = \dfrac{r}{s}, \cos R = \dfrac{t}{s}, \tan R = \dfrac{r}{t}, \cot R = \dfrac{t}{r}, \sec R = \dfrac{s}{t}, \csc R = \dfrac{s}{r}$

  b. $\sin T = \dfrac{t}{s}, \cos T = \dfrac{r}{s}, \tan T = \dfrac{t}{r}, \cot T = \dfrac{r}{t}, \sec T = \dfrac{s}{r}, \csc T = \dfrac{s}{t}$

7. $\sin \theta = \frac{12}{13}, \cos \theta = -\frac{5}{13}, \tan \theta = -\frac{12}{5}, \cot \theta = -\frac{5}{12}, \sec \theta = -\frac{13}{5}, \csc \theta = \frac{13}{12}$

9. $\sin \theta = \dfrac{2}{\sqrt{5}}, \cos \theta = -\dfrac{1}{\sqrt{5}}, \tan \theta = -2, \cot \theta = -\dfrac{1}{2}, \sec \theta = -\sqrt{5}, \csc \theta = \dfrac{\sqrt{5}}{2}$

11. $\sin \theta = -\frac{7}{25}, \cos \theta = \frac{24}{25}, \tan \theta = -\frac{7}{24}, \cot \theta = -\frac{24}{7}, \sec \theta = \frac{25}{24}, \csc \theta = -\frac{25}{7}$

13. $\sin 225° = -\dfrac{1}{\sqrt{2}}, \cos 225° = -\dfrac{1}{\sqrt{2}}, \tan 225° = 1, \cot 225° = 1, \sec 225° = -\sqrt{2}, \csc 225° = -\sqrt{2}$

15. $\sin \dfrac{4\pi}{3} = -\dfrac{\sqrt{3}}{2}, \cos \dfrac{4\pi}{3} = -\dfrac{1}{2}, \tan \dfrac{4\pi}{3} = \sqrt{3}, \cot \dfrac{4\pi}{3} = \dfrac{1}{\sqrt{3}}, \sec \dfrac{4\pi}{3} = -2, \csc \dfrac{4\pi}{3} = -\dfrac{2}{\sqrt{3}}$

17. $\sin \dfrac{3\pi}{4} = \dfrac{1}{\sqrt{2}}, \cos \dfrac{3\pi}{4} = -\dfrac{1}{\sqrt{2}}, \tan \dfrac{3\pi}{4} = -1, \cot \dfrac{3\pi}{4} = -1, \sec \dfrac{3\pi}{4} = -\sqrt{2}, \csc \dfrac{3\pi}{4} = \sqrt{2}$

19. $\sin\left(-\dfrac{\pi}{6}\right) = -\dfrac{1}{2}, \cos\left(-\dfrac{\pi}{6}\right) = \dfrac{\sqrt{3}}{2}, \tan\left(-\dfrac{\pi}{6}\right) = -\dfrac{1}{\sqrt{3}}, \cot\left(-\dfrac{\pi}{6}\right) = -\sqrt{3}, \sec\left(-\dfrac{\pi}{6}\right) = \dfrac{2}{\sqrt{3}}, \csc\left(-\dfrac{\pi}{6}\right) = -2$

21. $\sin 600° = -\dfrac{\sqrt{3}}{2}$, $\cos 600° = -\dfrac{1}{2}$, $\tan 600° = \sqrt{3}$, $\cot 600° = \dfrac{1}{\sqrt{3}}$, $\sec 600° = -2$, $\csc 600° = -\dfrac{2}{\sqrt{3}}$

23. $\sin 540° = 0$, $\cos 540° = -1$, $\tan 540° = 0$, $\cot 540°$ undefined, $\sec 540° = -1$, $\csc 540°$ undefined

25. **a.** $\dfrac{\sqrt{3}}{2}$ **b.** $\dfrac{1}{\sqrt{2}}$ **c.** $\dfrac{1}{\sqrt{3}}$ **d.** -2 **e.** -2 **27.** **a.** $-\sqrt{3}$ **b.** -1 **c.** 1 **d.** $-\dfrac{1}{2}$ **e.** 0

29. **a.** -2 **b.** $\sqrt{3}$ **c.** $-\dfrac{\sqrt{3}}{2}$ **d.** $-\dfrac{1}{\sqrt{2}}$ **e.** $-\dfrac{2}{\sqrt{3}}$

31. **a.** $-\sin 70°$ **b.** $-\cos 48°$ **c.** $-\tan 83°$ **d.** $\sec 80°$ **e.** $-\csc 50°$

33. **a.** $-\cos 56°20'$ **b.** $-\tan 47°45'$ **c.** $\csc 60°38'$ **d.** $\cot 83°24'$ **e.** $-\sin 21°47'$ **35.** 3

37. **a.** 0.9978 **b.** -0.3907 **c.** -2.125 **d.** 1.086

39. **a.** -5.759 **b.** 0.1411 **c.** 0.1287 **d.** -0.8391

41. $\cos \theta = -\dfrac{3}{5}$, $\tan \theta = \dfrac{4}{3}$, $\cot \theta = \dfrac{3}{4}$, $\sec \theta = -\dfrac{5}{3}$, $\csc \theta = -\dfrac{5}{4}$

43. 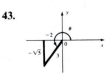 $\sin \theta = -\dfrac{\sqrt{5}}{3}$, $\cos \theta = -\dfrac{2}{3}$, $\tan \theta = \dfrac{\sqrt{5}}{2}$, $\cot \theta = \dfrac{2}{\sqrt{5}}$, $\csc \theta = -\dfrac{3}{\sqrt{5}}$

45. $\sin \theta = -\dfrac{1}{\sqrt{5}}$, $\cos \theta = \dfrac{-2}{\sqrt{5}}$, $\cot \theta = 2$, $\sec \theta = -\dfrac{\sqrt{5}}{2}$, $\csc \theta = -\sqrt{5}$

47. Solution 1. $\cos \theta = \sqrt{1-y^2}$, $\tan \theta = \dfrac{y}{\sqrt{1-y^2}}$, $\cot \theta = \dfrac{\sqrt{1-y^2}}{y}$, $\sec \theta = \dfrac{1}{\sqrt{1-y^2}}$, $\csc \theta = \dfrac{1}{y}$

47. Solution 2. 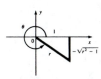 $\cos \theta = -\sqrt{1-y^2}$, $\tan \theta = -\dfrac{y}{\sqrt{1-y^2}}$, $\cot \theta = -\dfrac{\sqrt{1-y^2}}{y}$, $\sec \theta = -\dfrac{1}{\sqrt{1-y^2}}$, $\csc \theta = \dfrac{1}{y}$

49. Solution 1. $\sin \theta = \dfrac{\sqrt{r^2-1}}{r}$, $\cos \theta = \dfrac{1}{r}$, $\tan \theta = \sqrt{r^2-1}$, $\cot \theta = \dfrac{1}{\sqrt{r^2-1}}$, $\csc \theta = \dfrac{r}{\sqrt{r^2-1}}$

Solution 2. $\sin \theta = -\dfrac{\sqrt{r^2-1}}{r}$, $\cos \theta = \dfrac{1}{r}$, $\tan \theta = -\sqrt{r^2-1}$, $\cot \theta = -\dfrac{1}{\sqrt{r^2-1}}$, $\csc \theta = -\dfrac{r}{\sqrt{r^2-1}}$

51. For n even, $n\pi$ is coterminal with 0, so $\cos n\pi = 1 = (-1)^n$; for n odd, $n\pi$ is coterminal with π, so $\cos n\pi = -1 = (-1)^n$.

53. $\cos\dfrac{\pi}{3}(1 + 3k) = \cos\left(\dfrac{\pi}{3} + k\pi\right)$. If k is even, $\dfrac{\pi}{3} + k\pi$ is coterminal with $\dfrac{\pi}{3}$, so $\cos\left(\dfrac{\pi}{3} + k\pi\right) = \dfrac{1}{2} = \dfrac{(-1)^k}{2}$; if k is odd, $\dfrac{\pi}{3} + k\pi$

is coterminal with $\dfrac{4\pi}{3}$, so $\cos\left(\dfrac{\pi}{3} + k\pi\right) = -\dfrac{1}{2} = \dfrac{(-1)^k}{2}$.

55. $(\overline{AB})^2 = 65, (\overline{BC})^2 = 13, (\overline{AC})^2 = 52$, so $(\overline{AB})^2 = (\overline{BC})^2 + (\overline{AB})^2$. $\sin A = \dfrac{1}{\sqrt{5}}$, $\cos A = \dfrac{2}{\sqrt{5}}$, $\tan A = \dfrac{1}{2}$, $\cot A = 2$,

$\sec A = \dfrac{\sqrt{5}}{2}$, $\csc A = \sqrt{5}$.

Exercise Set 5, page 329

1. $\sin\theta = -2\sqrt{2}/3$, $\cos\theta = -\tfrac{1}{3}$, $\csc\theta = -3/2\sqrt{2}$, $\sec\theta = -3$, $\tan\theta = 2\sqrt{2}$, $\cot\theta = 1/2\sqrt{2}$

3. $\sin\theta = -\tfrac{12}{13}$, $\cos\theta = \tfrac{5}{13}$, $\csc\theta = -\tfrac{13}{12}$, $\sec\theta = \tfrac{13}{5}$, $\tan\theta = -\tfrac{12}{5}$, $\cot\theta = -\tfrac{5}{12}$

5. $\sin\theta = -\sqrt{15}/4$, $\cos\theta = \tfrac{1}{4}$, $\csc\theta = -4/\sqrt{15}$, $\sec\theta = 4$, $\tan\theta = -\sqrt{15}$, $\cot\theta = -1/\sqrt{15}$

7. $\theta = 4\pi/3$; $\sin\theta = -\sqrt{3}/2$, $\cos\theta = -\tfrac{1}{2}$, $\csc\theta = -2/\sqrt{3}$, $\sec\theta = -2$, $\tan\theta = \sqrt{3}$, $\cot\theta = 1/\sqrt{3}$

9. $\sin\theta = -2\sqrt{2}/3$, $\cos\theta = \tfrac{1}{3}$, $\csc\theta = -3/2\sqrt{2}$, $\sec\theta = 3$, $\tan\theta = -2\sqrt{2}$, $\cot\theta = -1/2\sqrt{2}$

11. a. $(0, 1)$ **b.** $(-1, 0)$ **c.** $(0, -1)$ **d.** $(-\sqrt{3}/2, \tfrac{1}{2})$ **e.** $(-\tfrac{1}{2}, -\sqrt{3}/2)$

13. a.   **b.**   **c.**   **d.**

15. a. $\pi/6$ **b.** $\pi/4$ **c.** $\pi/3$ **d.** $\pi/2$ **e.** 0

17. a. $\sin x = 0.5300, \cos x = -0.8480, \tan x = -0.6250$

    b. $\sin x = 0.3091, \cos x = 0.9510, \tan x = 0.3250$

    c. $\sin x = 0.9270, \cos x = 0.3750, \tan x = 2.472$

    d. $\sin x = -0.6672, \cos x = -0.7449, \tan x = 0.8957$

19. a. $\dfrac{\pi}{2}$ **b.** 0 **c.** $\dfrac{\pi}{4}, \dfrac{5\pi}{4}$ **d.** $0, \pi$ **e.** $\dfrac{\pi}{2}, \dfrac{3\pi}{2}$ **f.** $0, \pi$ **g.** $\dfrac{3\pi}{2}$ **h.** π **i.** $\dfrac{3\pi}{4}, \dfrac{7\pi}{4}$

21. Tan θ and sec θ are defined for all values of θ for which the x coordinate of $P(\theta)$ is not 0, that is, except for angles of $\theta = \dfrac{\pi}{2}$ or $\dfrac{3\pi}{2}$ radians, or angles coterminal with either of these, namely, $\dfrac{\pi}{2} + n\pi$, or $\left(\dfrac{2n+1}{2}\right)\pi$, for $n = 0, \pm 1, \pm 2, \ldots$.

23. By Definition 3, $\sec\theta = \dfrac{1}{x}(x \neq 0)$ and $\csc\theta = \dfrac{1}{y}(y \neq 0)$, and since $|x| \leq 1$ and $|y| \leq 1$, it follows that $|\sec\theta| \geq 1$ and $|\csc\theta| \geq 1$.

25. If f denotes any one of the trigonometric functions, $f(x) = f(x + 2\pi)$ for all x in the domain of f, so f is not 1–1.

27. From Problem 26 if $P(\theta) = (x, y)$, then $P(\theta + 2\pi) = (x, y)$ and $P(\theta + \pi) = (-x, -y)$. Thus $\sin(\theta + 2\pi) = y = \sin\theta$ and $\cos(\theta + 2\pi) = x = \cos\theta$. Similarly, $\sec(\theta + 2\pi) = \dfrac{1}{x} = \sec\theta$ and $\csc(\theta + 2\pi) = \dfrac{1}{y} = \csc\theta$. Also $\tan(\theta + \pi) = \dfrac{-y}{-x} = \dfrac{y}{x} = \tan\theta$, and $\cot(\theta + \pi) = \dfrac{-x}{-y} = \dfrac{x}{y} = \cot\theta$.

Exercise Set 6, page 337

1. Amplitude $= 1$, period $= \pi$

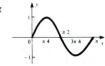

3. Amplitude $= 2$, period $= 2\pi$

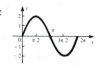

5. Amplitude = 3, period = 4

7. Amplitude = 2, period = $2\pi/3$

9. Amplitude = 1, period = 4π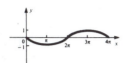

11. Amplitude = 1, period = π

13. Amplitude = 2, period = $2\pi/3$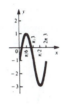

15. Period = 1

17. Amplitude = 1, period = 2π, phase shift = $\pi/3$

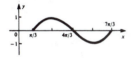

19. Amplitude = 1, period = π, phase shift = $-\pi/4$

21. Amplitude = 2, period = $2\pi/3$, phase shift = $-\frac{2}{3}$

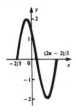

23. Amplitude = 2, period = 2, phase shift = $\frac{1}{4}$

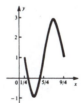

25. Amplitude = 2, period = $2\pi/3$, phase shift = $\pi/6$

27. Period $= 6\pi$

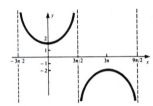

29. a.

Even function, asymptotic
to $y = 0$. "Amplitude" gets
smaller and smaller as $x \to \pm\infty$

b.

Odd function, asymptote $y = 0$,
"period" gets smaller and smaller
as $x \to 0$

c.

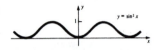

Even function, always
positive, period $= \pi$

d.

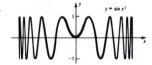

Even function, "period"
gets smaller and smaller
as $x \to \pm\infty$

e.

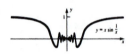

Even function, both "amplitude" and "period" get
smaller and smaller as $x \to 0$, asymptote $y = 1$

Exercise Set 7, page 345

1. a. $-\dfrac{\pi}{3}$ **b.** $\dfrac{5\pi}{6}$ **3. a.** $\dfrac{3\pi}{4}$ **b.** $\dfrac{\pi}{2}$ **5. a.** 0 **b.** $-\dfrac{\pi}{6}$ **7. a.** $\dfrac{\pi}{2}$ **b.** $\dfrac{7\pi}{6}$ **9.** $\dfrac{\pi}{2}$ **11.** 2π

13. $-\sqrt{15}$ **15.** $\frac{5}{4}$ **17.** $\dfrac{4}{\sqrt{15}}$ **19.** $\frac{3}{5}$ **21.** $\dfrac{3}{\sqrt{5}}$ **23.** $-\frac{7}{24}$ **25.** $\dfrac{\pi}{4}$ **27.** $-\frac{1}{2}$

29. a. -0.72973 **b.** -1.1071 **31. a.** 0.89115 **b.** 1.9705 **33. a.** 4.3726 **b.** 2.4981

35. Let $\alpha = \sin^{-1} x$ and $\beta = \sin^{-1}\sqrt{1 - x^2}$. From the figure $\alpha + \beta = \dfrac{\pi}{2}$.

37. a. Let $\theta = \sin^{-1} x$.
$\sin(\pi - \theta) = \sin\theta = x$
(See figure, for $x > 0$.)

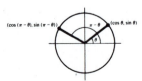

b. Let $\theta = \cos^{-1} x$
$\cos(\pi - \theta) = -\cos\theta = -x$
(See figure, for $x > 0$.)

39.

$y = \frac{1}{2}\sin^{-1} 2(x + 1)$

41.

$y = 2\cos^{-1}(x - 1) + \frac{\pi}{2}$

Review Exercise Set, page 347

1. a. $\dfrac{4\pi}{3}$ **b.** $\dfrac{7\pi}{4}$ **c.** $\dfrac{\pi}{12}$ **d.** 3π **e.** $\dfrac{8\pi}{9}$ **3. a.** $4\pi \approx 12.57$ **b.** $\dfrac{240°}{\pi} \approx 76.4°$

5. a. $\sin A = \dfrac{1}{2}, \cos A = \dfrac{\sqrt{3}}{2}, \tan A = \dfrac{1}{\sqrt{3}}, \cot A = \sqrt{3}, \sec A = \dfrac{2}{\sqrt{3}}, \csc A = 2$

  b. $\sin B = \frac{15}{17}, \cos B = \frac{8}{17}, \tan B = \frac{15}{8}, \cot B = \frac{8}{15}, \sec B = \frac{17}{8}, \csc B = \frac{17}{15}$

7. a. $\frac{1}{2}$ **b.** $-\frac{1}{2}$ **c.** $\dfrac{1}{\sqrt{3}}$ **d.** 0 **e.** $-\sqrt{2}$ **f.** 2 **g.** 1 **h.** -1 **i.** -2 **j.** $-\dfrac{\sqrt{3}}{2}$

9. a. $\sin \theta = \dfrac{\sqrt{5}}{3}, \cos \theta = -\dfrac{2}{3}, \tan \theta = -\dfrac{\sqrt{5}}{2}, \cot \theta = -\dfrac{2}{\sqrt{5}}, \sec \theta = -\dfrac{3}{2}, \csc \theta = \dfrac{3}{\sqrt{5}}$

  b. $\sin \theta = -\frac{5}{13}, \cos \theta = -\frac{12}{13}, \tan \theta = \frac{5}{12}, \cot \theta = \frac{12}{5}, \sec \theta = -\frac{13}{12}, \csc \theta = -\frac{13}{5}$

11.

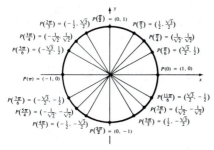

13. a. $\dfrac{2\pi}{3}$ **b.** $-\dfrac{\pi}{3}$ **c.** $-\dfrac{\pi}{3}$ **d.** π **e.** $-\dfrac{\pi}{4}$ **15. a.** $\sqrt{5}$ **b.** $\dfrac{4}{\sqrt{15}}$ **17. a.** $\dfrac{11\pi}{6}$ **b.** -2π

19. a. Period $= 4$ **b.** Amplitude $= 2$, period $= 4$

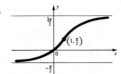

21.

Amplitude $= 2$
Period $= \pi$
Phase shift $= \frac{3}{2}$

23.

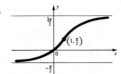

25. $\sin \theta = \sqrt{1 - x^2}, \ -1 \le x \le 1$

$\cos \theta = x, \ -1 \le x \le 1$

$\tan \theta = \dfrac{\sqrt{1 - x^2}}{x}, \ -1 \le x \le 1, \ x \ne 0$

$\cot \theta = \dfrac{x}{\sqrt{1 - x^2}}, \ -1 < x < 1$

$\sec \theta = \dfrac{1}{x}, \ -1 \le x \le 1, \ x \ne 0$

$\csc \theta = \dfrac{1}{\sqrt{1 - x^2}}, \ -1 < x < 1$

27. **a.** $-1 \le x \le 1$ **b.** $-\dfrac{\pi}{2} \le x \le \dfrac{\pi}{2}$ **c.** $-1 \le x \le 1$ **d.** $0 \le x \le \pi$ **e.** $-\infty < x < \infty$ **f.** $-\dfrac{\pi}{2} < x < \dfrac{\pi}{2}$

29. $\sin \theta = -\sqrt{1 - x^2}$

$\cos \theta = |x|$

$\tan \theta = -\dfrac{\sqrt{1 - x^2}}{|x|}, \ x \ne 0$

$\cot \theta = -\dfrac{|x|}{\sqrt{1 - x^2}}, \ |x| \ne 1$

$\sec \theta = \dfrac{1}{|x|}, \ x \ne 0$

$\csc \theta = -\dfrac{1}{\sqrt{1 - x^2}}, \ |x| \ne 1$

Chapter 8

Exercise Set 1, page 355

1. $\cos \theta = -\frac{4}{5}, \csc \theta = -\frac{5}{3}, \sec \theta = -\frac{5}{4}, \tan \theta = \frac{3}{4}, \cot \theta = \frac{4}{3}$

3. $\sin \theta = \frac{4}{5}, \cos \theta = -\frac{3}{5}, \csc \theta = \frac{5}{4}, \sec \theta = -\frac{5}{3}, \cot \theta = -\frac{3}{4}$

5. $\sin \theta = -1/\sqrt{5}, \csc \theta = -\sqrt{5}, \sec \theta = \sqrt{5}/2, \tan \theta = -\frac{1}{2}, \cot \theta = -2$

7. $\cos \theta = -2\sqrt{2}/3, \csc \theta = 3, \sec \theta = -3/2\sqrt{2}, \tan \theta = -1/2\sqrt{2}, \cot \theta = -2\sqrt{2}$

Exercise Set 2, page 362

1. $(\sqrt{2} + \sqrt{6})/4, (\sqrt{2} - \sqrt{6})/4$ **3.** **a.** $-(\sqrt{6} + \sqrt{2})/4$ **b.** $-(\sqrt{6} + \sqrt{2})/4$

5. **a.** $(\sqrt{2} - \sqrt{6})/4$ **b.** $-(\sqrt{6} + \sqrt{2})/4$ **7.** **a.** $(\sqrt{6} + \sqrt{2})/4$ **b.** $(\sqrt{6} + \sqrt{2})/4$ **9.** $-\frac{63}{65}$ **11.** $\frac{63}{65}$

13. **a.** $-\frac{56}{65}$ **b.** $-\frac{63}{65}$ **15.** **a.** $(2\sqrt{10} - 2)/9$ **b.** $(-\sqrt{5} + 4\sqrt{2})/9$ **17.** **a.** $31/17\sqrt{5}$ **b.** $38/17\sqrt{5}$

19. **a.** $-\frac{65}{33}$ **b.** $-\frac{65}{16}$

21. **a.** $\sin(\pi/2 + \theta) = \sin(\pi/2)\cos \theta + \cos(\pi/2)\sin \theta = \cos \theta$ **b.** $\cos(\pi/2 + \theta) = \cos(\pi/2)\cos \theta - \sin(\pi/2)\sin \theta = -\sin \theta$

23. **a.** $\sin(\pi - \theta) = \sin \pi \cos \theta - \cos \pi \sin \theta = \sin \theta$ **b.** $\cos(\pi - \theta) = \cos \pi \cos \theta + \sin \pi \sin \theta = -\cos \theta$ **35.** $0, \pi$

37. **a.** $\sin \alpha \cos \beta \cos \gamma + \cos \alpha \sin \beta \cos \gamma + \cos \alpha \cos \beta \sin \gamma - \sin \alpha \sin \beta \sin \gamma$

   b. $\cos \alpha \cos \beta \cos \gamma - \sin \alpha \sin \beta \cos \gamma - \sin \alpha \cos \beta \sin \gamma - \cos \alpha \sin \beta \sin \gamma$

Exercise Set 3, page 368

1. $\sin 2\theta = -\frac{24}{25}, \cos 2\theta = -\frac{7}{25}$ **3.** $\sin 2\theta = \frac{120}{169}, \cos 2\theta = \frac{119}{169}$ **5.** $\sin 2\theta = -\frac{4}{5}, \cos 2\theta = \frac{3}{5}$

7. $\sin 2\theta = 2x\sqrt{1 - x^2}, \cos 2\theta = 1 - 2x^2$ **9.** $\sin \theta = \sqrt{3}/3, \cos \theta = -\sqrt{6}/3$ **11.** $\sin \theta = -\frac{3}{5}, \cos \theta = \frac{4}{5}$

13. **a.** $\sqrt{2 - \sqrt{2}}/2$ **b.** $\sqrt{2 + \sqrt{3}}/2$ **c.** $\sqrt{2 + \sqrt{3}}/2$ **d.** $\sqrt{2 - \sqrt{2}}/2$

15. $\sin(\alpha/2) = 3/\sqrt{13}, \cos(\alpha/2) = -2/\sqrt{13}$ **17.** $\sin(\alpha/2) = -\sqrt{3}/3, \cos(\alpha/2) = \sqrt{6}/3$ **19.** $\frac{7}{9}$ **21.** $-\dfrac{12\sqrt{10}}{49}$

23. $\frac{3}{5}$ **25.** $\dfrac{2}{\sqrt{5}}$ **27.** $\dfrac{1}{5\sqrt{2}}$ **29.** $\dfrac{-4\sqrt{19}+3}{10\sqrt{5}}$ **31.** $P(\theta) = (\frac{3}{5}, \frac{4}{5})$

33. **a.** $-\sin\theta$ **b.** $-\sin\theta$ **c.** $\cos\theta$ **d.** $-\cos\theta$ **e.** $\cos\theta$
51. $\sin 2\theta = 2x/(1+x^2)$, $\cos 2\theta = (1-x^2)/(1+x^2)$ **53.** $\sin 3\theta = 3\sin\theta - 4\sin^3\theta$, $\cos 3\theta = 4\cos^3\theta - 3\cos\theta$

Exercise Set 4, page 372
1. **a.** $2+\sqrt{3}$ **b.** $2-\sqrt{3}$ **3.** $\frac{24}{7}$ **5.** $(9+5\sqrt{2})/2$
7. $\tan 2\alpha = \frac{120}{119}$, $\tan 2\beta = -\frac{240}{161}$, $\tan(\alpha/2) = -\frac{2}{3}$, $\tan(\beta/2) = 4$ **9.** $-\frac{7}{9}$ **11.** $-\frac{4}{3}$ **13.** $-\frac{2}{3}$ **15.** $-\frac{2}{11}$
17. $\tan 2\theta = -\frac{24}{7}$, $\tan(\theta/2) = 3$ **19.** **a.** $\sqrt{6}/2$ **b.** $-\sqrt{6}/2$ **21.** **a.** $\sqrt{2}/2$ **b.** $-\sqrt{2}/2$
23. **a.** $2\sin 4x\cos x$ **b.** $\frac{1}{2}(\sin 8x + \sin 2x)$

Exercise Set 5, page 381
1. $\{\pi/6, 11\pi/6\}$ **3.** $\{\pi/3, 5\pi/3\}$ **5.** $\{\pi/3, \pi/2, 3\pi/2, 5\pi/3\}$ **7.** $\{\pi/3, \pi, 5\pi/3\}$ **9.** $\{\pi/2, 3\pi/2\}$
11. $\{\pi/6, \pi/2, 5\pi/6, 3\pi/2\}$ **13.** $\{\pi/3, 2\pi/3, 4\pi/3, 5\pi/3\}$ **15.** No solution   **17.** $\{\pi/2, 3\pi/2\}$
19. $\{0, 2\pi/9, \pi/3, 4\pi/9, 2\pi/3, 8\pi/9, \pi, 10\pi/9, 4\pi/3, 14\pi/9, 5\pi/3, 16\pi/9\}$ **21.** $\{0, \pi/2\}$
23. $\{\pi/24, 11\pi/24, 13\pi/24, 23\pi/24, 25\pi/24, 35\pi/24, 37\pi/24, 47\pi/24\}$ **25.** No solution
27. $\{2\pi/9, \pi/4, 5\pi/9, 3\pi/4, 8\pi/9, 11\pi/9, 5\pi/4, 14\pi/9, 7\pi/4, 17\pi/9\}$ **29.** No solution
31. $\{0.8481, 2.2935, 3.8713, 5.5535\}$ **33.** $\{0.4636, 1.1071, 3.6052, 4.2487\}$ **35.** $\{0.2846, 2.8570\}$
37. $\{\pi/3, 2\pi/3, 3\pi/4, 4\pi/3, 5\pi/3, 7\pi/4\}$ **39.** $\{\pi/6, 5\pi/6, 3\pi/2\}$ **41.** $\{1.244, 4.028\}$ **43.** $\{1.587, 4.978\}$

Review Exercise Set, page 382
1. **a.** $-2-\sqrt{3}$ **b.** $\dfrac{\sqrt{2}-\sqrt{6}}{4}$ **c.** $\dfrac{\sqrt{2}+\sqrt{2}}{2}$ **d.** $-\dfrac{\sqrt{2}+\sqrt{6}}{4}$ **3.** **a.** $\frac{7}{25}$ **b.** $-\frac{24}{7}$ **c.** $-\dfrac{3}{\sqrt{10}}$

5. **a.** $\frac{63}{65}$ **b.** $-\frac{56}{65}$ **c.** $\frac{63}{16}$ **7.** $\sin\theta = \frac{4}{5}$, $\cos\theta = \frac{3}{5}$ **9.** **a.** $\frac{7}{25}$ **b.** $-\frac{120}{169}$ **25.** $\left\{0, \dfrac{\pi}{6}, \dfrac{5\pi}{6}, \pi\right\}$

27. $\left\{\dfrac{2\pi}{9}, \dfrac{4\pi}{9}, \dfrac{8\pi}{9}, \dfrac{10\pi}{9}, \dfrac{14\pi}{9}, \dfrac{16\pi}{9}\right\}$ **29.** **a.** $\frac{140}{221}$ **b.** $-\dfrac{\pi}{4}$ **37.** $\left\{\dfrac{\pi}{3}, \dfrac{5\pi}{3}\right\}$ **39.** $\left\{0, \dfrac{3\pi}{2}\right\}$

43.
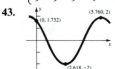

Chapter 9

Exercise Set 1, page 392
1. $B = 50.8°$, $b = 20.7$, $c = 26.7$ **3.** $A = 35.4°$, $a = 11.5$, $b = 16.1$ **5.** $B = 63°30'$, $a = 14.6$, $b = 29.3$
7. $A = 18.8°$, $b = 144$, $c = 153$ **9.** $A = 35.25°$, $B = 54.75°$, $c = 22.92$ **11.** $B = 47°47'$, $a = 184.8$, $c = 275.0$
13. $A = 77°26'$, $a = 13.92$, $c = 14.26$ **15.** $A = 55.11°$, $B = 34.89°$, $a = 31.30$ **17.** Approximately 359 ft
19. Approximately 137 ft   **21.** 45.1 ft, 49.9 ft   **23.** 19.2 mi, 51.3°   **25.** 840 meters from B, 926 meters from A
27. 6.142 mi and 8.331 mi   **29.** 249°10', 219 mph   **31.** 54.3 ft   **33.** 51.5 mi   **35.** 13.8 km

Exercise Set 2, page 400
1. $C = 115.2°$, $b = 2.07$, $c = 5.06$
3. Solution 1. $A = 76°41'$, $B = 67°39'$, $a = 145$ Solution 2. $A = 31°59'$, $B = 112°21'$, $a = 79.0$
5. $B = 14.4°$, $C = 138.3°$, $c = 30.3$ **7.** $A = 108.8°$, $a = 18.5$, $b = 14.1$ **9.** $C = 80°05'$, $a = 10.03$, $c = 16.59$
11. $A = 112.97°$, $C = 24.77°$, $a = 25.10$ **13.** $A = 35°28'$, $C = 39°05'$, $a = 66.52$
15. Solution 1. $A = 36°50'$, $B = 119°36'$, $b = 7.790$ Solution 2. $A = 143°10'$, $B = 13°16'$, $b = 2.056$ **17.** 108.8 in.
19. $x \approx 1{,}029$ ft   **21.** S 66°20' W, 37 min
23. Heading going $= 348°26'$; heading returning $= 191°34'$; total time $= 3$ hr 30 min

Exercise Set 3, page 404

1. $A = 45.6°, C = 74.4°, b = 18.3$ 3. $A = 51.1°, B = 35.1°, c = 6.56$ 5. $A = 53.2°, B = 33.9°, C = 92.9°$
7. $B = 41.9°, C = 27.8°, a = 346$ 9. $A = 24.4°, B = 44.1°, C = 111.5°$ 11. $A = 61°02', C = 32°23', b = 37.71$
13. $B = 42.89°, C = 11.71°, a = 4.143$ 15. $A = 56.59°, B = 90.00°, C = 33.41°$ 17. 13.56 cm   19. 58.27 ft
21. 1.04 and 3.86   23. 477.7 miles   25. 22.8 ft   27. 34.4 ft

Exercise Set 4, page 408

1. 90   3. 679   5. 1.29   7. 177   9. 168   11. 1.21   13. 73.5   15. 82.77   17. 233
19. 3.640 and 11.00

Exercise Set 5, page 416

1. $\langle 2, 5 \rangle$ 3. $\langle -3, 1 \rangle$ 5. $\langle -6, 1 \rangle$ 7. $\langle 1, -5 \rangle$

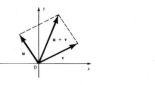   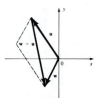

9. $\langle 0, 7 \rangle$ 11. $\langle 11, 2 \rangle$ 13. $\langle -4, -2 \rangle$ 15. $\langle -3, -4 \rangle$

17.   19.   21. 5   23. 17   25. $2\sqrt{10}$ 27. -6

29. -11 31. 126.9°   33. 143.1°   35. 164.7 lb; angle from horizontal = 24.4°   37. 210°, 138.6 mph
39. 57°51′ from horizontal, 82.7 ft/sec   41. $|F_1| = 156.3$ lb, $\theta = 68°00'$ 43. 54.2, 30.3°
45. a. $\theta = \pi/2$ if and only if $\cos \theta = 0$, which is true if and only if the dot product is 0.
    b. $\vec{AB} = \langle 10, -6 \rangle, \vec{CD} = \langle 3, 5 \rangle, \vec{AB} \cdot \vec{CD} = 0$
47. 84.5 lb, 61.7°   49. $\langle a, b \rangle = \langle a, 0 \rangle + \langle 0, b \rangle = a\langle 1, 0 \rangle + b\langle 0, 1 \rangle = ai + bj$ 51. 458 mph, 208.8°

Review Exercise Set, page 417

1. $B = 30°, c = 10, a = 5\sqrt{3}$ 3. $A = 30°, c = 8, b = 4\sqrt{3}$ 5. $A = 45°, a = 5\sqrt{2}, b = 5\sqrt{2}$
7. $A = 30°, B = 60°, b = 7\sqrt{3}$ 9. $B = 57°, a = 7.08, b = 10.90$ 11. $a = 111.34, A = 47.51°, B = 42.49°$
13. $A = 63.8°, a = 30.06, b = 14.79$ 15. $b = 10.72, c = 20.82, C = 118.9°$
17. **Solution 1.** $A = 64.49°, B = 83.51°, a = 13.62$
    Solution 2. $A = 51.51°, B = 96.49°, a = 11.82$
19. $A = 100.2°, B = 52.2°, C = 27.6°$ 21. No solution   23. $A = 37.36°, B = 43.05°, C = 99.59°$
25. **Solution 1.** $B = 104.21°, C = 50.59°, b = 34.38$
    Solution 2. $B = 25.39°, C = 129.41°, b = 15.21$

27. 365   **29.** 20.2   **31.** 61.2 ft   **33.** 103 ft, 4915 sq ft   **35.** $|V_2| = 15.0, |V_1 + V_2| = 25.0$ **37.** 80.6 lb, 53°38'
39. Let $A = (3, -1), B = (5, 4), C = (-5, 8), D = (-7, 3)$. Then $\overrightarrow{AB} \cdot \overrightarrow{BC} = 0$ and $\overrightarrow{BC} \cdot \overrightarrow{CD} = 0$. Area = 58.
41. 22.86 ft   **43.** 526.5 miles   **45.** 8.48 miles from first tower, 7.93 miles from second   **47.** 646 ft, S 10.6° E

Chapter 10

Exercise Set 1, page 430

1.
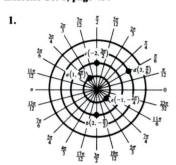

3. $(2, 8\pi/3), (-2, 5\pi/3),$
$(2, -4\pi/3), (-2, -\pi/3),$
$(2, 14\pi/3)$

5. a. $(2, 2)$ **b.** $(3\sqrt{3}, -3)$ **c.** $(-5\sqrt{2}, 5\sqrt{2})$ **d.** $(-4, 0)$ **e.** $(2, 2)$

7. a. $(4\sqrt{2}, 7\pi/4)$ **b.** $(4, 5\pi/6)$ **c.** $(2, 3\pi/2)$ **d.** $(1, 5\pi/3)$ **e.** $(2\sqrt{6}, 7\pi/6)$ **9.** $\theta = \dfrac{\pi}{4}$ **11.** $\theta = \dfrac{\pi}{3}$

13. $r \cos \theta = 2$ **15.** $r^2 \cos 2\theta = 1$ **17.** $r = 4 \cos \theta$ **19.** $r = -2 \cos \theta$ **21.** $r^2 = \tan 2\theta$ **23.** $y = x$
25. $y = 3$ **27.** $y + 4 = 0$ **29.** $2x + 5y = 7$ **31.** $x^2 + y^2 + 2y = 0$ **33.** $(x^2 + y^2)^2 = x^2 - y^2$
35. $x^2 - 3y^2 + 4y = 1$
37. Circle, radius 2

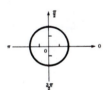

39. Line $x = 2$

41. Line $2x - 3y = 6$

43. Circle, radius 2

45. Cardioid

47. Limaçon

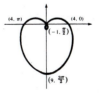

49. 4-leaf rose

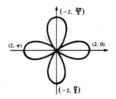

51. 3-leaf rose

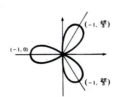

53. Lemniscate

55. Limaçon

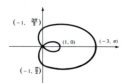

57. Cardioid

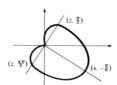

59. Lemniscate

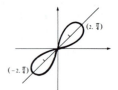

61.

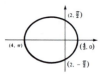

63. $r = 5\cos(\theta + \alpha)$, where $\alpha = \cos^{-1}\frac{3}{5}$

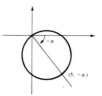

65.

67. $y^2 = -16(x - 4)$

69.

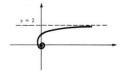

71.

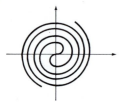

73.

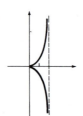

75. $(2, \pi/2), (2, 3\pi/2)$

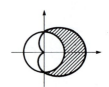

77. $(\frac{3}{2}, \pi/3), (\frac{3}{2}, 5\pi/3)$

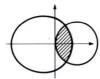

79. $(\frac{5}{2}, \pi/3), (\frac{5}{2}, 5\pi/3)$

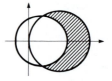

Exercise Set 2, page 437

1.

t	-2	-1	0	1	2
x	-5	-3	-1	1	3
y	0	1	2	3	4

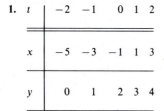

3.

t	0	1	2
x	2	1	0
y	0	2	4

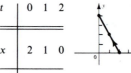

5.

t	0	$\dfrac{\pi}{3}$	$\dfrac{\pi}{2}$	$\dfrac{2\pi}{3}$	π
x	0	$\dfrac{3}{4}$	1	$\dfrac{3}{4}$	0
y	1	$\dfrac{1}{2}$	0	$-\dfrac{1}{2}$	-1

7. $x - 2y + 7 = 0$

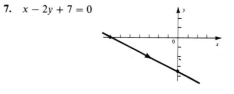

9. $(x + 1)^2 + (y - 2)^2 = 1$

11. $x = 2y^2 - 1, |x| \le 1, |y| \le 1$

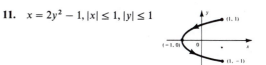

13. $y = x^2 - 1, x \ge 1, y \ge 0$

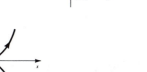

15. $xy = 1, x > 0, y > 0$

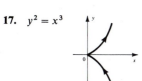

17. $y^2 = x^3$

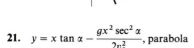

19. $(x + 1)^2 + (y - 2)^2 = 1$

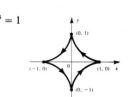

21. $y = x \tan \alpha - \dfrac{gx^2 \sec^2 \alpha}{2v_0^2}$, parabola   **23.** $x = a \cos^{-1}\left(\dfrac{a - y}{a}\right) - \sqrt{2ay - y^2}$ **25.** $x^{2/3} + y^{2/3} = 1$

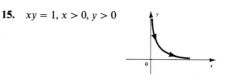

27.

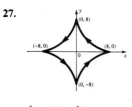

$\begin{cases} x = 8 \cos^3 t \\ y = 8 \sin^3 t \end{cases}$ $0 \le t \le 2\pi$

29.

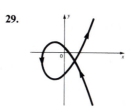

$\begin{cases} x = t^2 - 3 \\ y = t^3 - 4t - 1 \end{cases}$ $-3 \le t \le 3$

31.

$$\begin{cases} x = (1 + \cos\theta)\cos\theta \\ y = (1 + \cos\theta)\sin\theta \end{cases}$$

33.

$$\begin{cases} x = (4 - 2\cos\theta)\cos\theta \\ y = (4 - 2\cos\theta)\sin\theta \end{cases}$$

35.

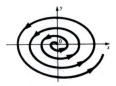

$$\begin{cases} x = \dfrac{\theta}{2}\cos\theta \\ y = \dfrac{\theta}{2}\sin\theta \end{cases}, \quad 0 \le \theta \le 25$$

Exercise Set 3, page 444

1. $\theta = 7\pi/4, r = \sqrt{2}$;
$1 - i = \sqrt{2}[\cos(7\pi/4) + i\sin(7\pi/4)]$

3. $\theta = 11\pi/6, r = 2$;
$\sqrt{3} - i = 2[\cos(11\pi/6) + i\sin(11\pi/6)]$

5. $\theta = \pi, r = 4$; $-4 = 4(\cos\pi + i\sin\pi)$

7. $\theta = 4\pi/3, r = 4$;
$-2 - 2i\sqrt{3} = 4[\cos(4\pi/3) + i\sin(4\pi/3)]$

9. $\theta = 0, r = 5$; $5 = 5(\cos 0 + i\sin 0)$

11. $-1 - i\sqrt{3}$ **13.** $-5i$ **15.** $3\sqrt{3} - 3i$ **17.** $1 - i$ **19.** $(-3\sqrt{3}/2) + (3i/2)$

21. $z_1 z_2 = 6[\cos(3\pi/2) + i\sin(3\pi/2)] = -6i; z_1/z_2 = \frac{2}{3}[\cos(-\pi) + i\sin(-\pi)] = -\frac{2}{3}$

23. $z_1 z_2 = 8[\cos(5\pi/4) + i\sin(5\pi/4)] = -4\sqrt{2} - 4i\sqrt{2}; z_1/z_2 = 2[\cos(-\pi/4) + i\sin(-\pi/4)] = \sqrt{2} - i\sqrt{2}$

25. $z_1 z_2 = 8[\cos(13\pi/6) + i\sin(13\pi/6)] = 4\sqrt{3} + 4i; z_1/z_2 = \frac{1}{2}[\cos(3\pi/2) + i\sin(3\pi/2)] = -i/2$

27. $z_1 z_2 = 24(\cos 170° + i\sin 170°); z_1/z_2 = \frac{2}{3}(\cos 30° + i\sin 30°) = (\sqrt{3}/3) + (i/3)$

29. $729[\cos(9\pi/2) + i\sin(9\pi/2)] = 729[\cos(\pi/2) + i\sin(\pi/2)] = 729i$

31. $256[\cos(44\pi/3) + i\sin(44\pi/3)] = 256[\cos(2\pi/3) + i\sin(2\pi/3)] = -128 + 128i\sqrt{3}$

33. $256[\cos(20\pi/3) + i\sin(20\pi/3)] = 256[\cos(2\pi/3) + i\sin(2\pi/3)] = -128 + 128i\sqrt{3}$

35. $w_1 = 2[\cos(\pi/6) + i\sin(\pi/6)] = \sqrt{3} + i; w_2 = 2[\cos(5\pi/6) + i\sin(5\pi/6)] = -\sqrt{3} + i;$
$w_3 = 2[\cos(3\pi/2) + i\sin(3\pi/2)] = -2i$

37. $w_1 = 1(\cos 0 + i\sin 0) = 1; w_2 = 1[\cos(\pi/3) + i\sin(\pi/3)] = \frac{1}{2} + i\sqrt{3}/2; w_3 = 1[\cos(2\pi/3) + i\sin(2\pi/3)] = -\frac{1}{2} + i\sqrt{3}/2;$
$w_4 = 1(\cos \pi + i\sin \pi) = -1; w_5 = 1[\cos(4\pi/3) + i\sin(4\pi/3)] = -\frac{1}{2} - i\sqrt{3}/2;$
$w_6 = 1[\cos(5\pi/3) + i\sin(5\pi/3)] = \frac{1}{2} - i\sqrt{3}/2$

39. $w_1 = 4[\cos(3\pi/4) + i\sin(3\pi/4)] = -2\sqrt{2} + 2i\sqrt{2}; w_2 = 4[\cos(7\pi/4) + i\sin(7\pi/4)] = 2\sqrt{2} - 2i\sqrt{2}$

41. $\{\sqrt{3} + i, 2i, -\sqrt{3} + i, -\sqrt{3} - i, -2i, \sqrt{3} - i\}$ **43.** $\{2, 1 + i\sqrt{3}, -1 + i\sqrt{3}, -2, -1 - i\sqrt{3}, 1 - i\sqrt{3}\}$

Review Exercise Set, page 445

1. a. $(0, -2)$ **b.** $\left(-\dfrac{3}{\sqrt{2}}, -\dfrac{3}{\sqrt{2}}\right)$ **c.** $\left(\dfrac{1}{\sqrt{2}}, \dfrac{1}{\sqrt{2}}\right)$ **d.** $(-2, -2\sqrt{3})$

3. a. $r(3\cos\theta - 5\sin\theta) = 7$ **b.** $r^2 = 2\sec\theta\csc\theta$ **5. a.** $r = \dfrac{\cos\theta}{1 - \sin 2\theta}$ **b.** $r = \dfrac{1}{1 + \sin\theta}$

7. a. $x + y = 2$ **b.** $12x - 5y = 0$ **9. a.** $(x^2 + y^2)^2 = 4xy$ **b.** $x^2 + y^2 - 2x + y = 0$

11. a. Line $2x + 4y = 3$ **b.** Line $x = -3$

13. a. Cardioid   **b.** Limaçon

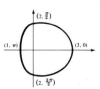

15. a. Cardioid   **b.** Lemniscate $r^2 = \cos 2\theta$

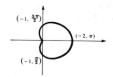

17. $(x - 1)^2 + y^2 = 1$

19. $(x - 1)^2 = -(y - 2), x \geq 2, y \leq 1$

21. a. $2\sqrt{2}\left(\cos\dfrac{3\pi}{4} + i\sin\dfrac{3\pi}{4}\right)$ **b.** $3\cos\left(\dfrac{3\pi}{2} + i\sin\dfrac{3\pi}{2}\right)$ **23. a.** $3i$ **b.** $-2\sqrt{3} - 2i$

25. $z_1z_2 = -8\sqrt{3} - 8i, \dfrac{z_1}{z_2} = 2\sqrt{3} + 2i$ **27.** $\dfrac{-27 + 27i}{\sqrt{2}}$ **29.** $2, 1 + \sqrt{3}i, -1 + \sqrt{3}i, -2, -1 - \sqrt{3}i, 1 - \sqrt{3}i$

31. $y^2 = 4(x + 1)$

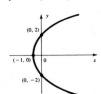

33. $\left(3 + 2\sqrt{3}, \pm\dfrac{5\pi}{6}\right)$, pole

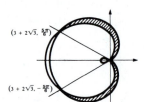

35.

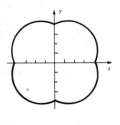

Cumulative Review Exercise Set III (Chapters 7–10), page 447

1. a. $B = 60°, c = 8\sqrt{3}, a = 4\sqrt{3}$ **b.** $A = 30°, c = 8, b = 4\sqrt{3}$ **c.** $B = 45°, a = b = 8\sqrt{2}$
   d. $A = 30°, B = 60°, b = 10\sqrt{3}$ **e.** $A = B = 45°, a = 6\sqrt{2}$

3. a. $-\frac{24}{25}$ **b.** $\frac{33}{65}$ **c.** $-1/\sqrt{5}$ **d.** $-\sqrt{5}$

5. a. $z_1z_2 = 8[\cos(3\pi/2) + i\sin(3\pi/2)] = -8i$ **b.** $z_1/z_2 = 2[\cos(7\pi/6) + i\sin(7\pi/6)] = -\sqrt{3} - i$
   c. $z_2^4 = 16[\cos(2\pi/3) + i\sin(2\pi/3)] = -8 + 8i\sqrt{3}$
   d. $w_1 = 2[\cos(2\pi/3) + i\sin(2\pi/3)] = -1 + i\sqrt{3}; w_2 = 2[\cos(5\pi/3) + i\sin(5\pi/3)] = 1 - i\sqrt{3}$

7. a. $\frac{7}{9}$ **b.** $-\frac{3}{5}$ **c.** $\frac{4}{3}$ **d.** $-1/\sqrt{5}$

9. a. 1.97 radians   **b.** 11.82 inches   **c.** 35.46 square inches   **d.** 16.58 square inches

11. a. $r = 1.194$ kilometers   **b.** 16.66 miles per hour   **15.** $\{4i, -2\sqrt{3} - 2i, 2\sqrt{3} - 2i\}$

17. a. Amplitude $= 2$, period $= 4\pi$, phase shift $= 2\pi/3$ **b.** Amplitude $= \frac{1}{2}$, period $= 2$, phase shift $= -\frac{1}{4}$

21. a. Period $= 2$

b. The given equation is equivalent to
$x = \frac{1}{2}\sin 2(y - \pi/3)$, with
$-\pi/2 \leq 2(y - \pi/3) \leq \pi/2$, or $\pi/12 \leq y \leq 7\pi/12$.

23. a. $\{0, 4\pi/3\}$ **b.** $\{\pi/6, 5\pi/6, 7\pi/6, 11\pi/6\}$

25. $\left(2, \pm\dfrac{2\pi}{3}\right), (1, \pi)$ is nonsimultaneous intersection.

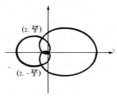

27. $h = 11.76, \theta = 35°21'$

29. $(x - 1)^2 = -\frac{1}{2}(y - 2), 0 \le x \le 2, 0 \le y \le 2$ **31.** $x^2 - y^2 = 4, x \ge 2$ **33.** $x = 1 - e^{-2y}$

    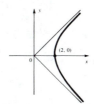   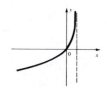

35. Vertical asymptotes: $x = \pm\dfrac{\pi}{3}, x = \pi$; x intercepts: $-\dfrac{\pi}{6}, -\dfrac{5\pi}{6}$; y intercept: $\dfrac{1}{2}$

Chapter 11

Exercise Set 1, page 461

1. $(4, 2)$ **3.** $(-1, -3)$ **5.** $(4, 3)$ **7.** $(\frac{17}{11}, \frac{2}{11})$ **9.** $(4, -2)$ **11.** $(-5, 6)$ **13.** $(-2, 3, 4)$

15. $(2, -1, 4)$ **17.** $(-\frac{11}{14}, -\frac{11}{24})$ **19.** $\left(\dfrac{ac + bd}{a^2 + b^2}, \dfrac{bc - ad}{a^2 + b^2}\right)$ **21.** $(8, 5)$

23. If both sides of the second equation are multiplied by $-\frac{1}{2}$ and then 5 is added to both sides, the result is the same as the first equation. So the equations are equivalent. The solution set is $\left\{\left(c, \dfrac{3c - 5}{4}\right):\ c \in R\right\}$.

25. $8x - 6y = 3$ This is impossible, so the equations are inconsistent.
    $\underline{-8x + 6y = 4}$
            $0 = 7$

27. The solution of the system consisting of the first two equations is $(-3, 2)$, and this pair also satisfies the third equation.

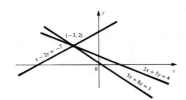

29. 95 and 32   **31.** 4 by 8   **33.** 3 miles per hour walking, 5 miles per hour jogging   **35.** 86

37. \$4,000 at 5%, \$7,000 at 7%   **39.** 14 cubic centimeters of 10% solution, 16 cubic centimeters of 25% solution

41. $10\frac{2}{3}$ pounds Indian tea, $9\frac{1}{3}$ pounds Ceylon tea   **43.** Coffee is \$4.25 per pound, eggs are 90¢ per dozen

45. $(\frac{1}{5}, \frac{1}{10} - \frac{7}{10})$ **47.** $(\frac{23}{10}, -\frac{8}{5}, -\frac{9}{2}, \frac{29}{5})$

51. The system is dependent. The solution set is $\{(2c - 2, 5c - 5, 3c - 2, c): \ c \in R\}$.
$\{(2c - 2, 5c - 5, 3c - 2, c): \ c \in R\}$.

53. $a = -1, b = 2, c = 5$ **55.** $x^2 + y^2 - 5x - y - 6 = 0$ **57.** $\frac{1}{2}$ hour, 15 miles

59. 30 hours for A, 24 hours for B, 40 hours for C

Exercise Set 2, page 471

1. $(-4, 3, 1)$ **3.** $(2, -3, 7)$ **5.** $(-3, 4, -2)$ **7.** $\{(2 - c, 1 - c, c): \ c \in R\}$ **9.** $(5, 2, -3)$ **11.** $(2, -1, 4)$

13. $\{(-3 + 7c, c, 4 - 3c): \ c \in R\}$ **15.** $\{(10c, 8c, c): \ c \in R\}$ **17.** $(\frac{10}{11}, -\frac{3}{11}, \frac{3}{11})$ **19.** Inconsistent

21. $(\frac{2}{3}, -\frac{22}{3}, -4)$ **23.** $\{(2, c - 5, 3c - 9, c): \ c \in R\}$ **25.** $\{(-5c, -2c, c): \ c \in R\}$ **27.** $(-\frac{4}{3}, \frac{2}{3}, \frac{7}{3})$

29. $(3, -1, 5, 6)$ **31.** $(-19, 16, -22, -9)$ **33.** $(2, 3, 0, -1)$ **35.** $(3, 4, -1, 2)$ **37.** Inconsistent

39. $\{(c, 2c, -c, 3c): \ c \in R\}$ **41.** $(-27, -21, 3, -6, 5)$

Exercise Set 3, page 478

1. a. 11   **b.** -78 **c.** -60 **3. a.** 0   **b.** 5   **c.** 0

5. a. -1 **b.** $\begin{vmatrix} 3 & 1 \\ -4 & -3 \end{vmatrix} = -5$ **c.** 5   **7. a.** -6 **b.** $\begin{vmatrix} -2 & 1 \\ 5 & -3 \end{vmatrix} = 1$ **c.** 1

9. a. $1\begin{vmatrix} -4 & 5 \\ -6 & -1 \end{vmatrix} - (-3)\begin{vmatrix} 3 & -2 \\ -6 & -1 \end{vmatrix} + 7\begin{vmatrix} 3 & -2 \\ -4 & 5 \end{vmatrix} = 34 + 3(-15) + 7(7) = 38$

    b. $(-6)\begin{vmatrix} -2 & 1 \\ 5 & -3 \end{vmatrix} - (-1)\begin{vmatrix} 3 & 1 \\ -4 & -3 \end{vmatrix} + 7\begin{vmatrix} 3 & -2 \\ -4 & 5 \end{vmatrix} = -6(1) + (-5) + 7(7) = 38$

11. 14   **13.** 118   **15.** -14 **17.** 62   **19.** $\{4, -1\}$

21. $\begin{vmatrix} a + kb & b \\ c + kd & d \end{vmatrix} = ad + kbd - bc - kbd = ad - bc = \begin{vmatrix} a & b \\ c & d \end{vmatrix}$

    $\begin{vmatrix} a + kc & b + kd \\ c & d \end{vmatrix} = ad + kcd - bc - kcd = ad - bc = \begin{vmatrix} a & b \\ c & d \end{vmatrix}$

23. 44   **25.** 38

29. If x is replaced by x_1 and y is replaced by y_1, the determinant is 0, since two rows are identical (see Problem 28). Similarly, if x is replaced by x_2 and y by y_2, the determinant is 0. Thus (x_1, y_1) and (x_2, y_2) satisfy the equation. Expanding the determinant shows that the equation is linear, and the conclusion follows.

31. $\{0, 1, -2, -\frac{3}{2}\}$

Exercise Set 4, page 484

1. $(\frac{37}{11}, -\frac{1}{11})$ **3.** $(-\frac{14}{13}, \frac{16}{13})$ **5.** $(\frac{13}{82}, -\frac{25}{41})$ **7.** $(2, \frac{1}{2})$ **9.** $(-\frac{5}{7}, \frac{13}{7})$ **11.** $(14, 8)$ **13.** $(\frac{4}{3}, \frac{22}{9}, \frac{5}{9})$

15. $(3, -2, -3)$ **17.** $(\frac{3}{2}, -\frac{1}{2}, 2)$ **19.** $(\frac{16}{11}, \frac{31}{11}, -\frac{6}{11})$ **21.** $\left(\dfrac{2a - 3}{1 - 5a}, \dfrac{a + 5}{5a - 1}\right)$ **23.** $\left(\dfrac{2x^2 - 1}{2x^2 + 1}, \dfrac{2x}{2x^2 + 1}\right)$

25. $\left(\dfrac{3}{4}, \dfrac{6 - 3m}{2}\right)$ **27.** $(\frac{7}{9}, \frac{26}{9}, \frac{19}{9})$

Exercise Set 5, page 489

1. $\begin{bmatrix} 2 & 6 & 3 \\ 10 & 4 & 16 \end{bmatrix}$ **3.** $\begin{bmatrix} -2 & 2 \\ -6 & 11 \end{bmatrix}$ **5.** $\begin{bmatrix} -2 & 14 \\ 0 & 0 \end{bmatrix}$ **7.** $\begin{bmatrix} -5 & 8 \\ 10 & -1 \\ 27 & 14 \end{bmatrix}$ **9.** $\begin{bmatrix} 21 & 14 \\ 17 & 26 \end{bmatrix}$ **11.** $\begin{bmatrix} 7 & -1 & 10 \\ 13 & 11 & 25 \end{bmatrix}$

13. $\begin{bmatrix} 1 & 9 & -7 & 7 \\ 10 & 6 & -28 & 0 \\ -7 & 15 & 10 & 16 \end{bmatrix}$ **15.** $x = 5, y = -2$

17. $\begin{bmatrix} 0 & 0 \\ 0 & 0 \end{bmatrix}$ Neither A nor B need be 0 in order for their product to be 0.   **21.** $B = \begin{bmatrix} -1 & 2 \\ -3 & 5 \end{bmatrix}$

25. $A^2 = \begin{bmatrix} 2 & 2 \\ 2 & 2 \end{bmatrix}$, $A^3 = \begin{bmatrix} 4 & 4 \\ 4 & 4 \end{bmatrix}$, $A^4 = \begin{bmatrix} 8 & 8 \\ 8 & 8 \end{bmatrix}$, $A^n = \begin{bmatrix} 2^{n-1} & 2^{n-1} \\ 2^{n-1} & 2^{n-1} \end{bmatrix}$ **27.** $(A + B)(A - B) = A^2 - AB + BA - B^2$

29. $A = \begin{bmatrix} a_{11} & a_{12} & a_{13} & \cdots & a_{1n} \\ a_{21} & a_{22} & a_{23} & \cdots & a_{2n} \\ a_{31} & a_{32} & a_{33} & \cdots & a_{3n} \\ \vdots & \vdots & \vdots & \vdots & \vdots \\ a_{m1} & a_{m2} & a_{m3} & \cdots & a_{mn} \end{bmatrix}$, $X = \begin{bmatrix} x_1 \\ x_2 \\ x_3 \\ \vdots \\ x_n \end{bmatrix}$, $B = \begin{bmatrix} b_1 \\ b_2 \\ b_3 \\ \vdots \\ b_m \end{bmatrix}$

Exercise Set 6, page 494

1. $\begin{bmatrix} 2 & -1 \\ 3 & -2 \end{bmatrix}$ **3.** $\begin{bmatrix} -1 & 2 \\ -\frac{3}{2} & \frac{5}{2} \end{bmatrix}$ **5.** $\begin{bmatrix} 1 & 0 & 0 \\ \frac{1}{2} & \frac{1}{2} & 0 \\ -1 & -\frac{2}{3} & \frac{1}{3} \end{bmatrix}$ **7.** $\begin{bmatrix} 0 & 1 & 2 \\ 1 & -2 & -4 \\ 2 & -2 & -5 \end{bmatrix}$ **9.** $\begin{bmatrix} -\frac{1}{2} & \frac{1}{2} & 0 \\ 0 & \frac{1}{2} & \frac{1}{2} \\ \frac{1}{2} & 0 & -\frac{1}{2} \end{bmatrix}$ **11.** Singular

13. $\begin{bmatrix} -80 & 67 & 4 \\ -19 & 16 & 1 \\ -98 & 82 & 5 \end{bmatrix}$ **15.** Singular   **17.** $\begin{bmatrix} 5 & -2 \\ 3 & -1 \end{bmatrix}$ **19.** $\begin{bmatrix} \frac{2}{5} & -\frac{3}{5} \\ \frac{3}{10} & -\frac{1}{5} \end{bmatrix}$ **21.** $\begin{bmatrix} 1 & -2 & 5 \\ 0 & 1 & -4 \\ 0 & 0 & 1 \end{bmatrix}$

23. $\begin{bmatrix} 3 & 3 & -8 \\ 6 & 7 & -18 \\ -1 & -1 & 3 \end{bmatrix}$ **25.** $\begin{bmatrix} 2 & -2 & -3 \\ -2 & 3 & 4 \\ -3 & 4 & 6 \end{bmatrix}$ **27.** $\begin{bmatrix} 23 & -30 & 12 \\ -6 & 8 & -3 \\ 19 & -25 & 10 \end{bmatrix}$ **29.** $\begin{bmatrix} -9 & -3 & -7 \\ -\frac{2}{3} & 0 & -\frac{1}{3} \\ -\frac{17}{3} & -2 & -\frac{13}{3} \end{bmatrix}$ **31.** $(2, 3, -4)$

33. $(\frac{1}{2}, -\frac{3}{2}, 2)$ **35.** $(4, -3, 5)$

39. $|A| = a_{11}a_{22}a_{33} \neq 0$ if, and only if, $a_{11} \neq 0$, $a_{22} \neq 0$, $a_{33} \neq 0$

$$A^{-1} = \frac{1}{a_{11}a_{22}a_{33}} \begin{bmatrix} a_{22}a_{33} & -a_{12}a_{33} & a_{12}a_{23} - a_{22}a_{13} \\ 0 & a_{11}a_{33} & -a_{11}a_{23} \\ 0 & 0 & a_{11}a_{22} \end{bmatrix}$$

41. $\begin{bmatrix} 0 & 0 & -1 & -1 \\ 1 & 4 & 1 & 3 \\ 1 & 2 & 1 & 2 \\ 0 & -1 & 0 & -1 \end{bmatrix}$ **43.** Singular   **49.** $\begin{bmatrix} 1.090 & 0.3789 & -0.4003 \\ 0.8379 & 0.4532 & -0.3122 \\ 0.4750 & 0.2408 & -0.2942 \end{bmatrix}$

Exercise Set 7, page 500

1. $\{(2, 1), (-1, -2)\}$ **3.** $\{(3, 0), (-1, -2)\}$ **5.** $\{(\frac{5}{3}, -\frac{11}{9}), (-1, -3)\}$ **7.** $\{(8, 2), (\frac{17}{4}, -\frac{1}{2})\}$

9. $\{(\frac{3}{4}, \frac{3}{2}), (-\frac{3}{4}, -\frac{3}{2})\}$ **11.** $\{(1, -1), (\frac{25}{4}, \frac{5}{2})\}$ **13.** $\{(-1, -1), (\frac{7}{4}, \frac{4}{7})\}$ **15.** $\{(1, \sqrt{2}), (1, -\sqrt{2})\}$ **17.** $\{(4, 3), (\frac{8}{3}, \frac{1}{3})\}$

19. $\{(9, 6), (9, -6)\}$ **21.** $\{(0, -4), (4, 4)\}$ **23.** $\{(2, 3), (2, -3), (-2, 3), (-2, -3)\}$ **25.** $\{(5, 0), (-\frac{2}{3}, \frac{17}{6})\}$

27. $\{(\frac{3}{2}, \frac{1}{2}), (\frac{3}{2}, -\frac{1}{2}), (-\frac{3}{2}, \frac{1}{2}), (-\frac{3}{2}, -\frac{1}{2})\}$ **29.** $\{(7, 2), (4, -1)\}$ **31.** $\{(2, 1), (\frac{2}{5}, -\frac{11}{5})\}$

33. $\left\{ \left(\frac{\sqrt{39}}{3}, \frac{1}{3} \right), \left(-\frac{\sqrt{39}}{3}, \frac{1}{3} \right), (\sqrt{2}, -2), (-\sqrt{2}, -2) \right\}$ **35.** 7 by 18   **37.** 12 and 8   **39.** 12 by 5

41. $2\sqrt{5}$ inches wide by $4\sqrt{5}$ inches high   **43.** \$3,600 at 5%   **45.** 3 cm by 4 cm

47. $\left\{ (3, -1), (-4, -8), \left(\frac{-1 + \sqrt{17}}{2}, \frac{7 + \sqrt{17}}{2} \right), \left(\frac{-1 - \sqrt{17}}{2}, \frac{7 - \sqrt{17}}{2} \right) \right\}$ **49.** $\{(1, 1), (\frac{1}{2}, 2)\}$ **51.** $\frac{5\pi}{2} - 5$

53. $r = 3, h = 8$; or $r = 2\sqrt{2 + \sqrt{13}} \approx 4.74$, $h = -4 + 2\sqrt{13} \approx 3.21$ **55.** $x + 2y - 8 = 0$ **57.** 8

59. 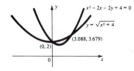 $x^2 - 2x - 2y + 4 = 0$, $y = \sqrt{x^2 + 4}$, $(3.088, 3.679)$, $(0, 2)$

61. 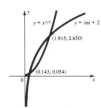 $y = x^{3/2}$, $y = \ln x + 2$, $(1.915, 2.650)$, $(0.143, 0.054)$

63. 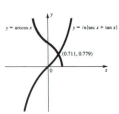 $y = \arccos x$, $y = \ln|\sec x + \tan x|$, $(0.711, 0.779)$

Exercise Set 8, page 506

1.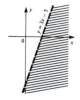

3.

5.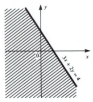

7.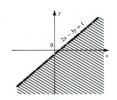

9.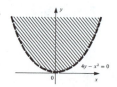

11.

13.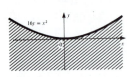

15.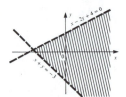

17.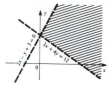

19.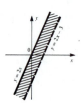

21.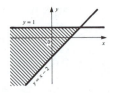

23.

25.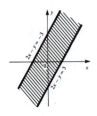

27.

29.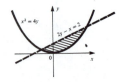

31.

33.

35.

37.

39.

41.

Exercise Set 9, page 510

1. Max $F = 11$ at $(4, 3)$ 3. Max $G = 24$ at $(0, 4)$ 5. Min $H = 16$ at $(2, 2)$ 7. Min $C = 0$ at $(0, 5)$
9. 30 cases of each kind   11. 7 hours for A, 5 hours for B 13. 500 gallons to each owner
15. 8 parts of type A, 10 parts of type B 17. 20 days for each plant
19. 1 pound of feed number 1, 4 pounds of feed number 2
21. 100 gallons from store 1 to A and 250 gallons from store 2 to A, 400 gallons from store 1 to B

Review Exercise Set, page 512

1. $(3, 2)$ 3. $(-8, 5)$ 5. $(9, 5, 3)$ 7. $(6, 5, -4, 3)$ 9. $(-6, 8, -5)$ 11. $(4, -6, -8)$
13. $\left\{\left(\dfrac{13 - c}{7}, \dfrac{5 + 5c}{7}, c\right): \ c \in R\right\}$ 15. $(10, 6, -4, -3)$ 17. $\{(-6 - c, 10, 2c + 3, c): \ c \in R\}$ 19. Inconsistent
21. **a.** 1   **b.** 167   23. $x = 1$ or $-\frac{4}{5}$ 25. **a.** $\left(-\frac{68}{3}, -43\right)$ **b.** $\left(\frac{19}{44}, \frac{12}{11}\right)$ 27. $(0, -1, 1)$
29. $\begin{bmatrix} -\frac{7}{5} & 2 & -\frac{8}{5} \\ -\frac{13}{5} & 3 & -\frac{12}{5} \\ 1 & -1 & 1 \end{bmatrix}$ 31. $(3, 7, 2)$ 33. $66\frac{2}{3}$ miles, $53\frac{1}{3}$ miles
35. 12 cubic centimeters 40% solution, 18 cubic centimeters 15% solution   37. 11 by 5
39. $18\frac{3}{4}$ gallons pure water, $31\frac{1}{4}$ gallons 40% salt solution   41. $1,800 at 6% interest
43. $4\frac{1}{6}$ hours for the larger pump, $6\frac{1}{4}$ hours for smaller one   45. $\{(-1, 2), (3, -\frac{2}{3})\}$ 47. $\{(8, -22), (-2, 8)\}$
49. $\{(2, 0), (-\frac{8}{5}, -\frac{12}{5})\}$ 51. $\{(\frac{3}{4}, \frac{3}{4}), (\frac{2}{3}, \frac{4}{5})\}$ 53. 15 and 7 or -7 and -15 55. 24 by 7 inches
57. **a.**

b.

59. **a.**

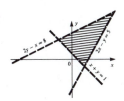

b.

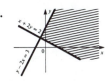

61. Max $F = 60$ at $(6, 0)$ **63.** Min $G = 7$ at $(\frac{4}{3}, 4)$ **65.** 250 acres corn, 50 acres oats   **67.** 9 short trips, 1 long trip
69. $\left\{ \left(\dfrac{50 + 13c}{12}, \dfrac{2 - 3c}{4}, \dfrac{2 - 5c}{12}, c \right): c \in R \right\}$ **71.** $\{(-3c, -4c, c, 2c): c \in R\}$ **73.** $(7, 3, 9, -6, -4)$
75. $(-2, 9, 8, 11)$ **77.** $\left(-\dfrac{6}{a + 2}, -\dfrac{2a + 1}{a + 2} \right), a \neq \pm 2$ **79.** $a = 2, b = -3, c = -5$ **81.** $\{(3, 4), (\frac{21}{34}, \frac{1}{34})\}$
83. $2\frac{2}{3}$ gallons 40% solution, $5\frac{1}{3}$ gallons 25% solution   **85.** $(3, 7)$ and $(-1, -1)$
87. 15 miles per hour riding, 3 miles per hour walking   **89.** 82 bags lawn fertilizer, 40 bags garden fertilizer

Chapter 12

Exercise Set 2, page 520

1. $x^2 + y^2 - 2x - 4y - 4 = 0$; circle of radius 3, with center $(1, 2)$ **3.** $3x^2 + 3y^2 - 18x + 14y + 10 = 0$
5. $x^2 + y^2 - 8x + 12y + 32 = 0$ **7.** $y^2 = 12x$ **9.** $x^2 - 2x + 8y + 1 = 0$ **11.** $5x^2 + 9y^2 = 45$
13. $x^2 = 4y$ **15.** $4x^2 + 3y^2 - 10y + 7 = 0$ **17.** $36x^2 - 64y^2 + 36x + 128y - 199 = 0$
19. $16x^2 - 24xy + 9y^2 - 142x - 106y + 76 = 0$

Exercise Set 3, page 523

1. a. $x^2 + y^2 - 2x + 6y + 6 = 0$ **b.** $x^2 + y^2 + 4x + 8y + 4 = 0$
3. $(x - 2)^2 + (y + 3)^2 = 9$; circle, center $(2, -3)$, radius 3   **5.** $(x + 1)^2 + (y - 4)^2 = 16$; circle, center $(-1, 4)$, radius 4

7. $(x + 4)^2 + (y - 2)^2 = 0$; degenerate circle—graph consists of point $(-4, 2)$ only
9. $(x + \frac{3}{4})^2 + (y - \frac{5}{4})^2 = \frac{29}{8}$; circle, center $(-\frac{3}{4}, \frac{5}{4})$, radius $\sqrt{\frac{29}{8}} \approx 1.9$

11. $x^2 + y^2 - 2x - 8y - 1 = 0$ **13.** $3x - 4y + 7 = 0$
15. $x^2 + y^2 - 12x - 2y + 1 = 0$ or $x^2 + y^2 + 12x + 10y + 25 = 0$
17. Distance between centers = sum of radii, so they are tangent; line joining centers, $12x + 5y - 16 = 0$; common tangent
line, $5x - 12y + 2 = 0$
19. $x^2 + y^2 + 2x + 4y - 20 = 0$
21. $(0, 1)$ and $(\frac{3}{2}, -\frac{1}{2})$; common chord, $x + y - 1 = 0$; perpendicular bisector, $2x - 2y - 1 = 0$

Exercise Set 4, page 530

1. a. $y^2 = 8x$ **b.** $x^2 = -12y$ **3. a.** $(x - 1)^2 = 8(y + 2)$ **b.** $(x + 3)^2 = -8(y - 1)$
5. a. $(x - 2)^2 = 12(y + 3)$ **b.** $(y + 1)^2 = 6(x + \frac{1}{2})$ **7.** $(y - 4)^2 = -12(x - 3)$ **9.** $(y + 2)^2 = -4(x - 3)$
11. $3x^2 - 4x - 4y + 5 = 0$
13. $(x - 2)^2 = 12(y + 1)$; focus $(2, 2)$ **15.** $(x - 1)^2 = y + 1$; focus $(1, -\frac{3}{4})$ **17.** $(y + \frac{1}{2})^2 = -6(x - \frac{3}{2})$; focus $(0, -\frac{1}{2})$

19. $f(x) = 4 - x^2$

21.

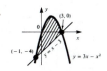

23.

25. $(x - 1)^2 + (y - 1)^2 = 4$

27.

29. $4x^2 + 4xy + y^2 + 28x - 16y + 4 = 0$

Exercise Set 5, page 537

1. **a.** $\dfrac{x^2}{16} + \dfrac{y^2}{4} = 1$

b. $\dfrac{x^2}{25} + \dfrac{y^2}{64} = 1$

3. $\dfrac{(x + 2)^2}{9} + \dfrac{(y - 4)^2}{25} = 1$

5. $\dfrac{(x - 4)^2}{20} + \dfrac{(y + 5)^2}{4} = 1$

7. $\dfrac{(x - 1)^2}{27} + \dfrac{(y + 3)^2}{36} = 1$

9. **a.** $\dfrac{x^2}{4} + \dfrac{y^2}{25} = 1$

b. $\dfrac{x^2}{5} + \dfrac{y^2}{4} = 1$

11. **a.** $\dfrac{x^2}{4} + \dfrac{y^2}{\frac{4}{9}} = 1$

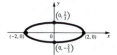

b. $\dfrac{x^2}{\frac{3}{5}} + \dfrac{y^2}{\frac{1}{5}} = 1$

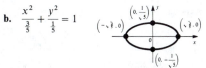

13. $\dfrac{(x - 1)^2}{4} + \dfrac{(y + 2)^2}{1} = 1$

15. $\dfrac{(x - \frac{5}{4})^2}{\frac{75}{16}} + \dfrac{(y + \frac{3}{4})^2}{\frac{75}{32}} = 1$

17. $\dfrac{(x - 3)^2}{16} + \dfrac{y^2}{9} = 1$

19. $\dfrac{(x+3)^2}{1} + \dfrac{\left(y-\frac{5}{2}\right)^2}{\frac{25}{4}} = 1$

21. $2\sqrt{3} \approx 3.46$

23. Circumscribed circle: $(x-2)^2 + (y+3)^2 = 25$; inscribed circle: $(x-2)^2 + (y+3)^2 = \frac{25}{2}$

25. $24x^2 + 4xy + 21y^2 - 500 = 0$

Exercise Set 6, page 544

1. **a.** $\dfrac{x^2}{4} - \dfrac{y^2}{5} = 1$

b. $\dfrac{y^2}{9} - \dfrac{x^2}{7} = 1$

3. $\dfrac{(y+4)^2}{9} - \dfrac{(x+2)^2}{16} = 1$

5. $\dfrac{(x-2)^2}{4} - \dfrac{(y-1)^2}{36} = 1$

7. $\dfrac{(x-3)^2}{16} - \dfrac{(y-2)^2}{9} = 1$

9. **a.** $\dfrac{x^2}{9} - \dfrac{y^2}{4} = 1$
Vertices: $(\pm 3, 0)$
Eccentricity: $\sqrt{13}/3$
Asymptotes: $y = \pm 2x/3$

9. **b.** $\dfrac{y^2}{4} - \dfrac{x^2}{9} = 1$
Vertices: $(0, \pm 2)$
Eccentricity: $\sqrt{13}/2$
Asymptotes: $y = \pm 2x/3$

11. **a.** $\dfrac{y^2}{9} - \dfrac{x^2}{4} = 1$
Vertices: $(0, \pm 3)$
Eccentricity: $\sqrt{13}/3$
Asymptotes: $y = \pm 3x/2$

11. **b.** $\dfrac{x^2}{9} - \dfrac{y^2}{25} = 1$
Vertices: $(\pm 3, 0)$
Eccentricity: $\sqrt{34}/3$
Asymptotes: $y = \pm 5x/3$

13. $\dfrac{(x-4)^2}{1} - \dfrac{(y-2)^2}{4} = 1$

Vertices: $(5, 2), (3, 2)$

Eccentricity: $\sqrt{5}/1$

Asymptotes: $y - 2 = \pm 2(x - 4)$

15. $\dfrac{(x+2)^2}{16} - \dfrac{(y-3)^2}{9} = 0$; degenerate hyperbola; graph consists of two lines: $y - 3 = \pm\frac{3}{4}(x + 2)$.

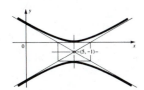

17. $\dfrac{(y+1)^2}{1} - \dfrac{(x-5)^2}{3} = 1$

Vertices: $(5, 0), (5, -2)$

Eccentricity: 2

Asymptotes: $y + 1 = \pm\frac{1}{\sqrt{3}}(x - 5)$

19. $\dfrac{(y-\frac{1}{2})^2}{36} - \dfrac{(x+\frac{3}{2})^2}{16} = 1$

Vertices: $\left(-\frac{3}{2}, -\frac{11}{2}\right), \left(-\frac{3}{2}, \frac{13}{2}\right)$

Eccentricity: $\dfrac{\sqrt{13}}{3}$

Asymptotes: $y - \frac{1}{2} = \pm\frac{3}{2}(x + \frac{3}{2})$

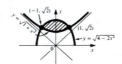

21.

23.

25. $\dfrac{(x-\frac{1}{2})^2}{\frac{4}{3}} - \dfrac{(y+\frac{3}{4})^2}{\frac{1}{3}} = 1$; hyperbola with horizontal transverse axis; center $(\frac{1}{2}, -\frac{3}{4})$;

eccentricity $\sqrt{5}/2$; asymptotes $y + \frac{3}{4} = \pm\frac{1}{2}(x - \frac{1}{2})$

27. $3y^2 - 4xy + 8x - 8 = 0$; vertices $\left(3 + \dfrac{4\sqrt{5}}{5}, 2 + \dfrac{2\sqrt{5}}{5}\right)$ and $\left(3 - \dfrac{4\sqrt{5}}{5}, 2 - \dfrac{2\sqrt{5}}{5}\right)$

end points of conjugate axis $(3 + \sqrt{5}/5, 2 - 2\sqrt{5}/5$ and $(3 - \sqrt{5}/5, 2 + 2\sqrt{5}/5)$

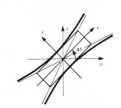

Exercise Set 7, page 555

1. $x'^2 = 2y'$

3. $\dfrac{y'^2}{10} - \dfrac{x^2}{15} = 1$

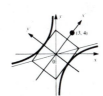

5. $\dfrac{y'^2}{4} - \dfrac{x'^2}{36} = 1$

7. $y'^2 = -4(x' - 1)$

9. $xy + 2 = 0$

11. $(x - 3)(y + 2) = 8$

13. $(x + 1)(y - 2) = -2$

15. $(x + 3)(y - 4) = -8$

17. a. Hyperbola   **b.** Hyperbola   **c.** Ellipse   **d.** Parabola   **e.** Ellipse
19. $(x' - 1)^2 = -4(y' + 1)$

Review Exercise Set, page 555

1. a. $x^2 + y^2 = 4$; circle

b. $\dfrac{y^2}{4} - \dfrac{x^2}{4} = 1$; hyperbola

c. $y^2 = -(x - 4)$; parabola

d. $x^2 = y - 4$; parabola

3. a. $(y - 1)^2 = x + 1$; parabola

b. $\dfrac{x^2}{1} - \dfrac{y^2}{4} = 1$; hyperbola

5. $\dfrac{(x + 3)^2}{4} + \dfrac{(y - 2)^2}{1} = 1$; ellipse

7. $\dfrac{(y + 1)^2}{4} - \dfrac{(x - 2)^2}{9} = 1$; hyperbola

9. $(x - 5)^2 + (y + 4)^2 = 0$; degenerate circle, single point $(5, -4)$ **13.** $(y + 2)^2 = -8(x - 3)$, or $y^2 + 8x + 4y - 20 = 0$
11. $(x - 3)^2 = \frac{15}{2}(y + 4)$; parabola

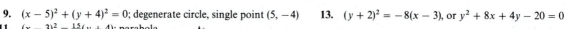

15. $\dfrac{(y-2)^2}{9} - \dfrac{(x-2)^2}{16} = 1$, or $9x^2 - 16y^2 - 36x + 64y + 116 = 0$ **17.** $\dfrac{y^2}{9} - \dfrac{x^2}{4} = 1$

19. $(x-3)^2 = -8(y+4)$, or $x^2 - 6x + 8y + 41 = 0$ **21.** $(x-3)^2 + (y-4)^2 = 64$, or $x^2 + y^2 - 6x - 8y - 39 = 0$

23. $x^2 - 2x + y - 4 = 0$ **25.** $x - y - 3 = 0$ **27. a.** Hyperbola   **b.** Hyperbola   **c.** Parabola   **d.** Ellipse

29. $y'^2 = -\tfrac{4}{5}x'$; parabola   **31.** $\dfrac{y'^2}{1} - \dfrac{x'^2}{3} = 1$; hyperbola

33. $5x^2 - 6xy - 3y^2 - 8x - 4 = 0$; hyperbola   **35.** $\dfrac{(x-4)^2}{\frac{16}{5}} - \dfrac{(y-1)^2}{4} = 1$, or

$$5x^2 - 4y^2 - 40x + 8y + 60 = 0$$

Chapter 13

Exercise Set 1, page 562

1. $\tfrac{2}{4}, \tfrac{4}{7}, \tfrac{6}{10}, \tfrac{8}{13}, \tfrac{10}{16}$ or $\tfrac{1}{2}, \tfrac{4}{7}, \tfrac{3}{5}, \tfrac{8}{13}, \tfrac{5}{8}$ **3.** $\dfrac{1}{1 \cdot 2}, \dfrac{-1}{2 \cdot 3}, \dfrac{1}{3 \cdot 4}, \dfrac{-1}{4 \cdot 5}, \dfrac{1}{5 \cdot 6}$ or $\tfrac{1}{2}, -\tfrac{1}{6}, \tfrac{1}{12}, -\tfrac{1}{20}, \tfrac{1}{30}$

5. $1, \tfrac{2}{3}, (\tfrac{2}{3})^2, (\tfrac{2}{3})^3, (\tfrac{2}{3})^4$ or $1, \tfrac{2}{3}, \tfrac{4}{9}, \tfrac{8}{27}, \tfrac{16}{81}$ **7.** $\tfrac{2}{4}, \tfrac{4}{10}, \tfrac{8}{28}, \tfrac{16}{82}, \tfrac{32}{244}$ or $\tfrac{1}{2}, \tfrac{2}{5}, \tfrac{2}{7}, \tfrac{8}{41}, \tfrac{8}{61}$ **9.** $\tfrac{2}{7}, \tfrac{2}{10}, \tfrac{2}{13}, \tfrac{2}{16}, \tfrac{2}{19}$ or $\tfrac{2}{7}, \tfrac{1}{5}, \tfrac{2}{13}, \tfrac{1}{8}, \tfrac{2}{19}$

11. $\tfrac{1}{243}, \tfrac{1}{729}, \tfrac{1}{2,187}$ **13.** $\tfrac{1}{11}, \tfrac{1}{13}, \tfrac{1}{15}$ **15.** $\tfrac{1}{32}, \tfrac{1}{64}, \tfrac{1}{128}$ **17.** $\tfrac{12}{37}, \tfrac{14}{50}, \tfrac{16}{65}$ **19.** $-\tfrac{6}{15}, \tfrac{7}{17}, -\tfrac{8}{19}$ **21.** $\dfrac{1}{2n-1}$

23. $\dfrac{1}{3n}$ **25.** $\dfrac{2^n - 1}{2^n}$ **27.** $\dfrac{1}{3n-2}$ **29.** $5(0.1)^n$ **31.** $2, \tfrac{4}{3}, \tfrac{8}{9}, \tfrac{16}{27}, \tfrac{32}{81}$ **33.** $4, -\tfrac{4}{3}, \tfrac{4}{9}, -\tfrac{4}{27}, \tfrac{4}{81}$

35. $2, -4, 12, -48, 240$ **37.** $\dfrac{n}{2n^2 - 3}$ **39.** $\dfrac{(-1)^{n-1} 2^{n-1}}{3^n + 1}$ **41.** $1, 1, 2, 3, 5$ **43.** $1, 2, 2, 1, \tfrac{1}{2}$

45. For $n = 10, 100,$ and $1,000$ the points are

$(\tfrac{7}{5}, \tfrac{20}{101}) = (1.4, 0.2)$

$(\tfrac{26}{25}, \tfrac{200}{10,001}) = (1.04, 0.02)$

$(\tfrac{251}{250}, \tfrac{2,000}{1,000,001}) = (1.004, 0.002)$

The limit point is $(1, 0)$.

Exercise Set 2, page 567

1. $a_{n+1} - a_n = 3(n+1) - 3n = 3$; so $d = 3$, and the sequence is arithmetic; $15, 18, 21, 24, 17$

3. $a_{n+1} - a_n = [8 - 3(n+1)] - (8 - 3n) = -3$; so $d = -3$, and the sequence is arithmetic; $-7, -10, -13, -16, -19$

5. $a_{n+1} - a_n = \dfrac{13 - 3(n+1)}{2} - \dfrac{13 - 3n}{2} = -\dfrac{3}{2}$; so $d = -\tfrac{3}{2}$, and the sequence is arithmetic; $-1, -\tfrac{5}{2}, -4, -\tfrac{11}{2}, -7$

7. $\dfrac{a_{n+1}}{a_n} = \dfrac{(-\tfrac{1}{2})^n}{(-\tfrac{1}{2})^{n-1}} = -\dfrac{1}{2}$; so $r = -\tfrac{1}{2}$, and the sequence is geometric; $-\tfrac{1}{8}, \tfrac{1}{16}, -\tfrac{1}{32}, \tfrac{1}{64}$

9. $\dfrac{a_{n+1}}{a_n} = \dfrac{3^n/9}{3^{n-1}/9} = 3$; so $r = 3$, and the sequence is geometric; $3, 9, 27, 81$

11. Arithmetic, $d = 3$; $a_{16} = 44$ **13.** Geometric, $r = -\frac{1}{3}$; $a_{10} = \frac{1}{6,561}$ **15.** Arithmetic, $d = 4$; $a_{21} = 70$

17. Geometric, $r = \frac{1}{2}$ **19.** Arithmetic, $d = 4$ **21.** Neither   **23.** Neither   **25.** Neither

27. Geometric, $r = \frac{2}{3}$ **29.** Neither   **31.** $a_1 = -25$ **33.** $a_1 = -14$, $d = 2$ **35.** $a_1 = -1$, $r = -\frac{2}{3}$

37. a. $x = \frac{25}{2}$ **b.** $x = \pm 10$ **39.** 14, 17, 20, 23, 26, 29

41. $x_{n+1} - x_n = (a_{n+1} + 5) - (a_n + 5) = a_{n+1} - a_n = $ a constant; so $x_1, x_2, x_3, \ldots$ is an arithmetic sequence.

43. Arithmetic sequence, $d = \ln 2$

45. Let $r = a_{n+1}/a_n$. Then $\log a_{n+1} - \log a_n = \log(a_{n+1}/a_n) = \log r$, which is a constant. So the sequence $\log a_1$, $\log a_2$, $\log a_3, \ldots$ is arithmetic, with common difference $d = \log r$.

47. 1,024

49. $A = P(1 + r)^n$; the sequence is $P(1 + r)$, $P(1 + r)^2$, $P(1 + r)^3, \ldots$, which is geometric, with common ratio $1 + r$.

51. \$7,864.32

Exercise Set 3, page 573

1. $4 + \frac{7}{3} + \frac{10}{5} + \frac{13}{7}$ **3.** $\dfrac{1}{1 \cdot 2} - \dfrac{1}{2 \cdot 3} + \dfrac{1}{3 \cdot 4} - \dfrac{1}{4 \cdot 5} + \dfrac{1}{5 \cdot 6} - \dfrac{1}{6 \cdot 7}$ **5.** $\frac{1}{2} - \frac{1}{5} + \frac{1}{10} - \frac{1}{17} + \frac{1}{26} - \frac{1}{37} + \frac{1}{50} - \frac{1}{65}$

7. $2 + 7 + 12 + 17 + 22 + 27 + 32 + 37 + 42 + 47$ **9.** $e^{-1} + 2e^{-2} + 3e^{-3} + 4e^{-4} + 5e^{-5} + 6e^{-6}$ **11.** 1,010

13. 9,300   **15.** 0.222222222222   **17.** 34.5   **19.** 60,700   **21.** 400   **23.** $6[1 - (\frac{2}{3})^8] = \frac{12,610}{2,187}$

25. $(\frac{12}{7})[1 - (-\frac{3}{4})^6] = \frac{1,443}{1,024}$ **27.** 5,050   **29.** 322   **31.** 120

33. First job sixth year salary = \$15,500, accumulated earnings = \$82,500; second job sixth year salary = \$14,693.28, accumulated earnings = \$73,359.29; second salary exceeds first in 8 years; accumulated earnings exceed first in 13 years.

35. Should take first offer; earnings are \$10,737,418.23 and \$465,000, respectively.

Exercise Set 4, page 579

1. Converges to $\frac{3}{2}$ **3.** Diverges   **5.** Converges to $\frac{10}{19}$ **7.** Converges to $\frac{30}{11}$ **9.** Diverges

11. Converges to 5   **13.** Converges to 27   **15.** Diverges   **17.** Converges to $1/(\sqrt{2} - 1)$

19. Converges to $\frac{10}{11}$ **21.** $\frac{5}{33}$ **23.** $\frac{17}{11}$ **25.** $\frac{117}{37}$ **27.** $\dfrac{x}{2 - x}$, $|x| < 2$

29. $a_1 = 2$, $a_n = \dfrac{-1}{n(n - 1)}$ for $n \geq 2$, so series is $2 - \dfrac{1}{2 \cdot 1} - \dfrac{1}{3 \cdot 2} - \dfrac{1}{4 \cdot 3} - \cdots$; $S = \lim_{n \to \infty} S_n = \lim_{n \to \infty} \left[1 + \left(\dfrac{1}{n} \right) \right] = 1.$

31. 10 meters   **33.** $\dfrac{x}{3 + x^2}$, valid for $|-x^2/3| < 1$ or $|x| < \sqrt{3}$ **35.** $\frac{2}{3}(2 + i)$

37. $1 + 2x + 4x^2 + 8x^3 + \cdots = \sum\limits_{n=0}^{\infty} (2x)^n$, valid for $|x| < \frac{1}{2}$ **39.** $\frac{1}{2} + \dfrac{x}{4} + \dfrac{x^2}{8} + \cdots = \sum\limits_{n=0}^{\infty} \dfrac{x^n}{2^{n+1}}$ valid for $|x| < 2$

41. The sum is $\lim_{n \to \infty} [4 - (1/3^n)] = 4$; $a_1 = \frac{11}{3}$ and for $n \geq 2$, $a_n = S_n - S_{n-1} = 2/3^n$. So the series is $\frac{11}{3} + \frac{2}{9} + \frac{2}{27} + \frac{2}{81} + \cdots$. Beginning with the second term, this is geometric, with $r = \frac{1}{3}$.

Exercise Set 6, page 589

1. a. 3   **b.** 20   **c.** 28   **d.** 120   **e.** 1   **f.** 252   **g.** 435   **h.** 4,368   **i.** 7   **j.** 28

3. a. $(k + 1)!$ **b.** $k + 1$ **c.** $n(n - 1)$ **5.** $\dbinom{r}{r} = \dfrac{r(r - 1)(r - 2) \cdots (r - r + 1)}{1 \cdot 2 \cdot 3 \cdots r} = \dfrac{r!}{r!} = 1$

7. $\dbinom{n}{n - k} = \dfrac{n!}{(n - k)![n - (n - k)]!} = \dfrac{n!}{(n - k)!k!} = \dbinom{n}{k}$ **9.** $(x - y)^5 = x^5 - 5x^4y + 10x^3y^2 - 10x^2y^3 + 5xy^4 - y^5$

11. $a^4 - 12a^3b + 54a^2b^2 - 108ab^3 + 81b^4$ **13.** $s^6 - 12s^5t + 60s^4t^2 - 160s^3t^3 + 240s^2t^4 - 192st^5 + 64t^6$

15. $x^{12} - 12x^{10}y + 60x^8y^2 - 160x^6y^3 + 240x^4y^4 - 192x^2y^5 + 64y^6$ **17.** $495x^8y^4$ **19.** $-2,835a^4b^3$ **21.** 1.126

23. $(1 + 1)^n = \dbinom{n}{0}1^n + \dbinom{n}{1}1^{n-1} \cdot 1 + \dbinom{n}{2}1^{n-2} \cdot 1^2 + \dbinom{n}{3}1^{n-3} \cdot 1^3 + \cdots + \dbinom{n}{n}$ or: $2^n = \dbinom{n}{0} + \dbinom{n}{1} + \dbinom{n}{2} + \cdots + \dbinom{n}{n}$

25. $\dfrac{n - r + 1}{r} \cdot \dbinom{n}{r - 1} = \dfrac{n - r + 1}{r} \cdot \dfrac{n!}{(r - 1)!(n - r + 1)!} = \dfrac{(n - r + 1)n!}{r \cdot (r - 1)!(n - r + 1)(n - r)!} = \dfrac{n!}{r!(n - r)!} = \dbinom{n}{r}$

27. $\sqrt{1 + x} = (1 + x)^{1/2} = 1 + \dfrac{x}{2} - \dfrac{x^2}{8} + \dfrac{x^3}{16} - \dfrac{5x^4}{128} + \cdots$ **29.** 0.99499

Review Exercise Set, page 590

1. **a.** $\frac{1}{2}, \frac{2}{3}, \frac{3}{4}, \frac{4}{5}$ **b.** 1, 5, 9, 13   **3. a.** $\frac{1}{4}, -\frac{3}{16}, \frac{9}{64}, -\frac{27}{256}$ **b.** $1/\log 2, -1/(2 \log 3), 1/(3 \log 4), -1/(4 \log 5)$

5. **a.** $\dfrac{1}{n(n+1)}$ **b.** $\dfrac{n}{n+1}$ **7.** Arithmetic, $\frac{21}{2}$ **9.** Geometric, $-\frac{1}{2,048}$

11. **a.** 2, 5, 11, 23, 47   **b.** 1, 2, 6, 24, 120   **13. a.** 9   **b.** $\pm 3\sqrt{5}$

15. **a.** $\frac{243}{64}$ inches   **b.** $\frac{3,367}{64}$ inches   **c.** 64 inches   **17.** 1,590   **19.** $-2,400$ **21.** $-1,200$ **23.** $\frac{33}{2}$

25. $\frac{3}{8}$ **27. a.** $\frac{6}{11}$ **b.** $\frac{4}{27}$ **c.** $\frac{24}{11}$ **d.** $\frac{1,004}{37}$ **e.** $\frac{101}{99}$ **29.** 182

39. **a.** 2,300   **b.** 84   **c.** 4   **d.** 1   **e.** 66

41. **a.** $x^7 - \dfrac{7x^6 y}{2} + \dfrac{21x^5 y^2}{4} - \dfrac{35x^4 y^3}{8} + \dfrac{35x^3 y^4}{16} - \dfrac{21x^2 y^5}{32} + \dfrac{7xy^6}{64} - \dfrac{y^7}{128}$

   b. $1 + 10x^2 + 45x^4 + 120x^6 + 210x^8 + 252x^{10} + 210x^{12} + 120x^{14} + 45x^{16} + 10x^{18} + x^{20}$

43. **a.** $-25,344x^5 y^7$ **b.** $-85,995,520x^{11}$ **45.** $-\frac{14,581}{729}$ **47.** $73,205$

49. $23\frac{1}{4}, 22\frac{1}{2}, 21\frac{3}{4}, 21, 20\frac{1}{4}, 19\frac{1}{2}, 18\frac{3}{4}, 18, 17\frac{1}{4}, 16\frac{1}{2}, 15\frac{3}{4}$

51. $\dbinom{n}{k} - \dbinom{n-1}{k} = \dfrac{n!}{k!(n-k)!} - \dfrac{(n-1)!}{k!(n-1-k)!} = \dfrac{n! - (n-1)!(n-k)}{k!(n-k)!} = \dfrac{(n-1)!(n-n+k)}{k!(n-k)!} = \dfrac{(n-1)!k}{k!(n-k)!}$

   $= \dfrac{(n-1)!}{(k-1)![(n-1)-(k-1)]!} = \dbinom{n-1}{k-1}$

53. **a.** $1 - x + x^2 - x^4 + \cdots = \displaystyle\sum_{n=1}^{\infty} (-1)^{n-1} x^{n-1}, |x| < 1$ **b.** $\dfrac{3}{2} + \dfrac{3x^2}{4} + \dfrac{3x^4}{8} + \dfrac{3x^6}{16} + \cdots = \displaystyle\sum_{n=1}^{\infty} \dfrac{3}{2}\left(\dfrac{x^2}{2}\right)^{n-1}, |x| < \sqrt{2}$

Cumulative Review Exercise Set IV (Chapters 11–13), page 592

1. $\left(\frac{5}{2}, -\frac{3}{4}, \frac{1}{2}\right)$

3. $y^2 + 4x + 2y - 11 = 0$
   Slope of $2x - y = 3$ is 2 and slope of $x + 2y = 4$ is $-\frac{1}{2}$, so these sides are perpendicular, and the triangle is therefore a right triangle.

5. $(x - 3)^2 + (y - 2 - \sqrt{10})^2 = 10$

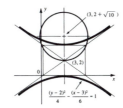

7. 80 standard, 60 deluxe

9. **a.** 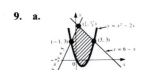   **b.**   **11.** $(x + 2)^2 = -8(y - 3)$, or $x^2 + 4x + 8y - 20 = 0$

13. Hyperbola: $\dfrac{y'^2}{14} - \dfrac{x'^2}{6} = 1$

   $\theta = \cos^{-1} \dfrac{1}{\sqrt{10}}$

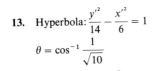
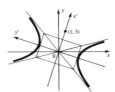

15. 27 type one parts and 15 type two parts. Minimum cost = $114.75

17. Solution set $\{(\frac{2}{3}, -\frac{5}{3}), (2, 1)\}$

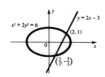

19. b. $\log 2^k = k \log 2$, so $\displaystyle\sum_{k=1}^{n} \log 2^k = \sum_{k=1}^{n} k \log 2 = \log 2 + 2 \log 2 + 3 \log 2 + \cdots + n \log 2$

$a = \log 2, d = \log 2, \quad S_n = \dfrac{n}{2}(\log 2 + n \log 2) = \dfrac{n}{2}(n+1)\log 2 = \dfrac{n^2 + n}{2}\log 2$

21. $a_1 = \sqrt{2}, a_2 = \sqrt{2 + \sqrt{2}}, a_3 = \sqrt{2 + \sqrt{2 + \sqrt{2}}}, a_4 = \sqrt{2 + \sqrt{2 + \sqrt{2 + \sqrt{2}}}}$

$a_5 = \sqrt{2 + \sqrt{2 + \sqrt{2 + \sqrt{2 + \sqrt{2}}}}}$

$a_1 = \sqrt{2} \le 2$. If $a_k \le 2$, then $a_{k+1} = \sqrt{2 + a_k} \le \sqrt{2 + 2} = 2$

23. a. $-4032 x^8 y^5$

b. $(1 + x)^{-1} = 1 - x + x^2 - x^3 + x^4 - \cdots + (-1)^{n-1} x^{n-1} + \cdots$; geometric series; ratio $-x$;

$\displaystyle\sum_{n=1}^{\infty} (-1)^{n-1} x^{n-1} = \dfrac{1}{1 + x} = (1 + x)^{-1}$

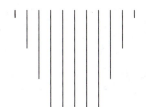

INDEX

Trigonometric Functions

For a right triangle

$$\sin A = \frac{a}{c} \qquad \csc A = \frac{c}{a}$$

$$\cos A = \frac{b}{c} \qquad \sec A = \frac{c}{b}$$

$$\tan A = \frac{a}{b} \qquad \cot A = \frac{b}{a}$$

For a general angle

$$\sin \theta = \frac{y}{r} \qquad \csc \theta = \frac{r}{y}$$

$$\cos \theta = \frac{x}{r} \qquad \sec \theta = \frac{r}{x}$$

$$\tan \theta = \frac{y}{x} \qquad \cot \theta = \frac{x}{y}$$

For a real number x

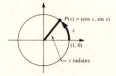

π radians $= 180°$

If f is any trigonometric function,
$f(\text{Real number } x) = f(\text{Angle of } x \text{ radians})$

Special Triangles

Law of Sines

$$\frac{a}{\sin A} = \frac{b}{\sin B} = \frac{c}{\sin C}$$

Trigonometric Form of a Complex Number

$z = a + bi = r(\cos \theta + i \sin \theta)$

Law of Cosines

$$a^2 = b^2 + c^2 - 2bc \cos A$$
$$b^2 = a^2 + c^2 - 2ac \cos B$$
$$c^2 = a^2 + b^2 - 2ab \cos C$$

Area of a Triangle

$\text{Area} = \frac{1}{2}ab \sin C = \frac{1}{2}ac \sin B = \frac{1}{2}bc \sin A$

De Moivre's Theorem: $\quad z^n = r^n(\cos n\theta + i \sin n\theta)$